Microelectronic Circuits

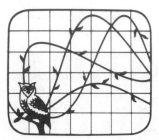

HRW
Series in
Electrical
Engineering

M. E. Van Valkenburg, Series Editor Electrical Engineering

SECOND EDITION

Microelectronic Circuits

Adel S. Sedra
University of Toronto

Kenneth C. Smith
University of Toronto

HOLT, RINEHART AND WINSTON
New York   Chicago   San Francisco   Philadelphia
Montreal   Toronto   London   Sydney   Tokyo
Mexico City   Rio de Janeiro   Madrid

TO
DORIS, PAUL AND MARK
LAURA AND PECAN
for their love,
patience, and understanding

Library of Congress Cataloging in Publication Data

Sedra, Adel S.
  Microelectronic circuits.

  Includes bibliographies and index.
  1. Electronic circuits. 2. Integrated circuits.
I. Smith, Kenneth Carless. II. Title.
TK7867.S39 1987   621.3815′3   86-25645
ISBN 0-03-007328-6

Printed in the United States of America

7 8 9 0   039   9 8 7 6 5 4 3

CBS COLLEGE PUBLISHING
Holt, Rinehart and Winston
The Dryden Press
Saunders College Publishing

Contents

Chapter 3

OPERATIONAL AMPLIFIERS   86

Chapter 4

DIODES   150

Chapter 8

BIPOLAR JUNCTION TRANSISTORS (BJTs)   398

Chapter 9

DIFFERENTIAL AND MULTISTAGE AMPLIFIERS   485

Chapter 10

OUTPUT STAGES AND POWER AMPLIFIERS   548

Chapter 11

FREQUENCY RESPONSE   599

Chapter 12

FEEDBACK   670

Chapter 13

ANALOG INTEGRATED CIRCUITS   743

Chapter 14

FILTERS, TUNED AMPLIFIERS AND OSCILLATORS   784

Chapter 15

MOS DIGITAL CIRCUITS   844

Chapter 16

BIPOLAR DIGITAL CIRCUITS   901

APPENDIXES   **follow page 962**

Preface

Microelectronic Circuits, Second Edition, is intended as a text for the core courses in electronic circuits taught to majors in electrical and computer engineering. It should also prove useful to engineers wishing to update their knowledge through self-study.

As was the case with the first edition, the objective of this book is to develop in the reader the ability to analyze and design electronic circuits, both analog and digital, discrete and integrated. While the application of integrated circuits is covered, emphasis is placed on transistor circuit design. This is done because of our belief that even if the majority of those studying the book were not to pursue a career in IC design, knowledge of what is inside the IC package would enable intelligent and innovative application of such chips. Furthermore, with the advances of VLSI technology and design methodology, IC design itself is becoming accessible to an increasing number of engineers.

Although the philosophy, pedagogical approach, and organization of the first edition have been retained, several changes have been made. Most importantly, coverage of MOS analog circuits has been greatly expanded. This reflects the developments of the last half-dozen years. Coverage of bipolar integrated circuits has also been expanded. A new chapter dealing with output stages and power amplifiers, including thermal considerations, has been added. New sections on tuned amplifiers, LC oscillators, and crystal oscillators are included in this edition. The digital electronics material has been updated and consolidated into two chapters. Here too, MOS circuits have been given the prominence they currently deserve.

To make room for various additions, we had to delete some of the introductory digital systems concepts. We felt this was acceptable since this material usually is included in a separate digital systems course. Also, coverage of digital memory has been considerably reduced; it is too specialized a topic for the core electronics course.

In keeping with our firmly held belief that good circuits are designed by people and not by computers, we have emphasized manual analysis and design techniques. Computer aids are, of course, useful at a later stage in the design process. We explain and illustrate this in a new appendix dealing with computer-aided design.

The exercise problems (with answers) that are sprinkled throughout the text were a well-received feature of the first edition. We have increased the variety and number (to about 400) of these exercise problems. Solving these exercises should enable the reader to gauge his or her grasp of the material immediately covered.

Also, the variety and number of the end-of-chapter problems has been increased: there are now about 800 problems, almost all of which are new to this edition. We have adopted a rating system to indicate the degree of challenge that each problem presents. While straightforward problems are not marked, moderately difficult problems are marked with an asterisk (*); more difficult problems are marked with two asterisks (**); and very difficult (and/or time consuming) problems are marked with three asterisks (***). We must admit, however, that this classification is by no means exact. Our rating no doubt has depended to some degree on our thinking (and mood!) at the time a particular problem was created. Answers to about half the problems are given in Appendix D. Complete solutions for all the exercises and the problems are included in the *Instructor's Manual*, which is available from the publisher for those who adopt the book.

To help the reader integrate the material covered in a chapter, a summary section has been added at the end of each chapter.

The prerequisite for studying the material in this book is a first course in circuit analysis. No prior knowledge of physical electronics is assumed. All required device physics is included and a brief appendix on IC fabrication is provided.

AN OUTLINE FOR THE READER

The book starts with an overview of electronic systems and signal processing in Chapter 1. This chapter discusses some electronic system examples and introduces the idea of a signal as an information container. This first chapter provides motivation and places the study of electronic circuits in its proper perspective.

Chapter 2 serves as a bridge between the material on circuit analysis, normally taught in a first course on this subject, and the electronic circuits topics treated in this book. Amplifiers are introduced as circuit building blocks and their various types and models are studied. Pertinent circuit analysis techniques are reviewed and the terminology and conventions used throughout the text are established.

Chapter 3 deals with operational amplifiers, their terminal characteristics, simple applications, and limitations. We have chosen to discuss the op amp as a circuit building block at this early stage simply because it is easy to deal with and because the student can experiment with op-amp circuits that perform nontrivial tasks with relative ease and with a sense of accomplishment. We have found this approach to be highly motivating to the student.

Chapter 4 introduces the ideal diode and the real *pn* junction diode. Here we concentrate on providing an understanding of the diode's terminal characteristics and its hierarchy of models. A qualitative description of the physical operation of the *pn* junction is also given.

Drawing on the knowledge of op amps and diodes acquired in Chapters 3 and 4, Chapter 5 presents many interesting and practical nonlinear circuit applications. These include various types of rectifiers, limiters and comparators, the bistable circuit, waveform generators, etc. Though it is customary in texts and courses to reserve this material for an advanced stage, we believe that it fits nicely here since no advanced mathematical methods or concepts are required. These circuits are exciting and have lots of action in them. They also perform useful and significant tasks. Of course the reader may skip some of the sections of this chapter and return to them at a later stage in her or his study, as desired.

Each of the next three chapters deals with a single type of transistor: the JFET in Chapter 6, the MOSFET in Chapter 7, and the BJT in Chapter 8. For each device we present a description of its physical operation, the different modes of operation, the terminal characteristics, the equivalent circuit models, and the basic circuit applications. Emphasis is placed on using simple models (both implicitly and explicitly) to perform rapid circuit analysis. Our hope is that by the end of each of these chapters the reader will be thoroughly familiar and intimately comfortable with the device treated.

By the end of Chapter 8 the reader will have learned about the basic building blocks of electronic circuits and will be ready to consider the more advanced topics covered in the second half of the book.

Chapter 9 is the first of a sequence of five chapters devoted to the study of analog circuits emphasizing amplifiers. The main topic of Chapter 9 is the differential amplifier, in both its bipolar and FET forms. Chapter 10 deals with the various types of amplifier output stages. Thermal design is studied and examples of IC power amplifiers are presented.

In Chapter 11 we study the frequency response of amplifiers. Here emphasis is placed on the choice of configuration to obtain wideband operation.

Chapter 12 deals with the important topic of feedback. Practical circuit applications of negative feedback are presented. We also discuss the stability problem in feedback amplifiers and treat frequency compensation in some detail.

Chapter 13 presents an introduction to analog integrated circuits. Both bipolar and CMOS op amps are discussed. This chapter ties together the ideas and methods presented in the previous chapters.

Chapter 14 is concerned with the design of filters, tuned amplifiers, and oscillators. It also includes an introduction to switched-capacitor filters.

The last two chapters in the book, 15 and 16, are concerned with digital circuits. With a few obvious exceptions, these two chapters do not rely on the material in Chapters 9 to 14 and, if desired, can be studied earlier. Chapters 15 and 16 present a concise, modern treatment of digital electronics and should serve as the basis for a more detailed study of digital circuits and systems and/or VLSI design.

TO THE INSTRUCTOR

The book contains sufficient material for a sequence of two single-semester courses (each of 40 to 50 lecture hours). The organization of the book provides considerable flexibility in course design. For example, one may choose to cover only part of the op-amp chapter (Chapter 3) in a first course. Also, the BJT can be covered before the FETs. Similarly, the digital-circuit material in Chapters 15 and 16 can be covered immediately after the study of the MOSFET and the BJT.

A first course in electronics may be based on Chapters 1 to 8 as follows:

Chapter 1   light treatment and/or reading assignment

Chapter 2   pertinent topics depending on student background

Chapter 3   all, except perhaps for Sections 3.9 to 3.13 which may be postponed to the second course

Chapter 4   all

Chapter 5   all, or a selection of topics; some sections may be postponed to the second course

Chapter 6   all, or Sections 6.1 to 6.8; the remainder of the chapter may be postponed to the second course

Chapter 7   all, or Sections 7.1 to 7.5; the remainder of the chapter can be postponed to the second course

Chapter 8   all

The second course may be based on Chapters 9 to 16 in addition to the sections of the preceding chapters which were not covered in the first course. The second course can be entirely analog (based on Chapters 9 to 14) or a mixture of analog and digital. In the latter case some of the more specialized analog topics (e.g., parts of Chapters 10, 13, and 14) may be postponed to a third, more advanced analog circuits course.

Mention has already been made of the *Instructor's Manual*, available from the publisher. In addition to complete solutions to all exercises and problems, the *Manual* includes problems and solutions for examinations and masters for overhead transparencies. Also available for students is a *Laboratory Manual*. It includes specially designed experiments that follow the sequence of the text material.

ACKNOWLEDGMENTS

Many of the changes in this edition were made in response to feedback received from some of the instructors who adopted the first edition. We are grateful to all those instructors and students who took the time to write to us. In particular, we wish to thank Professor Artice M. Davis of San Jose State University. In addition to supplying us with many helpful comments and suggestions, Professor Davis has written the *Laboratory Manual*. Thanks are also due to Professors D. Cooper of the University of Illinois, M. Ghorab of Ryerson Polytechnic Institute, A. Heckbert of San Jose State University, and W. Sansen of the Catholic University of Leuven, and to Mr. Robert Machado of the University of Arizona.

We are also indebted to Professor Martin Snelgrove, our colleague at the University of Toronto, who has been a source of many innovative ideas, and to our former colleague, Dr. Frank Holmes, who supplied us with Appendix A on IC fabrication. Our students Fred Gohh and Ed Nowicki helped with the preparation of Appendix C. Laura Fujino assisted with the preparation of the index.

Various portions of this edition were reviewed by several people. We thank the following reviewers for their helpful suggestions and hope that they will be pleased with the final result: Eugene Chenette of the University of Florida, Randall L. Geiger of Texas A & M University, Paul McGrath of Clarkson College of Technology, Alan B. MacNee of the University of Michigan, J. Alvin Connolly of Georgia Institute of Technology, Rolf Schaumann of the University of Minnesota, Yuh Sun of California State University, Douglas Brumm of Michigan Technical University, Jami Ramirez-Angulo of Texas A & M University, Glen Gerhard of the University of New Hampshire, Darrel Vines of Texas Tech University, William Sayle, III of Georgia Institute of Technology, Arnold Dippert of the University of Illinois, Richard Jaeger of Auburn University, Bernhard M. Schmidt of the University of Dayton, Doug Hamilton of the University of Arizona, Frank Barnes of the University of Colorado, and Artice M. Davis of San Jose State University.

The manuscript was skillfully typed by Penny Vardy and the production was most ably guided by Lila M. Gardner of Cobb/Dunlop Publisher Services, Inc. Last but not least, we wish to express our gratitude to our editor, Deborah L. Moore. It has been a great pleasure working with her.

Adel. S. Sedra
Kenneth C. Smith

Chapter 1

Electronic Systems

INTRODUCTION

The subject of this book is modern electronics, a field that has come to be known as *microelectronics*. Microelectronics refers to the integrated-circuit (IC) technology that at the time of this writing is capable of producing circuits that contain more than 1 million components in a small piece of silicon (known as a *silicon chip*) whose area is of the order of 60 mm². One such microelectronic circuit, for example, is a complete digital computer, which, accordingly, is known as a *microcomputer* or, more generally, a *microprocessor*.

In this book we shall study electronic devices that can be used singly (in the design of *discrete* circuits) or as components of an integrated-circuit chip. We shall study the design and analysis of interconnections of these devices, which form discrete and integrated circuits of varying complexity and perform a wide variety of functions. We shall also learn about available IC chips and their application in the design of electronic systems.

The purpose of this introductory chapter is to set the stage on which electronics will be seen to play a major part as interpreter of the vast body of work known broadly as signal and information processing. Although there are other interpreters of this art—mechanics, fluidics, and optics being examples—none has played center stage or is likely to without the strong support that electronics can provide. Our challenge then has a broad range: to equip the reader with some appreciation of the underlying themes of signal and information processing and to provide some of the ambience in which (and through which) the play unfolds. □

1.1 INFORMATION AND SIGNALS

In the broadest sense, *information* is simply knowledge of a situation, of an event, or of a trend. Likewise, *signals* are the means by which such knowledge is communicated. Thus the status of a mechanical system such as an automobile can be described by many pieces of information, the number of which depends on the need and the context of the situation. Thus whereas it is usually important for the observer of a parked car to know only its location, for a moving car information on velocity and even acceleration may be necessary. Such information is communicated to the observer by signals, components of the situation that the observer notices. Of course such signals often are not explicit: While the driver of an automobile can observe the speedometer directly, the pedestrian must derive a car's velocity in other ways; and although the driver is given a speedometer to read, he or she must sense acceleration by other means. In such cases an observer receives a collection of signals in which the desired information is *encoded:* The information itself is obtained only after some reflection, an activity that is really *signal processing.*

Electronic systems are employed to aid in the process of extracting the required information from a received set of signals. However, in order for a signal to be processed by an electronic system, it must first be converted into an electric signal, that is, a voltage or a current. This process is accomplished by devices known as *transducers.* A variety of transducers exist, each suitable for one of the various forms of physical signals. For instance, the sound waves generated by a human can be converted to electric signals using a microphone, which is in effect a pressure transducer. It is not our purpose here to study transducers; rather, we shall assume that the signals of interest already exist in the electrical domain. Our concern will be with the processing of such signals using electronic circuits.

From the discussion above it should be apparent that a signal is a time-varying quantity that can be represented by a graph such as that shown in Fig. 1.1. In fact, the

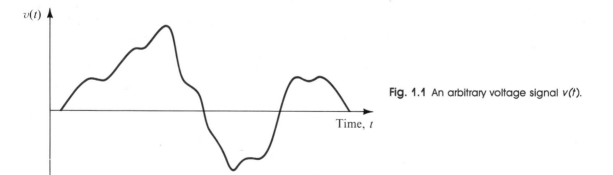

Fig. 1.1 An arbitrary voltage signal v(t).

information content of the signal is represented by the changes in its magnitude as time progresses; that is, the information is contained in the "wiggles" in the signal waveform. In general, such waveforms are difficult to characterize mathematically. In other words, it is not easy to describe succinctly an arbitrary looking waveform such as that of Fig. 1.1. Of course, such a description is of great importance for the purpose of designing appropriate signal-processing circuits that perform desired functions on the given signal.

1.2 FREQUENCY SPECTRUM OF SIGNALS

An extremely useful characterization of a signal, and for that matter of any arbitrary function of time, is in terms of its *frequency spectrum*. Such a description of signals is obtained through the mathematical tools of *Fourier series* and *Fourier transform*.[1] The Fourier series applies only to signal waveforms that are periodic functions of time. The Fourier transform, which is a limiting case of the Fourier series, can be used to obtain the frequency spectrum of nonperiodic functions.

The Fourier series allows us to express a given periodic function of time as the sum of an infinite number of sinusoids whose frequencies are harmonically related. For instance, the symmetrical square-wave signal in Fig. 1.2 can be expressed as

$$v(t) = \frac{4V}{\pi} \left(\sin \omega_0 t + \tfrac{1}{3} \sin 3\omega_0 t + \tfrac{1}{5} \sin 5\omega_0 t + \cdots \right) \tag{1.1}$$

where V is the amplitude of the square wave and $\omega_0 = 2\pi/T$ (T is the period of the square wave) is called the *fundamental frequency*. Note that because the amplitudes of the harmonics progressively decrease, the infinite series can be truncated, with the truncated series providing an approximation to the square waveform.

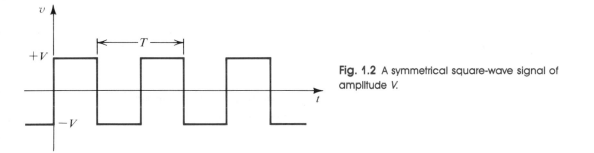

Fig. 1.2 A symmetrical square-wave signal of amplitude V.

The sinusoidal components in the series of Eq. (1.1) constitute the frequency spectrum of the square-wave signal. Such a spectrum can be graphically represented as in Fig. 1.3, where the horizontal axis represents the angular frequency ω in radians per second.

The Fourier transform can be applied to a nonperiodic function of time, such as that depicted in Fig. 1.1, and provides its frequency spectrum as a *continuous*

---

[1] The reader who has not yet studied these topics should not be alarmed. No detailed application of this material will be made until Chapter 11. Nevertheless, an understanding of Section 1.2 should be very helpful when studying earlier parts of the book.

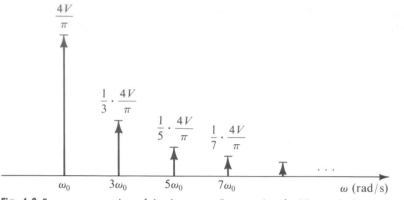

Fig. 1.3 Frequency spectrum (also known as line spectrum) of the periodic square wave of Fig. 1.2.

function of frequency, as indicated in Fig. 1.4. Unlike the case of periodic signals where the spectrum consists of discrete frequencies (at ω_0 and its harmonics), the spectrum of a nonperiodic signal contains in general all possible frequencies. Nevertheless, the essential parts of the spectra of practical signals are usually confined to relatively short segments of the frequency (ω) axis—an observation that is very useful in the processing of such signals. For instance, the spectrum of audible sounds such as speech and music extends from about 20 Hz to about 20 kHz—a frequency range known as the *audio* band. Here we should note that although some musical tones have frequencies above 20 kHz, the human ear is incapable of hearing frequencies that are much above 20 kHz.

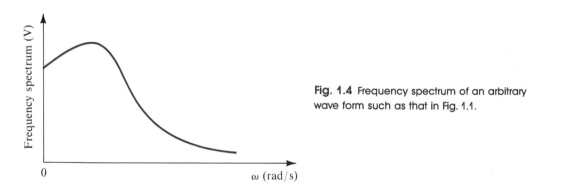

Fig. 1.4 Frequency spectrum of an arbitrary wave form such as that in Fig. 1.1.

The concept of frequency spectrum of signals is very important in the design of communication systems such as radio, TV, and telephone systems.

Exercise

1.1 When the square-wave signal of Fig. 1.2, whose Fourier series is given in Eq. (1.1), is applied to a resistor, the total power dissipated may be calculated directly using the relationship $P = 1/T \int_0^T (v^2/R) \, dt$, or indirectly by summing the contribution of each of the harmonic components, that is, $P = P_1 + P_3 + P_5 + \cdots$, which may be found directly from rms values. (*Hint:* $V_{rms} = V_{peak}/\sqrt{2}$ for a sine wave). Verify that the two approaches are equivalent. What fraction of the energy of a square wave is in its fundamental? In its first five harmonics? In its first seven? First nine? In what number of harmonics is 90% of the energy? (Note that in counting harmonics, the fundamental at ω_0 is the first, the one at $2\omega_0$ is the second, etc.)

Ans. 0.81; 0.93; 0.95; 0.96; 3

1.3 ANALOG AND DIGITAL SIGNALS

The voltage signal depicted in Fig. 1.1 is called an *analog signal*. The name derives from the fact that such a signal is analogous to the physical signal that it represents. The magnitude of an analog signal can take on any value; that is, the amplitude of an analog signal exhibits a continuous variation over its range of activity. The vast majority of signals in the world around us are analog. Electronic circuits that process such signals are known as *analog circuits*. A variety of analog circuits will be studied in this book.

An alternative form of signal representation is that of a sequence of numbers, each number representing the signal magnitude at an instant of time. The resulting signal is called a *digital signal*. To see how a signal can be represented in this form—that is, how signals can be converted from analog to digital form—consider Fig. 1.5a.

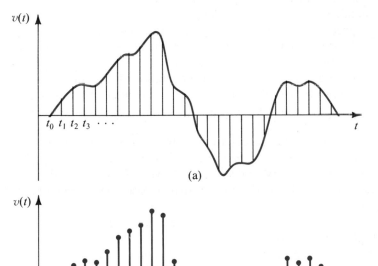

(a)

(b)

Fig. 1.5 Sampling the continuous-time analog signal in **(a)** results in the discrete time signal in **(b)**.

Here the curve represents a voltage signal, identical to that in Fig. 1.1. At equal intervals along the time axis we have marked the time instants t_0, t_1, t_2, and so on. At each of these time instants the magnitude of the signal is measured, a process known as *sampling*. Figure 1.5b shows a representation of the signal of Fig. 1.5a in terms of its samples. The signal of Fig. 1.5b is defined only at the sampling instants; it no longer is a continuous function of time, rather it is a *discrete-time* signal. However, since the magnitude of each sample can take any value in a continuous range, the signal in Fig. 1.5b is still an analog signal.

Now if we represent the magnitude of each of the signal samples in Fig. 1.5b by a number having a finite number of digits, then the signal amplitude will no longer be continuous, rather it is said to be *quantized*, *discretized*, or *digitized*. The resulting digital signal then is simply a sequence of numbers that represent the magnitudes of the successive signal samples. We will discuss the process of signal sampling at a later stage in this book.

Electronic circuits that process digital signals are called *digital circuits*. The digital computer is a system that is constructed of digital circuits. All the internal signals in a digital computer are digital signals. Digital processing of signals has become quite popular primarily because of the tremendous advances made in the design and fabrication of digital circuits. Another reason for the popularity of digital signal processing is that one generally prefers to deal with numbers. For instance, there is little doubt that the majority of us find the digital display of time (as in a digital watch) much more convenient that the analog display (hands moving relative to a graduated dial). While the latter form of display calls for interpretation on the part of the observer, the former is explicit, eliminating any subjective judgment. This is an important point that perhaps is better appreciated in the context of an instrumentation system, such as that for monitoring the status of a nuclear reactor. In such a system human interpretation of instrument readings and the associated inevitable lack of consistency could be hazardous. Furthermore, in such an instrumentation system the measurement results usually have to be fed to a digital computer for further analysis. It would be convenient therefore if the signals obtained by the measuring instruments were already in digital form.

Digital processing of signals is economical and reliable. Furthermore, it allows a wide variety of processing functions to be performed—functions that are either impossible or impractical to implement by analog means. Nevertheless, as already mentioned, most of the signals in the physical world are analog. Also, there remain many signal-processing tasks that are best performed by analog circuits. It follows that a good electronics engineer must be proficient in both forms of signal processing. Such is the philosophy adopted in this text.

Before leaving this discussion of signals we should point out that not all the signals with which electronic systems deal originate in the physical world. For instance, the electronic calculator and the digital computer perform mathematical and logical operations to solve problems. The internal digital signals represent the variables and parameters of these problems and are obviously not supplied directly from external physical signals.

Exercise

1.2 Consider the process of communicating information by samples that are too infrequent. For example, list the sequence of output samples of a system that samples at 5-ms intervals starting at time $t = 0$, when supplied with a 100-Hz, ± 1-V symmetrical square wave that goes positive 2.5 ms earlier. What if the input is a sine wave of ± 1-V amplitude at 100 Hz that has the same phase as the fundamental of the square wave? What if a square wave of 200 Hz starting at -1.25 ms is applied? What if a square wave of 400 Hz starting at -0.625 ms is applied?

Ans. $+1, -1, +1, -1, \ldots; +1, -1, +1, -1, \ldots; +1, +1, +1 \ldots; +1, +1, +1, \ldots$

Note: To convey useful information, sampling must be done at a frequency at least twice that of the highest-frequency component in the signal to be communicated. This is called *Shannon's sampling theorem.*

1.4 AMPLIFICATION AND FILTERING

In this section we shall introduce two fundamental signal-processing functions that are employed in some form in almost every electronic system. These functions are amplification and filtering.

Signal Amplification

From a conceptual point of view the simplest signal processing task is that of *signal amplification*. The need for amplification arises because transducers provide signals that are said to be "weak," that is, in the microvolt (μV) or millivolt (mV) range and possessing little energy. Such signals are too small for reliable processing and processing is much easier if the signal magnitude is made larger. The functional block which accomplishes this task is the *signal amplifier*.

It is appropriate at this point to discuss the need for *linearity* in amplifiers. When amplifying a signal, care must be exercised so that the information contained in the signal is not changed and no new information is introduced. Thus when feeding the signal shown in Fig. 1.1 to an amplifier we want the output signal of the amplifier to be an exact replica of that at the input, except of course for having larger magnitude. In other words, the "wiggles" in the output waveform must be identical to those in the input waveform. Any change in waveform is considered to be *distortion* and is obviously undesirable.

An amplifier that preserves the details of the signal waveform is characterized by the relationship

$$v_o(t) = Av_i(t) \tag{1.2}$$

where v_i and v_o are the input and output signals and A is a constant representing the magnitude of amplification, known as *amplifier gain*. Equation (1.2) is a linear relationship, hence the amplifier it describes is a *linear amplifier*. It should be easy to see that if the relationship between v_o and v_i contains higher powers of v_i then the waveform of v_o will no longer be identical to that of v_i. The amplifier is then said to exhibit *nonlinear distortion*.

The amplifiers discussed so far are primarily intended to operate on very small input signals. Their purpose is to make the signal magnitude larger and therefore are thought of as *voltage amplifiers*. The *preamplifier* in the home stereo system is an example of a voltage amplifier. However, it usually does more than just amplify the signal; specifically it performs some shaping of the frequency spectrum of the input signal. This topic, however, is beyond our need at this moment.

At this time we wish to mention another type of amplifier, namely, the power amplifier. Such an amplifier provides little or no voltage gain but substantial current gain. Thus while absorbing little power from the input signal source to which it is connected, often a preamplifier, it delivers large amounts of power to its load. An example is found in the power amplifier of the home stereo system, whose purpose is to provide sufficient power to drive the loudspeaker. Here we should note that the loudspeaker is the output transducer of the stereo system; it converts the electric output signal of the system into an acoustic signal. A further appreciation of the need for linearity can be acquired by reflecting on the power amplifier. A linear power amplifier causes both soft and loud music passages to be reproduced without distortion.

Signal Filtering

Another common signal-processing task is that of signal filtering. Filtering is the process by which the essential and useful part of a signal is separated from extraneous and undesirable components that are generally referred to as noise and that somehow get mixed with the signal.

Thus a general view of an electric filter is that it separates the signal from noise. Here noise is used to refer to two different things. The first is the extraneous signals (also known as *interference*) that might be picked up by the electronic system under consideration. An example of this is found in the interference caused to a radio or a TV set by the switching action in the mechanism of a home appliance. Such an interfering signal is usually coupled to the radio or TV set via transmission over the power supply wiring in the home. In other cases, however, the interference is coupled directly via electromagnetic radiation.

The second item included in the general category of noise refers to signals generated by the electronic components themselves. Such noise signals are due to the continuous movement of charged particles in the materials of the electronic components. A detailed study of noise in electronic components will not be carried out in this text; the interested reader is referred to Gray and Meyer (1984).[2]

A specific filter example will be discussed in Section 1.5, and the analysis and design of filters will be studied more completely in subsequent chapters. Finally, it should be noted that filtering functions can be implemented by either analog or digital circuits.

---

[2] See Bibliography at the end of the chapter.

1.5 COMMUNICATIONS

Historically, the most important application for electronics has been in the communication of information. It was the goal of information dissemination, exchange, manipulation, and storage that fostered developments in telegraphy, telephony, radio, television, and computers. Accordingly a brief overview of communication systems may serve to introduce concepts that will be found useful in the study of modern electronics.

A Radio Broadcast System

We shall first consider a simple radio broadcast system. Here the information is speech and music, originating from a human source, a record player, or a tape recorder. After transduction and amplification we obtain electric signals of reasonable strength that are ready for transmission to remotely located listeners. However, a problem arises if one attempts to transmit these signals directly. The problem is twofold: First, since the signals are in the audio-frequency range their wavelengths are quite long, and hence the transmitting antenna required to convert the signals to electromagnetic radiation would have to be very long. Second, if all radio stations were to transmit the same frequency band (audio), the remote listener would find it impossible to distinguish between the broadcasts received from the various stations.

A solution to this problem is provided by the process of *modulation*, which allows one to shift the frequency band to be transmitted from the audio range to a location at a much higher frequency [in a range known as the *radio frequency* (RF) range]. In this way the required antenna is of moderate length. Furthermore, since each of the radio stations in a given geographical area is assigned a different location in the radio-frequency band for its transmission, the listener can easily select a desired station by simply *tuning* the receiver to the frequency location at which this particular station is broadcasting.

Amplitude Modulation

To be specific let us consider the simplest form of modulation, namely, *amplitude modulation* (AM). In amplitude-modulation systems each station is assigned a separate frequency, known as the carrier frequency f_c, in the range 535 to 1605 kHz. The carrier wave is a sinusoid that can be described by

$$v_c(t) = \hat{V}_c \cos \omega_c t \tag{1.3}$$

where $\hat{V}_c$ is the carrier amplitude and ω_c is its frequency. Amplitude modulation consists of causing the carrier amplitude to vary in correspondence with the instantaneous magnitude of the audio-frequency signal (called the *modulating signal*). To illustrate, assume that the audio signal consists of a single sinusoidal tone given by

$$v_s(t) = \hat{V}_s \cos \omega_s t \tag{1.4}$$

We have to arrange that the instantaneous carrier amplitude becomes $\hat{V}_c + k\hat{V}_s \cos \omega_s t$, where k is a constant. This results in an amplitude-modulated wave given

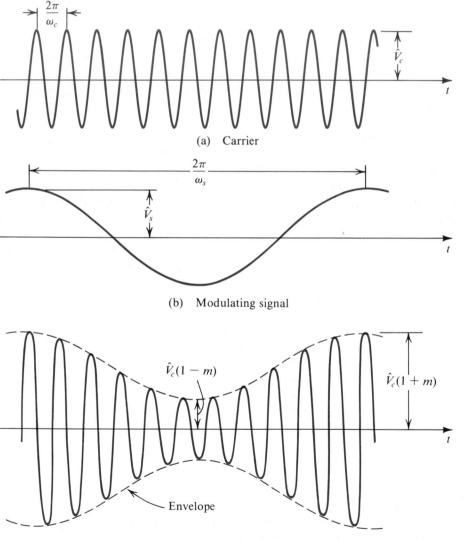

(a) Carrier

(b) Modulating signal

(c) Amplitude modulated (AM) wave

Fig. 1.6 Amplitude-modulating the carrier wave in **(a)** with the signal waveform in **(b)** results in the AM wave in **(c)**.

by

$$v_{AM}(t) = (\hat{V}_c + k\hat{V}_s \cos \omega_s t) \cos \omega_c t$$
$$= \hat{V}_c(1 + m \cos \omega_s t) \cos \omega_c t \qquad (1.5)$$

where $m \equiv k\hat{V}_s/\hat{V}_c$ is called the *modulation depth* or *modulation index*. Figure 1.6 illustrates the process. It shows that the information to be transmitted is contained in the *envelope* of the AM wave. Note that to avoid distortion the modulation depth m must be kept less than or equal to unity.

The AM wave in Eq. (1.5) can be expressed as

$$v_{AM}(t) = \hat{V}_c \cos \omega_c t + \tfrac{1}{2}m\hat{V}_c \cos(\omega_c + \omega_s)t + \tfrac{1}{2}m\hat{V}_c \cos(\omega_c - \omega_s)t \qquad (1.6)$$

Thus the spectrum of the AM wave consists of the carrier at ω_c, a component at $\omega_c + \omega_s$, and another component at $\omega_c - \omega_s$.

An actual audio-frequency signal has a spectrum that occupies a band of frequencies, say from ω_{s1} to ω_{s2}. Correspondingly, the spectrum of the resulting AM wave will consist of the carrier and two bands: the *upper sideband* occupying the range $\omega_c + \omega_{s1}$ to $\omega_c + \omega_{s2}$ and the *lower sideband* occupying the range $\omega_c - \omega_{s2}$ to $\omega_c - \omega_{s1}$. The spectrum therefore extends over the band from $\omega_c - \omega_{s2}$ to $\omega_c + \omega_{s2}$ and is centered around the carrier frequency ω_c, as illustrated in Fig. 1.7. This is the type of spectrum that an AM radio station transmits.

Since all the radio stations in a given geographical location have to share the AM broadcast band (535 to 1,605 kHz), the band allotted to each station is limited to 10 kHz. This means that the highest audio frequency transmitted must be only

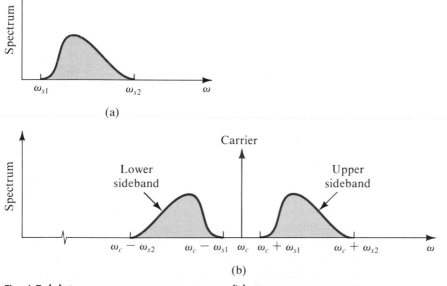

Fig. 1.7 (a) Spectrum of modulating signal. **(b)** Spectrum of AM wave.

5 kHz! This in turn implies that the higher-frequency notes present in music are suppressed, which explains the modest quality observed in music received on AM radio.

The AM Radio Receiver

We shall now discuss the AM radio receiver. The receiver antenna (which can be just a piece of wire), located in the field of the transmitting antenna, will have small induced voltage signals corresponding to the various radio transmissions in the geographical area. The antenna delivers these signals to the first stage of the receiver, whose function is twofold: first to select the signal of the desired station, and second to amplify the received signal to a level suitable for further processing. The first stage therefore consists of an amplifier and a *tunable bandpass filter*. A bandpass filter is a filter that passes signals whose spectrum occupies a specified band and rejects (stops or attenuates) other signals. To receive a particular station, the bandpass filter at the receiver input must be tuned so as to make its *passband* coincide with the spectrum of the AM wave emanating from the selected station.

In order to recover the audio signal, the received AM wave is fed to a *demodulator*, also known as a *detector*. A circuit for AM demodulation will be studied in Chapter 5. The audio output of the demodulator is then amplified and fed to the loudspeaker.

This completes our oversimplified description of one of the simplest and most common communications systems. The reader will note the variety of signal-processing functions embodied in such a system.

Exercise

1.4 An AM broadcast station operating at an assigned frequency of 1,000 kHz broadcasts a pure tone of 1,000 Hz at a level of one-tenth full modulation. Use Eq. (1.6) to find the amplitudes and frequencies of the three components of the AM wave.

Ans. V_c at 1,000 kHz, $0.05V_c$ at 1,001 kHz, and $0.05V_c$ at 999 kHz

Telephone Communications

Telephone communication systems have undergone tremendous development in the last few years. Many of these developments have been a direct result of advances in microelectronics. Currently, the telephone system is used to communicate not only speech but also television programs and digital data. Such *data communications* take place between computer terminals located at widely separated locations and a central computer, as well as between two or more computers. Familiar examples of data communication systems include systems used in banks and for airline reservations. Perhaps less glamorous but extremely useful is the ability to dial up a computer from one's home, connect the telephone via an *acoustic coupler* to a computer terminal, and then proceed to enter one's programs on the terminal. This arrangement was employed for typing and editing a good part of the manuscript for this book.

Advances in microelectronics are also allowing the addition of "intelligence" capabilities to the telephone set. As an example, "smart" telephones that "remember" frequently dialed numbers and that allow the user to reach any one of such numbers by pushing a single button are currently available.

Let us briefly examine the telephone communication system. For the sake of simplicity, let us restrict ourselves to speech communication. In order to make it possible to use a single telephone cable for the transmission of many separate conversations (separated in what is called *telephone channels*) the bandwidth allotted for each channel is limited to about 4 kHz (actually, from 300 to 3,400 Hz). It has been established that this relatively narrow bandwidth is sufficient for the transmission of speech signals and provides speech of reasonable quality. After transduction, the speech signal is fed to a *low-pass filter* whose passband extends to about 4 kHz. This low-pass filter eliminates the higher-frequency components, thus avoiding the possibility of the various channels carried on the same cable "running into one another."

To carry a number of telephone channels on a single cable, modulation is used to shift the spectrum of each channel to a different frequency band. Such a system is known as a *frequency division multiplexing* (FDM) system. At the receiving end the channels are separated using filtering and demodulation.

Another more recent approach to multiplexing a number of telephone channels on a single cable is the use of *time-division multiplexing* (TDM). In TDM the signals from the channels to be multiplexed (say 12 channels) are sampled in sequence. That is, a sample is taken from Channel 1, then a sample taken from Channel 2, and so on to the last channel, Channel 12. Then a new round of sampling is started and the process is periodically repeated. The samples are transmitted on the same cable in the same order in which they were obtained. Thus samples 1, 13, 25, etc. belong to the signal of Channel 1, and so on.

At the receiving end, the samples have to be separated, a process known as *demultiplexing*. The samples of each particular channel are then used to reconstruct a good approximation to the audio signal being transmitted.

TDM systems usually convert the signal samples to digital form, and the signals transmitted are actually digits. Due to the advances made in the digital circuits area and to the fact that the telephone network is carrying more and more digital data, the current trend is to perform a great deal of the processing digitally. It is probable therefore that in the not-too-distant future each telephone will contain its own analog-to-digital (A/D) and digital-to-analog (D/A) converters. Nevertheless, analog circuits will still be needed for processing the speech signals prior to converting to digital form and for reconstructing an analog signal from the received digital information.

Exercise

1.6 To preserve adequate information in a TDM telephone system, signal sampling must be done at least at twice the upper frequency of the 300- to 3,400-Hz channel. Available technology allows a sample amplitude to be transmitted in 10 μs. What is the greatest number of channels that can be multiplexed on a single line in such a system?

Ans. 14

1.6 COMPUTERS

The digital computer has already been mentioned on a number of occasions in this introductory chapter, especially in connection with the rapidly growing field of data communication systems. The digital computer will also be mentioned in Section 1.7

because of the key role it is playing in instrumentation and control systems. In fact, computers from microprocessors to large machines are playing a role in almost every aspect of our daily lives. It is therefore futile to attempt to list the variety of tasks in which computers are employed. Rather, we wish first to point out the close relationship between developments in microelectronics and those in computers. These two areas have greatly influenced each other and will likely continue to do so. Although the original digital computers used vacuum tubes, it was the invention of the transistor nearly 40 years ago and the silicon integrated circuit a decade later that made possible the digital computers that we know today. In turn, the demand for digital computers that are cheaper, smaller, faster, more powerful, and so on, has accelerated the development of microelectronics. It is also important to note that at the present time it would be almost impossible to design complex integrated circuits without the aid of the computer (see Appendix C).

The development of the microprocessor about 14 years ago has to be regarded as the beginning of a new era in the application of microelectronics. For the first time the concept of distributed computing or *distributed intelligence* has become feasible. By including microprocessors in a variety of devices, such as instruments and computer terminals, such devices acquire a degree of intelligence or "smartness" that enables them to do considerable processing. If desired, these local processors can be made to communicate among themselves as well as with a more powerful central computer for further processing and higher-level decision making.

We conclude this section with a discussion of the oversimplified block diagram of a digital computer, shown in Fig. 1.8. The digital computer consists of three basic

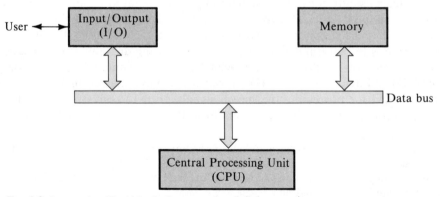

Fig. 1.8 An oversimplified block diagram of a digital computer.

blocks: The *input/output* (I/O) devices, which enable the user to interact with the machine, to enter programs and data, and to obtain results; the *memory*, in which the program and data are stored; and the *central processing unit* (CPU), which is the heart of the machine, where arithmetic and logical operations are performed and the various control signals are generated.

Under the control of a program whose instructions reside in memory, the CPU operates on the input data, which is also stored in memory. As a result of these arithmetic and logical operations, decisions are made regarding, for instance, the sequence of executing program instructions. Ultimately, output data are generated.

The required I/O devices such as video terminals, magnetic tape units, printers, plotters, and so on, as well as digital computers themselves and other digital systems, are constructed of two digital circuit types: logic gates and memory circuits. We shall study both logic circuits in detail in subsequent chapters.

1.7 INSTRUMENTATION AND CONTROL

Another area in which microelectronics is playing a key role is that of measurement and control. The measurement of a variety of phenomena is the key to the automatic control of variables and processes ranging from the temperature in our homes to the operation of industrial plants.

An instrumentation system can be as simple as just one instrument measuring a variable of interest and displaying the result as the reading of an analog or a digital meter. In such a system it is left to the human observer to react to the measurement result and make adjustments to the process being monitored, if such adjustments are called for.

A further level of sophistication is achieved by having the instrument compare the result of its measurement with a preset value or range of values. If the measured value is found to be outside the preset range, a "flag" is automatically raised; for instance, an alarm lamp is lit. Then again it is up to the human observer to take corrective action.

An even greater level of sophistication is obtained by having the instrumentation system determine what kind of corrective action is required, initiate such action, and supervise its completion. A simple example of such a system is found in the home thermostat, a device no doubt familiar to most readers. Such an *automatic control system* is based on a *negative-feedback loop*. The loop consists essentially of a *sensor* (transducer), a *processor* that today is most often electronic, and an *actuator*, which responds to the commands of the processor and causes the corrective action to take place physically.

Traditionally, the processor in an instrumentation system has been analog, with a growing use of separate digital logic and memory circuits. Only the most sophisticated systems employ a central computer to do the bulk of the processing. However, the advent of microprocessors and other advances in analog and digital microelectronic circuits is causing a major change to occur in instrumentation systems. Each sensor can now be equipped with sophisticated analog circuits as well as a microprocessor, thus enabling the sensor to do considerable processing locally. Such a "smart" sensor would also be capable of making local decisions. It would then need to communicate only infrequently with a central computer. The central computer would therefore be relieved of routine processing and be able to implement more extensive programs of automation. Some such automatic control systems are already available; witness the use of robots in the manufacture of automobiles.

Advances in microelectronics are also influencing such routine laboratory instruments as the oscilloscope, the function generator, the spectrum analyzer, the digital multimeter, and so on. Newer versions of these instruments feature such facilities as automatic adjustment, improved display, self-calibration, and others. The result is an instrument that is much easier to use and that relies less on the skill and attentiveness of the human operator. In addition, such features permit many more measurements to be made in a given length of time.

Finally, there is the area of *automatic testing*, an essential aspect of an assembly line. This area also has benefited greatly from advances in microelectronics and specifically those in digital computer technology. For further reading on the influence of microelectronics on instrumentation the reader is urged to consult Oliver (1977).

In conclusion, the area of instrumentation and control is a fertile one for the application of the latest developments in analog and digital microelectronic circuits.

1.8 CONCLUDING REMARKS

The focus of this chapter has been on signals and signal processing. The electronic system examples discussed are intended to illustrate the variety of signal-processing tasks with which electronics engineers deal. We have attempted also to illustrate the role played by microelectronics in at least three broad-ranging areas: communications, computers, and instrumentation and control.

It is not the purpose of this book to study the design of any particular kind of electronic system (such as a radio, a TV, or a computer). Rather, we shall study the principles of design and analysis of electronic circuits that realize many of the signal-processing functions needed in a variety of electronic systems. The circuits treated are both analog and digital, discrete and integrated. Although our approach is not to provide recipes for designing specific circuits, it is very down to earth. It is our goal that the reader develop the abilities and sharpen the skills required to innovate and create in this exciting field.

The reader who has not already done so is encouraged to read the overview of the text given in the Preface. Such an effort will prove useful before delving further into this book.

BIBLIOGRAPHY

P. R. Gray and R. G. Meyer, *Analysis and Design of Analog Integrated Circuits*, 2nd ed., Chap. 11, New York: Wiley, 1984.

M. E. Van Valkenburg, *Network Analysis*, 3rd ed., Englewood Cliffs, N.J.: Prentice-Hall, 1974. (The Fourier series and the Fourier transform are studied in Chaps. 15 and 16.)

B. M. Oliver, "The role of microelectronics in instrumentation and control," *Scientific American*, vol. 237, no. 3, pp. 180–190, Sept. 1977.

Scientific American, special issue on microelectronics, Sept. 1977.

PROBLEMS

1.1 Illustrate the composition of a square-wave signal by sketching the first four terms of the series given in Eq. (1.1) and then performing a graphical summation.

1.2 For a square-wave audio signal of 10 kHz, what fraction of the available energy is perceived by an average adult listener of age 40 whose hearing extends only to 16 kHz?

1.3 Numerically evaluate the series expansion for a square wave given in Eq. (1.1) truncated after four terms at $t = T/8$, $T/4$, $5T/8$, and $3T/4$.

1.4 What fraction of the energy contained in a square wave of frequency f and peak-to-peak amplitude 2 V is contained in the harmonic at frequency $9f$?

1.5 An amplifier, such as the one in Exercise 1.3, has a voltage gain of 1,000 and limits the output to ± 10 V. Sketch the output waveform when the input is a 1-V amplitude, 1-kHz frequency sine-wave signal. Convince yourself that the output is a good approximation to a square wave. Hence, with the aid of Eq. (1.1), find the "effective gain" of the amplifier for the component at 1 kHz.

1.6 In Problem 1.5 the components of the output at frequencies other than the fundamental constitute distortion whose rms value is the square root of the sum of the squares of the rms values of the components. Calculate the rms value of (a) the signal component at 1 kHz and (b) all other components up to 10 kHz. Find the ratio of (b) to (a) expressed as a percentage (it is called the percent total harmonic distortion).

1.7 The output voltage of an amplifier is given by

$$v_o = 100v_i + 50v_i^2$$

For $v_i = V_i \sin \omega t$ show that the output will consist of the desired component of frequency ω in addition to a dc component and a component of frequency 2ω. The ratio of the amplitude of the second-harmonic component to the amplitude of the fundamental, expressed as a percentage, is called the second-harmonic distortion.

(a) If it is desired to limit the second-harmonic distortion to 1%, find the largest allowable amplitude of v_i.

(b) Repeat for 10%.

(c) Find the second-harmonic distortion obtained when the desired output component has a 10-V amplitude.

(d) If the output of this amplifier is fed to the input of another amplifier which is perfectly linear and having a gain of 2, find the second-harmonic distortion obtained at the output of the new amplifier when the desired output component has a 10-V amplitude.

Chapter 2

Linear Circuits

INTRODUCTION

Electronic circuits can be broadly divided into two categories: *linear circuits* and *nonlinear circuits*. Filters and amplifiers are examples of linear circuits. Modulators and rectifiers (used to convert ac to dc) are examples of nonlinear circuits. A very important class of nonlinear circuits is that of *digital circuits*, which include logic and memory elements.

We begin this chapter by making the distinction between linear and nonlinear circuits. We then study the terminal characteristics of the various types of linear amplifiers. The rest of the chapter is devoted to the study of important circuit analysis techniques. Although a good part of this material is a review, the objective here is to demonstrate and illustrate specific applications of these techniques in electronics. As will become apparent throughout this book, an attribute of a "good" circuit designer is the ability to perform rapid circuit analysis.

In addition to reviewing relevant topics from linear circuit theory, this chapter introduces a number of new concepts and establishes terminology and conventions used throughout the text. □

2.1 LINEAR AND NONLINEAR ONE-PORT NETWORKS

Figure 2.1a shows a two-terminal, or one-port network labeled A. The *total instantaneous voltage* between the two terminals is denoted $v_A(t)$, and the *total instantaneous current* through the terminals is denoted $i_A(t)$. The static *i-v* characteristic of this one-port network can be measured point by point by first applying a constant

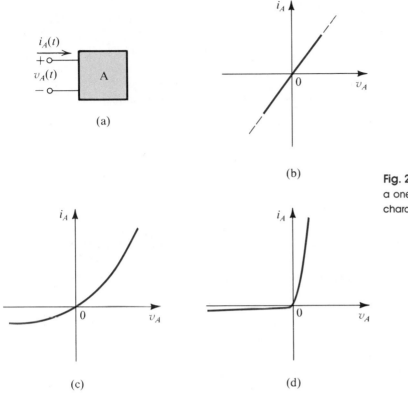

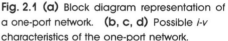

Fig. 2.1 (a) Block diagram representation of a one-port network. (b, c, d) Possible *i-v* characteristics of the one-port network.

(dc) voltage V_A and measuring the corresponding current I_A; then the value of V_A is changed and a new value of I_A is measured, and so on. The resulting characteristic may take one of the shapes shown in Figs. 2.1b to 2.1d.

The *i-v* characteristic in Fig. 2.1b is a straight line and thus the corresponding circuit is said to be linear. An ideal resistor has such an *i-v* relationship. On the other hand, the *i-v* characteristic in Fig. 2.1c is clearly nonlinear and thus belongs to a nonlinear circuit. The *i-v* characteristic in Fig. 2.1d also is nonlinear, exhibiting a very sharp change in slope at the origin. It will be seen in Chapter 4 that silicon diodes exhibit such a nonlinear *i-v* relationship.

Because the diode is a very important circuit element we shall use it here for further illustrating nonlinear circuits. This material may be considered a preview of the study of diodes in Chapter 4 and nonlinear circuit applications in Chapter 5. Figure 2.2a shows the circuit symbol of the diode and Fig. 2.2b shows an idealized approximation of its *i-v* characteristic of Fig. 2.1d. Note specifically that the characteristic of Fig. 2.1d indicates that the diode can conduct large positive currents while exhibiting a very small positive voltage drop. This is the *forward direction* of the diode and corresponds to the direction of its arrowlike circuit symbol in Fig. 2.2a. In the idealization of Fig. 2.2b this forward conduction property is exaggerated; the diode is shown to conduct any amount of positive current while exhibiting zero voltage drop.

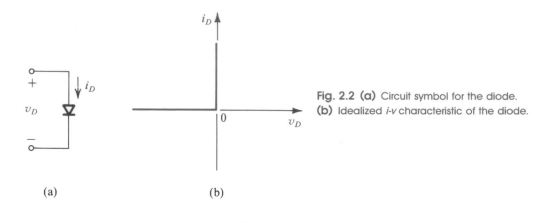

(a)   (b)

Fig. 2.2 (a) Circuit symbol for the diode.
(b) Idealized *i-v* characteristic of the diode.

Consider once more the diode *i-v* characteristic in Fig. 2.1d. Negative (reverse) voltages are shown to cause very little reverse current through the diode. This property also is exaggerated in the idealized characteristic in Fig. 2.2b; the diode is shown to conduct zero current in the *reverse direction*, independently of how large a reverse voltage is applied.

Nonlinear circuits have many varied and useful applications. As an example consider the circuit of Fig. 2.3a. It consists of the series connection of a diode D and a resistor R. Let the input voltage v_I be the sinusoid shown in Fig. 2.3b and assume the diode to be ideal. During the positive half-cycles of the input sinusoid, the positive v_I will cause current to flow through the diode in its forward direction. It follows that the diode voltage v_D will be very small—ideally zero. Thus the output voltage

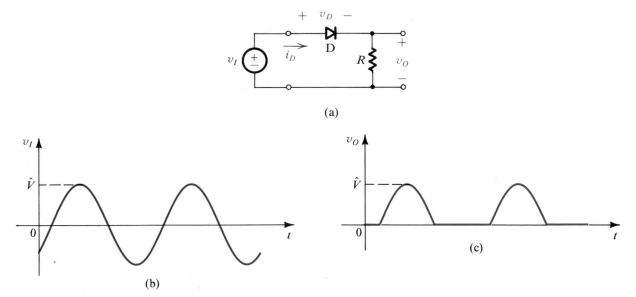

Fig. 2.3 (a) Rectifier circuit. **(b)** Input waveform. **(c)** Output waveform.

v_O will be equal to the input voltage v_I. On the other hand, during the negative half-cycles of v_I, the diode will not conduct and v_O will be zero. Thus, the output voltage will have the waveform shown in Fig. 2.3c. Note that while v_I alternates in polarity and has a zero average value, v_O is unidirectional and has a finite average value or a dc component. Thus the circuit of Fig. 2.3a *rectifies* the signal and hence is called a *rectifier*. It can be used to generate dc from ac. We will study rectifier circuits in detail in Chapter 5.

Most electronic devices exhibit nonlinearities of various forms. Such nonlinearities can be profitably utilized as illustrated in the example above. There are many occasions, however, where one desires to use these devices as linear circuit elements. In the next section we shall study a common technique for obtaining linear performance from devices that are basically nonlinear.

Exercises

2.1 For the circuit in Fig. 2.3a sketch the transfer characteristic v_O versus v_I.

Ans. See Fig. E2.1

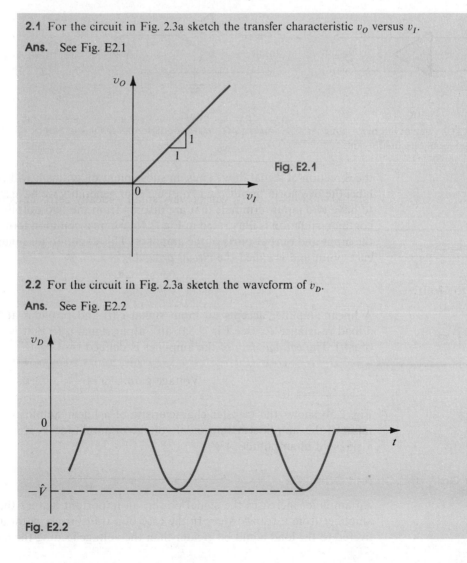

Fig. E2.1

2.2 For the circuit in Fig. 2.3a sketch the waveform of v_D.

Ans. See Fig. E2.2

Fig. E2.2

2.2 AMPLIFIERS

We now turn our attention to two-port networks, and in particular to a very important class of two-port networks: namely, amplifiers. A two-port circuit whose function is to amplify input signals is conveniently represented by the circuit symbol of Fig. 2.4a. This symbol clearly distinguishes the input and output ports and indicates

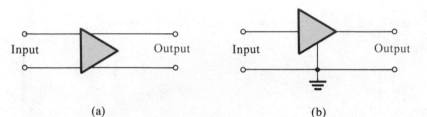

Input	Output	Input	Output
(a)		(b)	

Fig. 2.4 (a) Circuit symbol for amplifier. **(b)** An amplifier with a common terminal (ground) between the input and output ports.

the direction of signal flow. Thus, in subsequent diagrams it will not be necessary to label the two ports "input" and "output." For generality we have shown the amplifier to have two input terminals that are distinct from the two output terminals. A more common situation is illustrated in Fig. 2.4b, where a common terminal exists between the input and output ports of the amplifier. This common terminal is used as a reference point and is called the *circuit ground*.

Voltage Gain

A linear amplifier accepts an input signal $v_I(t)$ and provides at the output, across a load resistance R_L (see Fig. 2.5a), an output signal $v_O(t)$ that is a magnified replica of $v_I(t)$. The *voltage gain* of the amplifier is defined by

$$\text{Voltage gain } (A_v) \equiv \frac{v_O}{v_I} \tag{2.1}$$

Fig. 2.5b shows the transfer characteristic of a linear amplifier. If we apply to the input of this amplifier a sinusoidal voltage of amplitude $\hat{V}$, we obtain at the output a sinusoid of amplitude $A_v\hat{V}$.

Power Gain and Current Gain

An amplifier increases the signal power, an important feature that distinguishes an amplifier from a transformer. In the case of a transformer, although the voltage delivered to the load could be greater than the voltage feeding the input side (primary),

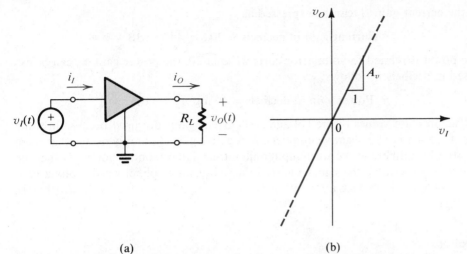

(a)   (b)

Fig. 2.5 (a) A voltage amplifier fed with a signal $v_I(t)$ and connected to a load resistance R_L.
(b) Transfer characteristic of a linear voltage amplifier with voltage gain A_v.

the power delivered to the load is less than or at most equal to the power supplied by the signal source. On the other hand, an amplifier provides the load with power greater than that obtained from the signal source. That is, amplifiers have power gain. The *power gain* of the amplifier in Fig. 2.5a is defined as

$$\text{Power gain } (A_p) \equiv \frac{\text{load power } (P_L)}{\text{input power } (P_I)} \tag{2.2}$$

$$= \frac{v_O i_O}{v_I i_I} \tag{2.3}$$

where i_O is the current that the amplifier delivers to the load (R_L), $i_O = v_O/R_L$, and i_I is the current the amplifier draws from the signal source. The *current gain* of the amplifier is defined as

$$\text{Current gain } (A_i) \equiv \frac{i_O}{i_I} \tag{2.4}$$

From Eqs. (2.1) to (2.4) we note that

$$A_p = A_v A_i \tag{2.5}$$

Expressing Gain in Decibels

The amplifier gains defined above are ratios of similarly dimensioned quantities. Thus they will be expressed either as dimensionless numbers or, for emphasis, as V/V for the voltage gain, A/A for the current gain, and W/W for the power gain. Alternately, for a number of reasons, some of them historic, electronics engineers express amplifier gain with a logarithmic measure. Specifically the voltage gain A_v can be expressed as

$$\text{Voltage gain in decibels} = 20 \log|A_v| \quad \text{dB}$$

and the current gain A_i can be expressed as

$$\text{Current gain in decibels} = 20 \log|A_i| \quad \text{dB}$$

Since power is related to voltage (or current) squared, the power gain A_p can be expressed in decibels as follows:

$$\text{Power gain in decibels} = 10 \log A_p \quad \text{dB}$$

The absolute values of the voltage and current gains are used because in some cases A_v or A_i may be negative numbers. A negative gain A_v simply means that there is a 180° phase difference between input and output signals; it does not imply that the amplifier is *attenuating* the signal. On the other hand, an amplifier whose voltage gain is, say, -20 dB is in fact attenuating the input signal by a factor of 10 (that is, $A_v = 0.1$).

The Amplifier Power Supplies

Since the power delivered to the load is greater than the power drawn from the signal source, the question arises as to the source of this additional power. The answer is found by observing that amplifiers need dc power supplies for their operation. These dc sources supply the extra power delivered to the load as well as any power that might be dissipated in the internal circuit of the amplifier (such power is converted to heat). In Fig. 2.5a we have not explicitly shown these dc sources.

Figure 2.6a shows an amplifier that requires two dc sources: one positive of value V_1 and one negative of value V_2. The amplifier has two terminals, labeled V^+ and V^-,

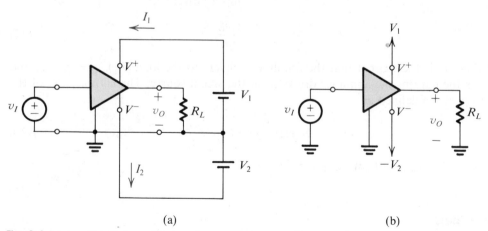

(a)   (b)

Fig. 2.6 An amplifier that requires two dc supplies for operation.

for connection to the dc supplies. For the amplifier to operate, the terminal labeled V^+ has to be connected to the positive side of a dc source whose voltage is V_1 and whose negative side is connected to the circuit ground. Also, the terminal labeled V^- has to be connected to the negative side of a dc source whose voltage is V_2 and whose positive side is connected to the circuit ground. Now, if the current drawn from the positive supply is denoted I_1 and that from the negative supply is I_2 (see Fig. 2.6a), then the dc power delivered to the amplifier is

$$P_{dc} = V_1 I_1 + V_2 I_2$$

If the power dissipated in the amplifier circuit is denoted $P_{dissipated}$, the power balance equation for the amplifier can be written as

$$P_{dc} + P_I = P_L + P_{dissipated}$$

Since the power drawn from the signal source is usually small, the amplifier *efficiency* is defined as

$$\eta \equiv \frac{P_L}{P_{dc}} \times 100 \qquad (2.6)$$

The power efficiency is an important performance parameter for amplifiers that handle large amounts of power. Such amplifiers, called power amplifiers, are used, for example, as output amplifiers of stereo systems.

In order to simplify circuit diagrams, we shall adopt the convention illustrated in Fig. 2.6b. Here the V^+ terminal is shown connected to an arrowhead pointing upward and the V^- terminal to an arrowhead pointing downward. Next to each arrowhead the corresponding voltage is indicated. Note that in many cases we will not explicitly show the connections of the amplifier to the dc power sources. Finally, we note that some amplifiers require only one power supply.

EXAMPLE 2.1

Consider an amplifier operating from ± 10-V power supplies. It is fed with a sinusoidal voltage having 1 V peak and delivers a sinusoidal voltage output of 5 V peak to a 1-kΩ load. The amplifier draws a current of 1 mA from the positive supply and 0.5 mA from the negative supply. The input current of the amplifier is found to be sinusoidal with 0.1 mA peak. Find the voltage gain, the current gain, the power gain, the power drawn from the dc supplies, the power dissipated in the amplifier, and the amplifier efficiency.

Solution

$$A_v = \frac{5}{1} = 5 \text{ V/V}$$

or

$$A_v = 20 \log 5 \simeq 14 \text{ dB}$$

$$\hat{I}_o = \frac{5 \text{ V}}{1 \text{ k}\Omega} = 5 \text{ mA}$$

$$A_i = \frac{\hat{I}_o}{\hat{I}_i} = \frac{5}{0.1} = 50 \text{ A/A}$$

or

$$A_i = 20 \log 50 = 34 \text{ dB}$$

$$P_L = V_{o_{\mathrm{rms}}} I_{o_{\mathrm{rms}}}$$

$$= \frac{5}{\sqrt{2}} \frac{5}{\sqrt{2}} = 12.5 \text{ mW}$$

$$P_I = V_{i_{\mathrm{rms}}} I_{i_{\mathrm{rms}}} = \frac{1}{\sqrt{2}} \frac{0.1}{\sqrt{2}} = 0.05 \text{ mW}$$

$$A_p = \frac{P_L}{P_I} = \frac{12.5}{0.05} = 250 \text{ W/W}$$

or

$$A_p = 10 \log 250 = 24 \text{ dB}$$

$$P_{\mathrm{dc}} = 10 \times 1 + 10 \times 0.5 = 15 \text{ mW}$$

$$P_{\mathrm{dissipated}} = P_{\mathrm{dc}} + P_I - P_L$$

$$= 15 + 0.05 - 12.5 = 2.55 \text{ mW}$$

$$\eta = \frac{P_L}{P_{\mathrm{dc}}} \times 100 = 83.3\%$$

Amplifier Saturation

The amplifier transfer characteristic remains linear over only a limited range of input and output voltages. For an amplifier operated from two power supplies the output voltage cannot exceed a specified positive limit and cannot decrease below a specified negative limit. The resulting transfer characteristic is shown in Fig. 2.7, with the positive and negative saturation levels denoted L_+ and L_-, respectively. Each of the two saturation levels is usually within 1 or 2 volts of the voltage of the corresponding power supply.

Obviously, in order to avoid distorting the output signal waveform, the input signal swing must be kept within the linear range of operation,

$$\frac{L_-}{A_v} \leq v_I \leq \frac{L_+}{A_v}$$

Figure 2.7 shows two input waveforms and the corresponding output waveforms. We note that the peaks of the larger waveform have been clipped off because of amplifier saturation.

Nonlinear Transfer Characteristics and Biasing

Except for the output saturation effect discussed above, the amplifier transfer characteristics have been assumed to be perfectly linear. In practical amplifiers the transfer characteristic may exhibit nonlinearities of various magnitudes, depending on how elaborate the amplifier circuit is and how much effort has been expended in the design to ensure linear operation. Consider as an example the transfer characteristic

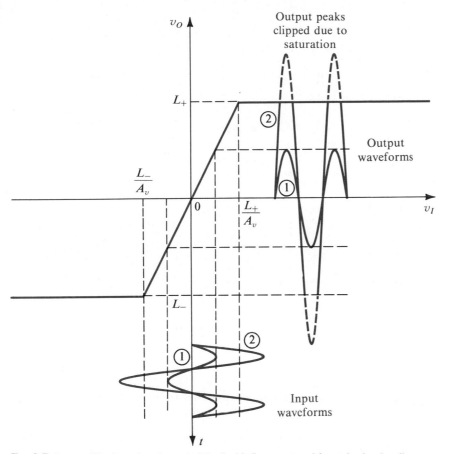

Fig. 2.7 An amplifier transfer characteristic that is linear except for output saturation.

depicted in Fig. 2.8. Such a characteristic is typical of simple amplifiers that are operated from a single (positive) power supply. The transfer characteristic is obviously nonlinear and, because of the single-supply operation, is not centered around the origin. Fortunately, a simple technique exists for obtaining linear amplification from an amplifier with such a nonlinear transfer characteristic.

The technique consists of first *biasing* the circuit to operate at a point near the middle of the transfer characteristic. This is achieved by applying a dc voltage V_I, as indicated in Fig. 2.8, where the operating point is labeled Q and the corresponding dc voltage at the output is V_O. The point Q is known as the quiescent point, the dc bias point, or simply the operating point. The time-varying signal to be amplified, $v_i(t)$, is then superimposed on the dc bias voltage V_I as indicated in Fig. 2.8. Now, as the total instantaneous input $v_I(t)$,

$$v_I(t) = V_I + v_i(t)$$

varies around V_I, the instantaneous operating point moves up and down the transfer curve around the operating point Q. In this way, one can determine the waveform

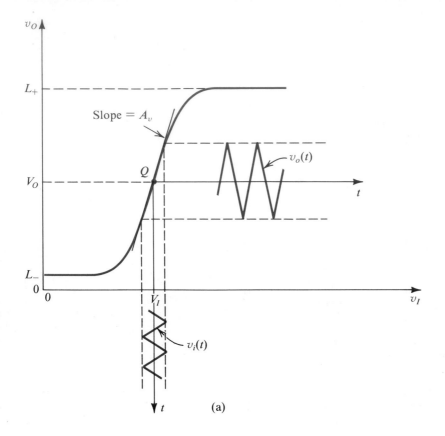

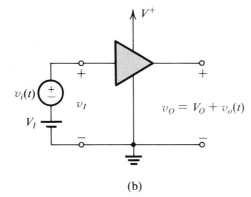

Fig. 2.8 (a) An amplifier transfer characteristic that shows considerable nonlinearity. **(b)** To obtain linear operation the amplifier is biased as shown and the signal amplitude is kept small.

of the total instantaneous output voltage $v_O(t)$. It can be seen that by keeping the amplitude of $v_i(t)$ sufficiently small, the instantaneous operating point can be confined to an almost linear segment of the transfer curve centered about Q. This in turn results in the time-varying portion of the output being proportional to $v_i(t)$; that is,

$$v_O(t) = V_O + v_o(t)$$

with

$$v_o(t) = A_v v_i(t)$$

where A_v is the slope of the almost linear segment of the transfer curve; that is,

$$A_v = \frac{dv_O}{dv_I}\bigg|_{\text{at } Q}$$

In this manner linear amplification is achieved. Of course, there is a limitation: the input signal must be kept sufficiently small. Increasing the amplitude of the input signal can cause the operation to be no longer restricted to an almost linear segment of the transfer curve. This in turn results in a distorted output signal waveform. Such nonlinear distortion is undesirable: the output signal contains additional spurious information that is not part of the input. We shall use this biasing technique and the associated small-signal approximation frequently in the design of transistor amplifiers.

EXAMPLE 2.2

A transistor amplifier has the transfer characteristic

$$v_O = 10 - 10^{-11} e^{40 v_I} \tag{2.7}$$

which applies for $v_I \geq 0$ V and $v_O \geq 0.3$ V. Find the limits L_- and L_+ and the corresponding values of v_I. Also, find the value of the dc bias voltage V_I that results in $V_O = 5V$ and the voltage gain at the corresponding operating point.

Solution
The limit L_- is obviously 0.3 V. The corresponding value of v_I is obtained by substituting $v_O = 0.3$ V in Eq. (2.7); that is,

$$v_I = 0.690 \text{ V}$$

The limit L_+ is determined by $v_I = 0$ and is thus given by

$$L_+ = 10 - 10^{-11} \simeq 10 \text{ V}$$

To bias the device so that $V_O = 5$ V we require a dc input V_I whose value is obtained by substituting $v_O = 5$ V in Eq. (2.7) as follows:

$$V_I = 0.673 \text{ V}$$

The gain at the operating point is obtained by evaluating the derivative dv_O/dv_I at $v_I = 0.673$ V. The result is

$$A_v = -200 \text{ V/V}$$

A sketch of the amplifier transfer characteristic (not to scale) is shown in Fig. 2.9.

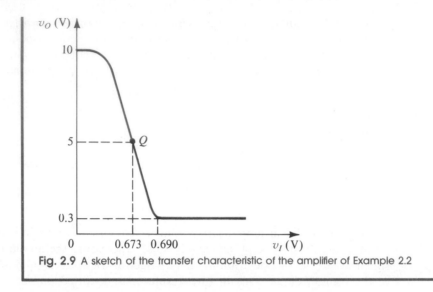

Fig. 2.9 A sketch of the transfer characteristic of the amplifier of Example 2.2

At this point we draw the reader's attention to the terminology used in the above. Total instantaneous quantities are denoted by a lowercase symbol with an uppercase subscript, for example, $i_A(t)$, $v_C(t)$. Direct-current (dc) quantities will be denoted by an uppercase symbol with an uppercase subscript, for example, I_A, V_C. Finally, incremental signal quantities will be denoted by a lowercase symbol with a lowercase subscript, for example $i_a(t)$, $v_c(t)$.

Once an amplifier is properly biased and the input signal is kept sufficiently small, the operation is assumed to be linear. We can then employ the techniques of linear circuit analysis to analyze the signal operation of the amplifier circuit. The remainder of this chapter provides a review and application of these analysis techniques.

Exercises

2.4 An amplifier has a voltage gain of 100 V/V and a current gain of 1,000 A/A. Express the voltage and current gains in decibels and find the power gain.

Ans. 40 dB; 60 dB; 50 dB

2.5 An amplifier operating from a single 15-V supply provides a 10-V rms signal to a 1-kΩ load, and draws negligible input current from the signal source. The dc current drawn from the 15-V supply is 9 mA. What is the power dissipated in the amplifier and what is the amplifier efficiency?

Ans. 35 mW; 74.1%

2.6 The object of this exercise is to investigate the limitation of the small-signal approximation. Consider the amplifier of Example 2.2 with a positive input signal of 2 mV superimposed on the dc bias voltage V_I. Find the corresponding signal at the output:
(a) Assume the amplifier is linear around the operating point, that is, use the value of gain evaluated in Example 2.2.

2.3 CIRCUIT MODELS FOR AMPLIFIERS

A good part of this book is concerned with the design of amplifier circuits using transistors of various types. Such circuits will vary in complexity from those using a single transistor to those with 20 or more devices. In order to be able to apply the resulting amplifier circuit as a building block in a system, one must be able to characterize, or model, its terminal behavior. In this section we study simple but effective amplifier models. These models apply irrespective of the complexity of the internal circuit of the amplifier. The values of the model parameters can be found either by analyzing the amplifier circuit or by performing measurements at the amplifier terminals.

Voltage Amplifiers

Figure 2.10a shows a circuit model for the voltage amplifier. The model consists of a voltage-controlled voltage source having a gain factor A_{vo}, an input resistance R_i that accounts for the fact that the amplifier draws an input current from the signal source, and an output resistance R_o that accounts for the change in output voltage as the amplifier is called upon to supply output current to a load. To be specific, we

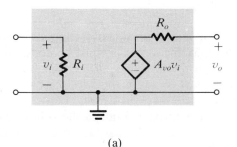

(a)

Fig. 2.10 (a) Circuit model for the voltage amplifier. **(b)** The voltage amplifier with input signal source and load.

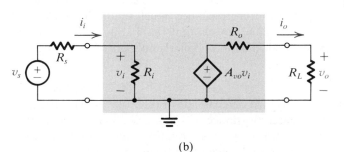

(b)

show in Fig. 2.10b the amplifier model fed with a signal voltage source v_s having a resistance R_s and connected at the output to a load resistance R_L. The non-zero output resistance R_o causes only a fraction of $A_{vo}v_i$ to appear across the output. Using the voltage divider rule we obtain

$$v_o = A_{vo}v_i \frac{R_L}{R_L + R_o}$$

Thus the voltage gain is given by

$$A_v \equiv \frac{v_o}{v_i} = A_{vo} \frac{R_L}{R_L + R_o} \tag{2.8}$$

It follows that in order not to lose gain in coupling the amplifier output to a load, the output resistance R_o should be much smaller than the load resistance R_L. In other words, for a given R_L one must design the amplifier so that its R_o is much smaller than R_L. An ideal voltage amplifier is one with $R_o = 0$. Equation (2.8) indicates also that for $R_L = \infty$, $A_v = A_{vo}$. Thus A_{vo} is the voltage gain of the unloaded amplifier or the *open-circuit voltage gain*. It should also be clear that in specifying the voltage gain of an amplifier, one must also specify the value of load resistance at which this gain is measured or calculated. If a load resistance is not specified, it is normally assumed that the given voltage gain is the open-circuit gain A_{vo}.

The finite input resistance R_i introduces another voltage-divider action at the input, with the result that only a fraction of the source signal v_s actually reaches the input terminals of the amplifier; that is,

$$v_i = v_s \frac{R_i}{R_i + R_s} \tag{2.9}$$

It follows that in order not to lose a significant portion of the input signal in coupling the signal source to the amplifier input, the amplifier must be designed to have an input resistance R_i much greater than the resistance of the signal source, $R_i \gg R_s$. An ideal voltage amplifier is one with $R_i = \infty$. In this ideal case both the current gain and power gain become infinite.

There are situations in which one is interested not in the voltage gain but in a significant power gain. For instance, the source signal can be of a respectable voltage but the source resistance can be much greater than the load resistance. Connecting the source directly to the load would result in significant signal attenuation. In such a case one requires an amplifier with a high input resistance (much greater than the source resistance) and a low output resistance (much smaller than the load resistance) but with a modest voltage gain (or even unity gain). Such an amplifier is referred to as a *buffer amplifier*. We shall encounter buffer amplifiers often throughout this text.

EXAMPLE 2.3

Figure 2.11 depicts an amplifier composed of a cascade of three stages. The amplifier is fed by a signal source with a source resistance of 100 kΩ and delivers its output into a load resistance of 100 Ω. The first stage has a relatively high input resistance and a modest gain factor of 10. The second stage has a higher gain factor but lower

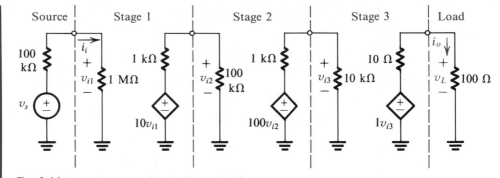

Fig. 2.11 Three-stage amplifier for Example 2.3.

input resistance. Finally, the last, or output, stage has unity gain but a low output resistance. We wish to evaluate the overall voltage gain, that is v_L/v_s, the current gain, and the power gain.

Solution
The fraction of source signal appearing at the input terminals of the amplifier is obtained using the voltage-divider rule at the input, as follows:

$$\frac{v_{i1}}{v_s} = \frac{1\ M\Omega}{1\ M\Omega + 100\ k\Omega} = 0.909$$

The voltage gain of the first stage is obtained by considering the input resistance of the second stage to be the load of the first stage; that is,

$$A_{v1} \equiv \frac{v_{i2}}{v_{i1}} = 10\ \frac{100\ k\Omega}{100\ k\Omega + 1\ k\Omega} = 9.9$$

Similarly, the voltage gain of the second stage is obtained by considering the input resistance of the third stage to be the load of the second stage.

$$A_{v3} \equiv \frac{v_{i3}}{v_{i2}} = 100\ \frac{10\ k\Omega}{10\ k\Omega + 1\ k\Omega} = 90.9$$

Finally, the voltage gain of the output stage is as follows:

$$A_{v3} \equiv \frac{v_L}{v_{i3}} = 1\ \frac{100\ \Omega}{100\ \Omega + 10\ \Omega} = 0.909$$

The total gain of the three stages in cascade can be now found from

$$A_v \equiv \frac{v_L}{v_{i1}} = A_{v1}A_{v2}A_{v3} = 818$$

or 58.26 dB.

To find the voltage from source to load we multiply A_v by the factor representing the loss of gain at the input; that is,

$$\frac{v_L}{v_s} = \frac{v_L}{v_{i1}} \frac{v_{i1}}{v_s} = A_v \frac{v_{i1}}{v_s}$$

$$= 818 \times 0.909 = 743.6 \text{ V/V}$$

or 57.4 dB.

The current gain is found as follows:

$$A_i \equiv \frac{i_o}{i_i} = \frac{v_L/100 \ \Omega}{v_{i1}/1 \ \text{M}\Omega}$$

$$= 10^4 \times A_v = 8.18 \times 10^6 \text{ A/A}$$

or 138.26 dB.

The power gain is found from

$$A_p \equiv \frac{P_L}{P_I} = \frac{v_L i_o}{v_{i1} i_i}$$

$$= A_v A_i = 818 \times 8.18 \times 10^6 = 66.9 \times 10^8 \text{ W/W}$$

or 98.25 dB. Note that

$$A_p \text{ (dB)} = \tfrac{1}{2}\big[A_v \text{ (dB)} + A_i \text{ (dB)}\big]$$

Exercises

2.7 A phonograph cartridge characterized by a voltage of 1 V rms and a resistance of 1 MΩ is available to drive a 10-Ω loudspeaker. If connected directly, what voltage and power levels result at the loudspeaker? If a unity-gain (that is, $A_{vo} = 1$) buffer amplifier with 1-MΩ input resistance and 10-Ω output resistance is interposed between source and load, what do the output voltage and power levels become? For the new arrangement find the voltage gain from source to load, and the power gain (both expressed in decibels).

Ans.   10 μV rms; 10^{-11} W; 0.25 V; 6.25 mW; −12 dB; 44 dB

2.8 The output voltage of a voltage amplifier has been found to decrease by 20% when a load resistance of 1 kΩ is connected. What is the value of the amplifier output resistance?

Ans.   250 Ω

2.9 An amplifier with a voltage gain of +40 dB, an input resistance of 10 kΩ, and an output resistance of 1 kΩ is used to drive a 1-kΩ load. What is the value of A_{vo}? Find the value of power gain in dB.

Ans.   100 V/V; 44 dB

Other Amplifier Types

In the design of an electronic system, the signal of interest—whether at the system input, at an intermediate stage, or at the output—can be either a voltage or a current. For instance, some transducers have very high output resistances and can be more appropriately modeled as current sources. Similarly, there are applications in which the output current rather than the voltage is of interest. Thus, although it is the most popular, the voltage amplifier considered above is just one of four possible amplifier types. The other three are the current amplifier, the transconductance amplifier, and the transresistance amplifier.

Figure 2.12a shows a circuit model for the current amplifier. It consists of a current-controlled current source with a current-gain factor A_{is}, an input resistance R_i, and an output resistance R_o. Figure 2.12b shows the current amplifier fed with

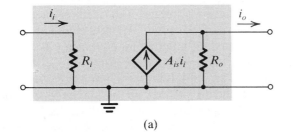

(a)

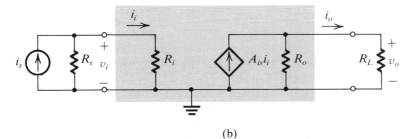

(b)

Fig. 2.12 (a) Circuit model for the current amplifier. **(b)** The current amplifier with input signal source and load.

a signal current source i_s having a resistance R_s, and with a load resistance R_L connected at the output. Using the current-divider rule at the output we find i_o as

$$i_o = A_{is}i_i \frac{R_o}{R_o + R_L}$$

Thus, the current gain of the loaded amplifier is given by

$$A_i \equiv \frac{i_o}{i_i} = A_{is} \frac{R_o}{R_o + R_L} \tag{2.10}$$

It follows that to avoid loss of gain in coupling the current amplifier to its load, the amplifier must be designed so that its output resistance R_o is much greater than the load resistance R_L. An ideal current amplifier has an infinite output resistance. Also note that with $R_L = 0$, the current gain is equal to A_{is}. Thus A_{is} is called the *short-circuit current gain*.

At the input side, the input resistance R_i causes current-divider action, with the result that only a fraction of i_s reaches the input of the amplifier; that is,

$$i_i = i_s \frac{R_s}{R_s + R_i} \tag{2.11}$$

To reduce the signal loss at the input side, the current amplifier must be designed so that $R_i \ll R_s$. The ideal current amplifier has $R_i = 0$.

Figure 2.13a shows a circuit model for the transconductance amplifier. This type of amplifier is intended to work with a voltage input signal and to provide an output current signal, as indicated in Fig. 2.13b. The gain parameter G_m is the ratio of the short-circuit output current to the input voltage. It is called the *short-circuit transconductance* and has the dimension of mhos or A/V. An ideal transconductance amplifier has an infinite input resistance and an infinite output resistance.

Finally, we show in Fig. 2.14a an equivalent circuit model for the transresistance amplifier. As indicated in Fig. 2.14b this type of amplifier is intended to operate with an input current signal and to provide an output voltage signal. The gain parameter R_m is the ratio of the open-circuit output voltage to the input current. It is called the *open-circuit transresistance* and has the dimension of ohms or V/A. An ideal transresistance amplifier has a zero input resistance and a zero output resistance.

Relationships Between the Four Amplifier Models

Although for a given amplifier a particular one of the four models above is most preferable, *any of the four can be used to model the amplifier*. In fact, simple relationships can be derived to relate the parameters of the various models. For instance, the open-circuit voltage gain A_{vo} can be related to the short-circuit current gain A_{is} as follows: The open-circuit output voltage given by the voltage amplifier model of Fig. 2.10a is $A_{vo}v_i$. The current amplifier model of Fig. 2.12a gives an open-circuit output voltage of $A_{is}i_iR_o$. Equating these two values and noting that $i_i = v_i/R_i$ gives

$$A_{vo}v_i = A_{is}\left(\frac{v_i}{R_i}\right)R_o$$

Thus

$$A_{vo} = A_{is}\left(\frac{R_o}{R_i}\right) \tag{2.12}$$

Similarly, we can show that

$$A_{vo} = G_m R_o \tag{2.13}$$

and

$$A_{vo} = \frac{R_m}{R_i} \tag{2.14}$$

The expressions in Eqs. (2.12) to (2.14) can be used to relate any two of the gain parameters A_{vo}, A_{is}, G_m, and R_m.

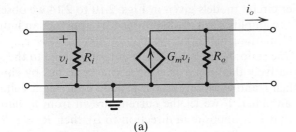

(a)

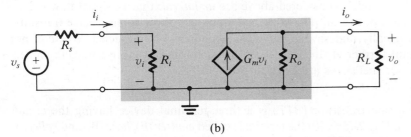

(b)

Fig. 2.13 (a) Circuit model for the transconductance amplifier.   **(b)** The transconductance amplifier with input signal source and load.

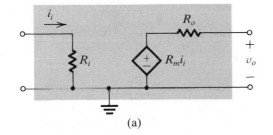

(a)

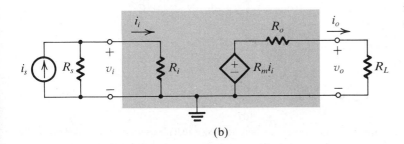

(b)

Fig. 2.14 (a) Circuit model for the transresistance amplifier.   **(b)** The transresistance amplifier with signal source and load.

From the amplifier circuit models given in Figs. 2.10 to 2.14 we observe that the input resistance R_i of the amplifier can be determined by applying an input voltage v_i and measuring (or calculating) the input current i_i; that is, $R_i = v_i/i_i$. The output resistance is found as the ratio of the open-circuit output voltage to the short-circuit output current. Alternatively the output resistance can be found by eliminating the input signal source (then i_i and v_i will both be zero) and applying a voltage signal v_x to the output of the amplifier. If we let the current drawn from v_x *into* the output terminals be i_x (note that i_x is opposite in direction to i_o), then $R_o = v_x/i_x$. Although these techniques are conceptually correct, in actual practice more refined methods are employed in measuring R_i and R_o.

The amplifier models considered above are *unilateral;* that is, signal flow is unidirectional, from input to output. Most real amplifiers show some reverse transmission, which is usually undesirable but must nonetheless be modeled. We shall not pursue this point further at this time except to mention that more complete models for linear two-port networks are given in Appendix B.

EXAMPLE 2.4

The *bipolar junction transistor (BJT)* is a three-terminal device having the circuit symbol shown in Fig. 2.15a with the terminals called *emitter* (E), *base* (B), and *collector* (C). The device is basically nonlinear; thus the relationships between the total instan-

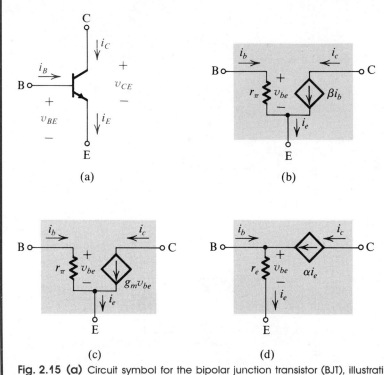

(a)　　　　　　　　　　　　(b)

(c)　　　　　　　　　　　　(d)

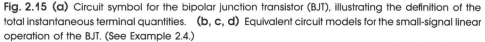

Fig. 2.15 (a) Circuit symbol for the bipolar junction transistor (BJT), illustrating the definition of the total instantaneous terminal quantities. **(b, c, d)** Equivalent circuit models for the small-signal linear operation of the BJT. (See Example 2.4.)

taneous terminal currents and voltages are generally nonlinear. Nevertheless, the linearization technique discussed in Section 2.2 can be used to provide linear operation for small signals—that is, incremental variations in currents and voltages around a bias or quiescent point. Thus the total instantaneous quantities can be expressed as the sums of dc or bias quantities and signal quantities as

$$v_{BE} = V_{BE} + v_{be} \qquad i_B = I_B + i_b$$

$$v_{CE} = V_{CE} + v_{ce} \qquad i_C = I_C + i_c$$

$$i_E = I_E + i_e$$

It will be shown in Chapter 8 that under the small-signal approximation the relationships between the signal quantities are linear and that the three-terminal transistor can be represented by one of the equivalent circuit models of Fig. 2.15b, c, and d. If these models are equivalent, we wish to find the relationships between their parameters.

Solution
For the model in Fig. 2.15b we have

$$i_b = \frac{v_{be}}{r_\pi} \tag{2.15}$$

$$i_c = \beta i_b \tag{2.16}$$

$$i_e = (\beta + 1)i_b \tag{2.17}$$

For the model in Fig. 2.15c we have

$$i_c = g_m v_{be} \tag{2.18}$$

Use of Eqs. (2.16), (2.17), and (2.18) gives

$$g_m = \frac{\beta}{r_\pi}$$

For the model in Fig. 2.15d we have

$$i_b = (1 - \alpha)i_e \tag{2.19}$$

$$i_c = \alpha i_e \tag{2.20}$$

$$i_e = \frac{v_{be}}{r_e} \tag{2.21}$$

Use of Eqs. (2.16) and (2.17) gives

$$\frac{i_c}{i_e} = \frac{\beta}{\beta + 1}$$

Comparing this result with Eq. (2.20), we get

$$\alpha = \frac{\beta}{\beta + 1}$$

Equations (2.15) and (2.17) can be combined to yield

$$i_e = \frac{\beta + 1}{r_\pi} v_{be}$$

Comparison of this result with Eq. (2.21) provides

$$r_e = \frac{r_\pi}{\beta + 1}$$

We have thus obtained expressions for the parameters of the models in Fig. 2.15c and d in terms of those for the model in Fig. 2.15b.

EXAMPLE 2.5

A *gyrator* is a two-port network that can be realized by connecting in parallel and back-to-back two voltage-controlled current sources of opposite polarities. Such an arrangement is shown in Fig. 2.16a (we assume ideal voltage-controlled current sources). It is required to show that if port 2 of the gyrator is terminated in a capacitance, as shown in Fig. 2.16b, then input port 1 will have the *i-v* relationship of an inductance.

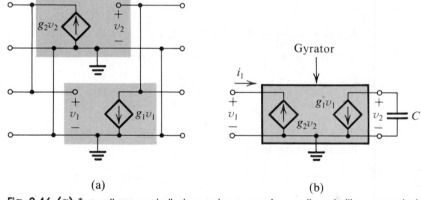

(a)   (b)

Fig. 2.16 (a) Two voltage-controlled current sources of opposite polarities connected back-to-back in parallel to form a gyrator. **(b)** The gyrator of (a) terminated at port 2 in a capacitance Example 2.5.)

Solution
We must show that in the circuit of Fig. 2.16b

$$v_1 = L \frac{di_1}{dt}$$

where L is a constant representing an inductance. To obtain the relationship between i_1 and v_1, consider first port 2, where we have

$$v_2 = -\frac{1}{C} \int_0^t g_1 v_1 \, dt$$

where it has been assumed that C was originally (at $t = 0$) uncharged. The current i_1 is given by

$$i_1 = -g_2 v_2 = \frac{g_1 g_2}{C} \int_0^t v_1 \, dt$$

which can be written in differential form as

$$v_1 = \frac{C}{g_1 g_2} \frac{di_1}{dt}$$

Thus the i_1-v_1 relationship is of the form found in an inductance, with the inductance L given by

$$L = \frac{C}{g_1 g_2}$$

Thus the gyrator can be used to realize an inductance using active elements (controlled sources) and a capacitance. This is a significant and useful result because physical inductors are almost impossible to realize using integrated circuit (IC) technology.

Exercises

2.10 A current amplifier with an input resistance of 100 Ω, an output resistance of 100 kΩ, and a current gain of 10,000 is connected between a signal current source with a 100-kΩ source resistance and a 100-ohm load. Find the voltage gain and the power gain.

Ans. 80 dB; 80 dB

2.11 A transconductance amplifier with an input resistance of 10 kΩ, an output resistance of 10 kΩ, and a transconductance of 1,000 mA/V is connected between a 10-kΩ voltage source and a 10-kΩ load. Find the voltage gain from source to load (that is, v_o/v_s).

Ans. 2,500 V/V or 68 dB

2.12 Consider a transistor modeled as in Fig. 2.15b to be connected between a voltage source with a 1-kΩ source resistance and a load resistance of 1 kΩ. Let $r_\pi = 2$ kΩ and $\beta = 90$. Calculate the voltage gain (v_c/v_s) and the current gain (i_c/i_b) as ratios. Also give the power gain in decibels.

Ans. −30 V/V; 90 A/A; 36.1 dB

2.13 Find the input resistance between terminals B and G in the circuit shown in Fig. E2.13.

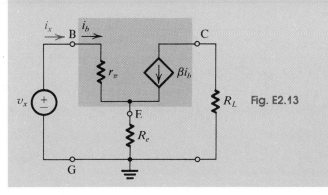

Fig. E2.13

The voltage v_x is a test voltage with the input resistance R_{in} defined as $R_{in} \equiv v_x/i_x$.

Ans. $R_{in} = r_\pi + (\beta + 1)R_e$

2.4 FREQUENCY RESPONSE OF AMPLIFIERS

The Sine-Wave Signal

The concept of the frequency spectrum of a signal was introduced in Chapter 1. To repeat, a signal that is an arbitrary function of time can be represented as the sum of sine waves through the mathematical devices of Fourier series and Fourier transform.[1] This fact has made the sine wave, or sinusoid, a standard signal waveform useful in the characterization of linear networks such as filters and amplifiers. Figure 2.17 shows a sine-wave voltage signal $v_A(t)$,

$$v_A(t) = V_p \sin(\omega t) \tag{2.22}$$

where V_p denotes the peak value or amplitude and ω denotes the angular frequency in radians per second, that is,

$$\omega = 2\pi f \quad \text{rad/s}$$

where f is the frequency in hertz,

$$f = \frac{1}{T} \quad \text{Hz}$$

and T is the period in seconds.

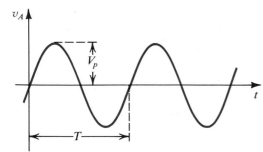

Fig. 2.17 Sine-wave voltage signal of amplitude V_p and frequency $f = 1/T$ Hz.

The sine-wave signal is completely characterized by its peak value V_p, its frequency ω, and its phase with respect to an arbitrary reference time. In this case the time origin has been chosen so that the phase angle is 0. It should be mentioned that it is common to express the amplitude of a sine-wave signal in terms of its root-mean-

---

[1] If the reader has not encountered these topics in the study of circuit theory, there is no need for concern; no significant use will be made of them until Chapter 11.

square value V_{rms} rather than the peak value V_p. The two quantities are related by

$$V_{rms} = V_p/\sqrt{2}$$

For instance, when we speak of the wall power supply in our homes as being 110 V, we mean that it has a sine waveform of $110\sqrt{2}$ peak value.

Because of the special role played by sinusoids in the characterization of linear networks we will use a special symbol to refer to them. The symbol $V_a(\omega)$ will refer to the sine-wave signal of Eq. (2.22), with the frequency ω indicated in the functional description for emphasis.

Measuring the Frequency Response

If the sinusoid $V_a(\omega)$ is applied at the input of an amplifier, the output will be a sinusoid of the same frequency. In fact the sine wave is the only signal that does not change in shape in passing through a linear network. The output sinusoid $V_b(\omega)$, however, could have a different amplitude and phase than the input $V_a(\omega)$. Thus an amplifier can be characterized in terms of the change it causes in the amplitude and phase of sinusoids of various frequencies applied at the input. To be more specific, the *frequency response* of an amplifier is measured as follows: A sinusoid of a given frequency and amplitude is applied at the input, and the amplitude of the output sinusoid and its phase angle relative to the input sinusoid are measured. Thus at this particular frequency we know the magnitude of the amplifier gain or transmission, which is the ratio of the output amplitude to the input amplitude, and we know the phase angle of the amplifier gain. Then the frequency of the input sinusoid is changed and the experiment is repeated, and so on. The end result will be a table or graph of gain magnitude versus frequency and a table or graph of phase angle versus frequency. Let us for the time being concentrate on the first, the graph of gain magnitude versus frequency. This is called the *magnitude response* or, loosely, the frequency response[2] of the amplifier.

Amplifier Bandwidth

Figure 2.18 shows the magnitude response of an amplifier. It indicates that the gain is almost constant over a wide frequency range, roughly between ω_1 and ω_2. Signals whose frequencies are below ω_1 or above ω_2 will experience lower gain, with the gain decreasing as we move farther away from ω_1 and ω_2. The band of frequencies over which the gain of the amplifier is almost constant, to within a certain number of decibels (usually 3 dB), is called the *amplifier bandwidth*. Normally the amplifier is designed so that its bandwidth coincides with the spectrum of the signals it is required to amplify. If this were not the case, the amplifier would *distort* the frequency spectrum of the input signal, with different components of the input signal being amplified by different amounts.

---

[2] Frequency response refers more properly to the combination of magnitude response and phase response.

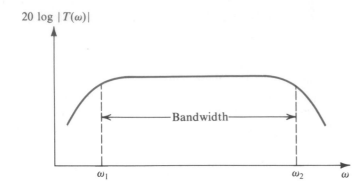

$20 \log |T(\omega)|$

Bandwidth

ω_1 ω_2 ω

Fig. 2.18 Typical magnitude response of an amplifier. $T(\omega)$ is the amplifier transfer function—that is, the ratio of the output $V_o(\omega)$ to the input $V_i(\omega)$.

Evaluating the Frequency Response of Amplifiers

Above, we described the method used to measure the frequency response of an amplifier. We now briefly discuss the method for analytically obtaining an expression for the frequency response. What we are about to say is just a preview of this important subject whose detailed study starts in Chapter 11.

To evaluate the frequency response of an amplifier one has to analyze the amplifier equivalent circuit model, taking into account all reactive components.[3] Circuit analysis proceeds in the usual fashion but with inductances and capacitances represented by their reactances. An inductance L has a reactance or impedance $j\omega L$, and a capacitance C has a reactance or impedance $1/j\omega C$, or equivalently a susceptance or admittance $j\omega C$. Thus in a *frequency-domain* analysis we deal with impedances and/or admittances. The result of the analysis is the amplifier transfer function $T(\omega)$,

$$T(\omega) = \frac{V_o(\omega)}{V_i(\omega)}$$

where $V_i(\omega)$ and $V_o(\omega)$ denote the input and output sinusoids, respectively. $T(\omega)$ is generally a complex function whose magnitude $|T(\omega)|$ gives the magnitude of transmission or the magnitude response of the amplifier. The phase of $T(\omega)$ gives the phase response of the amplifier.

In the analysis of a circuit to determine its frequency response, the algebraic manipulations can be considerably simplified by using the *complex frequency variable* s. In terms of s, the impedance of an inductance L is sL and that of a capacitor C is $1/sC$. Replacing the reactive elements with their impedances and performing standard circuit analysis we obtain the transfer function $T(s)$ as

$$T(s) \equiv \frac{V_o(s)}{V_i(s)}$$

Subsequently, we replace s by $j\omega$ to determine the network transfer function for *physical frequencies*, $T(j\omega)$. Note that $T(j\omega)$ is the same function we called $T(\omega)$ in the

---

[3] Note that in the models considered in previous sections no reactive components were included.

above;[4] the additional j is included in order to emphasize that $T(j\omega)$ is obtained from $T(s)$ by replacing s with $j\omega$.

EXAMPLE 2.6

Derive the transfer function of the two-port RC network shown in Fig. 2.19.

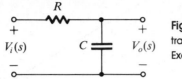

Fig. 2.19 A two-port RC network whose transfer function $T(s) = V_o(s)/V_i(s)$ is derived in Example 2.6.

Solution

Replacing the capacitor C by its impedance $1/sC$, we can use the voltage-divider rule to obtain the transfer function $T(s)$,

$$T(s) \equiv \frac{V_o(s)}{V_i(s)}$$

$$= \frac{1/sC}{R + 1/sC}$$

Thus

$$T(s) = \frac{1}{1 + sCR}$$

For physical frequencies we substitute $s = j\omega$,

$$T(j\omega) = \frac{1}{1 + j\omega CR}$$

Thus the magnitude response $|T(j\omega)|$ is given by

$$|T(j\omega)| = \frac{1}{\sqrt{1 + (\omega CR)^2}}$$

and the phase response $\phi(\omega)$ is

$$\phi(\omega) = -\tan^{-1}(\omega CR)$$

Figure 2.20 shows sketches of the magnitude and phase response using a logarithmic frequency axis with the frequency normalized with respect to $\omega_0 = 1/CR$ and using a decibel scale for the magnitude axis.[5] Note that this network *passes low frequencies* and attenuates high-frequency signals. The magnitude of transmission

---

[4] At this stage we are using s simply as a shorthand for $j\omega$. We shall not require detailed knowledge of s-plane concepts until Chapter 11.

[5] Such diagrams are known as *Bode plots*.

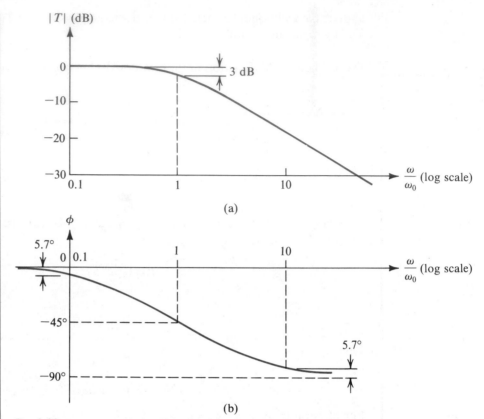

Fig. 2.20 The magnitude and phase responses of the two-port network in Fig. 2.19. Note that frequencies have been normalized to $\omega_0 = 1/RC$. At this frequency the transmission drops by 3 dB below the transmission at dc, and the phase is $-45°$.

decreases as the frequency increases. Thus this network is called a *low-pass filter*. At any specific frequency the phase response gives the phase angle by which the output sinusoid leads the input sinusoid. The significance of the phase response will be explained at a later stage in the text.

Finally, it should be noted that the reason for the falloff in magnitude of transmission of the network in Fig. 2.19 with frequency is that as the frequency increases, the impedance of the capacitor decreases; at very high frequencies the capacitor acts almost as a short circuit. Thus while it acts as an open circuit for dc, a capacitor acts as a short circuit for sinusoids with infinite frequency.

Classification of Amplifiers Based on Frequency Response

Amplifiers can be classified based on the shape of their magnitude-response curve. Figure 2.21 shows typical frequency response curves for various amplifier types. In Figure 2.21a the gain remains constant over a wide frequency range but falls off at

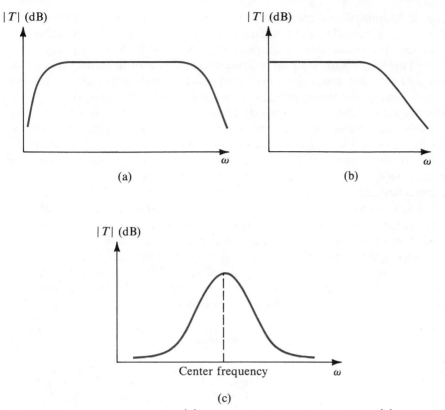

Fig. 2.21 Frequency response for **(a)** a capacitively coupled amplifier, **(b)** a direct-coupled amplifier, and **(c)** a tuned or bandpass amplifier.

low and high frequencies. This is a common type of frequency response found in audio amplifiers.

As will be shown in Chapter 11, *internal capacitances* in the device (transistor) cause the falloff of gain at high frequencies. On the other hand, the falloff of gain at low frequencies is usually caused by *coupling capacitors* used to connect one amplifier stage to another, as indicated in Fig. 2.22. This practice is usually adopted to simplify the design process of the different stages. The coupling capacitors are usually chosen

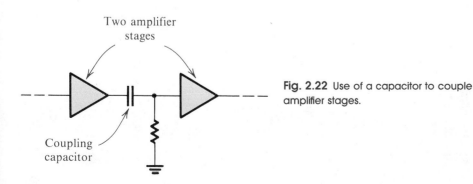

Fig. 2.22 Use of a capacitor to couple amplifier stages.

quite large (a fraction of a microfarad to a few tens of microfarads) so that their reactance (impedance) is small at the frequencies of interest. Nevertheless, at sufficiently low frequencies the reactance of a coupling capacitor will become large enough to cause part of the signal being coupled to appear as a voltage drop across the coupling capacitor and thus not reach the subsequent stage. Coupling capacitors will thus cause loss of gain at low frequencies and cause the gain to be zero at dc.

There are many applications in which it is important that the amplifier maintain its gain at low frequencies down to dc. Furthermore, monolithic integrated circuit (IC) technology does not allow the fabrication of large coupling capacitors. Thus IC amplifiers are usually designed as *directly coupled* or *dc amplifiers* (as opposed to *capacitively coupled* or *ac amplifiers*). Figure 2.21b shows the frequency response of a dc amplifier. Such a frequency response characterizes what is referred to as a *low-pass amplifier*. Although it is not very appropriate, the term low-pass amplifier is also usually used to refer to the amplifier whose response is shown in Fig. 2.21a.

In the discussion of a radio receiver in Chapter 1, the need for a tuned circuit became apparent. Amplifiers whose frequency response peaks around a certain frequency (called the *center frequency*) and falls off on both sides of this frequency, as shown in Fig. 2.21c, are called *tuned amplifiers*, *bandpass amplifiers*, or *bandpass filters*. The design of bandpass amplifiers is studied in Chapter 14.

Exercises

2.14 Find the frequencies f and ω of a sine-wave signal with a period of 1 ms.

Ans. $f = 1{,}000$ Hz; $\omega = 2\pi \times 10^3$ rad/s

2.15 For the low-pass circuit analyzed in Example 2.6, calculate the magnitude in decibels and the phase in degrees at one-tenth and ten times the corner frequency ω_0.

Ans. -0.04 dB; $-5.7°$; -20 dB; $-84.3°$

2.16 What is the period T of sine waveforms characterized by frequency
(a) $f = 60$ Hz, **(b)** $f = 10^{-3}$ Hz, and **(c)** $f = 1$ MHz.

Ans. 16.7 ms; 1,000 s; 1 μs

2.17 Consider the RC circuit analyzed in Example 2.6. Find the output $v_o(t)$ if the input is a sinusoid with frequency $\omega = 10^3$ rad/s and 10-V amplitude. The component values are $R = 1$ kΩ, $C = 1$ μF.

Ans. $v_o(t) = 7.07 \sin(10^3 t - \frac{1}{4}\pi)$

2.18 Derive an expression for the magnitude of transmission of the two-port RC network shown in Fig. E2.18.

Ans. $\left| \dfrac{V_o(j\omega)}{V_i(j\omega)} \right| = \dfrac{\omega CR}{\sqrt{1 + (\omega CR)^2}}$

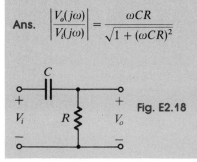

Fig. E2.18

2.5 SOME USEFUL NETWORK THEOREMS

Thévenin's Theorem

Thévenin's theorem is used to represent a part of a network by a voltage source V_t and a series impedance Z_t, as shown in Fig. 2.23. Figure 2.23a shows a network divided into two parts, A and B. In Fig. 2.23b part A of the network has been replaced by its Thévenin equivalent: a voltage source V_t and a series impedance Z_t. Figure 2.23c illustrates how V_t is to be determined: simply open-circuit the two terminals of network A and measure (or calculate) the voltage that appears between these two terminals. To determine Z_t we reduce all external (that is, independent) sources in network A to zero by short-circuiting voltage sources and open-circuiting current sources. The impedance Z_t will be equal to the input impedance of network A after this reduction has been performed, as illustrated in Fig. 2.23d.

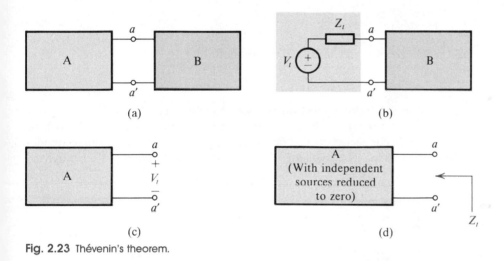

Fig. 2.23 Thévenin's theorem.

Norton's Theorem

Norton's theorem is the *dual* of Thévenin's theorem. It is used to represent a part of a network by a current source I_n and a parallel impedance Z_n, as shown in Fig. 2.24. Figure 2.24a shows a network divided into two parts, A and B. In Fig. 2.24b part A has been replaced by its Norton's equivalent: a current source I_n and a parallel impedance Z_n. The Norton's current source I_n can be measured (or calculated) as shown in Fig. 2.24c. The terminals of the network being reduced (network A) are shorted, and the current I_n will be equal simply to the short-circuit current. To determine the impedance Z_n we first reduce the external excitation in network A to zero, that is, short-circuit independent voltage sources and open-circuit independent current sources. The impedance Z_n will be equal to the input impedance of network A after this source elimination process has taken place. Thus the Norton impedance Z_n is

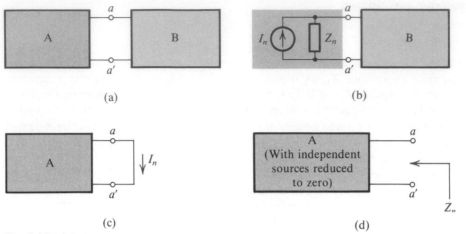

Fig. 2.24 Norton's theorem.

equal to the Thévenin impedance Z_t. Finally, note that $I_n = V_t/Z$, where $Z = Z_n = Z_t$.

EXAMPLE 2.7

Figure 2.25a shows a bipolar junction transistor circuit. The transistor is a three-terminal device with the terminals labeled E (emitter), B (base), and C (collector). As shown, the base is connected to the dc power supply V^+ via the voltage divider composed of R_1 and R_2. The collector is connected to the dc supply V^+ through R_3

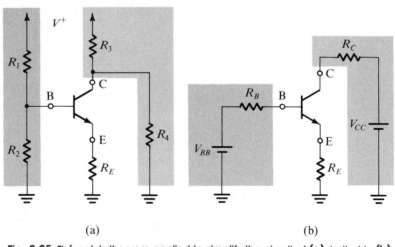

(a)   (b)

Fig. 2.25 Thévenin's theorem applied to simplify the circuit of **(a)** to that in **(b)**. (See Example 2.7.)

and to ground through R_4. To simplify the analysis we wish to reduce the circuit through application of Thévenin's theorem.

Solution

Thévenin's theorem can be used at the base side to reduce the network composed of V^+, R_1, and R_2 to a dc voltage source V_{BB},

$$V_{BB} = V^+ \frac{R_2}{R_1 + R_2}$$

and a resistance R_B,

$$R_B = R_1 \parallel R_2$$

where $\parallel$ denotes "in parallel with." At the collector side Thévenin's theorem can be applied to reduce the network composed of V^+, R_3, and R_4 to a dc voltage source V_{CC},

$$V_{CC} = V^+ \frac{R_4}{R_3 + R_4}$$

and a resistance R_C,

$$R_C = R_3 \parallel R_4$$

The reduced circuit is shown in Fig. 2.25b.

Source-Absorption Theorem

Consider the situation shown in Fig. 2.26. In the course of analyzing a network we find a controlled current source I_x appearing between two nodes whose voltage difference is the controlling voltage V_x. That is, $I_x = g_m V_x$ where g_m is a conductance. We can replace this controlled source by an impedance $Z_x = V_x/I_x = 1/g_m$, as shown in Fig. 2.26, because the current drawn by this impedance will be equal to the current of the controlled source that we have replaced.

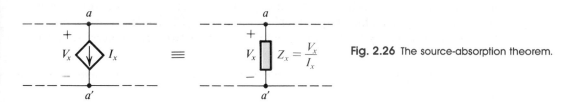

Fig. 2.26 The source-absorption theorem.

EXAMPLE 2.8

Figure 2.27a shows the small-signal equivalent circuit model of a transistor. We want to find the resistance R_{in} "looking into" the emitter terminal E—that is, between the emitter and ground—with the base B and collector C grounded.

Solution

From Fig. 2.27a we see that the voltage v_π will be equal to $-v_e$. Thus looking between E and ground we see a resistance r_π in parallel with a current source drawing a current

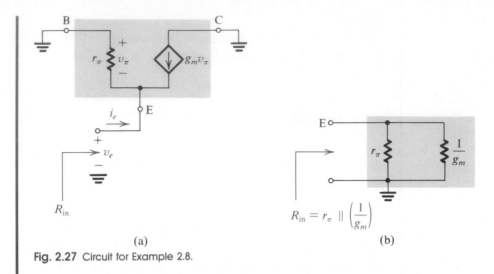

(a)

(b)

Fig. 2.27 Circuit for Example 2.8.

$g_m v_e$ away from terminal E. This latter source can be replaced by a resistance $(1/g_m)$, resulting in the input resistance R_{in} given by

$$R_{in} = r_\pi \,\|\, (1/g_m)$$

as illustrated in Fig. 2.27b.

Miller's Theorem

Consider the situation shown in Fig. 2.28a. We have identified two nodes, 1 and 2, together with the reference or ground terminal of a particular network. As shown, an admittance Y is connected between nodes 1 and 2. In addition, nodes 1 and 2 may be connected by other components to other nodes in the network, which is signified by the broken lines emanating from nodes 1 and 2. Miller's theorem provides the means for replacing the "bridging" admittance Y by two admittances: Y_1 between

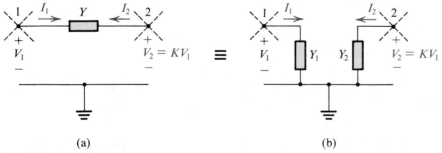

(a)

(b)

Fig. 2.28 Miller's theorem.

node 1 and ground and Y_2 between node 2 and ground, as shown in Fig. 2.28b. This replacement is usually quite helpful in simplifying circuit analysis.

The Miller equivalence we are about to derive is based on the premise that it is possible, by independent means, to determine the voltage gain from node 1 to node 2, denoted K, where $K \equiv V_2/V_1$. With the gain K known, the values of Y_1 and Y_2 can be determined as follows: It may be seen from Fig. 2.28a that the only way that node 1 "knows of the existence" of the admittance Y is through the current drawn by Y away from node 1. This current, I_1, is given by

$$
\begin{aligned}
I_1 &= Y(V_1 - V_2) \\
&= YV_1(1 - V_2/V_1) \\
&= YV_1(1 - K)
\end{aligned}
$$

For the circuit of Fig. 2.28b to be equivalent to that of Fig. 2.28a it is essential that the admittance Y_1 be of such value that the current it draws from node 1 is equal to I_1:

$$ Y_1 V_1 = I_1 $$

which leads to

$$ Y_1 = Y(1 - K) \tag{2.23} $$

Similarly, node 2 "feels" the existence of admittance Y only through the current I_2 drawn by Y away from node 2 (note that $I_2 = -I_1$):

$$
\begin{aligned}
I_2 &= Y(V_2 - V_1) \\
&= YV_2(1 - V_1/V_2)
\end{aligned}
$$

Thus

$$ I_2 = Y(1 - 1/K)V_2 $$

For the circuit of Fig. 2.28b to be equivalent to that of Fig. 2.28a it is essential that the value of Y_2 be such that the current it draws from node 2 is equal to I_2:

$$ Y_2 V_2 = I_2 $$

which leads to

$$ Y_2 = Y(1 - 1/K) \tag{2.24} $$

Equations (2.23) and (2.24) are the two necessary and sufficient conditions for the network in Fig. 2.28b to be equivalent to that of Fig. 2.28a. Note that in both networks the voltage gain from node 1 to node 2 is equal to K. We shall apply Miller's theorem quite frequently throughout this book.

Before we illustrate the application of Miller's theorem, an important caution is in order. *The Miller equivalent circuit of Fig. 2.28b is valid only as long as the conditions that existed in the network when K was determined are not changed.* It follows that the Miller equivalent circuit *cannot* be used directly to determine the output resistance of the amplifier. This is because in determining the output resistance in the conventional way the input signal source is eliminated and a test voltage is applied to the output terminals. This obviously changes the value of K and makes the Miller equivalence invalid. Similarly, the Miller equivalent cannot be used directly to determine the reverse (output to input) transmission of the amplifier. Note that the above statements assume that node 1 is at the input side of the amplifier and node 2 is at the output side. It turns out, in fact, that the Miller equivalent is useful only for determining the amplifier input impedance and its forward transmission or gain. These points are further illustrated by the following example.

EXAMPLE 2.9

Figure 2.29a shows a transconductance amplifier with an infinite input resistance, a 10-kΩ output resistance, and a transconductance $G_m = 0.1$ A/V. A 1-MΩ resistor R_f is connected from the output of the amplifier back to its input. (The resistor R_f applies *feedback* to the amplifier and is called a feedback resistor.) The amplifier input is fed with a voltage source v_s having a resistance $R_s = 1$ kΩ. It is required to calculate the input resistance of the amplifier (that is, the resistance R_{in} seen between terminals 1 and 1′), the voltage gain from source to load (v_o/v_s), and the output resistance of the overall amplifier (R_{out}).

Solution

We shall use Miller's theorem to replace the feedback resistance R_f with the two resistors, R_1 and R_2, as shown in Fig. 2.29b. Before we do that, however, we must calculate the voltage gain from node 1 to node 2—that is, v_o/v_i. This can be done by noting that the current through R_f is given by $(v_i - v_o)/R_f$. Now, writing a node equation at node 2,

$$\frac{v_i - v_o}{R_f} = G_m v_i + \frac{v_o}{R_o}$$

and rearranging terms we obtain

$$K \equiv \frac{v_o}{v_i} = \frac{-G_m + (1/R_f)}{(1/R_o) + (1/R_f)}$$

Substituting $G_m = 0.1$ A/V, $R_f = 10^6$ Ω and $R_o = 10^4$ Ω yields

$$K \simeq -1{,}000 \text{ V/V}$$

Now we can apply Miller's theorem and determine the values of R_1 and R_2, as follows:

$$R_1 = \frac{1}{(1/R_f)(1 - K)} = \frac{R_f}{1 - K} = \frac{10^6}{1 - (-1{,}000)} \simeq 10^3 \ \Omega$$

$$R_2 = \frac{1}{(1/R_f)(1 - 1/K)} = \frac{R_f}{1 - (-1/1{,}000)} \simeq R_f = 10^6 \ \Omega$$

We note that in the equivalent circuit of Fig. 2.29b $v_o = Kv_i$, where K is the value calculated above. This equivalent circuit can be used to determine the input resistance, R_{in}, which in this case is simply equal to R_1,

$$R_{in} = R_1 = 1 \text{ k}\Omega$$

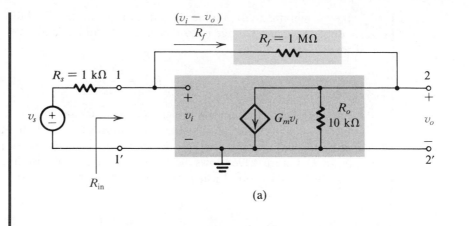

(a)

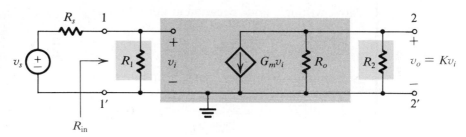

(b)

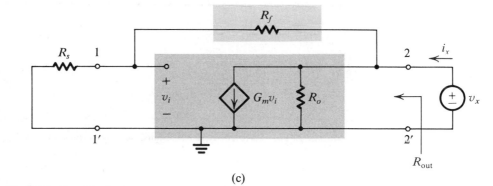

(c)

Fig. 2.29 Circuit for Example 2.9.

The equivalent circuit can also be used to determine the overall voltage gain, as follows:

$$\frac{v_o}{v_s} = \frac{v_o}{v_i} \times \frac{v_i}{v_s} = -1,000 \times \frac{R_1}{R_1 + R_s}$$

$$= -1,000 \times \frac{1}{1 + 1} = -500 \text{ V/V}$$

It now remains to determine the output resistance. This is done by reducing v_s to zero, applying a test voltage v_x to the output terminals and determining the current i_x drawn from v_x. Then

$$R_{\text{out}} \equiv \frac{v_x}{i_x}$$

If we attempt to do that on the Miller equivalent circuit in Fig. 2.29b, we change the conditions that prevailed during the calculation of K. This in turn invalidates the Miller equivalent circuit. Thus, to determine R_{out} we must use the original circuit in Fig. 2.29a. Figure 2.29c shows the circuit prepared for determining R_{out}. We observe that v_i can be easily found using the voltage-divider rule,

$$v_i = v_x \frac{R_s}{R_s + R_f} \tag{2.25}$$

Now, writing a node equation at node 2 gives

$$i_x = \frac{v_x}{R_o} + G_m v_i + \frac{v_x}{R_f + R_s}$$

Substituting for v_i from Eq. (2.25) and dividing both sides by v_x gives

$$\frac{1}{R_{\text{out}}} \equiv \frac{i_x}{v_x} = \frac{1}{R_o} + G_m \frac{R_s}{R_s + R_f} + \frac{1}{R_f + R_s}$$

Substituting the numerical values given results in

$$R_{\text{out}} \simeq 5 \text{ k}\Omega$$

Exercises

2.19 A source is measured to have a 10-V open-circuit voltage and to provide 1 mA into a short circuit. Calculate its Thévenin and Norton equivalent source parameters.

Ans. $V_t = 10$ V; $Z_t = Z_n = 10$ kΩ; $I_n = 1$ mA

2.20 In the circuit shown in Fig. E2.20 the diode has a voltage drop $V_D \simeq 0.7$ V. Use Thévenin's theorem to simplify the circuit and hence calculate the diode current I_D.

Ans. 1 mA

2.21 The two-terminal device M in the circuit of Fig. E2.21 has a current $I_M \simeq 1$ mA independent of the voltage V_M across it. Use Norton's theorem to simplify the circuit and hence calculate the voltage V_M.

Ans. 5 V

2.22 Consider an ideal voltage amplifier with a gain $\mu = -100$ and with a capacitor $C = 10$ pF connected from output to input. Use Miller's theorem to find the input capacitance of the amplifier.

Ans. 1,010 pF

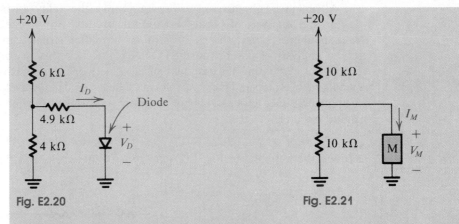

Fig. E2.20   **Fig. E2.21**

Note: This exercise illustrates an important effect in amplifiers with large negative gains: a small feedback capacitance gives rise to a large input capacitance. This capacitance multiplication is known as the *Miller effect*.

2.23 Consider an ideal voltage amplifier with gain $\mu = 0.95$ and a resistance $R = 100$ kΩ connected in the feedback path—that is, between the output and input terminals. Use Miller's theorem to find the input resistance of this circuit.

Ans. 2 MΩ

2.24 For the circuit in Fig. E2.24 find R_{in}, v_o/v_i, and R_{out}.

Ans. 2 Ω; 5,000 V/V; 5 kΩ

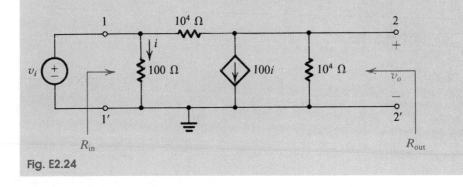

Fig. E2.24

2.6 SINGLE-TIME-CONSTANT NETWORKS

Single-time-constant (STC) networks are those networks that are composed of, or can be reduced to, one reactive component (inductance or capacitance) and one resistance. An STC network formed of an inductance L and a resistance R has a time constant $\tau = L/R$. The time constant τ of an STC network composed of a capacitance C and a resistance R is given by $\tau = CR$.

Although STC networks are quite simple, they play an important role in the design and analysis of linear and digital circuits. For instance, it will be shown in later chapters that the analysis of an amplifier circuit can usually be reduced to the analysis of one or more STC networks. For this reason we will review the process of evaluating the response of STC networks to sinusoidal and other input signals such as step and pulse waveforms. The latter signal waveforms are encountered in some amplifier applications but are more important in switching circuits, including digital circuits.

EXAMPLE 2.10

Reduce the network in Fig. 2.30a to an STC network, and find its time constant.

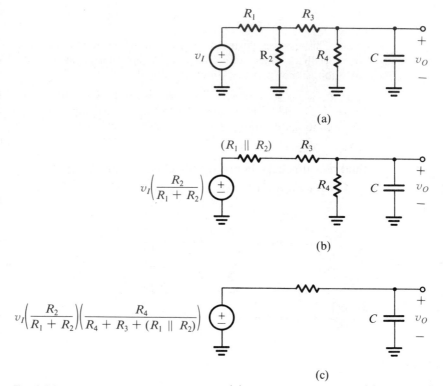

Fig. 2.30 The reduction of the network in **(a)** to the STC network in **(c)** through the repeated application of Thévenin's theorem. (See Example 2.10).

Solution

The reduction process is illustrated in Fig. 2.30 and consists of repeated applications of Thévenin's theorem. The final circuit is shown in Fig. 2.30c, from which we obtain the time constant as

$$\tau = C\{R_4 \| [R_3 + (R_2 \| R_1)]\}$$

Rapid Evaluation of τ

In many instances it will be important to be able to evaluate rapidly the time constant τ of a given STC network. A simple method for accomplishing this goal consists first of reducing the excitation to zero; that is, if the excitation is by a voltage source, short it, and if by a current source, open it. Then if the network has one reactive component and a number of resistances, "grab hold" of the two terminals of the reactive component (capacitance or inductance) and find the equivalent resistance R_{eq} seen by the component. The time constant is then either L/R_{eq} or CR_{eq}. As an example, in the circuit of Fig. 2.30a we find that the capacitor C "sees" a resistance R_4 in parallel with the series combination of R_3 and (R_2 in parallel with R_1). Thus

$$R_{eq} = R_4 \,\|\, [R_3 + (R_2 \,\|\, R_1)]$$

and the time constant is CR_{eq}.

In some cases it may be found that the network has one resistance and a number of capacitances or inductances. In such a case the procedure should be inverted; that is, "grab hold" of the resistance terminals and find the equivalent capacitance C_{eq}, or equivalent inductance L_{eq}, seen by this resistance. The time constant is then found as $C_{eq}R$ or L_{eq}/R. This illustrated in Example 2.11.

EXAMPLE 2.11

Find the time constant of the circuit in Fig. 2.31.

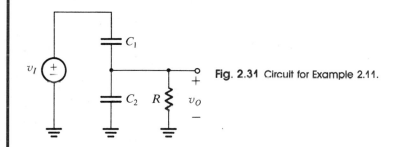

Fig. 2.31 Circuit for Example 2.11.

Solution
After reducing the excitation to zero by short-circuiting the voltage source, we see that the resistance R "sees" an equivalent capacitance $C_1 + C_2$. Thus the time constant τ is given by

$$\tau = (C_1 + C_2)R$$

Finally, there are cases where an STC network has more than one resistance and more than one capacitance (or more than one inductance). In such a case some initial work must be performed to simplify the network, as illustrated by Example 2.12.

EXAMPLE 2.12

Here we show that the response of the network in Fig. 2.32a can be obtained using the method of analysis of STC networks.

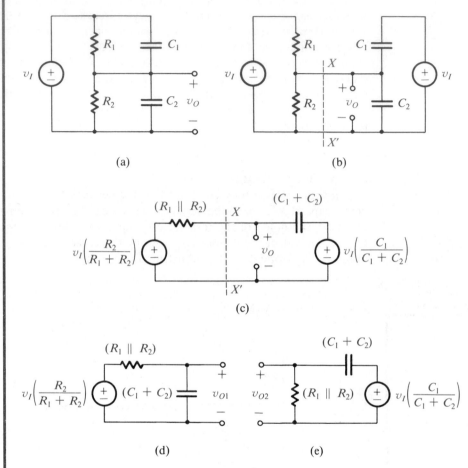

Fig. 2.32 Details for Example 2.12.

Solution

The analysis steps are illustrated in Fig. 2.32. In Fig. 2.32b we show the network excited by two separate but equal voltage sources. The reader should convince himself or herself of the equivalence of the networks in Fig. 2.32a and Fig. 2.32b. The "trick" employed to obtain the arrangement in Fig. 2.32b is a very useful one.

Application of Thévenin's theorem to the circuit to the left of the line XX' and then to the circuit to the right of that line results in the circuit of Fig. 2.32c. Since this is a linear circuit, the response may be obtained using the principle of superposition. Specifically, the output voltage v_O will be the sum of the two components v_{O1} and v_{O2}. The first component, v_{O1}, is the output due to the left-hand-side voltage source with the other voltage source reduced to zero. The circuit for calculating v_{O1} is shown in Fig. 2.32d. It is an STC network with a time constant given by

$$\tau = (C_1 + C_2)(R_1 \parallel R_2)$$

Similarly, the second component v_{O2} is the output obtained with the left-hand-side voltage source reduced to zero. It can be calculated from the circuit of Fig. 2.32e, which is an STC network with a time constant equal to that given above.

Finally, it should be observed that the fact that the circuit is an STC one can also be ascertained by setting the independent source v_I in Fig. 2.32a to zero. Also, the time constant is then immediately obvious.

Classification of STC Networks

STC networks can be classified into two categories, *low-pass* (LP) and *high-pass* (HP) types, with each of the two categories displaying distinctly different signal responses. The task of finding whether an STC network is of LP or HP type may be accomplished in a number of ways, the simplest of which uses the frequency-domain response. Specifically, low-pass networks pass dc (that is, signals with zero frequency) and attenuate high frequencies, with the transmission being zero at $\omega = \infty$. Thus we can test for the network identity either at $\omega = 0$ or at $\omega = \infty$. At $\omega = 0$ capacitors should be replaced by open circuits ($1/j\omega C = \infty$) and inductors should be replaced by short circuits ($j\omega L = 0$). Then if the output is zero, the circuit is of the high-pass type, while if the output is finite, the circuit is of the low-pass type. Alternatively, we may test at $\omega = \infty$ by replacing capacitors by short circuits ($1/j\omega C = 0$) and inductors by open circuits ($j\omega L = \infty$). Then if the output is finite, the circuit is of the HP type, whereas if the output is zero, the circuit is of the LP type. Table 2.1 provides a summary of these rules (s.c., short circuit; o.c., open circuit).

Table 2.1 RULES FOR FINDING THE TYPE OF STC NETWORK

Test at	Replace	Network is LP if	Network is HP if
$\omega = 0$	C by o.c. L by s.c.	Output is finite	Output is zero
$\omega = \infty$	C by s.c. L by o.c.	Output is zero	Output is finite

Figure 2.33 shows examples of low-pass STC networks, and Fig. 2.34 shows examples of high-pass STC networks. For each circuit we have indicated the input and output variables of interest. Note that a given network can be of either category, depending on the input and output variables. The reader is urged to verify, using the rules of Table 2.1, that the circuits of Figs. 2.33 and 2.34 are correctly classified.

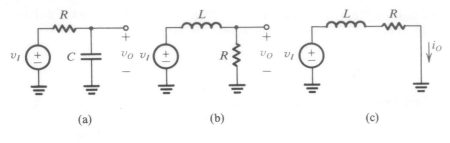

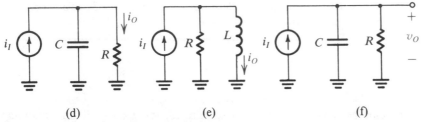

(d)   (e)   (f)

Fig. 2.33 STC networks of the low-pass type.

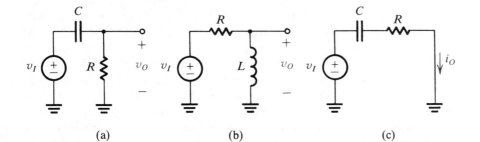

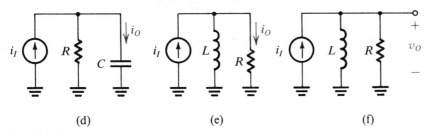

(d)   (e)   (f)

Fig. 2.34 STC networks of the high-pass type.

Exercises

2.25 Find the time constants for the circuits shown in Fig. E2.25.

Ans. (a) $\dfrac{(L_1 \| L_2)}{R}$;   (b) $\dfrac{(L_1 \| L_2)}{(R_1 \| R_2)}$

2.26 Classify the following circuits as STC high-pass or low-pass: Fig. 2.33a with output i_o in C to ground; Fig. 2.33b with output i_o in R to ground; Fig. 2.33d with output i_o in C to

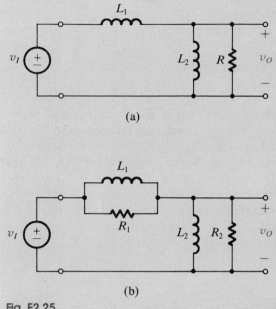

Fig. E2.25

ground; Fig. 2.33e with output i_O in R to ground; Fig. 2.34b with output i_O in L to ground; and Fig. 2.34d with output v_O across C.

Ans.   HP; LP; HP; HP; LP; LP

2.7 FREQUENCY RESPONSE OF STC NETWORKS

Low-Pass Networks

The transfer function $T(s)$ of an STC low-pass network always can be written in the form

$$T(s) = \frac{K}{1 + (s/\omega_0)} \tag{2.26}$$

which, for physical frequencies, where $s = j\omega$, becomes

$$T(j\omega) = \frac{K}{1 + j(\omega/\omega_0)} \tag{2.27}$$

where K is the magnitude of the transfer function at $\omega = 0$ (dc) and ω_0 is defined by

$$\omega_0 = \frac{1}{\tau}$$

with τ being the time constant. Thus the magnitude response is given by

$$|T(j\omega)| = \frac{K}{\sqrt{1 + (\omega/\omega_0)^2}} \tag{2.28}$$

and the phase response is given by

$$\phi(\omega) = -\tan^{-1}(\omega/\omega_0) \tag{2.29}$$

Figure 2.35 shows sketches of the magnitude and phase responses for an STC low-pass network. The magnitude response shown in Fig. 2.35a is simply a graph of the function in Eq. (2.28). The magnitude is normalized with respect to the dc gain K and is expressed in dB, that is, the plot is for $20 \log|T(j\omega)/K|$, with a logarithmic scale used for the frequency axis. Furthermore, the frequency variable has been normalized with respect to ω_0. As shown, the magnitude curve is closely defined by two straight-line asymptotes. The low-frequency asymptote is a horizontal straight line at 0 dB. To find the slope of the high-frequency asymptote consider Eq. (2.28) and

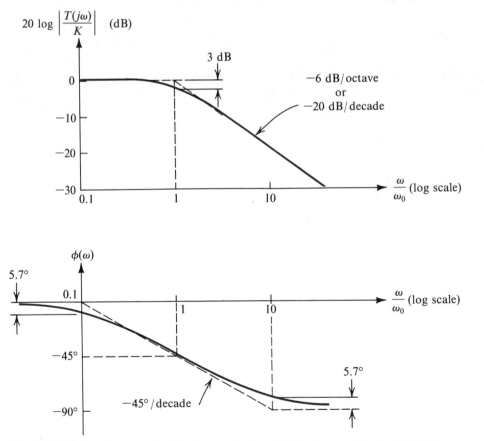

Fig. 2.35 (a) Magnitude and **(b)** phase response of STC networks of the lowpass type.

let $\omega/\omega_0 \gg 1$, resulting in

$$|T(j\omega)| \simeq K \frac{\omega_0}{\omega}$$

It follows that if ω doubles in value, the magnitude is halved. On a logarithmic frequency axis, doublings of ω represent equally spaced points, with each interval called an *octave*. Halving the magnitude function corresponds to a 6-dB reduction in transmission (20 log 0.5 = -6 dB). Thus the slope of the high-frequency asymptote is -6 dB/octave. This can be equivalently expressed as -20 dB/decade, where a decade refers to an increase in frequency by a factor of 10.

The two straight-line asymptotes of the magnitude-response curve meet at the "corner frequency" or "break frequency" ω_0. The difference between the actual magnitude-response curve and the asymptotic response is largest at the corner frequency, where its value is 3 dB. To verify that this value is correct, simply substitute $\omega = \omega_0$ in Eq. (2.28) to obtain

$$|T(j\omega_0)| = \frac{K}{\sqrt{2}}$$

Thus at $\omega - \omega_0$ the gain drops by a factor of $\sqrt{2}$ relative to the dc gain, which corresponds to a 3-dB reduction in gain. The corner frequency ω_0 is appropriately referred to as the 3-dB frequency.

Similar to the magnitude response, the phase-response curve, shown in Fig. 2.35b, is closely defined by straight-line asymptotes. Note that at the corner frequency the phase is $-45°$, and that for $\omega \gg \omega_0$ the phase approaches $-90°$. Also note that the $-45°$/decade straight line approximates the phase function, with a maximum error of 5.7°, over the frequency range $0.1\omega_0$ to $10\omega_0$.

EXAMPLE 2.13

Consider the circuit shown in Fig. 2.36a, where an ideal voltage amplifier of gain $\mu = -100$ has a small (10-pF) capacitance connected in its feedback path. The amplifier is fed by a voltage source having a source resistance of 100 kΩ. Show that the frequency response V_o/V_s of this amplifier is equivalent to that of an STC network, and sketch the magnitude response.

Solution
According to Miller's theorem the feedback capacitance C_f can be replaced by a capacitance C_1 at the input,

$$C_1 = C_f(1 - \mu) = 1,010 \text{ pF}$$

and a capacitance C_2 at the output,

$$C_2 = C_f(1 - 1/\mu) \simeq C_f = 10 \text{ pF}$$

Thus the equivalent circuit shown in Fig. 2.36b is obtained, from which we immediately recognize the low-pass RC network at the input. Thus the amplifier transfer

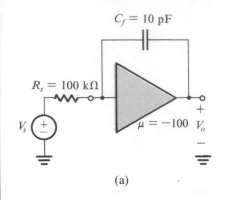

(a)

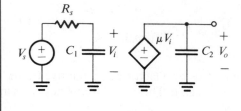

(b)

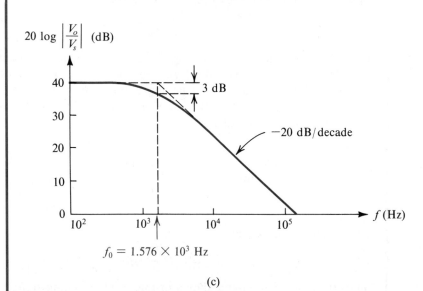

$f_0 = 1.576 \times 10^3$ Hz

(c)

Fig. 2.36 Details for Example 2.13.

function is given by

$$\frac{V_o}{V_s} = \frac{V_i}{V_s}\frac{V_o}{V_i} = \frac{V_i}{V_s}\mu$$

where V_i/V_s is the transfer function of the RC network formed by R_s and C_1. Note that since the output of the amplifier is an ideal voltage source, C_2 has no effect on the transfer function V_o/V_s. Now the input RC network has a dc transmission of unity; thus

$$\frac{V_o}{V_s} = \frac{\mu}{1 + j\omega/\omega_0}$$

where $\omega_0 = 1/C_1 R_s = 0.99 \times 10^4$ rad/s. It follows that the amplifier response is that of a low-pass STC network; thus its magnitude can be readily sketched, as shown in Fig. 2.36c.

High-Pass Networks

The transfer function $T(s)$ of an STC high-pass network always can be expressed in the form

$$T(s) = \frac{Ks}{s + \omega_0} \tag{2.30}$$

which for physical frequencies $s = j\omega$ becomes

$$T(j\omega) = \frac{K}{1 - j\omega_0/\omega} \tag{2.31}$$

where K denotes the gain as s or ω approaches infinity and ω_0 is the inverse of the time constant τ,

$$\omega_0 = 1/\tau$$

The magnitude response

$$|T(j\omega)| = \frac{K}{\sqrt{1 + (\omega_0/\omega)^2}} \tag{2.32}$$

and the phase response

$$\phi(\omega) = \tan^{-1}(\omega_0/\omega) \tag{2.33}$$

are sketched in Fig. 2.37. As in the low-pass case, the magnitude and phase curves are well defined by straight-line asymptotes. Because of the similarity (or more appropriately duality) with the low-pass case, no further explanation will be given.

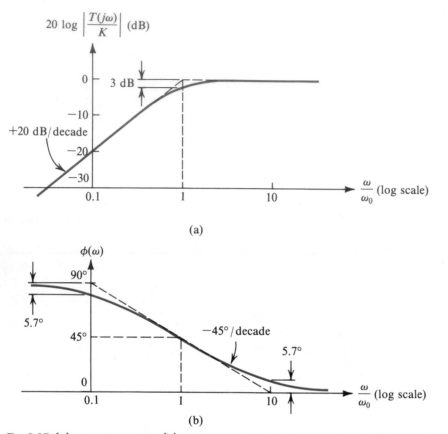

(a)

(b)

Fig. 2.37 (a) Magnitude and **(b)** phase response of STC networks of the high-pass type.

Exercises

2.27 Find the dc transmission, the corner frequency f_0, and the transmission at $f = 2\,\text{MHz}$ for the low-pass STC network shown in Fig. E2.27

Ans. $-6\,\text{dB}$; 318 kHz; $-22\,\text{dB}$

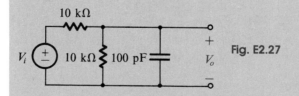

Fig. E2.27

2.28 Find the transfer function $T(s)$ of the circuit in Fig. 2.31. What type of STC network is it?

Ans. $T(s) = \dfrac{C_1}{C_1 + C_2} \dfrac{s}{s + [1/(C_1 + C_2)R]}$; HP

2.29 For the circuit of Exercise 2.28, if $R = 10$ kΩ, find the capacitor values that result in the circuit having a high-frequency transmission of 0.5 V/V and a corner frequency $\omega_0 = 10$ rad/s.

Ans. $C_1 = C_2 = 5$ μF

2.30 Find the high-frequency gain, the 3-dB frequency f_0, and the gain at $f = 1$ Hz of the capacitively coupled amplifier shown in Fig. E2.30. Assume the voltage amplifier to be ideal.

Ans.   40 dB; 15.9 Hz; 16 dB

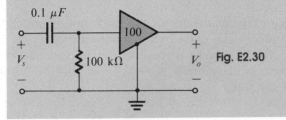

Fig. E2.30

2.8 STEP RESPONSE OF STC NETWORKS

In this section we consider the response of STC networks to the step-function signal shown in Fig. 2.38. Knowledge of the step response enables rapid evaluation of the response to other switching signal waveforms, such as pulses and square waves.

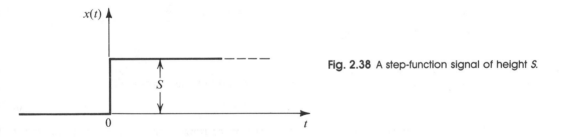

Fig. 2.38 A step-function signal of height S.

Low-Pass Networks

In response to an input step signal of height S, a low-pass STC network (with a dc gain $K = 1$) produces the waveform shown in Fig. 2.39. Note that while the input rises from 0 to S at $t = 0$, the output does not respond to this transient and simply begins to rise exponentially toward the *final* dc value of the input, S. In the long term—that is, for $t \gg \tau$—the output approaches the dc value S, a manifestation of the fact that low-pass networks faithfully pass dc.

The equation of the output waveform can be obtained from the expression

$$y(t) = Y_\infty - (Y_\infty - Y_{0+})e^{-t/\tau} \tag{2.34}$$

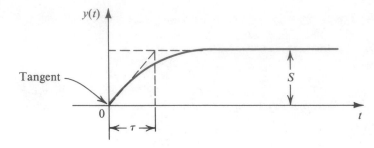

Fig. 2.39 The output $y(t)$ of a low-pass STC network excited by a step of height S.

where Y_∞ denotes the *final* value or the value toward which the output is heading and Y_{0+} denotes the value of the output immediately after $t = 0$. This equation states that the output at any time t is equal to the difference between the final value Y_∞ and a gap whose initial value is $Y_\infty - Y_{0+}$ and which is "shrinking" exponentially. In our case $Y_\infty = S$ and $Y_{0+} = 0$; thus

$$y(t) = S(1 - e^{-t/\tau}) \tag{2.35}$$

The reader's attention is drawn to the slope of the tangent to $y(t)$ at $t = 0$, which is indicated in Fig. 2.39.

High-Pass Networks

The response of an STC high-pass network (with a high-frequency gain $K = 1$) to an input step of height S is shown in Fig. 2.40. The high-pass network faithfully transmits the transient of the input signal (the step change) but blocks the dc. Thus the output at $t = 0$ follows the input,

$$Y_{0+} = S$$

and then it decays toward zero,

$$Y_\infty = 0$$

Substituting for Y_{0+} and Y_∞ in Eq. (2.34) results in the output $y(t)$,

$$y(t) = Se^{-t/\tau} \tag{2.36}$$

The reader's attention is drawn to the slope of the tangent to $y(t)$ at $t = 0$, indicated in Fig. 2.40.

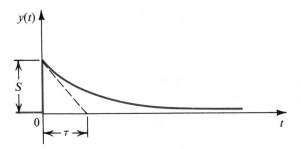

Fig. 2.40 The output $y(t)$ of a high-pass STC network excited by a step of height S.

EXAMPLE 2.14

This example is a continuation of the problem considered in Example 2.12. For an input v_I that is a 10-V step, find the condition under which the output v_O is a perfect step.

Solution

Following the analysis in Example 2.12, which is illustrated in Fig. 2.32, we have

$$v_{O1} = k_r[10(1 - e^{-t/\tau})]$$

where

$$k_r \equiv \frac{R_2}{R_1 + R_2}$$

and

$$v_{O2} = k_c(10e^{-t/\tau})$$

where

$$k_c \equiv \frac{C_1}{C_1 + C_2}$$

and

$$\tau = (C_1 + C_2)(R_1 \| R_2)$$

Thus

$$v_O = v_{O1} + v_{O2}$$
$$= 10k_r + 10e^{-t/\tau}(k_c - k_r)$$

It follows that the output can be made a perfect step of height $10k_r$ volts if we arrange that

$$k_c = k_r$$

that is, the resistive voltage-divider ratio is equal to the capacitive voltage-divider ratio.

This example illustrates an important technique, namely, that of the "compensated attenuator." An application of this technique is found in the design of the oscilloscope probe. The oscilloscope probe problem is investigated in Problem 2.43.

Exercises

2.31 For the circuit of Fig. 2.33f find v_O if i_I is a 3-mA step, $R = 1$ kΩ, and $C = 100$ pF.

Ans. $3(1 - e^{-10^7 t})$

2.32 In the circuit of Fig. 2.34f find $v_O(t)$ if i_I is a 2-mA step, $R = 2$ kΩ, and $L = 10$ μH.

Ans. $4e^{-2 \times 10^8 t}$

2.33 The amplifier circuit of Fig. E2.30 is fed with a signal source that delivers a 20-mV step. If the source resistance is 100 kΩ, find the time constant τ and $v_O(t)$.

Ans. $\tau = 2 \times 10^{-2}$ s; $v_O(t) = 1 \times e^{-50t}$

2.34 For the circuit in Fig. 2.31 with $C_1 = C_2 = 0.5 \ \mu\text{F}$, $R = 1 \ \text{M}\Omega$, find $v_O(t)$ if $v_I(t)$ is a 10-V step.

Ans. $5e^{-t}$

2.35 Show that the area under the exponential of Fig. 2.40 is equal to that of the rectangle of height S and width τ.

2.9 PULSE RESPONSE OF STC NETWORKS

Figure 2.41 shows a pulse signal whose height is P and whose width is T. We wish to find the response of STC networks to input signals of this form. Note at the outset that a pulse can be considered as the sum of two steps: a positive one of height P occurring at $t = 0$ and a negative one of height P occurring at $t = T$. Thus the response of a linear network to the pulse signal can be obtained by summing the responses to the two step signals.

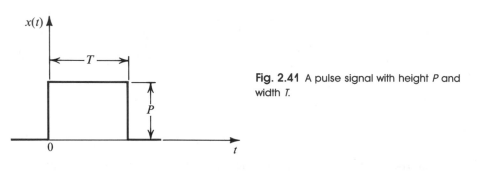

Fig. 2.41 A pulse signal with height P and width T.

Low-Pass Networks

Figure 2.42a shows the response of a low-pass STC network (having unity dc gain) to an input pulse of the form shown in Fig. 2.41. In this case we have assumed that the time constant τ is in the same range as the pulse width T. As shown, the LP network does not respond to the step change at the leading edge of the pulse; rather, the output starts to rise exponentially toward a final value of P. This exponential rise, however, will be stopped at time $t = T$, that is, at the trailing edge of the pulse when the input undergoes a negative step change. Again the output will respond by starting an exponential decay toward the final value of the input, which is zero. Finally, note that the area under the output waveform will be equal to the area under the input pulse waveform, since the LP network faithfully passes dc.

In connecting a pulse signal from one part of an electronic system to another, a low-pass effect usually occurs. The low-pass network in this case is formed by the output resistance (Thévenin's equivalent resistance) of the system part from which the signal originates and the input capacitance of the system part to which the signal is fed. This unavoidable low-pass filter will cause distortion—of the type shown in Fig. 2.42a—of the pulse signal. In a well-designed system such distortion is kept to a low value by arranging that the time constant τ be much smaller than the pulse width T.

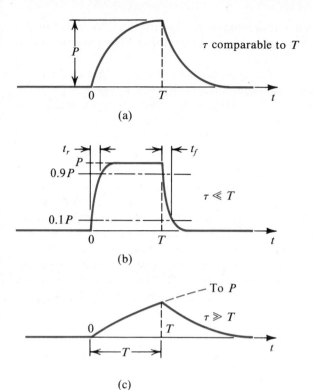

Fig. 2.42 Pulse response of STC low-pass networks.

In this case the result will be a slight rounding of the pulse edges, as shown in Fig. 2.42b. Note, however, that the edges are still exponential.

The distortion of a pulse signal by a parasitic (that is, unwanted) low-pass network is measured by its *rise time* and *fall time*. The rise time is the time taken by the amplitude to increase from 10% to 90% of the final value. Similarly, the fall time is the time during which the pulse amplitude falls from 90% to 10% of the maximum value. These definitions are illustrated in Fig. 2.42b. By use of the exponential equations of the rising and falling edges of the output waveform it can be easily shown that

$$t_r = t_f \simeq 2.2\tau \tag{2.37}$$

which can be also expressed in terms of $f_0 = \omega_0/2\pi = 1/2\pi\tau$ as

$$t_r = t_f \simeq \frac{0.35}{f_0} \tag{2.38}$$

Finally, we note that the effect of the parasitic low-pass networks that are always present in a system is to "slow down" the operation of the system: in order to keep the signal distortion within acceptable limits one has to use a relatively long pulse width (for a given time constant).

The other extreme case—namely, when τ is much larger than T—is illustrated in Fig. 2.42c. As shown, the output waveform rises exponentially toward the level P. However, since $\tau \gg T$, the value reached at $t = T$ will be much smaller than P. At

$t = T$ the output waveform starts its exponential decay toward zero. Note that in this case the output waveform bears little resemblance to the input pulse. Also note that because $\tau \gg T$ the portion of the exponential curve from $t = 0$ to $t = T$ is almost linear. Since the slope of this linear curve is proportional to the height of the input pulse, we see that the output waveform approximates the time integral of the input pulse. That is, a low-pass network with a large time constant approximates the operation of an *integrator*.

High-Pass Networks

Figure 2.43a shows the output of an STC HP network (with unity high-frequency gain) excited by the input pulse of Fig. 2.41, assuming that τ and T are comparable in value. As shown, the step transition at the leading edge of the input pulse is faithfully reproduced at the output of the HP network. However, since the HP network blocks dc, the output waveform immediately starts an exponential decay toward zero. This decay process is stopped at $t = T$ when the negative step transition of the input occurs and the HP network faithfully reproduces it. Thus at $t = T$ the output waveform exhibits an *undershoot*. Then it starts an exponential decay toward zero.

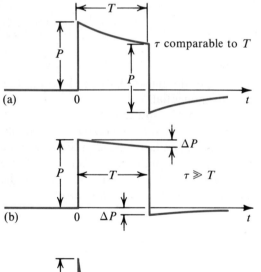

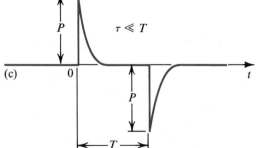

Fig. 2.43 Pulse response of STC high-pass networks.

Finally, note that the area of the output waveform above the zero axis will be equal to that below the axis for a total average area of zero, consistent with the fact that HP networks block dc.

In many applications an STC high-pass network is used to couple a pulse from one part of a system to another part. In such an application it is necessary to keep the distortion in the pulse shape as small as possible. This can be accomplished by selecting the time constant τ to be much longer than the pulse width T. If this is indeed the case, the loss in amplitude during the pulse period T will be very small, as shown in Fig. 2.43b. Nevertheless, the output waveform still swings negatively, and the area under the negative portion will be equal to that under the positive portion.

Consider the waveform in Fig. 2.43b. Since τ is much larger than T, it follows that the portion of the exponential curve from $t = 0$ to $t = T$ will be almost linear and that its slope will be equal to the slope of the exponential curve at $t = 0$, which is P/τ. We can use this value of the slope to determine the loss in amplitude ΔP as

$$\Delta P \simeq \frac{P}{\tau} T \qquad (2.39)$$

The distortion effect of the high-pass network on the input pulse is usually specified in terms of the per-unit or percentage loss in pulse height. This quantity is taken as an indication of the "sag" in the output pulse,

$$\text{Percentage sag} \equiv \frac{\Delta P}{P} \times 100 \qquad (2.40)$$

Thus

$$\text{Percentage sag} = \frac{T}{\tau} \times 100 \qquad (2.41)$$

Finally, note that the magnitude of the undershoot at $t = T$ is equal to ΔP.

The other extreme case—namely, $\tau \ll T$—is illustrated in Fig. 2.43c. In this case the exponential decay is quite rapid, resulting in the output becoming almost zero shortly beyond the leading edge of the pulse. At the trailing edge of the pulse the output swings negatively by an amount almost equal to the pulse height P. Then the waveform decays rapidly to zero. As seen from Fig. 2.43c, the output waveform bears no resemblance to the input pulse. It consists of two spikes: a positive one at the leading edge and a negative one at the trailing edge. Note that the output waveform is approximately equal to the time derivative of the input pulse. That is, for $\tau \ll T$ an STC high-pass network approximates a *differentiator*. However, the resulting differentiator is not an ideal one; an ideal differentiator would produce two impulses. Nevertheless, high-pass STC networks with short time constants are employed in some applications to produce sharp pulses or spikes at the transitions of an input waveform.

Exercises

2.36 Find the rise and fall times of a 1-μs pulse after it passes through a low-pass RC network with a corner frequency of 10 MHz.

Ans.   35 ns

2.37 Consider the pulse response of a low-pass STC circuit, shown in Fig. 2.42c. If $\tau = 100T$ find the output voltage at $t = T$. Also find the difference in the slope of the output waveform at $t = 0$ and $t = T$ (expressed as a percentage of the slope at $t = 0$).

Ans. $0.01P$; 1%

2.38 The output of an amplifier stage is connected to the input of another stage via a capacitance C. If the first stage has an output resistance of $10\text{ k}\Omega$ and the second stage has an input resistance of $40\text{ k}\Omega$, find the minimum value of C such that a 10-μs pulse exhibits less than 1% sag.

Ans. $0.02\ \mu\text{F}$

2.39 A high-pass STC network with a time constant of $100\ \mu$s is excited by a pulse of 1-V height and 100-μs width. Calculate the value of the undershoot in the output waveform.

Ans. 0.632 V

2.10 SUMMARY

● The transfer characteristic, v_O versus v_I, of a linear amplifier is a straight line with a slope equal to the voltage gain.

● Amplifiers increase the signal power and thus require dc power supplies for their operation.

● Linear amplification can be obtained from a device having a nonlinear transfer characteristic by employing dc biasing and keeping the input signal amplitude small.

● Depending on the signal to be amplified (voltage or current) and on the desired form of output signal (voltage or current), there are four basic amplifier types: voltage, current, transconductance, and transresistance amplifiers.

● Sinusoidal signals are used to measure the frequency response of amplifiers.

● The transfer function $T(s) \equiv V_o(s)/V_i(s)$ of a voltage amplifier can be determined from circuit analysis. Substituting $s = j\omega$ gives $T(j\omega)$, whose magnitude $|T(j\omega)|$ is the magnitude response, and whose phase $\phi(\omega)$ is the phase response, of the amplifier.

● Amplifiers are classified according to the shape of their frequency response, $|T(j\omega)|$.

● Application of Thévenin's and Norton's theorems can simplify circuit analysis considerably.

● Miller's theorem provides a convenient means for determining the input impedance of an amplifier that has a feedback impedance, provided that the amplifier gain is known or can be determined by other independent means.

● Single-time-constant (STC) networks are those networks that are composed of, or can be reduced to, one reactive component (L or C) and one resistance (R). The time constant τ is either L/R or CR.

● For an STC network, τ can be quickly determined by setting the external excitation to zero.

● STC networks can be classified into two categories: low-pass (LP) and high-pass (HP). LP networks pass dc and low frequencies and attenuate high frequencies. The opposite is true for HP networks.

● The gain of an LP (HP) STC circuit drops by 3 dB below the zero-frequency (infinite-frequency) value at a frequency $\omega_0 = 1/\tau$. At high frequencies (low frequencies) the gain falls off at the rate of 6 dB/octave.

● In response to an input step, an LP STC network produces an exponentially rising step. An HP STC network produces an initial step that decays exponentially to zero.

● Pulse signals undergo changes in shape as they are passed through linear circuits. The change can be minimized by ensuring that the time constant of the low-pass (high-pass) STC network is much smaller (greater) than the pulse width.

BIBLIOGRAPHY

E. F. Angelo, Jr., *Electronics: BJTs, FETs, and Microcircuits*, New York: McGraw-Hill, 1969.

L. S. Bobrow, *Elementary Linear Circuit Analysis*, 2nd ed., New York: Holt, Rinehart and Winston, 1987.

W. H. Hayt and J. E. Kemmerly, *Engineering Circuit Analysis*, 3rd ed., New York: McGraw-Hill, 1978.

R. Littauer, *Pulse Electronics*, New York: McGraw-Hill, 1965.

A. B. Macnee, "On the presentation of Miller's theorem," *IEEE Transactions on Education*, vol. E-28, no. 2, pp. 92-93, May 1985.

J. Millman and H. Taub, *Pulse, Digital, and Switching Waveforms*, New York: McGraw-Hill, 1965.

M. E. Van Valkenburg, *Network Analysis*, 3rd ed., Englewood Cliffs, N.J.: Prentice-Hall, 1974.

PROBLEMS

2.1 Consider the circuit of Fig. 2.3a with the diode reversed. Sketch the waveform of $v_O(t)$.

2.2 In the circuit of Fig. P2.2a the diode is ideal and the meter responds only to the dc (average) current flowing through it. It provides a full-scale reading when this current is 1 mA. Give the meter reading as a percentage of full scale when the input voltage applied is: (a) a sinusoid of 10 V peak; (b) the symmetrical triangular wave of Fig. P2.2b; (c) the symmetrical square wave of Fig. P2.2c.

2.3 As another illustration of the application of nonlinear devices consider the circuit shown in Fig. P2.3. Here the three one-port elements labeled N are identical and have the i-v relationship: $i = I_0 e^{v/V_0}$, where I_0 and V_0 are constants. The summing circuit is ideal in the sense that it draws zero input currents and provides at the output a voltage equal to the sum of the voltages applied at its two input ports (we shall study summing circuits in Chapter 3). Show that the circuit functions as a multiplier; that is, show that

$$i_3 = i_1 i_2 / I_0$$

2.4 A nonlinear one-port network has the i-v characteristic

$$i = 10 \sin\left(\frac{\pi}{20} v\right) \quad \text{mA} \qquad 10\,\text{V} \le v \le 10\,\text{V}$$

If v is a symmetrical triangular wave with a 10-V peak and 1-ms period, show how one can determine the waveform of i point by point. Give an expression for $i(t)$.

2.5 An amplifier fed by a sine-wave signal of 10-mV peak, delivers a sine wave output of 1-V peak to a load resistance of 1 kΩ. The input current of the amplifier is found to be a sine wave of 10-μA peak. Calculate the voltage gain, current gain, and power gain, as ratios and in decibels.

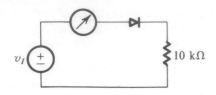

(a)

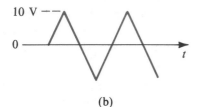

(b)

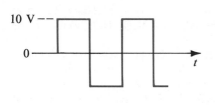

(c)

Fig. P2.2

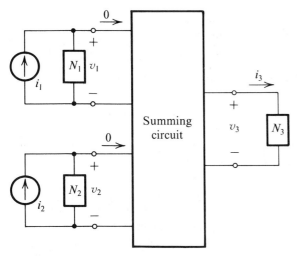

Fig. P2.3

2.6 An amplifier has a voltage gain of 60 dB and a power gain of 40 dB. Find its current gain.

2.7 An amplifier operating from ± 15-V power supplies delivers an output sine-wave signal of 20 V peak-to-peak to a 2-kΩ load resistance. The amplifier draws a current of 1.2 mA from each of its two power supplies and a negligible current from the input signal source. Find the power dissipated in the amplifier circuit and calculate the power efficiency of the amplifier.

***2.8** An amplifier designed using a single metal-oxide semiconductor (MOS) transistor has the transfer characteristic

$$v_O = 10 - 5(v_I - 2)^2$$

where v_I and v_O are in volts. This transfer characteristic applies for

$$2 \leq v_I \leq v_O + 2$$

and v_O is positive. At the limits of this region the amplifier saturates. Determine the output saturation levels L_+ and L_- and the corresponding values of v_I, and sketch the transfer characteristic. If the amplifier is biased with an input dc voltage $V_I = 2.5$ V, find the corresponding dc voltage at the output. Calculate the value of the small-signal gain at the bias point. Investigate the validity of the small-signal approximation by calculating the incremental changes in v_O corresponding to positive incremental changes in v_I (at the bias point) of 0.1 V, 0.2 V, and 0.5 V. In each case calculate Δv_O using the small-signal approximation and using the nonlinear transfer characteristic.

***2.9** The technique of biasing and small-signal approximation can also be applied to nonlinear one-port networks. Consider a silicon diode having the exponential i-v relationship

$$i_D = 10^{-15} e^{40 v_D}$$

Find the value of the dc bias voltage V_D that results in a dc current $I_D = 1$ mA. Now, if a signal v_d is superimposed on V_D and the amplitude of v_d is kept sufficiently small so that operation is restricted to an almost linear segment of the i_D-v_D curve, find the corresponding current signal i_d. (*Hint:* Remember that $e^x = 1 + x + x^2/2!$ $+ \cdots$, which for $x \ll 1$ is approximated by $e^x = 1 + x$.) Also calculate the value of the small signal or *incremental* resistance of the diode defined as $r_d \equiv v_d/i_d$. If the same diode is biased to operate at $I_D = 10$ mA, what is the corresponding value of r_d?

2.10 A signal source whose open-circuit voltage is 10 mV rms and whose short-circuit current is 1 μA rms, when connected to the input of an amplifier, supplies a current of 0.5 μA rms. The resulting output of the amplifier when connected to a 1-kΩ load is 1 V rms. Find the amplifier current gain, voltage gain, and power gain. Also find the amplifier input resistance. If, when the 1-kΩ load is disconnected from the output of the amplifier, the output voltage rises to 1.1 V rms, what is the output resistance of the amplifier? Also find its open-circuit voltage gain A_{vo}.

2.11 An amplifier utilizing ± 10-V power supplies provides an output voltage of 10 V rms to a 100-Ω load when a signal of 1 V rms is connected to its input. The input resistance of the amplifier is 100 Ω. If a current of 100 mA dc is provided by each supply, what is the total power lost (dissipated) in the amplifier? Calculate the power gain and efficiency.

2.12 An amplifier having an input resistance of 1 MΩ, an output resistance of 10 Ω, and unity voltage gain ($A_{vo} = 1$) is connected between a phonograph cartridge represented by a 1-V rms source having 1-MΩ source resistance and a small loudspeaker corresponding to a 10-Ω load. Find the output voltage, the output power, the overall voltage gain v_o/v_s, the current gain, and the power gain. Express all gains as ratios and in decibels.

*2.13 A multistage amplifier utilizes three stages, each of which has a ratio of input to output resistance of 10 to 1 and a voltage gain of 22. The last stage is designed for maximum power transfer to a load of 10 Ω (thus the output resistance is 10 Ω). Each of the other stages has an output resistance that is ten times the input resistance of the stage to which it is connected. The resistance of the source is 1 MΩ. Find the overall voltage gain (v_o/v_s) and the overall power gain.

2.14 A transconductance amplifier with input resistance of 10 kΩ, an output resistance of 10 kΩ, and a transconductance of 1 A/V is connected between a 10-kΩ source and a 10-kΩ load. Find the overall voltage gain v_o/v_s.

2.15 Figure P2.15 shows a transconductance amplifier whose output is *fed back* to its input. Find the input resistance R_{in} of the resulting one-port network. (*Hint:* Apply a test voltage v_x between the two input terminals and find the current i_x drawn from the source, then $R_{in} \equiv v_x/i_x$.)

2.16 Figure P2.16 shows a current amplifier whose output is fed back to the input. Find the input resistance R_{in} of the resulting one-port network. (See the hint given in Problem 2.15).

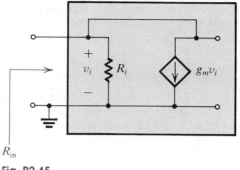

R_{in}

Fig. P2.15

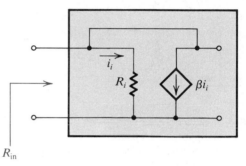

R_{in}

Fig. P2.16

2.17 A transconductance amplifier has $R_i = 2.5$ kΩ, $G_m = 40$ mA/V, and $R_o = \infty$. It feeds a load resistance of 10 kΩ. Calculate the voltage gain, the current gain, and the power gain.

2.18 Refer to Fig. 2.15c. A BJT, appropriately biased, is used as a so-called common (or grounded) emitter amplifier with E grounded, a source voltage v_s with resistance R_s connected to B and a load resistance R_L connected between C and ground. For $R_s = 10$ kΩ, $r_\pi = 5$ kΩ, $g_m = 20$ mA/V, and $R_L = 5$ kΩ, calculate the voltage gain v_o/v_s, where v_o is the voltage measured across R_L.

2.19 Consider the capacitively terminated gyrator of Fig. 2.16b. Apply an input test voltage V_x at port 1 and find the current I_x that is drawn from V_x. Thus show that the input impedance is given by

$$Z_{in}(s) \equiv \frac{V_x(s)}{I_x(s)} = \frac{sC}{g_1 g_2}$$

*2.20 In the gyrator circuit of Fig. 2.16a let each of the controlled sources have an input resistance R_i and an output resistance R_o. Now for the capacitively terminated gyrator of Fig. 2.16b with these input and output resistances included, show that the input admittance is given by

$$Y_{in}(s) = \frac{1}{R} + \frac{g_1 g_2 R}{sCR + 1}$$

where $R = R_i \| R_o$. Also show that this admittance can be represented by the circuit shown in Fig. P2.20 and find the values of L, R_1, and R_2 in terms of g_1, g_2, C, and R.

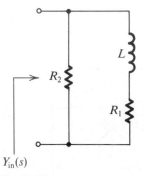

$Y_{in}(s)$

Fig. P2.20

2.21 With reference to the ideal gyrator of Fig. 2.16b, what value of inductance is obtained for $g_1 = g_2 = 5$ mA/V and $C = 0.01$ μF?

2.22 An amplifier with an input resistance of 10 kΩ, when driven by a current source of 1 μA and a source resistance of 1 MΩ, has a short-circuit output current of 10 mA and an open-circuit output voltage of 10 V. When driving a 2-kΩ load, what are the voltage gain, current gain, and power gain expressed as ratios and in dB?

2.23 For the BJT amplifier shown in Fig. E2.13 derive an expression for the voltage gain v_c/v_b, where v_c is the voltage between C and ground, and v_b is the voltage applied between b and ground (denoted v_x in Fig. E2.13). Evaluate the gain for the case $R_e = 1$ kΩ, $R_L = 10$ kΩ, $\beta = 100$, and $r_\pi = 2.5$ kΩ.

2.24 What is the angular frequency ω (rad/s) and frequency f(Hz) of a sine wave of period: **(a)** 1 s, **(b)** 1 μs, **(c)** 1 ns, **(d)** 1/60 s, **(e)** 16.6 ms?

2.25 Determine the period T (seconds) of sine waveforms characterized by the following frequencies.
(a) $\omega = 2\pi \times 10^3$ rad/s
(b) $f = 10^{-3}$ Hz
(c) $f = 60$ Hz
(d) $\omega = 376.8$ rad/s
(e) $\omega = 2\pi$ rad/s

(f) $f = 10$ GHz
(g) $f = 0.1$ MHz

***2.26 A linear amplifier has frequency response of the type shown in Fig. 2.18. Analytically it is specified as follows:

1. At intermediate frequencies, in the range 100 Hz to 10 kHz, $A_v = 100$.
2. At high frequencies, above 10 kHz,

$$A_v = \frac{100}{1 + jf/10^4}$$

3. At low frequencies, below 100 Hz,

$$A_v = \frac{100}{1 + 10^2/jf}$$

where f is the frequency in hertz.

A 10-mV peak-to-peak symmetrical square wave of frequency f_0 is applied to the amplifier. Using the Fourier series for a square wave given in Equation (1.1), calculate the magnitude and phase of the first six nonzero terms of the series for: $f_0 = 10$ Hz, 50 Hz, 200 Hz, 5 kHz, and 20 kHz. Sketch the output waveform in each case.

2.27 Derive expressions for the transfer function $V_o(s)/V_i(s)$ of each of the two-port RC networks shown in Fig.

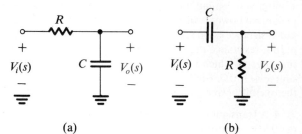

(a)   (b)

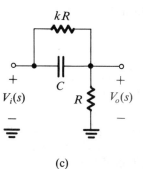

(c)

Fig. P2.27

P2.27. Evaluate the magnitude and phase responses for $R = 10\ \text{k}\Omega$, $C = 0.1\ \mu\text{F}$, and $k = 10$ at $\omega = 1{,}000\ \text{rad/s}$.

2.28 Using the transfer function derived in Problem 2.27 for the two-port STC circuit of Fig. P2.27c, construct its Bode plot—that is, a plot of the magnitude response in dB versus frequency on a log axis.

2.29 Consider the magnitude response of an STC low-pass circuit. Find the frequencies at which the response falls below the dc value by 1.5 dB, 1.0 dB, and 0.75 dB. These values can be used in estimating the 3-dB frequency of a cascade of n identical networks of this type for $n = 2$, 3, and 4, respectively, provided that the networks in the cascade are isolated from one another by buffer amplifiers as illustrated in the following problem.

2.30 Using the idea introduced in Problem 2.29 find the 3-dB frequency of the network shown in Fig. P2.30. The buffer amplifiers are ideal and have unity gain.

**2.31 Figure P2.31 shows a tuned, or bandpass, amplifier. Derive its transfer function $T(s) \equiv V_o(s)/V_i(s)$ and find $T(j\omega)$ and $|T(j\omega)|$. What is the value of gain at $\omega - 0$ and at $\omega = \infty$? At what frequency is the gain maximum? What is the value of the maximum gain? Find the two frequencies at which the gain drops by 3 dB below the maximum value. The 3-dB bandwidth of the amplifier is the difference between these two frequencies. Find its value.

2.32 Consider the circuit of Fig. E2.18, with $R = 1\ \text{k}\Omega$ and $C = 1\ \mu\text{F}$. If the input is a sinusoid of frequency $\omega = 10^3\ \text{rad/s}$ and 10-V amplitude, find the output voltage $v_o(t)$.

2.33 Consider the Thévenin equivalent circuit characterized by V_t and Z_t. Find the open-circuit voltage V_{oc} and the short-circuit current (that is, the current that flows when the terminals are shorted together) I_{sc}. Express Z_t in terms of V_{oc} and I_{sc}.

2.34 Repeat Problem 2.33 for a Norton equivalent characterized by I_n and Z_n.

2.35 A voltage divider consists of a 9-kΩ resistor connected to $+10$ V and a resistor of 1 kΩ connected to ground. What is the Thévenin equivalent of this voltage divider? What output voltage results if it is loaded with 1 kΩ? Calculate this two ways: directly and using your Thévenin equivalent.

2.36 Find the output voltage and output resistance of the circuit shown in Fig. P2.36 by considering a succession of Thévenin equivalent circuits.

2.37 Repeat Example 2.8 with a resistance R_B connected between B and ground in Fig. 2.27 (that is, rather than grounding the base B as indicated in Fig. 2.27).

2.38 Calculate the input capacitance of each of the circuits shown in Fig. P2.38. Assume the amplifiers to be ideal and to have the voltage gains indicated.

2.39 Figure P2.39a shows two transconductance amplifiers connected in a special configuration. Find v_o in terms of v_1 and v_2. Let $g_m = 100\ \text{mA/V}$ and $R = 5\ \text{k}\Omega$.

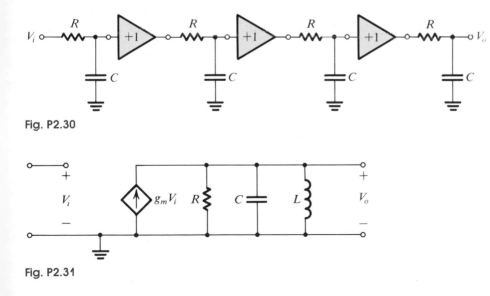

Fig. P2.30

Fig. P2.31

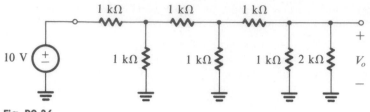

Fig. P2.36

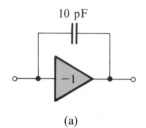 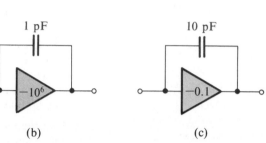

(a)   (b)   (c)

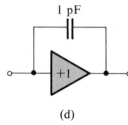 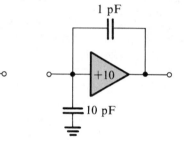

(d)   (e)

Fig. P2.38

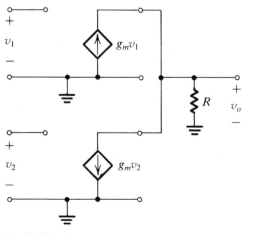 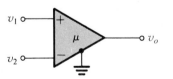

Fig. P2.39

If $v_1 = v_2 = 1$ V, find the value of v_o. Also find v_o for the case $v_1 = 1.01$ V and $v_2 = 0.99$ V. (*Note:* This circuit is called a *differential amplifier* and is given the symbol shown in Fig. P2.39b. A particular type of differential amplifier known as an operational amplifier will be studied in Chapter 3.)

*2.40 Figure P2.40a shows the circuit symbol of the junction field-effect transistor (JFET), which we shall study in Chapter 6. As indicated, the JFET has three terminals. When the gate terminal G is connected to the source terminal S, the two-terminal device shown in Fig. P2.40b is obtained. Its *i-v* characteristic is given by

$$i = I_{DSS}\left[2\frac{v}{V_P} - \left(\frac{v}{V_P}\right)^2\right] \quad \text{for } v \leq V_P$$

$$i = I_{DSS} \quad \text{for } v \geq V_P$$

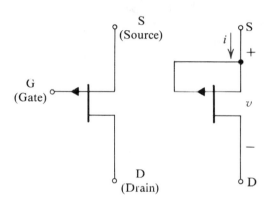

(a) (b)

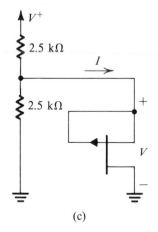

(c)

Fig. P2.40

where I_{DSS} and V_P are constants for the particular JFET. Now consider the circuit shown in Fig. P2.40c and let $V_P = 2$ V and $I_{DSS} = 2$ mA. For $V^+ = 10$ V show that the JFET is operating in the constant-current mode and find the voltage across it. What is the minimum value of V^+ for which this mode of operation is maintained? For $V^+ = 2$ V find the values of I and V.

2.41 Consider the circuit of Fig. 2.32a and the equivalent shown in (d) and (e). There the output, $v_O = v_{O1} + v_{O2}$, is the sum of outputs of a low-pass and a high-pass circuit, each with the time constant $\tau = (C_1 + C_2)(R_1 \| R_2)$. What is the condition that makes the contribution of the low-pass circuit at zero frequency equal to the contribution of the high-pass circuit at infinite frequency? Show that this condition can be expressed as $C_1 R_1 = C_2 R_2$. If this condition applies, sketch $|V_o/V_i|$ versus frequency for the case $R_1 = R_2$.

2.42 Use the voltage-divider rule to find the transfer function $V_o(s)/V_i(s)$ of the circuit in Fig. 2.32a. Show that the transfer function can be made independent of frequency if the condition $C_1 R_1 = C_2 R_2$ applies. Under such condition the circuit is called a *compensated attenuator*. Find the transmission of the compensated attenuator in terms of R_1 and R_2.

**2.43 The circuit of Fig. 2.32a is used as a compensated attenuator (see Problems 2.41 and 2.42) for the oscilloscope probe. The object is to reduce the signal voltage applied to the input amplifier of the oscilloscope, with the signal attenuation independent of frequency. The probe itself includes R_1 and C_1, while R_2 and C_2 model the oscilloscope input circuit. For an oscilloscope having an input resistance of 1 MΩ and an input capacitance of 30 pF design a compensated "10 to 1 probe"—that is, a probe that attenuates the input signal by a factor of 10. Find the input impedance of the oscilloscope together with the probe, which is the impedance seen by v_I in Fig. 2.32a. Show that this impedance is 10 times higher than that of the oscilloscope itself. This is a great advantage of the 10:1 probe.

2.44 In the circuits of Fig. 2.33 and 2.34 let $L = 10$ mH, $C = 0.01$ μF, and $R = 1$ kΩ. At what frequency does a phase angle of $45°$ occur?

*2.45 An amplifier with a frequency response of the type shown in Fig. 2.18 is specified to have a phase shift no greater than $11.4°$ over its bandwidth, which extends from 100 Hz to 1 kHz. It has been found that the gain falloff at the low-frequency end is determined by the response of a high-pass STC circuit, and that at the high-frequency end is determined by a low-pass STC circuit. What do you expect the corner frequencies of these two circuits to be? What is the drop in gain in decibels (relative to the maximum gain) at the two frequencies that define the amplifier bandwidth? What are the frequencies at which the drop in gain is 3 dB?

*2.46 Consider a voltage amplifier with $A_{vo} = -100$ V/V, $R_o = 0$, $R_i = 10\ k\Omega$, and an input capacitance C_i (in parallel with R_i) of 10 pF. The amplifier has a feedback capacitance (a capacitance connected between output and input) $C_f = 1\ pF$. The amplifier is fed with a voltage source V_s having a resistance $R_s = 10\ k\Omega$. Find the amplifier transfer function $V_o(s)/V_s(s)$ and sketch its magnitude response versus frequency (dB versus frequency on a log axis).

2.47 For the circuit in Fig. P2.47 assume the voltage amplifier to be ideal. Derive the transfer function $V_o(s)/V_i(s)$. What type of STC response is this? For $C = 0.01\ \mu F$ and $R = 100\ k\Omega$ find the corner frequency.

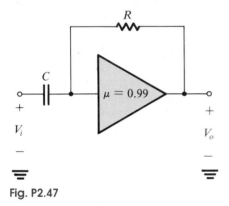

Fig. P2.47

2.48 What type of STC response does the circuit of Fig. P2.48 have? Find its corner frequency. Assume the voltage amplifier to be ideal.

*2.49 Assuming the voltage amplifier to be ideal, find the input resistance R_{in} of the circuit of Fig. P2.49.

2.50 For the circuits of Figs. 2.33b and 2.34b find $v_o(t)$ if v_I is a 10-V step, $R = 1\ k\Omega$, and $L = 1\ mH$.

2.51 Consider the exponential response of an STC low-pass network to a 10-V step input. In terms of the time constant τ find the time taken for the output to reach 5 V, 9 V, 9.9 V, and 9.99 V.

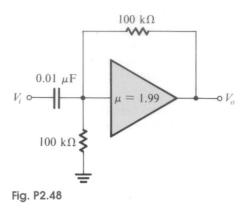

Fig. P2.48

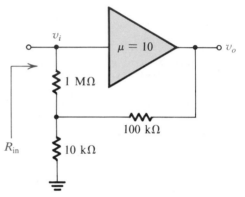

Fig. P2.49

2.52 The high-frequency response of an oscilloscope is specified to be like that of an STC LP network with a 100-MHz corner frequency. If this oscilloscope is used to display an ideal step waveform, what rise time (10% to 90%) would you expect to observe?

2.53 An oscilloscope whose step response is like that of a low-pass STC network has a rise time of t_s seconds. If an input signal having a rise time of t_w seconds is displayed, the waveform displayed will have a rise time t_d seconds, which can be found using the empirical formula $t_d = \sqrt{t_s^2 + t_w^2}$. If $t_s = 35\ ns$, what is the 3-dB frequency of the oscilloscope? What is the observed rise time for a waveform rising in 100 ns, 35 ns, and 10 ns? What is the actual rise time of a waveform whose displayed rise time is 49.5 ns?

2.54 A pulse of 10-ms width is transmitted through a system characterized as having an STC high-pass response with a corner frequency of 10 Hz. What undershoot would you expect to get with a 10-V pulse?

2.55 An RC differentiator having a time constant τ is used to implement a short-pulse detector. When a long pulse with $T \gg \tau$ is fed to the circuit, the positive and negative peak outputs are of equal magnitude. At what pulse width does the negative output peak differ from the positive one by 10%?

2.56 A high-pass STC network with a time constant of 1 ms is excited by a pulse of 10-V height and 1-ms width. Calculate the value of the undershoot in the output waveform. If an undershoot of 1 V or less is required, what is the time constant necessary?

2.57 A capacitor C is used to couple the output of an amplifier stage to the input of the next stage. If the first stage has an output resistance of 2 kΩ and the second stage has an input resistance of 3 kΩ, find the value of C so that a 1-ms pulse exhibits less than 1% sag. What is the associated 3-dB frequency?

2.58 An RC differentiator is used to convert a step voltage change V to a single pulse for a digital logic appli-cation. The logic circuit that the differentiator drives distinguishes signals above $V/2$ as "high" and below $V/2$ as "low." What must the time constant of the network be to convert a step input into a pulse that will be interpreted as "high" for 10 μs?

2.59 For the circuit in Fig. P2.59, $\mu = -100$, $C = 100$ pF, and the amplifier is ideal. Find the value of R so that the gain $|V_o/V_s|$ has a 3-dB frequency of 1 kHz.

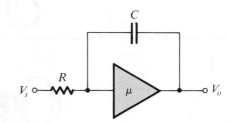

Fig. P2.59

Chapter

Operational Amplifiers

INTRODUCTION

Having learned basic amplifier concepts and terminology, we are now ready to undertake the study of a circuit building block of universal importance: the operational amplifier (op amp). Although op amps have been in use for a long time, their applications were initially in the areas of analog computation and instrumentation. Early op amps were constructed from discrete components (transistors and resistors), and their cost was prohibitively high (tens of dollars). In the mid-1960s the first integrated-circuit (IC) op amp was produced. This unit (the μA 709) was made up of a relatively large number of transistors and resistors all on the same silicon chip. Although its characteristics were poor (by today's standards) and its price was still quite high, its appearance signaled a new era in electronic circuit design. Electronics engineers started using op amps in large quantities, which caused their price to drop dramatically. They also demanded better-quality op amps. Semiconductor manufac-

turers responded quickly, and within the span of few years high-quality op amps became available at extremely low prices (tens of cents) from a large number of suppliers.

One of the reasons for the popularity of the op amp is its versatility. As we will shortly see, one can do almost anything with op amps! Equally important is the fact that the IC op amp has characteristics that closely approach the assumed ideal. This implies that it is quite easy to design circuits using the IC op amp. Also, op-amp circuits work at levels that are quite close to their predicted theoretical performance. It is for this reason that we are studying op amps at this early stage. It is expected that by the end of this chapter the reader should be able to successfully design nontrivial circuits using op amps.

As already implied, an IC op amp is made up of a large number of transistors, resistors, and (sometimes) one capacitor connected in a rather complex circuit. Since we have not yet studied transistor circuits, the circuit inside the op amp will not be discussed in this chapter. Rather, we will treat the op amp as a circuit building block and study its terminal characteristics and its applications. This approach is quite satisfactory in many op-amp applications. Nevertheless, for the more difficult and demanding applications it is quite useful to know what is inside the op-amp package. This topic will be studied in Chapter 13. Finally, it should be mentioned that more advanced applications of op amps will appear in later chapters. □

3.1 THE OP-AMP TERMINALS

From a signal point of view the op amp has three terminals: two input terminals and one output terminal. Figure 3.1 shows the symbol that we shall use to represent the op amp. Terminals 1 and 2 are input terminals, and terminal 3 is the output terminal. As explained in Section 2.2, amplifiers require dc power to operate. Most IC op amps require two dc power supplies, as shown in Fig. 3.2. Two terminals, 4 and 5, are brought out of the op-amp package and connected to a positive voltage V^+ and a negative voltage V^-, respectively. In Fig. 3.2b we explicitly show the two dc power supplies as batteries with a common ground. It is interesting to note that the reference grounding point in op-amp circuits is just the common terminal of the two power supplies; that is, no terminal of the op-amp package is physically connected to ground. In what follows we will not explicitly show the op-amp power supplies.

Fig. 3.1 Circuit symbol for the op amp.

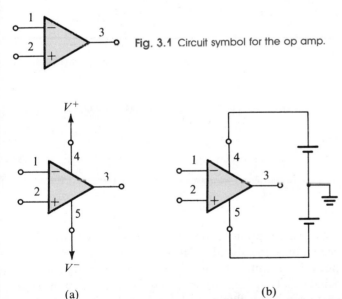

Fig. 3.2 The op amp shown connected to dc power supplies.

(a)   (b)

In addition to the three signal terminals and the two power-supply terminals, an op amp may have other terminals for specific purposes. These other terminals can include terminals for frequency compensation and terminals for offset nulling; both functions will be explained in later sections.

3.2 THE IDEAL OP AMP

We now consider the circuit function of the op amp. The op amp is supposed to sense the difference between the voltage signals applied at its two input terminals (that is, the quantity $v_2 - v_1$), multiply this by a number A, and cause the resulting voltage $A(v_2 - v_1)$ to appear at output terminal 3. Here it should be emphasized that when we talk about the voltage at a terminal we mean the voltage between that terminal and ground; thus v_1 means the voltage applied between terminal 1 and ground.

The ideal op amp is not supposed to draw any input current; that is, the signal current into terminal 1 and the signal current into terminal 2 are both zero. In other words, the input impedance of an ideal op amp is supposed to be infinite.

How about the output terminal 3? This terminal is supposed to act as the output terminal of an ideal voltage source. That is, the voltage between terminal 3 and ground will always be equal to $A(v_2 - v_1)$ and will be independent of the current that may be drawn from terminal 3 into a load impedance. In other words, the output impedance of an ideal op amp is supposed to be zero.

Putting together all of the above we arrive at the equivalent circuit model shown in Fig. 3.3. Note that the output is in phase with (has the same sign as) v_2 and out of phase with (has the opposite sign of) v_1. For this reason input terminal 1 is called the *inverting input terminal* and is distinguished by a "$-$" sign, while input terminal 2 is called the *noninverting input terminal* and is distinguished by a "$+$" sign.

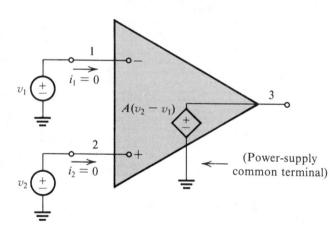

Fig. 3.3 Equivalent circuit of the ideal op amp.

As can be seen from the above description the op amp responds only to the *difference* signal $v_2 - v_1$ and hence ignores any signal *common* to both inputs. That is, if $v_1 = v_2 = 1$ V, then the output will—ideally—be zero. We call this property *common-mode rejection*, and we conclude that an ideal op amp has infinite common-mode rejection. We will have more to say about this point later. For the time being note that the op amp is a *differential-input*, *single-ended-output* amplifier, with the latter term referring to the fact that the output appears between terminal 3 and ground. Furthermore, gain A is called the *differential gain*, for obvious reasons. Perhaps not so obvious is another name that we will attach to A: the *open-loop gain*. The reason for this latter name will become obvious later on when we "close the loop" around the op amp and define another gain, the closed-loop gain.

An important characteristic of op amps is that they are direct-coupled devices or dc amplifiers, where dc stands for direct-coupled (it could equally well stand for direct current, since a direct-coupled amplifier is one that amplifies signals whose frequency is as low as zero). The fact that op amps are direct-coupled devices will allow us to use them in many important applications. Unfortunately, though, the direct-coupling property can cause some serious problems, as will be discussed in a later section.

How about bandwidth? The ideal op amp has a gain A that remains constant down to zero frequency and up to infinite frequency. That is, ideal op amps will amplify signals of any frequency with equal gain.

We have discussed all of the properties of the ideal op amp except for one, which in fact is the most important. This has to do with the value of A. The ideal op amp should have a gain A whose value is very large and ideally infinite. One may justifiably ask: If the gain A is infinite, how are we going to use the op amp? The answer is very simple: In almost all applications the op amp will *not* be used in an open-loop configuration. Rather, we will apply feedback to close the loop around the op amp, as will be illustrated in detail in Section 3.3.

Exercises

3.1 Consider an op amp that is ideal except that its open-loop gain $A = 10^3$. The op amp is used in a feedback circuit and the voltages appearing at two of its three signal terminals are measured. In each of the following cases, use the measured values to find the expected value of the voltage at the third terminal.
(a) $v_2 = 0$ V and $v_3 = 1$ V;   **(b)** $v_2 = +5$ V and $v_3 = -10$ V;   **(c)** $v_1 = 1.001$ V and $v_2 = 0.999$ V;   **(d)** $v_1 = -3.6$ V and $v_3 = -3.6$ V.
Ans. **(a)** $v_1 = -0.001$ V;   **(b)** $v_1 = +5.01$ V;   **(c)** $v_3 = -2$ V;   **(d)** $v_2 = -3.6036$ V

3.2 The internal circuit of a particular op amp can be modeled by the circuit shown in Fig. E3.2 (on the next page). Express v_3 as a function of v_1 and v_2. For the case $g_m = 20$ mA/V, $R = 10$ kΩ, and $\mu = 100$, find the value of the open-loop gain A.
Ans. $v_3 = \mu g_m R(v_2 - v_1)$; $A = 20,000$ V/V

3.3   ANALYSIS OF CIRCUITS CONTAINING IDEAL OP AMPS— THE INVERTING CONFIGURATION

Consider the circuit shown in Fig. 3.4, which consists of one op amp and two resistors R_1 and R_2. Resistor R_2 is connected from the output terminal of the op amp, terminal 3, *back* to the *inverting* or *negative* input terminal, terminal 1. We speak of R_2 as applying *negative feedback;* if R_2 were connected between terminals 3 and 2 we would have called this *positive feedback*. Note also that R_2 closes the loop around the op amp. In addition to adding R_2, we have grounded terminal 2 and connected a resistor R_1 between terminal 1 and an input signal source with a voltage v_I. The output of the overall circuit is taken at terminal 3 (that is, between terminal 3 and ground). Terminal 3 is, of course, a convenient point to take the output, since the impedance level there is ideally zero. Thus the voltage v_O will not

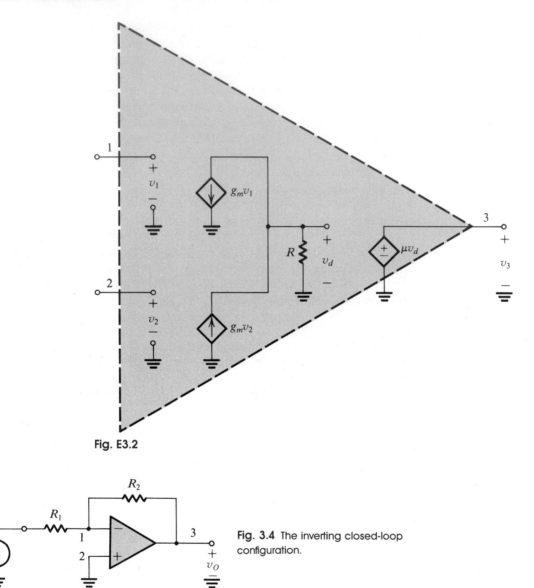

Fig. E3.2

Fig. 3.4 The inverting closed-loop configuration.

depend on the value of the current that might be supplied to a load impedance connected between terminal 3 and ground.

The Closed-Loop Gain

We now wish to analyze the circuit in Fig. 3.4 to determine the *closed-loop gain G,* defined as

$$G \equiv \frac{v_O}{v_I}$$

We will do so assuming the op amp to be ideal. Figure 3.5a shows the equivalent circuit, and the analysis proceeds as follows: The gain A is very large (ideally infinite). If we assume that the circuit is "working" and producing a finite output voltage at terminal 3, then the voltage between the op-amp input terminals should be negligibly small. Specifically, if we call the output voltage v_O, then, by definition,

$$v_2 - v_1 = \frac{v_O}{A} \simeq 0$$

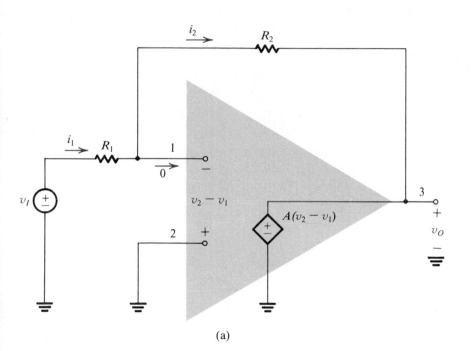

(a)

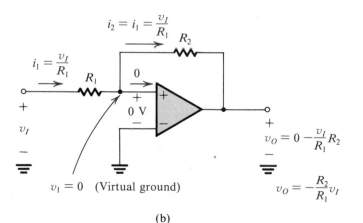

(b)

Fig. 3.5 Analysis of the inverting configuration.

It follows that the voltage at the inverting input terminal (v_1) is given by

$$v_1 \simeq v_2$$

That is, because the gain A approaches infinity, the voltage v_1 approaches v_2. We speak of this as the two input terminals "tracking each other in potential." We also speak of a "virtual short circuit" that exists between the two input terminals. Here the word *virtual* should be emphasized, and one should *not* make the mistake of physically shorting terminals 1 and 2 together while analyzing a circuit. A *virtual short circuit* means that whatever voltage is at 2 will automatically appear at 1 because of the infinite gain A. But terminal 2 happens to be connected to ground; thus $v_2 = 0$ and $v_1 \simeq 0$. We speak of terminal 1 as being a *virtual ground*—that is, having zero voltage but not physically connected to ground.

Now that we have determined v_1 we are in a position to apply Ohm's law and find the current i_1 through R_1 (see Fig. 3.5) as follows:

$$i_1 = \frac{v_I - v_1}{R_1} \simeq \frac{v_I}{R_1}$$

Where will this current go? It cannot go into the op amp, since the ideal op amp has an infinite input impedance and hence draws zero current. It follows that i_1 will have to flow through R_2 to the low-impedance terminal 3. We can then apply Ohm's law to R_2 and determine v_O; that is,

$$v_O = v_1 - i_1 R_2$$

$$= 0 - \frac{v_I}{R_1} R_2$$

Thus

$$\frac{v_O}{v_I} = -\frac{R_2}{R_1}$$

which is the required closed-loop gain. Figure 3.5b illustrates some of these analysis steps.

We thus see that the closed-loop gain is simply the ratio of the two resistances R_2 and R_1. The minus sign means that the closed-loop amplifier provides signal inversion. Thus if $R_2/R_1 = 10$ and we apply at the input (v_I) a sine-wave signal of 1 V peak-to-peak, then the output v_O will be a sine wave of 10 V peak-to-peak and phase-shifted $180°$ with respect to the input sine wave. Because of the minus sign associated with the closed-loop gain this configuration is called the *inverting configuration*.

The fact that the closed-loop gain depends entirely on external passive components (resistors R_1 and R_2) is a very interesting one. It means that we can make the closed-loop gain as accurate as we want by selecting passive components of appropriate accuracy. It also means that the closed-loop gain is (ideally) independent of the op-amp gain. This is a dramatic illustration of negative feedback: we started out with an amplifier having very large gain A, and through applying negative feedback we have obtained a closed-loop gain R_2/R_1 that is much smaller than A but is stable and predictable. That is, we are trading gain for accuracy.

Effect of Finite Open-Loop Gain

The points just made are more clearly illustrated by deriving an expression for the closed-loop gain under the assumption that the op-amp open-loop gain A is finite. Figure 3.6 shows the analysis. If we denote the output voltage v_O, then the voltage between the two input terminals of the op amp will be v_O/A. Since the positive input

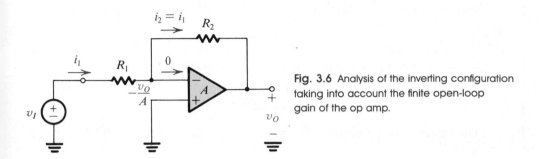

Fig. 3.6 Analysis of the inverting configuration taking into account the finite open-loop gain of the op amp.

terminal is grounded, the voltage at the negative input terminal must be $-v_O/A$. The current i_1 through R_1 can now be found from

$$i_1 = \frac{v_I - (-v_O/A)}{R_1} = \frac{v_I + v_O/A}{R_1}$$

The infinite input impedance of the op amp forces the current i_1 to flow entirely through R_2. The output voltage v_O can thus be determined from

$$v_O = -\frac{v_O}{A} - i_1 R_2$$

$$= -\frac{v_O}{A} - \left(\frac{v_I + v_O/A}{R_1}\right) R_2$$

Collecting terms, the closed-loop gain G is found as

$$G \equiv \frac{v_O}{v_I} = \frac{-R_2/R_1}{1 + (1 + R_2/R_1)/A} \tag{3.1}$$

We note that as A approaches ∞, G approaches the ideal value of $-R_2/R_1$. Also, from Fig. 3.6 we see that as A approaches ∞, the voltage at the inverting input terminal approaches zero. This is the virtual ground assumption we used in our earlier analysis when the op amp was assumed to be ideal. Finally note that Eq. (3.1) in fact indicates that to minimize the dependence of the closed-loop gain G on the value of the open-loop gain A, we should make

$$1 + \frac{R_2}{R_1} \ll A$$

EXAMPLE 3.1

Consider the inverting configuration with $R_1 = 1$ kΩ and $R_2 = 100$ kΩ.

(a) Find the closed-loop gain for the cases $A = 10^3$, 10^4, and 10^5. In each case determine the percentage error in the magnitude of G relative to the ideal value of R_2/R_1 (obtained with $A = \infty$). Also determine the voltage v_1 that appears at the inverting input terminal when $v_I = 0.1$ V.

(b) If the open-loop gain A changes from 100,000 to 50,000, what is the corresponding percentage change in the magnitude of the closed-loop gain G?

Solution

(a) Substituting the given values in Eq. (3.1) we obtain the values given in the following table where the percentage error ε is defined as

$$\varepsilon \equiv \frac{|G| - (R_2/R_1)}{(R_2/R_1)} \times 100$$

The values of v_1 are obtained from $v_1 = v_O/A = Gv_I/A$ with $v_I = 0.1$ V.

A	$|G|$	ε	v_1
10^3	90.83	-9.17%	-9.08 mV
10^4	99.00	-1.00%	-0.99 mV
10^5	99.90	-0.10%	-0.10 mV

(b) Using Eq. (3.1) we find that for $A = 50,000$, $|G| = 99.80$. Thus halving the open-loop gain results in a change of -0.1% in the closed-loop gain!

Input Resistance

The input resistance of the closed-loop inverting amplifier of Fig. 3.4 is simply equal to R_1. This can be seen from Fig. 3.5, where

$$R_{\text{in}} \equiv \frac{v_I}{i_1} = \frac{v_I}{v_I/R_1} = R_1$$

Thus to make R_{in} high we should select a high value for R_1. However, if the required gain R_2/R_1 is also high, then R_2 could become impractically large. We may conclude that the inverting configuration suffers from a low input resistance.

EXAMPLE 3.2

Assuming the op amp to be ideal, derive an expression for the closed-loop gain v_O/v_I of the circuit shown in Fig. 3.7. For $R_1 = 1$ kΩ, $R_2 = R_4 = 10$ kΩ, and $R_3 = 100$ Ω determine the value of the gain.

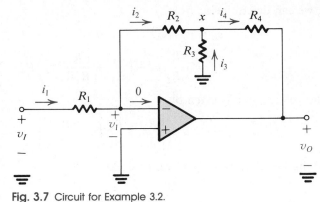

Fig. 3.7 Circuit for Example 3.2.

Solution

The analysis begins at the inverting input terminal of the op amp, where the voltage is

$$v_1 = \frac{-v_O}{A} = \frac{-v_O}{\infty} = 0$$

Here we have assumed that the circuit is "working" and producing a finite output voltage v_O. Knowing v_1 we can determine the current i_1, as follows:

$$i_1 = \frac{v_I - v_1}{R_1} = \frac{v_I - 0}{R_1} = \frac{v_I}{R_1}$$

Since zero current flows into the inverting input terminal, all of i_1 will flow through R_2, and thus

$$i_2 = i_1 = \frac{v_I}{R_1}$$

Now we can determine the voltage at node x:

$$v_x = v_1 - i_2 R_2 = 0 - \frac{v_I}{R_1} R_2 = -\frac{R_2}{R_1} v_I$$

This in turn enables us to find the current i_3,

$$i_3 = \frac{0 - v_x}{R_3} = \frac{R_2}{R_1 R_3} v_I$$

Next a node equation at x yields i_4,

$$i_4 = i_2 + i_3 = \frac{v_I}{R_1} + \frac{R_2}{R_1 R_3} v_I$$

Finally, we can determine v_O from

$$v_O = v_x - i_4 R_4$$

$$= -\frac{R_2}{R_1} v_I - \left(\frac{v_I}{R_1} + \frac{R_2}{R_1 R_3} v_I\right) R_4$$

Thus, the voltage gain is given by

$$\frac{v_O}{v_I} = -\left[\frac{R_2}{R_1} + \frac{R_4}{R_1}\left(1 + \frac{R_2}{R_3}\right)\right]$$

Substituting the given numerical values yields

$$\frac{v_O}{v_I} = -1{,}020 \text{ V/V}$$

We note that the same value of gain can be obtained by replacing the T network in the feedback with a single resistor of 1,020 kΩ. The T network, however, is a more practical arrangement because it uses relatively small resistances (that is, the maximum resistance required is 10 kΩ).

Exercises

3.3 Use the circuit of Fig. 3.4 to design an inverting amplifier having a gain of −10 and an input resistance of 10 kΩ. Give the values of R_1 and R_2.

Ans. $R_1 = 10 \text{ k}\Omega$; $R_2 = 100 \text{ k}\Omega$

3.4 The circuit shown in Fig. E3.4a can be used to implement a transresistance amplifier (see Section 2.3). Find the value of the input resistance R_i, the transresistance R_m, and the output resistance R_o of the transresistance amplifier. If the signal source shown in Fig. E3.4b is connected to the input of the transresistance amplifier, find its output voltage.

Ans. $R_i = 0$;. $R_m = -10 \text{ k}\Omega$; $R_o = 0$; $v_o = -5 \text{ V}$

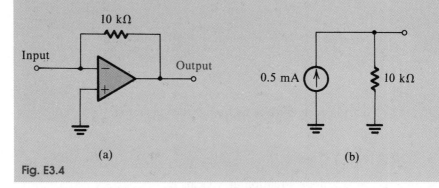

(a)   (b)

Fig. E3.4

3.4 OTHER APPLICATIONS OF THE INVERTING CONFIGURATION

Rather than using two resistors R_1 and R_2, we may use two impedances Z_1 and Z_2, as shown in Fig. 3.8. The closed-loop gain, or more appropriately the closed-loop

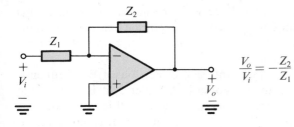

Fig. 3.8 The inverting configuration with general impedances in the feedback and at the input.

$$\frac{V_o}{V_i} = -\frac{Z_2}{Z_1}$$

transfer function, is given by

$$\frac{V_o}{V_i} = -\frac{Z_2}{Z_1}$$

As a special case consider first the following:

$$Z_1 = R \quad \text{and} \quad Z_2 = \frac{1}{sC}$$

$$\frac{V_o}{V_i} = -\frac{1}{sCR} \tag{3.2a}$$

which for physical frequencies, $s = j\omega$, becomes

$$\frac{V_o}{V_i} = -\frac{1}{j\omega CR} \tag{3.2b}$$

This transfer function can be shown to correspond to integration; that is, $v_o(t)$ will be the integral of $v_I(t)$. To see this in the time domain, consider the corresponding circuit shown in Fig. 3.9. It is easy to see that the current i_1 is given by

$$i_1 = \frac{v_I(t)}{R}$$

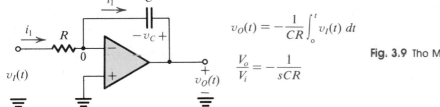

$$v_O(t) = -\frac{1}{CR}\int_0^t v_I(t)\, dt$$

$$\frac{V_o}{V_i} = -\frac{1}{sCR}$$

Fig. 3.9 The Miller or inverting integrator.

If at time $t = 0$ the voltage across the capacitor (measured in the direction indicated) is V_C, then

$$v_O(t) = V_C - \frac{1}{C}\int_0^t i_1(t)\, dt$$

$$= V_C - \frac{1}{CR}\int_0^t v_I(t)\, dt$$

Thus $v_O(t)$ is the time integral of $v_I(t)$, and voltage V_C is the initial condition of this integration process. The time constant CR is called the *integration time constant*. This integrator circuit is inverting because of the minus sign associated with its transfer function; it is known as the *Miller integrator*.

From the transfer function in Eq. (3.2) and our discussion of the frequency response of low-pass single-time-constant networks in Chapter 2, it is easy to see that the Miller integrator will have the magnitude plot shown in Fig. 3.10, which is identical to that of a low-pass network with zero break frequency. It is important to

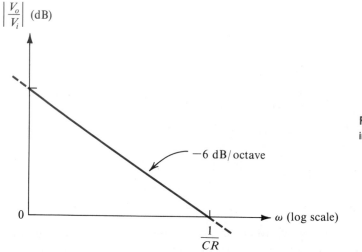

Fig. 3.10 Frequency response of an ideal integrator with a time constant CR.

note that at zero frequency the closed-loop gain is infinite. That is, at dc the op amp is operating as an open loop, which could be easily seen when we recall that capacitors behave as open circuits for dc. When op-amp imperfections are taken into account, we will find it necessary to modify the integrator circuit to make the closed-loop gain at dc finite. This will be illustrated in a later section. Another point to note is that the integrator magnitude response crosses the 0 dB (unity-gain) line at a frequency equal to the inverse of the time constant ($1/CR$). An important application of op-amp integrators is their use to convert square waveforms into triangular waves. This application is the subject of Exercise 3.5 below.

As another special case consider

$$Z_1 = \frac{1}{sC} \quad \text{and} \quad Z_2 = R$$

$$\frac{V_o}{V_i} = -sCR$$

or in terms of physical frequencies,

$$\frac{V_o}{V_i} = -j\omega CR$$

which corresponds to a differentiation operation, that is,

$$v_O(t) = -CR\frac{dv_I(t)}{dt}$$

The reader can easily verify that the circuit of Fig. 3.11 indeed implements this differentiation operation. Figure 3.12 shows a plot for the magnitude of the transfer function of the differentiator, which is identical to that of an STC high-pass network with an infinite-frequency break point. Note that the response crosses the 0 dB line at $\omega = 1/CR$.

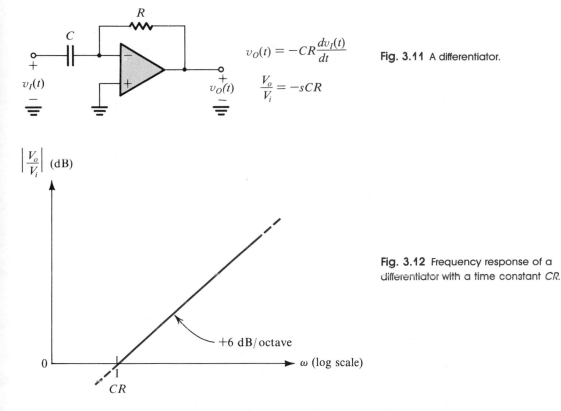

$$v_O(t) = -CR\frac{dv_I(t)}{dt}$$

$$\frac{V_o}{V_i} = -sCR$$

Fig. 3.11 A differentiator.

Fig. 3.12 Frequency response of a differentiator with a time constant CR.

The very nature of a differentiator circuit causes it to be a "noise magnifier." This is due to the spike introduced at the output every time there is a sharp change in $v_I(t)$; such a change could be a "picked up" interference. For this reason and because they suffer from stability problems (Chapter 12), differentiator circuits are generally avoided in practice.

As a final application of the inverting configuration consider the circuit shown in Fig. 3.13. Here we have a resistance R_f in the negative-feedback path (as before), but we have a number of input signals $v_1, v_2, \ldots, v_n$ each applied to a corresponding resistor $R_1, R_2, \ldots, R_n$, which are connected to the inverting terminal of the op amp. From our previous discussion, the ideal op amp will have a virtual ground appearing

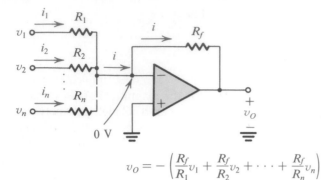

Fig. 3.13 A weighted summer.

$$v_O = -\left(\frac{R_f}{R_1}v_1 + \frac{R_f}{R_2}v_2 + \cdots + \frac{R_f}{R_n}v_n\right)$$

at its negative input terminal. Ohm's law then tells us that the currents $i_1, i_2, \ldots, i_n$ are given by

$$i_1 = \frac{v_1}{R_1}, \qquad i_2 = \frac{v_2}{R_2}, \qquad \cdots, \qquad i_n = \frac{v_n}{R_n}$$

All these currents sum together to produce the current i; that is,

$$i = i_1 + i_2 + \cdots + i_n$$

will be forced to flow through R_f (since no current flows into the input terminals of an ideal op amp). The output voltage v_O may now be determined by another application of Ohm's law,

$$v_O = 0 - iR_f = -iR_f$$

Thus

$$v_O = -\left(\frac{R_f}{R_1}v_1 + \frac{R_f}{R_2}v_2 + \cdots + \frac{R_f}{R_n}v_n\right)$$

That is, the output voltage is a weighted sum of the input signals $v_1, v_2, \ldots, v_n$. This circuit is therefore called a *weighted summer*. Note that each summing coefficient may be independently adjusted by adjusting the corresponding "feed-in" resistor (R_1 to R_n). This nice property, which greatly simplifies circuit adjustment, is a direct consequence of the virtual ground that exists at the inverting op-amp terminal. As the reader will soon come to appreciate, virtual grounds are extremely "handy."

In the above we have seen that op amps can be used to multiply a signal by a constant, integrate it, differentiate it, and sum a number of signals with prescribed weights. These are all mathematical operations—hence the name operational amplifier. The circuits above are functional building blocks needed to perform analog computation. For this reason the op amp has been the basic element of analog computers. Op amps, however, can do much more than just perform the mathematical operations required in analog computation. In this chapter we will get a taste of this versatility, with other applications presented in later chapters.

Exercises

3.5 Consider a symmetrical square wave of 20-V peak-to-peak, 0 average, and 1-ms period applied to a Miller integrator. Find the value of the time constant CR such that the triangular waveform at the output has a 20-V peak-to-peak amplitude.

Ans.   250 μs

3.6 Show that the circuit shown in Fig. E3.6 has an STC low-pass transfer function. For the case $R_1 = 1$ kΩ, $R_2 = 100$ kΩ, and $C_2 = 100$ pF, find the dc gain and the 3-dB frequency.

Ans. -100 V/V; 10^5 rad/s

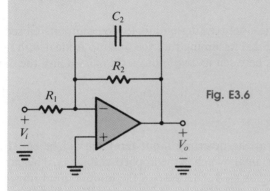

Fig. E3.6

3.7 Using an ideal op amp, design an inverting integrator with an input resistance of 10 kΩ and an integration time constant of 10^{-3} s. What is the gain magnitude and phase angle of this circuit at 10 rad/s and at 1 rad/s?

Ans. $R = 10$ kΩ, $C = 0.1$ μF; at $\omega = 10$ rad/s: $|V_o/V_i| = 100$ V/V and $\phi = +90°$; at $\omega = 1$ rad/s: $|V_o/V_i| = 1,000$ V/V and $\phi = +90°$

3.8 Design a differentiator to have a time constant of 10^{-2} s and an input capacitance of 0.01 μF. What is the gain magnitude and phase of this circuit at 10 rad/s, and at 10^3 rad/s? In order to limit the high-frequency gain of the differentiator circuit to 100, a resistor is added in series with the capacitor. Find the required resistor value.

Ans. $C = 0.01$ μF; $R = 1$ MΩ; at $\omega = 10$ rad/s: $|V_o/V_i| = 0.1$ and $\phi = -90°$; at $\omega = 1,000$ rad/s: $|V_o/V_i| = 10$ and $\phi = -90°$; 10 kΩ

3.9 Design an inverting op-amp circuit to form the weighted sum v_O of two inputs v_1 and v_2. It is required that $v_O = -(v_1 + 5v_2)$. Choose values for R_1, R_2, and R_f so that for a maximum output voltage of 10 V the current in the feedback resistor will not exceed 1 mA.

Ans.   A possible choice: $R_1 = 10$ kΩ, $R_2 = 2$ kΩ, and $R_f = 10$ kΩ

3.5 THE NONINVERTING CONFIGURATION

The second closed-loop configuration we shall study is shown in Fig. 3.14. Here the input signal v_I is applied directly to the positive input terminal of the op amp while one terminal of R_1 is connected to ground.

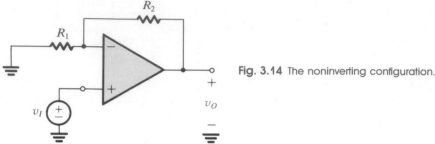

Fig. 3.14 The noninverting configuration.

Analysis of the noninverting circuit to determine its closed-loop gain (v_O/v_I) is illustrated in Fig. 3.15. Assuming that the op amp is ideal with infinite gain, a virtual short circuit exists between its two input terminals. Hence the difference input signal is

$$v_2 - v_1 = \frac{v_O}{A} = 0 \qquad \text{for } A = \infty$$

Thus the voltage at the inverting input terminal will be equal to that at the noninverting input terminal, which is the applied voltage v_I. The current through R_1 can then be determined as v_I/R_1. Because of the infinite input impedance of the op amp this current will flow through R_2, as shown in Fig. 3.15. Now the output voltage can be determined from

$$v_O = v_I + \left(\frac{v_I}{R_1}\right) R_2$$

which yields

$$\frac{v_O}{v_I} = 1 + \frac{R_2}{R_1} \tag{3.3}$$

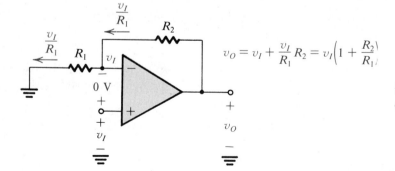

$$v_O = v_I + \frac{v_I}{R_1} R_2 = v_I \left(1 + \frac{R_2}{R_1}\right)$$

Fig. 3.15 Analysis of the noninverting circuit.

Further insight into the operation of the noninverting configuration can be obtained by considering the following: The voltage divider in the negative-feedback path causes a fraction of the output voltage to appear at the inverting input terminal of the op amp; that is,

$$v_1 = v_O\left(\frac{R_1}{R_1 + R_2}\right)$$

Then the infinite op-amp gain and the resulting virtual short circuit between the two input terminals of the op amp causes this voltage to be equal to that applied at the positive input terminal; thus

$$v_O\left(\frac{R_1}{R_1 + R_2}\right) = v_I$$

which yields the gain expression given in Eq. (3.3).

The gain of the noninverting configuration is positive—hence the name *non-inverting*. The input impedance of this closed-loop amplifier is ideally infinite, since no current flows into the positive input terminal of the op amp. This property of high input impedance is a very desirable feature of the noninverting configuration. It enables using this circuit as a buffer amplifier to connect a source with a high impedance to a low-impedance load. We have discussed the need for buffer amplifiers in Chapter 2. In many applications the buffer amplifier is not required to provide any voltage gain; rather it is used mainly as an impedance transformer or a power amplifier. In such cases we may make $R_2 = 0$ and $R_1 = \infty$ to obtain the unity-gain amplifier shown in Fig. 3.16. This circuit is commonly referred to as a *voltage follower*, since the output "follows" the input. In the ideal case, $v_O = v_I$, $R_{in} = \infty$, and $R_{out} = 0$.

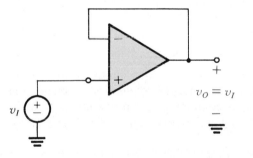

Fig. 3.16 The unity-gain buffer or follower amplifier.

Since the noninverting configuration has a gain greater than or equal to unity, depending on the choice of R_2/R_1, some prefer to call it "a follower with gain."

Exercises

3.10 Use the superposition principle to find the output voltage of the circuit shown in Fig. E3.10.

Ans. $v_O = 6v_1 + 4v_2$

3.11 If in the circuit of Fig. E3.10 the 1-kΩ resistor is disconnected from ground and connected to a third signal source v_3, use superposition to determine v_O in terms of v_1, v_2, and v_3.

Ans. $v_O = 6v_1 + 4v_2 - 9v_3$

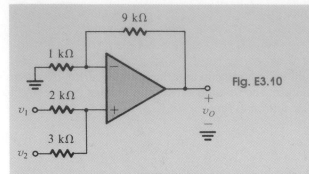

Fig. E3.10

3.12 Design a noninverting amplifier with a gain of 2. At the maximum output voltage of 10 V the current in the voltage divider is to be 10 μA.

Ans. $R_1 = R_2 = 0.5$ MΩ

3.13 (a) Show that if the op amp in the circuit of Fig. 3.14 has a finite open-loop gain A, then the closed-loop gain is given by

$$G \equiv \frac{v_O}{v_I} = \frac{1 + R_2/R_1}{1 + (1 + R_2/R_1)/A}$$

(b) For $R_1 = 1$ kΩ and $R_2 = 9$ kΩ find the percentage deviation ε of the closed-loop gain from the ideal value of $(1 + R_2/R_1)$ for the cases $A = 10^3$, 10^4, and 10^5. In each case find the voltage between the two input terminals of the op amp assuming that $v_I = 1$ V.

Ans. $\varepsilon = -1\%$, -0.1%; -0.01%; $v_2 - v_1 = 9.9$ mV, 1 mV, 0.1 mV

3.6 EXAMPLES OF OP-AMP CIRCUITS

Now that we have studied the two most common closed-loop configurations of op amps, we present a number of examples. Our objective is twofold: first, to enable the reader to gain experience in analyzing circuits containing op amps; second, to introduce the reader to some of the many interesting and exciting applications of op amps.

EXAMPLE 3.3

Figure 3.17 shows a circuit for an analog voltmeter of very high input impedance that uses an inexpensive moving-coil meter. As shown, the moving-coil meter is connected in the negative-feedback path of the op amp. The voltmeter measures the voltage v applied between the op-amp positive input terminal and ground. Assume that the moving coil produces full-scale deflection when the current passing through it is 100 μA; we wish to find the value of R such that the full-scale reading for v is $+10$ V.

Solution

The current in the moving-coil meter is v/R because of the virtual short circuit at the op-amp input and the infinite input impedance of the op amp. Thus we have to choose R such that

$$\frac{10}{R} = 100 \ \mu\text{A}$$

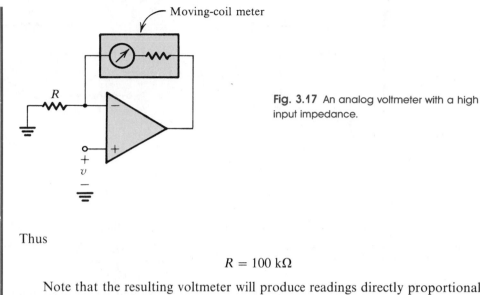

Fig. 3.17 An analog voltmeter with a high input impedance.

Thus

$$R = 100 \text{ k}\Omega$$

Note that the resulting voltmeter will produce readings directly proportional to the value of v, irrespective of the value of the internal resistance of the moving-coil meter—a very desirable property.

EXAMPLE 3.4

We need to find an expression for the output voltage v_O in terms of the input voltages v_1 and v_2 for the circuit in Fig. 3.18.

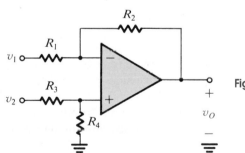

Fig. 3.18 A difference amplifier.

Solution

There are a number of ways to solve this problem; perhaps the easiest is using the principle of superposition. Obviously superposition may be employed here, since the network is linear. To apply superposition we first reduce v_2 to zero—that is, ground the terminal to which v_2 is applied—and then find the corresponding output voltage, which will be due entirely to v_1. We denote this output voltage v_{O1}. Its values may be found from the circuit in Fig. 3.19a, which we recognize as that of the inverting configuration. The existence of R_3 and R_4 does not affect the gain expression, since

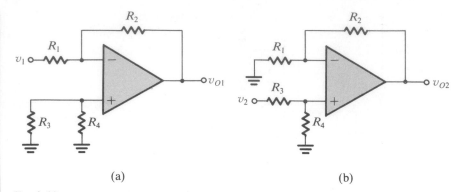

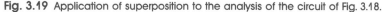

(a) (b)

Fig. 3.19 Application of superposition to the analysis of the circuit of Fig. 3.18.

no current flows through either of them. Thus

$$v_{O1} = -\frac{R_2}{R_1} v_1$$

Next, we reduce v_1 to zero and evaluate the corresponding output voltage v_{O2}. The circuit will now take the form shown in Fig. 3.19b, which we recognize as the noninverting configuration with an additional voltage divider, made up of R_3 and R_4, connected across the input v_2. The output voltage v_{O2} is therefore given by

$$v_{O2} = v_2 \frac{R_4}{R_3 + R_4} \left(1 + \frac{R_2}{R_1}\right)$$

The superposition principle tells us that the output voltage v_O is equal to the sum of v_{O1} and v_{O2}. Thus we have

$$v_O = -\frac{R_2}{R_1} v_1 + \frac{1 + R_2/R_1}{1 + R_3/R_4} v_2 \tag{3.4}$$

This completes the analysis of the circuit in Fig. 3.18. However, because of the practical importance of this circuit we shall pursue it further. We will ask, What is the condition under which this circuit will act as a differential amplifier? In other words, we wish to make the circuit respond (produce an output) in proportion to the difference signal $v_2 - v_1$ and reject common-mode signals (that is, produce zero output when $v_1 = v_2$). The answer can be obtained from the expression we have just derived [Eq. (3.4)]. Let us set $v_1 = v_2$ and require that $v_O = 0$. It is easy to see that this process leads to the condition

$$\frac{R_2}{R_1} = \frac{R_4}{R_3}$$

Substituting in Eq. (3.4) results in the output voltage

$$v_O = \frac{R_2}{R_1} (v_2 - v_1)$$

which is clearly that of a differential amplifier with a gain of R_2/R_1.

We next inquire about the input resistance seen between the two input terminals. The circuit is redrawn in Fig. 3.20 with the condition $R_2/R_1 = R_4/R_3$ imposed. In fact, to simplify matters and for other practical considerations we have made $R_3 = R_1$ and $R_4 = R_2$. We wish to evaluate the input differential resistance R_{in}, defined as

$$R_{in} \equiv \frac{v_2 - v_1}{i}$$

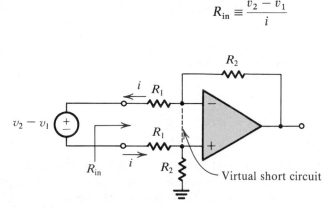

Fig. 3.20 Finding the input resistance of the differential amplifier.

Since the two input terminals of the op amp track each other in potential, we may write a loop equation and obtain

$$v_2 - v_1 = R_1 i + 0 + R_1 i$$

Thus

$$R_{in} = 2R_1$$

Note that if the amplifier is required to have a large differential gain, then R_1, of necessity, will be relatively small and the input resistance will be correspondingly small, a drawback of this circuit.

Differential amplifiers find application in many areas, most notably in the design of instrumentation systems. As an example, consider the case of a transducer that produces between its two output terminals a relatively small signal, say 1 mV. However, between each of the two wires (leading from the transducer to the instrumentation system) and ground there may be large picked-up interference, say 1 V. The required amplifier must reject this large interference signal, which is common to the two wires (a common-mode signal), and amplify the small difference (or differential) signal.

EXAMPLE 3.5

We desire to find the input resistance R_{in} of the circuit in Fig. 3.21.

Solution
To find R_{in} we apply an input voltage v and evaluate the input current i. Then R_{in} may be found from its definition, $R_{in} \equiv v/i$.

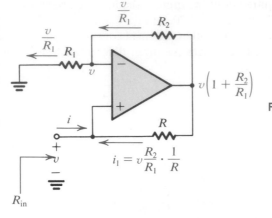

Fig. 3.21 Circuit for Example 3.5.

Owing to the virtual short circuit between the op-amp input terminals, the voltage at the inverting terminal will be equal to v. The current through R_1 will therefore be v/R_1. Owing to the infinite input impedance of the op amp, the current through R_2 will also be v/R_1. Thus the voltage at the op-amp output will be

$$v + \frac{v}{R_1} R_2 = \left(1 + \frac{R_2}{R_1} \right) v$$

We may now apply Ohm's law to R and obtain the current through it as

$$i_1 = \frac{v(1 + R_2/R_1) - v}{R} = v \frac{R_2}{R_1} \frac{1}{R}$$

Since no current flows into the positive input terminal of the op amp, we have

$$i = -i_1 = -\frac{v}{R} \frac{R_2}{R_1}$$

Thus

$$R_{\text{in}} = -R \frac{R_1}{R_2}$$

that is, the input resistance is negative with a magnitude equal to R, the resistance in the positive-feedback path, multiplied by the ratio R_1/R_2. This circuit is therefore called a *negative impedance converter* (NIC), where R may in general be replaced by an arbitrary impedance Z.

Let us investigate the application of this circuit further. Consider the case $R_1 = R_2 = r$, where r is an arbitrary value; it follows that $R_{\text{in}} = -R$. Let the input be fed with a voltage source V_s having a source resistance equal to R, as shown in Fig. 3.22a. We want to evaluate the current I_l that flows in an impedance Z_L connected as shown. In Fig. 3.22b we have utilized the information gained in the above and replaced the circuit in the dashed box by a resistance, $-R$. Figure 3.22c illustrates the conversion of the voltage source to its Norton's equivalent. Finally, the two parallel resistances R and $-R$ are combined to produce an infinite resistance, resulting in the circuit in

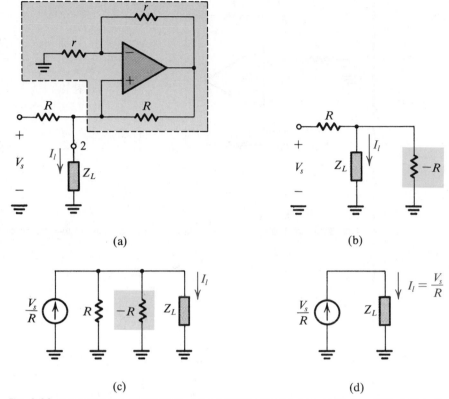

Fig. 3.22 Illustrating the application of the negative impedance converter of Fig. 3.21 in the design of a voltage-to-current converter or a controlled current source.

Fig. 3.22d, from which we see that the load current I_l is given by

$$I_l = \frac{V_s}{R}$$

independent of the value of Z_L! This is an interesting result; it tells us that the circuit of Fig. 3.22a acts as a *voltage-to-current converter*, providing a current I_l that is directly proportional to V_s ($I_l = V_s/R$) and is independent of the value of the load impedance. That is, terminal 2 acts as a current-source output, with the impedance looking back into 2 equal to infinity. Note that this infinite resistance is obtained via the cancellation of the positive source resistance R with the negative input resistance $-R$.

As it is, the circuit is quite useful and has practical applications where it is necessary to generate a current signal in proportion to a given voltage signal. A specific application is illustrated in Fig. 3.23, where a capacitor C is used as a load. From the above analysis we conclude that capacitor C will be supplied by a current $I = V_i/R$ and thus its voltage V_2 will be given by

$$V_2 = \frac{I}{sC} = \frac{V_i}{sCR}$$

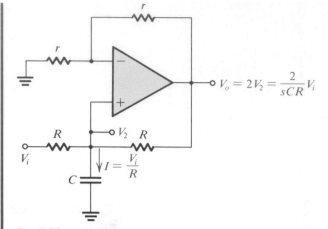

Fig. 3.23 Application of the voltage-to-current converter in the design of a noninverting integrator.

That is,

$$\frac{V_2}{V_i} = \frac{1}{sCR}$$

which is the transfer function of an integrator,

$$v_2 = \frac{1}{CR} \int_0^t v_i \, dt + V$$

where V is the voltage across the capacitor at $t = 0$. This integrator has some interesting properties. The transfer function does not have an associated negative sign, as is the case with the Miller integrator. Noninverting, or positive, integrators are required in many applications. Another useful property is the fact that one terminal of the capacitor is grounded. Among other things, this would simplify the initial charging of the capacitor, as may be necessary to simulate an initial condition in the solution of a differential equation on an analog computer.

The integrator circuit of Fig. 3.23 has a serious problem though. We cannot "take the output" at terminal 2 as indicated, since terminal 2 is a high-impedance point, which means that connecting any load resistance there will change the transfer function V_2/V_i. Fortunately, however, a low-impedance point exists where the signal is proportional to V_2. We are speaking of the op-amp output terminal, where, as the reader can easily verify,

$$V_o = 2V_2$$

Thus

$$\frac{V_o}{V_i} = \frac{2}{sCR}$$

EXAMPLE 3.6

We wish to analyze the circuit shown in Fig. 3.24 in order to derive an expression for the input impedance Z_{in}.

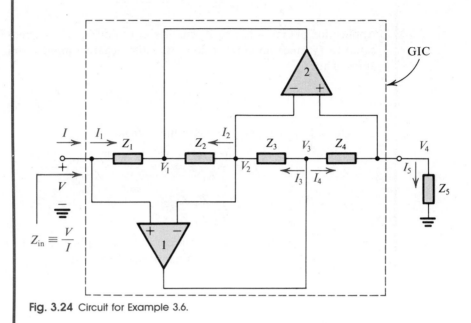

Fig. 3.24 Circuit for Example 3.6.

Solution

To find Z_{in} we apply a voltage V at the input and attempt to find an expression for the input current I. Then, the input impedance can be determined from

$$Z_{in} \equiv \frac{V}{I}$$

Because of the virtual short circuit between the input terminals of op amp 1, the voltage V_2 will be equal to V. Similarly, the virtual short circuit between the input terminals of op amp 2 will cause $V_4 = V_2 = V$. We now can evaluate the current in Z_5 as

$$I_5 = \frac{V_4}{Z_5} = \frac{V}{Z_5}$$

The current in Z_4 will be equal to I_5, since no current flows into the positive input terminal of op amp 2. Thus

$$I_4 = I_5 = \frac{V}{Z_5}$$

We are now able to determine V_3 as

$$V_3 = V_4 + I_4 Z_4 = V + \frac{V}{Z_5} Z_4 = \left(1 + \frac{Z_4}{Z_5}\right) V$$

Next the current in Z_3 can be determined as

$$I_3 = \frac{V_3 - V_2}{Z_3} = \frac{V(1 + Z_4/Z_5) - V}{Z_3} = \frac{V}{Z_3}\frac{Z_4}{Z_5}$$

Application of Ohm's law to Z_2 enables us to find V_1. The current in Z_2 is, of course, equal to I_3, since no current flows into the negative input terminals of the two op amps. Thus

$$V_1 = V_2 - I_2 Z_2 = V - I_3 Z_2 = V - \frac{V}{Z_3}\frac{Z_4}{Z_5}Z_2$$

The current I_1 in Z_1 may be determined from

$$I_1 = \frac{V - V_1}{Z_1} = \frac{V - V + (V/Z_3)(Z_4/Z_5)Z_2}{Z_1}$$

$$= \frac{V}{Z_3}\frac{Z_4}{Z_5}\frac{Z_2}{Z_1}$$

Finally, we recognize that the input current I should be equal to I_1, since no current flows into the positive input terminal of op amp 1. Thus

$$I = \frac{V}{Z_3}\frac{Z_4}{Z_5}\frac{Z_2}{Z_1}$$

from which we obtain

$$Z_{in} \equiv \frac{V}{I}$$

$$Z_{in} = \frac{Z_1 Z_3}{Z_2 Z_4}Z_5$$

This circuit is called the *generalized impedance converter* (GIC). The name follows from the above expression by considering the GIC to be the circuit within the box in Fig. 3.24 and Z_5 to be a terminating impedance. The GIC causes the impedance Z_5 to be multiplied by a general conversion factor $Z_1 Z_3/Z_2 Z_4$ and the resulting converted impedance to appear at the input terminals.

The GIC finds applications in the area of active inductorless filter design (Chapter 14). As a specific example of its use, consider the realization of an inductance. That is, let it be required to find suitable components (resistors and capacitors) for the impedances Z_1 to Z_5 such that

$$Z_{in} = sL$$

where L is the value of the required inductance. From the equation for Z_{in} derived above, we see that we have two choices if we wish to use only one capacitor. First, we have

$$Z_1 = R_1, \qquad Z_2 = \frac{1}{sC_2}, \qquad Z_3 = R_3, \qquad Z_4 = R_4, \qquad Z_5 = R_5$$

resulting in

$$Z_{in} = sC_2 \frac{R_1 R_3 R_5}{R_4}$$

Thus

$$L = C_2 \frac{R_1 R_3 R_5}{R_4}$$

Second, we have

$$Z_1 = R_1, \qquad Z_2 = R_2, \qquad Z_3 = R_3, \qquad Z_4 = \frac{1}{sC_4}, \qquad Z_5 = R_5$$

resulting in

$$Z_{in} = sC_4 \frac{R_1 R_3 R_5}{R_2}$$

Thus

$$L = C_4 \frac{R_1 R_3 R_5}{R_2}$$

Note that we have restricted ourselves to the use of a single capacitor, a practical constraint.

Exercises

3.14 Find values for the resistances in the circuit of Fig. 3.18 such that the circuit behaves as a differential amplifier with an input resistance of 4 kΩ and a gain of 100.

Ans. $R_1 = R_3 = 2$ kΩ; $R_2 = R_4 = 200$ kΩ

3.15 For the circuit shown in Fig. E3.15 derive an expression for the transfer function V_o/V_i. For the case $C = 0.01$ μF, $R = 20$ kΩ, and $R_1 = 10$ kΩ, find the magnitude and phase of the transfer function V_o/V_i at $\omega = 5{,}000$ rad/s.

Ans. $V_o/V_i = (s - 1/CR)/(s + 1/CR)$; $|V_o/V_i| = 1$; $\phi = +90°$

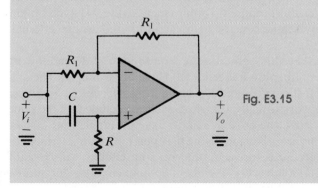

Fig. E3.15

3.16 This exercise illustrates the use of the negative impedance converter circuit of Fig. 3.21 as an amplifier. Let $R_1 = R_2 = 100$ kΩ and let the input be connected to a source $v_s = 10$ mV having a resistance $R_s = 0.9$ MΩ. Find the value of R so that a signal of 100 mV appears at the noninverting input terminal of the op amp. What is the value of the signal at the output terminal of the op amp?

Ans. 1 MΩ; 200 mV

3.17 It is required to use the GIC circuit studied in Example 3.6 to obtain an inductance of 100 mH for application at a frequency of 10^4 rad/s. A good design is one in which all resistors are equal. Furthermore, a detailed study of this circuit (which is beyond the scope of this book) has revealed that best performance is obtained if the resistor values are selected as equal to the magnitude of the reactance of the realized inductance at the frequency at which it is intended to be used. Find the component values.

Ans. $R_1 = R_2 = R_3 = R_5 = 1$ kΩ, $C_4 = 0.1$ μF

3.7 NONIDEAL PERFORMANCE OF OP AMPS

Above we defined the ideal op amp, and we presented a number of circuit applications of op amps. The analysis of these circuits assumed the op amps to be ideal. Although in many applications such an assumption is not a bad one, a circuit designer has to be thoroughly familiar with the characteristics of practical op amps and the effects of such characteristics on the performance of op-amp circuits. Only then will the designer be able to use the op amp intelligently, especially if the application at hand is not a straightforward one. The nonideal properties of op amps will, of course, limit the range of operation of the circuits analyzed in the previous examples.

In the following sections we consider the nonideal properties of the op amp. We do this by treating one parameter at a time.

3.8 FINITE OPEN-LOOP GAIN AND BANDWIDTH

The differential open-loop gain of an op amp is not infinite; rather it is finite and decreases with frequency. Figure 3.25 shows a plot for $|A|$, with the numbers typical of most general-purpose op amps (such as the 741-type op amp, which is available from many semiconductor manufacturers and whose internal circuit is studied in Chapter 13).

Note that although the gain is quite high at dc and low frequencies, it starts to fall off at a rather low frequency (10 Hz in our example). The uniform -20-dB/decade gain rolloff shown is typical of *internally compensated* op amps. These are units that have a network (usually a single capacitor) included on the same IC chip whose function is to cause the op-amp gain to have the single-time-constant low-pass response shown. This process of modifying the open-loop gain is termed *frequency compensation*, and its purpose is to ensure that op-amp circuits will be stable (as opposed to oscillatory). The subject of stability of op-amp circuits,—or, more generally, of feedback amplifiers—will be studied in Chapter 12.

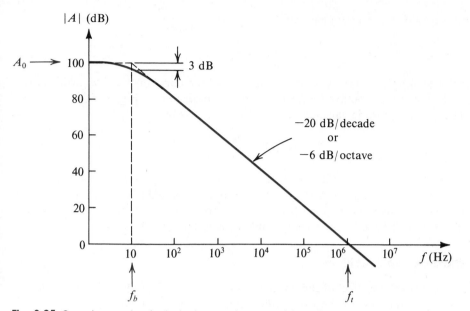

Fig. 3.25 Open-loop gain of a typical general-purpose internally compensated op amp.

By analogy to the response of low-pass STC circuits (Chapter 2), the gain $A(s)$ of an internally compensated op amp may be expressed as

$$A(s) = \frac{A_0}{1 + s/\omega_b} \qquad (3.5a)$$

which for physical frequencies, $s = j\omega$, becomes

$$A(j\omega) = \frac{A_0}{1 + j\omega/\omega_b} \qquad (3.5b)$$

where A_0 denotes the dc gain and ω_b is the 3-dB frequency (or "break" frequency). For the example shown in Fig. 3.25, $A_0 = 10^5$ and $\omega_b = 2\pi \times 10$ rad/s. For frequencies $\omega \gg \omega_b$ (about ten times and higher) Eq. (3.5b) may be approximated by

$$A(j\omega) \simeq \frac{(A_0\omega_b)}{j\omega} \qquad (3.6a)$$

from which it can be seen that the gain $|A|$ reaches unity (0 dB) at a frequency denoted by ω_t and given by

$$\omega_t = A_0\omega_b$$

Substituting in Eq. (3.6a) gives

$$A(j\omega) \simeq \frac{\omega_t}{j\omega} \qquad (3.6b)$$

where ω_t is called the *unity-gain bandwidth*. The unity-gain bandwidth $f_t = \omega_t/2\pi$ is usually specified on the data sheets of op amps. Also note that for $\omega \gg \omega_b$ the open-loop gain in Eq. (3.5a) becomes

$$A(s) \simeq \frac{\omega_t}{s} \tag{3.7}$$

Thus the op amp behaves as an integrator with time constant $\tau = 1/\omega_t$. This correlates with the -6-dB/octave frequency response indicated in Fig. 3.25.

The gain magnitude can be obtained from Eq. (3.6b) as

$$|A(j\omega)| \simeq \frac{\omega_t}{\omega} = \frac{f_t}{f} \tag{3.8}$$

Thus if f_t is known (10^6 Hz in our example), one can easily estimate the magnitude of the op-amp gain at a given frequency f. Finally, it should be mentioned that an op amp having this uniform -6 dB/octave gain rolloff is said to have a "single-pole" model. More will be said about poles and zeros in Chapter 11.

Frequency Response of Closed-Loop Amplifiers

We next consider the effect of the limited op-amp gain and bandwidth on the closed-loop transfer functions of the two basic configurations: the inverting circuit of Fig. 3.4 and the noninverting circuit of Fig. 3.14. The closed-loop gain of the inverting amplifier, assuming a finite op-amp open-loop gain A, was derived in Section 3.3 and given in Eq. (3.1), which we repeat here as

$$\frac{V_o}{V_i} = \frac{-R_2/R_1}{1 + (1 + R_2/R_1)/A} \tag{3.9}$$

Substituting for A from Eq. (3.5a) gives

$$\frac{V_o(s)}{V_i(s)} = \frac{-R_2/R_1}{1 + \dfrac{1}{A_0}\left(1 + \dfrac{R_2}{R_1}\right) + \dfrac{s}{\omega_t/(1 + R_2/R_1)}} \tag{3.10}$$

For $A_0 \gg 1 + R_2/R_1$, which is usually the case,

$$\frac{V_o(s)}{V_i(s)} \simeq \frac{-R_2/R_1}{1 + \dfrac{s}{\omega_t/(1 + R_2/R_1)}} \tag{3.11}$$

which is of the same form as Eq. (2.26). Thus, the inverting amplifier has an STC low-pass response with a dc gain of magnitude equal to R_2/R_1. The closed-loop gain rolls off at a uniform -20-dB/decade slope with a corner frequency (3-dB frequency) given by

$$\omega_{3dB} = \frac{\omega_t}{1 + R_2/R_1} \tag{3.12}$$

Similarly, analysis of the noninverting amplifier of Fig. 3.14, assuming a finite open-loop gain A, yields the closed-loop transfer function

$$\frac{V_o}{V_i} = \frac{1 + R_2/R_1}{1 + (1 + R_2/R_1)/A} \tag{3.13}$$

Substituting for A from Eq. (3.5a) and making the approximation $A_0 \gg 1 + R_2/R_1$ results in

$$\frac{V_o(s)}{V_i(s)} \simeq \frac{1 + R_2/R_1}{1 + \dfrac{s}{\omega_t/(1 + R_2/R_1)}} \qquad (3.14)$$

Thus the noninverting amplifier has an STC low-pass response with a dc gain of $(1 + R_2/R_1)$ and a 3-dB frequency given also by Eq. (3.12).

EXAMPLE 3.7

Consider an op amp with $f_t = 1$ MHz. Find the 3-dB frequency of closed-loop amplifiers with nominal gains of $+1{,}000$, $+100$, $+10$, $+1$, -1, -10, -100, and $-1{,}000$. Sketch the magnitude frequency response for the amplifiers with closed-loop gains of $+10$ and -10.

Solution
Using Eq. (3.12) we obtain the results given in the following table:

Closed-Loop Gain	$\dfrac{R_2}{R_1}$	$f_{3\text{dB}} = f_t/(1 + R_2/R_1)$
$+1{,}000$	999	1 kHz
$+100$	99	10 kHz
$+10$	9	100 kHz
$+1$	0	1 MHz
-1	1	0.5 MHz
-10	10	90.9 kHz
-100	100	9.9 kHz
$-1{,}000$	1,000	$\simeq 1$ kHz

Figure 3.26 shows the frequency response for the amplifier whose nominal dc gain is $+10$, and Fig. 3.27 shows the frequency response for the -10 case. An

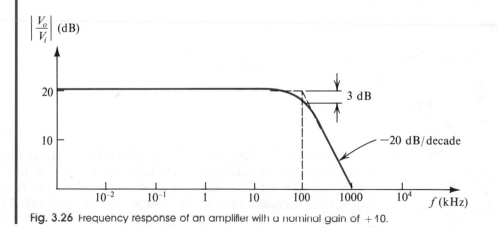

Fig. 3.26 Frequency response of an amplifier with a nominal gain of $+10$.

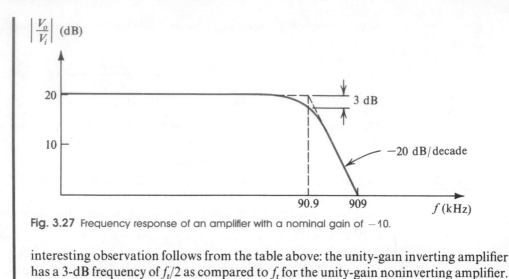

Fig. 3.27 Frequency response of an amplifier with a nominal gain of -10.

interesting observation follows from the table above: the unity-gain inverting amplifier has a 3-dB frequency of $f_t/2$ as compared to f_t for the unity-gain noninverting amplifier.

An Interpretation in Terms of Feedback

Example 3.7 clearly illustrates the trade-off between gain and bandwidth. For instance, the noninverting configuration exhibits a constant *gain–bandwidth product*. An interpretation of these results in terms of feedback theory will be given in Chapter 12. For the time being it should be mentioned that both the inverting and the noninverting configurations have identical "feedback loops." This can be seen by eliminating the excitation (that is, short-circuiting the input voltage source), resulting in both cases in the feedback loop shown in Fig. 3.28. Since their feedback loops are identical, the two configurations have the same dependence on the finite op-amp gain and bandwidth (for example, identical expressions for $f_{3\text{dB}}$).

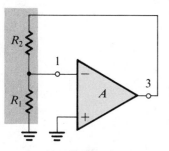

Fig. 3.28 The feedback loop of the inverting and the noninverting configurations.

Before leaving this section we wish to examine the feedback loop of Fig. 3.28 a bit more closely. The forward path of the loop (from node 1 to node 3) consists of an amplifier of gain $-A$. The feedback path (from node 3 to node 1) consists of a voltage divider of transmission ratio $R_1/(R_1 + R_2)$. This is usually called the feedback ratio (or feedback factor) and is denoted β; that is,

$$\beta = \frac{R_1}{R_1 + R_2} \tag{3.15}$$

The loop gain is given by

$$\text{Loop gain} = -A\beta \tag{3.16}$$

Note that the negative loop-gain signifies the fact that the feedback is negative. An important quantity in feedback amplifiers is the amount of feedback defined by

$$\text{Amount of feedback} \equiv 1 - \text{loop gain} = 1 + A\beta \tag{3.17}$$

Later on in this chapter we shall see that this quantity plays an important role.

Exercises

3.18 Consider an op amp having a 106-dB gain at dc and a single-pole frequency response with $f_t = 2$ MHz. Find the magnitude of gain at $f = 1$ kHz, 10 kHz, and 100 kHz.

Ans. 2,000; 200; 20

3.19 If the op amp in Exercise 3.18 is used to design a noninverting amplifier with nominal dc gain of 100, find the 3-dB frequency of the closed-loop gain. Also find the rise time of the output waveform if a voltage step is applied at the input.

Ans. 20 kHz; 17.5 μs

3.20 An op amp with a dc gain $A_0 = 10^4$ is connected in the noninverting configuration with $R_1 = 1$ kΩ and $R_2 = 99$ kΩ. Find the values of the feedback factor and the loop gain.

Ans. 0.01; -100

3.21 What is the value of the feedback ratio β for a unity-gain follower?

Ans. $\beta = 1$ (100% feedback)

3.22 Consider a unity-gain follower constructed using an op amp with infinite bandwidth but finite dc gain A_0. Find an expression for the closed-loop gain and evaluate its value for $A_0 = 10^3$ and 10^5.

Ans. $G = \dfrac{1}{1 + 1/A_0}$; 0.9990; 0.999990

3.9 THE INTERNAL STRUCTURE OF IC OP AMPS

Further insight into the frequency (and step) response of op amps can be gained by considering the internal circuit of the op amp. This will be done in detail in Chapter 13. For the time being let us consider Fig. 3.29, which shows the internal structure, in block diagram form, of most modern integrated-circuit (IC) op amps. The op amp is seen to consist of three stages: an input stage, which is basically a differential-input transconductance amplifier; an intermediate stage, which is a voltage amplifier with high negative voltage gain ($-\mu$) and with a feedback capacitor C; and an output stage, which is a unity-gain buffer whose purpose is to provide the op amp with a low output resistance. For the purpose of the present discussion we shall assume that the output buffer is an ideal unity-gain amplifier and consider it no further.

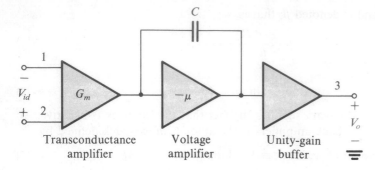

Fig. 3.29 The internal structure of modern IC op amps.

Figure 3.30a shows a simplified small-signal equivalent circuit model of the op amp with the output stage eliminated. The input stage is shown as having infinite input impedance. This stage senses the differential input voltage V_{id} ($V_{id} = V_2 - V_1$) and provides a proportionate current $G_m V_{id}$. It has an output resistance R_{o1}. The second stage is shown to have an input resistance R_{i2}, a voltage gain $-\mu$ and a zero output resistance. The feedback capacitor C is included for the purpose of ensuring that the op amp will be stable (does not oscillate and provide unwanted output

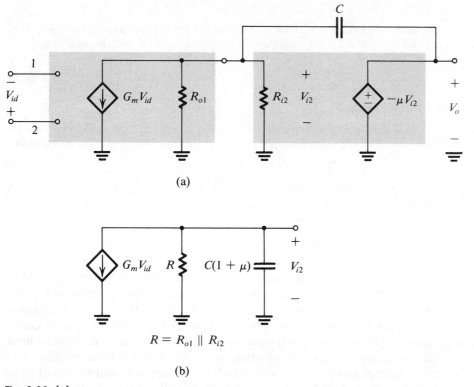

Fig. 3.30 (a) Small-signal model for the IC op amp whose internal structure is shown in Fig. 3.29. Here the output buffer is assumed ideal and is eliminated. **(b)** Equivalent circuit of the interconnection between the first and second stages.

signals) when it is connected in a feedback circuit. The topic of stability will be studied in Chapter 12. For the time being we will show that the capacitor C, known as the frequency-compensation capacitor, provides the op amp with the STC low-pass response studied in the previous section.

We wish to analyze the equivalent circuit of Fig. 3.30a to determine the open-loop gain $A \equiv V_o(s)/V_{id}(s)$. Toward that end we employ Miller's theorem and replace the bridging capacitor C by the equivalent capacitance $C(1 + \mu)$ at the input of the second stage, as shown in Fig. 3.30b. This figure shows the equivalent circuit at the interface between the first and second stages. Note that we have combined R_{o1} and R_{i2} into a single resistor R,

$$R = R_{o1} \parallel R_{i2}$$

For the circuit in Fig. 3.30b we can write

$$V_{i2} = \frac{-G_m V_{id}}{Y}$$

where

$$Y = \frac{1}{R} + sC(1 + \mu)$$

Thus

$$V_{i2} = -V_{id} \frac{G_m R}{1 + sC(1 + \mu)R}$$

Since

$$V_o = -\mu V_{i2}$$

the open-loop gain can now be found as

$$A(s) \equiv \frac{V_o(s)}{V_{id}(s)}$$

$$= \frac{\mu G_m R}{1 + sC(1 + \mu)R} \tag{3.18}$$

Comparing this expression to that in Eq. (3.5a), we find that

$$A_0 = \mu G_m R \tag{3.19}$$

$$\omega_b = \frac{1}{C(1 + \mu)R} \tag{3.20}$$

Now, since the unity-gain bandwidth ω_t is given by

$$\omega_t = A_0 \omega_b$$

it follows that

$$\omega_t = \frac{\mu G_m R}{C(1 + \mu)R}$$

Usually $\mu \gg 1$ and ω_t can be approximated by

$$\omega_t \simeq \frac{G_m}{C} \tag{3.21}$$

We also note that at frequencies much greater than the 3-dB frequency ω_b, the open-loop gain $A(s)$ in Eq. (3.18) can be approximated by

$$A(s) \simeq \frac{\mu G_m R}{sC(1 + \mu)R}$$

Again, for $\mu \gg 1$,

$$A(s) \simeq \frac{G_m}{sC} \tag{3.22}$$

At such frequencies the op-amp internal circuit can be represented by the structure in Fig. 3.31. Here we have assumed that μ approaches ∞ and thus the second stage

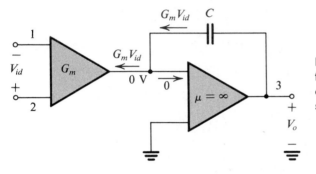

Fig. 3.31 Approximate equivalent circuit of the op amp at frequencies $f \gg f_b$ and assuming that the gain of the second stage is large.

together with the compensating capacitor acts as an integrator. A virtual ground appears at the input terminal of the second stage and all the current of the first stage $(G_m V_{id})$ flows through the feedback capacitor C. Thus

$$V_o = \frac{G_m V_{id}}{sC}$$

resulting in the open-loop gain given by Eq. (3.22). This simplified equivalent circuit will prove useful in our discussion of slew-rate limitation in the next section.

To conclude this section consider the popular 741-type op amp for which $G_m = 0.19$ mA/V, $R_{o1} = 6.7$ MΩ, $R_{i2} = 4$ MΩ, $\mu = 529$, $C = 30$ pF. Thus

$$A_0 = 529 \times 0.19 \times 10^{-3} \times (6.7 \parallel 4) \times 10^6$$

$$= 2.52 \times 10^5$$

$$\omega_t = \frac{G_m}{C} = \frac{0.19 \times 10^{-3}}{30 \times 10^{-12}} = 6.33 \text{ Mrad/s}$$

$$f_t = \frac{\omega_t}{2\pi} \simeq 1 \text{ MHz}$$

3.10 LARGE-SIGNAL OPERATION OF OP AMPS

In this section we study the limitations on the performance of op-amp circuits when large output signals are present.

Output Saturation

Similar to all other amplifiers, op amps operate linearly over a limited range of output voltages. Specifically, the op-amp output saturates in the manner shown in Fig. 2.7 with L_+ and L_- within 2 or 3 volts of the positive and negative power supplies respectively. Thus, an op amp that is operating from ± 15-V supplies will saturate when the output voltage reaches about $+12$ V in the positive direction and -12 V in the negative direction. For this op amp the *rated output voltage* is said to be ± 12 V. To avoid clipping off the peaks of the output waveform, and the resulting waveform distortion, the input signal must be kept correspondingly small.

Exercise

3.23 The rated output voltage of a given op amp is ± 10 V. If the op amp is used to design a noninverting amplifier with a gain of 100, what is the largest sine-wave input that can be handled without output clipping?

Ans. 0.2 V peak-to-peak

Slew Rate

Another phenomenon that can cause nonlinear distortion when large output signals are present is that of slew-rate limiting. We shall first describe slew-rate limiting and then explain the reason it occurs.

Consider the unity-gain follower shown in Fig. 3.32a. Let the input voltage v_I be the step of height V shown in Fig. 3.32b. Now the closed-loop gain of the amplifier is given by Eq. (3.14) with $R_2 = 0$ and $R_1 = \infty$; that is,

$$\frac{V_o}{V_i} = \frac{1}{1 + s/\omega_t} \tag{3.23}$$

which is a low-pass STC response. We would therefore expect the step input to produce the output waveform

$$v_O(t) = V(1 - e^{-t/\tau}) \tag{3.24}$$

where $\tau = 1/\omega_t$. Figure 3.32c shows a sketch of this exponentially rising waveform.

In practice, however, such a response is obtained only if the step size V is "small," where a more precise definition of "small" will follow shortly. For large step inputs (5 V, for example) the output waveform will be the linearly ramping signal shown in Fig. 3.32d. It is important to note that the slope of the linear ramp is smaller than the initial slope (at $t = 0$) of the exponentially rising waveform of the same magnitude V (shown in Fig. 3.32c), which is V/τ. The linear ramping response shown in Fig. 3.32d indicates that the op-amp output is unable to rise at the rate predicted by Eq. (3.23). When this occurs, the op amp is said to be slew-rate limited (or slewing), and the slope

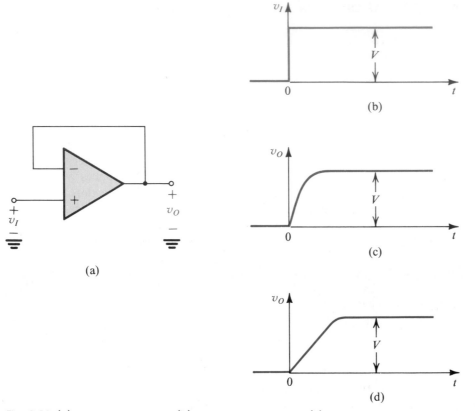

Fig. 3.32 (a) Unity-gain follower. **(b)** Input step waveform. **(c)** Exponentially rising output waveform obtained when V is small. **(d)** Linearly rising output waveform obtained when V is large (here the amplifier is slew-rate limited.)

of the linear ramp at the output is called the slew rate. The slew rate (SR) is the maximum possible rate of change of the op-amp output voltage,

$$SR = \frac{dv_O}{dt}\bigg|_{max} \tag{3.25}$$

and is usually specified on the op-amp data sheet in V/μs. It follows that the op amp in Fig. 3.32a will begin to slew for a signal V for which the initial slope of the exponentially rising ramp, V/τ, exceeds the op-amp slew rate.

We next investigate the origin of the slew-rate limitation. Consider once more the unity-gain follower in Fig. 3.32a with an input step of few volts. We see that at time $t = 0$, as the input rises to V volts, the output remains at zero volts. Thus the full size of the step appears between the two input terminals of the op amp. It follows that the input differential voltage V_{id} will be large and the input transconductance amplifier (See Fig. 3.31) will saturate in the manner indicated in Fig. 3.33. Under these conditions the transconductance amplifier supplies its maximum possible output current I_{max} to the second stage. (Note that I_{max} is smaller than $G_m V_{id}$ because the

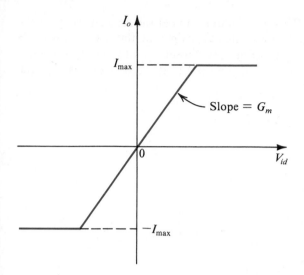

Slope $= G_m$

Fig. 3.33 Transfer characteristic of the input transconductance amplifier present in every modern IC op amp.

input stage has saturated.) The constant current I_{max} passes through the frequency-compensation capacitor C of the second stage and causes the output voltage to rise linearly with a slope equal to I_{max}/C. This, the highest possible rate of change of the output voltage, is the op-amp slew rate. Thus,

$$SR = \frac{I_{max}}{C} \tag{3.26}$$

Exercises

3.24 An op amp that has a slew rate of 1 V/μs and a unity-gain bandwidth f_t of 1 MHz is connected in the unity-gain follower configuration. Find the largest possible input voltage step for which the output waveform will still be given by the exponential ramp of Eq. (3.24). For this input voltage what is the 10% to 90% rise time of the output waveform? If an input step 10 times as large is applied, find the 10% to 90% rise time of the output waveform.

Ans. 0.16 V; 0.35 μs; 1.28 μs

3.25 For the 741-type op amp the maximum current that the first stage can supply is 19 μA and the compensation capacitor C is 30 pF. Find the slew rate.

Ans. 0.63 V/μs

Full-Power Bandwidth

Op-amp slew-rate limiting can cause nonlinear distortion in sinusoidal waveforms. Consider once more the unity-gain follower with a sine wave input given by

$$v_I = \hat{V}_i \sin \omega t$$

The rate of change of this waveform is given by

$$\frac{dv_I}{dt} = \omega \hat{V}_i \cos \omega t$$

and has a maximum value of $\omega \hat{V}_i$. This maximum occurs at the zero crossings of the input sinusoid. Now if $\omega \hat{V}_i$ exceeds the slew rate of the op amp, the output waveform will be distorted in the manner shown in Fig. 3.34. Observe that the output cannot keep up with the large rate of change of the sinusoid at its zero crossings, and the op amp slews.

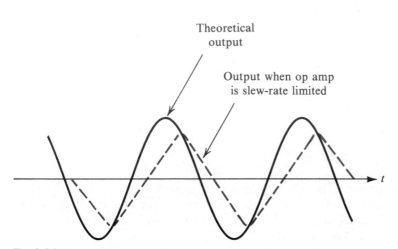

Theoretical
output

Output when op amp
is slew-rate limited

t

Fig. 3.34 Effect of slew-rate limiting on output sinusoidal waveforms.

The op-amp data sheets usually specify a frequency f_M called the *full-power bandwidth*. It is the frequency at which an output sinusoid with amplitude equal to the rated output voltage of the op amp begins to show distortion due to slew-rate limiting. If we denote the rated output voltage $V_{o\max}$, then f_M is related to SR as follows:

$$\omega_M V_{o\max} = \text{SR}$$

Thus,

$$f_M = \frac{\text{SR}}{2\pi V_{o\max}} \tag{3.27}$$

It should be obvious that output sinusoids of amplitudes smaller than $V_{o\max}$ will show slew-rate distortion at frequencies higher than ω_M. In fact, the maximum amplitude of the undistorted output sinusoid of frequency ω is given by:

$$V_o = V_{o\max}\left(\frac{\omega_M}{\omega}\right) \tag{3.28}$$

Finally, we note that slew-rate limiting is a phenomenon distinct from the small-signal frequency limitation studied in Section 3.8.

3.11 COMMON-MODE REJECTION

Practical op amps have finite nonzero *common-mode gain;* that is, if the two input terminals are tied together and a signal v_{Icm} is applied, the output will not be zero. The ratio of the output voltage v_O to the input voltage v_{Icm} is called the common-mode gain A_{cm}. Figure 3.35 illustrates this definition.

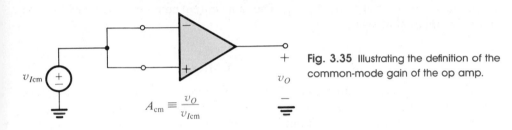

$$A_{cm} \equiv \frac{v_O}{v_{Icm}}$$

Fig. 3.35 Illustrating the definition of the common-mode gain of the op amp.

To be precise, consider an op amp with signals v_1 and v_2 applied to its inverting and noninverting input terminals, respectively. The difference between the two input signals is the differential-mode, or simply differential, input signal v_{id},

$$v_{id} = v_2 - v_1 \tag{3.29}$$

The average of the two input signals is the common-mode input signal v_{Icm},

$$v_{Icm} = \frac{v_1 + v_2}{2} \tag{3.30}$$

Now the output voltage v_O can be expressed as

$$v_O = Av_{id} + A_{cm}v_{Icm} \tag{3.31}$$

where A is the differential gain and A_{cm} is the common-mode gain.

The ability of an op amp to reject common-mode signals is specified in terms of the *common-mode rejection ratio* (CMRR), defined as

$$\text{CMRR} = \frac{|A|}{|A_{cm}|} \tag{3.32}$$

Usually the CMRR is expressed in decibels:

$$\text{CMRR} = 20 \log \frac{|A|}{|A_{cm}|} \tag{3.33}$$

The CMRR is a function of frequency, decreasing as the frequency is increased. Typical values of CMRR at low frequencies range from 80 to 100 dB.

The finite CMRR of op amps is unimportant in the case of the inverting configuration, since the positive input terminal is grounded and hence the common-mode input signal is approximately zero. On the other hand, in the noninverting configuration the common-mode input signal is nearly equal to the applied input signal, and thus the finite CMRR of the op amp may have to be taken into account in applications that demand high accuracy. The closed-loop configuration that is most adversely affected by the finite CMRR of the op amp is the differential amplifier of Fig. 3.18. We have found in Example 3.4 that by the appropriate selection of resistor values, the circuit can be made to respond only to differential input signals. This will no longer be true if the finite CMRR of the op amp is taken into account.

A simple method for taking into account the effect of the finite CMRR in calculating closed-loop gain is as follows: A common-mode input signal v_{Icm} gives rise to an output component of value $A_{cm}v_{Icm}$. The same output component can be obtained if a differential input signal

$$v_{error} = \frac{A_{cm}v_{Icm}}{A} = \frac{v_{Icm}}{CMRR} \tag{3.34}$$

is applied to an op amp with zero common-mode gain. Thus, in a given circuit, once the input common-mode signal is found, we simply add a signal generator v_{error} in series with one of the op-amp input leads and carry out the rest of the analysis assuming the op amp to be ideal.

As an example, Fig. 3.36 shows the analysis of the noninverting configuration taking into account the finite CMRR of the op amp. From Fig. 3.36a we observe that $v_{Icm} \simeq v_I$. Thus the op amp is replaced with an ideal one together with the $v_{error} = v_I/CMRR$ generator, as shown in Fig. 3.36b. Analysis of the latter circuit is straightforward, and the result is given in Fig. 3.36. Note that although the error-voltage generator is assigned a polarity, CMRR can be either positive or negative; its sign is generally not known.

Exercise

3.27 Use the method described above to calculate the common-mode gain of the difference amplifier circuit in Fig. 3.18 for the case of an op-amp CMRR = 80 dB and $R_2/R_1 = R_4/R_3 = 1,000$.

Ans.   CM gain = 0.1 V/V.

3.12 INPUT AND OUTPUT RESISTANCES

Figure 3.37 shows an equivalent circuit of the op amp, incorporating its finite input and output resistances. As shown, the op amp has a differential input resistance R_{id} seen between the two input terminals. In addition, if the two input terminals are tied together and the input resistance (to ground) is measured, the result is the common-mode input resistance R_{icm}. In the equivalent circuit we have split R_{icm} into two equal parts ($2R_{icm}$), each connected between one of the input terminals and ground.

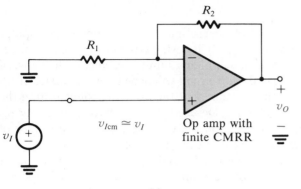

(a)

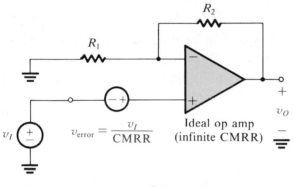

$$v_O = v_I\left(1 + \frac{1}{\text{CMRR}}\right)\left(1 + \frac{R_2}{R_1}\right)$$

$$\text{Gain error} = \frac{1}{\text{CMRR}} \times 100\%$$

(b)

Fig. 3.36 Analysis of the noninverting configuration with the finite CMRR of the op amp taken into account.

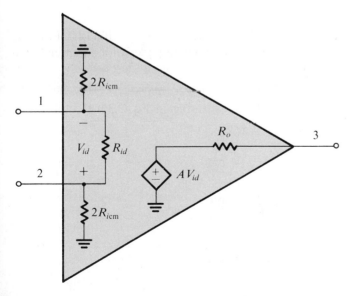

Fig. 3.37 Op-amp model with the input and output resistances shown.

Input Resistance

Typical values for the input resistances of general-purpose op amps using bipolar junction transistors are $R_{id} = 1$ MΩ and $R_{icm} = 100$ MΩ. Op amps that utilize field-effect transistors in the input stage have much higher input resistances. The value of the input resistance of a particular closed-loop circuit will depend on the values of R_{id} and R_{icm} as well as on the circuit configuration. For the inverting configuration the input resistance is approximately equal to R_1. Detailed analysis shows that taking R_{id} and R_{icm} into account has a negligible effect on the value of the input resistance of the inverting circuit. On the other hand, the input resistance of the noninverting configuration is strongly dependent on the values of R_{id} and R_{icm} as well as on the value of A and R_2/R_1. Straightforward analysis of the noninverting circuit using the op-amp model of Fig. 3.37 and assuming that

$$R_o = 0, \qquad R_1 \ll R_{icm}, \qquad \frac{R_2}{R_{id}} \ll A$$

results in the following approximate expression for the input resistance of the non-inverting circuit:

$$R_{in} \simeq \{2R_{icm}\} \,\|\, \{(1 + A\beta)R_{id}\} \tag{3.35}$$

where β is the feedback ratio, given by

$$\beta = \frac{R_1}{R_1 + R_2} \tag{3.36}$$

We observe that the input resistance consists of two components in parallel: $(2R_{icm})$, which is very large, and $(1 + A\beta)R_{id}$, which is also large because R_{id} is multiplied by the amount of feedback $(1 + A\beta)$. At low frequencies, $A = A_0$ and the amount of feedback $1 + A_0\beta$ is usually a large number. At higher frequencies, the dependence of A on frequency must be taken into account as illustrated in the following example.

EXAMPLE 3.8

Consider an op amp having $f_t = 1$ MHz, $R_{id} = 1$ MΩ, and $R_{icm} = 100$ MΩ. Find the components of the input impedance of a noninverting amplifier with a nominal gain of 100.

Solution
The input impedance can be obtained by substituting in Eq. (3.35) the following.

$$R_{icm} = 10^8 \ \Omega$$

$$R_{id} = 10^6 \ \Omega$$

Since

$$1 + \frac{R_2}{R_1} = 100$$

we find that $\beta = 0.01$. Also

$$A \simeq \frac{\omega_t}{s} = \frac{2\pi \times 10^6}{s}$$

The result is

$$Z_{in} = (2 \times 10^8) \left\| \left(10^6 + \frac{2\pi \times 10^6 \times 10^{-2} \times 10^6}{s} \right) \right.$$

Note that the second component of Z_{in} consists of a 10^6-Ω resistance in series with a capacitor of value $(1/2\pi \times 10^{-10})$ farads. Figure 3.38 shows an equivalent circuit of the input impedance obtained above.

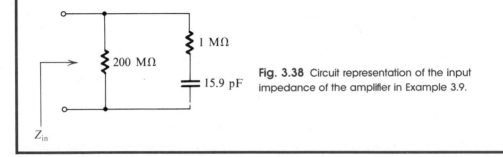

Z_{in}

Fig. 3.38 Circuit representation of the input impedance of the amplifier in Example 3.9.

Output Resistance

We now turn to the effect of the finite output resistance R_o shown in the model of Fig. 3.37. Typical values for the open-loop output resistance R_o are 75 to 100 Ω, although there exist amplifiers with much higher output resistances. We wish to find the output resistance of a closed-loop amplifier. To do this, we short the signal source, which makes the inverting and noninverting configurations identical, and apply a test voltage V_x to the output as shown in Fig. 3.39. Then, the output resistance $R_{out} \equiv V_x/I$ can be obtained by straightforward analysis of the circuit in Fig. 3.39 as follows:

$$V = -V_x \frac{R_1}{R_1 + R_2} = -\beta V_x$$

$$I = \frac{V_x}{R_1 + R_2} + \frac{V_x - AV}{R_o}$$

$$= \frac{V_x}{R_1 + R_2} + \frac{(1 + A\beta)V_x}{R_o}$$

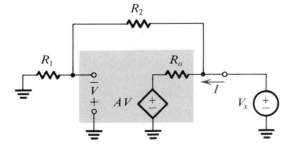

Fig. 3.39 Derivation of the closed-loop output resistance.

Thus

$$\frac{1}{R_{\text{out}}} \equiv \frac{I}{V_x} = \frac{1}{R_1 + R_2} + \frac{1 + A\beta}{R_o}$$

where the constant β is defined

$$\beta \equiv \frac{R_1}{R_1 + R_2}$$

This means that the closed-loop output resistance is composed of two parallel components,

$$R_{\text{out}} = [R_1 + R_2] \,\|\, [R_o/(1 + A\beta)] \qquad (3.37)$$

Normally R_o is much smaller than $R_1 + R_2$, resulting in

$$R_{\text{out}} \simeq \frac{R_o}{1 + A\beta} \qquad (3.38)$$

We note that the closed-loop output resistance is smaller than the op-amp open-loop output resistance by a factor equal to the amount of feedback, $1 + A\beta$. A further simplification of the expression for R_{out} is obtained by noting that normally $A\beta \gg 1$, which gives

$$R_{\text{out}} \simeq \frac{R_o}{A\beta} \qquad (3.39)$$

At very low frequencies A is real and large, resulting in a very small R_{out}. As an example, a voltage follower ($\beta = 1$) designed using an op amp with $R_o = 100 \ \Omega$ and $A_0 = 10^5$ will have

$$R_{\text{out}} = \frac{100}{10^5 \times 1} = 1 \ \text{m}\Omega$$

It is interesting to inquire about the effect of the finite op-amp bandwidth on the closed-loop output impedance. Substituting $A = \omega_t/s$ in Eq. (3.38) results in

$$Z_{\text{out}} = \frac{R_o}{1 + \beta\omega_t/s} \qquad (3.40)$$

Thus,

$$Y_{\text{out}} = \frac{1}{Z_{\text{out}}} = \frac{1}{R_o} + \frac{\beta\omega_t}{sR_o} \qquad (3.41)$$

which indicates that the output impedance consists of a resistance equal to R_o in parallel with an inductance of value $L = R_o/\beta\omega_t$. An equivalent circuit representation of the closed-loop output impedance is shown in Fig. 3.40a. Note that for this equivalent circuit we have used the more accurate gain expression

$$A = \frac{A_0}{1 + s/\omega_t}$$

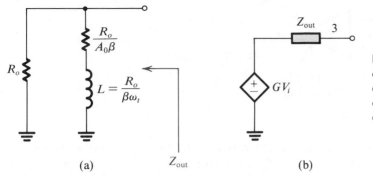

Fig. 3.40 (a) Equivalent circuit representation of the closed-loop output impedance of an op amp. **(b)** Thévenin equivalent of the output of a closed-loop circuit with gain G and output impedance Z_{out}.

Finally we remind the reader of the meaning of the closed-loop output impedance just evaluated by showing in Fig. 3.40b the Thévenin equivalent of the output of a closed-loop amplifier having a closed loop gain G and a closed-loop output impedance Z_{out}.

Exercises

3.28 Find the input resistance, at low frequencies, of a unity-gain buffer constructed using an op amp with $A_0 = 10^4$, $R_{icm} = 100$ MΩ and $R_{id} = 1$ MΩ.

Ans. $R_{in} \simeq 200$ MΩ

3.29 Repeat Exercise 3.28 for a noninverting amplifier with a nominal closed-loop gain of 100.

Ans. $R_{in} \simeq 67$ MΩ

3.30 Find the values of the various components in the output impedance (Fig. 3.40a) of a noninverting amplifier with a closed-loop gain of 100. Let the op amp have $R_o = 100$ Ω, $A_0 = 10^5$, and $f_t = 1$ MHz.

Ans. 100 Ω; 0.1 Ω; 1.59 mH

3.13 DC PROBLEMS

Offset Voltage

Because op amps are direct-coupled devices with large gains at dc, they are prone to dc problems. The first such problem is the dc offset voltage. To understand this problem consider the following conceptual experiment: If the two input terminals of the op amp are tied together and connected to ground, it will be found that a finite dc voltage exists at the output. This is the *output dc offset voltage*. To take this dc offset voltage into account in the analysis of closed-loop configurations, it has been found convenient to "refer it back" to the input. Specifically, if we divide the output dc offset voltage by the gain A_0 we obtain the *input offset voltage* V_{off}. The latter may be represented by a voltage source connected in series with one of the input leads of an ideal op amp—that is, one with zero offset voltage. This arrangement is shown in Fig. 3.41, where we note that V_{off} can be either positive or negative depending on the particular op amp, with the polarity of V_{off} not known a priori.

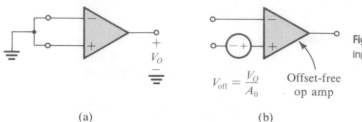

Fig. 3.41 Illustrating the definition of the input offset-voltage V_{off}.

$$V_{off} = \frac{V_O}{A_0}$$

Offset-free op amp

(a)   (b)

It should be obvious from Fig. 3.41b that if the two input terminals are tied together the output will be $A_0 V_{off} = V_O$, which correlates with our original definition of V_{off}. Another interpretation of V_{off} is that it is the magnitude of the voltage that, if applied between the two input terminals of a real op amp, would cause the output dc voltage to be reduced to zero. This interpretation assumes that the polarity of V_{off} is known, so the external source could be of opposite polarity.

General-purpose op amps have offset voltages V_{off} of the order of few millivolts (2 to 5 mV). The value of V_{off} and its temperature coefficient are usually specified in the op-amp data sheets. The latter specification is very important, especially if one is contemplating nulling V_{off} to zero. Many op amps are provided with two extra terminals and a prescribed technique for reducing V_{off} to zero. The question then becomes: Will V_{off} remain zero with changes in temperature? The answer can be found if we know the specification on *offset voltage drift* (usually given in $\mu V/°C$).

Let us now consider the effect of V_{off} on the performance of closed-loop amplifiers. In doing this we may simplify matters by grounding the signal source. In this manner both the inverting and the noninverting configurations will be identical, as shown in Fig. 3.42. It is easily seen that V_{off} will be amplified by the quantity $1 + R_2/R_1$, and the resulting dc voltage will appear at the output as an offset voltage. In the presence of an input signal this output dc voltage will be superimposed on the output signal. If the closed-loop gain is large, the output dc offset voltage could be high, thus reducing the maximum possible output signal swing.

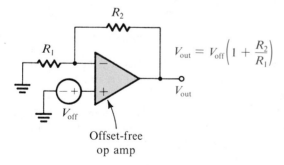

$$V_{out} = V_{off}\left(1 + \frac{R_2}{R_1}\right)$$

Fig. 3.42 Evaluating the output dc offset due to V_{off} in a closed-loop amplifier.

One way to overcome the dc offset problem is by capacitively coupling the amplifier. This, however, will be possible only in applications where the closed-loop amplifier is not required to amplify dc or very-low-frequency signals. Figure 3.43 shows a capacitively coupled inverting amplifier. The coupling capacitor will cause

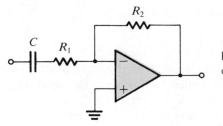

Fig. 3.43 A capacitively coupled inverting amplifier.

the gain to be zero at dc. In fact, the circuit will have an STC high-pass response with a 3-dB frequency $\omega_0 = 1/CR_1$, and the gain will be $-R_2/R_1$ for frequencies $\omega \gg \omega_0$. The advantage of this arrangement is that V_{off} will not be amplified. Thus the output dc voltage will be equal to V_{off} rather than $V_{\text{off}}(1 + R_2/R_1)$, which is the case without the coupling capacitor. The reader should convince himself or herself that the V_{off} generator indeed sees a unity-gain follower.

Input Bias Currents

The second dc problem encountered in op amps is illustrated in Fig. 3.44. In order for the op amp to operate, its two input terminals have to be supplied by finite dc currents, termed the *input bias currents*. In Fig. 3.44 these two currents are represented by two current sources, I_{B1} and I_{B2}, connected to the two input terminals. It should

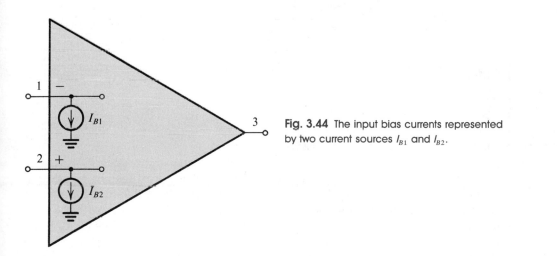

Fig. 3.44 The input bias currents represented by two current sources I_{B1} and I_{B2}.

be emphasized that the input bias currents are independent of the fact that the op amp has finite input resistance (not shown in Fig. 3.44). The op-amp manufacturer usually specifies the average value of I_{B1} and I_{B2} as well as their expected difference. The average value I_B is called the input bias current,

$$I_B = \frac{I_{B1} + I_{B2}}{2}$$

while the difference is called the *input offset current* and is given by

$$I_{\text{off}} = |I_{B1} - I_{B2}|$$

Typical values for general-purpose op amps that use bipolar transistors are $I_B = 100$ nA and $I_{\text{off}} = 10$ nA. Op amps that utilize field-effect transistors in the input stage have much smaller input bias current (of the order of picoamperes).

We now wish to find the dc output voltage of the closed-loop amplifier due to the input bias currents. To do this we ground the signal source and obtain the circuit shown in Fig. 3.45 for both the inverting and noninverting configurations. As shown in Fig. 3.45, the output dc voltage is given by

$$V_O = I_{B1}R_2 \simeq I_B R_2 \tag{3.42}$$

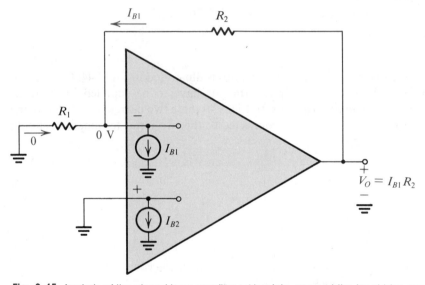

Fig. 3.45 Analysis of the closed-loop amplifier, taking into account the input bias currents.

This obviously places an upper limit on the value of R_2. Fortunately, however, a technique exists for reducing the value of the output dc voltage due to the input bias currents. The method consists of introducing a resistance R_3 in series with the non-inverting input lead, as shown in Fig. 3.46. From a signal point of view, R_3 has a negligible effect. The appropriate value for R_3 can be determined by analyzing the circuit in Fig. 3.46, where analysis details are shown and the output voltage is given by

$$V_O = -I_{B2}R_3 + R_2(I_{B1} - I_{B2}R_3/R_1) \tag{3.43}$$

Consider first the case $I_{B1} = I_{B2} = I_B$, which results in

$$V_O = I_B[R_2 - R_3(1 + R_2/R_1)]$$

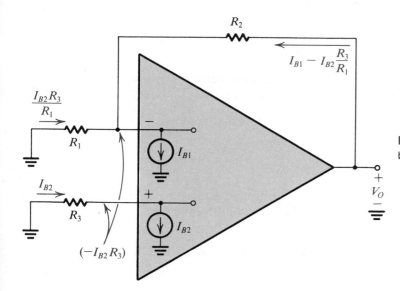

Fig. 3.46 Reducing the effect of the input bias currents by introducing a resistor R_3.

Thus we may reduce V_O to zero by selecting R_3 such that

$$R_3 = \frac{R_2}{1 + R_2/R_1} = \frac{R_1 R_2}{R_1 + R_2} \tag{3.44}$$

That is, R_3 should be made equal to the parallel equivalent of R_1 and R_2.

Having selected R_3 as in the above, let us evaluate the effect of a finite offset current I_{off}. Let $I_{B1} = I_B + I_{\text{off}}/2$ and $I_{B2} = I_B - I_{\text{off}}/2$, and substitute in Eq. (3.43). The result is

$$V_O = I_{\text{off}} R_2 \tag{3.45}$$

which is usually about an order of magnitude smaller than the value obtained without R_3 [Eq. (3.42)]. We conclude that to minimize the effect of the input bias currents one should place in the positive lead a resistance equal to the dc resistance seen by the inverting terminal. We should emphasize the word dc in the last statement; note that if the amplifier is ac-coupled, we should select $R_3 = R_2$, as shown in Fig. 3.47.

While we are on the subject of ac-coupled amplifiers we should note that one must always provide a continuous dc path between each of the input terminals of the op amp and ground. For this reason the ac-coupled noninverting amplifier of Fig. 3.48

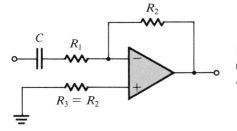

Fig. 3.47 In an ac-coupled amplifier the dc resistance seen by the inverting terminal is R_2; hence R_3 is chosen equal to R_2.

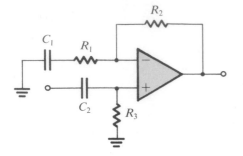

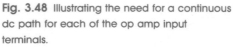

will *not* work without the resistance R_3 to ground. Unfortunately, including R_3 lowers considerably the input resistance of the closed-loop amplifier.

EXAMPLE 3.9

We wish to evaluate the effects of the input offset voltage and the input bias current on the performance of the Miller integrator shown in Fig. 3.49.

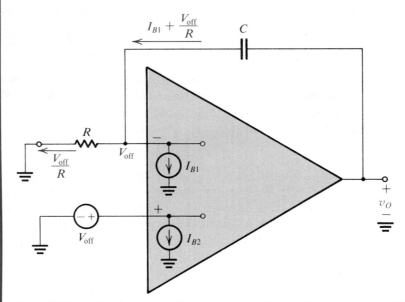

Fig. 3.49 Evaluating the effects of the dc bias and offsets on the performance of the Miller integrator.

Solution
Straightforward analysis shows that irrespective of the input signal there will be a dc current through the capacitor given by

$$I = I_{B1} + \frac{V_{off}}{R}$$

Correspondingly, there will be an output voltage given by

$$v_O = V_{off} + \frac{1}{C} \int I \, dt$$

$$= V_{off} + \frac{V_{off}}{CR} t + \frac{I_{B1}}{C} t \qquad (3.46)$$

which means that the output voltage will increase linearly with time until the amplifier leaves its linear range and saturates. Obviously this is an unacceptable situation! To overcome this problem we have to provide a dc path for the current I. This can be achieved by connecting a resistance in parallel with the capacitor. Such a resistance makes the closed-loop gain finite at dc. Unfortunately, however, the resulting circuit will no longer be an ideal integrator. Rather, it will have the response of a low-pass STC circuit. The performance of this "damped" or "leaky" integrator is investigated in Problem 3.59.

Exercises

3.31 A noninverting amplifier with a closed-loop gain of 1,000 is designed using an op amp having an input offset voltage of 4 mV and output saturation levels of ± 12 V. What is the maximum amplitude of the sine wave that can be applied at the input without the output clipping? If the amplifier is capacitively coupled in the manner indicated in Fig. 3.48, what would the maximum possible amplitude become?

Ans.   8 mV; 12 mV

3.32 Consider the differential amplifier circuit in Fig. 3.18. Let $R_1 = R_3 = 10$ kΩ and $R_2 = R_4 = 1$ MΩ. If the op amp has $V_{off} = 3$ mV, $I_B = 0.2$ μA, and $I_{off} = 50$ nA, find the dc offset voltage at the output.

Ans.   353 mV if V_{off} is positive; 253 mV if V_{off} is negative.

3.33 Consider the capacitively coupled amplifier of Fig. 3.43. Find values of C, R_1, and R_2 such that the corner frequency f_0 is 10 Hz and such that for $f \gg f_0$ the gain is approximately 100 and the input resistance is approximately 1 kΩ. Also find the magnitude and phase of the gain at $f = 100$ Hz.

Ans. $R_1 = 1$ kΩ; $R_2 = 100$ kΩ; $C = 15.9$ μF; $|G| = 99.5$; $\phi = 180° + 5.7°$

3.14 SUMMARY

● The IC op amp is a versatile circuit building block. It is easy to apply, and the performance of op-amp circuits closely matches theoretical predictions.

● Most op amps are differential amplifiers having a large differential gain (called open-loop gain, A) and a small common-mode gain. The ratio of the two gains is the CMRR.

● An ideal op amp has an infinite gain, an infinite input resistance, and a zero output resistance.

● In most applications, negative feedback is applied around the op amp, resulting in a closed-loop amplifier with gain determined almost entirely by the external components.

● The noninverting closed-loop configuration features a very high input resistance. A special case is the unity-gain follower, frequently employed as a buffer amplifier to connect a high-resistance source to a low-resistance load.

● In the analysis of op-amp circuits, if the op amp is assumed ideal, a virtual short circuit appears between its two input terminals. This, together with the fact that zero currents flow into the input terminals of an ideal op amp, simplify analysis considerably.

● For most internally compensated op amps, the open-loop gain falls off with frequency at a rate of -20 dB/decade, reaching unity at a frequency f_t (the unity-gain bandwidth).

● For both the inverting and the noninverting closed-loop configurations, the 3-dB frequency is equal to $f_t/(1 + R_2/R_1)$.

● The maximum rate at which the op-amp output voltage can change is called the slew rate. Op-amp slewing can result in nonlinear distortion of output signal waveforms.

● The full-power bandwidth, f_M, is the maximum frequency at which an output sinusoid with an amplitude equal to the op-amp rated output voltage, can be produced without distortion.

● The finite op-amp CMRR limits the performance of the difference-amplifier closed-loop configuration. The effect of finite CMRR can be taken into account in analysis, by including a signal source equal to ($v_{I\mathrm{cm}}$/CMRR) in series with the op-amp positive input lead.

● The input resistance of the noninverting configuration is approximately equal to $R_{id}(1 + A\beta)$, where β is the feedback factor.

● For both the inverting and the noninverting configurations the output resistance is equal to $R_o/(1 + A\beta)$.

● The input offset voltage, V_{off}, is the magnitude of dc voltage that when applied between the op amp input terminals, with appropriate polarity, reduces the dc offset voltage at the output to zero.

● The effect of V_{off} on performance can be evaluated by including, in the analysis, a dc source V_{off} in series with the op-amp positive input lead.

● Capacitively coupling an op amp reduces the dc offset voltage at the output considerably.

● The average of the two dc currents, I_{B1} and I_{B2}, that flow in the input terminals of the op amp, is called the input bias current, I_B. In a closed-loop amplifier, I_B gives rise to a dc offset voltage at the output of magnitude $I_B R_2$. This voltage can be reduced to $I_{off} R_2$ by connecting a resistance in series with the positive input terminal equal to the total dc resistance seen by the negative input terminal. I_{off} is the input offset current; that is, $I_{off} = |I_{B1} - I_{B2}|$.

● Connecting a large resistance in parallel with the capacitor of an op-amp integrator prevents op-amp saturation (due to the effect of V_{off} and I_B).

BIBLIOGRAPHY

G. B. Clayton, *Experimenting with Operational Amplifiers*, London: Macmillan, 1975.

G. B. Clayton, *Operational Amplifiers*, 2nd ed., London: Newnes-Butterworths, 1979.

J. G. Graeme, G. E. Tobey, and L. P. Huelsman, *Operational Amplifiers: Design and Applications*, New York: McGraw-Hill, 1971.

W. Jung, *IC Op Amp Cookbook*, Indianapolis: Howard Sams, 1974.

J. K. Roberge, *Operational Amplifiers: Theory and Practice*, New York: Wiley, 1975.

J. I. Smith, *Modern Operational Circuit Design*, New York: Wiley-Interscience, 1971.

J. E. Solomon, "The monolithic op amp: A tutorial study," *IEEE Journal of Solid-State Circuits*, vol. SC-9, no. 6, pp. 314–322, Dec. 1974.

J. V. Wait, L. P. Huelsman, and G. A. Korn, *Introduction to Operational Amplifier Theory and Applications*, New York: McGraw-Hill, 1975.

PROBLEMS

3.1 Assuming ideal op amps, find the voltage gain v_o/v_i and input resistance R_{in} of each of the circuits in Fig. P3.1.

3.2 Using an ideal op amp, design an inverting amplifier having a gain of -100 with input resistance of 10 kΩ.

3.3 Using the circuit of Fig. 3.4, design an amplifier with gain of -100 having the largest possible input resistance under the constraint of having to use resistors no larger than 10 MΩ. What is the input resistance of your design?

3.4 If when using the amplifier in Problem 3.3 an additional factor of 2 in gain is required, suggest a convenient means to achieve this using a resistor shunting (that is, connected in parallel with) one in the original design. Which resistor do you shunt? With what resistor do you shunt it? What does the input resistance become?

3.5 Using the circuit in Fig. 3.7 with resistors no larger than 1 MΩ, design an amplifier with a gain of -102 and an input resistance of 1 MΩ.

3.6 Find the transfer function of the circuit in Fig. P3.6

assuming the op amp to be ideal. Under what condition is the transfer function a constant independent of frequency? For a closed-loop gain of -10 V/V and an input impedance consisting of 1 MΩ shunted by 30 pF, what values of R_1, R_2, C_1, and C_2 must be used?

*__3.7__ For the circuit of Problem 3.6 sketch the frequency response for $R_1 = R_2 = 10$ kΩ and **(a)** $C_1 = 10C_2 = 0.1$ μF, **(b)** $C_2 = 10C_1 = 0.1$ μF. In each case what is the gain at low and high frequencies?

3.8 Using the circuit of Fig. E3.6, design a low-pass STC network with a dc gain of 20 dB, an input resistance of 10 kΩ, and a 3-dB frequency of 1000 rad/s. What is the gain in dB at $\omega = 10^4$ rad/s? Also give $v_o(t)$ if $v_I(t)$ is a $+1$-V step.

3.9 Consider a Miller integrator utilizing an ideal op amp and having an input resistor of 10 kΩ and a feedback capacitor of 0.1 μF. An input step rising from 0 to 1 V is applied at $t = 0$, at which time the voltage across the capacitor is zero. Describe the output voltage. At what time does it reach -10 V?

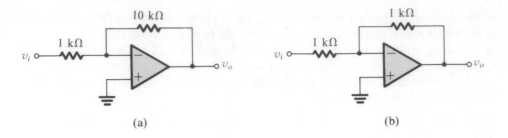

(a)   (b)

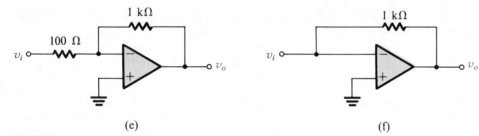

(c)   (d)

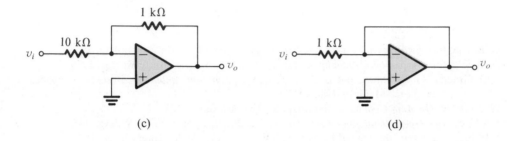

(e)   (f)

Fig. P3.1

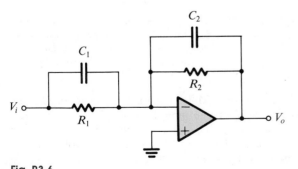

Fig. P3.6

3.10 Consider a Miller integrator with an ideal op amp, $R = 10$ kΩ, and $C = 0.1$ μF. A series of 1-V, 0.1-ms pulses occurring at intervals ranging randomly from 1 to 100 ms is applied at the input. With no pulse present, the input rests at 0 V. At $t = 0$ the output is $+10$ V. How many pulses does it take for the output to reach zero volts? (Such an integrator can be used as a pulse counter provided that the input pulses are of uniform amplitude and duration.) For the output initially at zero, what output level would correspond to the passage of 10 pulses?

3.11 Consider a differentiator consisting of an ideal op amp, a 0.01-μF capacitor and 10-kΩ resistor. Sketch the output waveform following application of a "slow-step" input, which rises linearly from 0 to 10 V in 1 ms. What happens if the step rises in 0.1 ms? What happens for an ideal step?

3.12 Design a two-op-amp circuit that accepts two independent voltages v_1 and v_2 and forms the output $v_O = v_1 - 5v_2$.

3.13 Derive the transfer function of the circuit shown in Fig. P3.13 assuming the op amp to be ideal. Sketch and clearly label the magnitude of the frequency response (dB versus ω on log scale). If the response crosses the 0-dB line at $\omega = 10$ rad/s and if $R = 100R_1$, find the magnitude of the transfer function at $\omega = 100$ rad/s and at $\omega = 10,000$ rad/s.

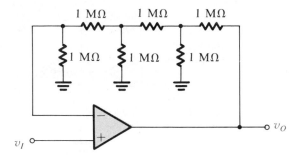

Fig. P3.14

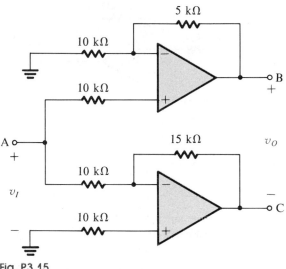

Fig. P3.15

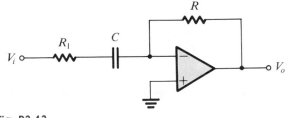

Fig. P3.13

3.14 Find the gain v_o/v_i of the amplifier circuit of Fig. P3.14 assuming that the op amp is ideal. *Hint:* Start at the left-hand side of the feedback circuit and work your way toward the output figuring out the various node voltages and branch currents in the feedback circuit.

3.15 The circuit shown in Fig. P3.15, called a bridge amplifier, is intended to supply current to floating loads (those for which both terminals are ungrounded) while

making greatest possible use of the available power supply. Assuming ideal op amps, sketch the voltage waveforms at B and C for a 1-V peak-to-peak sine wave applied at A. Also sketch v_O. What is the voltage gain v_O/v_I?

3.16 Consider the bridge amplifier of Problem 3.15 and assume that the op amps operate from ± 15-V power supplies and that their output saturates at ± 14-V (in the manner shown in Fig. 2.7). What is the largest sine wave output that can be accommodated? Specify both its peak-to-peak and rms values.

3.17 Find the gain v_o/v_i of the circuit shown in Fig. P3.17.

3.18 For the circuit in Fig. P3.18, express v_O as a function of v_1 and v_2.

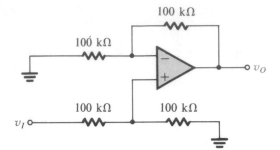

Fig. P3.17

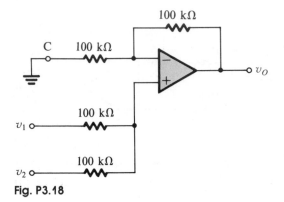

Fig. P3.18

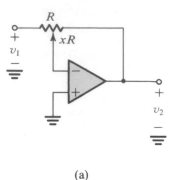

(a)

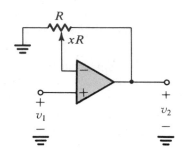

(b)

3.19 If node C in the circuit of Fig. P3.18 is removed from ground and connected to a third input v_3, express v_O as a function of v_1, v_2, and v_3.

*3.20 Use the circuit in Fig. 3.18 to design a difference amplifier with a gain of 10 and a differential input resistance of 20 kΩ. What is the resistance (to ground) seen at the v_1 terminal (with the v_2 terminal grounded)? What is the resistance (to ground) seen at the v_2 terminal (with the v_1 terminal grounded)? Revise your design to make these equal while maintaining other specifications.

*3.21 What is the gain v_2/v_1 as a function of x, the fraction of full rotation of the potentiometer R, for each of the circuits in Fig. P3.21?

3.22 For the circuits shown in Fig. P3.22 express i_o as a function of v.

3.23 The circuit shown in Fig. P3.23 is that of an "instrumentation amplifier," having high-impedance differential inputs and gain control by means of a single resistor (R_4). Assuming the op amps to be ideal, show that

$$v_O = -\frac{R_2}{R_1}\left(1 + \frac{2R_3}{R_4}\right)(v_2 - v_1)$$

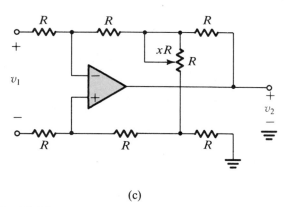

(c)

Fig. P3.21

and thus that the circuit amplifies differential input signals only.

3.24 A unity-gain voltage follower is designed using an op amp with an open-loop gain of 1,000. Find the closed-loop gain of the follower. What is the percentage change in closed-loop gain if the op amp is replaced with another having an open-loop gain of 2,000?

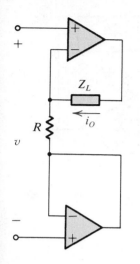

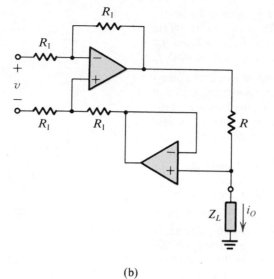

(a)   (b)

Fig. P3.22

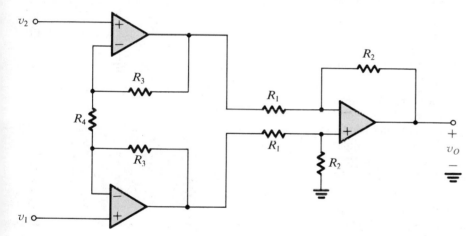

Fig. P3.23

3.25 Find the input impedance $Z_{in}(j\omega)$ of the GIC circuit in Fig. 3.24 for the case Z_1 and Z_5 is each a capacitor of value C and Z_2, Z_3, and Z_4 is each a resistor of value R.

3.26 Repeat Exercise 3.15 with the resistor R and the capacitor C interchanged.

3.27 For a number of internally compensated op amps characterized by a low-frequency gain A_0, a 3-dB frequency f_b, and a unity-gain bandwidth f_t, the following partial data are given. Provide the missing entries in the table at the right.

A_0		f_b		f_t	
V/V	dB	Hz	rad/s	Hz	rad/s
	100	10			
		10		10^6	
	100			10^6	
10^5			62.8		
10^6		1			
10^4				10^6	
10^5				10^7	

3.28 An internally compensated op amp is known to have an STC low-pass response. The gain is 10^3 at 100 Hz and 10 at 10^5 Hz. Determine the low-frequency gain A_0, the break frequency f_b, and the unity-gain bandwidth f_t.

*3.29 The designer of a new instrument requiring a unity-gain buffer amplifier considers three possibilities, each of which offers various advantages and disadvantages in the event that the design must later be modified. Evaluate the 3-dB frequency of each of his alternatives if the op amps he uses have $f_t = 2$ MHz:
(a) A noninverting op amp with direct feedback.
(b) A noninverting op amp having gain = 2 with an additional attenuator (voltage divider) of gain = 0.5.
(c) A cascade of two inverting op amps each of gain -1.

3.30 For the noninverting and inverting amplifier configurations evaluate the product of low-frequency gain and 3-dB frequency (called the gain–bandwidth product) in terms of R_1, R_2, and f_t. For which is it highest? For which is it constant (independent of closed-loop gain)? For what gain range is the difference most pronounced?

3.31 Consider a unity-gain buffer amplifier utilizing the noninverting op amp configuration with $R_2 = 0$. For what open-loop gain A does the closed-loop gain fall to 0.99? For this value of A what does the gain of an inverting amplifier with $R_1 = R_2$ become?

3.32 Two alternatives are being considered for implementing an amplifier with gain of 100: The first uses one op amp and two resistors in the noninverting configuration. The second employs a cascade of two amplifiers, each of gain of -10, using a total of two op amps and four resistors. For op-amp f_t of 1 MHz, what is the 3-dB frequency in each case?

3.33 Consider an op amp having a gain at dc of 112 dB and a single-pole STC rolloff with $f_t = 4$ MHz. Find the magnitude of the gain at 1 MHz, 100 kHz, 1 kHz, and 10 Hz.

3.34 The op amp in the previous problem is used to implement a noninverting amplifier with nominal dc gain of 400. Find the 3-dB frequency of the closed-loop gain. What is the rise time of the output if a 2.5-mV step is applied at the input?

3.35 Figure P3.35a shows a generalization of the noninverting configuration, where a resistive network having a transfer function $\beta \equiv V_1/V_o$ is connected in the negative feedback path. Show that if the open-loop gain is infinite, then the closed-loop gain $V_o/V_s = 1/\beta$. Find β and hence the closed-loop gain when the feedback network is im-

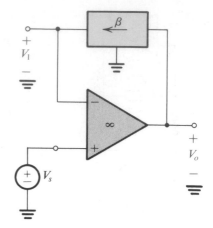

(a)

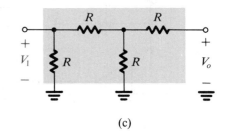

(b)

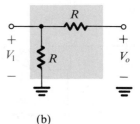

(c)

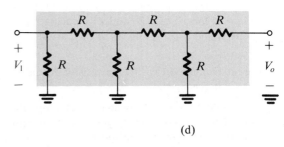

(d)

Fig. P3.35

plemented with the circuit in parts (b), (c), and (d) of the figure. [*Note:* Use of Thévenin's theorem greatly simplifies finding β for the networks in (c) and (d).]

3.36 If in the circuit of Fig. P3.35a the open-loop gain A is finite, derive an expression for the closed-loop gain V_o/V_s and find the condition under which the closed-loop gain can be approximated by $1/\beta$.

3.37 An op amp whose internal circuit can be modeled by the circuit shown in Fig. 3.30 has the following parameters: $G_m = 0.5$ mA/V, $R_{o1} = 200$ kΩ, $R_{i2} = 200$ kΩ, $\mu = 4{,}000$ and $C = 50$ pF. Find A_0, f_b, and f_t.

3.38 If the maximum current that the input stage of the op amp in the previous problem can supply is 0.1 mA, find the slew rate of the op amp. Also, if the rated output voltage is specified to be ±10 V, find the full-power bandwidth f_M.

3.39 The frequency of a 2-V peak-to-peak symmetrical square wave is raised until the output of a voltage follower to which it is applied becomes essentially triangular, with an amplitude of nearly 2 V peak-to-peak and a frequency of 500 kHz. What is the slew rate of this amplifier? If the amplifier is capable of linear operation to within 2 V of its supplies and supply voltages of ±15 V are used, what is the full-power bandwidth f_M?

3.40 For an op amp with a slew-rate limit of 10 V/μs what is the highest-frequency sine wave that can be reproduced at the output with **(a)** 1-V peak amplitude and **(b)** 10-V peak amplitude?

3.41 For an op amp with a full-power frequency of 50 kHz and rated output voltage of ±12 V, what is the largest possible undistorted sine wave output of 100 kHz?

3.42 A differential-input single-ended-output amplifier has a common-mode gain of 10^{-3} and a difference-mode (or differential) gain of 10^3. What is its common-mode rejection ratio (CMRR) in dB? If the output of this amplifier is connected to the input of another single-ended amplifier of gain 100, what is the CMRR of the combination? If the original amplifier is preceded by a differential-input differential-output amplifier having CMRR of 40 dB, what is the CMRR of the combination?

3.43 Consider the difference amplifier of Fig. 3.18, with all the resistors equal. Use the common-mode error approach to find the CMRR of the closed-loop amplifier assuming that the op amp is specified to have a CMRR of only 40 dB.

**3.44 Equation (3.4) gives the output voltage of the difference amplifier of Fig. 3.18 in terms of the two input voltages v_1 and v_2. Assume the op amp to have infinite CMRR. Expressing v_O as

$$v_O = G_d(v_2 - v_1) + G_{cm}\left(\frac{v_1 + v_2}{2}\right)$$

where G_d is the closed-loop differential gain and G_{cm} is the closed-loop common-mode gain, derive expressions for G_d and G_{cm} and for the closed-loop CMRR $\equiv G_d/G_{cm}$. If each of the two resistor ratios R_2/R_1 and R_4/R_3 is nominally 1,000 but accurate to $\pm1\%$, find the value of G_d, the largest expected value of G_{cm}, and the lowest expected value of CMRR.

3.45 Consider an op amp having $A_0 = 10^4$, $f_t = 10$ MHz, $R_{id} = 1$ MΩ, and $R_{icm} = 100$ MΩ used in a noninverting configuration with gain of 100. Evaluate the closed-loop input impedance at $f = 60$ Hz and 60 kHz.

*3.46 Consider the noninverting op-amp configuration with an external voltage v_i.
(a) Find v_o in terms of v_i, the open-loop gain A, and the feedback ratio $\beta \equiv R_1/(R_1 + R_2)$. Neglect the effects of the op-amp input and output resistances.
(b) Draw an equivalent circuit for the noninverting configuration, replacing the op amp with the equivalent circuit in Fig. 3.37 with R_o set to zero. Now, use the v_o obtained in (a) to find v_{id}.
(c) Use v_{id} obtained in (b) to help find the input current drawn from v_i and hence find the closed-loop input resistance R_{in}. Compare your result to that Eq. (3.35).

*3.47 An alternative derivation for the expression of the closed-loop output resistance R_{out} in Eq. (3.37) is possible. Consider the equivalent circuit in Fig. 3.39 and visualize this as that of an amplifier with input V_x and output AV and with R_o being a resistance between output and input. What is the gain of this amplifier? Use Miller's theorem to replace R_o with its equivalent parts and thus find the total resistance seen by the "input" voltage V_x. This is the output resistance of the original amplifier and should be identical to that in Eq. (3.37).

3.48 For an op amp having an output resistance $R_o = 10$ kΩ and an open-loop gain $A_0 = 10^4$, find the output resistance, at low frequencies, of **(a)** a unity-gain follower, **(b)** an inverting amplifier of gain 100 using $R_2 = 10$ kΩ, and **(c)** an inverting amplifier of gain 1,000 using $R_2 = 10$ kΩ. Note that although the approximate expression in Eq. (3.39) is adequate for cases (a) and (b), the more exact expression should be used in case (c).

3.49 Consider an op amp with $R_o = 1$ kΩ, $f_t = 1$ MHz, and $A_0 = 10^5$. What is the output impedance of an inverting amplifier of gain -100 designed using this op amp, at dc and at $f = 10$ kHz.

3.50 Consider an op amp with $f_t = 1$ MHz, a very large A_0, and $R_o = 10$ kΩ. This op amp is used with two resistors $R_1 = 1$ kΩ and $R_2 = 100$ kΩ to design an inverting amplifier. What is the output inductance of the closed-loop amplifier? What is the frequency at which resonance with a 0.1-μF load capacitor occurs?

3.51 An op amp having an input offset voltage of 1 mV is connected as a closed-loop amplifier with $R_2 = 100R_1$. What output dc offset results? Can anything be said about the polarity of this voltage?

3.52 Using offset-nulling facilities provided for the op amp, a closed-loop amplifier with gain of $+1,000$ is adjusted at 25°C to produce zero output with the input grounded. If the input offset-voltage drift of the op amp is specified to be 10 μV/°C, what output would you expect at 0°C and at 75°C? While nothing can be said separately about the polarity of the output offset at either 0 or 75°C, what would you expect their relative polarities to be?

3.53 A closed-loop amplifier of gain $-1,000$ uses an op amp whose output remains linear over a range of ± 12 V and whose input offset $\leq \pm 1$ mV at 25°C, with a temperature coefficient ≤ 20 μV/°C. What is the peak voltage of the largest possible sine wave that can be handled by this amplifier without limiting as the temperature varies from 0° to 75°C?

3.54 An op amp is connected in a closed-loop with gain of 100 utilizing a feedback resistor of 1 MΩ.
(a) If the input bias current is 100 nA, what output voltage results with the input grounded?
(b) If the input offset voltage is ± 1 mV, what is the largest possible output that can be observed with the input grounded?
(c) If the bias current compensation is used, what is the value of the required resistor? If the offset current is no more than one tenth the bias current, what is the resulting output offset voltage (due to offset current alone)?
(d) With the bias current compensation, as in (c) above, in place, what is the largest dc voltage at the output due to the combined effect of offset voltage and offset current?

3.55 An ac-coupled inverting amplifier with gain of -100 utilizes a 1-MΩ feedback resistor. For an input

bias current of 100 nA, what output offset voltage can be expected? If bias current compensation is used, what resistor should be employed? If the input offset current is 10 nA, what output offset voltage would you expect?

3.56 Using the topology indicated in Fig. 3.48 design an amplifier of gain $+100$ having an input resistance of 1 MΩ and input bias current compensation. Select the capacitor values such that the time constants of the two STC circuits involved are equal at 0.1 s.

3.57 Consider a Miller integrator with an input resistance of 10 kΩ and an integrating capacitor of 1 μF. Let the op amp have $V_{\text{off}} = 1$ mV and $I_B = 100$ nA. Refer to Fig. 3.49 and Eq. (3.46). What is the initial offset voltage at the output? Compare the values of the two contributors to the time-dependent output offset. If these contributions are of the same polarity, how long does it take for the amplifier to saturate at $+12$ V?

**3.58 A Miller integrator utilizing R_1 as an input resistor and C_2 as a feedback capacitor includes also resistor R_2 shunting C_2 to reduce the effects of amplifier imperfections—namely, an input offset voltage of 2 mV and a bias current of 5 nA. Design this integrator to have an input resistance of 100 kΩ, an integration time constant of 10^{-2} s and a maximum output dc offset of 100 mV. What is the duration of a positive input pulse of 10-V height that causes the output to fall to -10 V? If the input pulse is returned to zero when the output becomes -10 V, how long will it take the output to return to -1 V? Assume that the op-amp bias current flows into the amplifier and that the input voltage offset is of such a polarity as to reinforce the offset voltage due to the input bias current.

**3.59 Consider a Miller integrator with $R = 100$ kΩ and $C = 10$ nF. To reduce the effects on the dc offset voltage and dc bias current, C is shunted with a 10-MΩ resistor. Find the output of the resulting nonideal integrator when the input is a 25-Hz, 2-V peak-to-peak, zero-average, symmetrical square wave. Contrast this response to that of an ideal integrator having an equal integrating time constant.

**3.60 This problem is a challenging one. The objective is to find the effect of the finite CMRR of the op amps on the common-mode rejection of the instrumentation amplifier of Fig. P3.23. Use the technique discussed and illustrated in Fig. 3.36 to find an expression for v_O of the instrumentation amplifier in terms of v_1 and v_2 assuming that each op amp has a finite common-mode rejection ratio, denoted CMRR. Then let $v_1 = v_{icm}$ and $v_2 = v_{icm} + v_{id}$ and thus find the differential gain and the common-mode gain. Show that the common-mode rejection ratio of the instrumentation amplifier is approximately equal to $CMRR \times [1 + (2R_3/R_4)]$.

3.61 The op amp in Fig. P3.61 limits at ± 13 V and is otherwise ideal. What is the voltage at B with switch S closed? Sketch the waveform at B after S opens at time zero. How long after S opens does it take for B to reach $+11$ V?

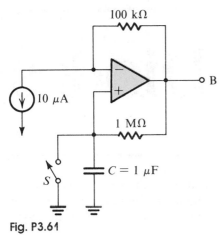

Fig. P3.61

*3.62 The op amp in Fig. P3.62 has an open-loop gain of 10^3 and an open-loop output resistance of 1 kΩ but is otherwise ideal. What is the output voltage at B with no load? If a current of 10 mA is extracted from the output, what does the output voltage become? What is the equivalent closed-loop output resistance?

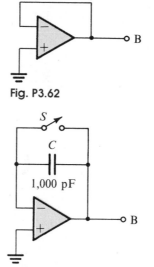

Fig. P3.62

Fig. P3.63

3.63 In the current of Fig. P3.63, when S is opened, the voltage at B falls at the rate of 10 mV/s. What is the input bias current of the op amp and in which direction does it flow?

*3.64 An inverting bandpass audio amplifier is constructed from an ideal op amp using an input network consisting of R_1 and C_1 in series and a feedback network of R_2 and C_2 in parallel. Design R_1, R_2, C_1, C_2 to meet the following specifications: an input resistance at high frequencies of 10 kΩ; a midband gain of -100; a lower 3-dB cutoff frequency of 100 Hz; an upper 3-dB cutoff frequency of 10 kHz. What is the minimum f_t of the op amp to ensure that the upper cutoff is relatively independent of the amplifier rolloff? Use a factor of 10 in separation of critical frequencies.

Chapter 4

Diodes

INTRODUCTION

In Section 2.1 the distinction was made between linear and nonlinear circuits. As an example of nonlinear devices, the diode was introduced and its application in the design of rectifiers was demonstrated. (The reader is encouraged to review this material before studying the present chapter.) The present chapter is devoted to the study of this fundamental nonlinear two-terminal device: the diode.

We begin by defining the characteristics of the *ideal diode* and consider the analysis of circuits containing ideal diodes. We then introduce the silicon junction diode, explain its terminal characteristics, and provide techniques for the analysis of diode circuits. The latter task involves the important subject of device modeling.

The last part of this chapter is concerned with the physical operation of the *pn junction*. In addition to being a diode, the *pn* junction is the basis of many other solid-state devices. Thus an understanding of the *pn* junction is essential to the study of modern electronics. Though simplistic, the qualitative view presented here provides sufficient background for our purpose—namely, the study of circuits. A more detailed study of the *pn* junction can be found in texts dealing with physical electronics.

Nonlinear circuits play a major role in signal processing. Examples include signal generators, modulators, demodulators, dc power supplies, and digital circuits. A number of these applications will be studied in the next chapter using circuits composed of diodes and op amps. □

4.1 THE IDEAL DIODE

The ideal diode may be considered the most fundamental nonlinear element. It is a two-terminal device having the circuit symbol of Fig. 4.1a and the *i-v* characteristic shown in Fig. 4.1b. The terminal characteristic of the ideal diode can

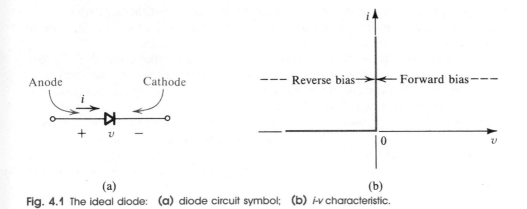

(a)

(b)

Fig. 4.1 The ideal diode: **(a)** diode circuit symbol; **(b)** *i-v* characteristic.

be interpreted as follows: If a negative voltage (relative to the reference direction indicated in Fig. 4.1a) is applied to the diode, no current flows and the diode behaves as an open circuit. Diodes operated in this mode are said to be *reverse-biased*, or operated in the reverse direction. An ideal diode has zero current when operated in the reverse direction.

On the other hand, if a positive current (relative to the reference direction indicated in Fig. 4.1a) is applied to the ideal diode, zero voltage drop appears across the diode. In other words, the ideal diode behaves as a short circuit in the *forward* direction; it passes any current with zero voltage drop.

From the above description it should be noted that the external circuit must be designed so as to limit the forward current through a conducting diode, and the reverse voltage across a cutoff diode, to predetermined values. Figure 4.2 shows two diode circuits that illustrate this point. In the circuit of Fig. 4.2a the diode is obviously conducting. Thus its voltage drop will be zero, and the current through it will be determined by the +10-V supply and the 1-kΩ resistor as 10 mA. The diode in the circuit of Fig. 4.2b is obviously cut off, and thus its current will be zero, which in turn means that the entire 10-V supply will appear as reverse bias across the diode.

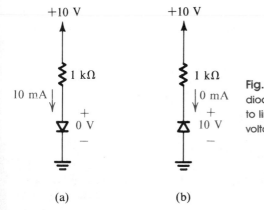

(a)

(b)

Fig. 4.2 The two modes of operation of ideal diodes and the use of an external circuit to limit the forward current and the reverse voltage.

The positive terminal of the diode is called the anode and the negative terminal the cathode, a carryover from the days of vacuum-tube diodes. The *i-v* characteristic of the ideal diode should explain the choice of its circuit symbol.

As should be evident from the above, the *i-v* characteristic of the ideal diode is highly nonlinear; it consists of two straight-line segments at 90° to one another. A nonlinear curve that consists of straight-line segments is said to be *piecewise linear*. If a device having a piecewise-linear characteristic is used in a particular application in such a way that the signal across its terminals swings only along one of the linear segments, then the device can be considered a linear circuit element as far as that particular circuit application is concerned. On the other hand, if signals swing past one or more of the break points in the characteristics, linear analysis is no longer possible.

EXAMPLE 4.1

Figure 4.3a shows a circuit for charging a 12-V battery. If v_S is a sinusoid with 24-V peak amplitude, find the fraction of each cycle during which the diode conducts. Also find the peak value of the diode current and the maximum reverse-bias voltage that appears across the diode.

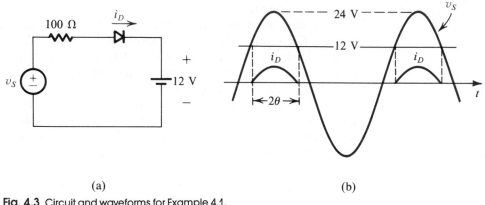

(a)   (b)

Fig. 4.3 Circuit and waveforms for Example 4.1.

Solution
The diode conducts when v_S exceeds 12 V, as shown in Fig. 4.3b. The conduction angle is 2θ, where θ is given by

$$24 \cos \theta = 12$$

Thus $\theta = 60°$ and the conduction angle is 120° or one-third of a cycle.

The peak value of the diode current is given by

$$I_d = \frac{24 - 12}{100} = 0.12 \text{ A}$$

The maximum reverse bias voltage across the diode occurs when v_S is at its negative peak and is equal to $24 + 12 = 36$ V.

EXAMPLE 4.2

Assuming the diodes to be ideal, find the values of I and V in the circuits of Fig. 4.4.

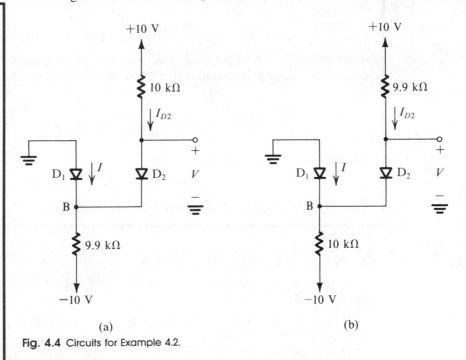

(a)   (b)

Fig. 4.4 Circuits for Example 4.2.

Solution

In these circuits it might not be obvious at first sight whether none, one, or both diodes are conducting. In such a case we make a plausible assumption, proceed with the analysis, and then check whether we end up with a consistent solution. For the circuit in Fig. 4.4a we shall assume that both diodes are conducting. It follows that $V_B = 0$ and $V = 0$. The current through D_2 can now be determined from

$$I_{D2} = \frac{10 - 0}{10} = 1 \text{ mA}$$

Writing a node equation at B,

$$I + 1 = \frac{0 - (-10)}{9.9}$$

results in

$$I = 0.01 \text{ mA}$$

Thus D_1 is conducting as originally assumed and the final result is $I = 0.01$ mA and $V = 0$ V.

For the circuit in Fig. 4.4b, if we assume that both diodes are conducting, then $V_B = 0$ and $V = 0$. The current in D_2 is obtained from

$$I_{D2} = \frac{10 - 0}{9.9} = 1.01 \text{ mA}$$

The node equation at B is

$$I + 1.01 = \frac{0 - (-10)}{10}$$

which yields $I = -0.01$ mA. Since this is not possible, our original assumption is not correct. We start again assuming that D_1 is off and D_2 is on. The current I_{D2} is given by

$$I_{D2} = \frac{10 - (-10)}{19.9} = 1.005 \text{ mA}$$

and the voltage at node B is

$$V_B = -10 + 10 \times 1.005 = +0.05 \text{ V}$$

Thus D_1 is reverse-biased as assumed, and the final result is $I = 0$ and $V = 0.05$ V.

Exercises

4.1 Find the values of I and V in the circuits shown in Fig. E4.1.

Ans. **(a)** 2 mA, 0 V;   **(b)** 0 mA, 5 V;   **(c)** 0 mA, 5 V;   **(d)** 2 mA, 0 V;   **(e)** 3 mA + 3 V;   **(f)** 4 mA + 1 V

4.2 Figure E4.2 shows a circuit for an ac voltmeter. It utilizes a moving-coil meter that gives a full-scale reading when the *average* current flowing through it is 1 mA. The moving-coil meter has a 50-Ω resistance. Find the value of R that results in the meter indicating a full-scale reading when the input sine-wave voltage v_I is 20 V peak-to-peak. (*Hint:* The average value of half-sine waves is V_p/π.)

Ans.   3.133 kΩ

4.2 TERMINAL CHARACTERISTICS OF REAL JUNCTION DIODES

In this section we study the characteristics of real diodes—specifically, semiconductor junction diodes made of silicon. The physical processes that give rise to the diode terminal characteristics, and to the name "junction diode," will be studied in the latter part of this chapter.

Figure 4.5 shows the *i-v* characteristic of a silicon junction diode. The same characteristic is shown in Fig. 4.6 with some scales expanded and others compressed, so as to portray details. Note that the scale changes have resulted in the apparent discontinuity at the origin.

As indicated, the characteristic curve consists of three distinct regions:

1. The forward-bias region, determined by $v > 0$

2. The reverse-bias region, determined by $v < 0$

3. The breakdown region, determined by $v < -V_{ZK}$

These three regions of operation are described in the following three sections.

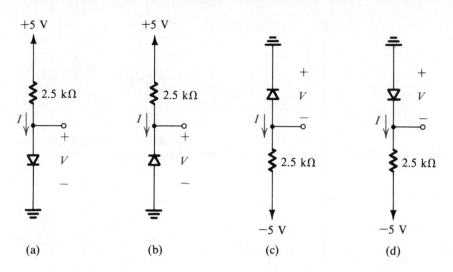

(a)   (b)   (c)   (d)

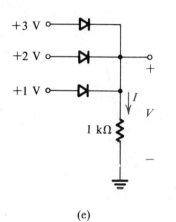

(e)

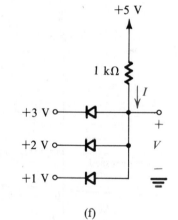

(f)

Fig. E4.1

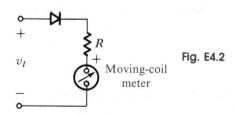

Fig. E4.2

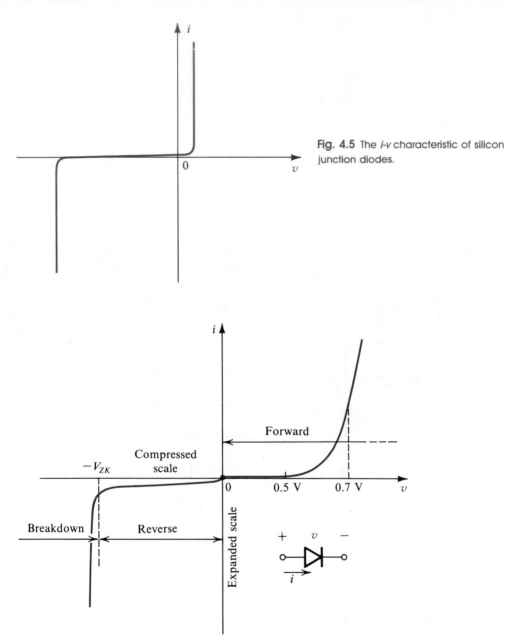

Fig. 4.5 The *i-v* characteristic of silicon junction diodes.

Fig. 4.6 The diode *i-v* relationship with some scales expanded and others compressed in order to reveal details.

4.3 THE FORWARD-BIAS REGION

The forward-bias—or, simply, forward—region of operation is entered when the terminal voltage v is positive. In the forward region the i-v relationship is closely approximated by

$$i = I_S \left(e^{v/nV_T} - 1 \right) \tag{4.1}$$

In this equation I_S is a constant for a given diode at a given temperature. The current I_S is usually called the *saturation current* (for reasons that will become apparent in the next section). A better name, one that we will use for I_S, is the *scale current*. This name arises from the fact that I_S is directly proportional to the cross-sectional area of the diode. Thus doubling of the junction area results in a diode with double the value of I_S and, as the diode equation indicates, double the value of current i for a given forward voltage v. For "small-signal" diodes, which are small-size diodes intended for low-power applications, I_S is of the order of 10^{-15} A. The value of I_S is, however, a very strong function of temperature. As a rule of thumb I_S doubles in value for every 5°C rise in temperature.[1]

The voltage V_T in Eq. (4.1) is a constant called the *thermal voltage*, given by

$$V_T = \frac{kT}{q} \tag{4.2}$$

where k = Boltzmann's constant = 1.38×10^{-23} joules/kelvin
$\quad T$ = the absolute temperature in kelvins = 273 + temperature in °C
$\quad q$ = the magnitude of electronic charge = 1.602×10^{-19} coulomb
At room temperature (20°C) the value of V_T is 25.2 mV. In rapid, approximate circuit analysis we shall use $V_T \simeq 25$ mV at room temperature.[2]

In the diode equation the constant n has a value between 1 and 2, depending on the material and the physical structure of the diode. Diodes made using the standard integrated-circuit fabrication process exhibit $n = 1$ when operated under normal conditions.[3] Diodes available as discrete two-terminal components generally exhibit $n = 2$.

For appreciable current i in the forward direction, specifically for $i \gg I_S$, Eq. (4.1) can be approximated by the exponential relationship

$$i \simeq I_S e^{v/nV_T} \tag{4.3}$$

---

[1] An excellent discussion of the temperature dependence of diode characteristics is given in Hodges and Jackson (1983), pp. 160–161. Also given is a derivation for the temperature coefficient of I_S.

[2] A slightly higher ambient temperature (25°C or so) is usually assumed for electronic equipment operating inside a cabinet. At this temperature, $V_T \simeq 25.8$ mV. Nevertheless, for the sake of simplicity and to promote rapid circuit analysis, we shall use $V_T \simeq 25$ mV throughout this book.

[3] On an integrated circuit, diodes are usually obtained by connecting a bipolar junction transistor (BJT) as a two-terminal device, as will be seen in Chapter 9.

This relationship can be expressed alternatively in the logarithmic form

$$v = nV_T \ln \frac{i}{I_S}$$ (4.4)

where ln denotes the natural (base e) logarithm.

The exponential relationship of the current i to the voltage v holds over many decades of current (a span of as many as seven decades—that is, a factor of 10^7—can be found). This is quite a remarkable property of junction diodes, one that is also found in bipolar junction transistors and one that has been exploited in many interesting applications.

Let us consider the forward i-v relationship in Eq. (4.3) and evaluate the current I_1 corresponding to a diode voltage V_1:

$$I_1 = I_S e^{V_1/nV_T}$$

Similarly, if the voltage is V_2, the diode current I_2 will be

$$I_2 = I_S e^{V_2/nV_T}$$

These two equations can be combined to produce

$$\frac{I_1}{I_2} = e^{(V_1-V_2)/nV_T}$$

which can be rewritten

$$V_1 - V_2 = nV_T \ln \frac{I_1}{I_2}$$

or, in terms of base 10 logarithms,

$$V_1 - V_2 = 2.3nV_T \log \frac{I_1}{I_2}$$ (4.5)

This equation simply states that for a decade (factor of 10) change in current the diode voltage drop changes by $2.3nV_T$, which is 60 mV for $n = 1$ and 120 mV for $n = 2$. This also suggests that the diode i-v relationship is most conveniently plotted on a semilog paper. Using the vertical, linear axis for v and the horizontal, log axis for i, one obtains a straight line with a slope of $2.3nV_T$ per decade of current. Finally, it should be mentioned that not knowing the exact value of n (which can be obtained from a simple experiment), circuit designers use the convenient approximate number of 0.1 V/decade for the slope of the diode logarithmic characteristic.

A glance at the i-v characteristic in the forward region (Fig. 4.6) reveals that the current is negligibly small for v smaller than about 0.5 V. This value is usually referred to as the *cut-in voltage*. It should be emphasized, however, that this apparent threshold in the characteristic is simply a consequence of the exponential relationship. Another consequence of this relationship is the rapid increase of i. Thus for a "fully conducting" diode the voltage drop lies in a narrow range, approximately 0.6 to 0.8 V. This gives rise to a simple "model" for the diode where it is assumed that a conducting diode has approximately a 0.7-V drop across it. Diodes with different current ratings (that is, different areas and correspondingly different I_S) will exhibit the 0.7-V drop at different currents. For instance, a small-signal diode may be considered to have a 0.7-V drop at $i = 1$ mA, while a higher-power diode may have a 0.7-V drop at $i = 1$ A. We will return to the topics of diode circuit analysis and diode models shortly.

EXAMPLE 4.3

A silicon diode said to be a 1-mA device displays a forward voltage of 0.7 V at a current of 1 mA. Evaluate the junction scaling constant I_S in the event that n is either 1 or 2. What scaling constants would apply for a 1-A diode of the same manufacture that conducts 1 A at 0.7 V?

Solution
Since

$$i = I_S e^{v/nV_T}$$

then

$$I_S = i e^{-v/nV_T}$$

For the 1-mA diode:

If $n = 1$: $I_S = 10^{-3} e^{-700/25} = 6.9 \times 10^{-16}$ A,   or about 10^{-15} A
If $n = 2$: $I_S = 10^{-3} e^{-700/50} = 8.3 \times 10^{-10}$ A,   or about 10^{-9} A

The diode conducting 1 A at 0.7 V corresponds to 1,000 1-mA diodes in parallel with a total junction area 1,000 times greater. Thus I_S is also 1,000 times greater, being 1 pA and 1 μA, respectively, for $n = 1$ and $n = 2$.

From this example it should be apparent that the value of n used can be quite important.

Temperature Dependence

The temperature dependence of the forward i-v characteristic is illustrated in Fig. 4.7. At a given constant diode current the voltage drop across the diode decreases by approximately 2 mV for every 1°C increase in temperature, owing to the dependence of

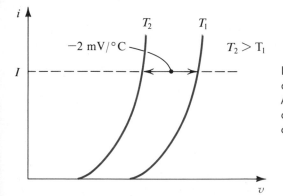

Fig. 4.7 Illustrating the temperature dependence of the diode forward characteristics. At a constant current *I* the voltage drop decreases by approximately 2 mV for every degree C increase in temperature.

both I_S and V_T on temperature. The change in diode voltage with temperature has been exploited in the design of electronic thermometers.

4.3 Consider a silicon junction diode with $n = 1.5$. Find the change in voltage if the current changes from 0.1 mA to 10 mA.

Ans.   172.5 mV

4.4 A silicon junction diode with $n = 1$ has $v = 0.7$ V at $i = 1$ mA. Find the voltage drop at $i = 0.1$ mA and $i = 10$ mA.

Ans.   0.64 V; 0.76 V

4.5 Using the fact that a silicon diode has $I_S = 10^{-14}$ A at 25°C and that I_S increases by 15% per °C rise in temperature, find the value of I_S at 125°C.

Ans. 1.17×10^{-8} A

4.4 THE REVERSE-BIAS REGION

The reverse-bias region of operation is entered when the diode voltage v is made negative. Equation (4.1) predicts that if v is negative and few times larger than V_T (25 mV) in magnitude, the exponential term becomes negligibly small compared to unity and the diode current becomes

$$i \simeq -I_S$$

that is, the current in the reverse direction is constant and equal to I_S. This is the reason behind the term *saturation current*.

Real diodes exhibit reverse currents that, though quite small, are much larger than I_S. For instance, a small-signal or a 1-mA diode whose I_S is of the order of 10^{-14} to 10^{-15} A could show a reverse current of the order of 1 nA. The reverse current also increases somewhat with the increase in magnitude of the reverse voltage. Note that because of the very small magnitude of current these details are not clearly evident on the diode *i-v* characteristic of Fig. 4.6.

A good part of the reverse current is due to leakage effects. These leakage currents are proportional to the junction area, just as I_S is. Finally, it should be mentioned that the reverse current is a strong function of temperature, with the rule of thumb being that it doubles for very 10°C rise in temperature.

4.6 The diode in the circuit of Fig. E4.6 is a large, high-current device whose reverse leakage is reasonably independent of voltage. If $V = 1$ V at 20°C, find the value of V at 40°C and at 0°C.

Ans.   4 V; 0.25 V

4.5 THE BREAKDOWN REGION AND ZENER DIODES

The third distinct region of diode operation is the breakdown region, which can be easily identified on the diode *i-v* characteristic in Fig. 4.6. For convenience the *i-v* characteristics in the reverse and breakdown regions are redrawn in Fig. 4.8. The

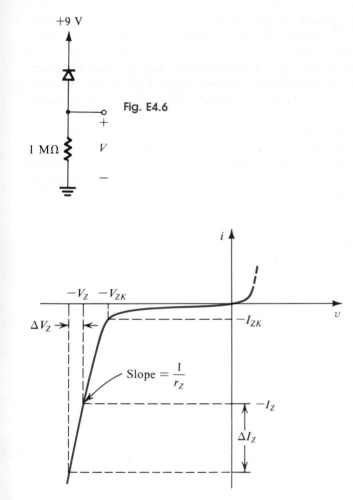

Fig. E4.6

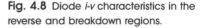

Fig. 4.8 Diode *i-v* characteristics in the reverse and breakdown regions.

breakdown region is entered when the magnitude of the reverse voltage exceeds a threshold value specific to the particular diode and called the *breakdown voltage*. This is the voltage at the "knee" of the *i-v* curve in Fig. 4.8 and is denoted V_{ZK}, where the subscript Z stands for zener (to be explained shortly) and K denotes knee.

As can be seen from Fig. 4.8, in the breakdown region the reverse current increases rapidly, with the associated increase in voltage drop being very small. Diode breakdown is normally not destructive provided that the power dissipated in the diode is limited by external circuitry to a "safe" level. This safe value is normally specified on the device data sheets. It therefore is necessary to limit the reverse current in the breakdown region to a value consistent with the permissible power dissipation.

Voltage Regulators

The fact that the diode *i-v* characteristic in breakdown is almost a vertical line enables it to be used in voltage regulation. A voltage regulator is a circuit whose purpose is to provide a constant dc voltage between its output terminals. This output voltage must remain as constant as possible in spite of changes in the load current drawn from the regulator and possible changes in the dc power supply that feeds the regulator circuit. The diode in the breakdown region exhibits a voltage drop that is almost constant and independent of the current through the diode. Special diodes have been manufactured to operate specifically in the breakdown region. Such diodes are called *breakdown diodes* or, more commonly, *zener diodes*, for reasons that will be explained shortly.

Figure 4.9 shows the circuit symbol of the zener diode. In normal application of zener diodes, current flows into the cathode, and the cathode is positive with respect to the anode; thus I and V in Fig. 4.9 have positive values. To illustrate the application of zener diodes, consider the circuit shown in Fig. 4.10. Here a zener diode,

Fig. 4.9 Circuit symbol for a zener diode.

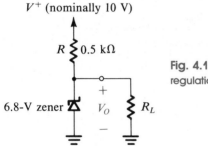

Fig. 4.10 Use of zener diodes in voltage regulation.

specified to have a voltage of 6.8 V, is shown being fed from a dc supply of $+10$ V through a resistance R of 0.5 kΩ. When no load resistance is connected, the current through the zener will be

$$I = \frac{10 - 6.8}{0.5} = 6.4 \text{ mA}$$

Now let a resistance $R_L = 2$ kΩ be connected across the zener diode. Assuming that the voltage across the zener will remain approximately constant (an assumption whose validity we will have to check shortly), we see that the load current I_L is given by

$$I_L = \frac{6.8}{2} = 3.4 \text{ mA}$$

Since the current through R is still approximately 6.4 mA, it follows that the current through the zener is now 3 mA. Thus the zener current has changed from 6.4 mA to 3 mA. Nevertheless, because of the sharp i-v curve the decrease in the voltage drop across the zener will be negligibly small. That is, the voltage remains almost constant in spite of changes in load current. Note, however, that the load current cannot exceed the current being fed to the zener through the resistance R. In fact, if the current in the zener decreases below a certain specified value (equal to the *knee current* I_{ZK} in Fig. 4.8), the voltage across the zener starts to decrease, and eventually the diode leaves the breakdown region and stops functioning as a voltage regulator. For instance, if R_L is 0.5 kΩ, the zener in Fig. 4.10 will be off and the load voltage will be equal to the value determined by the voltage divider composed of R and R_L,

$$V_O = 10\,\frac{0.5}{0.5 + 0.5} = 5 \text{ V}$$

which is less than the 6.8 V required for the zener to operate in the breakdown region. Finally, we note that regulator circuits of the type shown in Fig. 4.10 are known as *shunt regulators* because the regulator element (the zener diode) is in parallel (shunt) with the load.

Zener Resistance

Every zener diode has a specified value for its breakdown voltage, also called the *zener voltage* V_Z. This is the voltage drop across the zener when a specified nominal current I_Z is flowing through it (see Fig. 4.8). As the current through the zener deviates from the nominal value I_Z by a small amount ΔI_Z, the voltage across it will deviate from V_Z by an amount ΔV_Z given by

$$\Delta V_Z = r_Z\,\Delta I_Z$$

where r_Z is the incremental resistance of the zener diode at the operating point specified by V_Z and I_Z. The value of r_Z is equal to the inverse of the slope of the i-v curve at the operating point. (It is important to note that r_Z is *not* equal to V_Z/I_Z.) Typically, r_Z is a few tens of ohms in value, but it increases as the current decreases, reaching a very high value at the knee of the i-v curve. Obviously, as r_Z increases, the zener diode becomes less effective as a voltage regulator. The current at the knee, I_{ZK}, is also normally specified on the zener data sheets.

To illustrate the use of r_Z let us return to the voltage-regulator circuit of Fig. 4.10. Consider the no-load case. We have already found that when $V^+ = 10$ V, the zener has a current of 6.4 mA. Let us assume that at this current the zener resistance r_Z is 50 Ω. Now let V^+ change by 10%, to 11 V. The corresponding change in zener voltage can be calculated using the voltage-divider rule with the divider composed of R and r_Z,

$$\Delta V_Z = \Delta V^+\,\frac{r_Z}{r_Z + R}$$

Thus

$$\Delta V_Z = 1 \times \frac{50}{50 + 500} \simeq 91 \text{ mV}$$

Here the regulated voltage changes by 1.3%, corresponding to the 10% change in supply voltage; the 1.3% figure is usually called *voltage regulation*.

Avalanche Versus Zener Diodes

As will be explained in Section 4.12, there are two different mechanisms for diode breakdown: zener breakdown and avalanche breakdown. The zener effect, named in honor of an early worker in the area, is responsible for breakdown if it occurs at low voltages (2 to 5 V). If breakdown occurs at voltages about 7 V or greater, the mechanism responsible is avalanche breakdown. Avalanche breakdown results in diodes with sharper knees; these therefore make better voltage regulators. For diodes whose breakdown occurs between 5 and 7 V either or both mechanisms may be responsible. Irrespective of the breakdown mechanism, breakdown diodes are commonly called zener diodes.

Zener diodes are manufactured in a variety of standard voltages. The lower-voltage units, however, have rather weak knees. For this reason one may consider using a string of forward-biased diodes to provide low-voltage regulators. This application is based on the observation that a forward-conducting diode has a voltage drop of about 0.7 V and that this voltage drop changes very little with changes in current (about 0.1 V per decade of current change). Thus, for example, one can obtain a constant voltage drop of approximately 2 V by connecting three diodes in series. We will have more to say about this shortly.

Curves illustrating the sharpness of the breakdown knee are shown in Fig. 4.11. Here the attempt (though difficult to appreciate on a linear plot) is to show the very sharp knee of an avalanche zener of 6.8 V. A nominal 20-mA unit (normally rated at 400 mW) will hold its voltage within 0.3 V or so down to tens of microamperes, exhibiting a slope resistance of a few tens of ohms or less at high currents. Here it

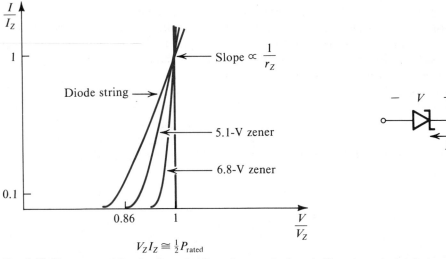

Fig. 4.11 Sharpness of the *i-v* characteristics of zener diodes of different nominal voltages and of a diode string. Note that I_z and V_z denote the nominal values of operating current and voltage.

should be mentioned that the reversed base–emitter junction of bipolar transistors (Chapter 8) suffers avalanche breakdown at about 7 V and may be used as a zener diode. Breakdown diodes of higher and lower voltages exhibit higher slope resistances and weaker knees, as shown by the characteristic of a 5.1-V unit, sketched in Fig. 4.11. Also indicated in Fig. 4.11, for comparison, is the forward characteristic of a silicon junction diode.

EXAMPLE 4.4

In this example we wish to contrast the use of a 1N5235 zener diode as a regulator in two applications. In each the supply voltage is nominally 9 V. The first application is one in which the supply of regulator current is no problem, but the source, the "raw" supply, changes ± 1 V (called *ripple* for reasons explained in Chapter 5). The second application is for a portable instrument where power is obtained from a rechargeable battery whose operating range extends from 10 V down to 8 V. For the purpose of this initial design, assume no load.

The 1N5235 unit is rated at $V_Z = 6.8$ V for $I_Z = 20$ mA with a typical zener resistance r_Z of 5 Ω. It is also specified that r_Z will be no greater than 750 Ω at 0.25 mA (nearer the knee). At other current values r_Z may be considered inversely proportional to the current.

Solution
In both cases, to obtain the security of using specified data and thus to avoid relying unduly on extrapolated assumptions, we will use the high and low values of current at which the zener resistance has been specified.

For the first case we will design for a zener current of 20 mA. Establishment of this current in the 6.8-V zener from a nominal 9-V supply in a circuit such as that in Fig. 4.10 requires a series resistance R of

$$R = \frac{9 - 6.8}{20} = 110 \ \Omega$$

The resulting regulator having $r_Z = 5 \ \Omega$ will reduce the ± 1-V ripple in the source to an output voltage ripple of

$$\Delta V_Z = \pm \frac{5}{110 + 5} \times 1 = \pm 43 \text{ mV} \qquad \text{for } 0.63\% \text{ regulation}$$

Because of the limited current capacity of the battery in the second application we will design for a zener current of 0.25 mA. To establish this current we need a resistance R of

$$R = \frac{9 - 6.8}{0.25} = 8.8 \text{ k}\Omega$$

The resulting regulator having $r_Z = 750 \ \Omega$ will reduce the ± 1-V change in the source to an output voltage change of

$$\Delta V_Z = \pm \frac{750}{750 + 8,800} \times 1 = \pm 79 \text{ mV} \qquad \text{for } 1.2\% \text{ regulation}$$

Example 4.4 indicates that although the equivalent reduction of disturbances is somewhat poorer at low currents, regulation is not dramatically different in the two designs. In fact, because the value of r_Z is approximately inversely proportional to the operating current, regulation will be almost independent of bias design for all other conditions being equal (for example, no load).

Temperature Effects

Zener diodes exhibit a temperature coefficient that depends on both voltage and current in a rather complex way, as indicated in Fig. 4.12. In particular, zeners with a nominal breakdown of 5.1 V exhibit a point of zero temperature coefficient at modest current levels. Such zeners, while exhibiting a relatively high zener impedance, can

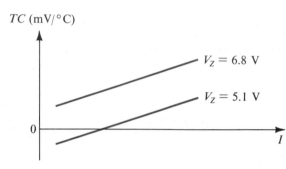

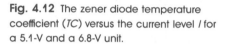

Fig. 4.12 The zener diode temperature coefficient (*TC*) versus the current level *I* for a 5.1-V and a 6.8-V unit.

be used in selected constant-current designs to establish temperature-independent voltages (particularly if the regulator output is buffered with an amplifier; see Problem 4.20). On the other hand, zeners of about 6.8 V exhibit a temperature coefficient of about $+2$ mV/°C, almost exactly the complement of that of a forward-conducting diode. Thus a series combination of a 6.8-V zener and a single forward junction exhibits a drop of about 7.5 V with a greatly reduced temperature coefficient. Other low-temperature-coefficient diodes (said also to have *low temco* or *low TC*) exist at somewhat higher voltages, utilizing other combinations of forward-conducting and breakdown diodes.

Exercises

4.7 Consider the regulator designed in Example 4.4. For both cases find the change in zener voltage when a 2-kΩ load resistance is connected.

Ans. -17 mV; -5.13 V

4.8 Consider a shunt regulator of the form shown in Fig. 4.10. Let the zener be a 10-V device with $r_Z = 100$ Ω and let $R = 1$ kΩ.
(a) For no load on the regulator find the output regulation corresponding to a $\pm 10\%$ variation in V^+.
(b) For a 2-mA load find the change in output voltage and hence the "load regulation" as the percent change in V_O per milliampere of load current.

Ans.   **(a)** $\pm 0.91\%$;   **(b)** -200 mV, -1%/mA

4.6 ANALYSIS OF DIODE CIRCUITS

Consider the circuit shown in Fig. 4.13 consisting of a dc source V_{DD}, a resistance R, and a diode. We wish to analyze this circuit to determine the diode current I and voltage V.

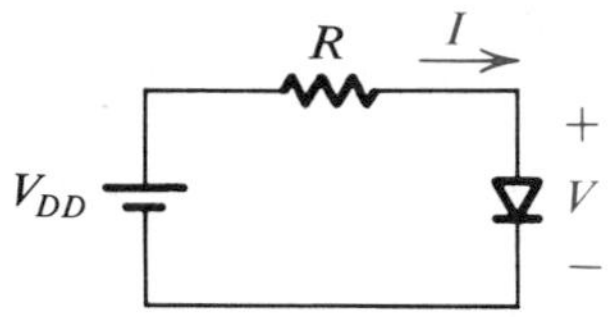

Fig. 4.13 A simple diode circuit.

Since the diode is biased in the forward direction, it will obviously be conducting (assuming that $V_{DD} > 0.5$ V), and we can write the two equations

$$I \simeq I_S e^{V/nV_T} \tag{4.6}$$

$$V = V_{DD} - RI \tag{4.7}$$

These are two equations in the two unknowns I and V, assuming that the diode parameters I_S and n are given. Although it is possible to solve these two equations to determine the values of I and V, the amount of work involved is usually not justified. In other words, one is usually interested in finding quick approximate answers.

Before presenting techniques of approximate analysis we will discuss a method for graphical analysis. However, it should be emphasized at the outset that one rarely uses graphical techniques to solve simple diode circuits. Nevertheless, it is important to understand the technique of graphical analysis, since similar techniques will be used in conjunction with transistor circuits to obtain insights not otherwise available.

Graphical Analysis

Graphical analysis consists simply of plotting the relationships of Eqs. (4.6) and (4.7) on the i-v plane. The solution can then be obtained as the coordinates of the point of intersection of the two graphs. Equation (4.6) represents the diode equation, and a sketch of it is given in Fig. 4.14. The other equation, Eq. (4.7), represents a straight line, called the *load line*, that intersects the voltage axis at $v = V_{DD}$ and has a slope of $-1/R$. As shown in Fig. 4.14, the load line intersects the diode curve at point Q, which represents the operating point (or quiescent point) of the diode and whose coordinates give the values of I and V.

Approximate Analysis

The method of approximate analysis is based on our earlier observation regarding the diode voltage drop. To repeat, a forward-conducting diode has a voltage drop that lies in a narrow range, approximately 0.6 to 0.8 V. Therefore as a gross approximation (one that in fact is not bad in many applications) we may assume that a 1-mA diode

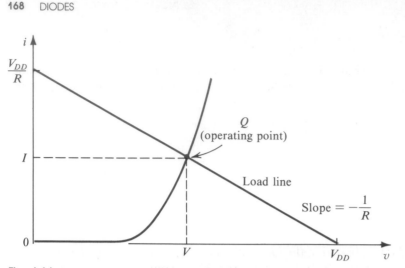

Fig. 4.14 Graphical analysis of the circuit in Fig. 4.13.

conducting current in the milliampere range will have a voltage drop of approximately 0.7 V.

Our approach to the problem depicted in Fig. 4.13 would then be as follows: First, we assume that the diode is conducting; thus its voltage drop is approximately 0.7 V. Then we proceed to use Kirchhoff's loop equation to determine the current I,

$$I = \frac{V_{DD} - 0.7}{R}$$

If we find that I is in the range assumed for the diode (that is, in the milliampere range for a 1-mA diode), then our original assumption is not a bad one and our work is completed. If, on the other hand, we find that the calculated value of I differs considerably from the assumed order of magnitude, then clearly our result needs further refinement, as will be explained shortly.

EXAMPLE 4.5

Find I and V in the circuit of Fig. 4.13; assume that the diode is a 1-mA diode, with $V_{DD} = 10$ V, and R = 10 kΩ.

Solution
Assuming $V \simeq 0.7$ V, then

$$I = \frac{10 - 0.7}{10} = 0.93 \text{ mA}$$

which is in the normal operating range of the diode. Thus our assumption of $V \simeq 0.7$ V is not a bad one.

Refining the Results

It is possible to use a simple iterative method to refine somewhat the results obtained by the approximate-analysis scheme above. The method consists of using the value of current obtained to calculate a new value for the diode voltage drop. Then we use this new value to calculate a better approximation for the value of the current. This completes one iteration. If better accuracy is desired, we can proceed with one or more further iterations. This process is best illustrated by numerical examples.

EXAMPLE 4.6

Find better approximations for the values of I and V in the circuit considered in Example 4.3. Assume that the diode is specified for 0.7 V at 1 mA and that $n = 1.8$.

Solution
The results of iteration 0, performed in Example 4.5, are $V = 0.7$ V, $I = 0.93$ mA. Since the diode is characterized by

$$I = I_S e^{V/nV_T}$$

then

$$10^{-3} = I_S e^{700/(1.8 \times 25)}$$

which results in

$$I_S = 10^{-3} e^{-700/(1.8 \times 25)}$$

Thus

$$I = 10^{-3} e^{(V - 700)/(1.8 \times 25)}$$

Substituting $I = 0.93$ mA gives

$$V = 696.7 \text{ mV}$$

Using Kirchhoff's voltage equation

$$I = \frac{V_{DD} - V}{R}$$

we obtain

$$I = \frac{10 - 0.6967}{10} = 0.9303 \text{ mA}$$

which is very close to the originally calculated value of 0.93 mA. Thus no further iterations are justified. In fact, because the result of our iteration 0 (0.93 mA) is so close to the current of 1 mA (corresponding to the 0.7-V drop), even the first iteration was not quite justified. The following example will better illustrate the need for iterating.

Finally, note that for this diode the value of n was given. If n is not known, we will have to assume a value (a good assumption is $n = 2$ for discrete silicon diodes or $n = 1$ for silicon diodes on an IC chip; otherwise 0.1 V per decade of current is often used for simplicity).

EXAMPLE 4.7

Calculate the values of V_D and I_D in the circuit of Fig. 4.15. Let the diode have $V_D = 0.7$ V at $I_D = 1$ mA, and let $n = 2$.

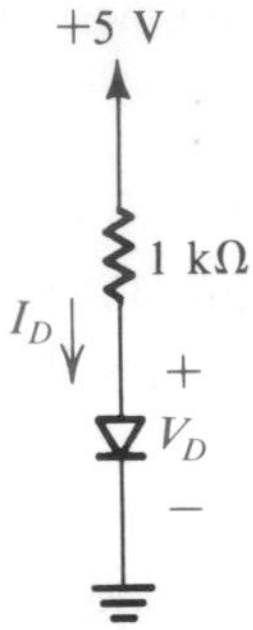

+5 V

1 kΩ

I_D

$+$

V_D

$-$

Fig. 4.15 Circuit for Example 4.7.

Solution

From the diode equation

$$i = I_s e^{v/nV_T}$$

it follows that corresponding to a voltage V_1 the current I_1 is

$$I_1 = I_s e^{V_1/nV_T}$$

and corresponding to a voltage V_2 the current I_2 is

$$I_2 = I_s e^{V_2/nV_T}$$

Thus

$$\frac{I_2}{I_1} = e^{(V_2 - V_1)/nV_T} \tag{4.8a}$$

or, alternatively,

$$V_2 - V_1 = nV_T \ln \frac{I_2}{I_1} \tag{4.8b}$$

These two relationships are useful in performing our iterative analysis.
Iteration 0

$$V_D = 0.7 \text{ V} \quad \text{and} \quad I_D = \frac{5 - 0.7}{1} = 4.3 \text{ mA}$$

Iteration 1   Use of Eq. (4.8) gives

$$V_2 - V_1 = 2 \times 25 \ln \frac{4.3}{1} = 72.9 \text{ mV}$$

Thus

$$V_D = 0.7729 \text{ V} \quad \text{and} \quad I_D = \frac{5 - 0.7729}{1} = 4.227 \text{ mA}$$

Iteration 2

$$V_2 - V_1 = 2 \times 25 \ln \frac{4.227}{4.3} = -0.85 \text{ mV}$$

Thus

$$V_D = 0.772 \text{ V} \quad \text{and} \quad I_D = \frac{5 - 0.772}{1} = 4.23 \text{ mA}$$

Obviously no further iterations are called for; in fact, even iteration 2 is hardly justified.

Exercises

4.9 Calculate the values of V and I in the circuits shown in Fig. E4.9. Assume that the diode has 0.7 V drop at a current of 1 mA, and let $n = 2$.

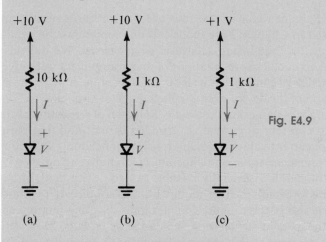

Fig. E4.9

(a)   (b)   (c)

Ans. **(a)** 696.4 mV, 0.930 mA; **(b)** 810.9 mV, 9.189 mA; **(c)** 647.6 mV, 0.352 mA

4.10 In the circuit shown in Fig. E4.10 assume that the three diodes are identical, with drops

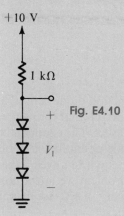

Fig. E4.10

of 0.7 V at currents of 1 mA and $n = 1$. Calculate the value of V_1. Also calculate the change in the value of V_1 when the dc supply voltage changes to $+15$ V and then to $+5$ V.

Ans.   2.254 V; 36.7 mV; -76 mV

4.7 MODELING THE DIODE FORWARD CHARACTERISTIC

Philosophy of Device Modeling

For the purpose of analyzing and designing circuits, each device must be represented by a model. Generally speaking, such a model takes the form of one or more relationship between the terminal voltages and currents. For instance, the model for a resistor is given by Ohm's law,

$$v = Ri$$

where R is the value of the resistance. Also, the various amplifier representations discussed in Chapter 2 are models. If one represents the equations of the model by a circuit, one obtains an *equivalent circuit model*. We did this in Chapter 2, where equivalent circuit representations of linear amplifiers were given.

Each device model has its limitations: the model will represent the actual physical operation of the device over a limited range of variables and only to a certain degree of accuracy. As an example, the $v = Ri$ model of a resistor assumes that the resistance remains constant as current is varied, an assumption that is valid only for a limited range of current values. Also, this resistor model does not account for the dependence of the resistance value on temperature, humidity, aging, etc.

Although it is generally possible to find device models that are extremely accurate, this is usually neither required nor desirable. As the model becomes more accurate it also becomes more complex, thus resulting in more effort being required in using the model for analyzing and designing circuits. More seriously, using complex models does not enable the circuit designer to obtain insight into the physical operation of the circuit. It is this insight that allows one to improve the design. Thus one should use as simple a device model as the application at hand permits. After the initial design is obtained one may analyze the circuit using a computer circuit-analysis program and employing more accurate device models (see Appendix C). As a result, the design may be further refined.

The Exponential Model

The exponential i-v relationship of the junction diode is itself a model of the physical processes that take place inside the pn junction that forms the diode. These physical processes will be described later in this chapter. The exponential relationship is a very good model in the forward region of diode operation. Use of this elaborate model, however, requires the solving of nonlinear equations even if the circuit being analyzed is a very simple one. To cope with this problem we introduced the iterative scheme of Section 4.6. Nevertheless, there are many diode circuit applications for which a much simpler diode model would be sufficient.

The "Ideal-Diode" Model

The simplest diode model is the ideal-diode model depicted in Fig. 4.16. Such a model assumes that the voltage drop across the diode in the forward direction is negligibly small. This model is quite appropriate for applications that make use of the gross nonlinearity exhibited by diodes, such as rectifiers of large voltages (much greater than 0.7 V). If the voltages involved are much larger than the diode voltage drop, we can safely neglect the latter and use the ideal diode as a model for the real junction diode.

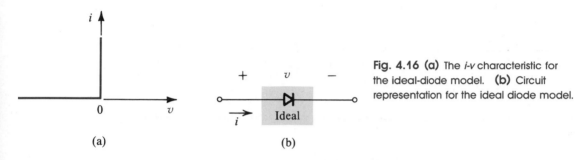

Fig. 4.16 (a) The *i-v* characteristic for the ideal-diode model. **(b)** Circuit representation for the ideal diode model.

(a)   (b)

Even in applications in which the diode voltage drop cannot be neglected, insights into circuit operation can be obtained by initially assuming that the diode is ideal. Subsequently the results can be refined with a more accurate diode model.

The Constant-Voltage-Drop Model

In the hierarchy of diode models the constant-voltage-drop model is the one next to the ideal-diode model. We have already alluded to the constant-voltage-drop model in the course of analyzing simple diode circuits. This model, depicted in Fig. 4.17, assumes that a conducting diode has a voltage drop that is almost constant and independent of the current flowing through the diode. Typically a value of 0.7 V is assumed for the constant voltage drop. Despite its simplicity, this model provides a reasonably accurate representation of the diode that is adequate for many applications. This fact can be appreciated by recalling that the voltage drop across an actual diode changes only by 60–120 mV when the current changes by an order of magnitude.

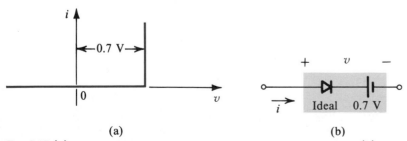

(a)   (b)

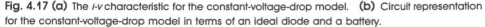

Fig. 4.17 (a) The *i-v* characteristic for the constant-voltage-drop model. **(b)** Circuit representation for the constant-voltage-drop model in terms of an ideal diode and a battery.

The "Battery-Plus-Resistance" Model

To account for the fact that the diode forward voltage drop is not constant but increases with the current flowing through the diode, a constant resistance may be included with the battery (or constant voltage) in the diode model. The result is the battery-plus-resistance model illustrated in Fig. 4.18. Obviously there is latitude in

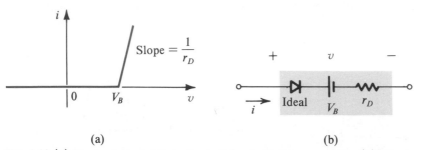

(a)   (b)

Fig. 4.18 (a) The *i-v* characteristic for the battery-plus-resistance model. **(b)** Circuit representation for the model.

selecting values for the battery voltage V_B and for the series resistance r_D. In doing this, one should attempt to select values that best fit the *i-v* characteristic over the expected range of operation. However, if the range of operation is not known, a simple scheme to be used is as follows: Since the diode conducts negligible currents when the forward voltage is less than about 0.5 V, we may take this voltage to be the battery voltage V_B. Then for a 1-mA diode—that is, a diode that conducts 1 mA when the voltage drop is about 0.7 V—we need a series resistance of

$$r_D = \frac{0.2 \text{ V}}{1 \text{ mA}} = 200 \text{ }\Omega$$

EXAMPLE 4.8

Find the current I_D in the circuit of Fig. 4.19 for two cases: (a) $V^+ = 10$ V and (b) $V^+ = 1$ V. In each case use (i) the ideal-diode model, (ii) the constant-voltage-drop model, (iii) the battery-plus-resistance model, and (iv) the exponential model, assuming the diode to be a 1-mA diode with $n = 2$.

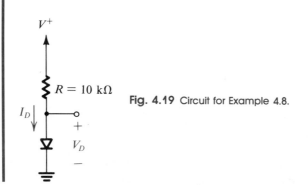

Fig. 4.19 Circuit for Example 4.8.

Solution

Case a: $V^+ = 10$ V
(i) Use of the ideal-diode model gives

$$V_D = 0 \text{ V} \quad \text{and} \quad I_D = \frac{V^+}{R} = \frac{10}{10} = 1 \text{ mA}$$

(ii) Use of the constant-voltage-drop model gives

$$V_D = 0.7 \text{ V} \quad \text{and} \quad I_D = \frac{V^+ - V_D}{R} = \frac{10 - 0.7}{10} = 0.93 \text{ mA}$$

(iii) Use of the battery-plus-resistance model with a battery voltage of 0.5 V and a series resistance r_D of 200 Ω gives

$$I_D = \frac{V^+ - 0.5}{R + r_D} = \frac{10 - 0.5}{10 + 0.2} = 0.931 \text{ mA} \quad \text{and} \quad V_D = 0.5 + I_D r_D = 0.69 \text{ V}$$

(iv) Use of the exponential model with $I_D = 1$ mA at $V_D = 0.7$ V and $n = 2$ together with the iterative scheme described in Section 4.7 gives

$$I_D = 0.930 \text{ mA} \quad \text{and} \quad V_D = 0.70 \text{ V}$$

Case b: $V^+ = 1$ V
(i) Use of the ideal-diode model gives

$$V_D = 0 \text{ V} \quad \text{and} \quad I_D = \frac{V^+}{R} = \frac{1}{10} = 0.1 \text{ mA}$$

(ii) Use of the constant-voltage-drop model gives

$$V_D = 0.7 \text{ V} \quad \text{and} \quad I_D = \frac{1 - 0.7}{10} = 0.03 \text{ mA}$$

(iii) Use of the battery-plus-resistance model gives

$$I_D = 0.049 \text{ mA} \quad \text{and} \quad V_D = 0.51 \text{ V}$$

(iv) Use of the exponential model with two iterations gives

$$I_D = 0.045 \text{ mA} \quad \text{and} \quad V_D = 0.55 \text{ V}$$

Example 4.8 shows that the constant-voltage-drop model provides good results when the voltages involved (V^+ in this case) are much greater than the diode voltage drop. When this is not the situation as in case b above, the slightly more elaborate battery-plus-resistance model provides a result much closer to the value obtained with the more elaborate exponential model. The choice of an appropriate device model is indeed an attribute of a "good" circuit designer.

The Small-Signal Model

There are applications in which a diode is biased to operate at a point on the forward *i-v* characteristic and a small ac signal is superimposed on the dc quantities. For this situation the diode is best modeled by a resistance equal to the inverse of the slope of the tangent to the *i-v* characteristic at the bias point. The concept of biasing a nonlinear device and restricting signal excursion to a short, almost-linear segment of its characteristic around the bias point was introduced in Section 2.2 for two-port networks. In the following section we develop such a small-signal model for the junction diode.

Exercise

4.11 Consider a voltage source v_S, $v_S = t$ volts, where t is the time in seconds, connected through a resistance of 10 kΩ to a 1-mA diode. Use the battery-plus-resistance diode model to find an expression for the current i flowing through the diode.

Ans. $i = 0$ $t \leq 0.5$ s

$$= \frac{t - 0.5}{10.2} \text{ mA} \quad t \geq 0.5 \text{ s}$$

4.8 THE SMALL-SIGNAL MODEL AND ITS APPLICATION

Consider the conceptual circuit in Fig. 4.20a and the corresponding graphical representation in Fig. 4.20b. A dc voltage V_D, represented by a battery, is applied to the diode and a time-varying signal $v_d(t)$, assumed (arbitrarily) to have a triangular waveform, is superimposed on the dc voltage V_D. In the absence of the signal $v_d(t)$ the diode voltage is equal to V_D, and correspondingly the diode will conduct a dc current I_D given by

$$I_D = I_S e^{V_D / nV_T} \tag{4.9}$$

When the signal $v_d(t)$ is applied, the total instantaneous diode voltage $v_D(t)$ will be given by

$$v_D(t) = V_D + v_d(t) \tag{4.10}$$

Correspondingly, the total instantaneous diode current $i_D(t)$ will be

$$i_D(t) = I_S e^{v_D / nV_T} \tag{4.11}$$

Substituting for v_D from Eq. (4.10) gives

$$i_D(t) = I_S e^{(V_D + v_d)/nV_T}$$

which can be rewritten

$$i_D(t) = I_S e^{V_D / nV_T} \, e^{v_d / nV_T}$$

Using Eq. (4.9) we obtain

$$i_D(t) = I_D e^{v_d / nV_T} \tag{4.12}$$

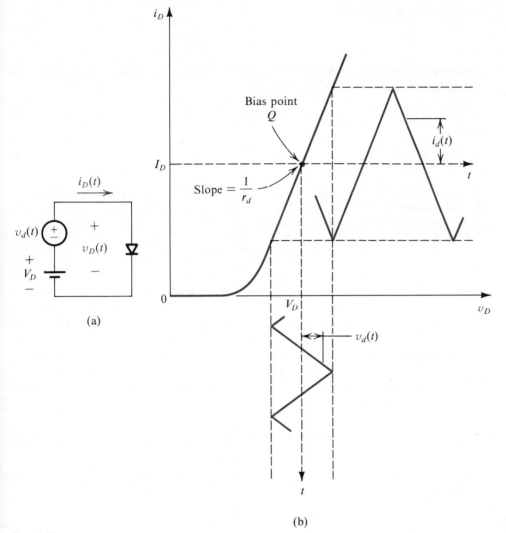

Fig. 4.20 Development of the diode small-signal model.

Now if the amplitude of the signal $v_d(t)$ is kept sufficiently small such that

$$\frac{v_d}{nV_T} \ll 1 \tag{4.13}$$

then we may expand the exponential of Eq. (4.12) in a series and truncate the series after the first two terms to obtain the approximate expression

$$i_D(t) \simeq I_D\left(1 + \frac{v_d}{nV_T}\right) \tag{4.14}$$

This is the *small-signal approximation*. It is valid for signals whose amplitudes are smaller than about 10 mV [see Eq. (4.13) and recall that $V_T = 25$ mV].

From Eq. (4.14) we have

$$i_D(t) = I_D + \frac{I_D}{nV_T} v_d \tag{4.15}$$

Thus superimposed on the dc current I_D we have a signal current component directly proportional to the signal voltage v_d. That is,

$$i_D = I_D + i_d \tag{4.16}$$

where

$$i_d = \frac{I_D}{nV_T} v_d \tag{4.17}$$

The quantity relating the signal current i_d to the signal voltage v_d has the dimensions of conductance, mhos ($\mho$), and is called the *diode small-signal conductance*. The inverse of this parameter is the *diode small-signal resistance*, or *incremental resistance*, r_d,

$$r_d = \frac{nV_T}{I_D} \tag{4.18}$$

Note that the value of r_d is inversely proportional to the bias current I_D.

Let us return to the graphical representation in Fig. 4.20b. It is easy to see that using the small-signal approximation is equivalent to assuming that the signal amplitude is sufficiently small such that the excursion along the i-v curve is limited to a short, almost linear segment. The slope of this segment, which is equal to the slope of the i-v curve at the operating point Q, is equal to the small-signal conductance. The reader is encouraged to prove that the slope of the i-v curve at $i = I_D$ is equal to I_D/nV_T, which is $1/r_d$.

EXAMPLE 4.9

Consider the circuit shown in Fig. 4.21. Let the nominal value of the power supply voltage V_{DD} be 10 V and let $R = 10$ kΩ. Calculate the value of the diode voltage, assuming that $n = 2$ and the diode is a 1-mA diode. If V_{DD} changes by ± 1 V, what are the corresponding changes in the diode voltage?

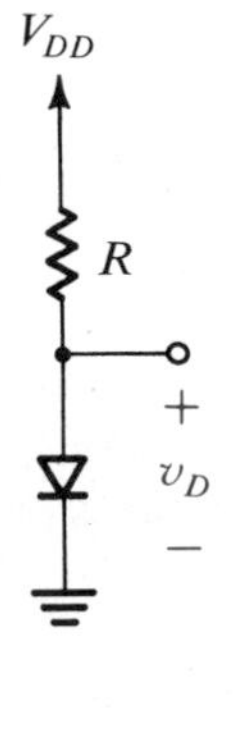

V_{DD}

R

Fig. 4.21 Circuit for Example 4.9.

$+$

v_D

$-$

Solution

When V_{DD} is at its nominal value of 10 V, one may assume that the diode voltage is

$$V_D \simeq 0.7 \text{ V}$$

Thus the diode current will be

$$I_D = \frac{10 - 0.7}{10} = 0.93 \text{ mA}$$

At this operating point the diode incremental resistance r_d is

$$r_d = \frac{nV_T}{I_D} = \frac{2 \times 25}{0.93} = 53.8 \ \Omega$$

The ± 1-V change in V_{DD} may be considered as a signal of 2 V peak-to-peak. Assuming the corresponding change in the diode voltage v_d to be small, we may find the peak-to-peak value of v_d using the ratio of the voltage divider composed of r_d and R, that is,

$$v_d \text{ (peak-to-peak)} = 2 \times \frac{r_d}{R + r_d}$$

$$= 2 \times \frac{0.0538}{10 + 0.0538} = 10.7 \text{ mV}$$

Thus the change in diode voltage is ± 5.35 mV. Since this value is quite small, our use of the small-signal model of the diode is justified.

Use of the Diode Forward Drop in Voltage Regulation

Example 4.9 suggests an important application of the diode forward characteristics, namely, the use of the diode in low-voltage (smaller than 3 or 4 V) regulators. Since the diode voltage drop remains almost constant at approximately 0.7 V independent of changes in the supply voltage, a string of forward-conducting diodes can be used to provide a constant voltage. For instance, the use of three diodes in series provides a voltage of about 2 V and would be equivalent to a low-voltage zener diode. Since low-voltage zeners do not have sharp knees, diode strings are quite competitive in generating constant (regulated) dc voltages lower than about 3 or 4 V. In such applications the small signal model of the diode can be used to calculate the changes in the regulated voltage due to changes in power supply voltage and/or changes in load current. This process is illustrated in Example 4.10.

EXAMPLE 4.10

Consider the circuit shown in Fig. 4.22. A string of three diodes is used to provide a constant voltage of about 2.1 V. We want to calculate the percentage change in this regulated voltage caused by (a) a $\pm 10\%$ change in the power supply voltage and (b) connection of a 1-kΩ load resistance. Assume $n = 2$.

Solution

With no load the nominal value of the current in the diode string is given by

$$I = \frac{10 - 2.1}{1} = 7.9 \text{ mA}$$

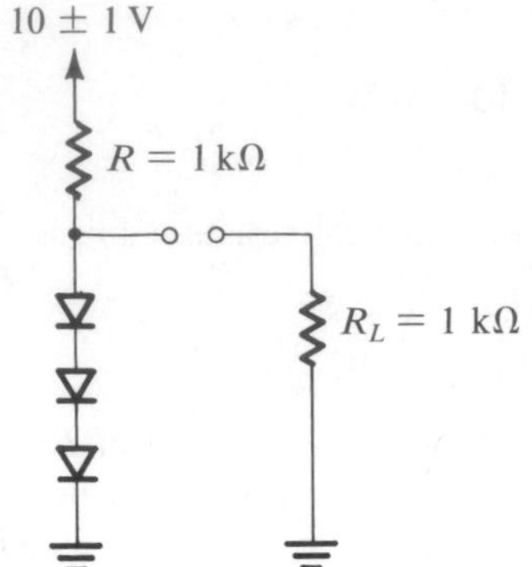

Fig. 4.22 Circuit for Example 4.10.

Thus each diode will have an incremental resistance of

$$r_d = \frac{nV_T}{I}$$

Using $n = 2$ gives

$$r_d = \frac{2 \times 25}{7.9} = 6.3\ \Omega$$

The three diodes in series will have a total incremental resistance of

$$r = 3r_d = 18.9\ \Omega$$

This resistance along with the resistance R forms a voltage divider whose ratio can be used to calculate the change in output voltage due to a $\pm 10\%$ (that is, ± 1 V) change in supply voltage. Thus the peak-to-peak change in output voltage will be

$$\Delta v_o = 2\,\frac{r}{r + R} = 2\,\frac{0.0189}{0.0189 + 1} = 37.1\ \text{mV}$$

that is, corresponding to the ± 1-V ($\pm 10\%$) change in supply voltage the output voltage will change by ± 18.5 mV or $\pm 0.9\%$. Since this implies a change of about ± 6.2 mV per diode, our use of the small-signal model is justified.

When a load resistance of 1 kΩ is connected across the diode string it draws a current of approximately 2.1 mA. Thus the current in the diodes decreases by 2.1 mA, resulting in a decrease in voltage across the diode string given by

$$\Delta v_o = -2.1 \times r = -2.1 \times 18.9 = -39.7\ \text{mV}$$

Since this implies that the voltage across each diode decreases by about 13.2 mV, our use of the small-signal model is not entirely justified. Nevertheless, a detailed calculation of the voltage change using the exponential model results in $\Delta v_o = -35.5$ mV, which is not too different from the approximate value obtained using the incremental model.

Exercises

4.12 For a diode with $n = 1$ find the percentage error involved in the small-signal approximation, that is, in approximating Eq. (4.12) by Eq. (4.14), for signals of 2 mV, 5 mV, 10 mV, and 25 mV.

Ans. 0.3%; 1.75%; 6.2%; 26.4%

4.13 Use the diode equation in Eq. (4.11) to find the small-signal resistance r_d,

$$r_d = \left[\frac{\partial i_D}{\partial v_D} \bigg|_{i_D = I_D} \right]^{-1}$$

Ans. $r_d = nV_T/I_D$

4.9 PHYSICAL OPERATION OF DIODES—BASIC SEMICONDUCTOR CONCEPTS

Having studied the terminal characteristics of junction diodes, we will now briefly consider the physical processes that give rise to these characteristics. The following treatment of device physics is qualitative; nevertheless, it should provide sufficient background for the design of diode and other semiconductor circuits.

The *pn* Junction

The semiconductor diode is basically a *pn* junction, as shown schematically in Fig. 4.23. As indicated, the *pn* junction consists of *p*-type semiconductor material (such as silicon) brought into close contact with *n*-type semiconductor material (silicon). In actual practice both the *p* and *n* regions are part of the same silicon crystal; that is, the *pn* junction is formed within a single silicon crystal by creating regions of different "dopings" (*p* and *n* regions). Appendix A provides a brief description of the process employed in the fabrication of *pn* junctions. As indicated in Fig. 4.23, external wire connections to the *p* and *n* regions (that is, diode terminals) are made through metal (aluminum) contacts.

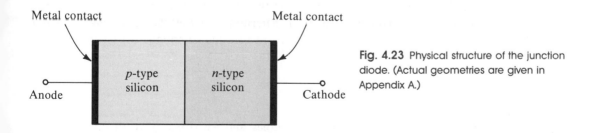

Fig. 4.23 Physical structure of the junction diode. (Actual geometries are given in Appendix A.)

In addition to being essentially a diode, the *pn* junction is the basic component of field-effect transistors (FETs) and bipolar-junction transistors (BJTs). Thus an understanding of the physical operation of *pn* junctions is important to the understanding of the operation and terminal characteristics of diodes and transistors.

Intrinsic Silicon

Although either silicon or germanium can be used to manufacture semiconductor devices—indeed earlier diodes and transistors were made of germanium—today's integrated-circuit technology is based entirely on silicon. For this reason we will deal exclusively with silicon devices throughout this book.

A crystal of pure or intrinsic silicon has a regular lattice structure where the atoms are held in their positions by bonds, called *covalent bonds*, formed by the four valence electrons associated with each silicon atom. At sufficiently low temperatures all covalent bonds are intact and no (or very few) *free electrons* are available to conduct electric current. However, at room temperature some of the bonds are broken by thermal ionization and some electrons are freed. When a covalent bond is broken an electron leaves its parent atom; thus a positive charge, equal to the magnitude of the electron charge, is left with the parent atom. An electron from a neighboring atom may be attracted to this positive charge, leaving its parent atom. This action fills up the "hole" that existed in the ionized atom but creates a new hole in the other atom. This process may repeat itself with the result that we effectively have a positively charged carrier, or hole, moving through the silicon crystal structure and being available to conduct electric current. The charge of a hole is equal in magnitude to the charge of an electron.

Thermal ionization results in free electrons and holes in equal numbers and hence equal concentrations. These free electrons and holes move randomly through the silicon crystal structure, and in the process some electrons may fill some of the holes. This process, called *recombination*, results in the disappearance of free electrons and holes. The recombination rate is proportional to the number of free electrons and holes, which is in turn determined by the ionization rate. The ionization rate is a strong function of temperature. In thermal equilibrium the recombination rate is equal to the ionization or thermal generation rate, and one can calculate the concentration of free electrons n, which is equal to the concentration of holes p,

$$n = p = n_i$$

where n_i denotes the concentration of free electrons or holes in intrinsic silicon at a given temperature.

Finally, it should be mentioned that the reason that silicon is called a semiconductor is that its conductivity, which is determined by the number of charge carriers available to conduct electric current, is between that of conductors (such as metals) and that of insulators (such as glass).

Diffusion and Drift

There are two mechanisms by which holes and electrons move through a silicon crystal—*diffusion* and *drift*. Diffusion is associated with random motion due to thermal agitation. In a piece of silicon with uniform concentrations of free electrons and holes this random motion does not result in a net flow of charge (that is, current). On the other hand, if by some mechanism the concentration of, say, free electrons is made higher in one part of the piece of silicon than in another, then electrons will diffuse from the region of high concentration to the region of low concentration. This diffusion process gives rise to a net flow of charge, or *diffusion current*. As an example, consider the bar of silicon shown in Fig. 4.24a, in which the hole *concentration profile* shown in Fig. 4.24b has been created along the x axis by some unspecified mechanism. The existence of such a concentration profile results in a hole diffusion current in the x direction, with the magnitude of the current at any point being proportional to the slope of the concentration curve, or the concentration gradient, at that point.

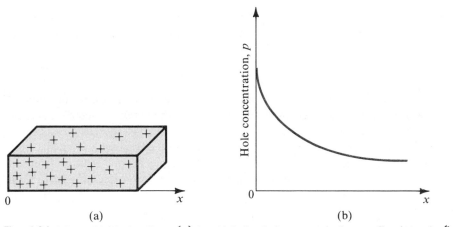

(a)   (b)

Fig. 4.24 A bar of intrinsic silicon **(a)** in which the hole concentration profile shown in **(b)** has been created along the x axis by some unspecified mechanism.

The other mechanism for carrier motion in semiconductors is drift. Carrier drift occurs when an electric field is applied across a piece of silicon. Free electrons and holes are accelerated by the electric field and acquire a velocity component (superimposed on the velocity of their thermal motion) called *drift velocity*. The resulting hole and electron current components are called *drift currents*. The relationship between the drift current and the applied electric field represents one form of Ohm's law.

Doped Semiconductors

The intrinsic silicon crystal described above has equal concentrations of free electrons and holes generated by thermal ionization. These concentrations, denoted n_i, are strongly dependent on temperature. Doped semiconductors are materials in which carriers of one kind (electrons or holes) predominate. Doped silicon in which the majority of charge carriers are the *negatively* charged electrons is called n type, while silicon doped so that the majority of charge carriers are the *positively* charged holes is called p type.

Doping of a silicon crystal to turn it into n type or p type is achieved by introducing a small number of impurity atoms. For instance, introducing impurity atoms of a pentavalent element such as phosphorus results in n-type silicon because the phosphorus atoms that replace some of the silicon atoms in the crystal structure have five valence electrons, four of which form bonds with the neighboring silicon atoms while the fifth becomes a free electron. Thus each phosphorus atom *donates* a free electron to the silicon crystal, and the phosphorus impurity is called a *donor*. It should be clear, though, that no holes are generated by this process; hence the majority of charge carriers in the phosphorus-doped silicon will be electrons. In fact, if the concentration of donor atoms (phosphorus) is N_D, in thermal equilibrium the concentration of free electrons in the n-type silicon, n_{n0}, will be

$$n_{n0} \simeq N_D$$

where the additional subscript 0 denotes thermal equilibrium. In this n-type silicon the concentration of holes, p_{n0}, that are generated by thermal ionization will be

$$p_{n0} \simeq \frac{n_i^2}{N_D}$$

Since n_i is a function of temperature, it follows that the concentration of the *minority* holes will be a function of temperature, whereas that of the *majority* electrons is independent of temperature.

To produce a p-type semiconductor, silicon has to be doped with a trivalent impurity such as boron. Each of the impurity boron atoms *accepts* one electron from the silicon crystal, so that they may form covalent bonds in the lattice structure. Thus each boron atom gives rise to a hole, and the concentration of the majority holes in p-type silicon, under thermal equilibrium, is approximately equal to the concentration N_A of the *acceptor* (boron) impurity,

$$p_{p0} \simeq N_A$$

In this p-type silicon the concentration of the minority electrons, which are generated by thermal ionization, will be

$$n_{p0} \simeq \frac{n_i^2}{N_A}$$

It should be mentioned that a piece of n-type or p-type silicon is electrically neutral; the majority free carriers (electrons in n-type silicon and holes in p-type silicon) are neutralized by *bound charges* associated with the impurity atoms.

4.10 THE *pn* JUNCTION UNDER OPEN-CIRCUIT CONDITIONS

Figure 4.25 shows a *pn* junction under open-circuit conditions—that is, the external terminals are left open. The "+" signs in the p-type material denote the majority holes. The charge of these holes is neutralized by an equal amount of bound negative charge associated with the acceptor atoms. For simplicity these bound charges are not shown in the diagram. Also not shown are the minority electrons generated in the p-type material by thermal ionization.

In the n-type material the majority electrons are indicated by "−" signs. Here also the bound positive charge, which neutralizes the charge of the majority electrons, is

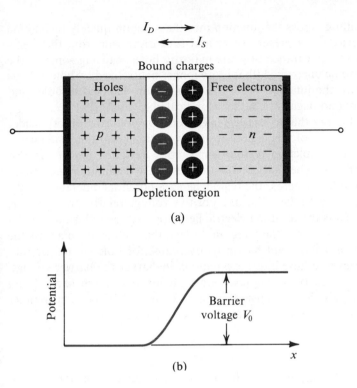

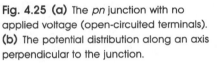

(a)

(b)

Fig. 4.25 (a) The *pn* junction with no applied voltage (open-circuited terminals). **(b)** The potential distribution along an axis perpendicular to the junction.

not shown in order to keep the diagram simple. The *n*-type material also contains minority holes generated by thermal ionization and not shown on the diagram.

The Diffusion Current I_D

Because the concentration of holes is high in the *p* region and low in the *n* region, holes diffuse across the junction from the *p* side to the *n* side; similarly, electrons diffuse across the junction from the *n* side to the *p* side. These two current components add together to form the diffusion current I_D, whose direction is from the *p* side to the *n* side, as indicated in Fig. 4.25.

The Depletion Region

The holes that diffuse across the junction into the *n* region quickly recombine with some of the majority electrons present there and thus disappear from the scene. This recombination process results in the disappearance of some free electrons from the *n*-type material. Thus some of the bound positive charge will no longer be neutralized by free electrons, and this charge is said to have been *uncovered*. Since recombination takes place close to the junction, there will be a region close to the junction that is depleted of free electrons and contains uncovered bound positive charge, as indicated in Fig. 4.25.

The electrons that diffuse across the junction into the *p* region quickly recombine with some of the majority holes present there and thus disappear from the scene. This results also in the disappearance of some majority holes, causing some of the bound negative charge to be uncovered (that is, no longer neutralized by holes). Thus in the *p* material close to the junction there will be a region depleted of holes and containing uncovered bound negative charge, as indicated in Fig. 4.25.

From the above it follows that a *carrier-depletion region* will exist on both sides of the junction, with the *n* side of this region positively charged and the *p* side negatively charged. This carrier-depletion region—or, simply, *depletion region*—is also called the *space-charge region*. The charge on both sides of the depletion region cause an electric field to be established across the region; hence a potential difference results across the depletion region, with the *n* side at a positive voltage relative to the *p* side, as shown in Fig. 4.25b. Thus the resulting electric field opposes the diffusion of holes into the *n* region and electrons into the *p* region. In fact, the voltage drop across the depletion region acts as a *barrier* that has to be overcome for holes to diffuse into the *n* region and electrons to diffuse into the *p* region. The larger the barrier voltage, the smaller the number of carriers that will be able to overcome the barrier, and hence the lower the magnitude of diffusion current. Thus the diffusion current I_D depends strongly on the voltage drop V_0 across the depletion region.

The Drift Current I_S and Equilibrium

In addition to the current component I_D due to majority carrier diffusion, a component due to minority carrier drift exists across the junction. Specifically, some of the thermally generated holes in the *n* material diffuse through the *n* material to the edge of the depletion region. There they experience the electric field in the depletion region, which sweeps them across that region into the *p* side. Similarly, some of the minority thermally generated electrons in the *p* material diffuse to the edge of the depletion region and get swept by the electric field in the depletion region across that region into the *n* side. These two current components—electrons moved by drift from *p* to *n* and holes moved by drift from *n* to *p*—add together to form the drift current I_S, whose direction is from the *n* side to the *p* side of the junction, as indicated in Fig. 4.25. Since the current I_S is carried by thermally generated minority carriers, its value is strongly dependent on temperature; however, it is independent of the value of the depletion layer voltage V_0.

Under open-circuit conditions (Fig. 4.25) no external current exists; thus the two opposite currents across the junction should be equal in magnitude:

$$I_D = I_S$$

This equilibrium condition is maintained by the barrier voltage V_0. Thus if for some reason I_D exceeds I_S, then more bound charge will be uncovered on both sides of the junction, the depletion layer will widen, and the voltage across it (V_0) will increase. This in turn causes I_D to decrease until equilibrium is achieved with $I_D = I_S$. On the other hand, if I_S exceeds I_D, then the amount of uncovered charge will decrease, the depletion layer will narrow, and the voltage across it will decrease. This causes I_D to increase until equilibrium is achieved with $I_D = I_S$.

The Terminal Voltage

When the *pn* junction terminals are left open-circuited the voltage measured between them will be zero. That is, the voltage V_0 across the depletion region *does not* appear between the diode terminals. This is because of the contact voltages existing at the metal–semiconductor junctions at the diode terminals, which counter and exactly balance the barrier voltage. If this were not the case, we would have been able to draw energy from the isolated *pn* junction, which would clearly violate the principle of conservation of energy.

Width of the Depletion Region

From the above it should be apparent that the depletion region exists in both the *p* and *n* materials and that equal amounts of charge exist on both sides. However, since usually the doping levels are not equal in the *p* and *n* materials, one can reason that the width of the depletion region will not be the same on the two sides. Rather, in order to uncover the same amount of charge the depletion layer will extend deeper into the more lightly doped material. In actual practice it is usual for one side of the junction to be much more lightly doped than the other, with the result that the depletion region exists almost entirely in one of the two semiconductor materials.

4.11 THE *pn* JUNCTION UNDER REVERSE-BIAS CONDITIONS

The behavior of the *pn* junction in the reverse direction is more easily explained on a microscopic scale if we consider exciting the junction with a constant-current source (rather than with a voltage source), as shown in Fig. 4.26. The current source I is obviously in the reverse direction. For the time being let the magnitude of I be less than I_S; if I is greater than I_S, breakdown will occur, as explained in Section 4.12.

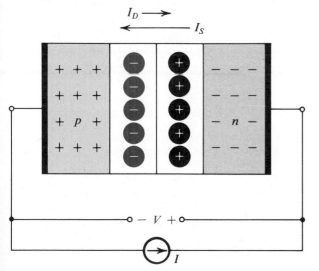

Fig. 4.26 The *pn* junction excited by a constant-current source I in the reverse direction. To avoid breakdown, I is kept smaller than I_S. Note that the depletion layer widens and the barrier voltage increases by V volts, which appears between the terminals as a reverse voltage.

The current I will be carried by electrons flowing in the external circuit from the n material to the p material (that is, in the direction opposite to that of I). This will cause electrons to leave the n material and holes to leave the p material. The free electrons leaving the n material cause the uncovered positive bound charge to increase. Similarly, the holes leaving the p material result in an increase in the uncovered negative bound charge. Thus the reverse current I will result in an increase in the width of, and the charge stored in, the depletion layer. This, in turn, will result in a higher voltage across the depletion region—that is, a greater barrier voltage V_0—which causes the diffusion current I_D to decrease. The drift current I_S, being independent of the barrier voltage, will remain constant. Finally, equilibrium (steady state) will be reached when

$$I_S - I_D = I$$

In equilibrium, the increase in depletion-layer voltage will appear as an external voltage between the diode terminals, with n being positive with respect to p.

We can now consider exciting the pn junction by a reverse voltage V, where V is less than the breakdown voltage V_{ZK}. When the voltage V is first applied, a reverse current flows in the external circuit from p to n. This current causes the increase in width and charge of the depletion layer. Eventually the voltage across the depletion layer will increase by the magnitude of the external voltage V, at which time an equilibrium is reached with the external reverse current I equal to $(I_S - I_D)$. Note, however, that initially the external current can be much greater than I_S. The purpose of this initial transient is to *charge* the depletion layer and increase the voltage across it by V volts.

From the above we observe the analogy between the depletion layer of a pn junction and a capacitor. As the voltage across the pn junction changes, the charge stored in the depletion layer changes accordingly. Figure 4.27 shows a sketch of typical charge-versus-external-voltage characteristic of a pn junction. Since this q-v characteristic is nonlinear, one has to be careful in speaking of the "depletion capacitance." As we did in Section 4.8 with the nonlinear i-v characteristic, we may consider

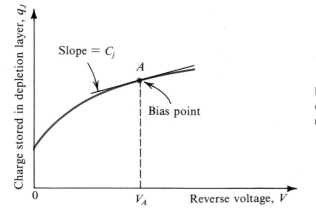

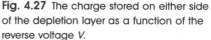

Fig. 4.27 The charge stored on either side of the depletion layer as a function of the reverse voltage *V*.

operation around a bias point on the q-v curve, such as point A in Fig. 4.27, and define the small-signal or incremental *depletion capacitance* C_j as the slope of the q-v curve at the operating point,

$$C_j = \frac{dq_J}{dv}\bigg|_{v=V_A}$$

It can be shown that

$$C_j = \frac{K}{(V_0 - V_D)^m}$$

where V_0 = the depletion-layer voltage with zero external voltage,

V_D = the voltage between the diode terminals (V_D is negative in the reverse direction),

K = a constant depending on the area of the junction and impurity concentrations, and

m = a constant depending on the distribution of impurity near the junction.
    The values of m ranges between $\frac{1}{3}$ and 3 or 4 for junctions of various types.

To recap, as a reverse bias voltage is applied to a *pn* junction, a transient occurs during which the depletion capacitance is charged to the new bias voltage. After the transient dies, the steady-state reverse current is simply equal to $I_S - I_D$. Usually I_D is very small when the diode is reverse-biased and the reverse current is approximately equal to I_S. This, however, is only a theoretical model that does not apply very well. In actual fact, currents as high as few nanoamperes (10^{-9} A) flow in the reverse direction, in devices for which I_S is of the order of 10^{-15} A. This large difference is due to leakage and other effects. Furthermore, the reverse current is dependent to a certain extent on the magnitude of the reverse voltage, contrary to the theoretical model, which states that $I \simeq I_S$ independent of the value of the reverse voltage applied. Nevertheless, because of the very low current involved, one is usually not interested in the details of the diode i-v characteristic in the reverse direction.

4.12 THE *pn* JUNCTION IN THE BREAKDOWN REGION

In considering diode operation in the reverse-bias region in Section 4.11 it was assumed that the reverse-current source I (Fig. 4.26) is smaller than I_S or, equivalently, that the reverse voltage V is smaller than the breakdown voltage V_{ZK}. We now wish to consider the breakdown mechanisms in *pn* junctions and explain the reasons behind the almost vertical line representing the i-v relationship in the breakdown region. For this purpose, let the *pn* junction be excited by a current source that causes a constant current I greater than I_S to flow in the reverse direction, as shown in Fig. 4.28. This current source will move holes from the p material through the external circuit[4] into

---

[4] The current in the external circuit will, of course, be carried entirely by electrons.

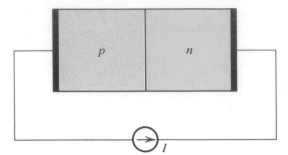

Fig. 4.28 The *pn* junction excited by a reverse-current source *I*, where $I > I_S$.

the n material and electrons from the n material through the external circuit into the p material. This action results in more and more of the bound charge being uncovered; hence the depletion layer widens and the barrier voltage rises. This latter effect causes the diffusion current to decrease; eventually it will be reduced to almost zero. Nevertheless, this is not sufficient to reach a steady state, since I is greater than I_S. Therefore the process leading to the widening of the depletion layer continues until a sufficiently high junction voltage develops, at which point a new mechanism sets in to supply the charge carriers needed to support the current I. As will be now explained, this mechanism for supplying reverse currents in excess of I_S can take one of two forms depending on the *pn* junction material, structure, and so on.

The two possible breakdown mechanisms are the *zener effect* and the *avalanche effect*. If a *pn* junction breaks down with $V_Z < 5$ V, the breakdown mechanism is usually the zener effect. Avalanche breakdown occurs when V_Z is greater than about 7 V. For junctions that breakdown between 5 and 7 V, the breakdown mechanism can be either the zener or the avalanche effect or a combination of the two.

Zener breakdown occurs when the electric field in the depletion layer increases to the point where it can break covalent bonds and generate electron–hole pairs. The electrons generated this way will be swept by the electric field into the n side and the holes into the p side. Thus these electrons and holes constitute a reverse current across the junction that helps support the external current I. Once the zener effect starts, a large number of carriers can be generated, with a negligible increase in the junction voltage. Thus the reverse current in the breakdown region will be determined by the external circuit, while the reverse voltage appearing between the diode terminals will remain close to the rated breakdown voltage V_Z.

The other breakdown mechanism is avalanche breakdown, which occurs when the minority carriers that cross the depletion region under the influence of the electric field gain sufficient kinetic energy to be able to break covalent bonds in atoms with which they collide. The carriers liberated by this process may have sufficiently high energy to be able to cause other carriers to be liberated in another ionizing collision. This process occurs in the fashion of an avalanche, with the result that many carriers are created that are able to support any value of reverse current, as determined by the external circuit, with a negligible change in the voltage drop across the junction.

As mentioned before, *pn* junction breakdown is not a destructive process, provided that the maximum specified power dissipation is not exceeded. This maximum power dissipation rating, in turn, implies a maximum value for the reverse current.

4.13 THE *pn* JUNCTION UNDER FORWARD-BIAS CONDITIONS

We next consider operation of the *pn* junction in the forward-bias region. Again it is easier to explain physical operation if we excite the junction by a constant-current source supplying a current I in the forward direction, as shown in Fig. 4.29. The electrons carrying the current I in the external circuit from the *p* material to the *n* material

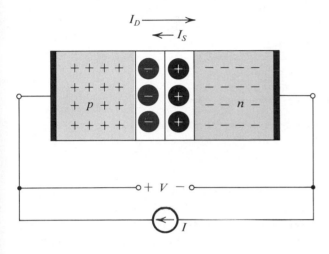

Fig. 4.29 The *pn* junction excited by a constant-current source supplying a current I in the forward direction. The depletion layer narrows and the barrier voltage decreases by V volts, which appears as an external voltage in the forward direction.

cause holes to be extracted from the *n* region and electrons to be extracted from the *p* region. This results in majority carriers being supplied to both sides of the junction by the external circuit: holes to the *p* material and electrons to the *n* material. These majority carriers will neutralize some of the uncovered bound charge, causing less charge to be stored in the depletion layer. Thus the depletion layer narrows and the depletion barrier voltage reduces. The reduction in barrier voltage enables more holes to cross the barrier from the *p* material into the *n* material and more electrons from the *n* side to cross into the *p* side. Thus the diffusion current I_D increases until equilibrium is achieved with $I_D - I_S = I$, the externally supplied forward current.

Let us now examine closely the current flow across the forward-biased *pn* junction in the steady state. The barrier voltage is now lower than V_0 by an amount V that appears between the diode terminals as a forward voltage drop (that is, the anode of the diode will be more positive than the cathode by V volts). Owing to the decrease in the barrier voltage or, alternatively, because of the forward voltage drop V, holes are *injected* across the junction into the *n* region and electrons are injected across the junction into the *p* region. The holes injected into the *n* region will cause the minority carrier concentration there, p_n, to exceed the thermal equilibrium value, p_{n0}. The *excess* concentration $p_n - p_{n0}$ will be highest near the edge of the depletion layer and will decrease (exponentially) as one moves away from the junction, eventually reaching zero. Figure 4.30 shows such a minority-carrier distribution.

In the steady state the concentration profile of *excess minority carriers* remains constant, and indeed it is such a distribution that gives rise to the increase of diffusion current I_D above the value I_S. This is because the distribution shown causes injected

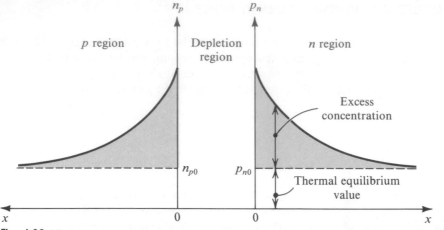

Fig. 4.30 Minority carrier distribution as a function of the distance from the edges of the depletion layer.

minority holes to diffuse away from the junction into the n region and disappear by recombination. To maintain equilibrium, an equal number of electrons will have to be supplied by the external circuit, thus replenishing the electron supply in the n material.

Similar statements can be made about the minority electrons in the p material. The diffusion current I_D is, of course, the sum of the electron and hole components.

Thus in the steady state, excess minority carrier distributions as shown in Fig. 4.30 exist in both the n and p materials. It can be shown that the hole current crossing the junction is proportional to the total excess hole charge stored in the n material. This charge is proportional to the area under the hole-concentration curve, which is shown in color in Fig. 4.30 and which, in turn, is proportional to the excess hole concentration at the edge of the junction. Similarly, it can be shown that the current component carried by electrons is proportional to the excess stored electron charge in the p material. Thus the electron-current component is proportional to the area under the minority carrier distribution curve and hence is proportional to the excess concentration of electrons at the edge of the depletion layer.

Finally, we note that if we change the value of the external current I or, alternatively, change the forward voltage drop V, the minority carrier charge stored in the p and n materials will have to be changed for a new steady state to be established (this, of course, is in addition to the change in charge stored in the depletion region). Thus the pn junction exhibits a capacitive effect—in addition to the depletion capacitance—related to the storage of minority carrier charges. Since these charges q_M are proportional to the current flowing across the junction I, it follows from the diode equation [Eq. (4.1)] that q_M is related to the forward voltage drop by a relationship of the form

$$q_M = q_0(e^{v/nV_T} - 1)$$

where q_0 is a constant charge proportional to the current I_S. Thus the q-v curve of this capacitive effect is clearly a nonlinear one. We may model this capacitive behavior, however, by a small-signal capacitance C_d,

$$C_d = \frac{dq_M}{dv}\Big|_{v=V_A}$$

where V_A is the dc diode voltage at the operating point A, around which the small-signal model is valid. The capacitance C_d is called the *diffusion capacitance*. From the above relationships one can easily establish that C_d is proportional to the value of $q_M + q_0$. Whereas in the reverse-bias region C_d is zero, its value in the forward-bias region is approximately proportional to the bias current I_A at the operating point.

4.14 THE COMPLETE SMALL-SIGNAL MODEL

From the above we conclude that an appropriate small-signal model for the *pn* junction consists of the diode resistance r_d in parallel with the depletion layer capacitance C_j, in parallel with the diffusion capacitance C_d. This model is depicted in Fig. 4.31.

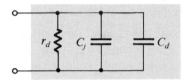

Fig. 4.31 Complete small-signal model of a *pn* junction.

If the diode is biased in the forward region at a point A on the *i-v* curve, then

$$r_d = \frac{nV_T}{I_A}$$

$$C_d = k_c I_A$$

$$C_j = \frac{K}{(V_0 - V_A)^m}$$

where k_c is a constant and V_A is a positive number. On the other hand, if the diode is reverse-biased with a reverse-bias voltage $v = -V_A$ (V_A is a positive number), then

$$r_d \simeq 0$$

$$C_d \simeq 0$$

$$C_j = \frac{K}{(V_0 + V_A)^m}$$

The values of C_d and C_j are, of course, dependent on the size of the junction, being directly proportional to the cross-sectional area. For a small-signal diode reverse-biased by few volts, C_j is typically of the order of 1 pF. The same diode, forward-biased with a current of few milliamperes, has a diffusion capacitance of the order of 10 pF.

4.15 SUMMARY

● In the forward direction, the ideal diode conducts any current forced by the external circuit while displaying a zero voltage drop. The ideal diode does not conduct in the reverse direction; any applied voltage appears as reverse bias across the diode.

● The unidirectional-current-flow property makes the diode useful in the design of rectifier circuits.

● The forward conduction of practical silicon diodes is accurately characterized by the relationship $i = I_S e^{v/nV_T}$.

● A silicon diode conducts a negligible current until the forward voltage is at least 0.5 V. Then the current increases rapidly, with the voltage drop increasing by 60 to 120 mV (depending on the value of n) for every decade of current change.

● In the reverse direction, a silicon diode conducts a current of the order of 10^{-9} A. This current is much greater than I_S and increases with the magnitude of reverse voltage.

● Beyond a certain value of reverse voltage (that depends on the diode) breakdown occurs, and current increases rapidly with a small corresponding increase in voltage.

● Diodes designed to operate in the breakdown region are called zener diodes. They are employed in the design of voltage regulators whose function is to provide a constant dc voltage that varies little with variations in power supply voltage and/or load current.

● A hierarchy of diode models exists, with the selection of an appropriate model dictated by the application.

● In many applications, a conducting diode is modeled as having a constant voltage drop, usually about 0.7 V.

● A diode biased to operate at a dc current I_D has a small-signal resistance $r_d = nV_T/I_D$.

● The silicon junction diode is basically a *pn* junction. Such a junction is formed in a single silicon crystal.

● In *p*-type silicon there is an overabundance of holes (positively charged carriers), while in *n*-type silicon electrons are abundant.

● A carrier-depletion region develops at the interface in a *pn* junction, with the *n* side positively charged and the *p* side negatively charged. The voltage difference resulting is called the barrier voltage.

● A diffusion current I_D flows in the forward direction (carried by holes from the *p* side and electrons from the *n* side), and a current I_S flows in the reverse direction (carried by thermally generated minority carriers). In an open-circuited junction, $I_D = I_S$ and the barrier voltage is denoted V_0.

● Applying a reverse-bias voltage $|V|$ to a *pn* junction causes the depletion region to widen, and the barrier voltage increases to $(V_0 + |V|)$. The diffusion current decreases and a net reverse current of $(I_S - I_D)$ flows.

• Applying a forward-bias voltage $|V|$ to a *pn* junction causes the depletion region to become narrower, and the barrier voltage decreases to $(V_0 - |V|)$. The diffusion current increases, and a net forward current of $(I_D - I_S)$ flows.

BIBLIOGRAPHY

E. J. Angelo, Jr., *Electronics: BJTs, FETs and Microcircuits*, New York: McGraw-Hill, 1969.

P. E. Gray and C. L. Searle, *Electronic Principles*, New York: Wiley, 1969.

D. A. Hodges and H. G. Jackson, *Analysis and Design of Digital Integrated Circuits*, New York: McGraw-Hill, 1983.

D. H. Navon, *Semiconductor Microdevices and Materials*, New York: Holt, Rinehart and Winston, 1986.

PROBLEMS

4.1 Find the values of I and V in the circuits shown in Fig. P4.1 assuming the diodes to be ideal.

4.2 For the circuits in Fig. P4.2 (page 196), each incorporating two ideal diodes, find the values of I and V.

4.3 For the circuits shown in Fig. P4.3 (page 197), assume that the diodes are ideal and find the values of I and V indicated.

4.4 Consider the rectifier circuit discussed in Section 2.1. Because this circuit utilizes only half the input waveform, it is known as a half-wave rectifier. The circuit is redrawn in Fig. P4.4 (page 198) and the diode is assumed ideal. Let the input be a sine wave of 60 Hz and 10-V rms. What is the peak output voltage across R_L? What is the interval during which the diode is in forward conduction? For what interval is the diode cut off? What is the average voltage across a load $R_L = 100\ \Omega$? What is the corresponding average diode current?

4.5 The half-wave rectifier circuit shown in Fig. P4.4 is driven by a source v_S having a Thévenin resistance R_s. If $R_s = R_L$, and the diode is assumed ideal, plot the transfer characteristic v_O versus v_S.

4.6 If the output in the half-wave rectifier circuit is taken across the diode, as shown in Fig. P4.6 (page 198), the circuit is called a clipper. Assuming the diode to be ideal, sketch the transfer characteristic of this circuit. What is the effect on the transfer characteristic of shunting the existing diode with a second ideal diode connected as follows: **(a)** anode to anode, cathode to cathode; **(b)** anode to cathode, cathode to anode.

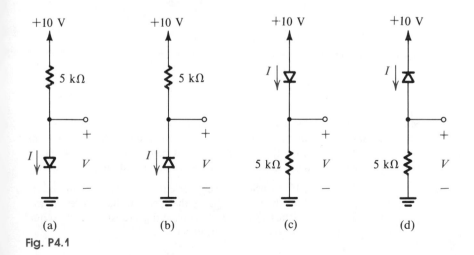

Fig. P4.1

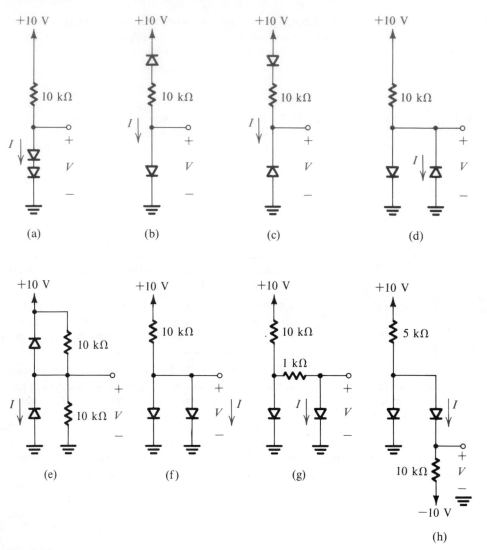

Fig. P4.2

4.7 Consider the battery charger described in Example 4.1. If the sinusoidal source is replaced by a square wave of the same amplitude, for what fraction of the cycle does the diode conduct? What is the peak diode current? What is the average diode current?

4.8 Repeat Problem 4.7 for a symmetrical triangular wave of 24-V amplitude.

4.9 Using the ideas introduced in Exercise 4.2, design an ac voltmeter having a full-scale reading of 15 V rms using the same 1-mA, 50-Ω meter movement. Assume the diode to be ideal. What value of R should be used?

4.10 Consider the clipper circuit of Fig. P4.6 and let the diode be ideal. What is the effect on the transfer characteristic v_O versus v_I, of connecting a resistor of value equal to R across the output terminals?

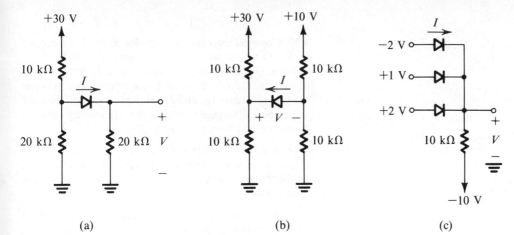

(a)   (b)   (c)

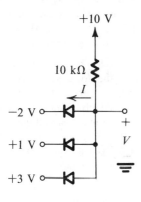

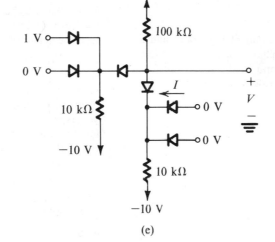

(d)   (e)

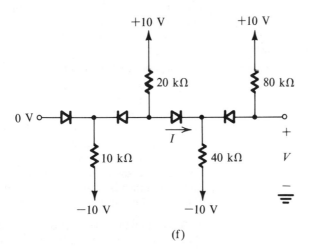

(f)

Fig. P4.3

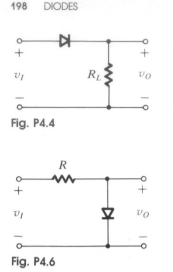

Fig. P4.4

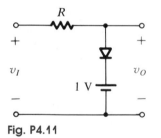

Fig. P4.6

4.11 Consider the clipper circuit of Fig. P4.11 and assume the diode to be ideal. What is the largest triangular-wave input for which the input and output waveforms are identical? Sketch the transfer characteristic.

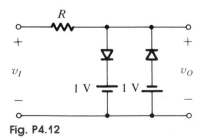

Fig. P4.11

4.12 Consider the clipper circuit of Fig. P4.12 assuming that the diodes are ideal. Sketch the transfer characteristic and the output waveform v_O obtained in response to $v_I = 2 \sin 100t$.

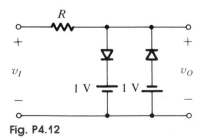

Fig. P4.12

4.13 Consider two junction diodes, each characterized by Eq. (4.1), and using the same fabrication technology, but one, diode A, having a junction area 1,000 times that of the other, diode B. What is the relationship between the scale currents I_{SA} and I_{SB} of the two diodes? For the same applied voltage v, what is the relationship between the two diode currents i_A and i_B?

4.14 A junction diode is operated in a circuit in which it is supplied with a constant current I. What is the effect on the forward voltage of shunting it by a second identical diode? Assume $n = 1$.

4.15 A silicon diode has $n = 1$ and $I_S = 10^{-15}$ A at a junction temperature of 20°C. If it is known that I_S increases by 15% for every °C rise in the junction temperature and that the thermal voltage V_T is given by Eq. (4.2), calculate the voltage drop across the diode when $i = 1$ mA, at 20°C and at 30°C. Hence find the temperature coefficient of the diode voltage drop.

4.16 Two silicon junction diodes, one intended for logic circuit applications and the other for use in power-supply rectification, are found to have a 0.7-V forward voltage drop at 1 mA and 10 A, respectively. What current would you expect each to conduct at the cut-in voltage of 0.5 V, for the cases in which the diodes are characterized by: **(a)** $n = 1$; **(b)** $n = 2$; **(c)** a logarithmic characteristic of 0.1 V/decade of current change.

4.17 A diode measured at two operating currents, 5 and 15 mA, is found to have corresponding voltage drops of 0.710 and 0.764 V. Find the values of n and I_S.

4.18 A small silicon junction diode for which the temperature coefficient is -2 mV/°C is operated at a current of 100 μA at which the junction voltage at 20°C is 700 mV. What junction voltage would result at 0°C? At 100°C?

4.19 A zener diode specified to have $V_Z = 6.8$ V and a low zener resistance has a knee current $I_{ZK} = 100$ μA. It is to be used in a shunt regulator connected to a power source whose lowest value is 10 V and to a load whose maximum current is 1 mA. For this situation what is the largest value of the regulator resistor R (see Fig. 4.10) that ensures that the zener will operate at a current that is at least twice the knee current? For this design, what does the zener diode current become if the load is removed and simultaneously the source voltage rises to 20 V? What is the power dissipated in the diode at this time?

4.20 To eliminate the effect of load variation on a zener shunt regulator, an op amp buffer may be added as shown in Fig. P4.20. If a voltage somewhat greater than that of the zener diode is required, a follower with gain may be used. Design a regulator circuit utilizing a 6.8-V zener and a supply of 15 V to provide a regulated output voltage of 10 V. Design for 1-mA zener current and a 0.1-mA current in the resistive network of the op amp.

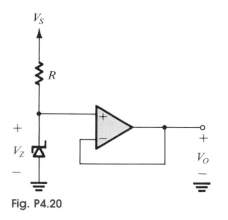

Fig. P4.20

4.21 Repeat Problem 4.20 for a +5-V output.

4.22 Repeat Problem 4.20 for a −5-V output.

**4.23 Reconsider the design of the 10-V regulator described in Problem 4.20. Although the op-amp follower solves the problem of the effect of load variation on the regulated voltage, it does nothing about the problem of output variation due to variation in V_S. To alleviate this problem consider the following. Since V_O is greater than V_Z, a resistor connected from the op-amp output to the cathode of the zener diode will provide the diode with a constant current independent of minor variations in the power supply V_S. Augment the design worked out in Problem 4.20 with such a resistor supplying a current of 1 mA to the zener. Raise the value of R to reduce its contribution to the zener current to 10 μA. Why is this resistor still necessary? (*Hint:* Think about start-up). If V_S rises to 20 V and if $r_Z = 100 \, \Omega$, and assuming the op amp to be ideal, find the change in V_O. Compare this result with the corresponding one for the unmodified circuit.

4.24 A 5.1-V zener diode is biased very nearly at the current at which its TC is zero (see Fig. 4.12). Its residual TC is + 0.1 mV/°C. Express the net variation as a fraction of the regulated output voltage in parts per million per °C (ppm/°C).

4.25 Consider a series circuit consisting of a 0.8-V battery polarized to cause the diode to conduct, a 1-kΩ

resistor, and a junction diode for which $n = 2$ and $I_S = 8.3 \times 10^{-10}$ A. Sketch the diode characteristic curve using data for currents at 0.1, 0.5, and 1.0 mA. Plot the load line corresponding to the circuit described. Use your sketch to estimate the diode current and voltage.

4.26 Attempt an alternative graphical solution for Problem 4.25, using semilog paper to plot the diode characteristic curve (a straight line) and the load line (not a straight line). Estimate the diode current and voltage.

4.27 Use the iterative procedure described in Section 4.6 to find the diode current and voltage for the circuit described in Problem 4.25. An acceptable solution is one in which the values of the current obtained in two successive iterations differ by less than 1%.

4.28 A 1-mA diode is connected in a series circuit with a 100-kΩ resistor and a 0.7-V battery polarized to cause the diode to conduct. If the diode exhibits a 0.1-V change of junction voltage for each decade change of current, what circuit current results? Use an iterative solution.

*4.29 A series circuit consists of a 1.5-V battery, a 1-KΩ resistor, a 1-A diode, and a 1-mA diode. For each of the diodes $n = 2$.
(a) Find the loop current and diode voltages using an iterative approach.
(b) If each of the two diodes is modeled as an ideal diode, what value of current results?
(c) If each diode is modeled by a constant 0.7-V drop, what value of current results?
(d) Find the battery voltage for which the approximate solutions in (b) and (c) above differ by no more than 10%.

***4.30 In the circuits of Figs. P4.1, P4.2, and P4.3 find the values of I and V as indicated. Assume all diodes to be 1-mA diodes—that is, conducting a current of 1 mA at a voltage drop of 0.7 V. Also assume that $n = 1$. (Note that this is a very long problem. The reader may consider solving only a few of the circuits shown.)

4.31 A particular junction diode has a voltage drop of 0.58 V at 1 mA and 0.64 V at 10 mA. Calculate the parameters of a battery-plus-resistance model that fits at the two data points given. What voltage would your model predict at 5.5 mA? At 19 mA?

4.32 For a diode with $n = 2$ find the percentage error involved in the small-signal approximation, that is, in approximating i_D in Eq. (4.12) by the expression in Eq. (4.14), for signals of 2 mV, 5 mV, 10 mV, and 25 mV. Compare your results with those obtained in Exercise 4.12 for the case $n = 1$.

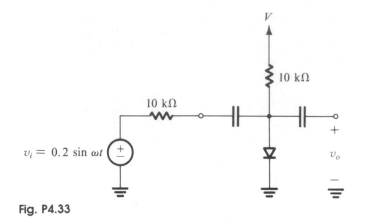

Fig. P4.33

Fig. P4.35

4.33 The capacitors in the circuit of Fig. P4.33 are for the purpose of blocking dc and transmitting the sine-wave signal. Assuming that the diode is a 1-mA device with $n = 2$, find the direct current through it for $V = 1$ V and $V = 10$ V. For each of these two cases use the diode small-signal model to find the sine-wave output signal v_o. Note that this circuit can be used as a voltage-controlled attenuator with V as the control voltage and v_i the signal being attenuated.

4.34 You are required to design a voltage regulator having an output of 2.34 V (at no load) utilizing a constant

current source of 10 mA feeding a network of 1-mA diodes for which $n = 2$. Show the circuit of such a regulator. (*Hint:* It uses six diodes.) What does the output voltage become if a load of 2.34 kΩ is connected to the regulator output? Use the diode small-signal model for this calculation. Then check the accuracy of your calculation using the exponential diode model.

**4.35 Using the constant-voltage-drop (0.7-V) model for each of the diodes (when conducting) find the transfer characteristic of the circuit shown in Fig. P4.35 for the range -10 V $\leq v_I \leq +10$ V.

Nonlinear Circuit Applications

INTRODUCTION

Having studied the op amp in Chapter 3 and the diode in Chapter 4, we are now in a position to consider combining them in the design of nonlinear circuits. The circuits considered in this chapter find a variety of applications, including dc power generation, measurement and instrumentation, generation of signals having a variety of waveforms, analog computation, and modulation and demodulation.

It will be seen that connecting the diode in the negative-feedback path of an op amp provides idealized behavior, with all the nonidealities of the diode being masked by the high gain of the op amp. This allows for the design of circuits having precise characteristics, as is usually needed in instrumentation and computation applications. This is another illustration of the utility of negative feedback, a concept that we encountered in the linear circuits of Chapter 3 and that we will study formally in Chapter 12.

In the current chapter we will also study some applications of *positive feedback*. As the name implies, positive feedback is exactly the opposite of negative feedback; rather than providing gain stability, as negative feedback does, positive feedback allows for the design of circuits that *oscillate* (that is, provide an output with no input) in a controlled manner. Such circuits can be made to generate signals having square, triangular, pulse, and other waveforms. Furthermore, we will show how circuits composed of diodes and resistors can be used for the purpose of shaping the waveform of a signal in a predetermined manner. The "sine shaper" discussed in Section 5.10 is one such circuit. It can be used to convert triangular waveforms into sinusoids.

This chapter has many objectives. One of these is to introduce the reader to interesting, challenging, and practical circuits that can be designed, assembled, and experimented with. Another objective of this chapter is to expose the reader to methods of analysis of nonlinear circuits. In keeping with the spirit of this book, our methods are usually approximate, but they "do the job." □

5.1 HALF-WAVE RECTIFICATION

In Section 2.1 we considered the simple diode circuit shown in Fig. 5.1a. With an ideal-diode model the transfer characteristic shown in Fig. 5.1b is obtained. For example, an input signal with triangular waveform (Fig. 5.1c) provides the output waveform shown in Fig. 5.1d. Since the output is unidirectional, the alternating-current (ac) signal at the input is said to have been *rectified*. Also, because only half-cycles of the input signal are utilized, the circuit is called a *half-wave rectifier*.

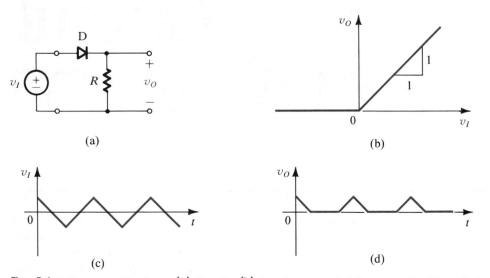

Fig. 5.1 Half-wave rectification. **(a)** Circuit. **(b)** Transfer characteristics using the ideal-diode model. **(c)** An input signal having a triangular waveform. **(d)** The corresponding output waveform.

Effect of the Diode Nonideal Characteristics

We will now consider the effect of the diode finite forward-voltage drop on the operation of the half-wave rectifier. In order for our discussion to be concrete we will consider the circuit operation with a specific input signal, namely, a triangular waveform with a 10-V peak amplitude. Figure 5.2a shows the positive half-cycle of this input waveform. To find the corresponding output waveform we first observe

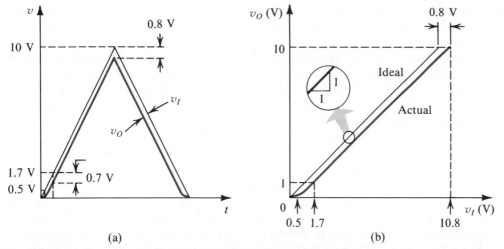

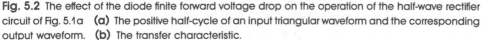

Fig. 5.2 The effect of the diode finite forward voltage drop on the operation of the half-wave rectifier circuit of Fig. 5.1a **(a)** The positive half-cycle of an input triangular waveform and the corresponding output waveform. **(b)** The transfer characteristic.

that the diode will not conduct appreciably until the input voltage exceeds about 0.5 V. Thus, until this point is reached, the output voltage will remain approximately zero. Then, as the diode conducts, the output voltage v_O will be less than that of the input by the diode drop. This diode drop, however, will change as the input voltage changes.

To obtain quantitative results, let $R = 1$ kΩ and assume that the diode is a 1-mA diode; that is, it displays a 0.7-V drop at a current of 1 mA. Furthermore, let us assume that we do not know the exact value of the constant n for this particular diode and therefore will use the 0.1-V/decade approximation. Under this approximation the 1-mA diode will exhibit a voltage drop of 0.5 V at 0.01 mA, 0.6 V at 0.1 mA, 0.8 V at 10 mA, and so on.

Now at the point where the output is 0.1 V the current is 0.1 V/1 kΩ = 0.1 mA and the input will be 0.7 V. When the output reaches 1 V, the current becomes 1 mA and the corresponding input voltage is 1.7 V. Finally, as the input reaches its peak value of 10 V, the output will lag by about 0.8 V, leading to a current of 9.2 mA, which indeed corresponds to a diode drop of almost 0.8 V.

Using the numbers obtained above we can sketch the output waveform shown in Fig. 5.2a and the transfer characteristic shown in Fig. 5.2b. Note that because of the continuous nature of the diode i-v characteristics, the corners of the resulting transfer characteristic and output waveform will be rounded, as shown in Fig. 5.2.

The simplicity of this analysis stems from the fact that we start at the output and work our way back to the input. Specifically, for a given value of v_O we first find the diode current $i_D = v_O/R$. Then the diode equation is used to find the corresponding diode voltage v_D. Finally, the corresponding value of v_I is found from $v_I = v_D + v_O$. The result is the coordinates of one point on the transfer characteristic. The process is then repeated to obtain the entire transfer curve. The latter can in turn be used to determine the waveform of v_O for any given waveform for v_I. Note that the above procedure can also be used to derive an expression for v_I in terms of v_O (see Exercise 5.1 below).

Although the process of finding the exact output waveform by taking into account the actual diode characteristics might be an interesting exercise, the amount of work involved is usually too great to be justified in practice. Depending on the particular application, one might just ignore the effects of the finite diode drop completely, investigate such effects qualitatively, or, if the application demands it, use a more elaborate circuit that "masks" these effects. For instance, if we are rectifying large input signals, it will most probably make little difference to the system operation whether the diode drop is 0.6 V or 0.8 V, and in some cases the assumption of an ideal diode is probably not a bad one. On the other hand, if the signals involved are reasonably small and if the application demands accuracy, we should either expend effort in accurately analyzing the circuit or, better yet, expend money in designing a more elaborate circuit, one in which the exact diode characteristics do not matter. As an illustration of the need for this latter alternative, consider the case of rectifying a signal only 100 mV in amplitude. Obviously our simple rectifier circuit will not do. As usual, the ubiquitous op amp will come to the rescue.

Exercise

5.1 Using the exponential diode characteristic show that for v_I and v_O both greater than zero the circuit of Fig. 5.1a has the transfer characteristic

$$v_I = v_O + v_D \, (\text{at } i_D = 1 \text{ mA}) + nV_T \ln\left(\frac{v_O}{R}\right)$$

where v_I and v_O are in volts and R is in kilohms.

5.2 PRECISION HALF-WAVE RECTIFIER—THE "SUPERDIODE"

Figure 5.3a shows a precision half-wave-rectifier circuit consisting of a diode placed in the negative-feedback path of an op amp, with R being the rectifier load resistance. The circuit works as follows: If v_I goes positive, the output voltage v_A of the op amp

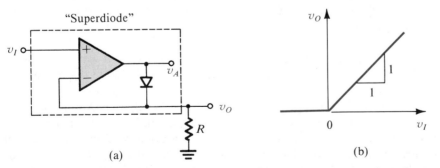

(a) (b)

Fig. 5.3 The "superdiode" precision half-wave rectifier and its almost ideal transfer characteristic. Note that when $v_I > 0$ and the diode conducts, the op amp supplies the load current and the source is conveniently buffered, an added advantage.

will go positive and the diode will conduct, thus establishing a closed feedback path between the op-amp output terminal and the negative input terminal. This negative-feedback path will cause the op amp to operate "normally," in the manner we studied in Chapter 3, causing a virtual short circuit to appear between the two input terminals. Thus the voltage at the negative input terminal, which is also the output voltage v_O, will equal (to within a few millivolts) that at the positive input terminal, which is the input voltage v_I,

$$v_O = v_I \qquad v_I \geq 0$$

Note that the offset voltage ($\simeq 0.5$ V) exhibited in the simple circuit of Fig. 5.1 is no longer present. For the op-amp circuit to start operation, v_I has to exceed only a negligibly small voltage equal to the diode drop divided by the op-amp open-loop gain. In other words, the straight-line transfer characteristic $v_O = v_I$ almost passes through the origin. This makes this circuit suitable for applications involving very small signals.

Consider now the case when v_I goes negative. The op-amp output voltage v_A will tend to follow and go negative. This will reverse-bias the diode, and no current will flow through resistance R, causing v_O to remain equal to 0 V. Thus for $v_I < 0$, $v_O = 0$. Since in this case the diode is off, the op amp will be operating in an open loop and its output will be at the negative saturation level.

The transfer characteristic of this circuit will be that shown in Fig. 5.3b, which is almost identical to the ideal characteristic of a half-wave rectifier. The nonideal diode characteristics have been almost completely masked by placing it in the negative-feedback path of an op amp. This is another dramatic application of negative feedback. The combination of diode and op amp, shown in the dotted box in Fig. 5.3a, is appropriately referred to as a "superdiode."

As usual, though, not all is well. The circuit of Fig. 5.3 has some disadvantages: When v_I goes negative and $v_O = 0$, the entire magnitude of v_I appears between the two input terminals of the op amp. If this magnitude is greater than few volts, the op amp may be damaged unless it is equipped with what is called "overvoltage protection" (a feature that most modern IC op amps have). Another disadvantage is that when v_I is negative, the op amp will be saturated. Although not harmful to the op amp, saturation should usually be avoided, since getting the op amp out of the saturation region and back into its linear region of operation requires some time. This time delay will obviously slow down circuit operation and limit the frequency of operation of the superdiode half-wave-rectifier circuit.

An Alternative Circuit

An alternative precision rectifier circuit that does not suffer from the disadvantages mentioned above is shown in Fig. 5.4. The circuit operates in the following manner: For positive v_I, diode D_2 conducts and closes the negative-feedback loop around the op amp. A virtual ground therefore will appear at the inverting input terminal, and the op-amp output will be *clamped* at one diode drop below ground. This negative voltage will keep diode D_1 off, and no current will flow in the feedback resistance R_2. It follows that the rectifier output voltage will be zero.

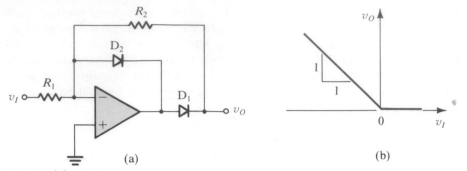

Fig. 5.4 (a) An improved version of the precision half-wave rectifier. Here diode D_2 is included to keep the feedback loop closed around the op amp during the off times of the rectifier diode D_1, thus preventing the op amp from saturating.   **(b)** The transfer characteristic for $R_2 = R_1$.

As v_I goes negative, the voltage at the inverting input terminal will tend to go negative, causing the voltage at the op-amp output terminal to go positive. This will cause D_2 to be reverse-biased and hence cut off. Diode D_1, however, will conduct through R_2, thus establishing a negative-feedback path around the op amp and forcing a virtual ground to appear at the inverting input terminal. The current through the feedback resistance R_2 will be equal to the current through the input resistance R_1. Thus for $R_1 = R_2$ the output voltage v_O will be

$$v_O = -v_I \qquad v_I \leq 0$$

The transfer characteristic of the circuit is shown in Fig. 5.4b. Note that unlike the previous circuit, here the slope of the characteristic can be set to any desired value, including unity, by selecting appropriate values for R_1 and R_2.

As mentioned before, the major advantage of this circuit is that the feedback loop around the op amp remains closed at all times. Hence the op amp remains in its linear operating region, avoiding the possibility of saturation and the associated time delay required to "get out" of saturation. Diode D_2 "catches" the output voltage as it goes negative and clamps it to one diode drop below ground; hence D_2 is called a "catching diode."

An Application: Measuring AC Voltages

As one of the many possible applications of the precision rectifier circuits discussed in this section, consider the basic ac voltmeter circuit shown in Fig. 5.5. The circuit consists of a half-wave rectifier—formed by op amp A_1, diodes D_1 and D_2, and resistors R_1 and R_2—and a first-order low-pass filter—formed by op amp A_2, resistors R_3 and R_4, and capacitor C. For an input sinusoid having a peak amplitude V_p the output v_1 of the rectifier will consist of a half sine wave having a peak amplitude of $V_p R_2/R_1$. It can be shown using Fourier series analysis that the waveform of v_1 has an average value of $(V_p/\pi)(R_2/R_1)$ in addition to harmonics of the frequency ω of the input signal. To reduce the amplitudes of all of these harmonics to negligible levels, the corner frequency of the low-pass filter should be chosen much smaller than the

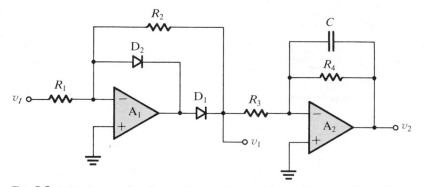

Fig. 5.5 A simple ac voltmeter consisting of a precision half-wave rectifier followed by a first-order low-pass filter.

lowest expected frequency ω_{min} of the input sine wave. This leads to

$$\frac{1}{CR_4} \ll \omega_{min}$$

Then the output voltage v_2 will be mostly dc, with a value

$$V_2 = -\frac{V_p}{\pi} \frac{R_2}{R_1} \frac{R_4}{R_3}$$

where (R_4/R_3) is the dc gain of the low-pass filter. Note that this voltmeter essentially measures the average value of the negative parts of the input signal but can be calibrated to provide root-mean-square (rms) readings for input sinusoids.

Exercises

5.2 Consider the operational rectifier or superdiode circuit of Fig. 5.3a, with $R = 1$ kΩ. For $v_I = 10$ mV, 1 V, and -1 V, what are the voltages that result at the rectifier output and at the output of the op amp? Assume that the op amp is ideal and that its output saturates at ± 12 V. The diode has a 0.7-V drop at 1-mA current, and the voltage drop changes by 0.1 V per decade of current change.

Ans. 10 mV, 0.51 V; 1 V, 1.7 V; 0 V, -12 V

5.3 If the diode in the circuit of Fig. 5.3a is reversed, find the transfer characteristic v_O as a function of v_I.

Ans. $v_O = 0$ for $v_I \geq 0$; $v_O = v_I$ for $v_I \leq 0$

5.4 Consider the circuit in Fig. 5.4a with $R_1 = 1$ kΩ and $R_2 = 10$ kΩ. Find v_O and the voltage at the amplifier output for $v_I = +1$ V, -10 mV, and -1 V. Assume the op amp to be ideal with saturation voltages of ± 12 V. The diodes have 0.7-V voltage drops at 1 mA, and the voltage drop changes by 0.1 V/decade of current change.

Ans. 0 V, -0.7 V; 0.1 V, 0.6 V; 10 V, 10.7 V

5.5 If the diodes in the circuit of Fig. 5.4a are reversed, find the transfer characteristic v_O as a function of v_I.

Ans. $v_O = -(R_2/R_1)v_I$ for $v_I \geq 0$; $v_O = 0$ for $v_I \leq 0$

5.6 Find the transfer characteristic for the circuit in Fig. E5.6.

Ans. $v_O = 0$ for $v_I \geq -5$ V; $v_O = -v_I - 5$ for $v_I \leq -5$ V

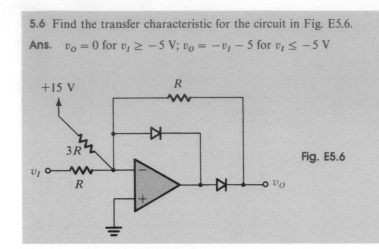

Fig. E5.6

5.3 FULL-WAVE RECTIFICATION

There are many applications in instrumentation (and in the design of power supplies) where the information (or the energy) provided by both halves of an ac signal is either useful or necessary. This need can be met by arranging that both the original signal and its inverse (or negative version) are made available and supplying each to a half-wave rectifier joined in turn to a common load. Such an arrangement, shown conceptually in Fig. 5.6, is called a *full-wave rectifier*. Here, when the voltage at A is positive,

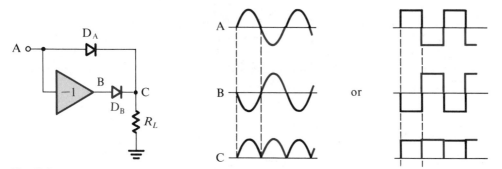

Fig. 5.6 Principle of full-wave rectification.

that at B is negative, causing D_A to conduct while D_B remains cut off. When the voltage at A goes negative, that at B goes positive and the diode roles interchange: D_B conducts while D_A is cut off. The output voltage resulting appears as a simple combination of alternate half-cycles of the input waveform, the negative halves having been inverted. The resulting waveforms, assuming ideal diodes, are depicted in Fig. 5.6. Note that these waveforms are more continuous and energetic than the corresponding half-wave ones. Also note that the average voltage at the output will be twice that obtained from a half-wave rectifier.

In practice there are many possible implementations of the full-wave rectifier. In the remainder of this section we develop implementations utilizing op amps. Circuits utilizing transformers for obtaining signal inversion will be presented in Section 5.4. Yet another alternative, the diode bridge rectifier, is presented in Section 5.5.

Amplifier-Coupled Full-Wave Rectifiers

Figure 5.7a shows a full-wave rectifier in which op amp A_1 and two equal resistors labeled R_1 and R_2 are used to implement the unity-gain inverting amplifier that provides the antiphase input to the second rectifier. Since for some applications an

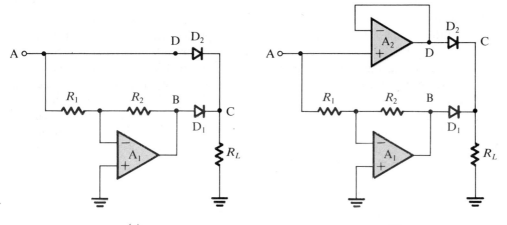

(a)   (b)

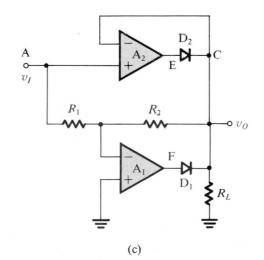

(c)

Fig. 5.7 Amplifier-coupled full-wave rectifiers. **(a)** Basic circuit, with an inverting amplifier used to provide the required inversion. **(b)** Same as (a) but with a buffer included in the in-phase path. **(c)** Same as (b) but with the diodes placed in the feedback loops.

asymmetry exists owing to second-order effects related to the amplifier, including its dynamics and its output resistance, it may be useful to introduce a unity-gain follower in the direct path, as shown in Fig. 5.7b. An incidental result that accrues, and is particularly useful in instrumentation applications, is that the input is not required to supply the output current directly; the output is said to be buffered. Note that if asymmetry is not a problem, then the input resistance can be made even higher by connecting the input resistor R_1 of the inverting amplifier to the output of the follower rather than directly to the input.

But there are other useful and more interesting results to be obtained by simple reconnections of the two amplifiers. Such a possibility is shown in Fig. 5.7c. An examination of this circuit reveals that it differs from its predecessor in Fig. 5.7b only in that diode D_1 has been placed in the negative-feedback loop of op amp A_1 and diode D_2 has been placed in the negative-feedback loop of op amp A_2. A further examination of this circuit reveals that it consists of two parallel circuits; the upper one, consisting of A_2 and D_2, may be recognized as the "superdiode" discussed in Section 5.2 that gives rise to the transfer characteristic shown in Fig. 5.8a. The lower circuit, composed of A_1, D_1, and the two resistors R_1 and R_2, may be recognized as the half-wave rectifier of Fig. 5.4a but without the catching diode. This lower circuit gives rise to the transfer characteristic shown in Fig. 5.8b. Connecting these two circuits in parallel implies combining their transfer characteristics to yield the characteristic shown in Fig. 5.8c, which may be recognized as that of a full-wave rectifier or, equivalently, an *absolute-value circuit*.

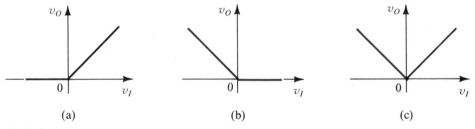

(a)   (b)   (c)

Fig. 5.8 Development of the transfer characteristic of the circuit in Fig. 5.7c.

A word of caution is in order. In connecting two circuits in parallel, one should be careful that nothing unusual results. For this reason it is prudent to reexamine the circuit of Fig. 5.7c to make sure that it indeed functions in the assumed manner. Toward that end, consider first the case where the input at A is positive. The output of A_2 will go positive, turning D_2 on, which will conduct through R_L and thus close the feedback loop around A_2. A virtual short circuit will thus be established between the two input terminals of A_2, and the voltage at the negative input terminal, which is the output voltage of the circuit, will become equal to the input. Thus no current will flow through R_1 and R_2, and the voltage at the inverting input of A_1 will be equal to the input and hence positive. Therefore the output terminal (F) of A_1 will go negative until A_1 saturates. This causes D_1 to be turned off.

Next consider the case when A goes negative. The tendency for a negative voltage at the negative input of A_1 causes F to rise, making D_1 conduct to supply R_L and allowing the feedback loop around A_1 to be closed. Thus a virtual ground appears at the negative input of A_1 and the two equal resistances R_1 and R_2 force the voltage at C, which is the output voltage, to be equal to the negative of the input voltage at A and thus positive. The combination of positive voltage at C and negative voltage at A causes the output of A_2 to saturate in the negative direction, thus keeping D_2 off.

The overall result is perfect full-wave rectification. This precision is, of course, a result of placing the diodes in op-amp feedback loops, thus masking their nonidealities. This circuit is one of many possible precision full-wave-rectifier or absolute-value circuits. Another related implementation of this function is examined in Exercise 5.7.

Exercises

5.7 In the full-wave rectifier circuit of Fig. 5.7c let $R_1 = R_2 = R_L = 10$ kΩ, and assume the op amp to be ideal except for output saturation at ± 12 V. When conducting a current of 1 mA each diode exhibits a voltage drop of 0.7 V, and this voltage changes by 0.1 V per decade of current change. Find v_O, v_E, and v_F corresponding to $v_I = +0.1, +1, +10, -0.1, -1$ and -10 volts.

Ans. $+0.1$ V, $+0.6$ V, -12 V; $+1$ V, $+1.6$ V, -12 V; $+10$ V, $+10.7$ V, -12 V; $+0.1$ V, -12 V, $+0.6$ V; $+1$ V, -12 V, $+1.6$ V; $+10$ V, -12 V, $+10.7$ V

5.8 The block diagram shown in Fig. E5.8a gives another possible arrangement for implementing the absolute-value or full-wave-rectifier operation depicted symbolically in Fig. E5.8b. As shown, the block diagram consists of two boxes: a half-wave rectifier, which can

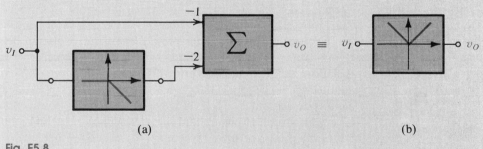

(a)   (b)

Fig. E5.8

be implemented by the circuit in Fig. 5.4a after reversing both diodes, and a weighted inverting summer. Convince yourself that this block diagram does in fact realize the absolute-value operation. Then draw a complete circuit diagram, giving reasonable values for all resistors.

5.4 TRANSFORMER COUPLING OF RECTIFIERS

Figure 5.9 illustrates the use of a transformer in generating the complementary signals required for full-wave rectification as well as in *isolating* the input and output circuits. Polarity marks on the windings indicate the direction of coupling—that is, the fact that all marked terminals will be polarized (relative to the corresponding unmarked terminals) in the same way at any moment. As Fig. 5.9 illustrates, the winding

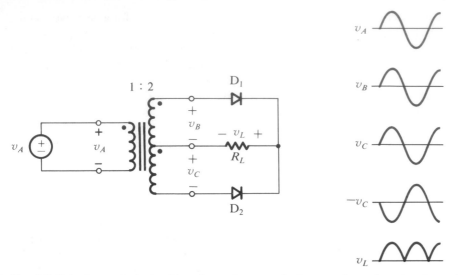

Fig. 5.9 Use of a center-tapped transformer to generate the complementary signals required for full-wave rectification.

providing v_B and v_C is continuous with a connection, at its midpoint, called a *center tap*.

Waveforms accompanying Fig. 5.9 illustrate the principle directly. Taking the center tap as a conceptual reference (in view of the symmetry of the circuit around it) the anode of D_1 will be positive when the anode of D_2 is negative, and vice versa. Thus diodes D_1 and D_2 will conduct alternately to provide full-wave operation.

Note that the transformer provides complete isolation between the input v_A and the output v_L by virtue of magnetic-flux coupling of otherwise separate windings. Isolation is a very important feature in power-supply systems, particularly in the context of connection to the power line. It prevents undesirable current flow between circuits so separated. Thus its use in a consumer product such as a stereo receiver is very important to ensure that the user, while connecting loudspeakers, for example, be totally protected from the possibility of shock from power-line voltages.[1]

Figure 5.10 illustrates, however, that an isolating transformer is not really essential for the full-wave function; only a center-tapped, well-coupled coil is required.

Exercise

5.9 Consider the full-wave rectifier circuit of Fig. 5.9 and assume the diodes to be ideal. If the input is a sinusoid with a 100-V peak amplitude and $R_L = 1 \text{ k}\Omega$, calculate
(a) The peak current in each diode.
(b) The peak reverse voltage appearing across the off diode.
(c) The average voltage across the load.
(d) The peak amplitude of the sinusoidal current supplied by the source.
(e) The power supplied by the source.

Ans: 100 mA; 200 V; 63.7 V; 100 mA; 5 W

---

[1] Here we are referring to the need for a power transformer, *not* for a transformer to couple the amplifier to the loudspeakers.

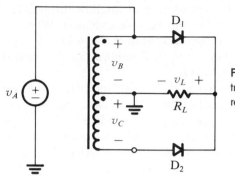

Fig. 5.10 If isolation is not required, the transformer in the circuit of Fig. 5.9 can be replaced by a center-tapped coil.

5.5 THE BRIDGE RECTIFIER

While the center-tapped transformer provides both isolation and a relatively simple circuit implementation of a full-wave rectifier, the effort required to tap the transformer is relatively great in practice. Moreover, each half of the secondary winding is used only half the time. Thus to provide a load voltage of V volts peak, the secondary must be capable of providing $2V$ volts total.

However, there is an alternative: At the modest expense of two extra diodes (and their voltage drops) and under the condition that the source and load do not share an essential common connection, a solution exists in what is known as the *diode-bridge configuration*. To emphasize the fact that the source and load should not have a common terminal, the circuit is first shown in Fig. 5.11a using a transformer. Note,

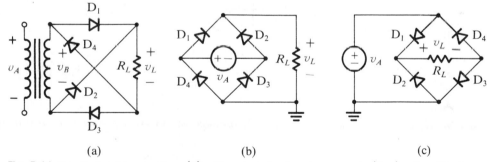

(a)   (b)   (c)

Fig. 5.11 The diode bridge rectifier **(a)** with an isolation transformer and **(b, c)** without the transformer. In (b) the source is floating and the load is grounded. In (c) the source is grounded and the load is floating.

however, that the secondary of this transformer is quite simple, having no tap and a voltage only slightly greater (to account for the one additional diode drop) than half that of the secondary in the circuit of Fig. 5.9. It should be obvious that one cannot have a common ground between the secondary winding and the load resistance, since this would result in shorting one of the two diodes D_1 and D_3. Of course with the transformer included, a common ground can exist between the source (primary winding) and the load (R_L).

Let us examine the operation: When v_B is positive, current will flow through D_1, R_L, and D_3, while at the same time D_2 and D_4 will be reverse-biased. When v_B reverses polarity and becomes negative, conduction will result through D_2, R_L, and D_4, with D_1 and D_3 reversed. Note that the current in R_L is unidirectional, supplied by either D_1 or D_2 and removed by either D_3 or D_4.

It is useful to reflect that there are many ways of looking at, and of drawing, the diode bridge. Two of these, emphasizing the traditional bridge shape, are shown in Figs. 5.11b and c with component labels corresponding to those in Fig. 5.11a. For simplicity and to imply other possibilities, the transformer has been eliminated and the source and load are, of course, assumed to share no common terminals. Figure 5.11b is most often used when the source is truly floating, isolated, and symmetrical, in which case the lower end of R_L may be grounded. Figure 5.11c is used when the source is asymmetric and ground-referenced while the load is floating. A classic example of the latter case is when the load is a dc meter movement and the source is an ac signal in a ground-referenced signal-processing or instrumentation system.

A Precision Bridge Rectifier for Instrumentation Applications

From Fig. 5.11c it should be apparent that the load current comes directly from, and is replicated in, the ac signal source. Consider next the circuit of Fig. 5.12a. Here the bridge load is assumed to be a moving-coil meter M, whose reading is proportional to the average value of the current flowing through it. We have also placed a unity-gain follower at the input. This follower will provide the circuit with a high input resistance while causing a replica of the input signal to appear at a low impedance level at

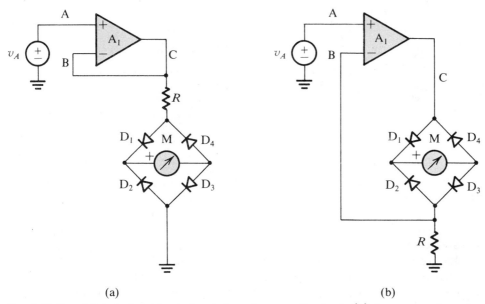

(a)   (b)

Fig. 5.12 Use of the diode bridge in the design of an ac voltmeter. In **(a)** the op amp is used as a buffer, and the resistance R determines the meter sensitivity. In **(b)** the bridge is included in the feedback loop, resulting in a precision instrument.

point C. Now comes the important point: In order to control the meter sensitivity we have included a resistance R in series with the bridge. We could, of course, have placed this resistance directly in series with meter M with identical results. However, placing it in series with the bridge emphasizes its role as being in effect a voltage-to-current converter. For a given value of the input voltage v_A the value of R, the resistance of the meter, and the diode voltage drops determine the average meter current and hence the meter reading.

To eliminate the effects of the usually unpredictable meter resistance and diode characteristics, as well as to make the circuit work for small voltages (less than the minimum of 1 V required in the circuit of Fig. 5.12a), a simple reconnection can be made. This is illustrated in Fig. 5.12b, where all that we have done is simply to place the diode bridge inside the negative-feedback loop of the op amp. If we assume that this feedback loop will be closed for both input polarities (which it will, as will be seen shortly), a virtual short circuit appears between the two input terminals of the op amp. This causes an almost perfect replica of the input voltage v_A to appear across resistance R. Thus the current through R will be alternating and directly proportional to the input voltage v_A (assuming an ideal op amp),

$$i_R = \frac{v_A}{R}$$

Now the positive half-cycles of this current will be coming from the op-amp output through diode D_1, meter M, and diode D_3. On the other hand, the negative half-cycles will be going into the op-amp output through diode D_4, the meter, and diode D_2. In both half-cycles the current in meter M will flow in the same direction.

The circuit in Fig. 5.12b thus provides a relatively accurate high-input-impedance ac voltmeter using an inexpensive moving-coil meter. Again, through the intelligent use of negative feedback, all nonidealities have been masked.

Exercises

5.10 In the circuit of Fig. 5.12b find the value of R that would cause the meter to provide a full-scale reading when the input voltage is a sine wave of 5 V rms. Let meter M have a 1-mA, 50-Ω movement (that is, its resistance is 50 Ω and it provides full-scale deflection when the average current through it is 1 mA). What are the approximate maximum and minimum voltages at the op-amp output? Assume that the diodes have constant 0.7-V drops when conducting.

Ans. 4.5 kΩ; +8.55 V; −8.55 V

5.11 Consider the full-wave bridge rectifier of Fig. 5.11c. Assume that the input is a sine wave with a 100-V peak and that the diodes are ideal. What is the maximum magnitude of reverse bias voltage that appears across the off diodes?

Ans. 100 V

5.6 THE PEAK RECTIFIER

While a simple rectifier circuit, whether half or full wave, is capable of supplying dc current or voltage, it does so with the addition of variable or ac components. For instance, an ideal half-wave rectifier fed by a sinusoidal waveform produces the output

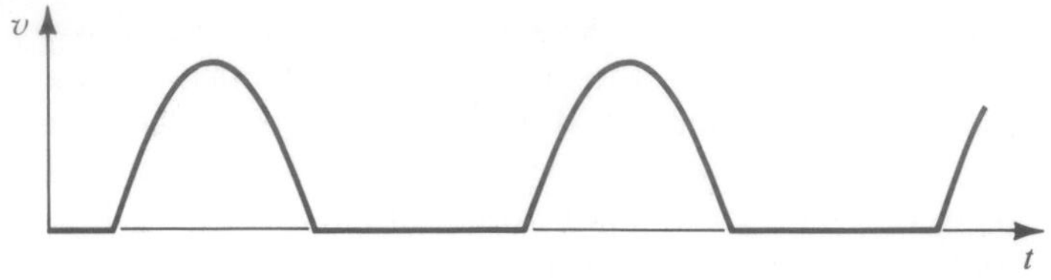

Fig. 5.13 The output waveform of an ideal half-wave rectifier fed by a sinusoid.

waveform shown in Fig. 5.13. This waveform can be decomposed into the sum of various components—a dc one, one at the fundamental frequency of the input sine wave, one at the third harmonic of the input sinusoid, and so on.

In a measurement application using a moving-coil meter, the ac components have no net effect provided that their frequency lies above the frequency of natural resonance of the mechanical meter; when the frequency is sufficiently high the needle cannot follow the pulse-to-pulse variation, and the ac components are in effect "filtered out" by mechanical inertia. In other measurement applications one may require a constant output voltage proportional to the average or dc component. In such a case the ac components may be removed by an electronic filter, as we have done in the circuit of Fig. 5.5. Both of these filtering processes fall into the class of linear filtering. We will deal with the subject of linear signal filtering in greater detail in Chapter 14.

In power-supply applications as well, one desires a pure dc output from the rectifier; hence some form of filtering is needed to remove the ac components. In this section we introduce an alternative approach to filtering that is useful in power-supply and other applications. As will be seen, the scheme is quite distinct from the linear filter circuits just mentioned. The process makes use of the interaction of the basic nonlinear element (diode) and the energy-storage element (capacitor). It will also be seen that from the point of view of power dissipation this is a much more efficient filter, and hence it is ideally suited to power-supply applications. Since the dc output voltage is almost equal to the peak of the input waveform, the circuit is called a *peak rectifier*.

The Basic Circuit

Figure 5.14a shows the basic peak rectifier circuit. Let the input signal be the triangular waveform shown in Fig. 5.14b and let the capacitor be initially discharged. To start with, let us also assume the diode to be ideal. As the input voltage increases above zero (at time t_0), the diode conducts and the capacitor charges up to the instantaneous value of the input voltage. In fact, if the diode is ideal, the voltage across the capacitor will be equal to the input voltage. This will continue until the time the positive peak is reached, time t_1. At $t = t_1$ the capacitor voltage will be equal to the peak of the input, V_p. Now at $t > t_1$ the input decreases, thus reverse-biasing the diode. The capacitor voltage therefore remains constant at the peak value V_p unless the value of a later positive peak of the input signal is larger. If this happens, the diode conducts for the time interval during which $v_I > V_p$ and the capacitor charges up to the new peak value.

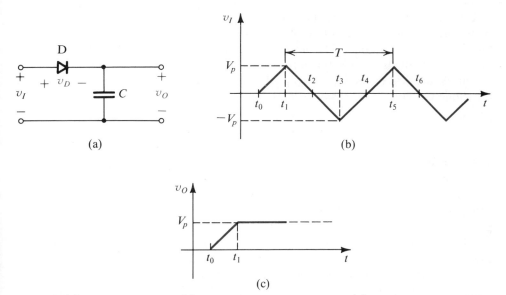

Fig. 5.14 (a) The peak rectifier. **(b)** An input triangular waveform. **(c)** The output that would be obtained if the diode were ideal.

From the above description we see that this circuit detects the positive peak of the input signal and provides an output dc voltage equal to the value of this peak. For this reason the circuit is called a *peak rectifier* or a *peak detector*. It now remains to consider the effect on circuit operation of the characteristics of real diodes and of connecting a load resistance R across the capacitor.

The Nonideal Case with No Load

Although it is not of immense practical interest, it is instructive to examine the operation of the peak-rectifier circuit taking nonideal diode characteristics into account. For simplicity we will still assume that the input signal is of triangular waveform, as shown in Fig. 5.14b. At the outset we note that one can write equations to describe the circuit operation accurately and solve for the output voltage waveform and for the diode current (Problem 5.19). For the time being, however, we wish to qualitatively describe circuit operation.

Assume that the capacitor was initially discharged, and consider the interval t_0 to t_1. As the input rises, the diode begins to conduct with a voltage drop related to current flow. After an initial transient during which the diode voltage exceeds 0.5 or 0.6 V (that is, as the diode current becomes much larger than I_S), it is easy to show that the output will be rising linearly with a slope equal to that of the input. To see that this indeed is the case we will assume that it is true and then verify our assumption. If the output is linearly increasing with time, then the current

$$i = C \frac{dv_O}{dt}$$

will be constant. This means that the diode will have a constant voltage drop v_D. Now since $v_O = v_I - v_D$, it follows that v_O will be increasing linearly with time with the same slope as that of v_I. The picture should be now clear: After an initial transient, the diode conducts a constant current and exhibits a constant voltage drop. The output is identical to the input except that it is lower by a diode drop. The value of the diode current can be easily estimated as

$$i = C\frac{dv_O}{dt} = C\frac{dv_I}{dt} = C\frac{V_p}{T/4}$$

The corresponding diode voltage can be found from the diode characteristics.

As a numerical example, consider $V_p = 10$ V, $T = 40$ ms, $C = 10$ μF. It follows that $i = 10$ mA. Assuming that the diode has a 0.7-V drop at 1 mA and that the drop changes by 0.1 V/decade of current, we see that $v_D = 0.8$ V. Thus at the time of the peak ($t = t_1$) the output voltage will be $10 - 0.8 = 9.2$ V.

Let us now consider what happens beyond the positive peak of the input. As the input voltage reverses slope and begins to decrease, the voltage drop across the diode will begin to decrease, causing the diode current to decrease. Nevertheless, current continues to flow beyond the peak point, and thus the capacitor voltage continues to increase as shown in Fig. 5.15. Shortly after the peak, however, the diode current will decrease to zero (that is, the diode cuts off), and the output voltage remains constant at a value slightly greater than 9.2 V (say, 9.3 V). The exact value reached by the output voltage can be found by solving the circuit equations using the exponential i-v model of the diode (Problem 5.19).

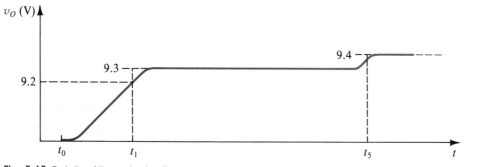

Fig. 5.15 Details of the output voltage waveform as a triangular wave with a 10-V peak is applied to the peak rectifier and the diode characteristics are taken into account.

Now during the interval t_1 to nearly t_5 the output remains constant at 9.3 V, since the diode is off and no current whatsoever can flow. Then when the input reaches 9.3 V again, the diode enters the forward-bias region, with the ultimate possibility of having as much as 0.7 V of forward bias. The result is that small currents will flow and the capacitor will charge up slightly more (say, to 9.4 V) before the diode is once again reversed. It is probably obvious that after a few cycles of this process the output will reach nearly 10 V. Thus in the steady state that is reached, the diode does not conduct and the output remains constant (pure dc) at a value equal to the peak of the input waveform.

The Peak Rectifier with Load

The no-load situation considered above is an idealization that cannot exist in practice; because of leakage effects the capacitor will always discharge slightly even if no actual load resistance is connected across it. Furthermore, in actual practical situations, such as the design of a power supply, a load resistance will be inherently present. In other applications of the peak rectifier, a load resistance may be essential in order to allow the capacitor to discharge in the event that the input positive peak decreases in value. Therefore we will now consider the operation of the peak rectifier with a load resistance R connected across the capacitor, as shown in Fig. 5.16a.

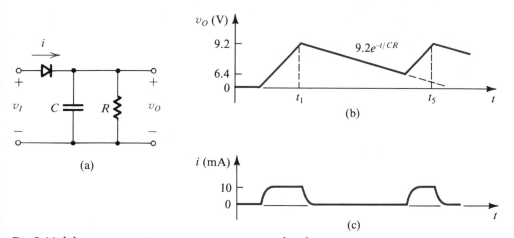

Fig. 5.16 (a) The peak rectifier with a load resistance.   **(b, c)** The output voltage and diode current for the case $C = 10 \, \mu F$, $R = 10 \, k\Omega$, and the input a triangular wave with a 10-V and 40-ms period.

At the very peak of the input, while the diode conducts, the load current, which is approximately equal to V_p/R, will be supplied directly by the input. However, between peaks—for example, in the interval t_1 to t_5—charge from the capacitor must supply the load current. This load current in turn will discharge the capacitor, lowering the output voltage in an exponential fashion (with time constant CR) for the greater part of the interval t_1 to t_5. At the point where the falling capacitor voltage intersects the rising edge of the input, the diode begins to conduct once more. Again conduction continues until a point just past the peak of the input. This process continues until a steady state is reached. In a steady state the diode conducts for a short interval near the peak to replenish the charge lost by the capacitor during the interval in which the diode is off (most of the peak-to-peak interval). Figures 5.16b and c show the waveforms of the output voltage and diode current for the same numerical example considered before but with a load resistance $R = 10 \, k\Omega$. Note that the output is no longer a dc voltage but may be considered as composed of a dc voltage on which an ac component, called *ripple*, is superimposed. We shall next show how the ripple voltage can be calculated in the important case of a peak rectifier fed with a sinusoidal input.

Calculation of the Ripple Voltage and Diode Current

Figure 5.17b shows the steady-state input and output voltage waveforms of the rectifier circuit of Fig. 5.17a under the assumption that $CR \gg T$, where T is the period of the input sinusoid. The waveforms of the load current

$$i_L = v_O/R_L \tag{5.1}$$

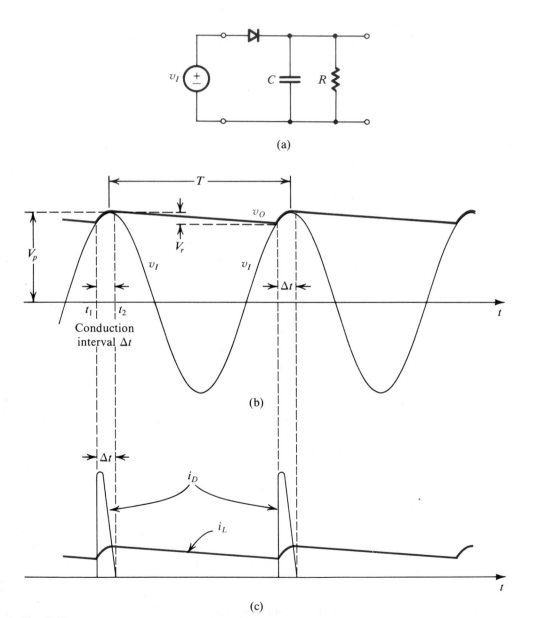

(a)

(b)

(c)

Fig. 5.17 Voltage and current waveforms in the peak rectifier circuit with $CR \gg T$.

and of the diode current

$$i_D = i_C + i_L \tag{5.2}$$

$$= C\frac{dv_I}{dt} + i_L \tag{5.3}$$

are shown in Fig. 5.17c. The following observations are in order:

1. The diode conducts for a brief interval, Δt, near the peak of the input sinusoid and supplies the capacitor with charge equal to that lost during the much longer discharge interval. The latter is approximately equal to the period T.

2. Assuming an ideal diode, the diode conduction begins at time t_1, at which the input v_I equals the exponentially decaying output v_O. Conduction stops at t_2 shortly after the peak of v_I; the exact value of t_2 can be determined by setting $i_D = 0$ in Eq. (5.2).

3. During the diode off interval the capacitor C discharges through R and thus v_O decays exponentially with a time constant CR. The discharge interval begins almost at the peak of v_I. At the end of the discharge interval, which lasts for almost the entire period T, $v_O = V_p - V_r$, where V_r is the peak-to-peak ripple voltage. When $CR \gg T$ the value of V_r is small.

4. When V_r is small, v_O is almost constant and equal to the peak value of v_I. Thus the dc output voltage is approximately equal to V_p. Similarly, the current i_L is almost constant and its dc component I_L is given by

$$I_L = \frac{V_p}{R} \tag{5.4}$$

A more accurate expression for the output dc voltage can be obtained by taking the average of the extreme values of v_O,

$$V_O = V_p - \tfrac{1}{2}V_r \tag{5.5}$$

With these observations in hand we now derive expressions for V_r and for the average and peak values of the diode current. During the diode off interval, v_O can be expressed as

$$v_O = V_p e^{-t/CR}$$

At the end of the discharge interval we have

$$V_p - V_r \simeq V_p e^{-T/CR}$$

Now since, $CR \gg T$, we can use the approximation $e^{-T/CR} \simeq 1 - T/CR$ to obtain

$$V_r \simeq V_p \frac{T}{CR} \tag{5.6}$$

We observe that to keep V_r small we must select a capacitance C so that $CR \gg T$. The ripple voltage V_r in Eq. (5.6) can be expressed in terms of the frequency

$f = 1/T$ as

$$V_r = \frac{V_p}{fCR} \tag{5.7}$$

Note that an alternative interpretation of the approximation made above is that the capacitor discharges by means of a constant current $I_L = V_p/R$. This approximation is valid as long as $V_r \ll V_p$.

Using Fig. 5.17b and assuming that diode conduction ceases almost at the peak of v_I, we can determine the conduction interval Δt from

$$V_p \cos(\omega \, \Delta t) = V_p - V_r$$

where $\omega = 2\pi f = 2\pi/T$ is the angular frequency of v_I. Since $(\omega \, \Delta t)$ is a small angle we can employ the approximation $\cos(\omega \, \Delta t) \simeq 1 - \frac{1}{2}(\omega \, \Delta t)^2$ to obtain

$$\omega \, \Delta t \simeq \sqrt{2V_r/V_p} \tag{5.8}$$

We note that when $V_r \ll V_p$, the conduction angle $\omega \, \Delta t$ will be small, as assumed.

To determine the average diode current during conduction, i_{Dav}, we equate the charge that the diode supplies the capacitor,

$$Q_{\text{supplied}} = i_{Cav} \, \Delta t$$

to the charge that the capacitor loses during the discharge interval,

$$Q_{\text{lost}} = CV_r$$

to obtain

$$i_{Dav} = I_L(1 + \pi\sqrt{2V_p/V_r}) \tag{5.9}$$

In deriving this expression we made use of Eq. (5.2) and assumed that i_{Lav} is given by Eq. (5.4). We also used Eqs. (5.7) and (5.8). Observe that when $V_r \ll V_p$, the average diode current during conduction is much greater than the dc load current. This is not surprising since the diode conducts for a very short interval and must replenish the charge lost by the capacitor during the much longer interval in which it is discharged by I_L.

The peak value of the diode current, i_{Dmax}, can be determined by evaluating the expression in Eq. (5.3) at the onset of diode conduction—that is, at $t = t_1 = -\Delta t$ (where $t = 0$ is at the peak). Assuming that i_L is almost constant at the value given by Eq. (5.4), we obtain

$$i_{Dmax} = I_L(1 + 2\pi\sqrt{2V_p/V_r}) \tag{5.10}$$

From Eqs. (5.9) and (5.10) we see that for $V_r \ll V_p$, $i_{Dmax} \simeq 2i_{Dav}$, which correlates with the fact that the waveform of i_D is almost a right-angle triangle (see Fig. 5.17c).

EXAMPLE 5.1

Consider a peak rectifier fed by a 60-Hz sinusoid having a peak value $V_p = 100$ V. Let the load resistance $R = 10$ kΩ. Find the value of the capacitance C that will result in a peak-to-peak ripple of 2 V. Also calculate the fraction of the cycle during which the diode is conducting and the average and peak values of the diode current.

Solution

From Eq. (5.7) we obtain the value of C as

$$C = \frac{V_p}{V_r f R} = \frac{100}{2 \times 60 \times 10 \times 10^3} = 83.3 \ \mu\text{F}$$

The conduction angle $\omega \, \Delta t$ is found from Eq. (5.8) as

$$\omega \, \Delta t = \sqrt{2 \times 2/100} = 0.2 \ \text{rad}$$

Thus the diode conducts for $(0.2/2\pi) \times 100 = 3.18\%$ of the cycle. The average diode current is obtained from Eq. (5.9), where $I_L = 100/10 = 10$ mA,

$$i_{Dav} = 10(1 + \pi\sqrt{2 \times 100/2}) = 324 \ \text{mA}$$

The peak diode current is found using Eq. (5.10),

$$i_{Dmax} = 10(1 + 2\pi\sqrt{2 \times 100/2}) = 638 \ \text{mA}$$

Full-Wave Peak Rectifiers

The full-wave rectifier circuits of Figs. 5.9, 5.10, and 5.11 can be converted to peak rectifiers by including a capacitor across the load resistor. As in the half-wave case the output dc voltage will be almost equal to the peak value of the input sinc wave (see Fig. 5.18). The ripple frequency, however, will be twice that of the input. The peak-to-peak ripple voltage, for this case, can be derived using a procedure identical to that above but with the discharge period T replaced by $T/2$, resulting in

$$V_r = \frac{V_p}{2fCR} \tag{5.11}$$

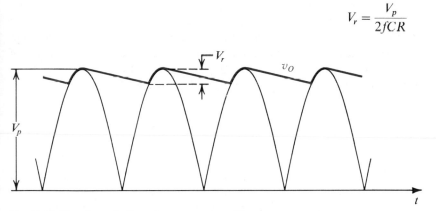

Fig. 5.18 Waveforms in the full-wave peak rectifier.

While the diode conduction interval, Δt, will still be given by Eq. (5.8), the average and peak currents in each of the diodes will be given by

$$i_{Dav} = I_L(1 + \pi\sqrt{V_p/2V_r}) \tag{5.12}$$

$$i_{Dmax} = I_L(1 + 2\pi\sqrt{V_p/2V_r}) \tag{5.13}$$

Comparing these expressions with the corresponding ones for the half-wave case we note that for the same values of V_p, f, R, and V_r (and thus the same I_L) we need a capacitor half the size of that required in the half-wave rectifier. Also the current in each diode in the full-wave rectifier is approximately half that which flows in the diode of the half-wave circuit.

Precision Peak Rectifiers

Including the diode of the peak rectifier inside the negative-feedback loop of an op amp, as shown in Fig. 5.19, results in a precision peak rectifier. The diode–op-amp combination will be recognized as the superdiode presented in Section 5.2 (Fig. 5.3).

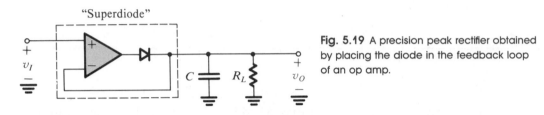

Fig. **5.19** A precision peak rectifier obtained by placing the diode in the feedback loop of an op amp.

Operation of the circuit in Fig. 5.19 is quite straightforward. For v_I greater than the output voltage the op amp will drive the diode on, thus closing the negative-feedback path and causing the op amp to act as a follower. The output voltage will therefore follow that of the input, with the op amp supplying the capacitor-charging current. This process continues until the input reaches its peak value. Beyond the positive peak the op amp will see a negative voltage between its input terminals. Thus its output will go negative to the saturation level and the diode will turn off. Except for possible discharge through the load resistance, the capacitor will retain a voltage equal to the positive peak of the input. Inclusion of a load resistance is essential if the circuit is required to detect reductions in the magnitude of the positive peak, as in the case of the amplitude-modulation (AM) detector discussed below.

A Buffered Precision Peak Detector

When the peak detector is required to hold the value of the peak for long times, the capacitor should be buffered, as shown in the circuit of Fig. 5.20. Here op amp A_2, which should have high input impedance and low input bias current, is connected as a voltage follower. The remainder of the circuit is quite similar to the half-wave-rectifier circuit of Figure 5.4. While diode D_1 is the essential diode for the peak rectification operation, diode D_2 acts as a catching diode to prevent negative saturation, and the associated delays, of op amp A_1. During the holding state, follower A_2 supplies D_2 with a small current through R. The output of op amp A_1 will then be clamped at one diode drop below the input voltage. Now if the input v_I increases above the value stored on C, which is equal to the output voltage v_O, op amp A_1 sees a net positive input which drives its output toward the positive saturation level, and turns diode D_2 off. Diode D_1 is then turned on and capacitor C is charged to

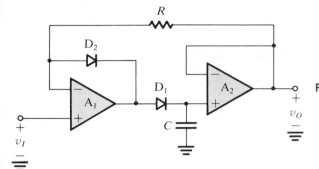

Fig. 5.20 A buffered precision peak rectifier.

the new positive peak of the input after which time the circuit returns to the holding state.

Finally, it should be noted that this circuit has a low-impedance output.

The Peak Detector as an AM Demodulator

The peak rectifier finds a great many applications, beyond the simple supply of power, in the areas of signal detection and processing. For example, the information contained in an amplitude-modulated (AM) signal can be extracted or *detected* using a diode peak rectifier. This is possible because of the ability of the diode peak rectifier to follow the peak value of the input signal from cycle to cycle. For this to happen, however, the time constant CR must be carefully chosen.

Figure 5.21 illustrates the process. Note that the output will consist of the modulating signal with some superimposed ripple at the carrier frequency. This high-frequency ripple can be removed by following the AM detector with a simple RC filter.

Obviously we have to select the detector time constant CR sufficiently small to be able to track every peak of the input, even at the steepest point on the envelope of the AM wave. On the other hand, selecting CR too small would result in an excessive amount of ripple on the detected output.

Let us consider quantitatively the question of selecting an appropriate value for the detector time constant. The AM waveform can be described by (see Chapter 1)

$$v_{AM} = V(1 + m \sin \omega_m t) \sin \omega_c t$$

where $V(1 + m \sin \omega_m t)$ represents the envelope. This envelope has a slope of $m\omega_m V \cos \omega_m t$, which has a maximum magnitude of $m\omega_m V$. This occurs at point A in Fig. 5.21a, where the value of the AM waveform is V. Now note that this envelope slope is proportional to the modulation depth m and the modulating frequency ω_m. Thus the highest possible slope occurs when $m = 1$ and $\omega_m = \hat{\omega}_m$ (where the caret "^" denotes maximum). For detection to be perfect, as measured by the ability to follow the most rapid variation of the input presented, the slope of the detector decay curve must exceed this. That is,

$$\frac{V}{CR} > \hat{\omega}_m V$$

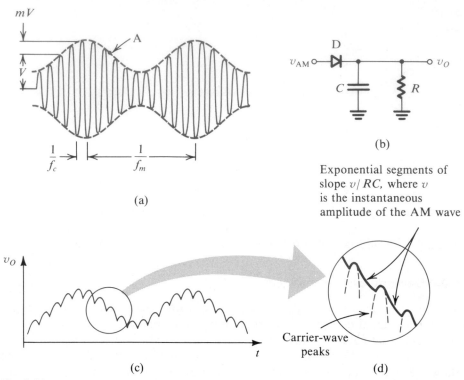

Exponential segments of slope v/RC, where v is the instantaneous amplitude of the AM wave

Carrier-wave peaks

(a)   (b)   (c)   (d)

Fig. 5.21 Use of the peak rectifier to detect the envelope of an amplitude-modulated wave.

which results in

$$CR < \frac{1}{\hat{\omega}_m} \qquad (5.14)$$

EXAMPLE 5.2

The highest-frequency signal components transmitted on the AM broadcast band are limited to 5 kHz. Design an AM detector to supply a resistive load of 10 kΩ, assuming that an ideal diode is available.

Solution

The appropriate circuit is the one shown in Fig. 5.21b, where D is an ideal diode and $R = 10$ kΩ. The challenge is to choose a value for the capacitance C. Clearly this follows directly from the previous result. At the limit,

$$C = \frac{1}{\hat{\omega}_m R}$$

$$= \frac{1}{2\pi \times 10^4 \times 5 \times 10^3} = 3{,}200 \text{ pF}$$

If the receiver cost and size are important, a somewhat smaller capacitor may be used.

Exercises

5.12 If in the rectifier circuit of Example 5.1 the output direct current I_L is doubled (by halving the value of R), find the value of C that will maintain the ripple voltage unchanged. Also, find the new values of average and peak diode currents.

Ans.   186.6 μF; 648 mA; 1.276 A

5.13 Repeat Example 5.1 with the half-wave peak rectifier replaced with a full-wave bridge rectifier with a capacitor C across the load resistor R. Find the value of C, the fraction of a cycle during which the diodes conduct, as well as i_{Dav} and i_{Dmax}.

Ans.   41.7 μF; 6.36%; 167.1 mA; 324.2 mA

5.14 In the circuit of Fig. 5.19 let $C = 1\ \mu F$ and $R_L = \infty$, and assume that the op amp is specified to have a slew rate of 0.1 V/μs and a maximum output current of 10 mA. What is the maximum rate of change of output voltage?

Ans.   10 V/ms

5.7 THE CLAMPED CAPACITOR OR DC RESTORER

If in the basic peak rectifier circuit the output is taken across the diode rather than across the capacitor, an interesting circuit with important applications results. The circuit, called a dc restorer, is shown in Fig. 5.22 fed with a square wave. Because of the polarity in which the diode is connected, the capacitor will charge to a voltage v_C (see Fig. 5.22) equal to the magnitude of the most negative peak of the input signal.

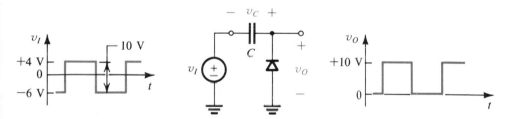

Fig. 5.22 The clamped capacitor or dc restorer with a square-wave input and no load.

Subsequently, the diode turns off and the capacitor retains its voltage indefinitely. If, for instance, the input square wave has the arbitrary levels -6 V and $+4$ V, then v_C will be equal to 6 V. Now, since the output voltage v_O is given by

$$v_O = v_I + v_C$$

it follows that the output waveform will be identical to that of the input, except that it is shifted upwards by v_C volts. In our example the output will be a square wave with levels of 0 V and $+10$ V.

Another way of visualizing the operation of the circuit in Fig. 5.22 is to note that because the diode is connected across the output with the polarity shown, it prevents the output voltage from going below 0 V (by conducting and charging up the capacitor, thus causing the output to rise to 0 V), but this connection will not constrain the positive excursion of v_O. The output waveform will therefore have its lowest peak *clamped* to 0 V, which is why the circuit is called a *clamped capacitor*. It should be obvious that reversing the diode polarity will provide an output waveform whose higest peak is clamped to 0 V. In either case the output waveform will have a finite average value or dc component. This dc component is entirely unrelated to the average value of the input waveform. As an application, consider a pulse signal being transmitted through a capacitively coupled or ac-coupled system. The capacitive coupling will cause the pulse train to lose whatever dc component it originally had. Feeding the resulting pulse waveform to a clamping circuit provides it with a well-determined dc component; a process known as *dc restoration*. This is why the circuit is also called a *dc restorer*.

Restoring dc is useful because the dc component of a pulse waveform is an effective measure of its duty cycle. The duty cycle of a pulse waveform can be modulated (in a process called pulsewidth modulation) and made to carry information. In such a system, detection or demodulation could be achieved simply by feeding the received pulse waveform to a dc restorer and then using a simple RC low-pass filter to separate the average of the output waveform from the superimposed pulses.

Effect of Load Resistance

When a load resistance R is connected across the diode in a clamping circuit, as shown in Fig. 5.23, the situation changes significantly. While the output is above ground, a net dc current must flow in R. Since at this time the diode is off, this current obviously comes from the capacitor, thus causing the capacitor to discharge

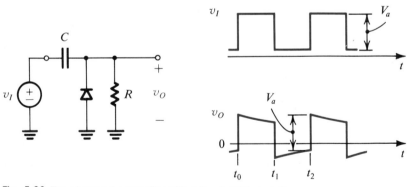

Fig. 5.23 The clamped capacitor with a load resistance R.

and the output voltage to fall. This is shown in Fig. 5.23 for a square-wave input. During the interval t_0 to t_1 the output voltage falls exponentially with time constant CR. At t_1 the input decreases by V_a volts and the output attempts to follow. This causes the diode to conduct heavily and to quickly charge the capacitor. At the end of the interval t_1 to t_2 the output voltage would normally be a few tenths of a volt negative (say, -0.5 V). Then as the input rises by V_a volts (at t_2) the output follows, and the cycle repeats itself. In a steady state the charge lost by the capacitor during the interval t_0 to t_1 is recovered during the interval t_1 to t_2. This charge equilibrium enables us to calculate the average diode current as well as the details of the output waveform.

A Precision Clamping Circuit

By replacement of the diode in the clamping circuit of Fig. 5.22 by a "superdiode," the precision clamp of Fig. 5.24 is obtained. Operation of this circuit should be self-explanatory.

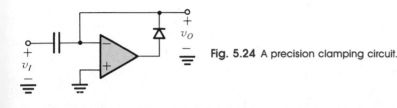

Fig. 5.24 A precision clamping circuit.

Voltage Doubler

Figure 5.25a shows a circuit composed of two sections in cascade: a clamp formed by C_1 and D_1, and a peak rectifier formed by D_2 and C_2. When excited by a sinusoid of amplitude V_p the clamping section provides the voltage waveform shown in

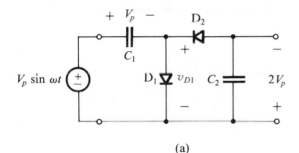

(a)

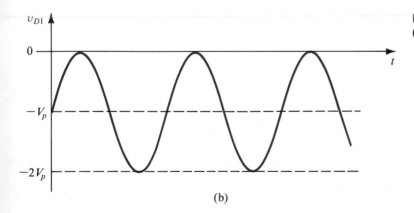

(b)

Fig. 5.25 Voltage doubler: (a) circuit; (b) waveform of the voltage across D_1.

Fig. 5.25b. Note that while the positive peaks are clamped to zero volts, the negative peak reaches $-2V_p$. In response to this waveform, the peak-detector section provides across capacitor C_2 a negative dc voltage of magnitude $2V_p$. Because the output voltage is double the input peak, the circuit is known as a voltage doubler. The technique can be extended to provide output dc voltages that are higher multiples of V_p.

<table>
<tr>
<td>Exercise</td>
<td>

5.15 Consider a precision clamping circuit such as the one in Fig. 5.24 but with the diode reversed. Let the input be a sinusoid with zero average value, 5 V peak, and 1 kHz frequency. If the output is fed to an RC low-pass filter with a corner frequency of 10 Hz, find the value of the dc component and the amplitude of the sinusoidal component at the output of the filter.

Ans. -5 V; 0.05 V

</td>
</tr>
</table>

5.8 LIMITERS AND COMPARATORS

We have studied rectifiers in some detail and now we shall consider other important nonlinear functional blocks or operators. Specifically, in this section we introduce two important nonlinear functions: the limiter operator and the comparator operator. Although both operations can be realized using diodes (including zener diodes) and resistors, to improve performance, add precision, and in fact simplify the design, op amps are usually used. Some circuit realizations will be discussed in Sections 5.9 and 5.10. Limiters and comparators are used in a variety of signal-processing applications, including analog computation and signal generation.

The Limiter Operator

The general transfer characteristic of the limiter operator is shown in Fig. 5.26. As indicated, the limiter acts as an amplifier of gain K, which can be either positive or negative, for inputs in a certain range; $L_-/K \le v_I \le L_+/K$. If v_I exceeds the upper *threshold* (L_+/K), the output voltage is *limited* or clamped to the upper limiting level

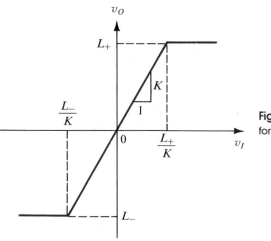

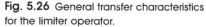

Fig. 5.26 General transfer characteristics for the limiter operator.

L_+. On the other hand, if v_I is reduced below the lower limiting threshold (L_-/K), the output voltage v_O is limited to the lower limiting level L_-.

The general transfer characteristic of Fig. 5.26 describes a *double limiter*—that is, a limiter that works on both the positive and negative peaks of an input waveform. *Single limiters*, of course, exist. Finally, note that if an input waveform such as that shown in Fig. 5.27 is fed to a double limiter its two peaks will be *clipped off*. Limiters therefore are sometimes referred to as *clippers*.

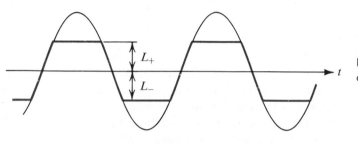

Fig. 5.27 Applying a sine wave to a limiter can result in clipping off its two peaks.

The limiter whose characteristics are depicted in Fig. 5.26 is described as a *hard limiter. Soft limiting* is characterized by smoother transitions between the linear region and the saturation regions and a slope greater than zero in the saturation regions, as illustrated in Fig. 5.28. Depending on the application, either hard or soft limiting may be preferred.

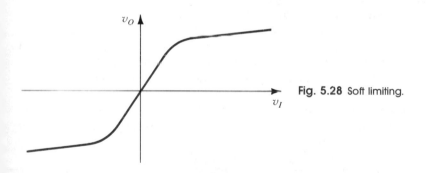

Fig. 5.28 Soft limiting.

The Comparator Operator

Figure 5.29 shows a block-diagram representation and transfer characteristic for the comparator operator. As indicated, the comparator compares the input v_I with a reference level V_R that constitutes the *comparator threshold*. If v_I is greater than the reference V_R, the comparator provides a high output L_+. Alternatively, for $v_I < V_R$ the comparator output will be the low level L_-.

Although special integrated circuits are manufactured to realize the comparator function, there are many instances where the comparator operator is more conveniently implemented using a general-purpose op amp. We will discuss some such arrangements in Section 5.9.

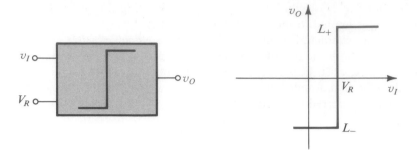

Fig. 5.29 Block-diagram representation and transfer characteristics for a comparator.

The comparator is a very useful circuit block. It finds application in such obvious operations as detecting the zero crossings of an arbitrary signal waveform and converting sine waves into square waves, as well as in more involved situations.

As an illustration of an important point that we wish to make, consider an application where it is required to count the instances of time at which a given arbitrary waveform crosses zero. This process can be implemented using a comparator that provides a step output every time a zero crossing occurs. These step changes at the output of the comparator can then be used to *trigger* a *monostable* (*one-shot*) circuit. This is a circuit that produces an output pulse of specified duration and height every time it is triggered. We will discuss monostables in Section 5.11. The standardized output pulses of the monostable can be then fed to a *counter*, which provides us with the required total number of zero crossings.

Imagine now what happens if the signal being processed has, as it usually does have, interference superimposed on it, say of a frequency much higher than that of the signal. It follows that the signal might cross the zero axis a number of times around each of the zero-crossing points we are trying to detect, as shown in Fig. 5.30. Correspondingly the comparator would change state a number of times at each of the zero crossings and our count would obviously be in error. If we have an idea of the expected peak-to-peak amplitude of the interference, the problem can be solved by introducing *hysteresis* in the comparator characteristics. Basically, this means that instead of having one threshold the comparator will have two thresholds, a low threshold V_{TL} and a high threshold V_{TH}. The two thresholds would be placed on either side of the reference level V_R and would differ by an amount of, say, 50 or 100 mV, depending on the magnitude of interference we wish to reject. This difference, $V_{TH} - V_{TL}$, is called the magnitude of hysteresis. Now if the input signal is increasing in magnitude, the comparator with hysteresis will remain in the low state until the input level exceeds the high threshold V_{TH}. Subsequently the comparator will remain in the high state even if, owing to interference, the signal decreases below V_{TH}. The comparator will switch to the low state only if the input signal is decreased below the low threshold V_{TL}. The situation is illustrated in Fig. 5.30, from which we see that including hysteresis in the comparator characteristics provides an effective means for rejecting interference (thus providing another form of filtering).

Figure 5.31 shows the comparator characteristic with hysteresis included. The arrows on the transfer characteristic illustrate the operation, as described above.

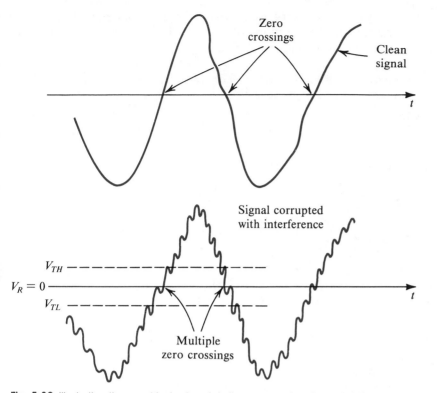

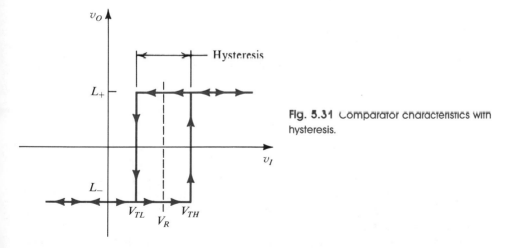

Fig. 5.30 Illustrating the need for hysteresis in the comparator characteristics as a means of rejecting interference.

Fig. 5.31 Comparator characteristics with hysteresis.

Comparators with hysteresis are very useful circuits because they provide a *memory* function, as will be explained in Section 5.10. Finally, it should be mentioned that the comparator with hysteresis is also known as a *Schmitt trigger*.

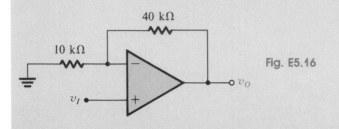
5.9 COMPARATOR AND LIMITER CIRCUITS

We will now present several circuit realizations of the two nonlinear operators introduced in Section 5.8.

Basic Circuits

First we note that the half-wave rectifier circuits of Sections 5.1 and 5.2 can be considered as limiters. The diode clamp or clipper obtained by taking the output in a half-wave rectifier circuit across the diode is also a limiter. Figure 5.32 shows a number of rudimentary realizations of the limiter operator, all based on the principle of the simple diode clipper. In each part of the figure both the circuit and a sketch of its transfer characteristic are given.

The circuits of Figs. 5.32a and b are identical except for the reversal of diode polarity. That in Fig. 5.32c uses two opposite-polarity diodes in parallel to provide double limiting. Thus feeding a sinusoid to the circuit of Fig. 5.32c results in a crude

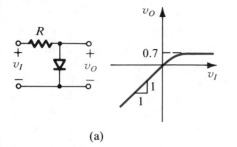

(a)

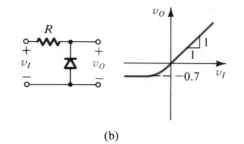

(b)

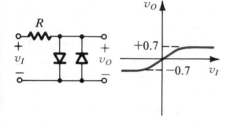

(c)

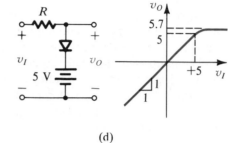

(d)

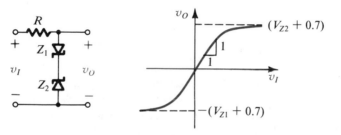

(e)

Fig. 5.32 A variety of basic limiting circuits.

approximation of a square wave, with about 1.4-V peak-to-peak amplitude. The limiting thresholds and levels can be controlled by using strings of diodes and/or by biasing the diode. The latter idea is illustrated in Fig. 5.32d. Finally, rather than strings of diodes we may use two zener diodes in series, as shown in Fig. 5.32e. In this circuit limiting occurs in the positive direction at a voltage of $V_{Z2} + 0.7$, where 0.7 V represents the voltage drop across zener diode Z_1 when conducting in the *forward* direction. For negative inputs, Z_1 acts as a zener, while Z_2 conducts in the forward direction. It should be mentioned that pairs of zener diodes connected in series are available commercially for applications of this type under the name *double-anode zener*.

Circuits Using Op Amps

All of the circuits in Fig. 5.32 suffer from the lack of flexibility in independently setting the linear gain, limiting thresholds, and limiting levels. In addition, all of them suffer from the essential inaccuracies that result from the nonideal characteristics of the diodes involved. Moreover, we have not yet been able to implement either a limiter with gain or a comparator. Better results can be obtained by combining diodes with op amps. As a first example, consider placing a zener diode in the negative-feedback path of an op amp, as shown in Fig. 5.33a. As v_I goes sufficiently positive to supply the zener with a current v_I/R greater than the knee current, the zener will enter the breakdown region and the output voltage will be limited to $-V_Z$. On the other hand, for negative v_I the zener functions as a forward-biased diode, limiting the output voltage to one diode drop or approximately $+0.7$ V. Thus it can be seen that the circuit functions as a comparator with the threshold set at approximately 0 V. This threshold, however, can be changed by biasing the circuit as shown in Fig. 5.33b.

The double-anode zener can be placed in the op-amp negative-feedback path as shown in Fig. 5.33c, thus providing a comparator with two symmetrical output levels and with a threshold controlled by the bias voltage.

Finally, consider the circuit of Fig. 5.33d. Here we have simply added a feedback resistance R_2 in parallel with the double-anode zener. The result of this simple change, though, is rather dramatic. The circuit becomes a limiter rather than a comparator. It is left to the reader to demonstrate that the transfer characteristic will be that shown in Fig. 5.33d.

A Comparator/Limiter Circuit with Adjustable Levels

As a last example, consider the circuit in Fig. 5.34a. This is a useful and practical circuit that functions as a comparator if the feedback resistance R_f is not included and as a limiter when R_f is included. In the following we will explain the circuit operation in the comparator configuration.

Assuming ideal diodes, the transfer characteristic of the circuit in Fig. 5.34a without the resistance R_f will take the shape shown in Fig. 5.34b. The circuit operates as follows: If v_I goes positive, diode D_1 conducts and clamps point A at zero voltage (assuming, for simplicity, an ideal diode). If the current through D_1 as it starts to conduct is negligibly small, the current through R_3 will be equal to that supplied from $+V$ through R_2. Thus

$$\frac{V - 0}{R_2} = \frac{0 - v_O}{R_3}$$

$$v_O = -V \frac{R_3}{R_2}$$

Since v_O is negative, diode D_2 will be off. As v_I increases in the positive direction diode D_1 conducts more and more current. Nevertheless, the voltage at point A remains at about one diode drop below ground. Thus the current through R_2 remains almost constant, and hence R_2 will have little effect on the operation of the circuit in this

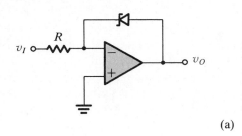
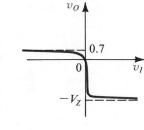

(a)

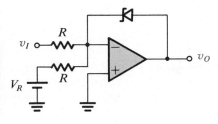
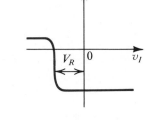

(b)

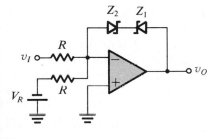
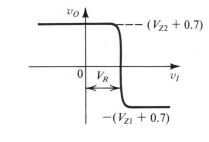

(c)

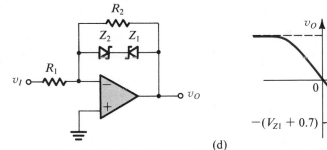
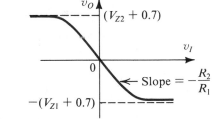

(d)

Fig. 5.33 Comparator and limiter circuits that employ op amps.

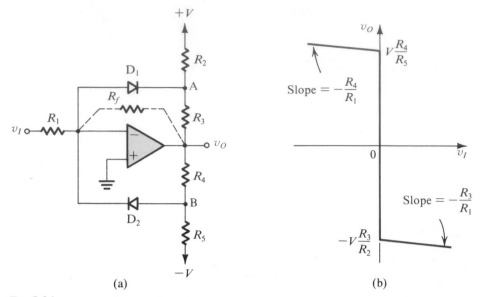

Fig. 5.34 A popular comparator circuit that functions as a limiter if R_f is included.

region. The extra current pushed through D_1 will therefore have to flow through R_3, thus causing the output voltage v_O to decrease. It may therefore be seen that in this region of operation the op amp operates as an inverting amplifier with a feedback resistance equal to R_3 plus the resistance presented by diode D_1. The latter resistance can be usually ignored in comparison with R_3, resulting in a linear transfer characteristic with a slope (gain) of $-R_3/R_1$. Normally, R_3 is chosen much smaller than R_1, rendering the slope quite small.

Consider next the situation as the input signal v_I goes negative. Diode D_1 will turn off and diode D_2 will turn on. This will cause the voltage at point B to become approximately zero. Neglecting the initial current through D_2, we see that the currents through R_4 and R_5 will be equal,

Thus

$$\frac{v_O - 0}{R_4} = \frac{0 + V}{R_5}$$

$$v_O = V\frac{R_4}{R_5}$$

As v_I is increased in the negative direction, the op amp operates as an inverting amplifier with a feedback resistance approximately equal to R_4 (neglecting the resistance presented by diode D_2). This results in a linear transfer characteristic with a slope of $-R_4/R_1$ that can be made small by selecting $R_4 \ll R_1$.

The nonideal diode characteristics will slightly modify the output levels of the comparator as well as making the transitions at the break points less sharp than those indicated in Fig. 5.34b. Thus as a limiter this circuit provides relatively soft limiting.

EXAMPLE 5.3

In the circuit of Fig. 5.34a let $V = 15$ V. Find suitable values for all resistances so that the comparator output levels are ± 5 V and that the slope of the limiting characteristics is 0.1. Also, indicate how to bias the comparator to cause the threshold to occur at $v_I = +5$ V.

Solution

To cause the threshold to occur at $v_I = +5$ V the comparator can be biased by connecting an additional resistance R_B from the inverting input terminal to a negative reference voltage. Selecting -15 V as the reference voltage, we use $R_B = 3R_1$.

To make the slope of the limiting characteristics equal to 0.1 we select

$$R_3 = R_4 = 0.1 \, R_1$$

Now for ± 5-V output levels we have

$$5 = 15 \frac{R_4}{R_5}, \qquad \text{leading to} \quad R_5 = 3R_4$$

$$5 = 15 \frac{R_3}{R_2}, \qquad \text{leading to} \quad R_2 = 3R_3$$

We may therefore select

$$R_3 = R_4 = 3 \text{ k}\Omega \qquad R_2 = R_5 = 9 \text{ k}\Omega \qquad R_1 = 30 \text{ k}\Omega \qquad R_B = 90 \text{ k}\Omega$$

Exercises

5.19 Assuming the diodes to be ideal, describe the transfer characteristic of the circuit shown in Fig. E5.19.

Ans. $v_O = v_I$ for $-5 \leq v_I \leq +5$
$v_O - \frac{1}{2}v_I - 2.5$ for $v_I \leq 5$
$v_O = \frac{1}{2}v_I + 2.5$ for $v_I \geq +5$

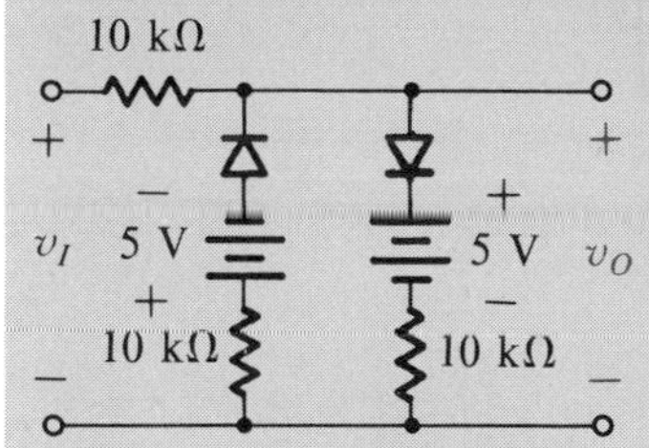

10 kΩ

Fig. E5.19

5.20 Consider the circuit of Fig. 5.34a with R_f included and with $V = 15$ V, $R_1 = 30$ kΩ, $R_f = 60$ kΩ, $R_2 = R_5 = 9$ kΩ, $R_3 = R_4 = 3$ kΩ. First convince yourself that the circuit operates as a limiter, and then find the value of the lower and upper threshold, the slope in the linear region, the output levels just as limiting starts, and the slope in the limiting regions (assume ideal diodes).

Ans. -2.5 V; $+2.5$ V; -2; ± 5 V; -0.095

5.10 COMPARATOR WITH HYSTERESIS—THE BISTABLE CIRCUIT

The advantages obtained by adding hysteresis to the comparator transfer characteristic were discussed in Section 5.8. We will now show how this may be accomplished. It will also be shown that the resulting circuit has *memory;* its usefulness therefore extends far beyond comparator applications.

Consider first the use of a single op amp as a comparator. If one of the op amp's input terminals is connected to a reference voltage V_R, then the open-loop op amp will obviously function as a comparator with the two output-limiting levels being the saturation voltages of the op amp. In view of the delay time required for an op amp to get out of saturation, it is sometimes desirable to fix, by external circuitry, the output voltages at levels lower than those of saturation. Except for this speed limitation, the op amp functions as an excellent comparator.

Next consider the circuit shown in Fig. 5.35a. Here we have connected the negative input terminal to ground while applying *positive feedback* through resistance R_2 and feeding the input v_I to the positive input terminal through another resistance R_1. First we note that because this circuit does *not* have *negative feedback*, there will be no virtual short circuit or virtual ground at the input. In fact, we will now show that this circuit functions as a comparator with hysteresis and that it has the transfer characteristics depicted in Fig. 5.35b, where L_+ and L_- denote the values of the op-amp saturation voltages (normally each is within about 2 V of the corresponding power-supply voltage).

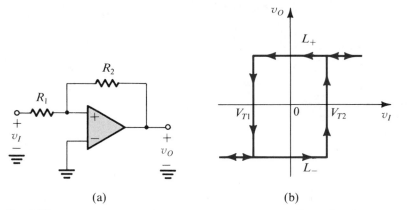

Fig. 5.35 A comparator circuit that employs an op amp with positive feedback to obtain transfer characteristics with hysteresis **(a)**, shown graphically in **(b)**.

By superposition, it follows that the voltage v_+ at the positive input terminal of the op amp is given by

$$v_+ = v_O \frac{R_1}{R_1 + R_2} + v_I \frac{R_2}{R_1 + R_2} \tag{5.15}$$

Consider now the case when the input v_I is at a high positive value (greater than the yet undetermined value V_{T2}). The transfer characteristic shows that the output should be L_+. Using these values in Eq. (5.15) shows that v_+ is positive, which indicates that the op amp is saturated in the positive direction and $v_O = L_+$ as originally assumed.

From the circuit we see that for the output to switch to the low level L_-, v_+ has to become slightly negative—that is, switching occurs when v_+ just passes zero in the negative direction. We can now use Eq. (5.15) to determine the value of v_I at which switching occurs, V_{T1}, by setting $v_+ = 0$ and $v_O = L_+$. The result is

$$V_{T1} = -L_+ \frac{R_1}{R_2}$$

If v_I is reduced below V_{T1}, the output remains at the low level L_-.

Next consider increasing v_I. Again we see from the circuit that for v_O to change from L_- to L_+ the voltage at the noninverting input terminal should become slightly positive. Thus the input threshold occurs at the value of v_I that results in v_+ being equal to zero. Setting $v_+ = 0$ and $v_O = L_-$ in Eq. (5.15) results in the upper threshold V_{T2} given by

$$V_{T2} = -L_- \frac{R_1}{R_2}$$

Increasing v_I beyond the positive value V_{T2} keeps v_+ positive and v_O equal to L_+.

We have thus shown that the circuit of Fig. 5.35a acts as a comparator with hysteresis. The reference level is 0 V, and the hysteresis width is $V_{T2} - V_{T1}$.

The comparison level can be changed by connecting the inverting input terminal to an appropriate reference voltage.

The circuit just described has more significance than just being a comparator with hysteresis. This circuit is *bistable;* that is, it has two stable states, one with the output equal to L_+, the other with the output equal to L_-. The circuit can remain in either stable state indefinitely. To change the circuit state we have to apply an input signal of appropriate value. Thus if the circuit is in the L_+ state we can *trigger* it to change state by applying an input $v_I < V_{T1}$. Here it is important to note that the word "trigger" is an excellent descriptor of the action caused by the input signal; the input signal merely starts the action of switching states. The switching process is completed by the action of the positive feedback. To see this consider the case when the output is high (at L_+) and an input signal $v_I < V_{T1}$ is applied. According to Eq. (5.15) the voltage v_+ at the positive input terminal goes negative. This causes the op-amp output to go negative, which in turn, according to Eq. (5.15), causes v_+ to become more negative. Once this *regenerative* process starts, we can remove the input signal and the process will continue on its own until the bistable switches completely to the other state.

Another important property of the bistable is that it has *memory.* This can be seen by noting that the response of the circuit at any moment is not determined solely by the value of the input signal at that moment but rather by the value of the input signal as well as the state in which the circuit is in. For instance, applying an input signal v_I in the range $V_{T1} < v_I < V_{T2}$ will result in output equal to L_+ if the circuit *was* in the L_+ state and in output equal to L_- if the circuit was in the L_- state.

Memory is a very useful property and will be exploited in Chapter 15 in the context of digital circuits.

Finally, we should note that the circuit of Fig. 5.35a is just one of many possible implementations of a class of circuits known as *bistable multivibrators*.

Exercises

5.21 In the circuit of Fig. 5.35a let $L_+ = -L_- = 10$ V, $R_1 = 10$ kΩ, and $R_2 = 20$ kΩ. Find the two threshold voltages V_{T1} and V_{T2}.

Ans. -5 V; $+5$ V

5.22 In the circuit of Fig. 5.35a let $L_+ = -L_- = 10$ V and $R_1 = 1$ kΩ. Find a value for R_2 that gives 100-mV hysteresis.

Ans. $R_2 = 200$ kΩ

5.11 WAVEFORM GENERATORS

Op amps, together with diodes, resistors, and capacitors, can be used to design circuits whose purpose is to generate signals having square, triangular, sine, and other waveforms. The subject of designing *waveform* or *function generators*, as they are normally called, is an important and challenging one. In this section we describe simple circuits for the generation of square and triangular waveforms. The resulting triangular waveform can be used as an input to a "sine-wave shaper," described in Section 5.12, to produce sinusoidal outputs. Other, more direct techniques for the generation of sinusoidal waveforms will be studied in Chapter 14. We will also describe a circuit for generating a single pulse. Other circuits for generating and processing pulse waveforms will be presented in Chapter 15.

A Single Op-Amp Square-Wave Generator—The Astable Multivibrator

Figure 5.36 shows a simple circuit that can be designed to provide a symmetrical square waveform of arbitrary frequency (limited normally by the speed of the op amp to the audio frequency range). The circuit operates in the following manner: Assume that for some reason a slight positive voltage develops between the op amp input terminals; that is, point C becomes more positive than point B. Because of the high open-loop gain of the op amp, its output will become positive at the upper saturation level L_+. This positive voltage L_+ at point A will, through the voltage divider composed of R_1 and R_2, cause the voltage at point C to become

$$v_C = v_A \frac{R_1}{R_1 + R_2} = \beta L_+ \tag{5.16}$$

where

$$\beta \equiv \frac{R_1}{R_1 + R_2}$$

Now B is lower than C in voltage, a situation that cannot continue because capacitor C_1 is connected to point A through R_3 and point A is at the higher positive voltage

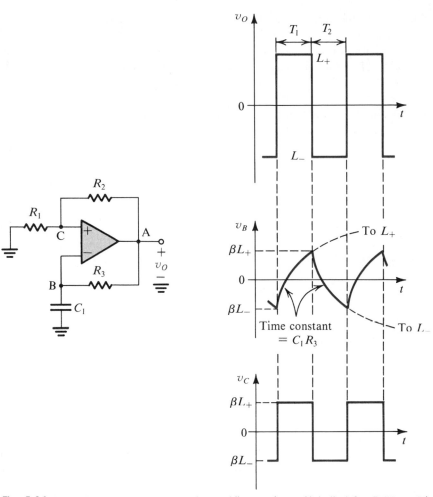

Fig. 5.36 A simple square-wave generator and its waveforms. Note that $\beta \equiv R_1/(R_1 + R_2)$.

L_+. Thus C_1 will charge and its voltage will rise exponentially, with a time constant C_1R_3, toward L_+. This process will be terminated when the voltage at B reaches and just exceeds the voltage at C, βL_+. At this moment the op amp will see a negative input voltage that, because of the high open-loop gain, will cause its output to change to the negative saturation level L_-. Now this negative voltage at A causes—through the R_1, R_2 divider—the voltage at C to change to the negative value βL_-. In turn, this negative voltage at C maintains the circuit for a time in its newly acquired state, the one with the output equal to L_-.

At this point we should note the *regenerative* action present: Once the voltage at B has risen above that at C, the switching action is started. Then the action is sustained and completed by the negative voltage fed back to point C through the R_1, R_2 divider. This regenerative action, which we have encountered before in the bistable, is a result of the judicious application of positive feedback.

Continuing with the description of circuit operation, we see that with the output at the negative voltage L_-, capacitor C_1 begins to discharge toward this negative level through R_3. The voltage v_B decreases exponentially, with a time constant C_1R_3, from the positive value βL_+. The discharge process will be terminated at the moment the magnitude of voltage at B just exceeds that at C (that is, βL_-), and so on. Thus the circuit oscillates, generating a square waveform at A.

Figure 5.36 shows the waveforms at various circuit points, from which the period T of the square wave can be found as follows: During the charging interval T_1 the voltage v_B across the capacitor at any time t, with $t = 0$ at the beginning of T_1, is given by (see Section 2.8)

$$v_B = L_+ - (L_+ - \beta L_-)e^{-t/\tau}$$

where

$$\tau = C_1 R_3$$

Substituting $v_B = \beta L_+$ at $t = T_1$ gives

$$T_1 = \tau \ln \frac{1 - \beta(L_-/L_+)}{1 - \beta} \tag{5.17}$$

Similarly, during the discharge interval T_2 the voltage v_B at any time t, with $t = 0$ at the beginning of T_2, is given by

$$v_B = L_- - (L_- - \beta L_+)e^{-t/\tau}$$

Substituting $v_B = \beta L_-$ at $t = T_2$ gives

$$T_2 = \tau \ln \frac{1 - \beta(L_+/L_-)}{1 - \beta} \tag{5.18}$$

Equations (5.17) and (5.18) can then be combined to obtain the period $T = T_1 + T_2$. Normally, $L_+ = -L_-$, resulting in symmetrical square waves of period T given by

$$T = 2\tau \ln \frac{1 + \beta}{1 - \beta} \tag{5.19}$$

Finally, we should mention that this square-wave generator can be made to have variable frequency by switching different capacitors C_1 (usually in decades) and by continuously adjusting R_3 (to obtain continuous frequency control within each decade of frequency). Also, although the waveform at B can be made almost triangular by using a small value for the parameter β, triangular waveforms of superior linearity can be easily generated using the scheme discussed next.

A General Scheme for Generating Triangular and Square Waveforms

Figure 5.37a shows a general scheme for generating triangular and square waveforms. The circuit consists of a feedback loop incorporating an inverting integrator and a bistable circuit. The latter is shown in block form, with the transfer characteristic

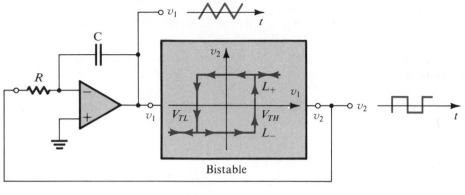

(a)

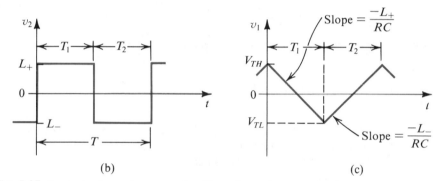

(b)   (c)

Fig. 5.37 General scheme for generating triangular and square waveforms.

shown inside the box. This bistable can be realized by a variety of circuits, a simple one of which is discussed in Section 5.10.

We now proceed to show how the feedback loop of Fig. 5.37a oscillates and generates a triangular waveform v_1 at the output of the integrator and a square waveform v_2 at the output of the bistable. Let the output of the bistable be at L_+. A current equal to L_+/R will flow into the resistor R and through capacitor C, causing the output of the integrator to *linearly* decrease with a slope of $-L_+/CR$, as shown in Fig. 5.37c. This will continue until the integrator output reaches the lower threshold V_{TL} of the bistable, at which point the bistable will switch states, its output becoming negative and equal to L_-. At this moment the current through R and C will reverse direction, and its value will become equal to $|L_-|/R$. It follows that the integrator output will start to increase linearly with a positive slope equal to $|L_-|/CR$. This will continue until the integrator output voltage reaches the positive threshold of the bistable, V_{TH}. At this point the bistable switches, its output becomes positive (L_+), the current into the integrator reverses direction, and the output of the integrator starts to decrease linearly, beginning a new cycle.

From the above discussion it is relatively easy to derive an expression for the period T of the square and triangular waveforms. During the interval T_1 we have, from Fig. 5.37c,

$$\frac{V_{TH} - V_{TL}}{T_1} = \frac{L_+}{CR}$$

from which we obtain

$$T_1 = CR \frac{V_{TH} - V_{TL}}{L_+}$$

Similarly, during T_2 we have

$$\frac{V_{TH} - V_{TL}}{T_2} = \frac{-L_-}{CR}$$

from which we obtain

$$T_2 = CR \frac{V_{TH} - V_{TL}}{-L_-}$$

Thus to obtain symmetrical square waves we design the bistable to have $L_+ = -L_-$.

Generation of a Standardized Pulse—The Monostable Multivibrator

In some applications the need arises for a pulse of known height and width, generated in response to a trigger signal. Because the width of the pulse is predictable, its trailing edge can be used for timing purposes—that is, to initiate a particular task at that specified time. A circuit that generates such a pulse is known as a *monostable multivibrator* or as a *one-shot*.

Figure 5.38a shows an op-amp monostable circuit. We observe that this circuit is an augmented form of the astable circuit of Fig. 5.36. Specifically, a clamping diode D_1 is added across the capacitor C_1, and a trigger circuit composed of capacitor C_2, resistor R_4, and diode D_2 is connected to the noninverting input terminal of the op amp. The circuit operates as follows: Being a monostable multivibrator, it has one stable state in which it can remain indefinitely. In the stable state, which prevails in the absence of the triggering signal, the output of the op amp is at L_+ and diode D_1 is conducting through R_3 and thus clamping the voltage v_B to one diode drop above ground. By selecting R_4 much larger than R_1, diode D_2 will be conducting a very small current and the voltage v_C will be very closely determined by the voltage divider R_1, R_2. Thus $v_C = \beta L_+$, where $\beta \equiv R_1/(R_1 + R_2)$. The stable state is maintained because βL_+ is greater than V_{D1}.

Now consider the application of a negative-going step at the trigger input and refer to the signal waveforms shown in Fig. 5.38b. The negative triggering edge will be coupled to the cathode of diode D_2 via capacitor C_2, and thus D_2 conducts heavily and pulls node C down. If the trigger signal is of sufficient height to cause v_C to go below v_B, the op amp will see a net negative input voltage and its output will switch to L_-. This in turn causes v_C to go negative to βL_- which keeps the op amp in its newly acquired state. Note that D_2 now cuts off, thus isolating the circuit from any further changes at the trigger input terminal.

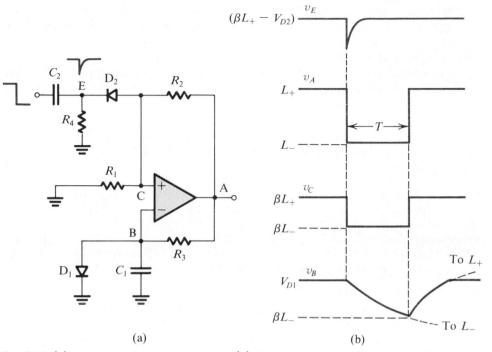

Fig. 5.38 (a) An op-amp monostable circuit. **(b)** Signal waveforms in the circuit of (a).

The negative voltage at A causes D_1 to cut off, and C_1 begins to discharge exponentially toward L_- with a time constant C_1R_3. The monostable is now in its *quasi-stable state*, which will prevail until the declining v_B goes below the voltage at node C, which is βL_-. At this instant the op amp output switches back to L_+ and the voltage at node C goes back to βL_+. Capacitor C_1 then charges toward L_+ until diode D_1 turns on and the circuit returns to its stable state.

From Fig. 5.38b we observe that a negative pulse is generated at the output during the quasi-stable state. The duration T of the output pulse is determined from the exponential waveform of v_B, as follows:

$$v_B(t) = L_- \quad (L_- \quad V_{D1})e^{-t/C_1R_3}$$

by substituting $v_B(T) = \beta L_-$,

$$\beta L_- = L_- - (L_- - V_{D1})e^{-T/C_1R_3}$$

which yields

$$T = C_1 R_3 \ln\left(\frac{V_{D1} - L_-}{\beta L_- - L_-}\right) \tag{5.20}$$

For $V_{D1} \ll |L_-|$, this equation can be approximated by

$$T \simeq C_1 R_3 \ln\left(\frac{1}{1-\beta}\right) \tag{5.21}$$

Finally, it should be noted that the monostable circuit should not be triggered again until capacitor C_1 has been recharged to V_{D1}, otherwise the resulting output pulse will be shorter than normal. This recharging time is known as the recovery period. Circuit techniques exist for shortening the recovery period.

We conclude this section by noting that special-purpose integrated circuits are available for implementing monostable and astable multivibrator functions. The most popular such circuit is the 555 IC Timer.

Exercises

5.23 Consider the circuit in Fig. 5.36a. Let the op-amp saturation voltages be ± 10 V. For $R_1 = 100$ kΩ, $R_2 = R_3 = 1$ MΩ, and $C_1 = 0.01$ μF, find the frequency of oscillation.

Ans.   274.2 Hz

5.24 Consider a modification of the circuit of Fig. 5.36 in which R_1 is replaced by a pair of diodes connected in parallel in opposite directions. For $L_+ = -L_- = 12$ V, $R_2 = R_3 = 10$ kΩ, $C_1 = 0.1$ μF, and the diode voltage a constant denoted V_D, find an expression for frequency as a function of V_D. If $V_D = 0.70$ V at 25°C with a TC of -2 mV/°C, find the frequency at 0°C, 25°C, 50°C, and 100°C. Note that the output of this circuit can be sent to a remotely connected frequency meter to provide a digital readout of temperature

Ans. $f = 500/\ln[(12 + V_D)/(12 - V_D)]$ Hz; 3,995 Hz, 4,281 Hz, 4,611 Hz, 5,451 Hz

5.25 Consider the circuit of Fig. 5.37a with the bistable realized by the circuit in Fig. 5.35a. If the op amps have saturation voltages of ± 10 V and a capacitor $C = 0.01$ μF and a resistor $R_1 = 10$ kΩ are used, find the values of R and R_2 (note that R_1 and R_2 are associated with the bistable of Fig. 5.35a) such that the frequency of oscillation is 1 kHz and the triangular waveform has a 10-V peak-to-peak amplitude.

Ans.   50 kΩ; 20 kΩ

5.26 For the monostable circuit of Fig. 5.38a find the value of R_3 that will result in a 100-μs output pulse for $C_1 = 0.1$ μF, $\beta = 0.1$, $V_D = 0.7$ V, and $L_+ = -L_- = 12$ V.

Ans.   6F171 Ω

5.12 OTHER APPLICATIONS

In this section we discuss three additional applications of diode–op-amp circuits: nonlinear wave shaping, logarithmic amplifiers, and diode logic gates.

Nonlinear Wave Shaping

Diodes together with resistors can be used to synthesize two-port networks having arbitrary nonlinear transfer characteristics. Such two-port networks can be employed in *waveform shaping*—that is, changing the waveform of an input signal in a prescribed manner to produce a waveform of a desired shape at the output. In this section we shall illustrate this application by a concrete example: the *sine-wave shaper*. This is a circuit whose purpose is to change the waveform of an input triangular wave signal to a sine wave. Though simple, this sine shaper is a practical building block used extensively in function generators. There are, of course, other, more direct techniques for generating sinusoidal signals; the general topic of sinusoidal oscillators will be discussed in Chapter 14. These direct techniques, however, are not very convenient at very low frequencies, and one might prefer the simple schemes discussed in this chapter—namely, generation of stable triangular waveforms and then the processing of them through the sine shaper discussed next.

Consider the circuit shown in Fig. 5.39a. It consists of a chain of resistors connected across the entire symmetric voltage supply $+V$, $-V$. The purpose of this voltage divider is to generate reference voltages that will serve to determine the breakpoints in the transfer characteristic. In our example these reference voltages are denoted $+V_2$, $+V_1$, $-V_1$, $-V_2$. Note that the entire circuit is symmetric, driven by a symmetric triangular wave and generating a symmetric sine wave. The circuit approximates each quarter-cycle of the sine wave by three straight-line segments, with the breakpoints between these segments determined by the reference voltages V_1 and V_2.

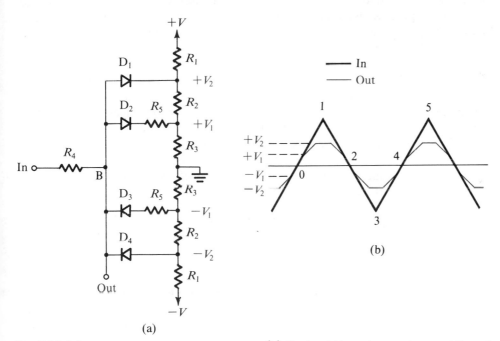

Fig. 5.39 (a) A three-segment sine-wave shaper.   **(b)** The input, triangular waveform and the output, approximately sinusoidal waveform.

The circuit works as follows: Let the input be the triangular wave shown in Fig. 5.39b, and consider first the quarter-cycle defined by the two points labeled 0 and 1. When the input signal is less in magnitude than V_1, none of the diodes conduct. Thus zero current flows through R_4, and the output voltage at B will be equal to the input voltage. But as the input rises to V_1 and above, D_2 (assumed ideal) begins to conduct.

Assuming that the conducting D_2 behaves as a short circuit, we see that, for $v_I > V_1$,

$$v_O = V_1 + (v_I - V_1)\frac{R_5}{R_4 + R_5}$$

This implies that as the input continues to rise above V_1 the output follows but with a reduced slope. This gives rise to the second segment in the output waveform, as shown in Fig. 5.39b. Note that in developing the above equation we have assumed that the resistances in the voltage divider are low-valued so as to cause the voltages V_1 and V_2 to be constant independent of the current coming from the input.

Next consider what happens as the voltage at point B reaches the second break-point determined by V_2. At this point D_1 conducts, thus limiting the output v_O to V_2 (plus, of course, the voltage drop across D_1 if it is not assumed ideal). This gives rise to the third segment, which is flat, in the output waveform. The result is to "bend" the waveform and shape it into an approximation of the first quarter-cycle of a sine wave. Then, beyond the peak of the input triangular wave, as the input voltage decreases, the process unfolds, the output becoming progressively more like the input. Finally, when the input goes sufficiently negative, the process begins to repeat at $-V_1$ and $-V_2$ for the negative half-cycle.

While the circuit is relatively simple, its performance is surprisingly good. A measure of goodness usually taken is to quantify the purity of the output sine wave by specifying the percentage *total harmonic distortion* (THD). This is the percentage ratio of the rms voltage of all harmonic components above the fundamental frequency (which is the frequency of the triangular wave) to the rms voltage of the fundamental. Interestingly, one reason for the good performance of the diode shaper is the beneficial effects produced by the nonideal i-v characteristics of the diodes—that is, the exponential knee of the junction diode as it goes into forward conduction. The consequence is a relatively smoothed transition from one line segment to the next.

The circuit discussed is an example of a family of circuits known as *diode function generators*. To enhance accuracy, such circuits usually employ op amps.

Logarithmic Amplifiers

The circuit shown in Fig. 5.40 provides an output voltage v_O given by

$$v_O = -nV_T \ln\left(\frac{v_I}{I_S R}\right) \qquad v_I > 0$$

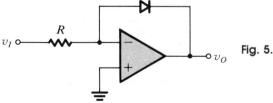

Fig. 5.40 A logarithmic amplifier.

where I_S is the scale current of the diode. Since the output voltage is proportional to the logarithm of the input voltage, the circuit is known as a logarithmic amplifier. Such amplifiers find application in situations where it is desired to compress the input signal range. A disadvantage of this simple implementation of the log amplifier is that its characteristics depend critically on the value of I_S, which is a strong function of temperature. More elaborate implementations of the log amplifier make use of the exponential characteristics of bipolar junction transistors (see Roberge, 1975).

Diode Logic Gates

Diodes together with resistors can be used to implement digital logic functions. Figure 5.41 shows two diode-logic gates. To see how these circuits function consider a positive logic system in which voltage values close to zero volts correspond to logic 0 (or low) and voltage values close to $+5$ V correspond to logic 1 (or high). The

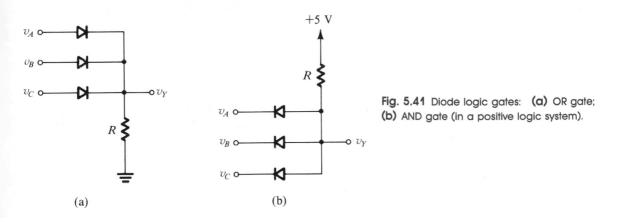

Fig. 5.41 Diode logic gates: **(a)** OR gate; **(b)** AND gate (in a positive logic system).

(a) (b)

circuit in Fig. 5.41a has three inputs v_A, v_B, and v_C. It is easy to see that diodes connected to $+5$-V inputs will conduct, thus clamping the output v_Y to a value equal to $+5$ less a diode drop. This positive voltage at the output will keep the diodes whose inputs are low (around zero volts) cut off. Thus the output will be high if one or more of the inputs are high. The circuit therefore implements the logic OR function, which in Boolean notation is expressed as

$$Y = A + B + C$$

A disadvantage of this diode gate is that the value of the logic 1 at the output is less than that at the input because of the finite diode drop. The problem is aggravated if the gate is cascaded with other similar gates (as is common in the realization of a digital system). In this case the value of the logic 1 deteriorates progressively until it becomes indistinguishable from that of logic 0. The problem can be solved by the addition of an amplifying device, for example a transistor, to restore the logic level. This will be discussed in Chapter 16.

Finally, the reader is encouraged to show that using the same logic system mentioned above, the circuit of Fig. 5.41b implements the logic AND function,

$$Y = A \cdot B \cdot C$$

5.27 The circuit in Fig. E5.27 is required to provide a three-segment approximation to the nonlinear i-v characteristic

$$i = 0.1v^2$$

where v is the voltage in volts and i is the current in milliamperes. Find the values of R_1, R_2, and R_3 such that the approximation is perfect at $v = 2, 4$, and 8 volts. Calculate the error in current value at $v = 3, 5, 7$, and 10 volts. Assume ideal diodes.

Ans. $5 \text{ k}\Omega$, $1.25 \text{ k}\Omega$, $1.25 \text{ k}\Omega$; -0.3 mA, $+0.1 \text{ mA}$, -0.3 mA, 0

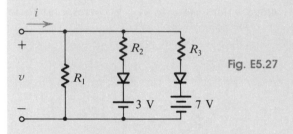

Fig. E5.27

5.28 If the diode in the circuit of Fig. 5.40 has 0.6-V drop at 1 mA and a 0.1 V/decade characteristic, find the value of R that will result in $v_O = -0.4$ V to -0.7 V corresponding to $v_I = 10$ mV to 10 V.

Ans. $1 \text{ k}\Omega$

5.13 SUMMARY

● In rectifier circuits the diode forward voltage drop can be neglected if the input signal is large.

● By judiciously incorporating op amps in diode circuits, the diode voltage drop can be masked, resulting in precision circuits that operate with very small input signals.

● Full-wave rectifiers invert either the negative or the positive parts of the input waveform, thus providing a unidirectional output. The signal inversion is performed using an amplifier (in small-signal applications) or a center-tapped transformer (in large-signal or power-supply applications). Alternatively, the diode bridge configuration can be employed, either alone or in conjunction with an op amp to add precision.

● Placing a capacitor across the load resistance of a rectifier results in a peak rectifier, also called a peak detector.

● For a sine-wave input, the output of the peak rectifier consists of a dc component approximately equal to the peak of the input sinusoid, superimposed on which is a small ripple waveform. The amplitude of the ripple voltage is inversely proportional to the rectifier capacitance and the load resistance.

● The peak rectifier can be employed as an AM demodulator.

● The dc restorer provides an output waveform whose positive or negative peaks are clamped to a predetermined level. The dc component of the output is independent of that of the input.

● The output of a limiter circuit is linearly related to the input over a specified range. Outside this range, the output is limited to predetermined values.

● The output of a comparator circuit can have one of two possible values. It changes from one value to the other as the input exceeds a reference value. A comparator with a zero reference can be used to detect the zero crossings of an input waveform.

● Limiters can be implemented using diodes and resistors. Incorporating op amps provides improved, more versatile implementations.

● There are three types of multivibrators: bistable, monostable, and astable. Op-amp circuit implementations of multivibrators are useful in analog circuit applications that require high precision. Implementations using digital logic gates will be studied in Chapter 15.

● The bistable multivibrator has two stable states and can remain in either state indefinitely. It changes state when triggered. A comparator with hysteresis is bistable.

● A monostable multivibrator has one stable state, in which it can remain indefinitely. When triggered, it goes into a quasi-stable state in which it remains for a predetermined interval, thus generating at the output a pulse of known width. It is also known as a one-shot.

● An astable multivibrator has no stable state. It oscillates between two quasi-stable states, remaining in each for a predetermined interval. It thus generates a periodic waveform at the output.

● A feedback loop consisting of an integrator and a bistable can be used to generate triangular and square waveforms.

● A sine wave can be generated by feeding a triangular waveform to a "sine shaper". The latter can be designed using diodes and resistors. Other means for generating sinusoidal waveforms will be studied in Chapter 14.

BIBLIOGRAPHY

G. B. Clayton, *Experimenting with Operational Amplifiers*, London: Macmillan, 1975.

G. B. Clayton, *Operational Amplifiers*, 2nd ed., London: Newnes-Butterworths, 1979.

J. G. Graeme, G. E. Tobey, and L. P. Huelsman, *Operational Amplifiers: Design and Applications*. New York: McGraw-Hill, 1971.

W. Jung, *IC Op Amp Cookbook*, Indianapolis: Howard Sams, 1974.

Nonlinear Circuits Handbook, Norwood, Mass.: Analog Devices, Inc., 1976.

J. K. Roberge, *Operational Amplifiers: Theory and Practice*, New York: Wiley, 1975.

J. I. Smith, *Modern Operational Circuit Design*, New York: Wiley-Interscience, 1971.

J. V. Wait, L. P. Huelsman, and G. A. Korn, *Introduction to Operational Amplifier Theory and Applications*, New York: McGraw-Hill, 1975

PROBLEMS

5.1 Using an ideal diode, design a rectifier circuit having the following transfer characteristic.

$$v_O = 0 \qquad \text{for } v_I \geq 0$$

$$v_O = v_I \qquad \text{for } v_I \leq 0$$

5.2 Using an ideal diode and two resistors, design a circuit having the following transfer characteristic.

$$v_O = 0 \qquad \text{for } v_I \geq 0$$

$$v_O = v_I/2 \qquad \text{for } v_I \leq 0$$

5.3 Using an ideal diode, a resistor, and a 2-V battery, design a circuit having the following transfer characteristic.

$$v_O = +2 \text{ V} \qquad \text{for } v_I \leq 2 \text{ V}$$

$$v_O = v_I \qquad \text{for } v_I \geq 2 \text{ V}$$

*5.4 Assuming that the diodes in the circuit of Fig. P5.4 are ideal, sketch the transfer characteristic v_O versus v_I.

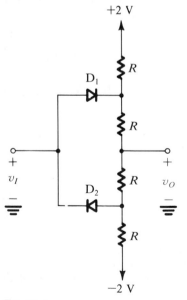

Fig. P5.4

As well describe the output analytically in terms of the input and critical (threshold) voltages.

*5.5 If in the circuit of Fig. P5.4 both diodes are reversed, what does v_O become, expressed as a function of v_I? Assume ideal diodes.

*5.6 Repeat Problem 5.4 assuming that the diodes are real junction diodes, each modeled as a constant 0.7-V battery (when conducting).

*5.7 Repeat Problem 5.5 assuming that the diodes are real junction diodes, each modeled as a constant 0.7-V battery (when conducting).

*5.8 Use the result of Exercise 5.1 to plot the transfer characteristic of the half-wave rectifier of Fig. 5.1a for the case in which $R = 1 \text{ k}\Omega$ and the diode is a 1 mA device with $n = 2$. Sketch the output waveform if

(a) v_I is a triangular wave with a 10-V peak.

(b) v_I is a triangular wave with a 1-V peak.

(c) v_I is a symmetric square wave with a 10-V amplitude.

5.9 Consider the superdiode circuit of Fig. 5.3a augmented by a second diode shunting the first but with the polarity reversed. Sketch the transfer characteristic of the revised circuit. On the same graph sketch the transfer characteristic from the input to the output of the op amp. Assume that the diodes have a constant 0.7-V drop when conducting.

5.10 Consider Fig. 5.3a. Show the addition of a second resistor of value R and a new output connection at which the output is $2v_I$ for $v_I \geq 0$.

5.11 Consider the superdiode circuit represented by the part enclosed in the broken-line box of Fig. 5.3a. Let two superdiodes be connected at their outputs to a common load resistance R. If the input to one superdiode is $+1$ V and that to the other is $+2$ V, what is the output voltage? Repeat with the diodes reversed and the inputs changed to -1 V and -2 V.

5.12 Consider the circuit of Fig. 5.4 modified by connecting the positive op amp terminal to a voltage V_1.

Sketch the transfer characteristic of the resulting circuit for **(a)** $V_1 = 1$ V and **(b)** $V_1 = -1$ V. Assume that $R_2 = R_1$.

5.13 (a) Sketch the transfer characteristic of the circuit in Fig. E5.6.

(b) Suppose that in this circuit both diodes are reversed and the $+15$-V source is replaced with a -15-V source. Sketch the transfer characteristic of the resulting circuit.

(c) Consider the circuit formed by combining the circuit of Fig. E5.6 with the circuit of (b) above as follows: Connect the two input terminals together and connect the two output terminals to an op-amp summer circuit having input resistors R and R, and a feedback resistor R. Sketch the transfer characteristic of the resulting composite circuit—that is, from the input to the output terminal of the summing amplifier.

5.14 Sketch the transfer characteristic of the circuit shown in Fig. P5.14.

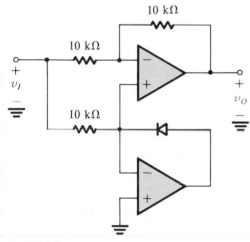

10 kΩ

10 kΩ

$+$
v_I

$+$
v_O

$-$

10 kΩ

Fig. P5.14

5.15 Consider the circuit of Fig. 5.7c with $R_1 = R_2$. For a sine wave input of amplitude V_p, what is the average (dc) output voltage? If by accident R_2 is made π times the value of R_1, what does the average output become?

*5.16 Augment the circuit of Fig. 5.10 with two additional diodes connected so as to produce a negative rectified output voltage across a second load resistance R_{L2}. (Think of the existing load as R_{L1}.) What is the rms value of v_A required to produce a peak output across R_{L2} of

-15 V assuming that the diodes have a constant 0.7-V drop when conducting? If the output voltage is taken between the ungrounded ends of R_{L1} and R_{L2}, find the maximum, minimum, and average values of the resulting voltage.

5.17 For the circuit in Fig. 5.11c, assuming that v_A is a sinusoid of peak voltage V_p, sketch the signal waveform (referred to ground) at each of the two ends of R_L. Then sketch the waveform for v_L. Assume ideal diodes.

5.18 Consider the circuit in Fig. 5.12b. For a sine-wave input of 5 V rms and $R = 1$ kΩ, what is the average current in M? What is the waveform across R? What is its peak value? For an op amp that saturates at ±12 V, a meter having 50-Ω resistance and diodes with 0.7-V drop (when conducting), what is the rms voltage of the largest possible sine wave for which the circuit operates as a voltmeter?

***5.19 It is required to analyze the circuit of Fig. 5.14a when excited with the triangular signal shown in Fig. 5.14b. Using the exponential diode model, show that shortly after time t_0 and for $t \le t_1$ the output is given by

$$v_O \simeq nV_T \ln\left(\frac{I_S T}{4CV_p}\right) + \left(\frac{V_p}{T/4}\right)t$$

Use this relationship to find the value reached by v_O at time t_1 for the case $T = 40$ ms, $C = 10$ μF, and $V_p = 10$ V. Assume the diode to be a 1-mA device whose voltage drop changes by 0.1 V per decade of current change. Continue the analysis to determine the value that v_O reaches when the diode cuts off beyond time t_1.

5.20 For the half-wave peak rectifier circuit with sine wave input analyzed in the text show that the dc output voltage is given by:

$$V_O = V_p - I_L\left(\frac{1}{2fC}\right)$$

Use this to find the Thévenin equivalent circuit of the rectifier as a dc source. Find the value of the Thévenin resistance when $C = 100$ μF and $f = 60$ Hz.

*5.21 A positive peak rectifier utilizing a 1-μF capacitor and an ideal diode and having an initial output of zero is driven by a series of pulses of 10-V amplitude and 10-μs duration, originating in a 1-kΩ source. What is the voltage on the capacitor after one pulse? After two pulses? After ten pulses? How many pulses are required for the output to reach 6.32 V? 9 V? 9.9 V?

***5.22 A positive-peak rectifier circuit utilizes a 1-μF capacitor and a diode for which the voltage drop at 1 mA is 0.7 V and for which $n = 2$. A series of pulses of 10-μs duration and 0.7-V amplitude is applied to the circuit. What is the voltage on the capacitor following the first pulse, approximately? Exactly? What is the voltage on the capacitor after 2 pulses? After 10 pulses? How many pulses does it take for the output to rise to 0.63 V?

*5.23 A positive-peak rectifier utilizing a fast op amp and a junction diode in a superdiode configuration, and a 10-μF capacitor initially uncharged, is driven by a series of 1-V pulses of 10 μs duration. If the maximum output current that the op amp can supply is 10 mA, what is the voltage on the capacitor following one pulse? Two pulses? Ten pulses? How many pulses are required to reach 0.5 V? 1.0 V? 2.0 V?

5.24 A peak rectifier utilizing an ideal diode and a 100-μF capacitor and loaded by a resistance of 10 kΩ is driven by a triangular wave of 20-V amplitude (peak-to-peak) and a frequency of 100 Hz. Find the dc output voltage, the peak-to-peak ripple voltage, and the average current in the diode during conduction. What is the length of time during which the diode conducts?

5.25 Consider a peak rectifier fed by a 60-Hz sinusoid having a peak value of 100 V and loaded by a resistance of 1 kΩ. If the diode drop can be neglected, what value of capacitance C must be used to ensure a peak-to-peak ripple of 2 V? For what fraction of the cycle does the diode conduct? Find the average value of diode current during conduction. Also find the peak diode current.

5.26 A full-wave diode rectifier operating at 60 Hz with a 10-kΩ load utilizes a filter capacitor to smooth the output. Assuming that the diode voltage drops are negligible, find the value of the capacitor required to reduce the ripple, expressed as a percentage of the output, to 10%? To 1%?

*5.27 Consider a circuit arrangement in which a positive-peak rectifier and a negative-peak rectifier each using a capacitor C are driven by a single sine-wave source of peak amplitude V_p and frequency f. A load resistance $2R$ is connected between the outputs of the two rectifiers. What is the average voltage across the load? What is the peak-to-peak ripple that results? Sketch and label the waveform of the ripple voltage across the load. Assume the diodes to be ideal.

5.28 Consider the buffered precision peak rectifier shown in Fig. 5.20 when connected to a triangular input of 1-V

peak-to-peak amplitude and 1,000-Hz frequency. It utilizes an op amp whose bias current (directed into A_2) is 10 nA and diodes whose reverse leakage current is 1 nA. What is the smallest capacitor that can be used to guarantee an output ripple less than 1%?

*5.29 Consider the circuit of Fig. 5.20 under the conditions that the bias current of A_2 flows out of the input terminal. What will happen to the voltage on C? The

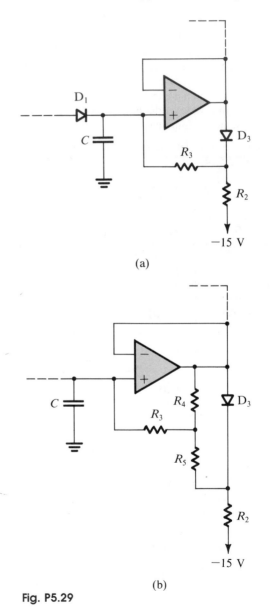

(a)

(b)

Fig. P5.29

circuit shown in Fig. P5.29a is one solution to this problem. If the diode voltage is 0.7 V and the op-amp bias current is 10 nA, what is the value of R_3 required? If the largest value of resistor conveniently available is 10 MΩ, use the modified arrangement shown in Fig. P5.29b. Design for a 0.1 mA in the voltage divider shunting D_3. Give values for all resistors.

5.30 Consider the bridge rectifier circuit of Fig. 5.11a with the addition of a filtering capacitor C across R_L. The secondary voltage v_B is a sinusoid of 10-V peak amplitude at 60 Hz. Diodes can be modeled as 0.7-V drops independent of current. For $R_L = 100\ \Omega$, find C for a peak-to-peak ripple that is 10% of the peak output. What is the average output voltage across R_L?

5.31 Consider the circuit of Fig. 5.20 with an input v_I applied, as follows:

$$v_I = 2 + 5 \sin \omega t$$

What are the peak reverse voltages appearing across diodes D_1 and D_2?

5.32 In medical instruments which utilize ultrasound to detect movement of some internal part of the body, such as a heart value, acoustic energy at a frequency of, say, 2 MHz is reflected from the moving body part in such a way as to produce a signal closely resembling an amplitude-modulated wave. For such an instrument, design a peak detector capable of following motions that correspond to a fall of signal amplitude of 50% in 5 ms. Using a load resistor of 10 kΩ, what capacitor C should be used? Consider the diode to be ideal. What is the percent carrier ripple for this design when a static object is being scanned?

**5.33 For the circuits in Fig. P5.33, each utilizing an ideal diode (or diodes), sketch the output for the input shown. Label the most positive and most negative output levels. Assume $CR \gg T$.

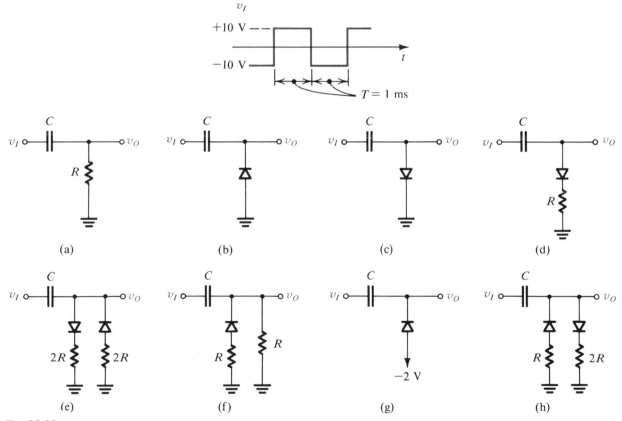

Fig. P5.33

5.34 Plot the transfer characteristic of the circuit in Fig. P5.34 by evaluating v_I corresponding to $v_O = 0.5, 0.6, 0.7, 0.8, 0, -0.5, -0.6, -0.7$, and -0.8 V. Assume that the diodes are 1-mA units having a 0.1-V/decade logarithmic chracteristic. Characterize the circuit as a hard or soft limiter. What is the value of K? Estimate L_+ and L_-.

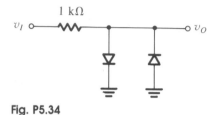

Fig. P5.34

5.35 A particular comparator with hysteresis has $L_+ = +12$ V, $L_- = -12$ V, $V_{TL} = -1$ V, and $V_{TH} = +1$ V. What is the amount of hysteresis? For a 0.5-V-amplitude sine-wave input having zero average, what is the output? Describe the output if a sinusoid of frequency f and amplitude 1.1 V is applied. By how much can the average of this sinusoid input shift before the output ceases to switch?

*5.36 In the limiter circuit shown in Fig. P5.36 the two diodes are identical 1-mA devices with a 0.1-V/decade characteristic. Find v_I corresponding to $v_O = 0.4$ V, 0.5 V, 0.52 V, 0.54 V, and 0.56 V. (*Hint:* At each value of v_O find $i_{D2}, v_{D1}, i_{D1}, i_I$, and v_I, in that order). Hence sketch the transfer characteristic v_O versus v_I. How does this limiter compare to another formed by R_1 and D_1 alone?

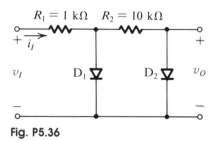

Fig. P5.36

5.37 Design limiter circuits using only diodes and 10-kΩ resistors to provide an output signal limited to the range: **(a)** -0.7 V and above; **(b)** -2.1 V and above; and **(c)** ± 1.4 V. Assume that each diode has a 0.7-V drop when conducting.

***5.38 Consider the limiter circuit shown in Fig. P5.38. The op amp saturates at ± 13 V. Note that the current in the voltage divider is much greater than the diode current.

(a) Assuming that the diode when conducting can be modeled as a constant 0.5-V voltage drop, find the transfer characteristic v_O versus v_I. Find the voltage gain in the various regions of operation.

(b) Assuming that the diode is a 1-mA device with a 0.1-V/decade characteristic, find the transfer characteristic. (*Hint:* The simplest approach is to start with a given diode current, in the range 0.1 μA to 8 μA, and then calculate the diode voltage, the voltage at the cathode of the diode, v_O, the current in the feedback resistor, the input current, and v_I in that order). Sketch the transfer characteristic and compare it to the approximate one obtained in (a) above.

(c) Show how another diode can be added to the circuit to obtain double limiting.

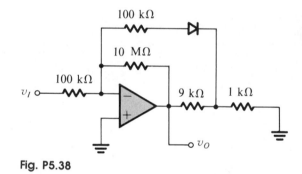

Fig. P5.38

*5.39 Using the ideas incorporated in the circuits of Fig. 5.33c and d and Fig. 3.7, design a limiter that hard limits at ± 7.5-V output, has a gain of -10, an input resistance of 10 MΩ, an input threshold voltage centered around -7.5 V, utilizing no resistor greater than 10 MΩ.

*5.40 Design a limiter based on the circuit of Fig. 5.34a, using ± 15-V supplies, which limits at ± 10 V and has a slope of 1 in the limiting regions. In the linear regions the gain is to be 100. The input resistance is to be 100 kΩ and the input thresholds should be centered around 10 V. Assume that the diodes are ideal.

5.41 Consider the Schmitt trigger comparator shown in Fig. 5.35. For an op amp that saturates at ± 13 V, and $R_1 = 1$ MΩ, find the value of R_2 that results in thresholds of ± 1.3 V.

*5.42 Consider the Schmitt trigger comparator shown in Fig. 5.35 under the condition that the negative op-amp input is connected to a reference voltage V_R. Using an op amp that saturates at $\pm V$ and $R_1 = 10$ kΩ, find the values of R_2 and V_R that will result in thresholds of 0 and $V/10$.

*5.43 Consider the circuit of Fig. 5.35 augmented by a resistor R_3 connected from the common node of R_1 and R_2 to the power supply ($+V$ or $-V$ as may be required). For an op amp powered by $\pm V$ that saturates at $(V-2)$ and $(-V+2)$ design a comparator with thresholds at 0 and $V/10$. Use $R_1 = 10$ kΩ and $V = 15$ V.

5.44 For the circuit in Fig. P5.44 sketch the transfer characteristic v_O versus v_I. The diodes are assumed to have a constant 0.7-V drop and the amplifier saturates at ± 12 V.

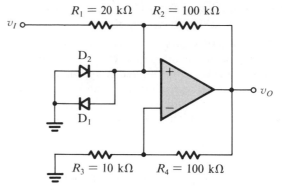

$R_1 = 20$ kΩ $R_2 = 100$ kΩ

$R_3 = 10$ kΩ $R_4 = 100$ kΩ

Fig. P5.47

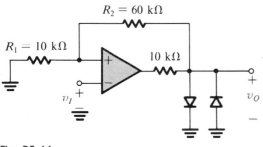

Fig. P5.44

*5.45 Consider the circuit of Fig. P5.44. Using a dc source of $+15$ V or -15 V and a resistor R_3, arrange to shift the thresholds such that the hysteresis is centered around $+2$ V. What is the required value of R_3? What is the width of the hysteresis region that results?

*5.46 Design a circuit resembling that in Problem 5.44 that has thresholds of ± 0.7 V and uses a single op amp, a single resistor, and two diodes.

**5.47 Show that the circuit in Fig. P5.47 functions as a bistable with thresholds and output levels that are independent of the op-amp characteristics. Assume that a conducting diode has 0.7-V drop. Find the thresholds and output levels. Replacing the op amp together with R_2, R_3, and R_4 with the equivalent negative input resistance (see Example 3.5), find what happens if R_1 is reduced toward 10 kΩ, if $R_1 = 10$ kΩ, and if $R_1 < 10$ kΩ.

*5.48 Assume that the saturation voltages of the op amp in Fig. P5.48 are ± 9 V. Show that the circuit operates

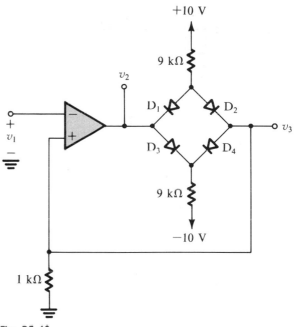

Fig. P5.48

as a bistable. Sketch v_2 versus v_1. Give the threshold voltages. Assume that a conducting diode has 0.7-V drop.

5.49 Using the circuit of Fig. 5.36 design an oscillator that generates a square wave of 1-kHz frequency employing an op amp for which $L_+ = -L_-$. Select $\beta = 0.1$ and use resistors no smaller than 10 kΩ but as small as possible otherwise. As a measure of linearity of the waveform at node B, evaluate the difference in slope of v_B at the beginning and end of each half-period. Express this slope difference as a percentage of the maximum slope.

5.50 Repeat Exercise 5.25 using a capacitor of 1,000 pF and $R_1 = 100$ kΩ for a frequency of oscillation of 10 kHz.

**5.51 Figure P5.51 shows a monostable multivibrator circuit. In the stable state, $v_O = L_+$, $v_A = 0$, and $v_B = -V_{ref}$. The circuit can be triggered by applying a positive input pulse of height greater than V_{ref}. Show the resulting waveforms of v_O and v_A. Also, show that the pulse generated at the output will have a width T given by

$$T = CR \ln\left(\frac{L_+ - L_-}{V_{ref}}\right)$$

Note that this circuit has the interesting property that the pulse width can be controlled by changing V_{ref}.

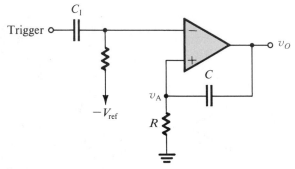

Fig. P5.51

**5.52 The two-diode circuit shown in Fig. P5.52 can provide a crude approximation to a sine-wave output when driven by a triangular waveform. To obtain a good approximation, the peak of the triangular waveform, V, is selected so that the slope of the desired sine wave at the zero crossings is equal to that of the triangular wave. Also, the value of R is selected so that when v_I is at its peak the output voltage is equal to the desired peak of the sine wave. If the diodes are 1-mA devices with 0.1-V/decade i-v characteristics, find the values of V and R that will yield an approximation to a sine waveform of 0.7-V peak amplitude. Then find the angles θ (where $\theta = 90°$ when v_I is at its peak) at which the output of the circuit is 0.7, 0.65, 0.6, 0.55, 0.5, 0.4, 0.3, 0.2, 0.1, and 0 V. Use the angle values obtained to determine the values of the exact sine wave (that is, $0.7 \sin \theta$). Plot both the exact sine wave and the approximate version provided by the circuit over half a cycle.

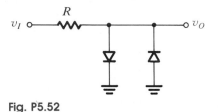

Fig. P5.52

5.53 Design a two-segment sine-wave shaper using a 10-kΩ input resistor, two diodes, and two clamping voltages. The circuit, fed by a 10-V peak-to-peak triangular wave, should limit the amplitude of the output signal via a 0.7-V diode to a value corresponding to that of a sine wave whose zero crossing slope matches that of the triangle. What are the clamping voltages you have chosen?

Junction Field-Effect Transistors (JFETs)

INTRODUCTION

In this chapter we begin our study of *three-terminal* semiconductor devices. Three-terminal devices are far more useful than two-terminal ones because they can be used in a multitude of applications ranging from signal amplification to the realization of logic and memory functions. The basic principle involved is the use of the voltage between two terminals to control the current flowing in the third terminal. In this way a three-terminal device can be used to realize a controlled source, which is the basis for amplifier design. Also, in the extreme the control signal can be used to cause the current in the third terminal to change from zero to a large value, thus allowing the device to act as a switch.

The device studied in this chapter is a special type of field-effect transistor (FET) called the *junc-tion field-effect transistor* (JFET). The other type of FET—namely, the metal-oxide-semiconductor FET, or MOSFET—will be studied in the next chapter. The name field-effect transistor arises from the fact that current flow between two of the device terminals is controlled by an electric field, which in turn is established by a voltage applied to the third terminal, as will be explained below. FETs are also called *unipolar transistors* because current is conducted by charge carriers (electrons or holes) flowing through one type of semiconductor (*n*-type in *n*-channel FETs and *p*-type in *p*-channel FETs). This is in contrast to *bipolar transistors* (which we will study in Chapter 8), where current passes through both *n*-type and *p*-type semiconductor materials in series.

JFETs are useful in the design of special amplifier circuits, especially those with very high input impedances. For instance, op amps with very high input impedances usually have a first stage made up of JFETs. JFETs can be also combined with bipolar transistors to provide high-performance linear circuits (called BIFET circuits). The JFET structure employing a metal-semiconductor (Schottky) junction is used with the semiconductor gallium arsenide to form the MESFET, a JFET-like device suitable for application in amplifiers and logic circuits in the gigahertz range. The JFET is also used as an analog switch and in a variety of other analog circuit applications. □

6.1 PHYSICAL OPERATION

There are two types of JFET: the *n*-channel device and the *p*-channel device. In the following we will explain in detail the operation of the *n*-channel FET. The *p*-channel FET works in a similar manner except for a reversal of polarities of all currents and voltages, as will be briefly explained.

Figure 6.1 shows the basic structure[1] of the *n*-channel JFET. It consists of a slab of *n*-type silicon with *p*-type silicon diffused on both sides. The *n* region is called the *channel*, while the *p*-type regions are electrically connected together and form the *gate*. Metal contacts are made to both ends of the channel, with the terminals called the *source* (S) and the *drain* (D). Similarly, a metal contact is made to the *p*-type region to provide the *gate* terminal (G). We will always assume that the gate regions

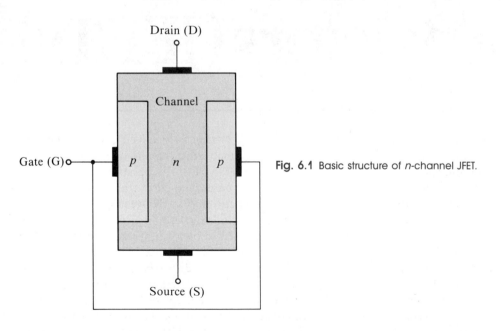

Fig. 6.1 Basic structure of *n*-channel JFET.

---

[1]For actual geometries and fabrication methods the reader is referred to Appendix A. The structure discussed here is a simplified and idealized one intended only to illustrate the basic principles of operation.

are electrically connected together, but we will not explicitly show such connections in order to keep the diagrams simple.

Figure 6.2 shows the circuit symbol for the *n*-channel JFET. Note that the gate line has an arrowhead whose direction indicates the type of the device (that is, *n*-channel or *p*-channel JFET). For the *n*-channel device we are considering, the arrow points toward the *n* channel, that is, in the forward direction of the gate-to-channel junction. Although JFETs are usually symmetric (that is, the drain and source are interchangeable), it is convenient in analyzing and designing FET circuits to indicate which terminal is the source. For this reason we will distinguish the source by drawing the gate line closer to it than to the drain.

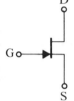

D

G

S

Fig. 6.2 Circuit symbol for *n*-channel JFET.

It can be seen from Fig. 6.1 that the JFET has one *pn* junction, the gate-to-channel junction. In almost all applications this junction will be reverse-biased and hence only a very small leakage current (of the order of 10^{-9} A) will flow in the gate terminal. This also means that the input impedance looking into the gate will be very high.

Although only one *pn* junction exists, it will be seen that the operation of the JFET is dependent on establishing different values of reverse bias at the two ends (source and drain) of this junction. For this reason one is sometimes tempted to speak of the gate-to-drain junction and the gate-to-source junction.

Operation When v_{DS} Is Small

Consider first the operation of the device when a small positive voltage (a fraction of a volt) v_{DS} is applied between the drain and source, as shown in Fig. 6.3. If $v_{GS} = 0$, there will be a narrow depletion region and a current i_D will flow in the channel. The value of i_D will be determined by the value of v_{DS} and the channel resistance r_{DS}. As v_{GS} is made negative, the depletion region widens and the channel narrows. Note that since v_{DS} is small, the reverse-bias voltage will be approximately the same at both ends of the channel, and the channel width will be uniform. The narrowing of the channel causes its resistance to increase, and the i_D-v_{DS} characteristic remains a straight line but with a smaller slope, as shown in Fig. 6.4.

If we keep increasing the magnitude of v_{GS} in the negative direction, a point will be reached at which the depletion region occupies the entire width of the channel. In other words, the channel will be completely depleted of charge carriers (electrons for the *n*-channel device), and hence no current will flow. This condition, called

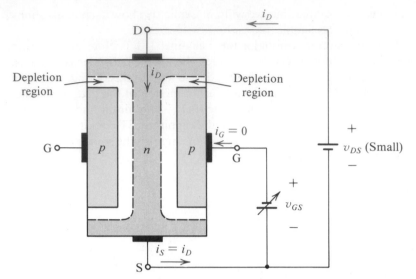

Fig. 6.3 Physical operation of the *n*-channel JFET for small v_{DS}. Although not indicated, the gate terminals are joined together.

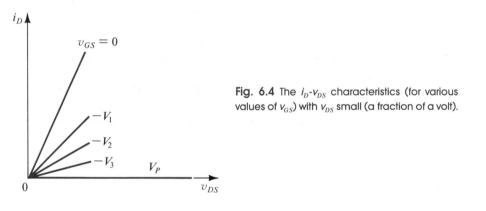

Fig. 6.4 The i_D-v_{DS} characteristics (for various values of v_{GS}) with v_{DS} small (a fraction of a volt).

pinch-off, is illustrated in Fig. 6.5. The voltage v_{GS} at which pinch-off occurs is called the *pinch-off voltage* and is denoted by V_P,

$$V_P = v_{GS}|_{i_D = 0, \, v_{DS} = \text{small}} \tag{6.1}$$

Thus for an *n*-channel device, V_P is a negative voltage and is a parameter of the particular FET. On the characteristics of Fig. 6.4, pinch-off is represented by a horizontal line at zero current level.

The JFET characteristics in Fig. 6.4 suggest that for small v_{DS} the device acts as a linear resistance r_{DS} whose value is controlled by the voltage v_{GS}. In fact, the JFET is used as a *voltage-controlled resistance* (VCR) in some applications, such as in *automatic gain control* (AGC) circuits which are employed in communications receivers. As will be shown, this region of operation of the JFET is not useful for linear-amplifier applications.

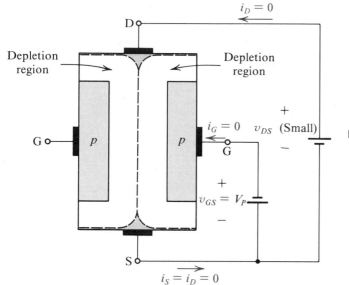

Fig. 6.5 Pinch-off when v_{DS} is small.

Operation as v_{DS} Is Increased

Consider now the characteristics as the value of v_{DS} is increased. Let v_{GS} be kept constant—say, at 0 V to begin with. The gate-to-channel junction will have zero voltage across it at the source end and (because of the voltage drop along the channel) a progressively increasing reverse bias as we move toward the drain. At the drain end the reverse bias voltage v_{DG} will be equal in magnitude to v_{DS}. Since the width of the depletion region depends on the magnitude of reverse bias, the depletion region will have the tapered shape shown in Fig. 6.6, with the result that the channel will be narrowest at the drain end. Thus as v_{DS} is increased, the channel resistance will increase, causing the i_D-v_{DS} characteristic to "bend" and thus become nonlinear, as shown in Fig. 6.7 by the curve corresponding to $v_{GS} = 0$. If we keep increasing v_{DS}, a value will be reached at which the channel is pinched off at the drain end. This should happen at the value that results in the reverse-bias voltage at the drain end being equal to the pinch-off voltage; that is,

$$v_{DG} = -V_P \tag{6.2}$$

(recall that V_P is a negative voltage). Since we are now considering the case $v_{GS} = 0$, Eq. (6.2) implies that pinch-off at the drain end will happen when v_{DS} is given by

$$v_{DS} = -V_P$$

Any further increase in v_{DS} will not alter the channel shape, and hence the current i_D will remain constant at the value reached for $v_{DS} = -V_P$. This value of *saturated drain-to-source current* (I_{DSS}) is specified on the data sheets of the JFET. It is defined as follows:

$$I_{DSS} = i_D\big|_{v_{GS}=0,\, v_{DS}=-V_P} \tag{6.3}$$

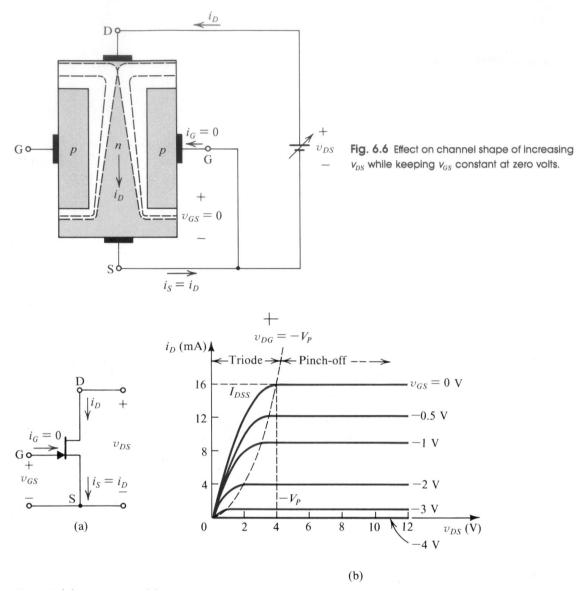

Fig. 6.6 Effect on channel shape of increasing v_{DS} while keeping v_{GS} constant at zero volts.

Fig. 6.7 (a) Circuit and **(b)** complete family of $i_D - v_{DS}$ characteristics for an n-channel JFET with $V_P = -4$ V and $I_{DSS} = 16$ mA.

Here it is important to note the difference between the case when the channel is completely pinched off and the case when pinch-off occurs only at the drain end. In the former case the channel is entirely depleted of charge carriers (electrons for the n-channel device); hence no current will flow ($i_D = 0$). In the latter case, current will continue to flow through the channel; the electrons that flow through the channel will simply drift through the pinched-off region at the drain end of the channel and reach the drain terminal.[2] The voltage across the channel and the current through it remain constant. The difference between the applied voltage v_{DS} and the value $-V_P$ appears across the depletion region at the drain end of the channel.

Consider next the case $v_{GS} = -V$, where V is a positive voltage smaller than $|V_P|$. For small v_{DS} the channel will be of uniform width, since the reverse-bias voltage at the drain end is almost equal to that at the source end. Note that in this case the channel will be narrower than for the previous case where $v_{GS} = 0$. Now, although as v_{DS} is increased the reverse-bias voltage at the source end remains constant, it increases at all other points along the channel. The largest magnitude of reverse bias will be at the drain end. Thus the channel will again have a tapered shape, and its resistance will increase as v_{DS} is increased. Eventually, the channel will be pinched off at the drain end. This will happen at the value of the drain voltage corresponding to

$$v_{DG} = -V_P$$

which can be written as

$$v_{DS} - v_{GS} = -V_P$$

Since in this case $v_{GS} = -V$, it follows that pinch-off will occur at

$$v_{DS} = -V_P - V$$

which is smaller than the value for the $v_{GS} = 0$ case by the magnitude of the reverse bias voltage v_{GS}. Since the channel in this case is narrower than it was for the $v_{GS} = 0$ case, the saturated value of drain current will be smaller than I_{DSS}.

Continuing in this manner for other values of v_{GS} down to $v_{GS} = V_P$, we obtain the complete family of i_D-v_{DS} characteristics shown in Fig. 6.7.

6.2 STATIC CHARACTERISTICS

The static characteristics of an n-channel JFET are shown in Fig. 6.7. The reason for the word "static" is that these characteristics are measured with dc or low-frequency signals; thus effects of internal capacitances are not observed.

As shown in Fig. 6.7, there are two distinct regions of operation: the *triode* region [also known as the voltage-controlled-resistance (VCR) region] and the *pinch-off* region. The two regions are separated by a parabolic boundary, represented by the broken-line curve in Fig. 6.7.

The Triode Region

In the triode region the FET acts as a resistance (r_{DS}) whose value is controlled by the gate-to-source voltage v_{GS}. This resistance is linear for small values of v_{DS}. The

---

[2] In practice, the channel width at the pinched-off end is not zero; rather, it approaches some small limiting value and the drain current i_D flows through this narrow neck.

i_D-v_{DS} relationship in the triode region is parabolic and is described by

$$i_D = I_{DSS}\left[2\left(1 - \frac{v_{GS}}{V_P}\right)\frac{v_{DS}}{-V_P} - \left(\frac{v_{DS}}{V_P}\right)^2 \right] \tag{6.4}$$

where V_P and I_{DSS} are parameters of the JFET whose values are usually given on the device data sheets. For n-channel FETs V_P is a negative voltage. Since v_{GS} is always negative, the gate-to-channel junction remains reverse-biased at all times. This results in the gate current being almost zero. In fact, the gate current will consist mostly of the leakage current of the reverse-biased pn junction. For silicon devices this current is of the order of a few nanoamperes, but it increases with temperature. Temperature effects will be discussed later.

For small values of v_{DS}, Eq. (6.4) can be approximated

$$i_D \simeq \frac{2I_{DSS}}{-V_P}\left(1 - \frac{v_{GS}}{V_P}\right)v_{DS} \tag{6.5}$$

This linear relationship represents the i_D-v_{DS} characteristics near the origin. The linear resistance r_{DS} is therefore given by

$$r_{DS} = \left.\frac{v_{DS}}{i_D}\right|_{v_{DS}\text{ small}}$$

Thus

$$r_{DS} = \left[\frac{2I_{DSS}}{-V_P}\left(1 - \frac{v_{GS}}{V_P}\right)\right]^{-1} \tag{6.6}$$

The Boundary Between the Triode and Pinch-Off Regions

Pinch-off is reached when the reverse-bias voltage at the *drain end* is equal to the pinch-off voltage; that is,

$$v_{DG} = -V_P \tag{6.7}$$

This equation represents the boundary between the triode region and the pinch-off region. It can be rewritten

$$v_{DS} = v_{GS} - V_P \tag{6.8}$$

Substituting in Eq. (6.4) yields

$$i_D = I_{DSS}\left(\frac{v_{DS}}{V_P}\right)^2 \tag{6.9}$$

which is the equation of the broken-line parabola in Fig. 6.7.

Thus to operate in the triode region, for which Eq. (6.4) applies, the drain-to-gate voltage should be less than $-V_P$,

$$v_{DG} < -V_P \qquad \text{(Triode region)} \tag{6.10}$$

which can be rewritten

$$v_{DS} < v_{GS} - V_P \qquad \text{(Triode region)} \tag{6.10a}$$

On the other hand, operation in the pinch-off region is obtained for

$$v_{DG} \geq -V_P \qquad \text{(Pinch-off region)} \qquad (6.11)$$

which implies that *the drain voltage should be higher than the gate voltage by at least* $|V_P|$. The condition in Eq. (6.11) can be alternately expressed

$$v_{DS} \geq v_{GS} - V_P \qquad \text{(Pinch-off region)} \qquad (6.11a)$$

The Pinch-Off Region

In the pinch-off region the i_D-v_{DS} characteristics are (ideally) horizontal straight lines whose heights are determined by the value of v_{GS}. It follows that in pinch-off the JFET operates as a constant-current source with the value of the current controlled by v_{GS}. Furthermore, this constant-current source has ideally an infinite resistance (that is, looking back into the drain terminal one sees an infinite resistance). Also, the input impedance of this controlled source (looking between the control terminals G and S) is ideally infinite. The control relationship is given approximately by the square law,

$$i_D = I_{DSS}\left(1 - \frac{v_{GS}}{V_P}\right)^2 \qquad (6.12)$$

This characteristic, which applies only in pinch-off, is sketched in Fig. 6.8. An equivalent circuit representation of the operation of the JFET in pinch-off is shown in Fig. 6.9. This is a large-signal model for the JFET for which $v_{DG} \geq |V_P|$. Since in

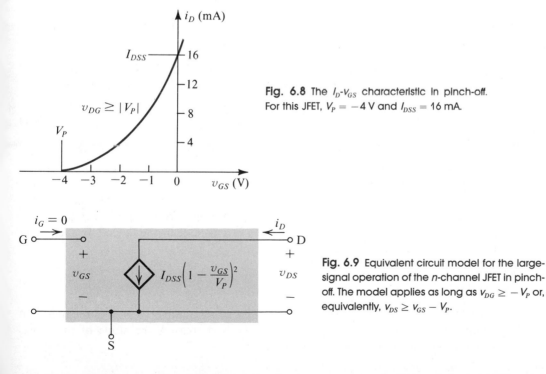

Fig. 6.8 The I_D-V_{GS} characteristic in pinch-off. For this JFET, $V_P = -4$ V and $I_{DSS} = 16$ mA.

Fig. 6.9 Equivalent circuit model for the large-signal operation of the *n*-channel JFET in pinch-off. The model applies as long as $v_{DG} \geq -V_P$ or, equivalently, $v_{DS} \geq v_{GS} - V_P$.

pinch-off the JFET operates as a voltage-controlled current source, the pinch-off region is useful for applications that involve using the JFET as an amplifier. For this reason the pinch-off region is also called the *active region*. JFET amplifiers will be studied in a later section.

Finite Output Resistance in Pinch-Off

The i_D-v_{DS} characteristic curves of practical JFETs show finite, nonzero slope in pinch-off, as illustrated in Fig. 6.10. In fact, as indicated in Fig. 6.10, the characteristic curves in pinch-off are straight lines that, when extrapolated, intercept the v_{DS} axis at a single point. If we denote the voltage corresponding to this point $-V_A$, where

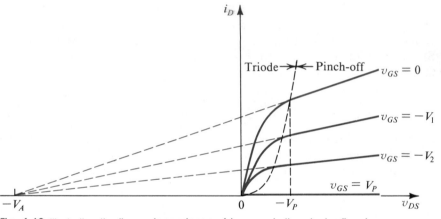

Fig. 6.10 Illustrating the linear dependence of i_D on v_{DS} in the pinch-off region.

V_A is a positive voltage whose value for JFETs fabricated using the standard IC process is typically about 100 V, we can express the pinch-off characteristics by

$$i_D = I_{DSS}\left(1 - \frac{v_{GS}}{V_P}\right)^2\left(1 + \frac{v_{DS}}{V_A}\right) \tag{6.13}$$

where we have included a factor that represents the linear dependence of i_D on v_{DS}. At a given value of v_{GS}, the linear dependence of i_D on v_{DS} in pinch-off can be represented by including a resistance r_o in parallel with the controlled current source in the model of Fig. 6.9. The resistance r_o is equal to the inverse of the slope of the i_D-v_{DS} characteristic lines. Its value can be determined from Eq. (6.13) and can be shown to be inversely proportional to the current level in the device. In fact, from Fig. (6.10) we see that if $V_A \gg |V_P|$, then

$$r_o \simeq \frac{V_A}{I_D} \tag{6.14}$$

where I_D is the value of the drain current, for a given v_{GS}, at the onset of the pinch-off region. Figure 6.10 clearly shows that as I_D is decreased, the slope of the i_D-v_{DS} straight lines decreases, and thus r_o increases.

Junction Breakdown

As the drain voltage is increased, a value is reached at which the reverse-biased *pn* junction of the JFET breaks down in an avalanche manner. Breakdown occurs at the point of maximum reverse bias, which is at the drain end of the junction, when v_{DG} exceeds the junction breakdown voltage. At $v_{GS} = 0$ this breakdown voltage is denoted BV_{DG0}. As v_{GS} is made more negative, the breakdown occurs at correspondingly lower values of v_{DS}, as indicated in Fig. 6.11. Note that, just as in a diode, in breakdown i_D increases rapidly and must be limited by the external circuit.

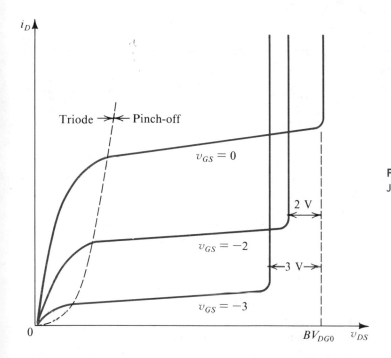

Fig. 6.11 Illustrating junction breakdown in the JFET.

Temperature Effects

Circuit designers are usually concerned that their circuit meets specifications over a specified range of operating temperatures. Therefore the change in device parameters with temperature is of interest. For the JFET we have already mentioned one consequence of changing temperature—namely, that the gate leakage current increases with temperature, roughly doubling for every 10°C rise in temperature. In addition, both the conductivity of the channel and the barrier voltage V_0 of the gate–channel junction are functions of temperature. The result of this can be expressed in terms of the change in gate-to-source voltage v_{GS} when temperature is changed while the drain current is held constant. It can be shown, however, (see Hayt and Neudeck, 1984), that there exists a particular value of drain current at which the temperature coefficient of v_{GS} is zero. Thus with proper circuit design one can arrange that the changes with temperature are very small. We shall not pursue this point any further here.

Exercises

In the following problems, let the *n*-channel JFET have $V_P = -4$ V and $I_{DSS} = 10$ mA, and unless otherwise specified assume that in pinch-off the output resistance is infinite.

6.1 For $v_{GS} = -2$ V, find the minimum v_{DS} for the device to operate in pinch-off. Calculate i_D for $v_{GS} = -2$ V and $v_{DS} = 3$ V.

Ans. 2 V; 2.5 mA

6.2 For $v_{DS} = 3$ V, find the change in i_D corresponding to a change in v_{GS} from -2 to -1.6 V.

Ans. 1.1 mA

6.3 For small v_{DS} calculate the value of r_{DS} at $v_{GS} = 0$ V and at $v_{GS} = -3$ V.

Ans. 200 Ω; 800 Ω

6.4 If $V_A = 1,000$ V, find the JFET output resistance when operating in pinch-off at a current of 1 mA, 2.5 mA, and 10 mA.

Ans. 1,000 kΩ; 400 kΩ; 100 kΩ

6.3 THE *p*-CHANNEL JFET

The circuit symbol and the static characteristics of a *p*-channel JFET are shown in Fig. 6.12. The circuit symbol differs from that of the *n*-channel device only in the direction of the arrowhead on the gate line. Indeed, it is the direction of the arrowhead that indicates the polarity of the device (that is, *n*-channel or *p*-channel JFET). For the *p*-channel device the arrow is pointing "outward," in the forward direction of the channel (*p*-type)-to-gate (*n*-type) junction.

Another difference in the circuit symbol is that it is drawn "upside down." This is done so that currents always flow from the top of the page down. This drawing convention, which is adopted throughout this book, enhances the ability of the reader to rapidly recognize the functional parts of a complex circuit. Needless to say, the value of this convention will be appreciated only at later stages.

We have indicated on the circuit symbol of the *p*-channel FET in Fig. 6.12 the reference directions for currents and voltages where the correlation between these reference directions and the symbols used should be noted. For instance, v_{SD} means, "how high v_S is with respect to v_D." In normal applications v_{SD} will be positive, which means that the source is more positive than the drain. On the other hand, v_{SG} will always be negative because the gate-to-channel junction has to be reverse-biased. Finally, i_D is shown flowing from source to drain, which is the actual direction of drain-current flow in a *p*-channel device. As in the *n*-channel device, the gate current is very small and ideally is zero; thus the source and drain currents are equal.

The i_D-v_{SD} characteristics look identical to those of the *n*-channel device. For operation in the pinch-off region, which is also called the active region, the drain has to be lower in potential than the gate by at least V_P volts. Note that for *p*-channel JFETs, V_P is positive, by convention. Thus the JFET will be in pinch-off when the following condition is satisfied:

$$v_{GD} \geq V_P \qquad \text{(Pinch-off region)} \qquad (6.15)$$

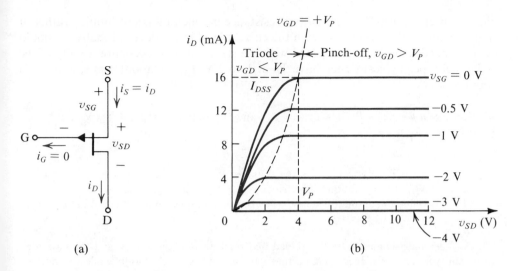

(a)   (b)

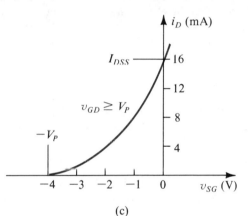

(c)

Fig. 6.12 Circuit symbol and static characteristics for a p-channel JFET whose $V_P = +4$ V and $I_{DSS} = 16$ mA.

and will be in the triode region under the condition

$$v_{GD} < V_P \qquad \text{(Triode region)} \tag{6.16}$$

The boundary between the two regions is characterized by

$$v_{GD} = V_P \tag{6.17}$$

Stated in words, for the p-channel JFET to operate in pinch-off the drain should be lower than the gate by at least V_P volts.

The relationships governing i_D, v_{GS}, and v_{DS} are identical to those for the n-channel device [Eqs. (6.4), (6.12), and (6.13)] except that here V_P is positive, v_{GS} is normally positive (because v_{SG} is negative), v_{DS} is normally negative, and V_A is negative. As in the case of n-channel FETs, the i_D-v_{SD} characteristics in pinch-off display a finite, nonzero slope. Thus the output resistance in pinch-off is not infinite; rather, it is finite with a value decreasing with the current level in the device. Finally, it should be mentioned that JFETs fabricated with the standard IC process are usually of the p-channel type. Typically they have V_P of 1 to 2 V, I_{DSS} of about 1 mA, and V_A of about -100 V.

Exercises

6.5 Consider a p-channel JFET with $V_P = 5$ V and $I_{DSS} = 10$ mA. If $v_{SG} = -3$ V, find i_D for $v_{SD} = 1$ V and for $v_{SD} = 2$ V.

Ans. 1.2 mA; 1.6 mA

6.6 A p-channel JFET fabricated with the standard IC process has $V_P = 1$ V, $I_{DSS} = 1$ mA, and $V_A = -100$ V. For operation in the triode region near the origin find the value of r_{DS} at $v_{SG} = 0$ V. Also find the output resistance when the device is operated in pinch-off at $v_{SG} = 0$.

Ans. 500 Ω; 100 kΩ

6.7 Consider the p-channel JFET specified in Exercise 6.6. If the source and the gate are grounded, what is the highest voltage that can be applied to the drain while the device operates in pinch-off? What current flows in the drain at this voltage? If the drain voltage is decreased by 4 volts, find the change in drain current.

Ans. -1 V; 1 mA; $+0.04$ mA

6.4 JFET CIRCUITS AT DC

In this section we consider the analysis of some simple JFET circuits fed by dc sources. Our goal is to familiarize the reader with these devices. For simplicity we shall neglect the effect of v_{DS} on i_D in pinch-off.

EXAMPLE 6.1

We want to analyze the circuit shown in Fig. 6.13 to determine the value of I_D and V_D. The FET has $V_P = -4$ V and $I_{DSS} = 10$ mA.

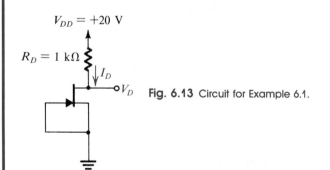

Fig. 6.13 Circuit for Example 6.1.

Solution
Since the gate is connected to the source, it follows that $V_{GS} = 0$. We do not know, however, whether the device is in the pinch-off region or in the triode region. To find out, we assume that it is in pinch-off, carry out the analysis, and then check whether

our original assumption is correct or not. If the device is in pinch-off, then $I_D = I_{DSS}$. Thus

$$I_D = 10 \text{ mA}$$

The drain voltage can then be calculated from

$$V_D = V_{DD} - R_D I_D$$
$$= 20 - 1 \times 10$$

Thus

$$V_D = +10 \text{ V}$$

Since the gate is at zero volts, the drain is higher in potential than the gate by 10 V. This is greater than $|V_P|$, and hence the device is indeed in pinch-off, as originally assumed.

The above example illustrates an interesting application of the JFET as a two-terminal constant-current device. Specifically, a JFET whose gate is connected to its source conducts a constant current equal to I_{DSS} provided that the external circuit ensures that the voltage across the two-terminal device (see Fig. 6.14) is always maintained greater than $|V_P|$. This constant-current two-terminal device is the dual of the zener diode discussed in Chapter 4. The latter provides a constant voltage as long as a certain minimum current is forced through it by the external circuit. Note, of course, that the current in the JFET will not be exactly constant, since the i_D-v_{DS} characteristic curve in pinch-off exhibits some finite slope. In other words, the current of the two-terminal device of Fig. 6.14 will be slightly dependent on the voltage that the external circuit forces across it. This again is parallel to the case of the zener diode.

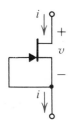

n-channel JFET

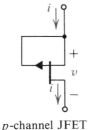

p-channel JFET

Fig. 6.14 The application of the JFET as a constant-current device. Ideally, i is constant and equal to I_{DSS} provided that $v \geq |V_P|$.

EXAMPLE 6.2

We wish to analyze the circuit of Fig. 6.15 to determine I_D and V_D. The JFET is specified to have $V_P = -4$ V and $I_{DSS} = 10$ mA.

Solution

This circuit has a topology identical to that of Example 6.1. The difference is that here $R_D = 1.8$ kΩ. Assuming operation in pinch-off, we obtain

$$I_D = I_{DSS} = 10 \text{ mA}$$

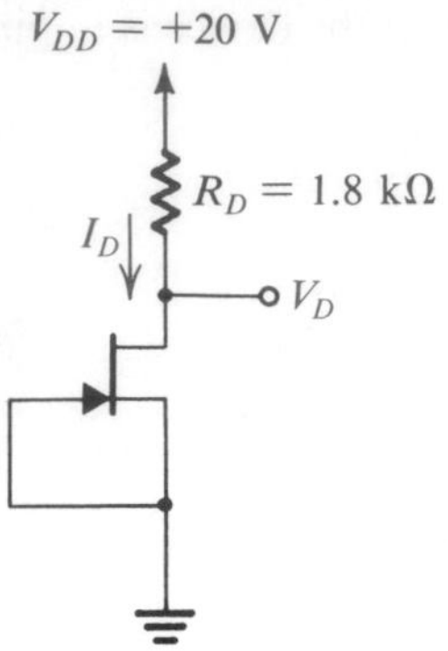

$V_{DD} = +20$ V

$R_D = 1.8$ kΩ

I_D

V_D

Fig. 6.15 Circuit for Example 6.2.

which results in

$$V_D = 20 - 1.8 \times 10 = 2 \text{ V}$$

Thus the drain is higher than the gate by only 2 V. Since we require a minimum of 4 V for pinch-off operation, we conclude that the device is in the triode region and thus our original assumption is incorrect. We now have to repeat the analysis assuming operation in the triode region. The I_D-V_{DS} relationship is that given in Eq. (6.4); that is,

$$I_D = I_{DSS}\left[2\left(1 - \frac{V_{GS}}{V_P}\right)\frac{V_{DS}}{-V_P} - \left(\frac{V_{DS}}{V_P}\right)^2\right]$$

Substituting $V_{GS} = 0$, $I_{DSS} = 10$ mA, and $V_P = -4$ V gives

$$I_D = 5V_{DS}(1 - \tfrac{1}{8}V_{DS}) \tag{6.18}$$

The other equation that governs circuit operation is

$$V_{DS} = V_{DD} - R_D I_D$$

in which we substitute $V_{DD} = 20$ V and $R_D = 1.8$ kΩ to obtain

$$I_D = \frac{20}{1.8} - \frac{V_{DS}}{1.8} \tag{6.19}$$

Equations (6.18) and (6.19) can now be solved to obtain V_{DS} and I_D. The results are

$$V_{DS} = 3 \text{ V} \quad \text{or} \quad 5.8 \text{ V}$$

The second answer is obviously inappropriate since it implies operation in the pinch-off region, which we have already ruled out. Thus

$$V_{DS} = 3 \text{ V}$$

and the corresponding current I_D is

$$I_D = 9.4 \text{ mA}$$

EXAMPLE 6.3

We wish to analyze the circuit in Fig. 6.16 to find the voltages and current. The FET has $V_P = -2$ V and $I_{DSS} = 4$ mA.

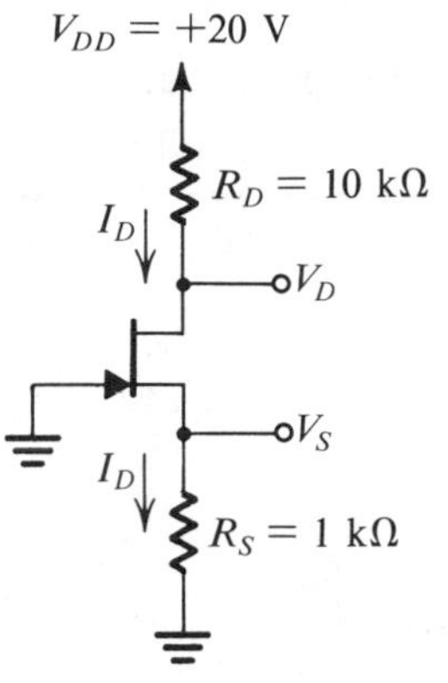

$V_{DD} = +20$ V

$R_D = 10$ kΩ

I_D

V_D

V_S

I_D

$R_S = 1$ kΩ

Fig. 6.16 Circuit for Example 6.3.

Solution

Assuming that the device is conducting, we see that the current I_D creates a voltage drop $I_D R_S$ across the source resistance R_S. Since the gate is at ground voltage, it follows that

$$V_{GS} = -I_D R_S \tag{6.20}$$

which is negative. It therefore is possible that the FET is operating either in pinch-off or in the triode region. Which region will depend on the voltage at the drain, which we do not know. Thus to begin the analysis we have to make an assumption. We shall assume that the FET is in the pinch-off region, carry out the analysis, and then check the validity of our original assumption.

Assuming operation in pinch-off, the I_D-V_{GS} relationship is given by

$$I_D = I_{DSS}\left(1 - \frac{V_{GS}}{V_P}\right)^2 \tag{6.21}$$

Equations (6.20) and (6.21) can be used to determine I_D and V_{GS}. Substituting $R_S = 1$ kΩ, $I_{DSS} = 4$ mA, and $V_P = -2$ V and combining the two equations, results in the quadratic equation

$$I_D^2 - 5I_D + 4 = 0 \tag{6.22}$$

This equation has two roots, $I_D = 4$ mA and $I_D = 1$ mA. The first solution is obviously inappropriate, since it equals I_{DSS}, which implies that $V_{GS} = 0$. This latter value is inconsistent with the value obtained by substituting in Eq. (6.20). It follows that

$$I_D = 1 \text{ mA}$$

and the corresponding V_{GS} is

$$V_{GS} = -1 \text{ V}$$

Thus

$$V_S = +1 \text{ V}$$

$$V_D = 20 - 1 \times 10 = +10 \text{ V}$$

Since the drain is higher than the gate by more than $|V_P|$, the device is indeed in pinch-off, as originally assumed.

Exercises

6.8 In the circuit of Fig. E6.8 let the FET have $V_P = 3$ V and $I_{DSS} = 9$ mA. Find the voltages V_G, V_S, and V_D.

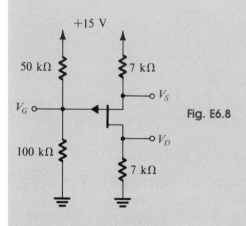

Fig. E6.8

Ans. $+10$ V; $+8$ V; $+7$ V

6.9 For the circuit shown in Fig. E6.9, find the values of R_S and R_D so that the JFET operates at $I_D = 0.5$ mA and $V_{DG} = 5$ V. The JFET parameters are $I_{DSS} = 2$ mA and $V_P = -2$ V.

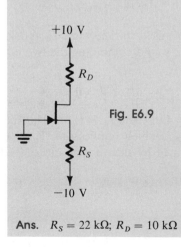

Fig. E6.9

Ans. $R_S = 22$ kΩ; $R_D = 10$ kΩ

6.5 Graphical Analysis

Since the JFET is well-characterized by equations, it is rarely necessary to apply graphical techniques in the analysis of JFET circuits. Nevertheless, it is illustrative and instructive to consider the graphical analysis of the basic circuit in Fig. 6.17. Here we assume that the voltage v_{GS} takes on negative values with a maximum of zero.

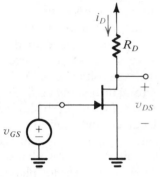

Fig. 6.17 Basic grounded-source JFET circuit.

Thus the gate-to-channel junction will always be reverse-biased, ensuring proper JFET operation. The operating point will be located on the i_D-v_{DS} curve corresponding to the specific value of v_{GS} applied. Where it will be on that curve will be determined by V_{DD} and R_D from

$$v_{DS} = V_{DD} - R_D i_D$$

which may be rewritten

$$i_D = \frac{V_{DD}}{R_D} - \frac{1}{R_D} v_{DS} \tag{6.23}$$

This is a linear equation in the variables i_D and v_{DS} and can be represented by a straight line on the i_D-v_{DS} plane. Such a line intercepts the v_{DS} axis at V_{DD} and has a slope equal to $-1/R_D$. Since when this circuit is used as an amplifier, R_D represents the *load resistance*, the straight line representing Eq. (6.23) is called the *load line*. The operating point of the JFET will lie at the intersection of the load line and the i_D-v_{DS} curve corresponding to the applied value of v_{GS}.

Amplifier Operation

We shall illustrate the graphical analysis outlined above by considering the operation of the circuit of Fig. 6.17 as an amplifier. Since the JFET is a nonlinear device, linear amplification is obtained by using the biasing technique introduced in Section 2.2. Thus the voltage v_{GS} will consist of a dc bias voltage V_{GS}, on which will be superimposed the time-varying signal v_{gs} that we wish to amplify; thus

$$v_{GS} = V_{GS} + v_{gs}$$

As an example, let the JFET have the characteristics depicted in Fig. 6.7. Figure 6.18a shows the circuit with a dc bias voltage $V_{GS} = -1$ V and an input signal v_{gs} that is triangular in shape with 1-V peak-to-peak amplitude, a load resistance $R_D = 1.33$ kΩ, and a dc supply $V_{DD} = 20$ V. Figure 6.18b shows the i_D-v_{DS} characteristics of the JFET, together with the load line corresponding to $R_D = 1.33$ kΩ. In the absence of the input signal v_{gs}, the JFET will operate at the point labeled Q, which is at the intersection of the curve for $v_{GS} = -1$ V and the load line. The coordinates of the dc bias point Q determine the dc current in the drain, $I_D = 9$ mA, and the dc voltage at the drain, $V_{DS} = 8$ V.

When the triangular-wave signal v_{gs} is applied, the instantaneous operating point will move along the load line in correspondence with the total instantaneous voltage v_{GS}. This is shown in Fig. 6.18b, from which we observe that, for instance, at the positive peak of the input signal, $v_{gs} = 0.5$ V, $v_{GS} = -1 + 0.5 = -0.5$ V, the corresponding drain current is 12.25 mA, and the corresponding drain voltage is 3.7 V. It follows that the total instantaneous drain current i_D and the total instantaneous drain voltage v_{DS} can be determined in this manner, point-by-point. Figure 6.18b shows the resulting waveforms. We note that superimposed on the dc current I_D we obtain a time-varying component i_d that is almost perfectly triangular in shape. Also, superimposed on the dc voltage V_{DS} we obtain a signal component that also is almost perfectly triangular in shape. This is the output voltage signal and, except for phase inversion, is an amplified replica of the input signal: It has a peak-to peak amplitude of 8 V, and thus the gain of the amplifier is -8 V/V.

In the example above we were able to obtain almost linear amplification from the nonlinear JFET by properly choosing the dc bias point Q, and by keeping the input signal amplitude small. In the following sections we will study the analysis and design of JFET amplifiers in detail. At this stage, however, it is important to note that the instantaneous operating point was confined to the pinch-off region. This is done so that the JFET operates as a current source whose magnitude is controlled by v_{gs}. If the instantaneous operating point is allowed to leave the pinch-off region, the FET no longer operates as a controlled current source and severe nonlinear distortion may result. This is illustrated in Fig. 6.18c where the JFET is still biased at $V_{GS} = -1$ V but a larger resistance $R_D = 1.78$ kΩ is used. As a result of using the larger load resistance the instantaneous operating point enters the triode region during most of the positive halves of v_{gs}. As can be seen, the signal component of i_D and the output signal voltage v_{ds} are severely distorted.

In summary, for the JFET to operate as a linear amplifier it must be biased at a point in the middle of the active (pinch-off) region, the instantaneous operating point must at all times be confined to the active region, and the input signal must be kept sufficiently small. The last point will be elaborated on in a later section.

Operation as a Switch

Another important application of the JFET is its use as a switch, with v_{GS} as the control signal and the drain and source as the switch terminals. This mode of operation is illustrated in Fig. 6.19. As indicated in Fig. 6.19a the control signal v_{GS} is a pulse

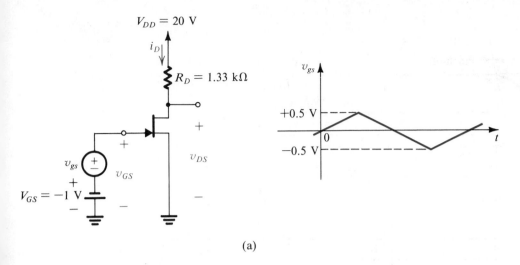

(a)

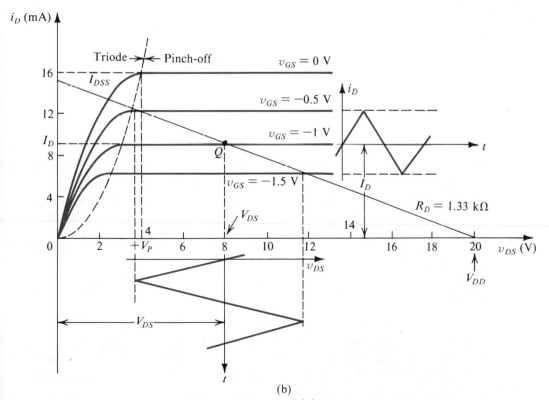

(b)

Fig. 6.18 (a) A grounded-source JFET amplifier circuit. (b) Graphical analysis of this circuit.

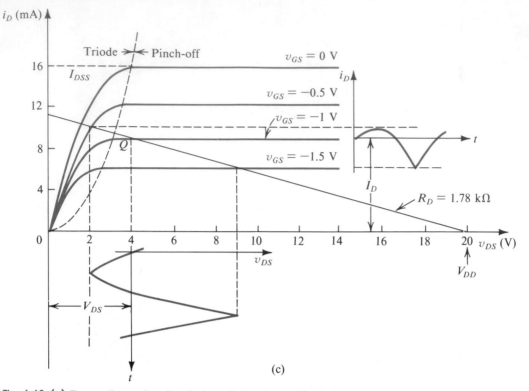

Fig. 6.18 (c) The nonlinear distortion that results by allowing the instantaneous operating point to enter the triode region of operation (here $R_D = 1.78$ kΩ).

having the two levels, V_P and zero. Figure 6.19b shows the graphical construction from which the two operating points corresponding to the two values of v_{GS} can be determined. When $v_{GS} = V_P$, operation is at point B and the JFET is cut off. It has the equivalent circuit shown in Fig. 6.19c, which corresponds to the open position of the FET switch. When $v_{GS} = 0$, operation is at point C in the triode region. Here v_{DS} is very small (0.25 V in the example depicted in Fig. 6.19), and the output resistance r_{DS} is also small (1.27 Ω for our example). In this state the JFET switch is closed and is represented by the equivalent circuit in Fig. 6.19d.

From the above we see that the waveform at the drain will be a pulse whose polarity is the inverse of the input pulse, as indicated in Fig. 6.19a. For this reason the circuit under consideration can be thought of as a logic inverter (see Chapter 15). Applications of the JFET switch will be studied in a later section. At this point, however, it is important to note that for the JFET to operate as a switch, the levels of v_{GS} as well as the value of R_D must be selected so that in one extreme the JFET is off and in the other it is operating in the triode region with a small closure voltage V_{ON} and a small "on" resistance r_{DS}.

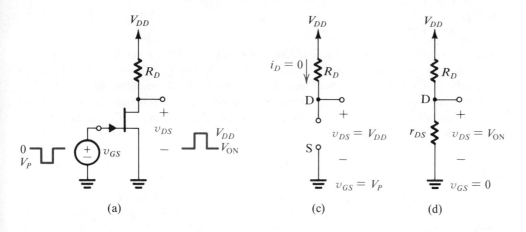

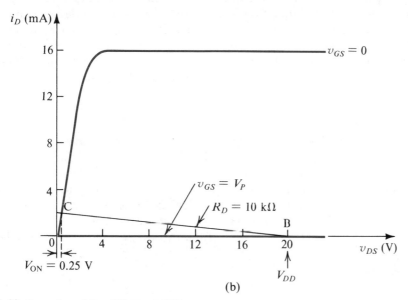

Fig. 6.19 Operation of the JFET as a switch.

6.10 Consider the circuit shown in Fig. 6.17 and let $V_{DD} = 15$ V, $R_D = 3$ kΩ, $I_{DSS} = 12$ mA, $V_P = -4$ V, and $v_{GS} = -2$ V. Calculate the value of I_D and V_D.

Ans. 3 mA; +6 V

6.11 For the circuit in Fig. 6.17 find the value of R_D so that when $v_{GS} = 0$, $v_{DS} = 0.1$ V. Let $V_{DD} = 15$ V, $I_{DSS} = 12$ mA, and $V_P = -4$ V. Also, calculate r_{DS} at the resulting operating point.

Ans. 25 kΩ; 168 Ω

6.6 BIASING

The first step in the design of a JFET amplifier is to establish a stable and predictable dc operating point. As explained in the previous section, this operating point should be in the active region (pinch-off region) and should allow for sufficient signal swing without the device entering the triode region or the cutoff region. Later we shall discuss the signal-swing problem in some detail. The present section is devoted to techniques for establishing a stable dc operating point.

Stable Operating Point

A stable operating point is one that is almost independent of variations in the device parameters V_P and I_{DSS}. These parameters vary with temperature; more important, they vary considerably between different units belonging to the same device type. It is usual to see on the data sheets of a JFET a wide range specified for V_P and I_{DSS} (say, V_P from -2 to -8 V and I_{DSS} from 4 to 16 mA). A good bias design is one that ensures that I_D and V_{DS} will always be within a certain range of their nominal value (say, $\pm 20\%$) independent of the value of V_P and I_{DSS}. In other words, we wish to find a biasing arrangement by which the value of I_D (and hence V_{DS}) will not change substantially if we replace the particular FET by another of the same type.

Extreme Devices

Although the problem of biasing JFETs is difficult because of the wide variation in device parameters, there is a simplifying factor. Devices whose $|V_P|$ is large tend also to have high values of I_{DSS}, whereas those with low $|V_P|$ also have low I_{DSS}. Thus if a device is specified to have V_P in the range -2 to -8 V and I_{DSS} in the range 4 to 16 mA, we may assume that the two extreme devices are as follows: the "low" device, with $V_P = -2$ V and $I_{DSS} = 4$ mA, and the "high" device, with $V_P = -8$ V and $I_{DSS} = 16$ mA. The i_D-v_{GS} characteristics of the two extreme devices are sketched in Fig. 6.20.

Fixed Bias

The simplest approach to biasing the JFET is to apply a constant dc voltage between gate and source; however, this results in very unsatisfactory performance. For instance, from the characteristics shown in Fig. 6.20 we find that a fixed voltage V_{GS} of, say, -3 V will result in the "high" device conducting a current I_D of 6.25 mA while the "low" device will be cut off! For $V_{GS} = -1$ V the "high" device will have $I_D = 12.25$ mA, while the "low" device will have $I_D = 1$ mA, a very large difference [see line (a) in Fig. 6.20].

Self-Bias

Better results are obtained with the self-bias circuit shown in Fig. 6.21. Here we have included a resistance R_S in the source lead. The voltage $I_D R_S$ developed across this

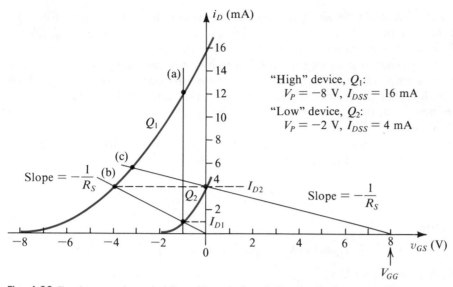

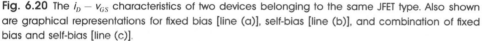

Fig. 6.20 The $i_D - v_{GS}$ characteristics of two devices belonging to the same JFET type. Also shown are graphical representations for fixed bias [line (a)], self-bias [line (b)], and combination of fixed bias and self-bias [line (c)].

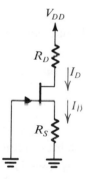

Fig. 6.21 JFET with self-bias.

resistance will constitute the reverse bias V_{GS},

$$V_{GS} = -I_D R_S \tag{6.24}$$

The term *self-bias* arises because no external dc source is used in the gate–source circuit. Self-bias is possible only in *depletion-type* devices such as JFETs, some MOSFETs, and vacuum tubes. As will be seen in Chapter 8, bipolar transistors are *enhancement-type* devices, and hence self-bias is not possible.

Let us continue with the analysis of the circuit in Fig. 6.21. We may use Eq. (6.24) together with the i_D-v_{GS} square-law relationship to write

$$I_D = I_{DSS}\left(1 + \frac{I_D R_S}{V_P}\right)^2 \tag{6.25}$$

Graphically, Eq. (6.24) can be represented by the straight line labeled (b) in Fig. 6.20. As can be seen, the value of I_D will vary from I_{D1} for the "low" device to I_{D2} for the "high" device, a less dramatic spread than in the case of fixed bias. Obviously, to minimize the variance in I_D we should select as high a value for R_S as possible (since a high value for R_S results in a straight line with a small slope). However, using a high value for R_S could result in a very low current in the "low" device. In Section 6.7 it will be shown that the gain is proportional to $\sqrt{I_D}$; thus one should attempt to use as high a value for I_D as possible.

As an example, let $R_S = 1\ \text{k}\Omega$ and consider the device whose two extreme characteristics are depicted in Fig. 6.20. For the "low" device we have

$$I_D = 4\left(1 - \frac{I_D}{2}\right)^2 \tag{6.26}$$

which results in

$$I_D = 1\ \text{mA} \qquad \text{and} \qquad V_{GS} = -1\ \text{V}$$

For the high device we have

$$I_D = 16\left(1 - \frac{I_D}{8}\right)^2 \tag{6.27}$$

which results in

$$I_D = 4\ \text{mA} \qquad \text{and} \qquad V_{GS} = -4\ \text{V}$$

Although the spread is smaller than in the case of fixed bias, it is still unacceptably large.

Combination of Fixed Bias and Self-Bias

From Fig. 6.20 it may be observed that to minimize the variance in I_D we should intersect the i_D-v_{GS} curves at a very small slope. This suggests the use of a bias circuit that results in a straight line such as that labeled (c) in Fig. 6.20. This line intercepts

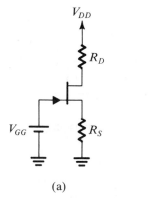

(a)

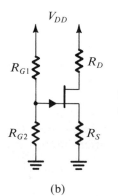

(b)

Fig. 6.22 JFET with combination of self-bias and fixed bias.

the v_{GS} axis at a positive voltage V_{GG} and has the equation

$$v_{GS} = V_{GG} - i_D R_S \tag{6.28}$$

Straightforward reasoning suggests the circuit shown in Fig. 6.22a. In a practical situation the voltage V_{GG} would be obtained as a fraction of V_{DD} through the use of a voltage divider, as shown in Fig. 6.22b. Finally, we should remind the reader that the value of R_D should be such that at all times the drain voltage is higher than the gate voltage by at least $|V_P|$, this being the condition for active-mode operation.

EXAMPLE 6.4

We desire to design the bias circuit of Fig. 6.22b for the JFET whose characteristics are depicted in Fig. 6.20. Assume that the available power supply $V_{DD} = 20$ V.

Solution
To obtain a small variance in I_D and still keep the current relatively high, we should use as high a value for V_{GG} and for R_S as possible. However, the higher the value we use for V_{GG}, the smaller the voltage drop that will be available across R_D (because we have to keep D higher than G by at least $|V_P|$). For a given I_D, a smaller voltage drop across R_D implies a lower value for R_D and hence (as will be shown in Section 6.7) a lower gain. Let us choose $V_{GG} = 8$ V. For the "low" device this value implies that the minimum drain voltage should be 10 V, leaving a maximum of 10 V for the drop across R_D. For the "high" device the minimum drain voltage is 16 V, which leaves a maximum of 4 V for the drop across R_D.

We next have to select a value for R_S. In order to maximize the value of I_D we select a value for R_S that results in the maximum possible I_D for the "low" device. To avoid forward-biasing the gate-to-channel junction, we use $I_D = I_{DSS} = 4$ mA. This corresponds to $V_{GS} = 0$ for the "low" device. It is important here to note that the applied signal will cause the gate-to-source voltage to become positive. This is allowed as long as the resulting forward-bias voltage is less than about 0.5 V.

For the bias line to pass through the point $V_{GS} = 0$, $I_D = 4$ mA, it follows that

$$R_S = \frac{V_{GG}}{4 \text{ mA}} = \frac{8}{4} = 2 \text{ k}\Omega$$

We can now calculate the current in the "high" device from

$$I_D = 16\left(1 - \frac{8 - 2I_D}{-8}\right)^2$$

which results in

$$I_D = 5.63 \text{ mA} \qquad V_{GS} = -3.26 \text{ V}$$

Thus the spread is from 4 mA to 5.63 mA, which is much less than what we have been able to achieve with self-bias alone.

Exercises

6.12 For the circuit considered in Example 6.4 find the maximum value of R_D that will ensure that both the "low" and "high" devices are in pinch-off. Also calculate V_D for these extreme devices.

Ans. 710 Ω; 17.2 V, 16 V

6.13 For the circuit in Fig. 6.22b find the values of R_D, R_S, R_{G1}, and R_{G2} that will result in $I_D = I_{DSS}/2$ and that a third of the power-supply voltage appears across each of R_D, R_S, and the FET (that is, V_{DS}). Let $V_{DD} = 15$ V, $I_{DSS} = 8$ mA, and $V_P = -2$ V. Use 1 μA in the voltage-divider network that feeds the gate.

Ans. $R_D = 1.25$ kΩ; $R_S = 1.25$ kΩ; $R_{G1} = 10.6$ MΩ; $R_{G2} = 4.4$ MΩ

6.14 For the circuit shown in Fig. E6.14 find the values of R_D and R_S that will result in $I_D = 1$ mA and $V_D = +6$ V. The FET has $I_{DSS} = 4$ mA and $V_P = -2$ V. Determine the percentage change in I_D that will result when the FET is replaced with another having **(a)** the same V_P but double the value I_{DSS} and **(b)** double V_p and double I_{DSS}.

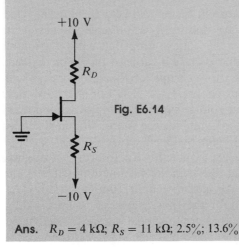

Fig. E6.14

Ans. $R_D = 4$ kΩ; $R_S = 11$ kΩ; 2.5%; 13.6%

6.7 THE JFET AS AN AMPLIFIER

Basic Amplifier Arrangement

The operation of the JFET as an amplifier was introduced in Section 6.5, where we graphically analyzed the basic circuit shown in Fig. 6.23. In this section we develop analytic models for the JFET amplifier. Note that in the circuit of Fig. 6.23 the gate-to-source reverse bias is established by a separate battery, which is not a good biasing arrangement, but our intention is to focus on the concepts involved. Superimposed on the bias voltage V_{GS} we have a signal v_{gs}; thus the total instantaneous gate-to-source voltage is given by

$$v_{GS} = V_{GS} + v_{gs} \tag{6.29}$$

Assuming that the FET will remain in pinch-off at all times, which is achieved by

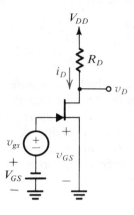

Fig. 6.23 Circuit for establishing the basis for the operation of the JFET as an amplifier.

keeping v_D higher than v_G by at least $|V_P|$; that is,

$$v_D \geq v_{GS} + |V_P| \tag{6.30}$$

it follows that

$$i_D = I_{DSS}\left(1 - \frac{v_{GS}}{V_P}\right)^2 \tag{6.31}$$

Substituting for v_{GS} from Eq. (6.29) into Eq. (6.31) yields

$$i_D = I_{DSS}\left(1 - \frac{V_{GS}}{V_P} - \frac{v_{gs}}{V_P}\right)^2$$

$$= I_{DSS}\left(1 - \frac{V_{GS}}{V_P}\right)^2 - \frac{2I_{DSS}}{V_P}\left(1 - \frac{V_{GS}}{V_P}\right)v_{gs} + I_{DSS}\left(\frac{v_{gs}}{V_P}\right)^2 \tag{6.32}$$

Setting $v_{qs} = 0$ in Eq. (6.32) results in $i_D = I_D$, which is the dc bias current,

$$I_D = I_{DSS}\left(1 - \frac{V_{GS}}{V_P}\right)^2 \tag{6.33}$$

Substituting in Eq. (6.32) results in

$$i_D = I_D + \frac{2I_{DSS}}{-V_P}\left(1 - \frac{V_{GS}}{V_P}\right)v_{gs} + I_{DSS}\left(\frac{v_{gs}}{V_P}\right)^2 \tag{6.34}$$

If we restrict ourselves to small signals satisfying the constraint

$$\left|\frac{v_{gs}}{V_P}\right| \ll 1 \tag{6.35}$$

we can neglect the last term in Eq. (6.34) and obtain

$$i_D \simeq I_D + \frac{2I_{DSS}}{-V_P}\left(1 - \frac{V_{GS}}{V_P}\right)v_{gs} \tag{6.36}$$

Thus the total drain current is composed of two components: the dc bias I_D and a signal component i_d given by

$$i_d = \frac{2I_{DSS}}{-V_P} \left(1 - \frac{V_{GS}}{V_P} \right) v_{gs} \tag{6.37}$$

Thus the signal current is linearly related to the signal voltage v_{gs}, which is a requirement in a linear amplifier. This linear relationship is based on the premise that the signal v_{gs} is much smaller than $|V_P|$, which is referred to as the *small-signal approximation*. Note that if the small-signal assumption is not valid, the term involving v_{gs}^2 in Eq. (6.34) has to be taken into account. This obviously results in the current i_d having components harmonically related to the input signal. Such nonlinear distortion is undesirable and should be minimized in the design of linear amplifiers.

Transconductance

Assuming that the small-signal assumption is adhered to, the signal current i_d is proportional to v_{gs} with the proportionality constant, called *transconductance* and denoted by g_m, given by

$$g_m \equiv \frac{i_d}{v_{gs}} = \frac{2I_{DSS}}{-V_P} \left(1 - \frac{V_{GS}}{V_P} \right) \tag{6.38}$$

Recalling that for n-channel FETs V_P is a negative number and that V_{GS} is also negative, we see that g_m is positive. A relationship for g_m that applies for both n- and p-channel devices can be written

$$g_m = \frac{2I_{DSS}}{|V_P|} \left(1 - \frac{V_{GS}}{V_P} \right) \tag{6.39}$$

Note that g_m is determined by the FET parameters I_{DSS} and V_P as well as by the dc operating point. We may use the square-law relationship of the JFET together with Eq. (6.39) to express g_m in the alternative form

$$g_m = \frac{2I_{DSS}}{|V_P|} \sqrt{\frac{I_D}{I_{DSS}}} \tag{6.40}$$

from which it is evident that g_m is proportional to the square root of the bias current I_D. It follows that g_m will be highest if the FET is biased at $V_{GS} = 0$ (or, equivalently, $I_D = I_{DSS}$). This maximum value of g_m is denoted g_{m0} and is given by

$$g_{m0} = \frac{2I_{DSS}}{|V_P|} \tag{6.41}$$

A graphical interpretation of the above results is shown in Fig. 6.24, which displays the i_D-v_{GS} characteristic in the pinch-off region. A triangular waveform is used for the input signal v_{gs}. Note that g_m is equal to the slope of the parabolic characteristic at the bias point Q,

$$g_m = \left. \frac{\partial i_D}{\partial v_{GS}} \right|_{i_D = I_D} \tag{6.42}$$

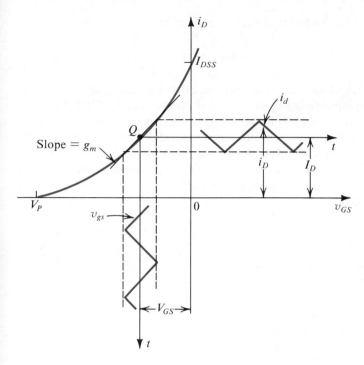

Fig. 6.24 Operation of the JFET as an amplifier.

Also note that the small-signal approximation implies that we are operating on a segment of the graph (around Q) that is sufficiently short to approximate a straight line of slope g_m. Finally, note that the reason that one does not usually bias the device at $V_{GS} = 0$ is to allow for sufficient input-signal swing without forward-biasing the gate-to-channel junction.

Voltage Gain

Returning to the circuit in Fig. 6.24, we can now find the total voltage at the drain, v_D, as

$$v_D = V_{DD} - i_D R_D$$
$$= V_{DD} - I_D R_D - i_d R_D$$
$$= V_D - i_d R_D \qquad (6.43)$$

Thus superimposed on the dc drain voltage V_D we have a signal component v_d given by

$$v_d = -i_d R_D$$

which can be written

$$v_d = -g_m v_{gs} R_D$$

The voltage gain of the FET amplifier is therefore given by

$$\frac{v_d}{v_{gs}} = -g_m R_D \tag{6.44}$$

The minus sign in Eq. (6.44) indicates that the output signal v_d is 180° out of phase with respect to the input signal v_{gs}. This is illustrated in Fig. 6.25, which shows v_{GS} and v_D. The input signal is assumed to have a triangular waveform with an amplitude much smaller than $|V_P|$ in order to ensure linear operation. For operation in the pinch-off region the minimum value of v_D should be greater than the corresponding value of v_G by at least $|V_P|$. Also, the maximum value of v_D should be smaller than V_{DD}; otherwise the FET will enter the cutoff region and the peaks of the output signal waveform will be clipped off.

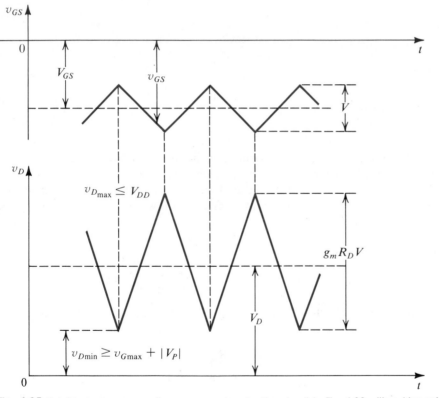

Fig. 6.25 Total instantaneous voltages v_{GS} and v_D for the circuit in Fig. 6.23 with a triangular-wave input signal.

Nonlinear Distortion

To quantify the magnitude of the nonlinear distortion produced by a JFET amplifier when the input signal is not kept sufficiently small, consider the case

$$v_{gs} = \hat{V} \sin \omega t$$

and substitute in Eq. (6.34). The total drain current i_D will be given by

$$i_D = I_D + \left(\frac{2I_{DSS}}{-V_P}\right)\left(1 - \frac{V_{GS}}{V_P}\right)\hat{V}\sin\omega t + I_{DSS}\left(\frac{\hat{V}}{V_P}\right)^2\sin^2\omega t$$

Now replacing $(1 - V_{GS}/V_P)$ by $\sqrt{I_D/I_{DSS}}$ and using the identity $\sin^2\omega t = \frac{1}{2} - \frac{1}{2}\cos 2\omega t$ yields

$$i_D = I_D + \frac{1}{2}I_{DSS}\left(\frac{\hat{V}}{V_P}\right)^2 + \left(\frac{2I_{DSS}}{-V_P}\right)\sqrt{\frac{I_D}{I_{DSS}}}\,\hat{V}\sin\omega t - \frac{1}{2}I_{DSS}\left(\frac{\hat{V}}{V_P}\right)^2\cos 2\omega t$$

We note that there is a slight shift in the dc drain current. More seriously, however, is the component of drain current at twice the input signal frequency. A measure of this second-harmonic distortion can be obtained as follows:

$$\text{Percent second-harmonic distortion} \equiv \frac{\text{Amplitude of second harmonic}}{\text{Amplitude of fundamental}} \times 100\%$$

$$= \frac{1}{4}\left(\frac{\hat{V}}{-V_P}\right)\sqrt{\frac{I_{DSS}}{I_D}} \times 100\%$$

As an example, consider the JFET amplifier that is analyzed graphically in Fig. 6.18b. It has $I_{DSS} = 16$ mA, $V_P = -4$ V, $V_{GS} = -1$ V, $I_D = 9$ mA, and a small-signal voltage gain of 8. If the input is a sinusoid of 0.5-V amplitude, the second-harmonic distortion will be about 4%.

Separating the DC Analysis and the Signal Analysis

From the above analysis we see that under the small-signal approximation, signal quantities are superimposed on dc quantities. For instance, the total drain current i_D equals the dc current I_D plus the signal current i_d, the total drain voltage $v_D = V_D + v_d$, and so on. It follows that the analysis and design can be greatly simplified by separating dc or bias calculations from small-signal calculations. That is, once a stable dc operating point has been established and all dc quantities calculated, we may then perform signal analysis ignoring dc quantities.

Small-Signal Equivalent Circuit Models

From a signal point of view the JFET behaves as a voltage-controlled current source. It accepts a signal v_{gs} between gate and source and provides a current $g_m v_{gs}$ at the drain terminal. The input resistance of this controlled source is vey high—ideally, infinite. The output resistance—that is, the resistance looking into the drain—also is high, and we have assumed it to be infinite thus far. Putting all of this together we arrive at the circuit in Fig. 6.26a, which represents the small-signal operation of the JFET and is thus a small-signal model or a small-signal equivalent circuit.

In the analysis of a JFET amplifier circuit, the JFET can be replaced by the equivalent circuit model shown in Fig. 6.26a. The rest of the circuit remains unchanged except that *dc voltage sources are replaced by short circuits*. This is a result of the fact that the voltage across an ideal dc voltage source does not change and

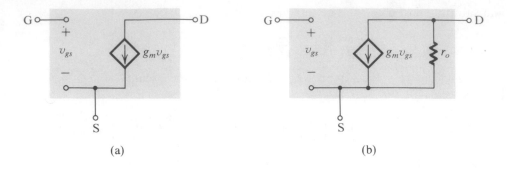

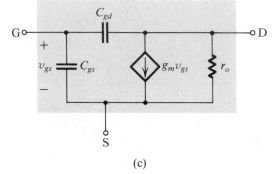

(c)

Fig. 6.26 Small-signal models for the JFET.

thus there will always be a zero voltage signal across a dc voltage source. The circuit resulting can then be used to perform any required signal analysis, such as calculating voltage gain.

The most serious shortcoming of the small-signal model of Fig. 6.26a is that it assumes that the drain current in pinch-off is independent of the drain voltage. From our study of the JFET characteristics in pinch-off in Section 6.2 we know that the drain current does in fact depend on v_{DS} in a linear manner [see Fig. 6.10 and Eq. (6.13)]. Such dependence was modeled by a finite resistance r_o between drain and source, whose value is given approximately by

$$r_o \simeq \frac{|V_A|}{I_D}$$

where V_A is a JFET parameter that is either specified or can be measured. Typically, r_o is in the range 10 to 1,000 kΩ. It follows that the accuracy of the small-signal model in Fig. 6.26a can be improved by including r_o in parallel with the controlled source, as shown in Fig. 6.26b.

It is important to note that the small-signal model parameters g_m and r_o depend on the dc bias point of the JFET.

Although the model of Fig. 6.26b is quite adequate for determining small-signal gain at low frequencies, it fails at high frequencies. This is not surprising since the model does not account for the capacitive effects present in the JFET. Specifically, recall from the physical description of JFET operation that the gate-to-channel junction is reverse-biased, and thus a depletion capacitance (see Section 4.11) exists between gate and source and another between gate and drain. Including these two capacitances results in the model of Fig. 6.26c, which can be used to predict the gain of JFET amplifiers at high frequencies, as will be demonstrated in Chapter 11. Typically C_{gs} is 1–3 pF and C_{gd} is 0.5–1 pF.

We conclude this discussion by noting that the equivalent circuit models of Fig. 6.26 apply for both n- and p-channel JFETs. The use of equivalent circuit models in the analysis of JFET amplifiers will be demonstrated in the following sections.

The Equivalent Resistance Between Source and Gate

Figure 6.27 shows a JFET that is part of a circuit the details of which are not shown. The small signal between gate and source is v_{gs} and if we ignore r_o, then $i_d = g_m v_{gs}$. Since the gate current is zero, the resistance between gate and source looking into the gate is infinite, as expected. The resistance between gate and source looking into

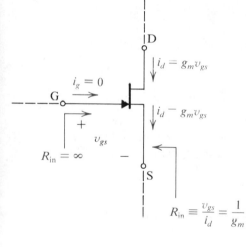

Fig. 6.27 An alternate view of the JFET small-signal operation. The resistance between gate and source, looking into the source, is equal to $1/g_m$.

the source can be obtained by dividing v_{gs} by i_d. The result is $1/g_m$. This can be also shown using the equivalent circuit of Fig. 6.26a. This alternate view of JFET operation can in certain situations lead to considerable simplification in analysis. It should be pointed out, however, that the analysis can always be performed by explicitly using the models in Fig. 6.26.

Exercises

6.15 Consider the JFET amplifier analyzed graphically in Fig. 6.18b. Find the value of g_m at the bias point indicated in the figure. Calculate the voltage gain $A_v \equiv v_{ds}/v_{gs}$.

Ans. 6 mA/V; -8 V/V

6.16 Show that if the finite output resistance r_o is taken into account, the voltage-gain expression in Eq. (6.44) becomes

$$A_v \equiv \frac{v_d}{v_{gs}} = -g_m(R_D \parallel r_o)$$

6.17 A p-channel JFET fabricated using the standard IC process has $I_{DSS} = 0.5$ mA, $V_P = 1$ V, and $|V_A| = 100$ V. What is the value of g_{m0}, its maximum g_m? If the device is biased to operate at $I_D = 0.25$ mA, find g_m and r_o at the operating point. Also, if $R_D = 10$ kΩ, find the voltage gain v_{ds}/v_{gs} with and without r_o.

Ans.   1 mA/V; 0.707 mA/V; 400 kΩ; -6.90 V/V; -7.07 V/V

6.18 Show that Eq. (6.42) leads to the expression for g_m given in Eq. (6.38).

6.19 In the circuit of Fig. 6.23 let the FET have $V_P = -3$ V, $I_{DSS} = 9$ mA, and $r_o = \infty$. For $V_{GS} = -2$ V find the value of g_m. Calculate the voltage gain obtained when $R_D = 10$ kΩ. If $V_{DD} = 15$ V and the input signal is a triangular waveform of 0.2 V peak-to-peak, find the minimum and maximum voltages at the drain.

Ans.   2 mA/V; -20 V/V; 3 V; 7 V

6.20 In the circuit specified in Exercise 6.19 let the input be a sinusoid of peak amplitude $\hat{V}$. If it is desired to limit the second-harmonic distortion to 1%, find the maximum allowed value of $\hat{V}$.

Ans.   0.04 V

6.8 THE JFET COMMON-SOURCE AMPLIFIER

Biasing

Figure 6.28 shows one of the most traditional configurations of a discrete-circuit JFET amplifier. Here the JFET is biased using a combination of fixed bias and self-bias, as explained in Section 6.6. Resistors R_{G1} and R_{G2} form a voltage divider across

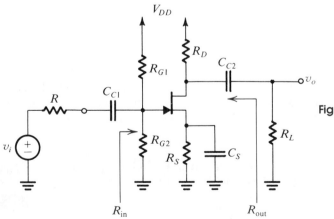

Fig. 6.28 JFET common-source amplifier.

the power supply V_{DD} and thus establish a voltage V_{GG} at the gate, as follows:

$$V_{GG} = V_{DD} \frac{R_{G2}}{R_{G1} + R_{G2}} \tag{6.45}$$

In order to keep the input resistance of the FET amplifier high, one should choose R_{G1} and R_{G2} as high as possible. Typically, R_{G1} and R_{G2} can be in the range 100 kΩ to a few megohms with no difficulty. Note, though, that the small gate current (a few nanoamperes) flows through the parallel equivalent of R_{G1} and R_{G2}. Since this current increases with temperature, its value imposes an upper limit on the values of R_{G1} and R_{G2}. Otherwise, the bias point of the FET becomes temperature-dependent.

Self-bias is obtained by connecting a resistance R_S in the source lead. As discussed in Section 6.6, the values of the biasing elements, including R_D, should be chosen to obtain a stable and predictable dc operating point. In the following we shall assume that this has already been accomplished, and we will concern ourselves with signal performance only.

Coupling the Source and the Load

The amplifier circuit of Fig. 6.28 is fed by a signal source with voltage v_i and resistance R. The signal is coupled to the gate of the FET through a capacitor C_{C1}. The purpose of this capacitor is to block dc so as not to disturb the bias conditions already established. The value of C_{C1} should be chosen sufficiently large so that at the signal frequencies of interest it acts as a short circuit. Obviously, C_{C1} will cause the amplifier gain to drop at low frequencies and will cause the gain to be zero at dc. As mentioned in Section 2.4, amplifiers that use coupling capacitors are called ac amplifiers in order to distinguish them from direct-coupled (dc) amplifiers, which are essential in some applications. Operational amplifiers are examples of dc amplifiers.

The output signal voltage at the drain is coupled to a load resistance R_L through another coupling capacitor C_{C2}. Again, C_{C2} should be chosen large in order to act as a perfect coupler at all frequencies of interest. The load resistance R_L represents either an actual load or the input resistance of another part of the system.

Bypassing R_S

As shown in Fig. 6.28, a capacitance C_S is connected across the bias resistance R_S. The purpose of C_S is to cause the total impedance between source and ground to be very small (ideally zero) at all frequencies of interest. In other words, C_S is chosen sufficiently large so that it acts as a short circuit for signals. For this reason C_S is called a *bypass capacitor*—signal currents pass through it, thus "bypassing" the resistance R_S. As will be shown below, inclusion of bypass capacitor C_S causes the gain of the amplifier to increase considerably. Alternatively, one can say that without C_S the gain would be quite small. In Chapter 11 we shall study the effects of the capacitance C_S on the frequency response of the amplifier. It can be easily seen that at low frequencies C_S will no longer act as a perfect short circuit.

Assume for now that C_S is acting as a perfect short circuit. It follows that the source of the FET will be at signal ground. Thus, viewed as a two-port network, the amplifier circuit of Fig. 6.28 has a common ground between input and output, which is the source terminal. For this reason the circuit is called a *common-source amplifier*.

Small-Signal Analysis

In the remainder of this section we shall concern ourselves with analysis of the common-source amplifier, assuming that the frequencies of interest are sufficiently high so that coupling and bypass capacitors behave as perfect short circuits. We shall also assume that the frequencies of interest are sufficiently low so that the FET internal capacitances C_{gs} and C_{gd} can be considered as open circuits. We speak of the frequency band at which both the above conditions are satisfied as *the midband*.

The first step in the analysis consists of evaluating the input resistance R_{in}, indicated in Fig. 6.28. This is the resistance seen between the gate terminal and ground. Since the input resistance of the JFET is infinite (the gate current is zero), R_{in} consists of R_{G2} in parallel with R_{G1}. (The terminal of R_{G1} connected to V_{DD} is at signal ground.) Thus

$$R_{in} = (R_{G1} \parallel R_{G2}) \qquad (6.46)$$

Having evaluated R_{in}, we now can find the fraction of the input signal voltage that appears between gate and ground. This can be done by recognizing that R_{in} and R form a voltage divider; thus

$$v_g = v_i \frac{R_{in}}{R_{in} + R} \qquad (6.47)$$

Since the source is bypassed to ground through C_S, it follows that

$$v_{gs} = v_g \qquad (6.48)$$

The drain-current signal i_d is given by

$$i_d = g_m v_{gs} \qquad (6.49)$$

To obtain the output voltage signal v_o, which is equal to v_d, we multiply i_d by the total resistance between drain and ground. This resistance consists of R_D in parallel with R_L and in parallel with the FET output resistance r_o,

$$v_o = -i_d(R_D \parallel R_L \parallel r_o) \qquad (6.50)$$

In most cases r_o will have only a small effect on gain calculations. Combining the above equations yields the voltage gain v_o/v_i of the common-source amplifier as

$$\frac{v_o}{v_i} = -\frac{R_{in}}{R_{in} + R} g_m(R_D \parallel R_L \parallel r_o) \qquad (6.51)$$

The output resistance R_{out} of this amplifier, exclusive of the load (see Fig. 6.28), is given by

$$R_{out} = (R_D \parallel r_o) \qquad (6.52)$$

EXAMPLE 6.5

We want to evaluate the voltage gain and the maximum allowable input signal swing of the common-source amplifier of Fig. 6.29. Ignore all capacitive effects; that is, perform the analysis at midband frequencies. The FET is specified to have $I_{DSS} = 12$ mA and $V_P = -4$ V. At $I_D = 12$ mA the output resistance $r_o = 25$ kΩ.

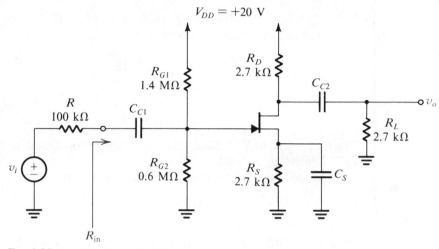

Fig. 6.29 Circuit for Example 6.5.

Solution

First we determine the dc operating point as follows:

$$V_{GG} = V_{DD} \frac{R_{G2}}{R_{G1} + R_{G2}}$$

$$= 20 \frac{0.6}{2} = 6 \text{ V}.$$

$$V_{GS} = V_{GG} - I_D R_S$$
$$= 6 - 2.7I_D$$

$$I_D = I_{DSS}\left(1 - \frac{V_{GS}}{V_P}\right)^2$$

$$= 12\left(1 - \frac{6 - 2.7I_D}{-4}\right)^2$$

This leads to the quadratic equation

$$I_D^2 - 7.59I_D + 13.7 = 0$$

The appropriate solution of this equation is

$$I_D = 2.96 \text{ mA}$$

The corresponding voltage V_{GS} is

$$V_{GS} = 6 - 2.7 \times 2.96 \simeq -2 \text{ V}$$

The drain voltage will be

$$V_D = V_{DD} - I_D R_D = 20 - 2.96 \times 2.7 = 12 \text{ V}$$

Since $V_D > V_G + |V_P|$, the device is indeed in pinch-off.

The value of g_m can be determined as

$$g_m = \frac{2I_{DSS}}{-V_P} \sqrt{\frac{I_D}{I_{DSS}}}$$

$$1 = \frac{2 \times 12}{4} \sqrt{\frac{2.96}{12}}$$

Thus $g_m = 2.98$ mA/V. Since r_o is inversely proportional to I_D, it follows that at $I_D = 2.96$ mA, $r_o \simeq 100$ kΩ.

The input resistance of the amplifier is given by

$$R_{\text{in}} = (R_{G1} \| R_{G2}) = \frac{1.4 \times 0.6}{1.4 + 0.6}$$

Thus $R_{\text{in}} = 0.42$ MΩ $= 420$ kΩ. The voltage gain can now be evaluated as

$$\frac{v_o}{v_i} = \frac{R_{\text{in}}}{R_{\text{in}} + R} (-g_m)(R_D \| R_L \| r_o)$$

$$= -\frac{420}{420 + 100} \times 2.98(2.7 \| 2.7 \| 100)$$

Thus $v_o/v_i = -3.2$.

To investigate the question of signal swing, we first consider the maximum amplitude of input signal for which the FET remains in pinch-off. Denote the peak value of the gate signal by $\hat{V}$. The condition for remaining in pinch-off can be expressed as

$$V_D - A_v \hat{V} \geq V_G + \hat{V} + |V_P|$$

where A_v denotes the magnitude of voltage gain between gate and drain,

$$A_v = g_m(R_L \| R_D \| r_o) = 3.97$$

The maximum allowable value of $\hat{V}$ can be obtained as

$$\hat{V} = \frac{V_D - V_G - |V_P|}{A_v + 1} = \frac{12 - 6 - 4}{4.97} = 0.4 \text{ V}$$

Since this signal is much smaller than $|V_P|$, it can be considered as a "small signal." In general, however, one can use the square-law relationship of the JFET to determine the second-harmonic distortion corresponding to this signal amplitude. Letting

$$v_{gs} = \hat{V} \sin \omega t$$

we have

$$i_D = I_{DSS}\left(1 - \frac{v_{GS}}{V_P}\right)^2$$

$$i_D = 12\left(1 - \frac{-2 + \hat{V}\sin\omega t}{-4}\right)^2$$

$$= 3 + \frac{3\hat{V}^2}{8} + 3\hat{V}\sin\omega t - \frac{3\hat{V}^2}{8}\cos 2\omega t$$

The second component on the right-hand side represents a very slight shift in the dc drain current. The last term represents the second-harmonic component of drain current. The percentage harmonic distortion is given by

$$\text{Percentage harmonic distortion} = \frac{3\hat{V}^2/8}{3\hat{V}} \times 100 \simeq 12.5\hat{V}$$

For $\hat{V} = 0.4$ V the harmonic distortion is 5%. This is quite small. If in a certain application the value of harmonic distortion is found to be unacceptably high, then the value of $\hat{V}$ should be reduced. In other words, the maximum allowable input signal swing is often determined by linearity considerations rather than by the condition that the device remain in the active mode.

Corresponding to v_{gs} having a peak value of 0.4 V, the input signal v_i will have a peak value of

$$\hat{V}_i = \hat{V}\frac{R_{\text{in}} + R}{R_{\text{in}}} \simeq 0.5 \text{ V}$$

Analysis Using Equivalent Circuit Models

In all of the discussion above, small-signal analysis was performed directly on the circuit diagram. Of course, the equivalent circuit model of the FET was used, but only implicitly. What we did was possible because the model at midband frequencies is quite simple. If more exact—and hence more complicated—models are used, a different approach to signal analysis should be adopted. Specifically, the FET should be explicitly replaced by the equivalent circuit model, and all analysis should be performed on the resulting circuit.

For the common-source amplifier of Fig. 6.28, which is redrawn in Fig. 6.30a, the equivalent circuit at midband frequencies is shown in Fig. 6.30b. Analysis of this latter circuit yields results identical to those obtained above.

The Common-Source Amplifier with Unbypassed Source Resistance

Figure 6.31 shows an amplifier circuit similar to that of Fig. 6.28, except that part of the source resistance R_S has been left unbypassed. An unbypassed resistance in the source lead results in a reduction in gain, but it has other positive attributes,

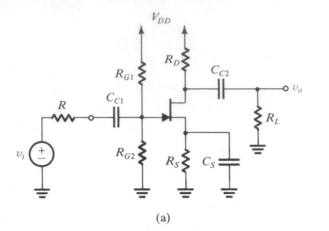

(a)

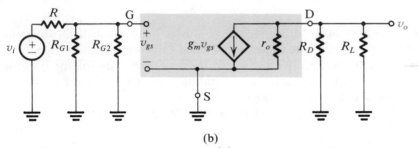

(b)

Fig. 6.30 (a) Common-source amplifier. **(b)** Equivalent circuit of the common-source amplifier at midband frequencies. Note that the model used is that of Fig. 6.26b.

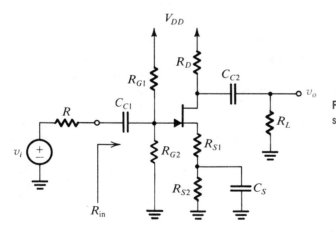

Fig. 6.31 JFET amplifier with an unbypassed source resistance R_{S1}.

such as increasing the amplifier bandwidth (the frequency range over which the amplifier provides gain close in value to the midband gain) and increasing the maximum allowable input signal swing. The effect on bandwidth will be discussed in Chapter 11.

To evaluate the voltage gain of the circuit in Fig. 6.31 we first observe that the input resistance R_{in} remains equal to $(R_{G1} \| R_{G2})$ and thus the fraction of v_i that appears between gate and ground (v_g) remains unchanged. However, unlike the situation in the common-source amplifier, this signal does not appear between gate and source. To find the latter signal, v_{gs}, it is easiest to use the fact illustrated in Fig. 6.27—namely, that the resistance between gate and source, looking into the source, is equal to $1/g_m$. This resistance appears in series with R_{S1}. Thus we can use the voltage-divider rule to obtain v_{gs} as

$$v_{gs} = v_g \frac{1/g_m}{R_{S1} + 1/g_m} \qquad (6.53)$$

Once v_{gs} is determined, the remainder of the analysis is straightforward; the final result is

$$\frac{v_o}{v_i} = -\frac{R_{in}}{R_{in} + R} \frac{(R_D \| R_L \| r_o)}{1/g_m + R_{S1}} \qquad (6.54)$$

This expression[3] is quite interesting. It consists of two factors: the first represents the gain between the signal generator and the gate terminal, and the second represents the gain between the gate and drain. This second factor is simply *the ratio between the total resistance in the drain lead and the total resistance in the source lead.* This observation enables one to write the gain expression directly without having to go through a step-by-step procedure.

The gain expression of Eq. (6.54) can be written in the alternative form

$$\frac{v_o}{v_i} = -\frac{R_{in}}{R_{in} + R} g_m \frac{(R_D \| R_L \| r_o)}{1 + g_m R_{S1}} \qquad (6.55)$$

Comparison of this expression with that obtained for the grounded-source amplifier in Eq. (6.51) reveals that including an unbypassed resistance R_{S1} reduces the gain by the factor $1 + g_m R_{S1}$.

Since only a fraction of v_g appears between gate and source [Eq. (6.53)], it follows that one can allow v_g and hence v_i to be larger (as compared to the grounded-source case) without violating the small-signal approximation.

Power Gain

Before leaving this section we should point out that although the amplifiers considered in the examples do not exhibit large voltage gains, their power gains are quite impressive. To illustrate, let us calculate the power gain for the amplifier of Example 6.5. The input power is given by

$$P_i = \frac{v_g^2}{R_{in}}$$

---

[3] In writing this expression it has been assumed that r_o appears in parallel with R_D. This is not exactly correct, since r_o appears between drain and source and the source is no longer grounded. Nevertheless, the approximation involved is good enough for our purposes at this stage.

while the output power is given by

$$P_o = \frac{v_o^2}{R_L}$$

Thus the power gain is

$$\frac{P_o}{P_i} = \frac{R_{\text{in}}}{R_L} \left(\frac{v_o}{v_g}\right)^2$$

Substitution of the values from the results of Example 6.5 gives

$$\frac{P_o}{P_i} = \frac{420}{2.7}(3.97)^2 = 2,452$$

In decibels the power gain is

$$10 \log \frac{P_o}{P_i} = 33.9 \text{ dB}$$

Exercises

6.21 Consider the circuit of Fig. 6.29 (Example 6.5). What does the voltage gain become if R_L is removed? What value of R_L reduces the gain to -1?

Ans. -6.33 V/V; 493 Ω

6.22 Using the JFET model of Fig. 6.26a sketch the complete equivalent circuit of the amplifier in Fig. 6.31. Analyze the circuit to obtain the voltage gain v_o/v_i. The result should be identical to that given by Eq. (6.55) with r_o set to ∞.

6.23 Consider the amplifier circuit of Example 6.5, but let 300 Ω of the source resistance R_S be left unbypassed. Calculate the voltage gain of the modified circuit. Also calculate the maximum allowable input signal swing that corresponds to a maximum v_{gs} of 0.4 V.

Ans. -1.7 V/V; 0.94 V

6.9 THE SOURCE FOLLOWER

Because of its very high input resistance the JFET is an excellent choice for the design of buffer or isolation amplifiers. These are amplifiers whose purpose is to allow the coupling of a high-impedance source to a low-impedance load with little loss in signal level. Normally a buffer amplifier is not required to provide large voltage gain; the main objective is to provide large current and power gains.

Figure 6.32 shows a simple buffer circuit using the JFET. Here the generator signal is capacitively coupled to the gate, while the output signal is taken from the source terminal and is capacitively coupled to the load. The drain terminal is connected to the dc supply and hence is at signal ground. For this reason this circuit is referred to as the *grounded-drain* or *common-drain configuration*. Finally, note that the FET is self-biased using a source resistance R_S. The resistance R_G provides dc continuity for the gate terminal. Since the gate current is very small (a few nanoamperes), we can safely assume that the dc voltage at the gate is zero. This is true even though one normally uses a large resistance R_G.

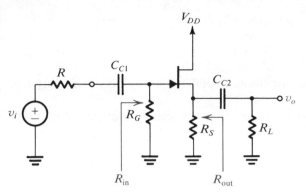

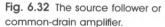

Fig. 6.32 The source follower or common-drain amplifier.

Input Resistance

The input resistance R_{in} of the circuit in Fig. 6.32 is simply equal to the bias resistance R_G. Since a buffer amplifier is required to have a high input resistance, one should choose as large a value for R_G as practicable. However, too high a value for R_G will cause an appreciable dc voltage drop across it. This voltage drop is due to the input leakage current of the FET and hence will increase with temperature, causing bias change.

Gain at Midband

The midband gain of the buffer circuit of Fig. 6.32 can be evaluated in a step-by-step manner starting from the signal source and proceeding toward the load. The fraction of the input signal v_i that appears at the gate is given by

$$v_g = v_i \frac{R_{in}}{R_{in} + R} \tag{6.56}$$

If R_{in} is much larger than the generator resistance R, then

$$v_g \simeq v_i \tag{6.57}$$

The signal v_g appears across the series combination of the gate-to-source resistance $1/g_m$ and the resistance $(R_S \| R_L \| r_o)$. Thus to obtain the output voltage v_o we simply use the voltage-divider rule:

$$v_o = \frac{(R_S \| R_L \| r_o)}{(R_S \| R_L \| r_o) + (1/g_m)} v_g \tag{6.58}$$

If $1/g_m \ll (R_S \| R_L \| r_o)$, then

$$v_o \simeq v_g \tag{6.59}$$

Combining Eqs. (6.56) and (6.58) yields the voltage gain of the buffer as

$$\frac{v_o}{v_i} = \frac{R_{in}}{R_{in} + R} \frac{(R_S \| R_L \| r_o)}{(R_S \| R_L \| r_o) + 1/g_m} \tag{6.60}$$

and, provided that $R_{\text{in}} \gg R$ and $1/g_m \ll (R_S \| R_L \| r_o)$

$$\frac{v_o}{v_i} \simeq 1 \tag{6.61}$$

Thus the voltage gain of the common-drain amplifier is always less than unity and at the limit approaches unity.

Because the signal at the source terminal is almost equal to that at the gate, this circuit is also called a *source follower*.

Output Resistance

To find the output resistance R_{out} (see Fig. 6.32) we look back into the circuit between the source terminal and ground while the generator is shorted. Since the gate current is zero, the gate will be at signal ground. Thus it can be seen that R_{out} is made up of R_S in parallel with the gate-to-source resistance $1/g_m$ in parallel with r_o,

$$R_{\text{out}} = [R_S \| (1/g_m) \| r_o] \tag{6.62}$$

It follows that the source follower exhibits a low output resistance.

Analysis Using the FET Equivalent Circuit Model

An alternate approach to obtaining the results above is to analyze the circuit obtained by replacing the JFET with its small-signal model. Figure 6.33 shows the source-follower circuit with this replacement. The reader is encouraged to analyze this circuit to determine the voltage gain and output resistance. Although this analysis is straight-forward, it is slightly more difficult than that used above.

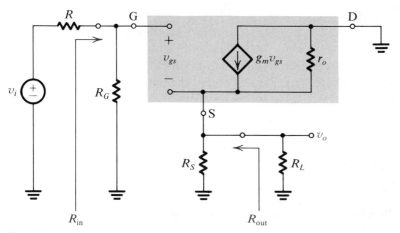

Fig. 6.33 Small-signal equivalent circuit of the source-follower circuit of Fig. 6.32.

Exercises

6.23 As an amplifier, the source-follower circuit of Fig. 6.32 can be modeled by its input resistance R_{in}, its open-circuit voltage gain v_o/v_g [obtained from Eq. (6.58) by setting $R_L = \infty$] and its output resistance R_{out} (see Section 2.3). Use this model to find the overall voltage gain v_o/v_i and show that it is identical to that in Eq. (6.60).

6.24 Consider the source-follower circuit with $R = 100$ kΩ, $R_G = 1$ MΩ, $R_S = 4$ kΩ, $R_L = 4$ kΩ, $V_{DD} = +10$ V, and with the lower end of R_S returned to another supply voltage of -10 V (instead of ground). If the JFET has $V_P = -4$ V, $I_{DSS} = 12$ mA and $r_o = \infty$, find the input resistance, the voltage gain from generator to load, and the output resistance.

Ans. 1 MΩ; 0.78 V/V; 307 Ω

6.25 A source follower, utilizing a resistance $R_S = 10$ kΩ and having $r_o = 100$ kΩ, has an unloaded (open-circuit) voltage gain from gate to source of 0.9 V/V. Find g_m and R_{out}. Also, find the voltage gain when the output of the source follower is capacitively coupled to a load resistance of 910 Ω.

Ans. 0.99 mA/V; 909 Ω; 0.5 V/V

6.10 DIRECT-COUPLED AND MULTISTAGE AMPLIFIERS

The JFET amplifier circuits considered thus far use coupling and bypass capacitors, which limit performance at low frequencies, as has already been mentioned. There exist applications in which gain is required down to zero frequency (dc). For such applications one requires a direct-coupled (dc) amplifier. Also, one has to seek an alternative to the biasing scheme that employs a bypass capacitor.

Another motivation to seek direct-coupled amplifier circuits is that integrated-circuit (IC) technology precludes the fabrication of large coupling and bypass capacitors. A different philosophy is thus required for the design of IC FET amplifiers. In Chapter 9 we shall study the most common configuration of dc JFET amplifier—namely, the differential pair.

Many amplifier applications require a number of stages connected in cascade. For instance, to obtain a low output resistance for the overall amplifier one might use a source follower as the last stage in the cascade. Multistage amplifiers will be encountered throughout this book. To illustrate the analysis of such amplifiers we consider a simple three-stage example.

EXAMPLE 6.6

Consider the three-stage amplifier shown in Fig. 6.34. The first stage uses a JFET Q_1 in the common-source configuration. The output of the first stage is directly coupled to the gate of the second-stage transistor Q_2, which is a p-channel device. Transistor Q_2 also is connected as a common-source amplifier. To obtain a low output resistance a third stage consisting of transistor Q_3 connected in the source-follower configuration is used. Although direct coupling is used between stages, this amplifier is classified as an ac amplifier (why?). In the following we shall carry out small-signal analysis at midband. Assume all FETs to have $|V_P| = 2$ V and $I_{DSS} = 8$ mA.

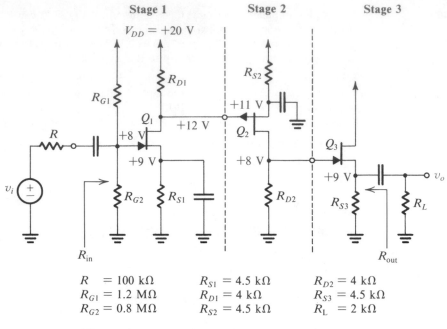

R $= 100$ kΩ	$R_{S1} = 4.5$ kΩ	$R_{D2} = 4$ kΩ
$R_{G1} = 1.2$ MΩ	$R_{D1} = 4$ kΩ	$R_{S3} = 4.5$ kΩ
$R_{G2} = 0.8$ MΩ	$R_{S2} = 4.5$ kΩ	$R_L = 2$ kΩ

All capacitances very large

Fig. 6.34 Three-stage amplifier for Example 6.6.

Bias Calculations

$$V_{G1} = V_{DD} \frac{R_{G2}}{R_{G1} + R_{G2}} = 20 \frac{0.8}{1.2 + 0.8} = +8 \text{ V}$$

$$V_{GS1} = 8 - I_{D1}R_{S1} = 8 - 4.5I_{D1}$$

Using this relationship together with the FET square-law equation gives $I_{D1} = 2$ mA and $V_{GS1} = -1$ V. Thus $V_{S1} = +9$ V and $V_{D1} = 20 - 4 \times 2 = +12$ V.

For transistor Q_2 we have

$$V_{SG2} = V_{DD} - I_{D2}R_{S2} - V_{D1} = 20 - I_{D2} \times 4.5 - 12$$

Thus $V_{SG2} = 8 - 4.5I_{D2}$. Substituting in the square-law relationship for Q_2,

$$I_D = I_{DSS}\left(1 - \frac{V_{GS2}}{V_P}\right)^2$$

where $I_{DSS} = 8$ mA and $V_P = 2$ V, and solving, gives $I_{D2} = 2$ mA, $V_{SG2} = -1$ V. Thus $V_{S2} = +11$ V, $V_{D2} = I_{D2}R_{D2} = +8$ V.

For Q_3 we have

$$V_{GS3} = V_{D2} - I_{D3}R_{S3} = 8 - 4.5I_{D3}$$

Using this together with the square-law relationship gives $I_{D3} = 2$ mA and $V_{GS3} = -1$ V. Thus $V_{S3} = +9$ V.

The various dc voltages evaluated are indicated on the circuit diagram. Note that all devices are well into the pinch-off region.

Input Resistance

The amplifier input resistance is determined by the parallel equivalent of R_{G1} and R_{G2},

$$R_{in} = \frac{1.2 \times 0.8}{1.2 + 0.8} = 480 \text{ k}\Omega$$

Voltage Gain

The transmission through the input coupling circuit is given by

$$\frac{v_{g1}}{v_i} = \frac{R_{in}}{R_{in} + R} = \frac{480}{480 + 100} = 0.83 \tag{6.63}$$

To evaluate the gain of the first stage we note that its load is the input impedance of the second stage, which is very high. Thus

$$\frac{v_{d1}}{v_{g1}} = -g_{m1} R_{D1}$$

where

$$g_{m1} = \frac{2 \times 8}{2} \sqrt{\frac{2}{8}} = 4 \text{ mA/V}$$

and where we have assumed $r_o \gg R_{D1}$. Thus the gain of the first stage is given by

$$\frac{v_{d1}}{v_{g1}} = -4 \times 4 = -16 \tag{6.64}$$

The second stage is a common-source amplifier with a load resistance equal to the input resistance of Q_3, which is very high. Thus the voltage gain of the second stage will be given by

$$\frac{v_{d2}}{v_{g2}} = \frac{v_{d2}}{v_{d1}} = -g_{m2} R_{D2}$$

where

$$g_{m2} = \frac{2 \times 8}{2} \sqrt{\frac{2}{8}} = 4 \text{ mA/V}$$

Thus

$$\frac{v_{d2}}{v_{d1}} = -4 \times 4 = -16 \tag{6.65}$$

The gain of the third stage can be evaluated from

$$\frac{v_o}{v_{d2}} = \frac{(R_{S3} \parallel R_L)}{(R_{S3} \parallel R_L) + 1/g_{m3}}$$

where $g_{m3} = 4$ mA/V and where r_{o3} has been neglected. Thus

$$\frac{v_o}{v_{d2}} = \frac{1.38}{1.38 + 0.25} = 0.84 \tag{6.66}$$

The overall voltage gain of the amplifier can now be obtained by combining Eqs. (6.63) through (6.66):

$$\frac{v_o}{v_i} = 0.83 \times -16 \times -16 \times 0.84 = 178.5 \tag{6.67}$$

or 45 dB.

Output Resistance

The amplifier output resistance, excluding the load resistance R_L, is equal to the output resistance of the source-follower stage,

$$R_{\text{out}} = [R_{S3} \| (1/g_{m3})]$$

Thus

$$R_{\text{out}} = (4.5 \text{ k}\Omega \| 0.25 \text{ k}\Omega) = 237 \ \Omega$$

Signal Swing

To determine the maximum allowable output signal swing we have to investigate each stage to find out the limitations it imposes. In this particular design it can be shown that the second stage is the one that limits the amplifier signal swing. Specifically, for Q_2 to remain in pinch-off the maximum drain voltage is limited to $+12 - |V_P| = +10$ V. Thus at the drain of Q_2 we can have a maximum signal swing of 2 V (that is, 4 V peak-to-peak). This means that at the output the maximum signal swing is

$$2 \times 0.84 = 1.68 \text{ V}$$

that is, 3.36 V peak-to-peak.

6.11 THE JFET AS A SWITCH

Basis of Operation

For small signals the i_D-v_{DS} characteristics of the JFET are straight lines that pass through the origin, as illustrated in Fig. 6.35. Note that for small v_{SD} the characteristics extend symmetrically into the third quadrant of the i_D-v_{SD} plane and that at $i_D = 0$ the voltage $v_{SD} = 0$. These properties make the JFET quite suitable for switching applications.

As an illustration of the operation of the JFET as a switch we consider the circuit in Fig. 6.36. The switching or control signal is applied to the gate of the FET. When this signal is equal to 0 V the FET is on and the switch is closed. The value of the signal v and the resistance R_D should be such that the device is in the triode region.

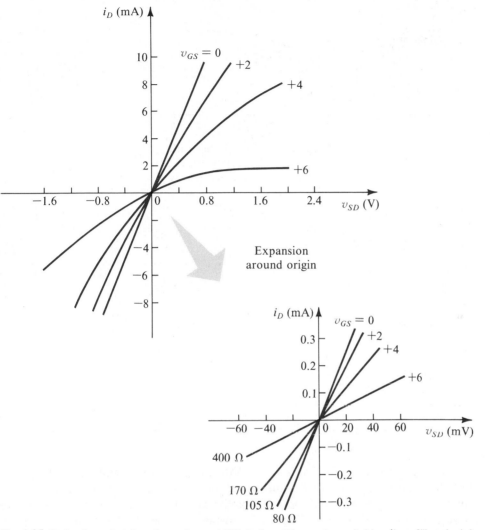

Fig. 6.35 Static characteristics of a p-channel JFET designed for analog switching (from Siliconix data sheets, 2N3386).

Figure 6.36b shows the i_D-v_{DS} curve for $v_{GS} = 0$ and a number of load lines corresponding to different values of v, positive and negative. Note that one can arrange that the operating point Q remain on an almost linear segment of the i_D-v_{DS} curve. In this case the FET can be considered as a resistance r_{DS} whose value is equal to the inverse of the slope of the i_D-v_{DS} line (see Fig. 6.36c). It should be mentioned that the "on" resistance can be as small as a few tens of ohms. Although bipolar transistors (Chapter 8) can be made to have smaller "on" resistances, they exhibit an undesirable offset voltage (that is, at zero current the voltage across the device is not zero as is the case with the JFET).

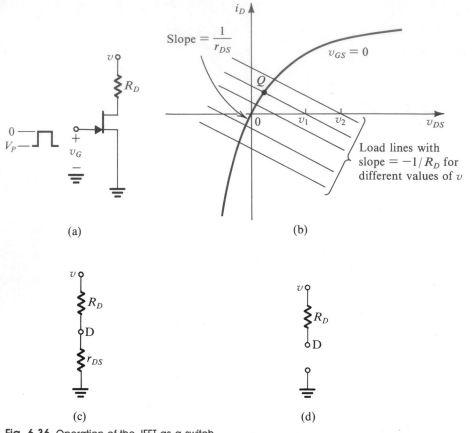

Fig. 6.36 Operation of the JFET as a switch.

To turn the JFET off—that is, to open the switch—the control signal at the gate should be made negative with a value higher than $|V_P|$. The current through the FET will be very close to zero, and the drain can be assumed to be disconnected from the source, as shown in Fig. 6.36. The "off" resistance of the JFET switch is very high.

Before considering specific applications, we wish to distinguish between analog and digital switches. In a digital switch, such as that used to implement a logic inverter, one can generally tolerate an offset voltage and inaccuracies in the value of the "on" resistance. In an analog switch, such as those used in the design of D/A converters, such nonideal properties are detrimental.

A Sawtooth-Waveform Generator

As an application of the JFET switch consider the circuit of Fig. 6.37a, whose purpose is to generate a sawtooth waveform. The input terminal I is fed with a square wave whose two levels are 0 and $-|V_P|$ volts. When the input goes negative to $-|V_P|$, the FET turns off and the capacitor C charges exponentially through R_D toward the V_{DD} level (see Fig. 6.37b). If the input remains negative for a period T_1 much smaller than

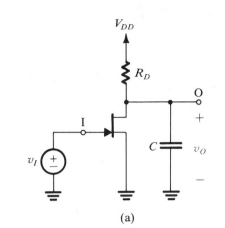

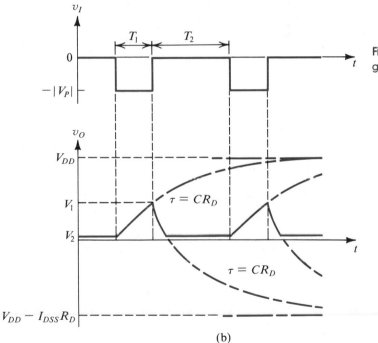

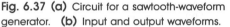

Fig. 6.37 (a) Circuit for a sawtooth-waveform generator. **(b)** Input and output waveforms.

the time constant $\tau = CR_D$, the curve of output voltage versus time will be almost linear.

At the end of interval T_1 the input goes to 0 V and the FET turns on. Since the voltage at the drain, which is the voltage across the capacitor, cannot change instantaneously, the FET will be in the pinch-off region operating at point B in Fig. 6.38. The drain terminal will carry a constant current equal to I_{DSS} that rapidly discharges the capacitor. The equivalent circuit is shown in Fig. 6.39a, which can be simplified using Thévenin's theorem to that in Fig. 6.39b. From the latter circuit we see that the voltage across the capacitor will decrease exponentially toward a value of $V_{DD} - I_{DSS}R_D$,

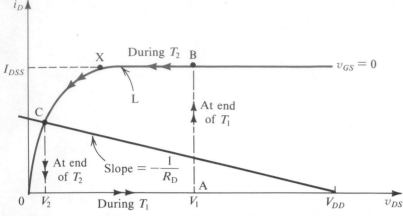

Fig. 6.38 Trajectory of the operating point of the JFET in the circuit of Fig. 6.37a.

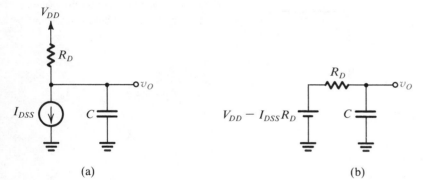

(a)   (b)

Fig. 6.39 Equivalent circuit that applies during the discharge time of C in the sawtooth-waveform generator of Fig. 6.37a.

which is usually a large negative number. The discharge time constant is still CR_D. As the capacitor discharges, the operating point will move along line L in Fig. 6.38. The discharge current will slightly decrease until point X, where the device leaves the pinch-off region. In the triode region the discharge current will be smaller than I_{DSS}. Finally, the device reaches operating point C and stays there for the remainder of interval T_2. It should be noted that the equivalent circuit in Fig. 6.39 assumes that the discharge current remains constant at the value I_{DSS}, which is only approximately true.

Although the discharge time constant is still CR_D, the time taken by the output voltage to reach the low value V_2 is quite short because the exponential discharge curve is heading toward a large negative value (see Fig. 6.37b). As a result we end up with an output signal that rises almost linearly from a low value V_2 to a higher value

V_1 and then quickly falls to the low value V_2. Thus the output voltage waveform is a good approximation of a sawtooth waveform.

The JFET Chopper

As another application of the JFET as a switch consider the circuit shown in Fig. 6.40a. Here v_I is a time-varying signal much smaller in magnitude than $|V_P|$. The control signal is a square wave v_G applied to the gate terminal (that is, between gate and ground). The square wave has the two levels 0 and $-|V_P|$ and has a frequency normally much higher than that of the input signal v_I.

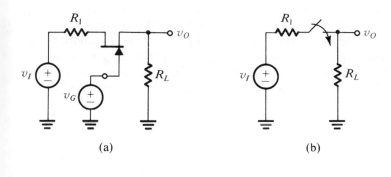

(a)   (b)

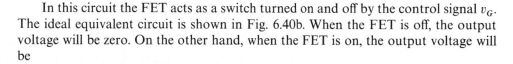

(c)

Fig. 6.40 Operation of the JFET chopper, with **(b)** the ideal equivalent circuit and **(c)** the various waveforms.

In this circuit the FET acts as a switch turned on and off by the control signal v_G. The ideal equivalent circuit is shown in Fig. 6.40b. When the FET is off, the output voltage will be zero. On the other hand, when the FET is on, the output voltage will be

$$v_O = v_I \frac{R_L}{R_L + r_{DS} + R_1}$$

Now if R_L is much greater than the "on" resistance r_{DS} and the generator resistance R_1, v_O will be approximately equal to v_I. Note that because v_I is small, v_O will also be small, and the gate-to-source voltage will be approximately equal to the square-wave signal v_G.

Figure 6.40c shows the output waveform v_O, which is almost equal to the input v_I during the "on" intervals and is zero during the "off" intervals. We speak of the input signal as being "chopped" in passing through this circuit and correspondingly refer to the circuit as a *chopper*. Alternatively, we may consider this circuit as a modulator in which the carrier is the square wave and the modulating signal is v_I. Choppers and modulators find extensive use in electronic systems (see Chapter 1).

A Sample-and-Hold Circuit

If analog signals are to be processed by digital circuitry, *samples* of the analog signals are taken at regular intervals. The value of each signal sample is then *held* (that is, stored) until the time comes for taking the next sample. During the hold time, the signal sample is converted to digital form (that is, a number) using an analog-to-digital converter (see Chapter 13). It is our purpose here to discuss the precision sample-and-hold circuit shown in Fig. 6.41. This circuit uses an op amp for two

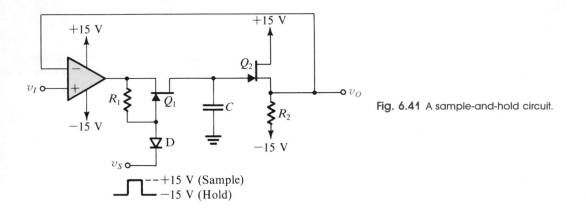

Fig. 6.41 A sample-and-hold circuit.

purposes: to buffer the input signal v_I and to enhance precision. The latter objective is achieved by placing both the JFET switch Q_1 and the JFET source follower Q_2 in the negative-feedback loop of the op amp. When the control signal v_S is at the $+15$-V level, diode D is off and thus resistance R_1 causes the gate-to-source voltage of Q_1 to be zero. Thus Q_1 turns on and presents a low series resistance between the op-amp output and capacitor C. The feedback loop around the op amp will thus be closed, and capacitor C charges to the value that results in the output voltage v_O being exactly (except for the op-amp offset-voltage error) equal to the sampled signal v_I (because of the virtual short circuit between the two input terminals of the op amp).

In the hold mode, v_S is at -15 V, which turns diode D on. Thus the gate of JFET Q_1 will be at about -14.3 V, which turns Q_1 off. Capacitor C will thus retain its charge and voltage. Here we note the need for the source follower Q_2, which buffers C and reduces considerably the droop in output voltage. The output voltage v_O will therefore remain at the value reached just before v_S was switched. When it is properly designed this circuit can achieve an output droop of less than 1 mV/s.

Exercises

6.26 Use the data given in the expanded plot of Fig. 6.35 to determine V_P and I_{DSS} for the Siliconix 2N3386 JFET.

Ans. 7.5 V; 46.9 mA

6.27 The circuit in Fig. E6.27 uses a JFET for which $r_{DS} = 100\ \Omega$ at $v_{GS} = 0$ and small v_{DS}. When v_C is low and $v_I = 1 \sin \omega t$ volts, find v_O for $R = 100$ kΩ and 1 kΩ. If the FET has a V_P of 5 V and the diode has a 0.7-V drop, what are the extremes of v_C required for proper switch operation?

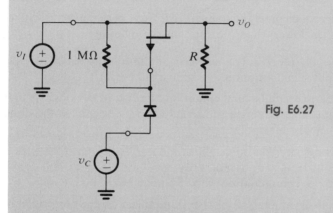

Fig. E6.27

Ans. 0.999 sin ωt; 0.909 sin ωt; -1 V and $+6.7$ V

6.28 For the sawtooth-waveform generator of Fig. 6.37 let $V_{DD} = 10$ V, $R_D = 10$ kΩ, $I_{DSS} = 10$ mA, $V_P = -2$ V, $T_1 - 1$ ms, and $T_2 = 5$ ms. Find the value of V_2. Also, find the value of C that will result in a sawtooth waveform of 1 V peak-to-peak.

Ans. 0.1 V; 0.94 μF

6.29 For the sample-and-hold circuit of Fig. 6.41 with $C = 0.1\ \mu$F find the maximum allowed total leakage current of Q_1 and Q_2 so that v_O droops by no more than 1 mV/s.

Ans. 100 pA

6.12 SUMMARY

● The essence of JFET operation is that the reverse-bias voltage across the gate–channel junction controls the width of the depletion region and hence the channel width. This in turn controls the current flow through the channel (the drain current).

● When the gate-to-source reverse bias exceeds the magnitude of the pinch-off voltage, $|V_P|$, the channel is fully depleted of charge carriers and no current flows.

● The drain-to-source voltage causes the reverse bias between gate and channel to increase progressively as one moves along the channel from source to drain. This results in a tapered channel that is narrowest at the drain end.

● When the gate-to-channel reverse bias *at the drain end* exceeds $|V_P|$, the JFET enters the pinch-off region of operation. Otherwise it operates in the triode region.

● For a given v_{GS} the channel geometry remains constant once the JFET enters pinch-off. Thus the current remains constant at the value reached at the onset of pinch-off.

● In the pinch-off region, the JFET operates as a voltage-controlled current source having the control relationship $i_D = I_{DSS}(1 - v_{GS}/V_P)^2$, where I_{DSS} and V_P are FET parameters. The current i_D depends also slightly on v_{DS}, thus giving the controlled source a finite output resistance r_o.

● V_P is negative for n-channel devices and positive for p-channel devices.

● For an n-channel (p-channel) JFET to operate in pinch-off, the drain voltage must be higher (lower) than the gate voltage by at least $|V_P|$.

● To apply the JFET as a linear amplifier, it is biased to operate in pinch-off and the input signal amplitude is kept small ($\ll |V_P|$).

● Under the small-signal condition, the JFET operates as a linear voltage-controlled current source, having a transconductance g_m equal to the slope of the i_D-v_{GS} characteristic at the bias point.

● Bias design seeks to establish a dc drain current that is as insensitive as possible to the wide variations normally encountered in V_P and I_{DSS} among JFETs of the same type. A combination of fixed bias and self-bias is best for discrete JFET amplifiers.

● Small-signal analysis of JFET amplifiers can be performed by replacing the JFET with its small-signal equivalent circuit model and by replacing the dc voltage sources with short circuits.

● In the common-source configuration the source is at signal ground, the input signal is applied to the gate, and the output is taken at the drain.

● In the source-follower or common-drain configuration the drain is at signal ground, the input signal is applied to the gate, and the output is taken at the source. The voltage gain is less than unity, but the output resistance is low.

● Switching applications make use of both the cutoff region and the triode region.

● In the triode region, for small v_{DS}, the JFET has a linear resistance (the "on" resistance of the switch) and is capable of bidirectional current flow, making it a good switch for analog signals.

BIBLIOGRAPHY

Analog Switches and Their Applications, Santa Clara, Calif.: Siliconix.

E. J. Angelo, Jr., *Electronics—BJTs, FETs, and Microcircuits*, New York: McGraw-Hill, 1969.

R. S. C. Cobbold, *Theory and Applications of Field-Effect Transistors*, New York: Wiley, 1969.

A. D. Evans (Ed.), *Designing with Field-Effect Transistors*, New York: McGraw-Hill, 1981.

W. H. Hayt and G. W. Neudeck, *Electronic Circuit Analysis and Design*, Second Edition, Boston: Houghton Mifflin, 1984.

PROBLEMS

6.1 A JFET of unknown type, but for which the source and drain connections are known, has the gate left floating and an ohmmeter connected between source and drain. When the gate is joined to the negative (red) lead of the ohmmeter, the apparent resistance does not change. When the gate is joined to the positive (black) lead of the ohmmeter, the apparent resistance decreases greatly. Is the JFET of the *n*- or *p*-channel type?

6.2 An *n*-channel JFET with characteristics as shown in Fig. 6.7 with a pinch-off voltage of -4 V is operated with a gate-to-source voltage of -1 V and a drain-to-source voltage of 0. What is the region of operation of the FET under these conditions?

As the drain voltage is raised, at what drain-to-gate voltage does the drain pinch off? What is the corresponding drain-to-source voltage? A second JFET having a pinch-off voltage of -5 V is operated with a gate-to-source voltage of -1 V. How high must its drain-to-source voltage be raised to ensure operation in the pinch-off region with the channel pinched off at the drain end?

6.3 Consider the JFET triode-region relationship presented in Eq. (6.4). What is the value of v_{DS} for which the squared term is less than 10% of the linear term? For which it is less than 1% of the linear term? What do these limits become for operation at $v_{GS} = 0$? At $v_{GS} = V_P/2$?

6.4 Consider Eq. (6.6). What does the linear resistance r_{DS} become for $v_{GS} = 0$? For $v_{GS} = V_P/2$? For $v_{GS} = 0.9$ V_P?

6.5 Consider the control characteristic shown in Fig. 6.8. What is the gate-to-source voltage for which the drain current is $I_{DSS}/2$? What fraction of I_{DSS} is the drain current for a gate-to-source voltage of $V_P/2$?

6.6 The gate-to-channel junction of a particular JFET for which $V_P = -4$ V is found to break down at 22 V. If the FET is operated with a gate signal varying ± 1 V around pinch-off, what is the highest value of drain-to-source voltage that can be used without encountering breakdown?

6.7 In a particular application a JFET whose gate-to-channel breakdown is 22 V is operated with a high-resistance gate supply of -4 V. By accident a supply of 20 V is connected to the drain (the source is grounded). Assuming the result to be nondestructive, at least on a short-term basis, what happens? What does v_{GS} become?

6.8 A JFET with $V_P = -1$ V and $I_{DSS} = 1$ mA shows an output resistance of 100 kΩ when operated in pinch-off with $v_{GS} = 0$. What is the value of output resistance when the device is operated in pinch-off with $v_{GS} = -0.5$ V?

6.9 For a particular JFET, $V_P = -2$ V and $I_{DSS} = 8$ mA. What is the minimum v_{DS} for which the device will operate in pinch-off with $v_{GS} = -1$ V? Find i_D for $v_{DS} = +2$ V and $v_{GS} = -1$ V.

6.10 For a JFET having $V_P = -2$ V and $I_{DSS} = 8$ mA and operating at $v_{GS} = -1$ V, find the change in v_{GS} required to increase i_D by 1 mA. What change in v_{GS} is required to decrease i_D by 1 mA from the same original condition?

6.11 For a JFET having $V_P = -2$ V and $I_{DSS} = 8$ mA operating at $v_{GS} = -1$ V and $v_{DS} = 0$ V, what is the value of r_{DS}? Find the value of v_{GS} at which r_{DS} becomes half this value.

6.12 A *p*-channel JFET operating with $v_{SD} = 0$ has a channel resistance of 100 Ω for $v_{SG} = 0$ and of 400 Ω for $v_{SG} = -1$ V. What must I_{DSS} and V_P be for this FET?

6.13 A *p*-channel JFET for which $V_P = 1$ V operates with $v_{SG} = 0$ V. For $v_{SD} = 1$ V, $i_D = 1$ mA. For $v_{SD} = 11$ V, $i_D = 1.1$ mA. Calculate I_{DSS} and r_o, the output resistance in pinch-off. If the gate-to-channel junction breaks down at 20 V and the source is connected to a $+10$-V supply, what is the lowest voltage that can be applied to the drain without junction breakdown?

6.14 A *p*-channel JFET operating at $v_{SG} = 0$ is connected in series with a power supply V and a resistor $R = 10$ kΩ. It has been found that for $V \geq 12$ V the voltage across R is constant at 10 V. What are I_{DSS} and V_P for this FET?

6.15 If, for the JFET described in Problem 6.14, the voltage V is reduced to 5 V and $R = 10$ kΩ, what does v_{SD} of the FET become? If, for $V = 5$ V, R is raised to 100 kΩ, what does v_{SD} become?

6.16 A *p*-channel JFET for which $I_{DSS} = 8$ mA and $V_P = 2$ V operates with the gate grounded, the source connected to ground through a 1-kΩ resistor and the drain connected to a -10-V supply via a 5-kΩ resistor. What drain current flows? What are the voltages on the source and drain?

6.17 Repeat Problem 6.16 for the case in which the drain resistor is increased to 20 kΩ.

***6.18** In the circuits shown in Fig. P6.18 each FET has $I_{DSS} = 2$ mA, $|V_P| = 2$ V, and the output resistance in pinch-off, r_o, is 50 kΩ. Find the currents and voltages indicated.

****6.19** For the circuits in Fig. P6.19 find the voltages labeled. For each FET, $I_{DSS} = 2$ mA, $|V_P| = 2$ V, and $r_o = \infty$.

***6.20** Using Eq. (6.4), explore the behavior of the $v_{GS} = 0$ curve of Fig. 6.7 at low values of v_{DS}. Find an expression

for the slope of the curve at $v_{DS} = 0$. If this tangent were extended, at what value of v_{DS} would it intersect the line $i_D = I_{DSS}$? From Eq. (6.4), what does i_D become for $v_{GS} = 0$ and $v_{DS} = -V_P$? Using the numerical values of I_{DSS} and V_P that apply to Fig. 6.7, calculate v_{DS} at $i_D = I_{DSS}/2$ and $0.75\ I_{DSS}$, for $v_{GS} = 0$. Compare these with the corresponding points on the extended tangent.

6.21 Using the characteristic curves of the JFET shown in Fig. 6.7, draw load lines for a grounded-source JFET having $V_{DD} = 12$ V and $R_D = 250$ Ω, 500 Ω, 1 kΩ, 2 kΩ, and 4 kΩ. In each case what would be the operating (quiescent) points for $v_{GS} = 0$, $V_P/2$, and V_P?

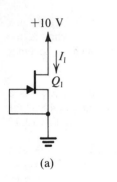

(a)

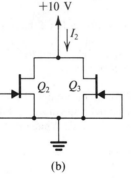

(b)

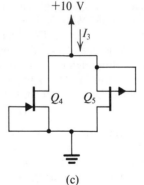

(c)

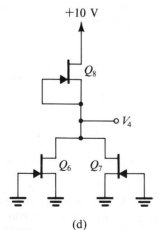

(d)

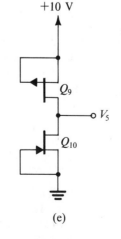

(e)

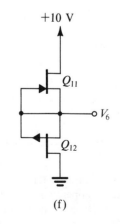

(f)

Fig. P6.18

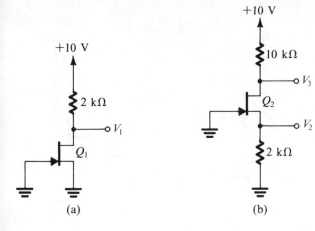

+10 V

2 kΩ

V_1

Q_1

(a)

+10 V

10 kΩ

V_3

Q_2

V_2

2 kΩ

(b)

+10 V

20 kΩ

V_5

Q_3

V_4

2 kΩ

(c)

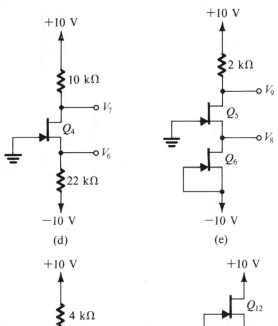

+10 V

10 kΩ

V_7

Q_4

V_6

22 kΩ

−10 V

(d)

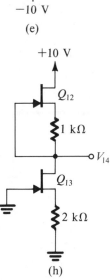

+10 V

2 kΩ

V_9

Q_5

V_8

Q_6

−10 V

(e)

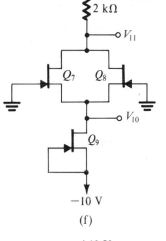

+10 V

2 kΩ

V_{11}

Q_7 Q_8

V_{10}

Q_9

−10 V

(f)

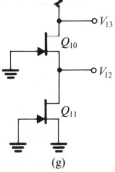

+10 V

4 kΩ

V_{13}

Q_{10}

V_{12}

Q_{11}

(g)

+10 V

Q_{12}

1 kΩ

V_{14}

Q_{13}

2 kΩ

(h)

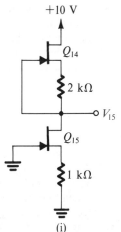

+10 V

Q_{14}

2 kΩ

V_{15}

Q_{15}

1 kΩ

(i)

Fig. P6.19

321

*6.22 A FET switch is connected with two load resistors as shown in Fig. P6.22. The intent is to provide somewhat complementary signals at X and Y; that is, when one rises, the other falls. For the FET, $I_{DSS} = 10$ mA and $V_P = -2$ V. For the diode, when conducting, $V_D = 0.7$ V. When the diode is cut off, what are the voltages at X and Y? What voltage is required at A to ensure that the diode is barely cut off (diode voltage is zero)? What voltage on A is required to cause the JFET to cut off? What voltages on X and Y result?

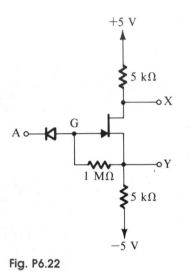

Fig. P6.22

6.23 Consider the JFET devices described by the curves in Fig. 6.20 operated using a constant-current-source supply. What is the largest current that can be used for the "low" device while maintaining the gate-to-channel junction totally cut off?

6.24 For JFETs specified as follows, find the value of the self-bias resistor R_S that results in $I_D = I_{DSS}/4$: [I_{DSS} (mA), V_P (V)] = (16, −8); (8, −4); (4, −2); (16, −4); (8, −2); (16, −2); (4, −1).

***6.25 Each of the circuits shown in Fig. P6.25 is designed to operate using an FET for which $I_{DSS} = 4$ mA and $V_P = -2$ V at $V_{GS} = -1$ V, $I_D = 1$ mA, and $V_D = +6$ V. As a challenge you are asked to evaluate the behavior of these circuits for devices for which I_{DSS} varies from 2 to 16 mA and V_P varies from −1 to −4 V. To do so, analyze the behavior of each circuit for the four devices representing the extreme values of I_{DSS} and V_P.

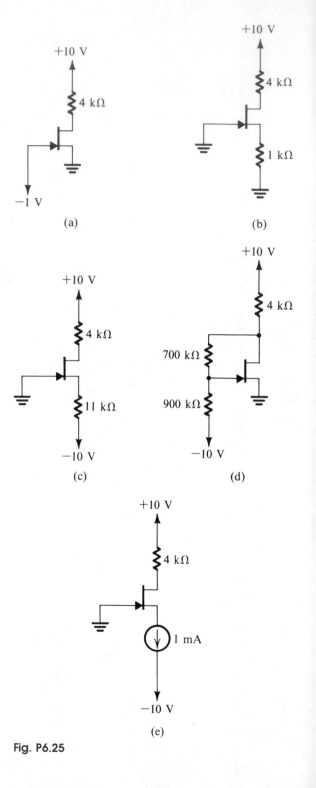

Fig. P6.25

(Note that here we assume that there are four rather than just two extreme devices.) For each (where appropriate), calculate I_D, V_D, V_{GS}, and

$$g_m = \frac{2I_{DSS}}{|V_P|}\left(1 - \frac{V_{GS}}{V_P}\right)$$

Prepare a table including the following data:

    (i) Whether the bias design results in pinch-off operation.

    (ii) The range of output signal swing measured as the smaller of: the difference between V_D and the supply voltage, and the difference between V_D and the boundary of the triode region.

    (iii) The range of g_m expressed as fractions of g_m at the design point.

*6.26 Show that for the circuit in Fig. P6.26 to operate in pinch-off the following two conditions must be satisfied.

$$I_D R_S \geq 0.5|V_P|$$
$$V_{DD} - I_D R_D \geq 1.5|V_P|$$

Assume that the two JFETs are matched. For $V_{DD} = 10$ V, $|V_P| = 2$ V, and $I_{DSS} = 4$ mA, design R_S and R_D so that:
(a) $V_{DG1} = |V_P|$ and $V_{DG2} = 2|V_P|$.
(b) $V_{DG1} = 1.5|V_P|$ and $V_{DG2} = 1.5|V_P|$.

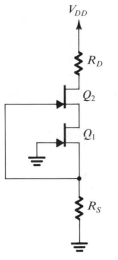

Fig. P6.26

*6.27 In the circuit shown in Fig. P6.27 calculate I_D for $I_{DSS} = 4$ mA if the diode is a 1-mA unit (has a drop of

0.7 V at a current of 1 mA) for which $n = 2$, under the conditions that **(a)** $V_P = -2$ V and **(b)** $V_P = -1$ V.

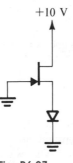

Fig. P6.27

6.28 Two JFETs having identical characteristics are connected in parallel. How are V_P, I_{DSS}, and g_{m0} of the combination and of a single transistor related?

6.29 For what change in bias voltage V_{GS} does the g_m of a JFET change by 10%?

6.30 A JFET circuit using a constant-current-source bias current I is equipped with a new FET type having the same V_P but an I_{DSS} that is 4 times as large as that for which the design was made. What increase in g_m and voltage gain should be expected?

6.31 Use the JFET model of Fig. 6.26a to find an expression for the input resistance seen looking into the source terminal with the gate grounded. Evaluate this input resistance for a JFET for which $I_{DSS} = 4$ mA and $V_P = -2$ V when operating at $I_D = 1$ mA.

*6.32 Design a two-resistor gate-bias network such as that shown in Fig. 6.28 to satisfy the following constraints: $V_{DD} = 15$ V; the gate should be biased at one-third the supply; the input resistance should exceed 1 MΩ; resistors of value greater than 10 MΩ are not available; the bias network should use less than 10 μA from the supply; the bias point should shift by less than 10% at high temperatures where the gate leakage current increases to 100 nA.

6.33 Consider the circuit of Fig. 6.29 modified to leave a small part of R_S, denoted R_{S1}, unbypassed. Find the value of R_{S1} that will make the gain v_o/v_i exactly -1.

6.34 A JFET for which $I_{DSS} = 8$ mA and $V_P = -2$ V operates at $V_{GS} = -1$ V with an input signal $v_{gs} = V \sin \omega t$. Find the percentage harmonic distortion for $V = 0.5$ V. For what input signal is the harmonic distortion 1%?

6.35 In Example 6.5, for what rms value of sine-wave input is the harmonic distortion 1%?

6.36 For what value of unbypassed source resistance is the effect of a reduction in g_m by one-half reduced to a change of 5% in amplifier gain?

6.37 What is the power gain of an amplifier with the topology of Fig. 6.31 for which $R_{in} = 1\ M\Omega$, $R_{S1} = 100\ \Omega$, $g_m = 10\ mA/V$, $R_D = 4\ k\Omega$, and $R_L = 8\ k\Omega$? If R_L is reduced to 4 kΩ, what does the power gain become? Express as ratios and in decibels.

6.38 A source follower utilizing a JFET for which $I_{DSS} = 10\ mA$, and $V_P = -5\ V$ is biased by a constant-current source (feeding the source terminal) of 5 mA. What is the output resistance of such a follower? For what value of capacitor-coupled load is the voltage gain reduced to one-half the open-circuit voltage gain?

6.39 A source follower can be biased at I_{DSS} using the circuit arrangement of Fig. P6.39. Here Q_2 is the biasing transistor and V_{SS} is greater than $|V_P|$.
(a) Analyze the circuit taking into account the source-to-drain resistance of each of the two FETs to show that the voltage gain is given by

$$\frac{v_o}{v_i} = \frac{g_{m0}r_o/2}{1 + (g_{m0}r_o/2)}$$

and that the output resistance is $(r_o/2)\|(1/g_{m0})$.
(b) Find the voltage gain and output resistance for the case $I_{DSS} = 1\ mA$, $V_P = -2\ V$, and $|V_A| = 100\ V$.

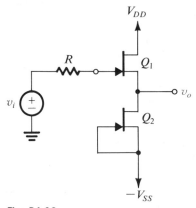

V_{DD}

R

Q_1

v_i

v_o

Q_2

$-V_{SS}$

Fig. P6.39

6.40 Consider the circuit of Fig. P6.39 with $R = 10\ M\Omega$. If at low temperatures the leakage current of the FET gate is 1 nA and it is assumed that the dc component

of v_i is zero, find the dc voltage at the output. If the leakage current increases to 100 nA, what does the output dc voltage become?

*6.41 Consider the JFET source follower of Fig. 6.32 with R set to zero. The most negative output voltage, v_{Omin}, occurs when the load current equals the current in R_S, leaving the JFET with zero source current. At this point $v_{GS} = V_P$ and the JFET just cuts off. Show that v_{Omin} and the corresponding value of v_I, denoted v_{Imin}, are given by

$$v_{Omin} = \frac{-I_D R_S}{1 + (R_S/R_L)}$$

$$v_{Imin} = (I_D R_S)\frac{(R_S/R_L)}{1 + (R_S/R_L)} - |V_P|$$

**6.42 Consider the circuit of Fig. 6.32 modified by connecting the lower end of R_G to the upper end of R_L. Note that the dc bias voltage on the gate is unaffected, but that the signal voltage across R_G is reduced. This circuit is said to have a *bootstrapped input resistor*.
(a) Using either Miller's theorem (see Section 2.5) or direct-circuit analysis employing the FET equivalent circuit model, show that

$$\frac{v_o}{v_g} = \frac{g_m + (1/R_G)}{g_m + \dfrac{1}{R_S} + \dfrac{1}{R_L} + \dfrac{1}{r_o} + \dfrac{1}{R_G}}$$

and

$$R_{in} \simeq R_G\left[1 + \frac{g_m}{\dfrac{1}{R_S} + \dfrac{1}{R_L} + \dfrac{1}{r_o}}\right]$$

(b) For the component values of Exercise 6.24 calculate v_o/v_g, R_{in}, and v_o/v_i.

**6.43 Consider the three-stage amplifier shown in Fig. 6.34. Scale the circuit to use approximately one-fourth of the present current by raising all resistor values by a factor of 4. What is the total supply current required? What is the voltage gain for $R_L = 2\ k\Omega$?

**6.44 Redesign the circuit in Fig. 6.34 to use a 10-V supply while maintaining the voltages V_{DS} and V_{GS} and the current I_D for all FETs unchanged. As a convenient basis for design, reduce the voltage on the drain of Q_1 to 6 V. What is the voltage gain of the modified amplifier for a load $R_L = 2\ k\Omega$?

***6.45 In the direct-coupled amplifier circuit of Fig. P6.45, Q_1, Q_2, Q_3, and Q_4 are identical JFETs for which

$I_{DSS} = 8$ mA, $V_P = -2$ V, and r_o (at a current of 8 mA) is 125 kΩ. Q_1 operates as a common-source amplifier with Q_2 providing a constant-current bias, and also acting as the load for Q_1 (that is, the load resistance for Q_1 is equal to r_o of Q_2). Q_3 operates as a source follower. R_0 provides an output-level shift using the constant current supplied by Q_4. The circuit from B to the output terminal E forms an op amp (single-ended) with B as its inverting input terminal, and R_1 and R_2 form a resistive feedback network that determines the closed-loop gain v_o/v_i.
(a) With the input A grounded, what are the dc voltages at B, E, D, and C? (For this calculation, assume $r_o = \infty$.)
(b) Find approximate values for the gain from B to C, from C to E, and hence from B to E. Use this value to estimate the closed-loop gain, v_o/v_i.

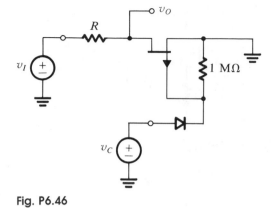

Fig. P6.46

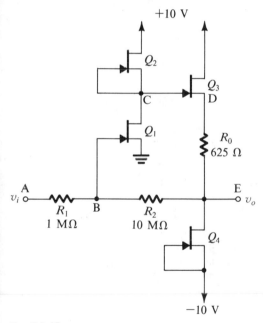

Fig. P6.45

6.46 The p-channel FET in the circuit of Fig. P 6.46 is operated as a switch in order to ground signal v_I via a resistor R when the gate control voltage v_C is low. If $r_{DS} = 100$ Ω at $v_{GS} = 0$ and v_{DS} small, find $v_0(t)$ for $v_I(t) = 5 \sin \omega t$ and $R = 10$ kΩ when v_C is low. If $V_P = 2$ V and the diode voltage drop is 0.7 V, find the minimum value of v_C that will turn the JFET off at all times for the given input signal.

6.47 If, in the sample-and-hold circuit of Fig. 6.41, the droop of v_O is to be less than 0.1 mV/s and a net leakage current of 10 nA is expected, how large must C be?

6.48 Consider the circuit of Fig. 6.41, with hold level as stated. For an op-amp saturation voltage of -12 V and a diode drop of 0.7 V, what is the largest allowed value of pinch-off voltage, $|V_P|$, for Q_1?

***6.49** The circuit shown in Fig. P6.49a is a *complementary source follower*. For matched JFETs with $I_{DSS} = 16$ mA and $|V_P| = 2$ V, what is the voltage v_O when $v_I = 0$ V, $+8$ V, -8 V, $+9$ V, and -9 V? For $v_I = 0$ V, what is the output resistance? For $v_I = 0$ V and the output terminal connected to the current-source load shown in Fig. P6.49b, what does the output voltage become?

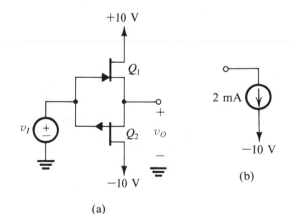

Fig. P6.49

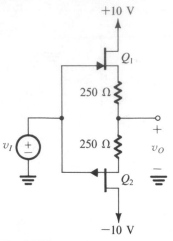

Fig. P6.50

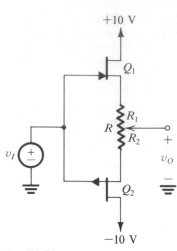

Fig. P6.51

***6.50** Repeat Problem 6.49 for the follower circuit shown in Fig. P6.50.

***6.51** For the follower circuit shown in Fig. P6.51 let the JFETs have equal $|V_P|$ of 2 V but let Q_1 have $I_{DSS} = 4$ mA and Q_2 have $I_{DSS} = 16$ mA. What is the minimum value of R that will make $v_O = v_I$? If a resistor R of twice the minimum required value is used, find the position of the potentiometer wiper (that is, find R_1 and R_2) to obtain $v_O = v_I$. Also find the current in both FETs.

****6.52** In the circuit of Fig. P6.52, both FETs have $I_{DSS} = 4$ mA and $|V_P| = 2$ V. For $v_I = 0$ V, what is v_O? Find the small-signal voltage gain v_o/v_i.

***6.53** Using ± 10-V supplies, a supply current of 0.5 mA, a drain-resistor load of 10 kΩ to $+10$ V, a direct-coupled input at 0 V, a source resistor R_S to -10 V, and a large source-bypass capacitor C_S, design a JFET amplifier with a voltage gain of -10. It should be biased at $V_{GS} = V_P/2$ and use a JFET with $|V_P|$ no less than 2 V and a minimum I_{DSS}. Specify your choice of V_P and I_{DSS} and give the value of R_S. For I_{DSS} twice your design value, find the value of the resistor that, when placed in series with C_S, restores the gain to -10.

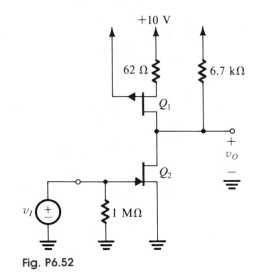

Fig. P6.52

****6.54** The JFETs in the circuits of Fig. P6.54 have $I_{DSS} = 4$ mA and $V_P = -2$ V. Find the amplifier transconductance $[I(s)/V(s)]$ as a function of the complex frequency variable s.

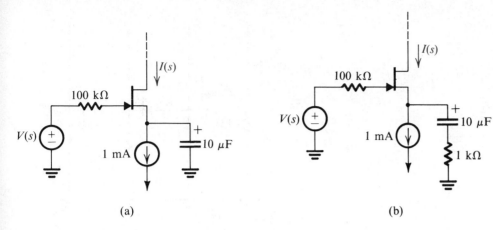

(a)

(b)

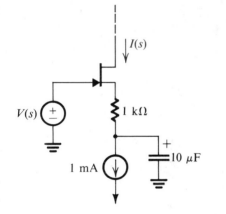

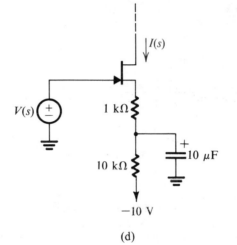

(c)

(d)

Fig. P6.54

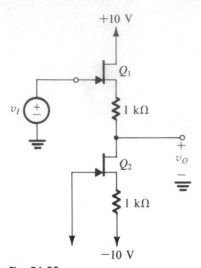

+10 V

Q_1

1 kΩ

v_I

v_O

Q_2

1 kΩ

−10 V

Fig. P6.55

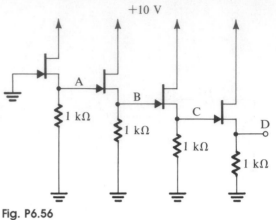

+10 V

A

1 kΩ

B

1 kΩ

C

1 kΩ

D

1 kΩ

Fig. P6.56

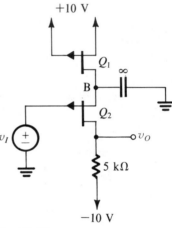

+10 V

Q_1

B

∞

Q_2

v_I

v_O

5 kΩ

−10 V

Fig. P6.57

**6.55 The JFETs in the circuit of Fig. P6.55 have $I_{DSS} = 4\,\text{mA}$ and $V_P = -2\,\text{V}$. Find v_O for $v_I = 0\,\text{V}$, $+1\,\text{V}$, and $-1\,\text{V}$. Also find R_{out}.

**6.56 The JFETs in the circuit of Fig. P6.56 have $I_{DSS} = 4\,\text{mA}$ and $V_P = -2\,\text{V}$. Find V_A, V_B, V_C, and V_D.

6.57 The JFETs in the circuit of Fig. P6.57 have $I_{DSS} = 1\,\text{mA}$ and $V_P = 1\,\text{V}$. For $v_I = 0\,\text{V}$ find v_B and v_O. Also, find the small-signal voltage gain v_o/v_i.

Chapter 7

Metal-Oxide-Semiconductor Field-Effect Transistors (MOSFETs)

INTRODUCTION

In this chapter we study a very popular class of field-effect transistor, the metal-oxide-semiconductor FET or, simply, MOSFET. As will be seen in Chapter 15, digital logic and memory functions can be implemented with circuits that exclusively use MOSFETs (that is, no resistors or diodes are needed). For this reason, because MOS devices can be made quite small (that is, occupying a small silicon area on the IC chip), and because their manufacturing process is relatively simple (as compared to that of bipolar transistors; see Appendix A), most very-large-scale-integrated (VLSI) circuits are currently made using MOS technology. Examples include microprocessor and memory packages. MOS technology has also been applied to the design of analog integrated circuits.

MOS transistors are either of the *p*-channel or *n*-channel type. Initially, *p*-channel devices (PMOS) were quite popular in digital applications, but more recently the NMOS (*n*-channel) technology has been perfected. The latter technology results in faster devices, which occupy smaller silicon area. Therefore the PMOS technology has practically disappeared and NMOS has become one of two currently dominant VLSI technologies. The other is *complementary-symmetry MOS* (COS MOS or CMOS). As the name implies, CMOS circuits use both *n*-channel and *p*-channel devices on the same IC chip. CMOS logic circuits are currently quite popular and are sometimes referred to as the "ideal" logic family. Moreover, CMOS is capable of im-

plementing a variety of analog circuit functions. At the time of this writing CMOS is the technology of choice whenever analog and digital functions are to be implemented on the same VLSI circuit chip.

MOS transistors have very high input impedance and consume little static power. This makes them quite useful in the design of micropower circuits, both digital and linear. Needless to say, the MOS transistor is also very useful in the design of amplifiers with extremely high input impedance. MOS transistors can also be used as analog switches. This together with the ability to manufacture capacitors whose ratios are very accurate using MOS technology, has motivated its application in the design of analog signal-processing circuits (see Chapter 14). □

7.1 THE DEPLETION-TYPE MOSFET

Physical Structure

Unlike JFETs, which are all of the depletion type, MOSFETs can be either depletion or enhancement type (the exact meaning of these words will become clearer shortly). The operation of the depletion-type MOSFET is very similar to that of the JFET. Let us start by considering the depletion-type MOSFET shown in Fig. 7.1. As shown, the *n*-channel device is formed on a *p*-type silicon substrate. The two heavily doped n^+ wells form low-resistance connections between the ends of the *n* channel and the metal contacts of the source (S) and the drain (D). A thin oxide layer is grown on the surface of the channel, and metal (aluminum) is deposited on it to

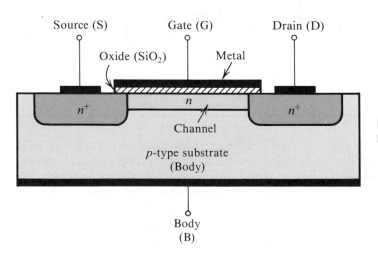

Fig. 7.1 Physical structure of *n*-channel MOSFET of the depletion type.

form the gate (G). At this point it should be clear that the name MOS is derived from the structure of the device.[1]

Whereas the JFET is controlled by the gate-to-channel *pn* junction, this is not the case in the MOSFET. Here the oxide layer acts as an insulator that causes the gate current to be negligibly small (10^{-12} to 10^{-15} A). This gives the MOS transistor its extremely high input resistance under all conditions.

Circuit Symbol

Figure 7.2 shows the circuit symbol of the *n*-channel depletion-type MOSFET. Note that the space between the line representing the gate electrode and the line representing the channel denotes the insulating oxide layer. As in the JFET symbol the gate terminal is drawn closer to the source than to the drain. The arrow on the substrate (also known as the body) line points in the forward direction of the substrate-to-channel *pn* junction and hence indicates the type (or polarity) of the device (*n*- or *p*-channel device).

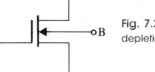

Fig. 7.2 Circuit symbol for the *n*-channel depletion-type MOSFET.

In many applications the substrate is electrically connected to the source. Since the drain voltage will be positive with respect to the source, the substrate-to-channel junction will always be reverse-biased and hence the substrate current will be almost zero. In the remainder of this section we will ignore the role of the substrate in our description of MOSFET operation. In later sections we will describe the role that the substrate plays if it is not connected to the source.

A simplified symbol for the *n*-channel depletion-type MOSFET is shown in Fig. 7.3. Note that the polarity of the device is indicated by the direction of the arrowhead

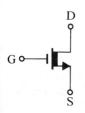

Fig. 7.3 Simplified circuit symbol for the depletion-type *n*-channel MOSFET.

---

[1] Modern IC MOS transistors utilize polysilicon for the gate electrode, in a technology known as silicon-gate technology (see Appendix A).

on the source line; the arrowhead points in the normal direction of current flow in the source lead. Since this simplified circuit symbol does not show the substrate terminal, its use will be confined to situations in which the effect of substrate on device operation is not important.

Physical Operation

As mentioned before, the depletion-type MOSFET operates much like the JFET. The fundamental difference is that in the JFET the channel is depleted by reverse-biasing the gate-to-channel junction. In the MOSFET control of the channel width is effected by an electric field created by the voltage on the gate.

Consider first the triode—or voltage-controlled resistance—region of operation. Let v_{DS} be small while v_{GS} is either negative or positive (note the difference between this case and the JFET case; for the JFET, v_{GS} has to be negative or at most 0.5 V positive, otherwise the gate-to-channel junction becomes forward-biased). If v_{GS} is negative, some of the electrons in the n-channel area will be repelled from the channel, and thus a depletion region will be created below the oxide layer. The result will be a narrower channel, as shown in Fig. 7.4. From Fig. 7.4 we also note that the channel has a uniform width because $v_{GS} \simeq v_{GD}$. The narrower channel will have a higher resistance r_{DS}, giving rise to the set of straight-line characteristics shown in Fig. 7.5.

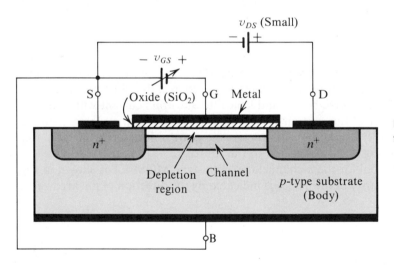

Fig. **7.4** Physical operation of the depletion-type MOSFET for small v_{DS}.

As v_{GS} is made more negative, a point will be reached at which the channel is completely depleted of charge carriers. The value of v_{GS} at which this happens is called the pinch-off voltage V_P,

$$V_P = v_{GS}|_{i_D = 0, \, v_{DS} = \text{small}}$$

Obviously, for an n-channel device V_P is negative.

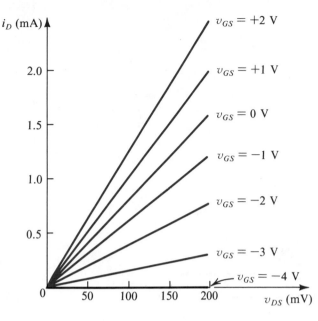

i_D (mA)

$v_{GS} = +2$ V

$v_{GS} = +1$ V

$v_{GS} = 0$ V

$v_{GS} = -1$ V

$v_{GS} = -2$ V

$v_{GS} = -3$ V

$v_{GS} = -4$ V

v_{DS} (mV)

Fig. 7.5 The i_D-v_{DS} characteristics of the depletion-type MOSFET for small v_{DS} (for this device $I_{DSS} = 16$ mA and $V_p = -4$ V).

Unlike the JFET case, v_{GS} can be made positive, with the result that the channel becomes wider and its resistance decreases, as illustrated in Fig. 7.5. The mechanism by which this occurs will be discussed when we consider enhancement devices.

Next consider the operation as v_{DS} is increased. If v_{GS} is kept constant while v_{DS} is increased, the depletion region will have a constant width at the source end, but it will be progressively wider as we move toward the drain. Thus the channel will have a tapered shape, and its width will be narrowest at the drain end. It follows that as v_{DS} is increased the channel resistance increases, giving rise to a nonlinear (parabolic) i_D-v_{DS} curve. Finally, if v_{DS} is increased to the value at which $v_{GD} = V_P$, the channel will be pinched off at the drain end. Any further increase in v_{DS} does not alter the channel shape, and hence the current in the channel remains constant at the value reached for $v_{GD} = V_P$ (or, correspondingly, $v_{DS} = v_{GS} - V_P$.) The result is the i_D-v_{DS} characteristic curves shown in Fig. 7.6.

Static Characteristics

The static characteristics of the n-channel depletion-type MOSFET take the form shown in Fig. 7.6. These characteristics are identical to those of the JFET except that in the MOSFET positive values for v_{GS} are allowed. A positive v_{GS} "enhances" the channel by attracting more electrons into it, thus increasing its width and reducing its resistance. A depletion-type MOSFET can therefore be applied in the *enhancement mode*. Figure 7.7 illustrates this point further by showing the i_D-v_{GS} characteristic curve for a device operating in the pinch-off region ($v_{DG} \geq |V_P|$).

The equations describing the characteristic curves of Figs. 7.6 and 7.7 are identical to those for the JFET. Specifically, in the triode region,

$$v_{DG} < -V_P$$

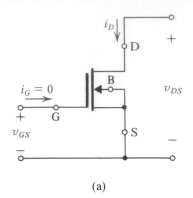

(a)

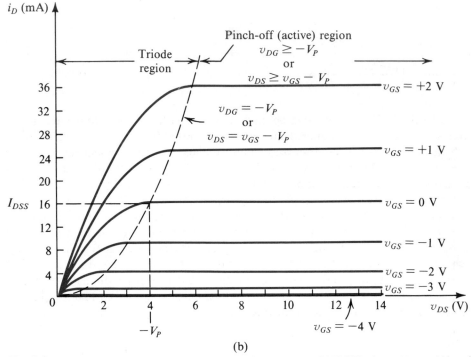

(b)

Fig. 7.6 The i_D-v_{DS} characteristics for a depletion-type n-channel MOSFET whose $V_P = -4$ V and $I_{DSS} = 16$ mA.

or, equivalently,

$$v_{DS} < v_{GS} - V_P$$

$$i_D = I_{DSS}\left[2\left(1 - \frac{v_{GS}}{V_P}\right)\left(\frac{v_{DS}}{-V_P}\right) - \left(\frac{v_{DS}}{V_P}\right)^2\right] \quad (7.1)$$

In the pinch-off region,

$$v_{DG} \geq -V_P,$$

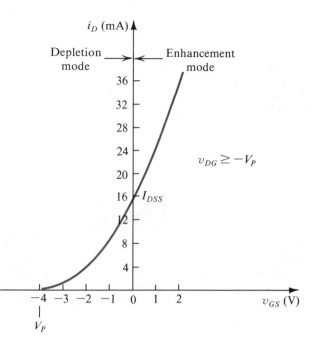

Fig. 7.7 The i_D-v_{GS} characteristic of an n-channel depletion-type MOSFET operating in the pinch-off region. Note that the device can be operated in both the depletion ($v_{GS} < 0$) and enhancement ($v_{GS} > 0$) modes.

or equivalently,

$$v_{DS} \geq v_{GS} - V_P$$

$$i_D = I_{DSS}\left(1 - \frac{v_{GS}}{V_P}\right)^2 \tag{7.2}$$

Equation (7.2) is based on the assumption that in pinch-off i_D is independent of v_{DS}—in other words, that the MOSFET operates as an ideal current source having an infinite output resistance. Practical MOSFETs, however, have a finite output resistance in pinch-off, as in the JFET case depicted in Fig. 6.10.

The p-Channel Device

The depletion-type p-channel MOSFET operates in the same manner as the n-channel device except that all voltages and currents are reversed. It has the circuit symbols shown in Fig. 7.8, and its operation is described by Eqs. (7.1) and (7.2) above, where the pinch-off voltage V_P is positive.

We conclude this section by noting that with the exceptions mentioned above, the depletion-type MOSFET behaves in much the same way as a JFET and can be applied using circuit techniques similar to those discussed for the JFET. We shall not repeat this material here; our only further concern with the depletion MOSFET will be its use as a load device in NMOS integrated circuits (see Sections 7.9 and 15.3).

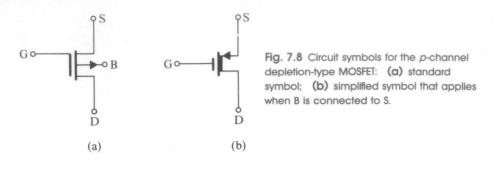

Fig. 7.8 Circuit symbols for the *p*-channel depletion-type MOSFET: **(a)** standard symbol; **(b)** simplified symbol that applies when B is connected to S.

Exercises

7.1 Consider an NMOS transistor with $V_P = -2$ V and $I_{DSS} = 8$ mA. Find the minimum v_{DS} required for the device to operate in pinch-off when $v_{GS} = +1$ V. What is the corresponding value of i_D?

Ans. 3 V; 18 mA

7.2 Find the *i-v* relationship (for both positive and negative *v*) of the depletion mode NMOS connected as a two-terminal device as shown in Fig. E7.2.

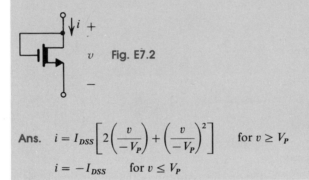

Fig. E7.2

Ans. $i = I_{DSS}\left[2\left(\dfrac{v}{-V_P}\right) + \left(\dfrac{v}{-V_P}\right)^2\right] \quad$ for $v \geq V_P$

$i = -I_{DSS} \quad$ for $v \leq V_P$

7.2 THE ENHANCEMENT-TYPE MOSFET

The enhancement-type MOSFET is the most widely used device in the design of MOS integrated circuits. In this section we study its physical operation and terminal characteristics.

Structure and Physical Operation

Figure 7.9 shows the physical structure of an enhancement-type MOSFET. As seen, the structure looks quite similar to that of the depletion device in Fig. 7.1, with one major exception: there is no channel. It follows that if the gate is left floating, or if $v_{GS} = 0$, the path from drain to source includes two series diodes back-to-back, which means that no drain current can flow. To cause current to flow from drain to source we first have to create an *n* channel. This can be done by applying a positive voltage v_{GS}. Such a positive voltage at the gate will attract electrons from the substrate and

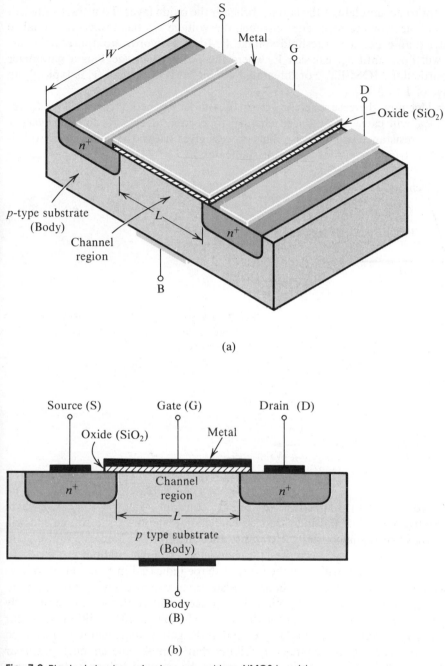

(a)

(b)

Fig. 7.9 Physical structure of enhancement-type NMOS transistor.

cause them to accumulate at the surface beneath the oxide layer. To attract sufficient numbers of electrons to form an n channel, the voltage v_{GS} has to become equal to or greater than a positive threshold voltage V_t. In other words, no appreciable current i_D will flow until v_{GS} exceeds V_t, where the positive voltage V_t is a parameter of the particular MOSFET. For integrated-circuit NMOS devices, V_t is typically in the range of 1 to 3 V.

Consider now the case where v_{DS} is small, and let v_{GS} be increased above V_t. Increasing v_{GS} will cause the induced channel to become deeper (become *enhanced*), and thus its resistance will decrease. This process gives rise to the voltage-controlled-resistance or triode region of operation, indicated in Fig. 7.10.

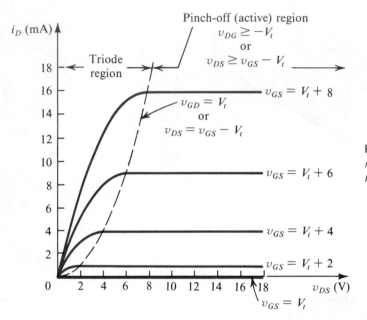

Fig. 7.10 Ideal $i_D - v_{DS}$ characteristics for an n-channel enhancement-type MOSFET whose $K = 0.25$ mA/V².

Next consider what happens if we keep v_{GS} constant—say, $v_{GS} = V_t + V$, where V is a positive voltage—and increase v_{DS}. The channel depth at the source end will not change, since v_{GS} is constant. However, since increasing v_{DS} means that v_{DG} increases or, equivalently, v_{GD} decreases, the channel will become shallower at the drain end. In fact, the channel will have the tapered shape indicated in Fig. 7.11. It follows that increasing v_{DS} will cause the channel resistance to increase, giving rise to a nonlinear i_D-v_{DS} curve in the triode region. This process will continue until eventually the channel depth becomes zero at the drain end. This pinch-off condition will occur when $v_{GD} \leq V_t$. To appreciate this fact, recall that the gate voltage had to be increased to V_t in order to create the channel. It follows that to make the channel disappear at the drain end the gate-to-drain voltage has to be reduced below V_t. In our case $v_{GS} = V_t + V$; thus pinch-off will occur when $v_{GD} \leq V_t$, which corresponds to $v_{DS} \geq V$.

Increasing the drain voltage above the value for pinch-off will not change the shape of the channel; hence the current i_D remains constant at the value reached at the onset of pinch-off.

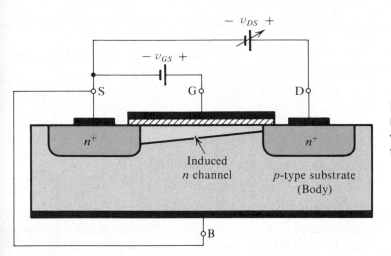

Fig. 7.11 Operation of the enhancement-type NMOS transistor, with v_{GS} kept constant at a value greater than V_t, and v_{DS} increased. Note the tapered shape of the induced channel.

Static Characteristics

The description above leads to the i_D-v_{DS} characteristics shown in Fig. 7.10, from which we see that there are two regions of operation: the triode region and the pinch-off region. Operation in the triode region (also called the *ohmic region*) is obtained for

$$v_{GS} \geq V_t \quad \text{(Induced channel)} \tag{7.3}$$

and

$$v_{GD} > V_t \quad \text{(Continuous channel)} \tag{7.4}$$

The latter condition can be expressed in the alternate form

$$v_{DS} < v_{GS} - V_t \quad \text{(Continuous channel)} \tag{7.5}$$

The i_D-v_{DS} characteristics in the triode region are described by

$$i_D = K[2(v_{GS} - V_t)v_{DS} - v_{DS}^2] \tag{7.6}$$

where the constant K is given by

$$K = \frac{1}{2} \mu_n C_{\text{ox}} \left(\frac{W}{L} \right) \tag{7.7}$$

where μ_n is the mobility of the electrons in the induced n channel, C_{ox} is the capacitance per unit area of the gate-to-channel capacitor for which the oxide layer serves as a dielectric, L is the length of the channel and W is its width (see Fig. 7.9). Since for a given IC fabrication process the quantity $(\frac{1}{2}\mu_n C_{\text{ox}})$ is a constant (approximately 10 $\mu\text{A/V}^2$ for the standard NMOS process with a 0.1-μm oxide thickness[2]), the aspect

---

[2] A micrometer (μm), or micron, is 10^{-6} meter.

ratio of the device, W/L, determines its conductivity parameter K. Note from Eq. (7.6) that K has the units A/V^2.

Operation in the pinch-off region (also called the *saturation region* because the drain current becomes constant for a given v_{GS}) is obtained for

$$v_{GS} \geq V_t \qquad \text{(Induced channel)} \tag{7.8}$$

and,

$$v_{GD} \leq V_t \qquad \text{(Channel pinched off at drain)} \tag{7.9}$$

The latter condition can be expressed in the alternate form

$$v_{DS} \geq v_{GS} - V_t \qquad \text{(Pinched-off channel)} \tag{7.10}$$

from which we observe that pinch-off operation is maintained even when the drain voltage falls below the gate voltage by as much as V_t volts. This should be contrasted with the case of the depletion MOSFET, where to operate in pinch-off the drain voltage must be greater than the gate voltage by at least $|V_P|$ volts. From the condition in Eq. (7.10) we find that the boundary between the triode region and the pinch-off region is determined from

$$v_{DS} = v_{GS} - V_t \qquad \text{(Boundary)} \tag{7.11}$$

Substituting for v_{DS} from Eq. (7.11) into Eq. (7.6) gives the i_D-v_{DS} characteristics in pinch-off,

$$i_D = K(v_{GS} - V_t)^2 \tag{7.12}$$

Thus, in pinch-off the enhancement MOSFET acts as a voltage-controlled current source having the square-law control characteristic sketched in Fig. 7.12. This figure

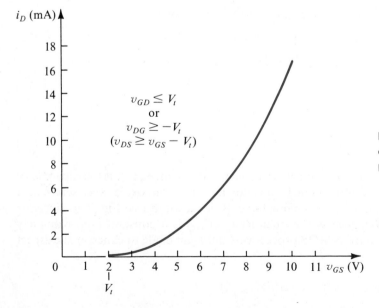

Fig. 7.12 The i_D-v_{GS} characteristic for an enhancement-type NMOS transistor in pinch-off ($V_t = 2$ V, $K = 0.25$ mA/V^2).

clearly illustrates the enhancement nature of the device: At $v_{GS} = 0$ the current is almost zero; for the device to conduct, v_{GS} has to exceed the positive threshold voltage V_t. In contrast, the depletion-type MOSFET (see Fig. 7.7) conducts the relatively large current I_{DSS} at $v_{GS} = 0$; then the current can be reduced by making v_{GS} negative and thus depleting the channel of charge carriers. However, depletion-type MOSFETs also operate in the enhancement mode by making v_{GS} positive.

The equation describing the boundary between the triode region and the pinch-off region, shown by the broken-line curve in Fig. 7.10, is obtained by substituting for v_{GS} from Eq. (7.11) into Eq. (7.6),

$$i_D = Kv_{DS}^2 \tag{7.13}$$

The characteristics of Fig. 7.10 indicate that in the pinch-off region the current i_D is a constant (for a given v_{GS}) independent of v_{DS}. In other words, the device acts as a current source with an infinite output resistance. However, as with JFETs and depletion-type MOSFETs, practical enhancement-type MOSFETs have finite output resistances as evident from the characteristics shown in Fig. 7.13. Note that in pinch-off the i_D-v_{DS} characteristic curves are straight lines that, when extrapolated, intercept

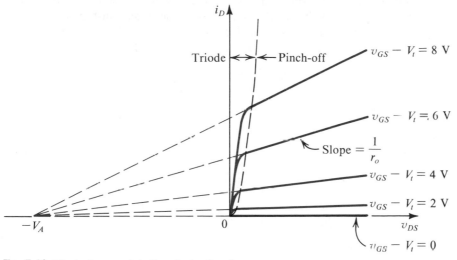

Fig. 7.13 Effect of v_{DS} on i_D in the pinch-off region.

the v_{DS} axis at the point $v_{DS} = -V_A$, where V_A is a positive voltage that is a parameter for the particular MOSFET. For devices fabricated using the standard IC process, $V_A = 30$ to 200 V. The linear dependence of i_D on v_{DS} in the pinch-off region (known as the *channel length modulation* effect) can be accounted for by incorporating the factor $(1 + v_{DS}/V_A)$ in Eq. (7.12) to obtain

$$i_D = K(v_{GS} - V_t)^2 \left(1 + \frac{v_{DS}}{V_A}\right) \tag{7.14}$$

This equation can be used to find the output resistance in pinch-off, r_o, which is the inverse of the slope of the i_D-v_{DS} characteristic,

$$r_o \equiv \left[\frac{\partial i_D}{\partial v_{DS}} \right]^{-1}_{v_{GS} = \text{constant}}$$

$$= [K(V_{GS} - V_t)^2 / V_A]^{-1} \tag{7.15}$$

Since the second term in Eq. (7.14) is usually small, the current I_D corresponding to the constant v_{GS} (that is, V_{GS}) is given approximately by

$$I_D \simeq K(V_{GS} - V_t)^2 \tag{7.16}$$

Substituting in Eq. (7.15) gives r_o as follows:

$$r_o \simeq \frac{V_A}{I_D} \tag{7.17}$$

Thus the output resistance is inversely proportional to the dc bias current I_D.

Circuit Symbol

Figure 7.14 shows the circuit symbol of the n-channel enhancement-type MOSFET. The only difference between this symbol and that of the depletion-type device (Fig. 7.2) is that here the channel is represented by a broken line. This signifies that no channel exists in the enhancement device (until v_{GS} exceeds V_t).

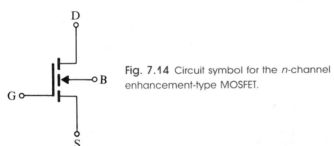

Fig. 7.14 Circuit symbol for the n-channel enhancement-type MOSFET.

The p-Channel Device

The operation of the p-channel MOSFET (or PMOS transistor) parallels that of the n-channel device described above. The characteristics are also similar except for a reversal of polarity of all currents and voltages. Also, V_t and V_A for p-channel devices are negative.

The characteristics of the p-channel MOSFET are described by the same equations as those for the n-channel device except for replacing v_{GS} by v_{SG}, v_{DS} by v_{SD}, V_t by $|V_t|$ and V_A by $|V_A|$. Also, μ_n in Eq. (7.7) should be replaced by μ_p, the mobility of holes in the induced p channel. Typically, $\mu_p \simeq \frac{1}{2}\mu_n$, with the result that for the same W/L ratio a PMOS transistor has half the value of K as the NMOS device. The circuit symbol for the enhancement-type PMOS transistor is shown in Fig. 7.15.

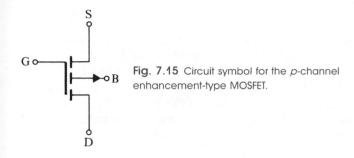

Fig. 7.15 Circuit symbol for the *p*-channel enhancement-type MOSFET.

The Role of the Substrate—The Body Effect

In many applications the substrate (or body) terminal B is connected to the source terminal, which results in the *pn* junction between the substrate and the induced channel (see Fig. 7.11) having a constant reverse bias. In such a case the substrate does not play any role in circuit operation and its existence can be ignored altogether.

In integrated circuits, however, the substrate is usually common to many MOS transistors. In order to maintain the reverse-bias condition on the substrate-to-channel junction, the substrate is usually connected to the most negative power supply in an NMOS circuit (the most positive in a PMOS circuit). The resulting reverse-bias voltage between source and body (V_{SB}) will have an effect on device operation. To appreciate this fact, observe that V_{SB} can control the channel depth in the same manner as the gate voltage in a JFET controls its channel; increasing V_{SB} depletes the channel of charge carriers. The body terminal, in effect, acts as a second gate for the MOSFET. The effect of the reverse-bias voltage V_{SB} on device operation can be described via the dependence of V_t on V_{SB}, given approximately by

$$V_t \simeq \text{constant} + \gamma\sqrt{V_{SB}} \tag{7.18}$$

where the constant part of V_t is independent of V_{SB} and γ is a device parameter that depends, among other things, on the doping of the substrate and is typically equal to 0.5 $V^{1/2}$.

Since the substrate is almost always connected to a dc supply, it will be at signal ground. Nevertheless, the source terminal may have a signal voltage on it with the result that a signal voltage appears between body and source. Now since the body acts as a second gate, such a signal voltage v_{bs} will give rise to a drain current component. This effect, known as the *body effect*, can cause considerable degradation in circuit performance, as will be shown in a later section.

Temperature Effects

Both V_t and K are temperature-sensitive. The magnitude of V_t decreases by about 2 mV for every 1°C rise in temperature. This decrease in $|V_t|$ gives rise to a corresponding increase in drain current as temperature is increased. However, because K decreases with temperature and its effect is a dominant one, the overall observed effect of a temperature increase is a *decrease* in drain current. This very interesting result is put to use in applying the MOSFET in power circuits (Chapter 10).

Breakdown and Input Protection

As the voltage on the drain is increased, a value is reached at which the substrate-to-channel junction breaks down. Such a breakdown is of the avalanche type and causes the current to increase rapidly, just as in the case of the JFET (see Fig. 6.11).

In the MOSFET another kind of breakdown occurs when the gate-to-source voltage exceeds about 50 V. This is the breakdown of the gate oxide and it results in permanent damage to the device. Although 50 V is high, it must be remembered that the MOSFET has a very high input impedance, and thus small amounts of static charge accumulating on the input capacitor (between gate and source) can cause this breakdown voltage to be exceeded.

To prevent the accumulation of static charge on the input capacitor of a MOSFET, gate protection devices are usually included at the input of MOS integrated circuits (such as the input terminals of a CMOS logic gate). The protection mechanism invariably makes use of clamping diodes (see Fig. 15.44).

Simplified Circuit Symbols

When the substrate is connected to the source, the simplified circuit symbols of Fig. 7.16 will be used to represent the enhancement-type MOSFET. The difference between these circuit symbols and those for the depletion-type MOSFETs in Fig. 7.3 and 7.8b should be noted. The extra bar in the depletion-device symbol is used to signify the existing channel.

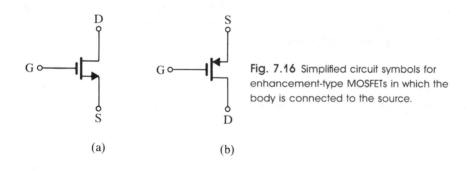

Fig. 7.16 Simplified circuit symbols for enhancement-type MOSFETs in which the body is connected to the source.

(a)   (b)

An Alternative View of the Depletion MOSFET

The depletion MOSFET studied in Section 7.1 can be considered as an enhancement device with a negative (for NMOS) threshold voltage. The negative threshold voltage, which is equal to the pinch-off voltage, is a result of the implanted channel. This view of the depletion device can be appreciated by comparing its i_D-v_{GS} characteristic in Fig. 7.7 with that of the enhancement device in Fig. 7.12. By adopting this viewpoint, the depletion-type device can be represented by the same equations presented above for the enhancement-type MOSFET with the parameter I_{DSS} given by

$$I_{DSS} = KV_t^2 \tag{7.19}$$

Recapitulation

Consider Fig. 7.17. For the n-channel enhancement device shown in the figure to conduct, v_{GS} has to be positive and greater than V_t. The device will operate in the pinch-off or active region if the drain voltage v_D is more positive than the gate voltage v_G by at least $-V_t$. That is, the device will still be in pinch-off even if the drain voltage is lower than that of the gate by V_t volts. If the drain voltage is further reduced, the device gets out of pinch-off and goes into the triode region.

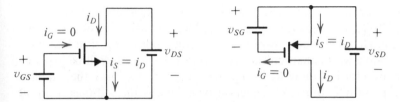

Fig. 7.17 Normal current-flow directions and voltage polarities in enhancement MOS transistors.

For the p-channel device to conduct, the source has to be made more positive than the gate by at least $|V_t|$ volts; that is, $v_{SG} \geq |V_t|$. The device will be in pinch-off as long as the drain voltage is lower than that of the gate, or even if it is higher than that of the gate by at most $|V_t|$. If v_D is increased more than $|V_t|$ volts above v_G, the device leaves pinch-off and enters the triode region.

Since in the pinch-off region the MOSFET operates as a voltage-controlled current source, it is the pinch-off region that is utilized for amplifier applications. On the other hand, switching applications make use of both the cut-off region and the triode region.

Exercises

7.3 Consider an enhancement NMOS transistor with $V_t = 2$ V that conducts a current $i_D = 1$ mA when $v_{GS} = v_{DS} = 3$ V. Neglecting the dependence of i_D on v_{DS} in pinch-off, find the value of i_D for $v_{GS} = 4$ V and $v_{DS} = 5$ V. Also, calculate the value of the drain-to-source resistance r_{DS} for small v_{DS} and $v_{GS} = 4$ V.

Ans.   4 mA; 250 Ω

7.4 An IC NMOS transistor has $W = 100$ μm, $L = 10$ μm, $(\mu_n C_{ox}) = 20$ μA/V^2, $V_A = 100$ V, $\gamma = \frac{1}{2} V^{1/2}$, and $V_t = 1$ V at $V_{SB} = 0$ V. Calculate the values of K and of V_t at $V_{SB} = 4$ V. Also, for $V_{GS} = 3$ V and $V_{DS} = 5$ V, calculate I_D for $V_{SB} = 0$ V and for $V_{SB} = 4$ V. What is the output resistance r_o for each of the two cases?

Ans.   0.1 mA/V^2; 2 V; 0.420 mA; 0.105 mA; 238 kΩ; 952 Ω

7.5 An enhancement NMOS transistor for which $V_t = 2$ V and $K = 0.25$ mA/V^2 is to be operated at $I_D = 4$ mA. What is the least drain-to-source voltage that can be used if the device is to be maintained in pinch-off?

Ans.   4 V

7.3 BIASING THE ENHANCEMENT MOSFET IN DISCRETE CIRCUITS

As mentioned before, the first step in the design of a transistor amplifier involves establishing a stable and predictable dc operating point inside the active region of operation. In the following we shall study two popular biasing arrangements for enhancement MOSFETs in discrete circuits. Integrated-circuit biasing techniques will be studied in a later section.

A First Biasing Scheme

Figure 7.18 shows our first biasing arrangement; although it looks identical to that used for biasing JFETs and depletion-type MOSFETs, the principle of operation is somewhat different. Here the source resistance R_S is not a self-bias resistance. In fact, self-bias is not possible with enhancement-type devices. Rather, as will be explained below, the sole reason for including R_S is to provide negative feedback that stabilizes the dc operating point. Of course, in depletion-type devices the self-bias resistance provides negative feedback also.

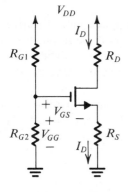

Fig. 7.18 A popular biasing arrangement for enhancement MOS discrete-circuit amplifiers.

The voltage divider R_{G1}-R_{G2} supplies the gate with a constant dc voltage V_{GG},

$$V_{GG} = V_{DD} \frac{R_{G2}}{R_{G1} + R_{G2}}$$

In the absence of R_S this voltage appears directly between gate and source, and the corresponding current I_D will be highly dependent on the exact value of V_{GG} and on the device parameters K and V_t. This point is illustrated in Fig. 7.19, where we show the i_D-v_{GS} characteristic curve for two extreme devices of the same type. The large difference in I_D between the two devices should be evident.

Including the resistance R_S results in the following describing equation:

$$V_{GG} = V_{GS} + I_D R_S$$

which can be rewritten

$$I_D = \frac{V_{GG}}{R_S} - \frac{1}{R_S} V_{GS}$$

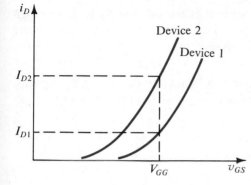

Fig. 7.19 High dependence of the bias current on device parameters if the resistance R_s is not used. Devices 1 and 2 represent two extremes among units of the same type.

This is the equation of the straight line shown in Fig. 7.20, where again we have shown the characteristics of two extreme devices. Note that the difference in the value of I_D between the two devices is much less than that observed with fixed bias alone.

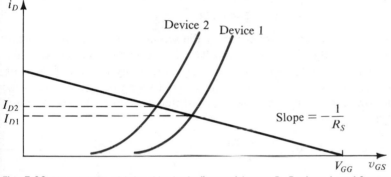

Fig. 7.20 Bias stability obtained by including resistance R_s. Devices 1 and 2 represent two extremes among units of the same type.

To gain more insight into the stabilizing action of R_S, consider once more the circuit of Fig. 7.18. Assume that for some reason (such as a change in temperature) the drain current increases by an amount Δi_D. This incremental change in the bias current results in an incremental increase in the voltage at the source, Δv_S,

$$\Delta v_S = R_S \, \Delta i_D$$

Since the gate is held at a constant voltage V_{GG}, an increase in the source voltage results in an equal decrease in V_{GS},

$$\Delta v_{GS} = -\Delta v_S = -R_S \, \Delta i_D$$

Since a decrease in V_{GS} results in a decrease in I_D, the net increase in I_D will be less than the original value Δi_D, indicating the presence of a negative-feedback mechanism. It should be mentioned again that in the JFET case also the self-bias resistance R_S provides a stabilizing negative-feedback action.

In the above it has been assumed implicitly that the value of R_D is chosen such that the device operates in the pinch-off region. This is accomplished by maintaining the drain voltage greater than $v_G - V_t$ at all times.

EXAMPLE 7.1

Consider an enhancement MOSFET with $K = 0.25$ mA/V^2 and $V_t = 2$ V. We wish to bias the device at $I_D = 1$ mA using the biasing arrangement of Fig. 7.18 with $V_{DD} = 20$ V.

Solution

To determine the required value of V_{GS} we use the relationship

$$I_D = K(V_{GS} - V_t)^2$$

For $K = 0.25$ mA/V^2, $V_t = 2$ V, and $I_D = 1$ mA, the value of V_{GS} should be $V_{GS} = 4$ V. If we choose a 4-V drop across R_S, then the gate voltage should be $V_{GG} = 8$ V. This voltage can be established by choosing $R_{G1} = 1.2$ MΩ and $R_{G2} = 0.8$ MΩ. Of course we should choose values for R_{G1} and for R_{G2} as large as practicable in order to keep the amplifier input resistance as high as possible. The value of R_S is given by

$$R_S = \tfrac{4}{1} = 4 \text{ k}\Omega$$

The choice of a value for R_D is governed by the required gain and signal swing. The higher the value of R_D the higher the gain will be. However, we should ensure that the drain voltage will at no time fall below the gate voltage by more than V_t volts. For this example let us assume that a maximum signal swing of ± 4 V is required at the drain. It follows that we may choose R_D such that $V_D = +10$ V. The required value of R_D will be

$$R_D = \frac{20 - 10}{1} = 10 \text{ k}\Omega$$

A Second Biasing Scheme

The second biasing arrangement we shall study is depicted in Fig. 7.21. As shown, a resistance R_G, usually quite large, is connected between drain and gate. Since the gate current is almost zero, the dc gate voltage will be equal to the dc drain voltage. This condition means that the device is still operating in the active (pinch-off) region. It should be obvious, however, that this biasing arrangement would not work with depletion-type devices.

To evaluate the dc operating point of the circuit of Fig. 7.21 consider Fig. 7.22, which shows the i_D-v_{DS} characteristics of the MOSFET. The parabolic boundary between the triode region and the active region is shown as a broken line. This boundary curve is the locus of the points at which $v_{DG} = -V_t$ or equivalently $v_{DS} = v_{GS} - V_t$. If one assumes ideal characteristics, then shifting this curve laterally by V_t volts gives the locus of the points for which $v_{DS} = v_{GS}$. Clearly the operating point Q lies on this latter curve, which is represented by the solid line in Fig. 7.22. From the circuit

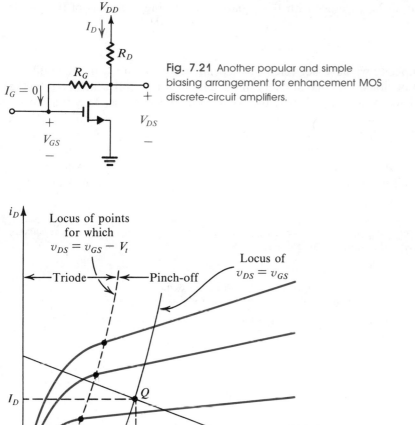

Fig. 7.21 Another popular and simple biasing arrangement for enhancement MOS discrete-circuit amplifiers.

Fig. 7.22 Evaluation of the dc operating point Q of the circuit in Fig. 7.21.

in Fig. 7.21 we can write

$$v_{DS} = V_{DD} - R_D i_D$$

or, equivalently,

$$i_D = \frac{V_{DD}}{R_D} - \frac{1}{R_D} v_{DS} \qquad (7.20)$$

which is the equation of the straight line shown in Fig. 7.22, referred to as the *load line*. The operating point Q will lie at the intersection of the load line and the solid-line parabola. Though quite illustrative, the above graphical procedure is seldom used in practice. It is far more expedient to determine i_D and v_{DS} at the operating point Q

by solving Eq. (7.20) together with the equation describing the locus of the points $v_{DS} = v_{GS}$,

$$i_D = K(v_{DS} - V_t)^2 \tag{7.21}$$

where for simplicity we have neglected the dependence of i_D on v_{DS} in pinch-off.

It should be noted that bias stability in the circuit of Fig. 7.21 is achieved by the negative-feedback action provided by the connection of R_G. To see how this works, assume that for some reason the drain current increases by an increment Δi_D. From the circuit we see that the drain voltage will decrease by $R_D \Delta i_D$. Since the current in R_G is nearly zero, the gate voltage, and hence V_{GS}, will decrease by an equal amount, namely, $R_D \Delta i_D$. As a result of the decrease in V_{GS} the drain current will decrease. Thus the overall increase in drain current will be much smaller than the value originally assumed, Δi_D.

Exercises

7.6 In the circuit of Example 7.1 find the percentage change in I_D if the FET is replaced by another having the same K value but $V_t = 3$ V.

Ans. -20%

7.7 Consider the bias arrangement of Fig. 7.21. Find the value of R_D required to establish a drain current of 1 mA provided that $V_{DD} = 20$ V, $V_t = 2$ V, and $K = 0.25$ mA/V^2. What is the percentage change in I_D if the device is replaced by another having the same K value but $V_t = 3$ V?

Ans. 16 kΩ; -6%

7.4 SMALL-SIGNAL OPERATION OF THE ENHANCEMENT MOSFET AMPLIFIER

Equivalent Circuit Model

Figure 7.23 shows the small-signal equivalent circuit model of the enhancement MOSFET for which the substrate (body) is connected to the source and thus there is no body effect. This model is identical in form to that of the JFET and the depletion-type MOSFET. The gate-to-source and gate-to-drain capacitances C_{gs} and C_{gd} range from a fraction of a picofarad to 3 pF. Capacitance C_{ds} is the drain-to-substrate

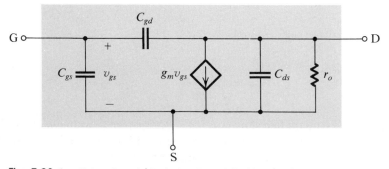

Fig. 7.23 Small-signal equivalent circuit model of the MOSFET.

capacitance and is usually about 1 pF or less. From Section 2.5 we can see that the feedback capacitance C_{gd} will take part in a Miller effect, resulting in a high input capacitance in parallel with C_{gs}. The total effective input capacitance together with the resistance of the signal source feeding the amplifier constitutes a first-order low-pass filter (see Chapter 2), which reduces the gain of the amplifier at high frequencies. The high-frequency response of transistor amplifiers will be studied in detail in Chapter 11. For the time being it is sufficient to note that in calculating the gain at low frequencies, the capacitances C_{gs}, C_{gd}, and C_{ds} can be ignored (that is, assumed to act as an open circuit). In this case the equivalent circuit model reduces to that shown in Fig. 7.24. The resistance r_o, which represents the finite output resistance of the MOSFET in the pinch-off region, is in the range of 10 to 100 kΩ and is inversely proportional to the dc bias current [Eq. (7.17)]. An expression for the transconductance g_m will be derived below.

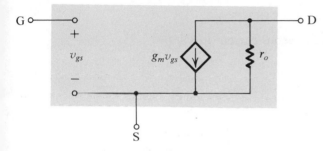

Fig. 7.24 Small-signal low-frequency MOSFET model.

The equivalent circuit model of Fig. 7.24 can be used for evaluating amplifier gain in the same manner as described for JFETs. Furthermore, it is sometimes expeditious to employ the model *implicitly* and carry out the small-signal analysis directly on the circuit diagram (rather than drawing a new diagram with each transistor explicitly replaced by its equivalent circuit model). On other occasions the model will be implicitly employed in a yet different way: the MOSFET will be treated as if a small-signal resistance equal to $1/g_m$ exists between gate and source, looking into the source. (Recall that we utilized this approach in the analysis of the JFET source follower.)

The Transconductance g_m

Consider the conceptual amplifier circuit in Fig. 7.25. Here the enhancement MOSFET is shown biased by a fixed dc voltage V_{GS}. A resistance R_D is connected between the drain and the power supply V_{DD} to establish a dc voltage V_D at the drain such that the device is in the active mode. The dc quantities I_D and V_D are given by

$$I_D = K(V_{GS} - V_t)^2 \tag{7.22}$$

$$V_D = V_{DD} - R_D I_D \tag{7.23}$$

As shown in Fig. 7.25, a voltage signal v_{gs} is superimposed on the dc voltage V_{GS}. Thus the total instantaneous gate-to-source voltage v_{GS} is given by

$$v_{GS} = V_{GS} + v_{gs} \tag{7.24}$$

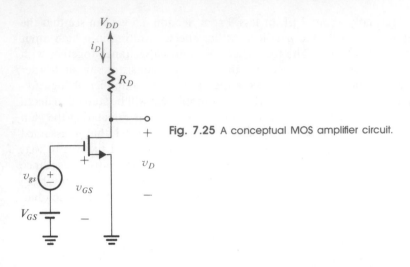

Fig. 7.25 A conceptual MOS amplifier circuit.

Correspondingly, the total instantaneous current i_D will be

$$i_D = K(v_{GS} - V_t)^2$$

$$= K(V_{GS} + v_{gs} - V_t)^2$$

$$= K(V_{GS} - V_t)^2 + 2K(V_{GS} - V_t)v_{gs} + Kv_{gs}^2 \tag{7.25}$$

The first term on the right-hand side of Eq. (7.25) can be recognized as the dc or quiescent current I_D [Eq. (7.22)]. The second term represents a current component that is directly proportional to the input signal v_{gs}. The last term is a current component that is proportional to the square of the input signal. This last component is undesirable because it represents nonlinear distortion. To reduce the nonlinear distortion introduced by the MOSFET, the input signal should be kept small,

$$v_{gs} \ll 2(V_{GS} - V_t) \tag{7.26}$$

If this small-signal condition is satisfied, we may neglect the last term in Eq. (7.25) and express i_D as

$$i_D \simeq I_D + i_d \tag{7.27}$$

where the signal current i_d is given by

$$i_d = 2K(V_{GS} - V_t)v_{gs}$$

The constant relating i_d and v_{gs} is the transconductance g_m,

$$g_m = 2K(V_{GS} - V_t) \tag{7.28}$$

Figure 7.26 presents a graphical interpretation of the small-signal operation of the enhancement MOSFET amplifier. Note that g_m is equal to the slope of the i_D-v_{GS} characteristic at the operating point,

$$g_m = \left. \frac{\partial i_D}{\partial v_{GS}} \right|_{v_{DS} = \text{constant}} \tag{7.29}$$

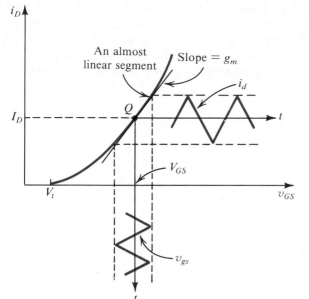

Fig. 7.26 Small-signal operation of the enhancement MOSFET amplifier.

The reason for maintaining v_{DS} constant is to eliminate the change in i_D due to changes in v_{DS} (which is modeled by r_o).

We shall now take a closer look at MOSFET transconductance. Substituting for K from Eq. (7.7) into Eq. (7.28) gives

$$g_m = (\mu_n C_{\text{OX}})(W/L)(V_{GS} - V_t) \qquad (7.30a)$$

This relationship indicates that g_m depends on the W/L ratio of the MOS transistor; hence to obtain relatively large transconductance the device must be short and wide. We also observe that for a given device the transconductance is proportional to $\Delta V = V_{GS} - V_t$, the amount by which the bias voltage V_{GS} exceeds the threshold voltage V_t.

Another useful expression for g_m can be obtained by substituting for $(V_{GS} - V_t)$ in Eq. (7.30a) by $\sqrt{I_D/K}$ [from Eq. (7.22)] and again substituting for K by $\frac{1}{2}\mu_n C_{\text{OX}}(W/L)$. The result is

$$g_m = \sqrt{2\mu_n C_{\text{OX}}}\sqrt{W/L}\sqrt{I_D} \qquad (7.30b)$$

This expression shows that

1. For a given MOSFET, g_m is proportional to the square root of the dc bias current.

2. At a given bias current, g_m is proportional to $\sqrt{W/L}$.

In contrast, the transconductance of the bipolar junction transistor (BJT) studied in Chapter 8 is proportional to the bias current and is independent of the physical size and geometry of the device.

To gain some insight into the values of g_m obtained in MOSFETs consider a device operating at $I_D = 1\,\text{mA}$ and having $\mu_n C_{\text{OX}} = 20\ \mu\text{A/V}^2$. Equation (7.30b) shows that for $W/L = 1$, $g_m = 0.2\,\text{mA/V}$, whereas a device for which $W/L = 100$ has $g_m = 2\,\text{mA/V}$. In contrast, a BJT operating at a collector current of 1 mA has $g_m = 40\,\text{mA/V}$. However, in spite of their low g_m, MOSFETs have many other advantages, including high input impedance, small size, low power dissipation, and ease of fabrication.

Voltage Gain

Returning to the circuit of Fig. 7.25, we can express the total instantaneous drain voltage v_D as follows:

$$v_D = V_{DD} - R_D i_D$$

Under the small-signal condition we have

$$v_D = V_{DD} - R_D(I_D + i_d)$$

which can be rewritten

$$v_D = V_D - R_D i_d$$

Thus the signal component of the drain voltage is

$$v_d = -i_d R_D$$
$$= -g_m R_D v_{gs}$$

which indicates that the voltage gain is given by

$$\frac{v_d}{v_{gs}} = -g_m R_D$$

It should be noted that in the above analysis we have assumed that the i_D-v_{DS} characteristics in the pinch-off region are horizontal straight lines. In other words, we have neglected the finite output resistance r_o of the voltage-controlled current source. For the amplifier circuit in Fig. 7.25 this resistance appears directly in parallel with R_D. Thus the voltage gain will be reduced to

$$\frac{v_o}{v_{gs}} = -g_m(R_D \,\|\, r_o) \tag{7.31}$$

All of the above analysis has been predicated on the assumption that the MOSFET remains in the active mode at all times, which is achieved by ensuring that the condition

$$v_D \geq v_G - V_t$$

is satisfied.

Although the small-signal analysis of the amplifier circuit of Fig. 7.25 was performed directly on the circuit, the same results can be obtained by replacing the MOSFET with its small-signal model of Fig. 7.24. The result is the circuit shown in Fig. 7.27, which can be easily analyzed to obtain the voltage-gain expression in

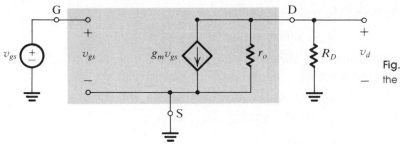

Fig. 7.27 Small-signal equivalent circuit of the amplifier in Fig. 7.25.

Eq. (7.31). Note that in the small-signal equivalent circuit, dc voltage sources are represented by short circuits; since their voltage is constant, they exhibit zero incremental or signal voltage. The explicit use of equivalent circuit models provides a systematic analysis approach.

EXAMPLE 7.2

Figure 7.28 shows a capacitively coupled grounded-source amplifier using an enhancement MOSFET with $V_t = 1.5$ V and $K = 0.125$ mA/V^2. We desire to evaluate the midband gain and input resistance. For this MOS transistor $V_A = 50$ V.

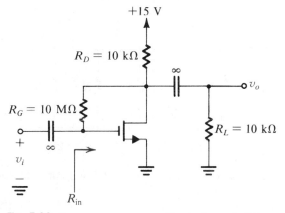

Fig. 7.28 Common-source amplifier for Example 7.2.

Solution

We first evaluate the dc operating point as follows:

$$I_D = 0.125(V_{GS} - 1.5)^2$$

But $V_{GS} = V_D$, and thus

$$I_D = 0.125(V_D - 1.5)^2 \tag{7.32}$$

Also,

$$V_D = 15 - R_D I_D$$
$$= 15 - 10I_D \tag{7.33}$$

Solving Eqs. (7.32) and (7.33) gives

$$I_D = 1.06 \text{ mA} \quad \text{and} \quad V_D = 4.4 \text{ V}$$

(Note that the other solution to the quadratic equation is not physically meaningful.)
The value of g_m is given by

$$g_m = 2K(V_{GS} - V_t)$$
$$= 0.25(4.4 - 1.5) = 0.725 \text{ mA/V}$$

The output resistance r_o is given by

$$r_o = \frac{50}{I_D} = \frac{50}{1.06} = 47 \text{ k}\Omega$$

Figure 7.29 shows the small-signal equivalent circuit of the amplifier. The midband voltage gain can be evaluated to a good approximation by neglecting the effect of the large resistance R_G:

$$\frac{v_o}{v_i} \simeq -g_m(R_D \parallel R_L \parallel r_o)$$

$$= -0.725(10 \parallel 10 \parallel 47) = -3.3 \text{ V/V}$$

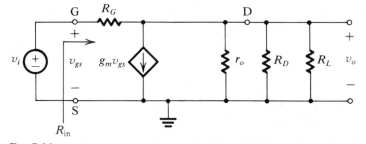

Fig. 7.29 Small-signal equivalent circuit of the MOSFET amplifier in Fig. 7.28.

To evaluate the input resistance we apply Miller's theorem (see Section 2.5) as follows:

$$R_{\text{in}} = \frac{R_G}{1 - A}$$

where

$$A = \frac{v_d}{v_g} = \frac{v_o}{v_i} = -3.3$$

Thus

$$R_{\text{in}} = \frac{10}{1 + 3.3} = 2.33 \text{ M}\Omega$$

Modeling the Body Effect

When the substrate (body) terminal B is not connected to the source S, a signal voltage v_{bs} may develop between body and source. Recalling that the body terminal acts as another gate for the MOSFET, the signal v_{bs} gives rise to a drain current component which we shall write as $g_{mb}v_{bs}$ where g_{mb} is the *body transconductance*. Figure 7.30 shows the low-frequency equivalent circuit of a MOSFET in which B is not connected to S. The body transconductance g_{mb} is defined as

$$g_{mb} \equiv \frac{\partial i_D}{\partial v_{BS}}\bigg|_{\substack{v_{GS} = \text{constant} \\ v_{DS} = \text{constant}}} \tag{7.34a}$$

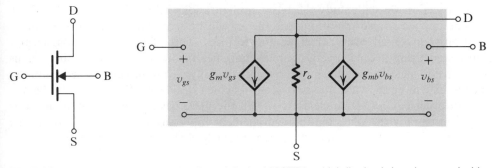

Fig. 7.30 Small-signal equivalent circuit model of a MOSFET in which the body is not connected to the source.

Recalling that i_D depends on v_{BS} through the dependence of V_t on V_{BS}, Eqs. (7.12), (7.18), and (7.28) can be used to obtain

$$g_{mb} = \chi g_m \tag{7.34b}$$

where

$$\chi \equiv \frac{\partial V_t}{\partial V_{SB}} \tag{7.35}$$

The value of χ lies in the range 0.1 to 0.3. The equivalent circuit of Fig. 7.30 will be employed to predict the gain of IC MOS amplifiers in Section 7.8.

Exercises

7.8 An enhancement NMOS transistor has $(\mu_n C_{OX}) = 20 \ \mu\text{A/V}^2$, $W/L = 64$, $V_t = 1$ V, and $V_A = 100$ V. Find g_m and r_o when
(a) the bias voltage $V_{GS} = 2$ V and
(b) the bias current $I_D = 1$ mA.

Ans.   (a) 1.28 mA/V, 156.25 kΩ; **(b)** 1.6 mA/V, 100 kΩ

7.9 For an NMOS transistor with $\gamma = 0.5 \ \text{V}^{1/2}$ and $V_{SB} = 4$ V, find the value of χ.

Ans. 0.125

7.10 Consider the circuit designed in Example 7.1. Let an input signal source be capacitively coupled to the gate and let the drain be capacitively coupled to a 10-kΩ load resistance. Also let the resistance in the source lead (R_S) be bypassed to ground by a large capacitance (as was done with the JFET amplifier in Fig. 6.28). Find the input resistance and the voltage gain assuming that for the MOSFET, $V_A = 50$ V.

Ans. 480 kΩ; -4.54 V/V

7.5 BASIC CONFIGURATIONS OF SINGLE-STAGE MOSFET AMPLIFIERS

In this section we shall study the basic configurations in which a single MOSFET can be applied to provide amplification. As will be seen, each of the three configurations considered offers some unique features. These features are most clearly illustrated using a capacitively coupled amplifier in order to separate the signals from the dc bias. The results, however, apply equally well to direct-coupled amplifier stages, as will be shown in the next section, which deals with IC MOS amplifiers.

In order to distinguish clearly between the three basic amplifier configurations we shall employ the same biasing arrangement in all cases. Figure 7.31 shows the basic circuit that will be used to implement each of the three basic configurations.

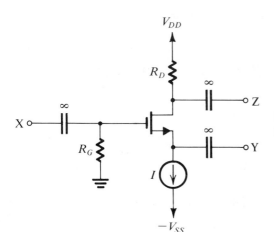

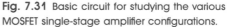

Fig. 7.31 Basic circuit for studying the various MOSFET single-stage amplifier configurations.

The MOSFET is biased by a dc current source connected to the negative supply. Although the device could be biased by connecting a resistor R_S to the negative supply, constant-current biasing is employed in order to simplify the analysis and thus focus attention on the salient features of the various amplifier configurations. Moreover, using current sources for biasing is a common practice in the design of integrated circuits. A resistor R_G connects the gate to ground, thus establishing dc continuity and fixing the dc voltage at the gate at zero volts. Because the gate current is extremely small, a large resistor R_G (in the megohm range) can be easily employed. A resistor R_D connects the drain to the positive supply voltage, V_{DD}, and thus establishes the dc drain voltage at a value that ensures pinch-off operation at all times while allowing for the required signal swing at the drain. Finally, three large-valued capacitors are used to couple the gate, source, and drain to signal source, load resistance, or ground, as required to configure the circuit in one of the three amplifier configurations. In the analysis to follow we assume that these capacitors act as perfect short circuits.

The Common-Source Amplifier

The common-source amplifier configuration is obtained by connecting terminal Y to ground, thus establishing a signal ground at the source. The input signal is connected to the gate and the load resistance to the drain, resulting in the configuration of Fig. 7.32a. Replacing the MOSFET with its small-signal equivalent circuit model leads to the circuit shown in Fig. 7.32b. (Note that the bias current source is replaced by an open circuit.)

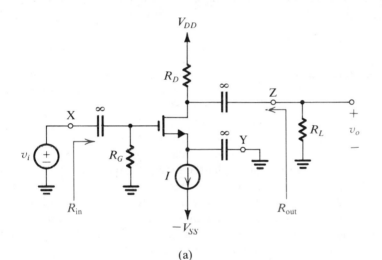

(a)

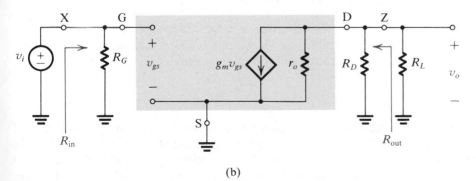

(b)

Fig. 7.32 (a) A MOSFET common-source amplifier. **(b)** Its small-signal equivalent circuit.

The amplifier input resistance R_{in}, output resistance R_{out}, and voltage gain A_v can be found by inspection of the circuit in Fig. 7.32b, as follows:

$$R_{in} = R_G \tag{7.36}$$

$$R_{out} = R_D \| r_o \tag{7.37}$$

$$A_v \equiv \frac{v_o}{v_i} = -g_m(R_L \| R_D \| r_o) \tag{7.38}$$

Note that the voltage gain in Eq. (7.38) includes the effect of the load resistance R_L. Alternatively, the amplifier can be characterized by its output resistance R_{out} in Eq. (7.37) and its open-circuit voltage gain A_{vo} obtained by setting R_L in Eq. (7.38) to ∞; that is,

$$A_{vo} \equiv \frac{v_o}{v_i}\bigg|_{R_L = \infty} = -g_m[R_D \| r_o]$$

The gain A_v for a particular R_L can then be evaluated using the voltage-divider rule:

$$A_v = A_{vo} \frac{R_L}{R_L + R_{out}}$$

From the above we note that the common-source amplifier provides a high input resistance, limited only by the value of the biasing resistor R_G, a large negative voltage gain, and a large output resistance. The last property is of course not a desirable one for voltage amplifiers (see Chapter 2).

A major drawback of the common-source configuration is its limited high-frequency response. This, of course, is not evident from the analysis above since we have not included in the MOSFET model the internal capacitances C_{gs}, C_{gd}, and C_{ds} (see Fig. 7.23). A detailed analysis of the frequency response of transistor amplifiers will be undertaken in Chapter 11. For the time being we note that of all the three internal capacitances, C_{gd} is the one that most seriously limits the high-frequency response of the common-source amplifier. To appreciate this fact refer to the amplifier circuit in Fig. 7.32 and imagine a capacitor C_{gd} connected between gate and drain. Since the voltage at the drain end is A_v times the voltage at the gate end, we can employ the Miller equivalence (Section 2.5) to replace C_{gd} with a capacitor $C_{gd}(1 - A_v)$ between gate and ground. Now, since A_v is a large negative number, this equivalent capacitance will be large. This capacitance, which adds to C_{gs}, together with the resistance of the signal generator (not shown in the figure) creates an STC low-pass circuit that causes the gain to roll off at high frequencies.

The Common-Gate Amplifier

The common-gate amplifier configuration is obtained by connecting terminal X of the circuit in Fig. 7.31 to ground. This establishes a signal ground at the gate. The input signal is then applied to the source by connecting the v_i generator to terminal Y, and the output is taken at the drain by connecting the load resistance R_L to terminal Z. The result is the circuit in Fig. 7.33a, whose small-signal equivalent is

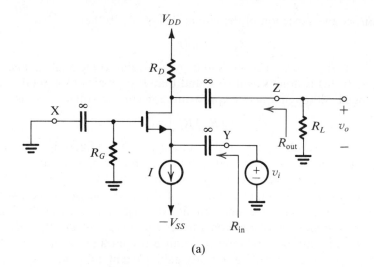

(a)

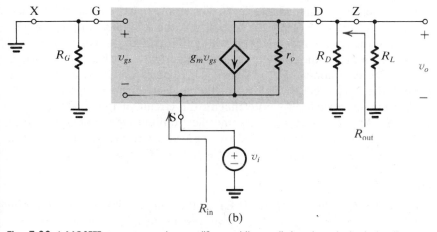

(b)

Fig. 7.33 A MOSFET common-gate amplifier and its small-signal equivalent circuit.

shown in Fig. 7.33b. To find the amplifier characteristics let us neglect for the moment the existence of r_o. We can see that $v_{gs} = -v_i$; thus at the input terminal the signal generator v_i sees a current $g_m v_i$ drawn away from v_i. Therefore the input resistance is simply $1/g_m$ (see the source absorption theorem in Section 2.5). This is consistent with the fact that the resistance between gate and source, looking into the source is equal to $1/g_m$. Thus

$$R_{\text{in}} \simeq 1/g_m \tag{7.39}$$

The voltage gain is found by noting that the drain current is (approximately, because r_o is neglected) $g_m v_{gs} = -g_m v_i$. Thus, $v_o = g_m v_i (R_L \| R_D)$ and

$$A_v \equiv \frac{v_o}{v_i} \simeq g_m[R_L \| R_D] \tag{7.40}$$

Finally, we can see by inspection of the circuit in Fig. 7.33b that

$$R_{\text{out}} = R_D \| r_o \qquad (7.41)$$

The effect of r_o on R_{in} and A_v can be found by analyzing the circuit in Fig. 7.33b with r_o in place. If this is done it will be found that R_{in} remains to a good approximation unchanged, although a better approximation for the voltage gain is

$$A_v = g_m[R_L \| R_D \| r_o] \qquad (7.42)$$

From the above we observe that the common-gate configuration provides a voltage gain that is almost equal to that obtained in the common-source amplifier except that here there is no signal inversion. An important difference between the two configurations is that the input resistance of the common-gate circuit is much smaller than that of the common-source circuit. Although this is a drawback in the case of a voltage amplifier, the common-gate circuit is almost always fed with a current signal. In this case the low input resistance becomes an advantage and the common-gate circuit acts simply as a unity-gain current amplifier or a *current follower*. It provides a drain signal current equal to the signal current fed to the source but at a much higher impedance level. The drain signal current is then fed to the parallel equivalent of R_L and R_D to produce the amplifier output voltage. This application of the common-gate circuit will be illustrated at a later stage.

The major advantage of the common-gate amplifier is that it has a much wider bandwidth than the common-source amplifier. This comes about because it does not suffer from the Miller effect that limits the high-frequency response of the common-source circuit. Specifically, note that in the circuit of Fig. 7.33a one side of C_{gd} is grounded. Also, although C_{gs} has some effect on limiting the high-frequency response, such effect is small because of the low input resistance $(1/g_m)$ at the source.

The Common-Drain Amplifier or Source Follower

In Section 6.9 we studied the use of the JFET in the source-follower configuration to provide a buffer amplifier having a high input resistance, a low output resistance, and a voltage gain close to unity. The MOSFET can be applied in exactly the same manner, as shown in Fig. 7.34a, with the equivalent circuit shown in Fig. 7.34b. For this circuit we see that

$$R_{\text{in}} = R_G \qquad (7.43)$$

To find R_{out} we reduce v_i to zero, which makes $v_s = -v_{gs}$. Invoking the source-absorption theorem (Section 2.5) we replace the current source $g_m v_{gs}$ by a resistance $1/g_m$. Thus the output resistance is given by

$$R_{\text{out}} = [(1/g_m) \| r_o] \qquad (7.44)$$

$$\simeq 1/g_m \qquad (7.44a)$$

The open-circuit voltage gain can be found by setting $R_L = \infty$ in the circuit of Fig. 7.34b. Thus

$$v_o = g_m v_{gs} r_o \qquad (7.45)$$

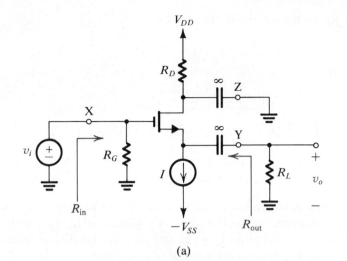

(a)

Fig. 7.34 (a) A MOSFET common-drain or source-follower amplifier.   **(b)** Its small-signal equivalent circuit.

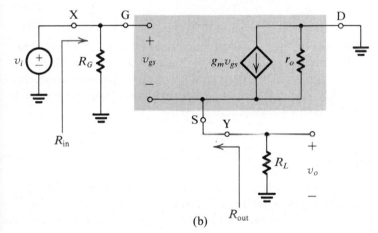

(b)

and

$$v_i = v_{gs} + v_o$$

$$= v_{gs} + g_m r_o v_{gs} \qquad (7.46)$$

Equations (7.45) and (7.46) can be combined to obtain the open-circuit voltage gain A_{vo}, as follows:

$$A_{vo} \equiv \left. \frac{v_o}{v_i} \right|_{R_L = \infty}$$

$$= \frac{g_m r_o}{1 + g_m r_o} \qquad (7.47)$$

The voltage gain with R_L connected can be found from

$$A_v \equiv \frac{v_o}{v_i}$$

$$= A_{vo} \frac{R_L}{R_L + R_{\text{out}}} \tag{7.48}$$

Substituting for R_{out} from Eq. (7.44) and for A_{vo} from Eq. (7.47) gives

$$A_v = \frac{g_m r_o R_L}{g_m r_o R_L + R_L + r_o} \tag{7.49}$$

A simpler derivation for this expression is available by simply inspecting the circuit of Fig. 7.34a. The total resistance between source and ground is $(R_L \| r_o)$ and v_o appears across this resistance. From input to output we have a voltage divider formed by the resistance $1/g_m$ (between source and gate) and $(R_L \| r_o)$; thus

$$A_v = \frac{(R_L \| r_o)}{(R_L \| r_o) + (1/g_m)} \tag{7.50}$$

which, as the reader can easily verify, is identical to that given in Eq. (7.49). This illustrates the fact that there are many occasions where it is quicker to perform the analysis directly on the circuit itself (rather than explicitly using the equivalent circuit). Finally, we note from Eq. (7.50) that the gain of the source follower approaches unity when $(R_L \| r_o)$ is made much greater than $(1/g_m)$.

Area of Applicability of the Various Configurations

A high-gain MOSFET amplifier usually consists of two or three stages connected in cascade. The bulk of the voltage gain is usually realized in one or two common-source amplifier stages. However, in order to reduce the Miller capacitance multiplication effect, which severely limits the high-frequency response of the common-source amplifier, a common-gate stage is often connected in cascade with the common-source transistor. The result is the composite amplifier circuit shown in Fig. 7.35 and known as the *cascode configuration*. We shall briefly explain here the operation of this important circuit, but will postpone the detailed analysis of its frequency response to Chapter 11.

In the cascode circuit, transistor Q_1 operates in the common-source configuration. It provides in its drain a current signal of value $g_m v_i$. Transistor Q_2 operates in the common-gate configuration, and thus the resistance looking into its source is $(1/g_{m2})$. Thus the total resistance between the drain of Q_1 and ground is $[r_{o1} \| (1/g_{m2})]$, which, since r_o is usually much greater than $1/g_m$, is approximately equal to $1/g_{m2}$. The voltage gain of Q_1 can now be determined as

$$\frac{v_{o1}}{v_i} = -g_{m1}(1/g_{m2}) \tag{7.51}$$

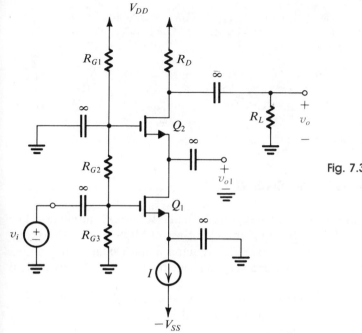

Fig. 7.35 The cascode configuration.

If Q_1 and Q_2 are identical, then $g_{m1} = g_{m2}$ and

$$\frac{v_{o1}}{v_i} = -1$$

It follows that the Miller multiplication effect in Q_1 will be considerably reduced and its bandwidth increased.

Transistor Q_1 supplies the source terminal of Q_2 with the current signal $(g_{m1}v_i)$. In turn, transistor Q_2 acting as a current follower passes this current on to its drain terminal. Thus the output voltage v_o will be given by

$$v_o = -g_{m1}v_i(R_L \parallel R_D)$$

where we have neglected the effect of r_{o2}. Finally, the voltage gain of the cascode stage can be obtained as

$$A_v \equiv \frac{v_o}{v_i} = -g_{m1}(R_L \parallel R_D) \tag{7.52}$$

which is almost equal to the gain of a simple common-source amplifier. The addition of the common-gate transistor Q_2 has, however, reduced the Miller effect in Q_1 with the attendant improvement in high-frequency performance.

Finally, we note that the source follower is usually used as the last (or output) stage of a multistage amplifier in order to provide a low output resistance. Other output-stage configurations will be studied in Chapter 10.

Exercise

7.11 Consider the circuit of Fig. 7.31 with $\mu_n C_{OX} = 20\mu A/V^2$, $W/L = 100$, $V_t = 1$ V, $V_A = 50$ V, $I = 1$ mA, $V_{DD} = V_{SS} = 5$ V, $R_G = 1$ MΩ, and $R_D = 3$ kΩ. Find the following.
(a) The dc voltages at the gate, source, and drain.
(b) g_m.
(c) R_{in}, R_{out}, and A_v for the common-source amplifier with $R_L = 3$ kΩ.
(d) R_{in}, R_{out}, and A_v for the common-gate amplifier with $R_L = 3$ kΩ.
(e) R_{in}, R_{out}, and A_v for the source follower with $R_L = 3$ kΩ.

Ans. (a) 0 V, -2 V, $+2$ V; (b) 2 mA/V; (c) 1 MΩ, 2.83 kΩ, -2.91 V/V; (d) 0.5 kΩ, 2.83 kΩ, $+2.91$ V/V; (e) 1 MΩ, 495 Ω, 0.85 V/V

7.6 INTEGRATED-CIRCUIT MOS AMPLIFIERS—AN OVERVIEW[3]

In this section we begin the study of integrated-circuit (IC) MOS amplifiers. The most important feature of IC MOS amplifiers is their use of MOS transistors as load elements, in place of resistors. Since a MOSFET requires a much smaller silicon area than a resistor, the resulting circuits are very efficient in their use of silicon "real estate."

At the present time there are two different MOS integrated-circuit technologies: NMOS and CMOS. NMOS refers to MOS integrated circuits that are based entirely on *n*-channel MOS transistors. The majority of these transistors are of the enhancement type; depletion-type transistors are used only as load devices, as will be explained below. By contrast, CMOS technology is based on using both *n*-channel and *p*-channel devices, all of which are of the enhancement type. The availability of both device polarities makes it easier to design high-quality circuits in CMOS. In fact, at the present time CMOS is by far the most popular technology for digital integrated circuits, and is rivaling bipolar technology (Chapter 8) for analog applications. The NMOS technology, though not as convenient for the circuit designer, currently offers the highest possible functional density (highest number of devices, and hence circuit functions, per chip) and it requires fewer processing steps than CMOS. Thus NMOS allows very high levels of integration. Both CMOS and NMOS are used extensively in the design of very-large-scale integrated (VLSI) circuits.

In the following sections we shall study some of the circuit techniques employed in the design of NMOS and CMOS amplifiers. More advanced MOS analog circuit design techniques will be studied in Chapters 9 and 13. Digital NMOS and CMOS circuits are studied in Chapter 15.

7.7 NMOS LOAD DEVICES

In NMOS technology two types of load elements are used: The enhancement MOSFET with the drain connected to the gate, and the depletion MOSFET with the gate connected to the source. Figure 7.36 shows the "diode-connected" enhance-

---

[3] Study of Sections 7.6–7.12 may be deferred and undertaken later in conjunction with the material on IC MOS amplifiers in Chapters 9 and 13.

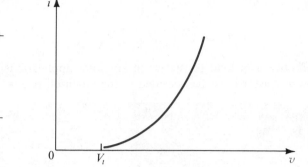

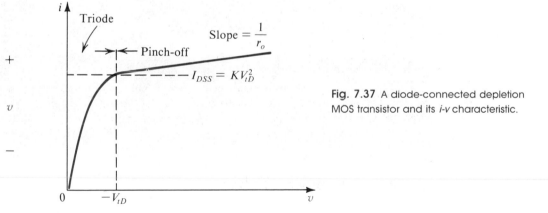

Fig. 7.36 A diode-connected enhancement MOS transistor and its *i-v* characteristic.

ment transistor, together with its *i-v* characteristic, which is described by

$$i = K(v - V_t)^2 \tag{7.53}$$

Observe that the transistor will always be operating in the pinch-off region. If the diode-connected transistor is biased at a voltage V, then its incremental or small-signal resistance will be equal to $1/g_m$, with the value of g_m evaluated at the bias point.

The diode-connected depletion MOSFET is shown in Fig. 7.37, together with its *i-v* characteristic. To operate in the pinch-off region, the voltage across the two-terminal device must exceed $-V_{tD}$, where V_{tD} is the threshold voltage of the depletion

Fig. 7.37 A diode-connected depletion MOS transistor and its *i-v* characteristic.

device (the pinch-off voltage V_P) and is negative, typically -1 to -4 V. In the triode region the *i-v* characteristic is described by

$$i = K(-2V_{tD}v - v^2) \tag{7.54}$$

At the onset of pinch-off, $v = -V_{tD}$ and

$$i = KV_{tD}^2 = I_{DSS} \tag{7.55}$$

Ideally, in pinch-off, the *i-v* characteristic is a horizontal line. However, because of channel-length modulation, the *i-v* characteristic is a straight line with a finite slope and is described by

$$i = KV_{tD}^2\left(1 + \frac{v}{V_A}\right) \tag{7.56}$$

Obviously, to be effective as a load resistance in amplifier applications, the diode-connected depletion transistor must be operated in the pinch-off region.

Exercises

7.12 Figure E7.12 shows a voltage divider composed of three diode-connected enhancement MOSFETs. Utilizing a current $I = 90$ μA, find the W/L ratios of the three transistors so that the divider provides $V_1 = +1$ V and $V_2 = -1$ V. Let $V_t = 1$ V and $\mu_n C_{OX} = 20$ μA/V^2. Neglect the small effect of r_o of each of the three devices.

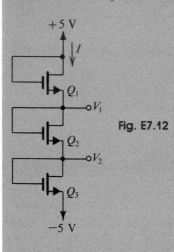

Fig. E7.12

Ans. $W_1/L_1 = W_3/L_3 = 1$; $W_2/L_2 = 9$

7.13 For the diode-connected enhancement MOSFET of Fig. 7.36 find an expression for its incremental (small-signal) resistance at $v = 2V_t$. Neglect the small effect of r_o. Evaluate r for the case $V_t = 1$ V, $\mu_n C_{OX} = 20$ μA/V^2, $W = 6$ μm, and $L = 30$ μm.

Ans. $r = 1/[(\mu_n C_{ox})(W/L)V_t]$; 250 k$\Omega$

7.8 NMOS AMPLIFIER WITH ENHANCEMENT LOAD

Figure 7.38a shows an enhancement MOSFET amplifier with an enhancement load. This circuit represents the simplest way of implementing an amplifier and a logic inverter in NMOS technology. We wish to derive the transfer characteristic v_O versus v_I. This can be done graphically, as illustrated in Fig. 7.38b, which shows a plot of the i_D-v_{DS} characteristics of the driving (amplifying) transistor Q_1. Note that the current i_{D1} is the same current that flows in the load device Q_2. Also note that $v_{DS1} = v_O$, that each of the characteristic curves corresponds to a constant value of v_{GS1}, and that $v_{GS1} = v_I$.

Superimposed on the static characteristics of Q_1 is the *load curve*, which is drawn in the same manner used to draw a load line; namely, we locate the V_{DD} point on

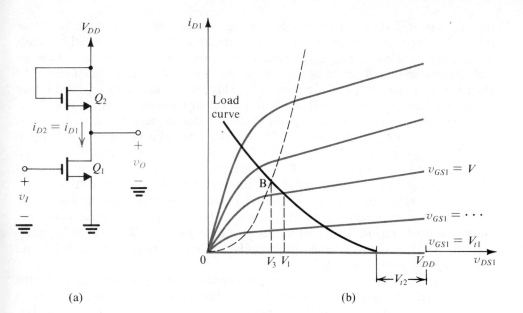

(a)

(b)

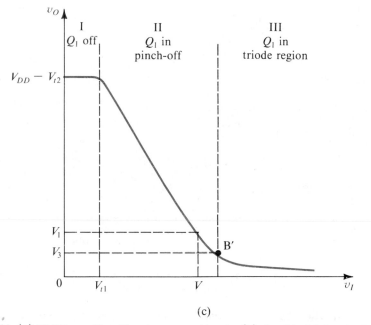

(c)

Fig. 7.38 (a) NMOS amplifier with enhancement load.   **(b)** Graphical determination of the transfer characteristic.   **(c)** Transfer characteristic.

the v_{DS1} axis and draw a mirror image of the i-v characteristic of the load device (Fig. 7.36). Now the transfer characteristic is determined by the intersection points of the load curve and the i_{D1}-v_{DS1} characteristic curves. For instance, for $v_I = V$ we find the intersection of the curve corresponding to $v_{GS1} = V$ and the load curve. As shown, at this point $v_{DS1} = V_1$; thus $v_O = V_1$. This process should be repeated for all possible values of v_I. The result is the transfer characteristic shown in Fig. 7.38c.

The transfer characteristic displays three well-defined regions. In region I, the driving transistor Q_1 is off, since $v_1 < V_{t1}$. Nevertheless, Q_2 is in the pinch-off region and is conducting a negligible current; thus the voltage across Q_2 is equal to V_{t2}, and hence the output voltage is $V_{DD} - V_{t2}$. (Note that, in fact, Q_2 is always in pinch-off.) In region II, Q_1 is conducting and is operating in pinch-off, and, as will be shown analytically, the transfer curve in region II is linear. Therefore, this region is very useful for amplifier operation. Finally, in region III, Q_1 leaves the active mode and enters the triode region. The onset of this region, point B′, corresponds to the intersection of the load curve and the boundary curve between the pinch-off and triode regions (point B in Fig. 7.38b).

We shall now derive the equation describing the transfer characteristic under the assumption that both devices have infinite resistance (that is, horizontal characteristic lines) in pinch-off. Furthermore, the two devices will be assumed to have equal threshold voltages V_t but different values of K (K_1, and K_2), a situation that corresponds to actual practice.

When Q_1 is in pinch-off we have

$$i_{D1} = K_1(v_{GS1} - V_t)^2 \tag{7.57}$$

Since $i_{D1} = i_{D2} = i_D$ and $v_{GS1} = v_I$, this equation can be rewritten

$$i_D = K_1(v_I - V_t)^2 \tag{7.58}$$

The operation of Q_2 is described by

$$i_D = K_2(v_{GS2} - V_t)^2$$

Since $v_{GS2} = V_{DD} - v_O$, this equation can be written

$$I_D = K_2(V_{DD} - v_O - V_t)^2 \tag{7.59}$$

Combining Eqs. (7.58) and (7.59) and with some simple manipulations we obtain

$$v_O = \left(V_{DD} - V_t + \sqrt{\frac{K_1}{K_2}}\, V_t \right) - \sqrt{\frac{K_1}{K_2}}\, v_I \tag{7.60}$$

which is a linear equation between v_O and v_I. This is obviously the equation of the straight-line portion of the transfer characteristic (region II) of Fig. 7.38c.

From Eq. (7.60) we see that the circuit behaves as a linear amplifier for large signals. The gain of the amplifier is

$$A_v = -\sqrt{\frac{K_1}{K_2}} \tag{7.61}$$

Substituting for K_1 and K_2 using the expression in Eq. (7.7) gives

$$A_v = -\sqrt{\frac{(W/L)_1}{(W/L)_2}} \tag{7.62}$$

Thus the gain is determined by the geometries of the two devices and is fixed for given devices. To obtain relatively large gains, $(W/L)_2$ must be made smaller than $(W/L)_1$. Thus the usual practice is to make the amplifier transistor Q_1 short and wide and the load transistor Q_2 long and narrow. Nevertheless, it is difficult to realize gains greater than about 10.

We shall next consider a small-signal analysis of the amplifier circuit of Fig. 7.38a, assuming that the circuit is biased to operate somewhere in region II of the transfer characteristic. Figure 7.39 shows the amplifier equivalent circuit obtained by replacing each of the two transistors by its equivalent circuit model. Since the voltage across the

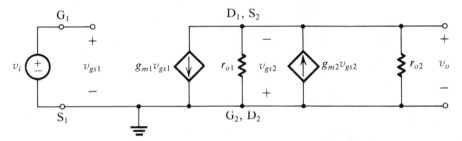

Fig. 7.39 Small-signal equivalent circuit of the enhancement-load amplifier of Fig. 7.38a.

controlled current source $g_{m2}v_{gs2}$ is v_{gs2}, the source-absorption theorem (Section 2.5) can be employed to replace the source with a resistance $1/g_{m2}$. Then for v_o we can write

$$v_o = -g_{m1}v_{gs1}[(1/g_{m2}) \| r_{o1} \| r_{o2}]$$

Substituting $v_{gs1} = v_i$, we obtain the voltage gain as follows:

$$A_v = \frac{v_o}{v_i} = \frac{-g_{m1}}{g_{m2} + 1/r_{o1} + 1/r_{o2}} \tag{7.63}$$

Now if r_{o1} and r_{o2} are much larger than $(1/g_{m2})$, the gain expression in Eq. (7.63) reduces to

$$A_v \simeq -\frac{g_{m1}}{g_{m2}} \tag{7.63a}$$

which can easily be shown to be identical to the expression in Eq. (7.61).

The analysis above neglected the body effect in the load device Q_2. Figure 7.40a shows the enhancement-load amplifier with the substrate connections explicitly indicated. Replacing each of the two transistors with the small-signal equivalent circuit of Fig. 7.30 gives rise to the equivalent circuit of Fig. 7.40b. The only difference between this equivalent circuit and that in Fig. 7.39 is the inclusion of the body effect of Q_2, modeled by the controlled source $g_{mb2}v_{bs2}$. Observing that the voltage across

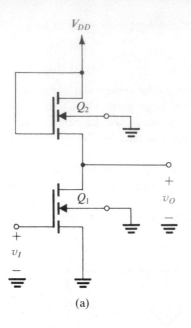

(a)

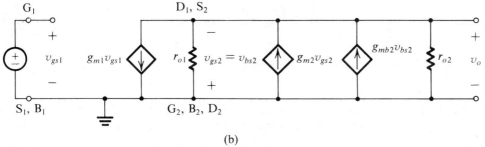

(b)

Fig. 7.40 (a) The enhancement-load amplifier with the substrate connections explicitly shown. **(b)** Small-signal equivalent circuit of the amplifier in (a), including the body effect of Q_2.

this current source is v_{bs2}, it can be replaced by a resistance equal to $1/g_{mb2}$. We thus obtain for v_o

$$v_o = -g_{m1}v_{gs1}\left[\left(\frac{1}{g_{m2}}\right)\middle\|\left(\frac{1}{g_{mb2}}\right)\middle\| r_{o1}\middle\| r_{o2}\right]$$

Substituting $v_{gs1} = v_i$ results in the voltage-gain expression

$$A_v = -\frac{g_{m1}}{g_{m2} + g_{mb2} + 1/r_{o1} + 1/r_{o2}} \qquad (7.64)$$

Assuming that r_{o1} and r_{o2} are large in comparison to $1/g_{m2}$, we can approximate Eq. (7.64) by

$$A_v \simeq \frac{-g_{m1}}{g_{m2} + g_{mb2}} \qquad (7.65)$$

Substituting for g_{mb2} from Eq. (7.34) gives

$$A_v = -\frac{g_{m1}}{g_{m2}} \frac{1}{1 + \chi} \tag{7.66}$$

Comparison of this expression with that in Eq. (7.63a) reveals that the body effect in the load device results in a reduction in gain by a factor $[1/(1 + \chi)]$.

A drawback of the enhancement-load amplifier is its rather limited signal swing. Specifically, it can be seen from Fig. 7.38c that the output voltage cannot exceed $V_{DD} - V_t$.

As a final point we note that the Miller effect will limit the high-frequency response of the enhancement-load amplifier in exactly the same manner as in a resistively loaded common-source amplifier. This limitation can be considerably reduced by adding another MOSFET connected as a common-gate amplifier. The resulting cascode configuration is shown in Fig. 7.41. V_{BIAS} denotes the gate bias voltage of

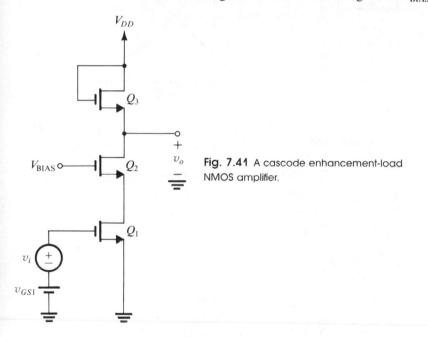

Fig. 7.41 A cascode enhancement-load NMOS amplifier.

Q_2 and must be chosen to keep Q_1 and Q_2 in pinch-off over the required signal swing. It is usual to design this circuit so that

$$(W/L)_2 \simeq (W/L)_1$$

$$(W/L)_1 \gg (W/L)_3$$

The voltage gain (neglecting the body effect) is given by

$$A_v \equiv \frac{v_o}{v_i} = -\sqrt{\frac{(W/L)_1}{(W/L)_3}} \tag{7.67}$$

The cascode circuit unfortunately suffers from a further reduction in signal swing.

EXAMPLE 7.3

The enhancement-load MOSFET amplifier can be also used in discrete circuits to design a linear amplifier for large input signals. As an example, consider the capacitively coupled amplifier shown in Fig. 7.42a. Resistance R_G establishes a dc operating point on the linear segment of the transfer curve. Figure 7.42b illustrates the process of determining the bias point Q, where the 45° straight line represents the constraint that R_G imposes: $V_{D1} = V_{G1}$. Thus the operating point lies at the

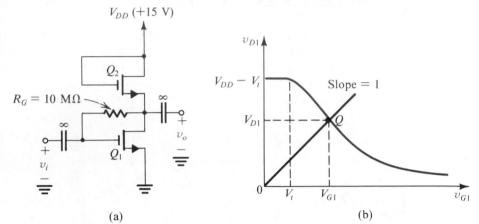

(a)   (b)

Fig. 7.42 (a) A capacitively coupled MOS amplifier with enhancement load.   **(b)** Illustrating the determination of the dc operating point Q.

intersection of this straight line with the transfer curve. Analytically, the straight-line portion of the transfer curve is described by Eq. (7.60). Thus

$$V_{D1} = \left(V_{DD} - V_t + \sqrt{\frac{K_1}{K_2}} V_t \right) - \sqrt{\frac{K_1}{K_2}} V_{G1}$$

Substitution of $V_{G1} = V_{D1}$ gives

$$V_{D1} = \frac{V_{DD} - V_t + V_t\sqrt{K_1/K_2}}{1 + \sqrt{K_1/K_2}}$$

Consider the case where $V_t = 2$ V, $V_{DD} = 15$ V, $K_1 = 270\ \mu A/V^2$, and $K_2 = 30\ \mu A/V^2$. The value of V_{D1} will be 4.75 V, and the dc drain current I_D will be given by

$$I_D = 0.27(4.75 - 2)^2 \simeq 2 \text{ mA}$$

At this operating point the voltage gain will be

$$\frac{v_o}{v_i} = -\sqrt{\frac{K_1}{K_2}} = -3$$

The actual voltage gain will be slightly lower than this value because of the finite output resistance r_o of each of the two MOSFETs. Also, if a load resistance is connected to the output, the voltage gain will be further reduced.

Exercises

7.14 For the enhancement-load amplifier of Fig. 7.38a let $W_1 = 100\ \mu$m, $L_1 = 6\ \mu$m, $W_2 = 6\ \mu$m, and $L_2 = 30\ \mu$m. If the body-effect parameter $\chi = 0.1$, find the voltage gain without and with the body effect taken into account. Neglect the effect of r_o.

Ans. -9.12 V/V; -8.3 V/V

7.15 For the enhancement-load amplifier of Fig. 7.38a let $W_1 = 100\ \mu$m, $L_1 = 6\ \mu$m, $W_2 = 6\ \mu$m, $L_2 = 30\ \mu$m, $V_{DD} = 10$ V, and $V_t = 1$ V. Find the coordinates of points A and B that define the linear region of the transfer characteristic in Fig. 7.38c.

Ans. $V_{IA} = 1$ V, $V_{OA} = 9$ V; $V_{IB} = 1.9$ V, $V_{OB} = 0.9$ V

7.9 NMOS AMPLIFIER WITH DEPLETION LOAD

Modern NMOS technology allows the fabrication of both enhancement and depletion devices on the same chip. The latter device is realized by implanting an n channel with the result that it exhibits a negative threshold voltage. As will be now shown, using the depletion MOSFET as a load device results in an amplifier with performance superior to that of the enhancement-load circuit. The same holds true if the circuit is to be used as a logic inverter (Chapter 15).

The depletion-load amplifier is shown in Fig. 7.43a. Figure 7.43b shows the i-v characteristic of the depletion load. The transfer characteristic of the amplifier can be determined using the graphical technique illustrated in Fig. 7.43c. Here the i-v load curve is superimposed on the i_D-v_{DS} characteristics of the enhancement transistor Q_1. The transfer characteristic can be determined point-by-point in the same manner used for the enhancement-load amplifier. The resulting characteristic is shown in Fig. 7.43d. We observe four distinct regions. For $v_I < V_{tE}$, the threshold voltage of the enhancement transistor Q_1, transistor Q_1 is off and $v_O = V_{DD}$. Here we note an important difference from the enhancement-load case where the maximum output is one threshold voltage lower than V_{DD}. Region II is obtained when v_I exceeds V_{tE} and Q_1 turns on; but because the output voltage is high, Q_2 is in the triode region. In fact, Q_2 remains in the triode region until v_O becomes lower than V_{DD} by $|V_{tD}|$. At this point the amplifier enters region III of its transfer characteristic. Here both Q_1 and Q_2 are operating in the pinch-off region and thus have a large output resistance, which gives rise to the large gain indicated by the sharp transfer curve in region III. Region III is the one of interest for amplifier operation; that is, the amplifier will be biased to operate in region III. Finally, region IV is entered when v_O becomes V_{tE} volts lower than v_I, at which point Q_1 enters the triode region.

If the depletion-load amplifier is biased to operate in region III, then the small-signal gain will be given by

$$A_v \equiv \frac{v_o}{v_i} = -g_{m1}[r_{o1} \,\|\, r_{o2}] \tag{7.68}$$

In practice, however, a gain much lower than this is realized due to the body effect on transistor Q_2. Specifically, note that because the substrate of Q_2 will be connected to ground, a voltage signal equal to $-v_o$ appears between body and source. The resulting

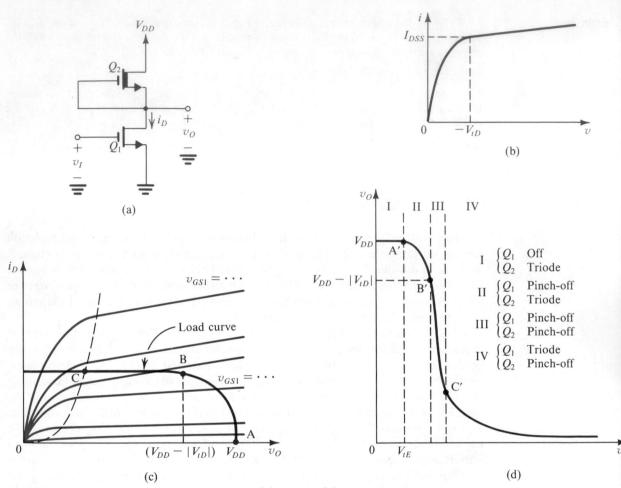

Fig. 7.43 The NMOS amplifier with depletion load: **(a)** circuit; **(b)** i-v characteristic of the depletion load; **(c)** graphical construction to determine the transfer characteristic; and **(d)** transfer characteristic.

small-signal equivalent circuit is shown in Fig. 7.44 from which we see that the controlled current source $g_{mb2}v_{bs2}$ can be replaced by a resistance $1/g_{mb2}$. Then the output voltage can be obtained as

$$v_o = -g_{m1}v_{gs1}\left[\left(\frac{1}{g_{mb2}}\right)\middle\|r_{o1}\middle\|r_{o2}\right]$$

Since $v_{gs1} = v_i$, the voltage gain is given by

$$A_v \equiv \frac{v_o}{v_i} = -g_{m1}\left[\left(\frac{1}{g_{mb2}}\right)\middle\|r_{o1}\middle\|r_{o2}\right] \tag{7.69}$$

It is usually the case that $1/g_{mb2}$ is much smaller than r_{o1} and r_{o2}, resulting in the

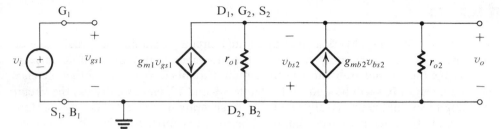

Fig. 7.44 Small-signal equivalent circuit of the depletion-load amplifier of Fig. 7.43a, incorporating the body effect on Q_2.

approximate gain expression

$$A_v \simeq -\frac{g_{m1}}{g_{mb2}}$$ (7.70)

Expressing g_{mb2} as $\chi\, g_{m2}$ gives

$$A_v = -\frac{g_{m1}}{g_{m2}}\left(\frac{1}{\chi}\right)$$ (7.71)

or, alternatively,

$$A_v = -\sqrt{\frac{(W/L)_1}{(W/L)_2}}\left(\frac{1}{\chi}\right)$$ (7.72)

Comparison of Eq. (7.71) with the gain expression for the enhancement-load amplifier [Eq. (7.66)] reveals that the gain of the depletion-load amplifier is a factor $(1 + \chi)/\chi$ greater. Since χ is typically 0.1 to 0.3, the gain increase can be as much as a factor 10.

Finally we observe that the dc bias current of the depletion-load amplifier is approximately equal to I_{DSS} of the depletion load which is given by Eq. (7.55) as

$$I_D \simeq I_{DSS} = K_D V_{tD}^2$$ (7.73)

Thus the bias current is determined by the technology and by the device geometry and cannot be changed by the circuit designer.

Exercise

7.16 For the depletion-load amplifier of Fig. 7.43a let $W_1 = 100\ \mu\mathrm{m}$, $L_1 = 6\ \mu\mathrm{m}$, $W_2 = 6\ \mu\mathrm{m}$, $L_2 = 30\ \mu\mathrm{m}$, $V_{tE} = 1.5$ V, $V_{tD} = -3$ V, $\mu_n C_{OX} = 100\ \mu\mathrm{A/V}^2$, $V_{DD} = 10$ V, $|V_A| = 90$ V, and $\chi = 0.1$.
(a) Calculate I_{DSS} for the depletion device.
(b) Neglecting the finite output resistance in pinch-off and the body effect, find the coordinates of points A', B', C', and D' that define the transfer characteristic of Fig. 7.43d. (Note that in this case the segment B'C' is a vertical straight line).
(c) For operation in region III, calculate g_{m1}, g_{m2}, g_{mb2}, r_{o1}, and r_{o2}.
(d) Taking into account the body effect and the finite output resistance, find the small-signal voltage gain in region III.

Ans. (a) 90 μA; **(b)** (1.5 V, 10 V),   (1.83 V, 7 V),   (1.83 V, 0.33 V); **(c)** 0.55 mA/V, 60 μA/V, 6 μA/V, 1 MΩ, 1 MΩ; **(d)** -68.8 V/V

7.10 THE CURRENT MIRROR

In both NMOS and CMOS analog integrated circuits a stable and predictable dc reference current is generated and is then used to generate proportional dc currents for biasing the various transistors in the circuit. Circuits for generating the reference current in MOS ICs will be studied in Chapters 9 and 13. Here we discuss the circuit building block that is universally employed to generate dc currents bearing constant multiples to the reference current source. The circuit is appropriately called a *current mirror* and is shown in its simplest form in Fig. 7.45a.

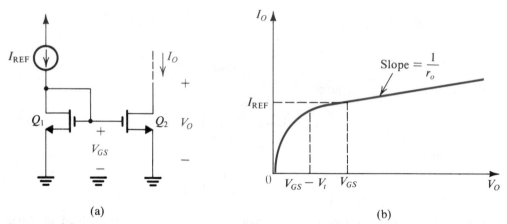

(a)   (b)

Fig. 7.45 (a) The basic current mirror circuit.   **(b)** The output characteristic of the current mirror.

The current mirror consists of two enhancement MOSFETs, Q_1 and Q_2, having equal threshold voltages V_t, but possibly different (W/L) ratios. Transistor Q_1 is fed with the reference current I_{REF}. The output current I_O is taken at the drain of Q_2, which must be operated in the pinch-off region. For Q_1 we can write

$$I_{REF} = K_1'(V_{GS} - V_t)^2 \tag{7.74}$$

where V_{GS} is the gate-to-source voltage corresponding to a drain current of I_{REF}. Since Q_2 is connected in parallel with Q_1, it will have the same V_{GS}; thus

$$I_O = K_2(V_{GS} - V_t)^2 \tag{7.75}$$

where we have neglected the finite output resistance of Q_2. Equations (7.74) and (7.75) can be combined to obtain

$$I_O = I_{REF}\left(\frac{K_2}{K_1}\right) \tag{7.76}$$

Expressing K_1 and K_2 in terms of the devices' (W/L) ratios gives

$$I_O = I_{REF}\frac{(W/L)_2}{(W/L)_1} \tag{7.77}$$

Thus, ideally, I_O will be a multiple of I_{REF} whose value is determined by device geometry. In practice this value of I_O will be obtained only when the voltage at the drain of Q_2 is equal to V_{GS}. Variation of the drain voltage will result in corresponding changes in I_O due to the finite output resistance r_o of Q_2. Figure 7.45b shows I_O as a function of V_O. This is simply the i_D-v_{DS} characteristic curve for Q_2 corresponding to the value of V_{GS} established by passing I_{REF} through Q_1. More elaborate current-mirror circuits will be presented in Chapter 9.

7.11 THE CMOS AMPLIFIER

In CMOS technology both n-channel and p-channel devices are available, thus making possible a greater variety of circuit design techniques. Furthermore, the devices are usually fabricated in a way that eliminates the body effect, which we have found to cause considerable degradation in the performance of NMOS circuits. The basic CMOS amplifier is shown in Fig. 7.46a. Here Q_2 and Q_3 are a matched pair of p-channel devices connected as a current mirror that is fed with the reference dc current I_{REF}. Thus Q_2 behaves as a current source and has the i-v characteristic shown in Fig. 7.46b. Note that Q_2 will be operating in the pinch-off region when the voltage at its drain is lower than that at its source (V_{DD}) by at least ($V_{SG} - |V_{tp}|$), where V_{SG} is the dc bias voltage corresponding to a drain current of I_{REF}. When in pinch-off, Q_2 has a high output resistance r_{o2},

$$r_{o2} = \frac{|V_A|}{I_{REF}} \tag{7.78}$$

Transistor Q_2 is used as the load resistance for the amplifying transistor Q_1, and is called an *active load*. It follows that when Q_1 is operating in pinch-off, the small-signal voltage gain will be equal to g_{m1} multiplied by the total resistance between the output and ground, which is ($r_{o1} \| r_{o2}$). Thus a large voltage gain is obtained in the CMOS amplifier.

Before we consider the voltage gain in more detail we wish to examine the transfer characteristic of the CMOS amplifier. Figure 7.46c shows the i_D-v_{DS} characteristics of Q_1, with the load curve corresponding to the active load device Q_2 superimposed. Since $v_{GS1} = v_I$, the transfer characteristic can be determined point-by-point by finding the intersections of the Q_1 characteristic curves, corresponding to different values of v_I, and the load curve. The horizontal coordinate of each intersection point gives the value of v_{DS1} that is equal to v_O. The resulting transfer characteristic

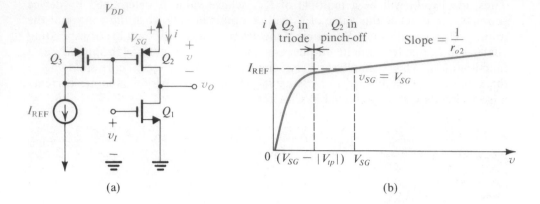

(a)

(b)

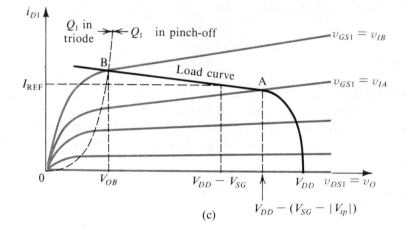

(c)

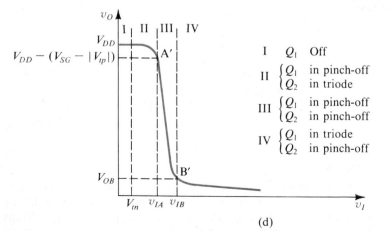

(d)

Fig. 7.46 The basic CMOS amplifier: **(a)** circuit; **(b)** *i-v* characteristic of the active load Q_2; **(c)** graphical construction to determine the transfer characteristic; and **(d)** transfer characteristic.

is sketched in Fig. 7.46d and its four distinct regions are indicated. For amplifier operation, region III is the one of interest. It can be shown (see Example 7.4) that the transfer characteristic in region III is almost linear and, because of the high gain, very sharp.

As already mentioned, in region III of the transfer characteristic the small-signal voltage gain is given by

$$A_v \equiv \frac{v_o}{v_i} = -g_{m1}[r_{o1} \| r_{o2}] \tag{7.79}$$

Since Q_1 is operating at a dc bias current equal to I_{REF}, g_{m1} can be expressed using Eq. (7.30b) as

$$g_{m1} = \sqrt{2(\mu_n C_{OX})(W/L)_1 I_{REF}} \tag{7.80}$$

Substituting in Eq. (7.79) for g_{m1} from Eq. (7.80) and using $r_{o1} = r_{o2} = |V_A|/I_{REF}$ we obtain

$$A_v = -\frac{\sqrt{K_n}|V_A|}{\sqrt{I_{REF}}} \tag{7.81}$$

Thus the voltage gain is inversely proportional to the square root of the bias current.

EXAMPLE 7.4

Consider a CMOS amplifier for which $V_{DD} = 10$ V, $V_{tn} = |V_{tp}| = 1$ V, $\mu_n C_{OX} = 2\mu_p C_{OX} = 20$ μA/V^2, $W = 100$ μm, and $L = 10$ μm for both n and p devices, $|V_A| = 100$ V and $I_{REF} = 100$ μA. Find the small-signal voltage gain. Also find the coordinates of the extremities of the amplifier region of the transfer characteristic, that is points A' and B'.

Solution

$$K_n = \tfrac{1}{2}\mu_n C_{OX}(W/L)$$
$$= \tfrac{1}{2} \times 20 \times (100/10) = 100 \ \mu\text{A/V}^2$$

From Eq. (7.81) we obtain

$$A_v = -\frac{\sqrt{100 \times 10^{-6}} \times 100}{\sqrt{10^{-4}}} = -100 \text{ V/V}$$

We observe that the gain is much greater than the values obtained in NMOS amplifiers.

The extremities of the amplifier region of the transfer characteristic are found as follows (refer to Fig. 7.46): First we determine V_{SG} of Q_2 and Q_3 corresponding to $I_D = I_{REF} = 100$ μA using

$$I_D = K_p(V_{SG} - |V_{tp}|)^2 \left(1 + \frac{V_{SD}}{|V_A|}\right)$$

Substituting $K_p = \tfrac{1}{2}\mu_p C_{OX}(W/L)$ and $V_{SD} = V_{SG}$ and neglecting for simplicity the factor $(1 + V_{SG}/|V_A|)$ we obtain

$$V_{SG} \simeq 2.414 \text{ V}$$

Thus for point A' we have

$$V_{OA} = V_{DD} - (V_{SG} - |V_{tp}|) = 8.586 \text{ V}$$

To find the corresponding value of v_I, V_{IA}, we equate the drain currents of Q_1 and Q_2,

$$i_{D1} = K_n(v_I - V_{tn})^2 \left(1 + \frac{v_O}{|V_A|}\right)$$

$$i_{D2} = K_p(V_{SG} - |V_{tp}|)^2 \left(1 + \frac{V_{DD} - v_O}{|V_A|}\right)$$

and substitute for $K_p(V_{SG} - |V_{tp}|)^2 \simeq I_{\text{REF}}$ to obtain

$$K_n(v_I - V_{tn})^2 = I_{\text{REF}} \frac{1 + (V_{DD} - v_O)/|V_A|}{1 + (v_O/|V_A|)}$$

$$\simeq I_{\text{REF}} \left(1 + \frac{V_{DD}}{|V_A|} - \frac{2v_O}{|V_A|}\right)$$

which yields

$$v_O = \frac{|V_A|}{2I_{\text{REF}}} \left[I_{\text{REF}}\left(1 + \frac{V_{DD}}{|V_A|}\right) - K_n(v_I - V_{tn})^2\right] \tag{7.82}$$

Substituting $v_O = V_{OA} = 8.586$ V gives the corresponding value of v_I; that is,

$$V_{IA} = 1.963 \text{ V}$$

Since the width of region III is narrow, we may assume that $V_{OB} = V_{IB} - V_{tn} \simeq 2 - 1 = 1$ V. Substituting this value in Equation (7.82), we obtain $V_{IB} = 2.039$ V. Thus a more exact value for V_{OB} is 1.039 V. The width of the amplifier region is therefore

$$\Delta V_I = V_{IB} - V_{IA} = 0.076 \text{ V}$$

The corresponding output signal swing is

$$\Delta V_O = V_{OA} - V_{OB} = 7.547 \text{ V}$$

The large signal voltage gain is

$$\frac{\Delta V_O}{\Delta V_I} = \frac{7.547}{0.076} = 99.3$$

which is very close to the small-signal value of 100, indicating that the transfer characteristic is quite linear.

Exercises

7.18 For the CMOS amplifier of Example 7.4, find the small-signal voltage gain for $I_{\text{REF}} = 25$ μA and 400 μA.

Ans. -200 V/V; -50 V/V

7.19 For small-signal operation, the CMOS amplifier of Fig. 7.46a can be represented as a transconductance amplifier (Section 2.3). Sketch the amplifier equivalent circuit.

Ans. See Fig. E7.19

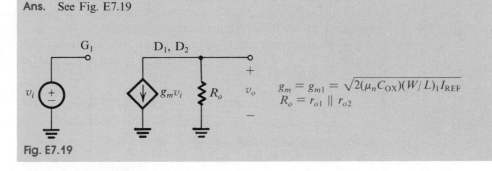

$$g_m = g_{m1} = \sqrt{2(\mu_n C_{OX})(W/L)_1 I_{REF}}$$
$$R_o = r_{o1} \parallel r_{o2}$$

Fig. E7.19

7.12 THE SOURCE FOLLOWER

The source-follower configuration was studied in Section 7.5. In the design of MOS IC amplifiers, the source follower is used as a buffer to obtain a low output resistance. Figure 7.47a shows a source follower as it is commonly connected in an IC amplifier. To calculate the small-signal voltage gain and output resistance, we show in Fig. 7.47b the circuit with the dc voltage sources replaced with grounds and the dc current source replaced with an open circuit. Also shown are the resistance $1/g_m$, which is the equivalent resistance seen between source and gate, looking into the source; the

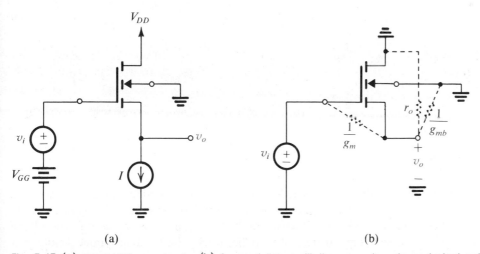

(a)   (b)

Fig. 7.47 (a) Source-follower circuit. **(b)** Source follower with the source-to-gate equivalent resistance (looking into the source), the source-to-body equivalent resistance (looking into the source), and the drain-to-source resistance shown.

resistance $1/g_{mb}$, which is the equivalent resistance between source and body, looking into the source; and the source-to-drain resistance r_o. Extreme caution should be exercised in using this equivalent circuit: $1/g_m$ in the resistance looking into the source. The resistance looking into the gate is infinite, since the gate current is zero. Thus the input resistance of the source follower is infinite.

The output voltage v_o appears across the total resistance between source and ground. This is the parallel equivalent of r_o and $1/g_{mb}$. To find the voltage gain we use the voltage-divider rule

$$\frac{v_o}{v_i} = \frac{[(1/g_{mb})\,\|\,r_o]}{(1/g_m) + [(1/g_{mb})\,\|\,r_o]} \tag{7.83}$$

If $r_o \gg 1/g_{mb}$, we obtain

$$\frac{v_o}{v_i} \simeq \frac{g_m}{g_m + g_{mb}} \tag{7.84}$$

Substituting $g_{mb} = \chi g_m$ gives

$$\frac{v_o}{v_i} = \frac{1}{1 + \chi} \tag{7.85}$$

Thus the body effect reduces the gain from approximately unity to the value given by Eq. (7.85). Note that this gain value is obtained with no load; it is the open-circuit voltage gain. It can be used together with the output resistance R_o of the source follower, to obtain the gain when a load is connected. The output resistance is the resistance between the source and ground with v_i reduced to zero. Short-circuiting the signal source v_i in Fig. 7.47b, we see that

$$R_o = (1/g_m)\,\|\,(1/g_{mb})\,\|\,r_o \tag{7.86}$$

Exercise

7.20 For the source-follower circuit of Fig. 7.47a, let $W = 100\ \mu m$, $L = 8\ \mu m$, $\mu_n C_{OX} = 100\ \mu A/V^2$, $V_t = 1\ V$, $V_A = 100\ V$, $I = 0.5\ mA$, and $\chi = 0.1$. Calculate the open-circuit voltage gain and the output resistance.

Ans.   0.91 V/V; 810 Ω

7.13 MOS ANALOG SWITCHES

Like the JFET, the MOSFET is well suited for analog-switching applications because of the almost linear i_D-v_{DS} characteristics in the region around the origin. Furthermore, these characteristic lines pass through the origin—that is, there is no voltage offset as in the case of bipolar transistors (Chapter 8)—and extend symmetrically into the third quadrant. Analog switches are useful in many applications such as sample-and-hold circuits, chopper circuits, and digital-to-analog converters. A very important application of MOS switches is to be found in the design of "switched-capacitor filters," discussed in Chapter 14.

An NMOS Switch

When the MOSFET is used to switch or gate analog signals, a specific configuration using CMOS circuitry is the most convenient. To appreciate this fact, consider first the analog switch formed by a single enhancement NMOS transistor, as shown in Fig. 7.48. Here v_A is the analog input signal and is assumed to be in the range -5 V to $+5$ V. The load to which this analog signal is to be connected is assumed to be a resistance R_L in parallel with a capacitance C_L. In order to keep the substrate-to-source and substrate-to-drain *pn* junctions reverse-biased at all times, the substrate terminal is connected to -5 V.

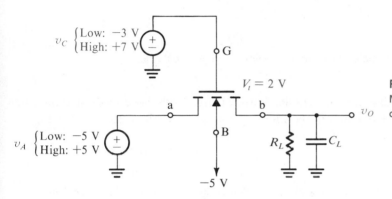

Fig. 7.48 Application of the enhancement MOSFET as an analog switch. The parallel combination of R_L and C_L represents the load.

The purpose of the control signal v_C is to turn the switch on and off. Assume that the device has a threshold voltage $V_t = 2$ V. It follows that in order to turn the transistor on for all possible input signal levels the high value of v_C should be at least $+7$ V. Similarly, to turn the transistor off for all possible input signal levels the low value of v_C should be a maximum of -3 V. Note, however, that these levels are not sufficient in practice, since the transistor will be barely on and barely off at the limits. In any case we see that the range of the control voltage has to be at least equal to the range of the analog input signal being switched. Unfortunately, though, the "on" resistance of the switch will depend to a large extent on the value of the analog input signal. Thus the transient response of the switch will depend on the value of input signal. Another obvious disadvantage is the rather inconvenient levels required for v_C.

The reader will observe that we have not indicated in Fig. 7.48 which terminal is the source and which is the drain. We have avoided labeling them first of all because the MOSFET is a symmetric device, with the source and the drain interchangeable. More important, the operation of the device as a switch is based on this interchangeability of roles. Specifically, if the analog input signal is positive, say, $+4$ V, then it is most convenient to think of terminal *a* as the drain and of terminal *b* as the source. For this case the circuit (when v_C is high) takes the familiar form shown in Fig. 7.49a. It is easy to see that the device will be operating in the triode region and that v_O will be very close to the input analog signal level of $+4$ V. On the other hand, if the input signal is negative, say, -4 V, then it is most convenient to think of terminal *a* as the source and of terminal *b* as the drain. In this case the circuit (in the case of

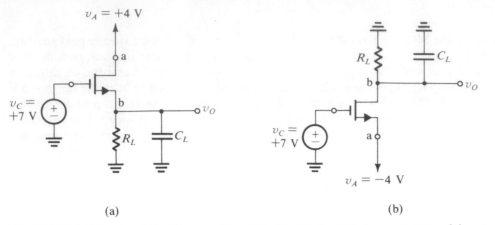

(a)   (b)

Fig. 7.49 Method for visualizing the operation of the circuit in Fig. 7.48 when v_C is high: **(a)** v_A is positive; **(b)** v_A is negative.

v_C high) can be redrawn in the familiar form of Fig. 7.49b. Here again it should be clear that the device operates in the triode region, and v_O will be only slightly higher than the input analog signal level of -4 V.

The CMOS Transmission Gate

Let us consider next the more elaborate CMOS analog switch shown in Fig. 7.50. This circuit is commonly known as a *transmission gate*. As in the previous case, we are assuming that the analog signal being switched lies in the range -5 V to $+5$ V. In order to prevent the substrate junctions from becoming foward-biased at any time, the substrate of the p-channel device is connected to the most positive voltage level ($+5$ V) and that of the n-channel device is connected to the most negative voltage level (-5 V). The transistor gates are controlled by two complementary signals denoted v_C and $\overline{v_C}$. Unlike the single NMOS switch, here the levels of v_C can be the same as the extremes of the analog signal, $+5$ V and -5 V. When v_C is at the low level, the gate of the n-channel device will be at -5 V, thus preventing the n-channel device from conducting for any value of v_A (in the range -5 V to $+5$ V). Simultaneously, the gate of the p-channel device will be at $+5$ V, which prevents that device from conducting for any value of v_A (in the range -5 V to $+5$ V). Thus with v_C low the switch is open.

In order for us to close the switch, we have to raise the control signal v_C to the high level of $+5$ V. Correspondingly, the n-channel device will have its gate at $+5$ V and will thus conduct for any value of v_A in the range of -5 V to $+3$ V. Simultaneously, the p-channel device will have its gate at -5 V and will thus conduct for any value of v_A in the range -3 V to $+5$ V. We thus see that for v_A less than -3 V only the n-channel device will be conducting, while for v_A greater than $+3$ V only the p-channel device will be conducting. For the range $v_A = -3$ V to $+3$ V both devices will be conducting. Furthermore, we can see that as one device conducts more heavily, conduction in the other device is reduced. Thus as the resistance r_{DS} of one device decreases, the resistance of the other device increases, with the parallel

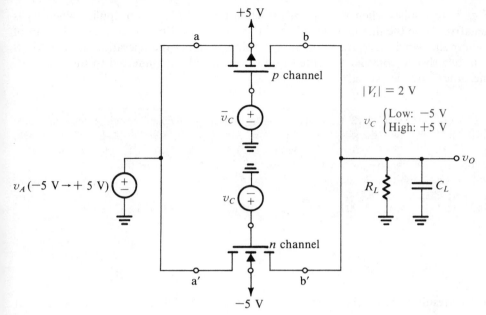

Fig. 7.50 The CMOS transmission gate.

equivalent, which is the "on" resistance of the switch, remaining approximately constant. This is clearly an advantage of the CMOS transmission gate over the single NMOS switch previously considered. This advantage operates in addition to the obviously more convenient levels required for the control signal in the CMOS switch.

The operation of the transmission gate (in the closed position) can be better understood from the two equivalent circuits shown in Fig. 7.51. Here the circuit in

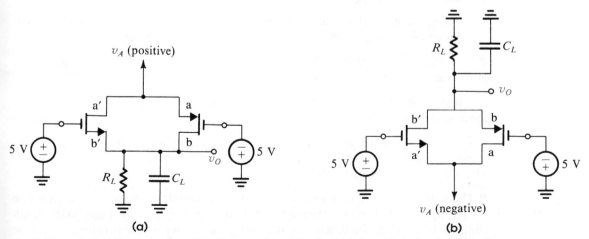

Fig. 7.51 Equivalent circuits for visualizing the operation of the transmission gate in the closed (on) position: **(a)** v_A is positive; **(b)** v_A is negative.

Fig. 7.51a applies when v_A is positive, while that in Fig. 7.51b applies when v_A is negative. Note the interchangeability of the roles played by the source and drain of each of the two devices. Also note that in thinking about the operation of the CMOS gate one should consider that the drain of one device is connected to the source of the other, and vice versa.

Exercise

7.21 Consider the CMOS transmission gate and its equivalent circuit shown in Fig. 7.51b. Let the two devices have $|V_t| = 2$ V and $K = 50\,\mu\text{A/V}^2$, and let $R_L = 50$ kΩ. For **(a)** $v_A = -5$ V, **(b)** $v_A = -2$ V, and **(c)** $v_A = 0$ V, calculate v_O and the total resistance of the switch. [*Hint:* Since both devices will be operating in the triode region and $|v_{DS}|$ will be small, use $i_D \simeq 2K(v_{GS} - V_t)v_{DS}$ for the *n*-channel device and $i_D \simeq 2K(v_{SG} - |V_t|)v_{SD}$ *for* the *p*-channel device.]
Ans. **(a)** −4.878 V, 1.25 kΩ; **(b)** −1.963 V, 1.649 kΩ; **(c)** 0 V, 1.667 kΩ

7.14 SUMMARY

● As a result of the insulated-gate structure, the MOSFET features a very-high input resistance.

● The depletion-type MOSFET operates in a manner similar to that of the JFET with one exception: the depletion MOSFET can also operate in the enhancement mode.

● To operate in the pinch-off region the drain voltage of the *n*-channel (*p*-channel) depletion MOSFET must be higher (lower) than the gate voltage by at least $|V_P|$.

● For the *n*-channel enhancement-type MOSFET to conduct, $v_{GS} \geq V_t$, where the threshold voltage V_t is positive. The device operates in pinch-off for $v_{DS} \geq v_{GS} - V_t$. In pinch-off, $i_D = K(v_{GS} - V_t)^2$.

● For the *p*-channel enhancement MOSFET to conduct, $v_{SG} \geq |V_t|$. Here V_t is negative. The device operates in pinch-off for $v_{SD} \geq v_{SG} - |V_t|$. In pinch-off, $i_D = K(v_{SG} - |V_t|)^2$.

● In pinch-off, i_D depends slightly on v_{DS}. This dependence is modeled by an output resistance $r_o = |V_A|/I_D$.

● For small signals, a MOSFET, biased in the pinch-off region, functions as a voltage-controlled current source with transconductance $g_m = \partial i_D/\partial v_{GS}$, at the bias point.

● The substrate (body) terminal B acts as a second gate for the MOSFET, an effect known as the body effect. This effect should be taken into account in situations for which B is not connected to S.

● The common-source configuration provides a reasonably high voltage gain and a high input resistance but a limited high-frequency response. A much wider bandwidth is achieved in the common-gate configuration but its input resistance is low. The cascode configuration combines the advantages of both common-source and common-gate configurations.

● The source follower provides a voltage gain less than unity but features a low output resistance.

● Integrated-circuit MOS amplifiers utilize MOS transistors both as amplifying and as load devices.

● In NMOS technology, the load devices can be either enhancement- or depletion-type MOSFETs connected as two-terminal devices. The voltage gain of a depletion-load amplifier is an order of magnitude greater than that of an enhancement-load amplifier.

● In CMOS technology, both n- and p-channel enhancement MOSFETs are used, thus providing the circuit designer with considerable flexibility.

● In the basic CMOS amplifier, a complementary MOSFET, operated as a constant-current source, is used as a load in an arrangement called an active load.

● The voltage gain of the basic CMOS amplifier is approximately $(g_m r_o/2)$.

● The CMOS transmission gate is widely used in the switching of analog signals.

BIBLIOGRAPHY

R. S. C. Cobbold, *Theory and Applications of Field-Effect* Transistors, New York: Wiley, 1969.

P. R. Gray, D. A. Hodges, and R. W. Brodersen, *Analog MOS Integrated Circuits*, New York: IEEE Press, 1980.

P. R. Gray and R. G. Meyer, *Analysis and Design of Analog Integrated Cicuits*, 2nd ed; New York: Wiley, 1984.

A. B. Grebene, *Bipolar and MOS Analog Integrated Circuit Design*, New York: Wiley, 1984.

D. J. Hamilton and W. G. Howard, *Basic Integrated Circuit Engineering*, New York: McGraw-Hill, 1975.

Y. Tsividis, "Design considerations in single-channel MOS analog integrated circuits—A tutorial," *IEEE Journal of Solid-State Circuits*, vol. SC-13, pp. 383 391, June 1978.

PROBLEMS

7.1 A depletion NMOS transistor has $I_{DSS} - 2$ mA and $V_P = -1$ V. What is the minimum value of v_{DS} for the device to operate in pinch-off **(a)** with $v_{GS} = +1$ V and **(b)** with $v_{GS} = +5$ V? What are the corresponding values of i_D?

7.2 For the NMOS transistor described by Fig. 7.6 calculate g_m at $V_{GS} = -1, 0,$ and 1 V.

7.3 For the NMOS transistor described by Fig. 7.6, for which $I_{DSS} = 16$ mA and $V_P = -4$ V, what is the gate-to-source bias required to establish $I_D = 32$ mA?

7.4 A depletion PMOS transistor has $I_{DSS} = 8$ mA and $V_P = 4$ V. Plot the i-v characteristic of this device when "diode-connected" as shown in Fig. P7.4, over the range $v = \pm 8$ V.

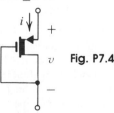

Fig. P7.4

7.5 An enhancement NMOS transistor for which $V_t = 2$ V and $K = 0.25$ mA/V^2 is operated as a diode with gate and drain joined. Plot the characteristic i_D versus v_{DS} over the range $v_{DS} = \pm 10$ V. Noting that for this connection $v_{GS} = v_{DS}$, compare your plotted characteristic with that in Fig. 7.12. Plot the locus of operation for the diode-connected device on the characteristics of Fig. 7.10. Compare this also with your results.

7.6 For an enhancement NMOS transistor, show that for small v_{DS} the resistance r_{DS} is given by $r_{DS} \simeq 1/2K(v_{GS} - V_t)$, for $v_{DS} \ll V_t$. For a device having $\mu_n C_{OX} = 10$ μA/V^2, $W/L = 10$, and $V_t = 1$ V, calculate r_{DS} when the device is operated at $v_{GS} = 2$ V and 5 V.

7.7 A number of enhancement NMOS transistors whose threshold voltage V_t is 1 V (with the source and substrate joined) are operated with a 10-V supply and a common substrate connected to the negative end of the supply. In operation the dc voltage levels on the source of the devices range from 0 to 9 V. For $\gamma = 1$ what is the expected range of thresholds that will be observed?

7.8 A static discharge of 10 mA occurs for 100 ns to the gate of a MOSFET whose capacitance is 1 pF. Calculate the voltage to which an ideal gate would be charged. What is likely to happen?

7.9 An enhancement PMOS transistor with $V_t = -1.5$ V conducts a current i_D of 5 mA when $v_{SG} = v_{SD} = 3$ V. For an ideal device (that is, one in which i_D is independent of v_{SD} in pinch-off) what is the drain current at $v_{SG} = v_{SD} = 4.5$ V? What is the source-to-drain resistance for v_{SD} small and $v_{SG} = 3$ V? 4.5 V?

7.10 An n-channel enhancement MOSFET operating at $v_{GS} = 4$ V and $v_{DS} = 4$ V has $i_D = 4$ mA. With v_{GS} held constant and v_{DS} raised to 8 V, i_D increases to 4.4 mA. What is r_o at i_D in the vicinity of 4 mA for this device? What is the value of the parameter V_A?

7.11 Consider the circuit of Fig. 7.18, where $R_{G1} = R_{G2}$, $R_S = R_D = 10$ kΩ, and $V_{DD} = 10$ V. If $V_S = 3$ V and $K = 0.3$ mA/V^2, what is V_t? Find the drain current and drain-to-source voltage. If the transistor is replaced with another having the same V_t but with K 10 times as large (that is, 3 mA/V^2), what do I_D and V_{DS} become?

7.12 Consider the circuit of Fig. 7.18 in which $V_{GG} = 5$ V and $R_S = 10$ kΩ operating in pinch-off with an FET for which $V_t = 2$ V and $K = 50$ μA/V^2. What I_D results? If the FET is replaced with one for which K is double, what I_D results? Compare the percentage change in I_D to the percentage change in K. The reduction is due to negative feedback.

*7.13 Use the biasing arrangement of Fig. 7.18 with an enhancement NMOS transistor for which $K = 0.5$ mA/V^2 and $V_t = 1.5$ V to establish operation at $V_{GS} = V_{DS} = 2V_t$ with equal values for R_S and R_D, a 12-V supply and the largest resistor available of 2.2 MΩ. For your design what do I_D and V_{DS} become if $V_t = 3$ V? If V_t remains at 1.5 V while K is increased to 1 mA/V^2, what do I_D and V_{DS} become?

*7.14 Repeat Problem 7.13 using the biasing arrangement shown in Fig. 7.21.

7.15 For the MOSFET represented by the equivalent circuit in Fig. 7.23, operating in a common-source amplifier having a voltage gain of -10, find the equivalent total input capacitance. All model capacitances are 1 pF.

7.16 Calculate g_m of enhancement MOSFETs under the following conditions.
(a) $K = 0.05$ mA/V^2, $V_t = 2$ V, $V_{GS} = 3$ V.
(b) $K = 0.5$ mA/V^2, $V_t = -2$ V, $V_{GS} = -2.5$ V.

7.17 For an enhancement NMOS amplifier for which $V_t = 2$ V biased at $V_{GS} = 4$ V, what is the peak value of the sine wave signal v_{gs} that maintains operation in the linear region to within 1% of second-harmonic distortion?

7.18 A grounded-source MOSFET operates with $R_D = 10$ kΩ and has $g_m = 1$ mA/V and $r_o = 50$ kΩ. Find the voltage gain. If the amplifier is biased with a resistance $R_G = 10$ MΩ connected between drain and gate, what is the value of the input resistance?

7.19 Consider an enhancement NMOS transistor having $W = 100$ μm, $L = 6$ μm, $\mu_n C_{OX} = 20$ μA/V^2, $V_t = 1.5$ V, $V_A = 50$ V and $\chi = 0.1$. Find V_{GS} to bias the device at $I_D = 100$ μA and $V_{DS} = 5$ V. Also find g_m, g_{mb}, and r_o at the bias point.

7.20 Show that g_m of an enhancement MOSFET biased at a gate-to-source voltage V_{GS} and a drain current I_D can be expressed as

$$g_m = \frac{2I_D}{V_{GS} - V_t}$$

**7.21 The MOSFET in the circuit of Fig. P7.21 has $V_t = 1$ V, $K = 0.4$ mA/V^2, and $V_A = 40$ V.
(a) Find the values of R_S, R_D, and R_G so that $I_D = 0.1$ mA, the largest possible value for R_D is used while a maximum signal swing at the drain of ± 1 V is possible, and the input resistance at the gate is 10 MΩ.
(b) Find the values of g_m and r_o at the bias point.
(c) If terminal Z is grounded and terminal X is connected to a signal source having a resistance of 1 MΩ while

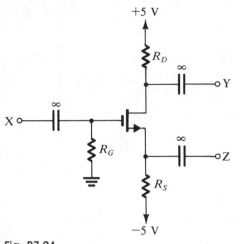

Fig. P7.21

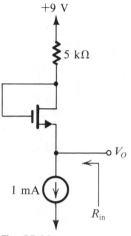

Fig. P7.23

terminal Y is connected to a load resistance of 40 kΩ, find the voltage gain from signal source to load.

(d) If terminal Y is grounded, find the voltage gain from X to Z with Z open-circuited. What is the output resistance of the source follower?

(e) If terminal X is grounded and terminal Z is connected to a current source delivering a signal current of 10 μA and having a resistance of 100 kΩ, find the voltage signal that can be measured at Y. For simplicity, neglect the effect of r_o.

7.22 In Fig. P7.22, $V_t = 2$ V, $K = 0.25$ mA/V², and $r_o = \infty$. Find the dc voltage at the source and the small-signal resistance looking into the source, R_{in}.

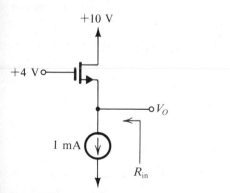

Fig. P7.22

7.23 In Fig. P7.23, $V_t = 2$ V, $K = 0.25$ mA/V², and $r_o = \infty$. Find the dc voltage V_O and the small-signal resistance R_{in}.

7.24 In Fig. P7.24, $V_t = 2$ V, $K = 0.25$ mA/V², and $r_o = 100$ kΩ. What is the dc voltage at the output? Use the small-signal equivalent circuit to determine the voltage gain v_o/v_i and the input resistance R_{in}.

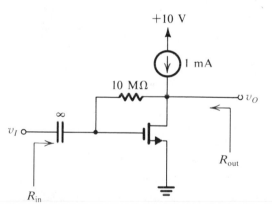

Fig. P7.24

*7.25 For the cascode amplifier in Fig. 7.35 let $I = 0.1$ mA, $V_{DD} = 5$ V, $V_t = 1$ V, $K = 0.4$ mA/V², $r_o = \infty$, $R_{G1} = 2.5$ MΩ, $R_{G2} = 1.5$ MΩ, $R_{G3} = 1$ MΩ, $R_D = 25$ kΩ, and $R_L = \infty$. Find the dc voltages V_{G1}, V_{S1}, V_{D1}, V_{G2}, and V_{D2}. Also find the voltage gains v_{o1}/v_i and v_o/v_i. What is the value of the input resistance?

**7.26 For the circuits in Fig. P7.26 calculate the drain current and drain-to-source voltage. For all devices $K = 0.5$ mA/V² and $|V_t| = 1$ V. The effect of r_o is to be neglected.

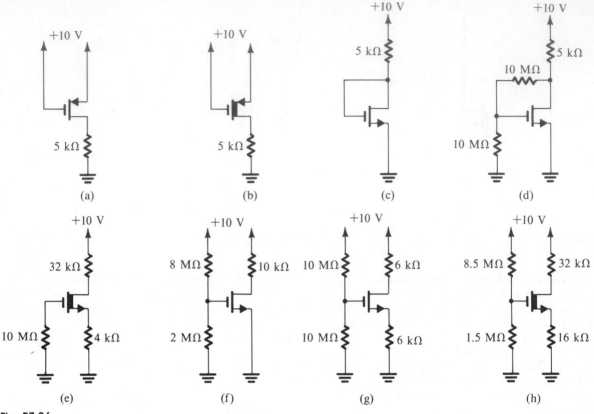

Fig. P7.26

*7.27 For the circuits in Fig. P7.27 calculate the labeled currents and voltages. For all devices $K = 0.5$ mA/V^2 and $|V_t| = 1$ V.

*7.28 The MOSFET in the circuit shown in Fig. P7.28 has $V_t = 1$ V. Find V_O and R_{out} for **(a)** $K = 0.05$ mA/V^2, **(b)** $K = 0.5$ mA/V^2, and **(c)** $K = 5$ mA/V^2.

*7.29 The MOSFET in the circuit shown in Fig. P7.29 has $V_t = 1$ V. Find V_O and R_{out} for **(a)** $K = 0.05$ mA/V^2, **(b)** $K = 0.5$ mA/V^2, and **(c)** $K = 5$ mA/V^2.

7.30 The MOSFETs in the circuit of Fig. P7.30 have $|V_t| = 1$ V and $K = 0.05$ mA/V^2. Find V_O and I for $V = 5$, 10, and 15 volts.

7.31 The enhancement-MOS load of an enhancement-MOS amplifier has a channel width 1/10 and a channel length 10 times that of the amplifier transistor. Ignoring the effects of r_o and the body effect, find the expected voltage gain.

*7.32 Reconsider the amplifier in Example 7.3. Draw its small-signal equivalent circuit including the finite output resistance r_o of each of the two transistors. Assuming that $V_A = 50$ V, evaluate the voltage gain.

7.33 **(a)** In the MOS voltage divider shown in Fig. P7.33, all devices are identical, with $K = 5$ μA/V^2 and $V_t = 2$ V. Find V_1, V_2, and I.

(b) If Q_2 is replaced by a device with $K_2 = 1$ μA/V^2 what are the values of V_1, V_2, and I.

7.34. For the circuit in Fig. P7.34 let Q_1 and Q_3 have $|V_t| = 2$ V and $K = 0.5$ mA/V^2, and let Q_2 have $I_{DSS} = 1$ mA and $V_t = -2$ V. Find V_1, V_2, and I.

7.35 The driver transistor in an enhancement-load amplifier has a W/L ratio of 9. Find the W/L ratio of the load to obtain a small-signal voltage gain of -10. Ignore the effect of r_o and assume that $\chi = 0.2$.

7.36 A depletion-load MOSFET amplifier has $K_1/K_2 = 4$. Find the small-signal voltage gain for $\chi = 0.2$.

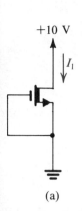

(a)

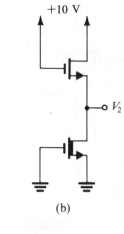

(b)

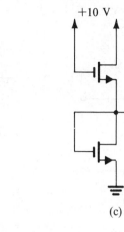

(c)

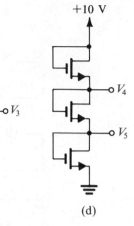

(d)

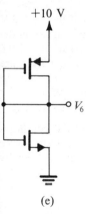

(e)

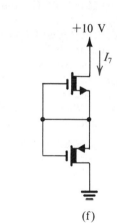

(f)

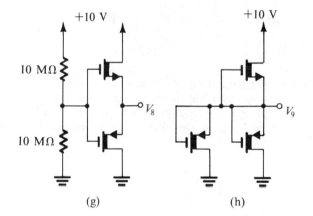

(g)

(h)

Fig. P7.27

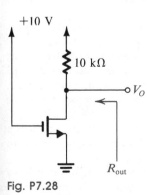

Fig. P7.28

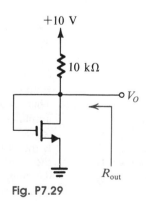

Fig. P7.29

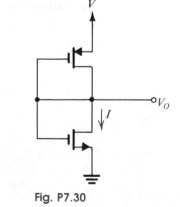

Fig. P7.30

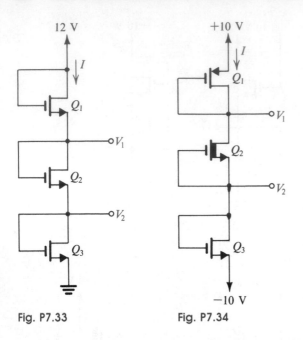

Fig. P7.33

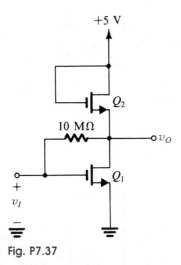

Fig. P7.34

7.37 For the circuit in Fig. P7.37 let $V_t = 2$ V. For each of the following cases find the dc voltage at the output and the small-signal voltage gain: **(a)** $K_2 = K_1$, **(b)** $K_2 = 0.1K_1$ and **(c)** $K_2 = 0.01K_1$. Neglect the effect of r_o and the body effect.

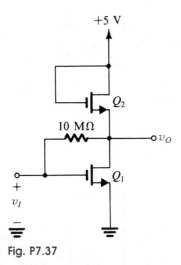

Fig. P7.37

7.38 For the circuit in the Fig. P7.38 let $|V_t| = 2$ V. For each of the cases **(a) $K_2 = K_1$, **(b)** $K_2 = 0.1K_1$, and **(c)** $K_2 = 0.01K_1$, find v_O corresponding to $v_I = 0$ V, 3 V, and 6 V.

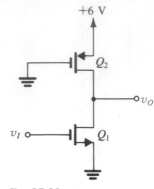

Fig. P7.38

**7.39 (a) The MOSFET in Fig. P7.39a has $V_t = 2$ V and $K = 0.025$ mA/V^2. Use the FET small-signal model to find R_{in}.
    (b) The MOSFET in Fig. P7.39b has $V_t = 2$ V and $K = 0.025$ mA/V^2. Use the FET small-signal model to find R_{in} in the two cases: (a) $r_o = \infty$ and (b) $r_o = \frac{1}{3}$ MΩ.
    (c) The FETs in the circuit of Fig. P7.39c are matched, with $g_m = 5$ mA/V. The gain v_o/v_i was found to be -25 V/V. What is r_o for each device? What is the output resistance R_{out}? What is the input resistance R_{in}?

7.40 (a) Figure P7.40a shows a MOSFET common-source amplifier biased by a constant drain current source I. Show that the small-signal voltage gain is given by

$$\frac{v_o}{v_i} = -g_m r_o$$

    (b) Figure P7.40b shows a cascade of two common-source amplifiers, each biased by a constant drain-current source I. Show that the small-signal voltage gain is given by

$$\frac{v_o}{v_i} = (g_{m1}r_{o1})(g_{m2}r_{o2})$$

*7.41 Figure P7.41 shows an IC MOS cascode amplifier. Transistor Q_1 operates in the common-source configuration and Q_2, the cascode transistor, operates in the common-gate configuration; V_{BIAS} is a dc bias voltage. Both devices are identical and operate at a dc current I. This amplifier can be modeled as a transconductance amplifier.
(a) Show that the short-circuit transconductance, which is the ratio of the short-circuit output signal current to the input voltage, is equal to g_{m1}.
(b) Short circuit v_i and use equivalent circuit models for Q_1 and Q_2 to show that the output resistance is given

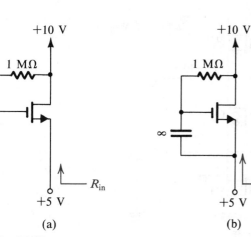

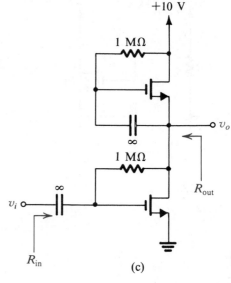

Fig. P7.39

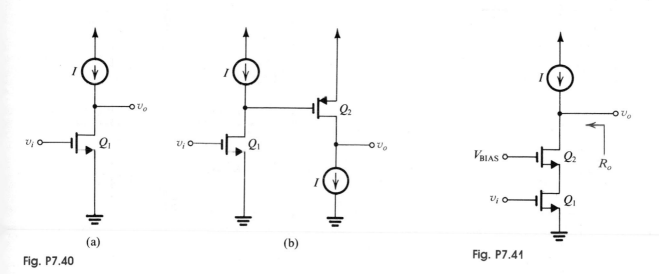

Fig. P7.40

Fig. P7.41

by

$$R_o = r_{o1} + r_{o2} + g_{m2}r_{o2}r_{o1} \simeq g_{m2}r_{o2}r_{o1}$$

(c) Use the results of (a) and (b) to obtain the open-circuit voltage gain v_o/v_i.

7.42 For the MOSFETs in the circuit of Fig. P7.42, $K = 5 \ \mu A/V^2$ and $V_t = 2V$. Select R to obtain $V_O = 6$ V. Neglect the effect of r_o.

7.43 For the MOSFETs in the circuit of Fig. P7.43, $K = 50 \ \mu A/V^2$ and $V_t = 2$ V. Find V_O. Neglect the effect of r_o.

7.44 The MOSFETs in the two-output current mirror of Fig. P7.44 have equal V_t and equal channel lengths. Transistor Q_1 has $W = 10 \ \mu m$. Neglecting the effects of r_o find the channel width of Q_2 and of Q_3 so that $I_2 = 20 \ \mu A$ and $I_3 = 100 \ \mu A$.

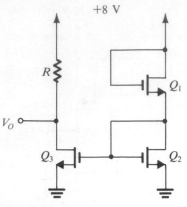

+8 V

Fig. P7.42

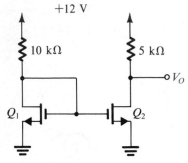

Fig. P7.43

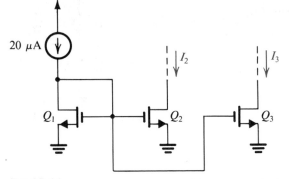

Fig. P7.44

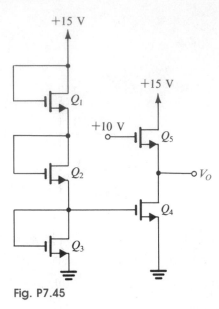

Fig. P7.45

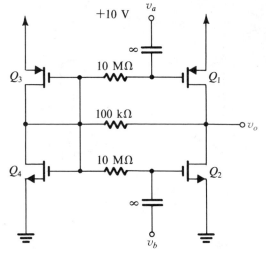

Fig. P7.46

7.45 In the circuit of Fig. P7.45 all devices are matched. Find the value of V_O.

***7.46 All the devices in the circuit of Fig. P7.46 are matched, with $K = 25 \ \mu\text{A/V}^2$ and $V_t = 2$ V. The small signal voltages applied are: $v_a = 10 \sin \omega t$ mV and $v_b =$ 10 $\sin(\omega t + \phi)$ mV. Find v_o for **(a)** $\phi = 0°$, **(b)** $\phi = 90°$, and **(c)** $\phi = 180°$.

***7.47 The MOSFETs in the circuit of Fig. P7.47 are matched, with $K = 25 \ \mu\text{A/V}^2$ and $|V_t| = 2$ V. The resistance $R_2 = 10$ MΩ. For G and D open, what are the drain currents I_{D1} and I_{D2}? For $r_o = \infty$, what is the voltage gain of the amplifier from G to D? For finite r_o ($r_o = |V_A|/I_D$, $|V_A| = 180$ V), what is the voltage gain from G to D and the input resistance at G? If G is driven

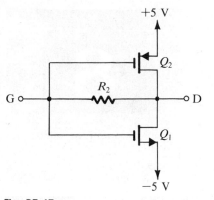

G o———/\/\/\———o D

R_2

Fig. P7.47

from a source v_i having a resistance of 1 MΩ, find the voltage gain v_d/v_i. For what range of output signals do Q_1 and Q_2 remain in the pinch-off region?

7.48 Repeat Exercise 7.20 for the case $I = 0.1$ mA. Also find the voltage gain when the source follower is loaded by a 10-kΩ resistor.

7.49 An NMOS switch has a signal input that varies over the range ± 5 V and a control input of ± 10 V. When the control input is high, turning the switch on, calculate the channel resistance at the two signal extremes. If the switch is to be used in a circuit whose loop resistance is R, how large must R be to ensure at most a 1% signal loss in the switch? For the transistor, $K = 25\ \mu A/V^2$ and $V_t = 2$ V.

7.50 A CMOS transmission gate for which $K = 25\ \mu A/V^2$ and $|V_t| = 2$ V utilizes control signals of ± 5 V for signals over the range ± 5 V. Calculate the switch resistance at the signal extremes and for signals at 0 V. If the switch is to be used in a circuit whose loop resistance is R, how large must R be to ensure at most a 1% signal loss in the switch?

7.51 A CMOS switch is used to connect a sinusoidal source $0.1 \sin \omega t$ to a load capacitance C. For ± 5-V control signals and $K_p = K_n = 25\ \mu A/V^2$, $V_{tn} = |V_{tp}| = 2$ V, what is the cut-off frequency introduced by the switch if $C = 1,000$ pF.

Chapter

Bipolar Junction Transistors (BJTs)

INTRODUCTION

In this chapter we shall study the characteristics and basic applications of another three-terminal solid-state device, the bipolar junction transistor (BJT). The BJT, which is often referred to simply as "the transistor," is widely used in discrete circuits as well as in integrated circuits (ICs), both analog and digital. The device characteristics are so well understood that one is able easily to design transistor circuits whose performance is remarkably predictable and quite insensitive to variations in device parameters.

We shall start by presenting a simple qualitative description of the operation of the transistor. Though simple, this physical description provides considerable insight into the performance of the transistor as a circuit element. We will quickly move from describing current flow in terms of holes and electrons to a study of transistor terminal characteristics. First-order models for transistor operation in different modes will be developed and utilized in the analysis of transistor circuits. One of the main objectives of this chapter is to develop in the reader a high degree of familiarity with the transistor. Thus by the end of the chapter the reader should be able to perform rapid first-order analysis of transistor circuits, as well as design single-stage transistor amplifiers. ☐

8.1 PHYSICAL STRUCTURE AND MODES OF OPERATION

Figure 8.1 shows a simplified structure for a BJT. A practical transistor structure will be shown later (see also Appendix A, which deals with fabrication technology).

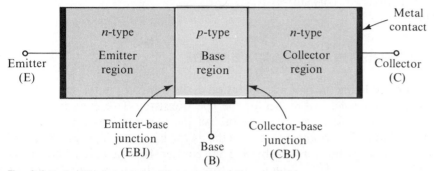

Fig. 8.1 A simplified structure of the *npn* transistor.

As shown in Fig. 8.1, the BJT consists of three semiconductor regions: the emitter region (*n* type), the base region (*p* type), and the collector region (*n* type). Such a transistor is called an *npn* transistor. Another transistor, a dual of the *npn* as shown in Fig. 8.2, has a *p*-type emitter, an *n*-type base, and a *p*-type collector and is appropriately called a *pnp* transistor.

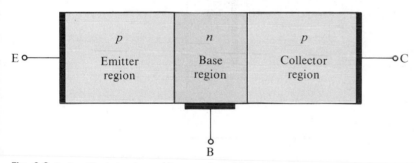

Fig. 8.2 A simplified structure of the *pnp* transistor.

A terminal is connected to each of the three semiconductor regions of a transistor, with the terminals labeled *emitter* (E), *base* (B), and *collector* (C).

The transistor consists of two *pn* junctions, the emitter–base junction (EBJ) and the collector–base junction (CBJ). Depending on the bias condition (forward or reverse) of each of these junctions, different modes of operation of the BJT are obtained, as shown in Table 8.1.

The active mode is the one used if the transistor is to operate as an amplifier. Switching applications (for example, logic circuits) utilize both the cutoff and the saturation modes.

Table 8.1 BJT MODES OF OPERATION

Mode	EBJ	CBJ
Cutoff	Reverse	Reverse
Active	Forward	Reverse
Saturation	Forward	Forward

8.2 OPERATION OF THE *npn* TRANSISTOR IN THE ACTIVE MODE

Let us start by considering the physical operation of the transistor in the active mode. This situation is illustrated in Fig. 8.3 for the *npn* transistor. Two external voltage sources (shown as batteries) are used to establish the required bias conditions for active mode operation. The voltage V_{BE} causes the *p*-type base to be higher in potential than the *n*-type emitter, thus forward-biasing the emitter–base junction. The collector–base voltage V_{CB} causes the *n*-type collector to be higher in potential than the *p*-type base, thus reverse-biasing the collector–base junction.

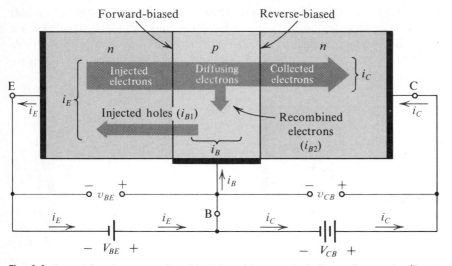

Fig. 8.3 Current flow in an *npn* transistor biased to operate in the active mode. (Reverse current components due to drift of thermally generated minority carriers are not shown.

Current Flow

In the following description of current flow only diffusion-current components are considered. Drift currents due to thermally generated minority carriers are usually very small and can be neglected. We will have more to say about these reverse-current components at a later stage.

The forward bias on the emitter–base junction will cause current to flow across this junction. Current will consist of two components: electrons injected from the emitter into the base, and holes injected from the base into the emitter. As will become apparent shortly, it is highly desirable to have the first component (electrons from emitter to base) at a much higher level than the second component (holes from base to emitter). This can be accomplished by fabricating the device with a heavily doped emitter and a lightly doped base; that is, the device is designed to have a high density of electrons in the emitter and a low density of holes in the base.

The current that flows across the emitter–base junction will constitute the emitter current i_E, as indicated in Fig. 8.3. The direction of i_E is "out of" the emitter lead, which is in the direction of the hole current and opposite to the direction of the electron current, with the emitter current i_E being equal to the sum of these two components. However, since the electron component is much larger than the hole component, the emitter current will be dominated by the electron component.

Let us now consider the electrons injected from the emitter into the base. These electrons will be *minority carriers* in the p-type base region. Because the base is usually very thin, in the steady state the excess minority carrier (electron) concentration in the base will have an almost straight-line profile as indicated by the solid straight line in Fig. 8.4. The concentration will be highest [denoted by $n_b(0)$] at the emitter side and lowest (zero) at the collector side.[1] As in the case of any forward-biased *pn* junction (Chapter 4), the concentration $n_b(0)$ will be proportional to e^{v_{BE}/V_T}, where v_{BE} is the forward-bias voltage and V_T is the thermal voltage, which is equal to approximately 25 mV at room temperature. The reason for the zero concentration at the collector side of the base is that the positive collector voltage v_{CB} causes the electrons at that end to be swept across the CBJ depletion region.

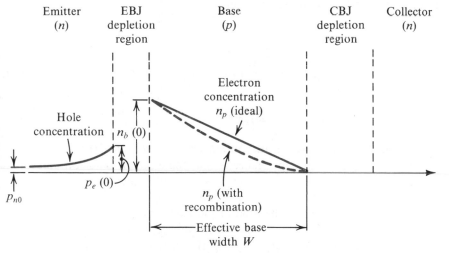

Fig. 8.4 Profiles of minority carrier concentrations in the base and emitter of an *npn* transistor operating in the active mode; $v_{BE} > 0$ and $v_{CB} \geq 0$.

---

[1] This minority carrier distribution in the base results from the boundary conditions imposed by the two junctions. It is not a "natural" diffusion-based distribution, which would result if the base region were infinitely thick.

The tapered minority carrier concentration profile (Fig. 8.4) causes the electrons injected into the base to diffuse through the base region toward the collector. This diffusion current is directly proportional to the slope of the straight-line concentration profile. Thus the diffusion current will be proportional to the concentration n_b (0) and inversely proportional to the base width W.

Some of the electrons that are diffusing through the base region will combine with holes, which are the majority carriers in the base. However, since the base is usually very thin, the percentage of electrons "lost" through this recombination process will be quite small. Nevertheless, the recombination in the base region causes the excess minority carrier concentration profile to deviate from a straight line and take the slightly concave shape indicated by the broken line in Fig 8.4. The slope of the concentration profile at the EBJ is slightly higher than that at the CBJ, with the difference accounting for the small number of electrons lost in the base region through recombination.

From the above we see that most of the diffusing electrons will reach the boundary of the collector–base depletion region. Because the collector is more positive than the base (by v_{CB} volts), these successful electrons will be swept across the CBJ depletion region into the collector. They will thus get "collected" to constitute the collector current i_C. By convention the direction of i_C will be opposite to that of electron flow; thus i_C will flow *into* the collector terminal.

Another important observation to make here is that the magnitude of i_C is independent of v_{CB}. That is, as long as the collector is positive with respect to the base, the electrons that reach the collector side of the base region will be swept into the collector and register as collector current.

The Collector Current

From the above discussion we can see that the collector current i_C may be expressed as

$$i_C = I_S e^{v_{BE}/V_T} \tag{8.1}$$

where I_S is a constant called the saturation current and V_T is the thermal voltage. The reason for the exponential dependence is that the electron diffusion current is proportional to the minority carrier concentration n_b (0), which in turn is proportional to e^{v_{BE}/V_T}. The saturation current I_S is inversely proportional to the base width W and is directly proportional to the area of the EBJ. Typically I_S is in the range of 10^{-12} to 10^{-15} A (depending on the size of the device) and is a function of temperature, approximately doubling for every 5°C rise in temperature.

Because I_S is directly proportional to the junction area (that is, the device size) it will also be referred to as the *current scale factor*. Two transistors that are identical except that one has an EBJ area, say, twice that of the other will have saturation currents with that same ratio (2). Thus for the same value of v_{BE} the larger device will have a collector current twice that in the smaller device. This concept is frequently employed in integrated-circuit design.

The Base Current

The base current i_B is composed of two components. The first and dominant component i_{B1} is due to the holes injected from the base region into the emitter region. This current component is proportional to e^{v_{BE}/V_T}. The second component of base current, i_{B2}, is due to holes that have to be supplied by the external circuit in order to replace the holes lost from the base through the recombination process. The number of electrons (and hence the number of holes) taking part in the recombination process is proportional to the concentration $n_b(0)$ and to the base width W. Thus i_{B2} will be proportional to e^{v_{BE}/V_T} and to W.

From the above discussion we conclude that the total base current $i_B(= i_{B1} + i_{B2})$ will be proportional to $e^{(v_{BE}/V_T)}$. We may therefore express i_B as a fraction of i_C, as follows:

$$i_B = \frac{i_C}{\beta} \tag{8.2}$$

Thus

$$i_B = \frac{I_S}{\beta} e^{v_{BE}/V_T} \tag{8.3}$$

where β is a constant for the particular transistor. For modern *npn* transistors β is in the range 100 to 200, but it can be as high as 1,000 for special devices. For reasons that will become clear later, the constant β is called the *common-emitter current gain*.

As may be seen from the above, the value of β is highly influenced by two factors: the width of the base region and the relative dopings of the emitter region and the base region. To obtain a high β (which is highly desirable since β represents a gain parameter) the base should be thin and lightly doped and the emitter heavily doped. The discussion thus far assumes an idealized situation, where β is independent of the current level in the device.

The Emitter Current

Since the current that enters a transistor should leave it, it can be seen from Fig. 8.3 that the emitter current i_E is equal to the sum of the collector current i_C and the base current i_B,

$$i_E = i_C + i_B \tag{8.4}$$

Use of Eqs. (8.2) and (8.4) gives

$$i_E = \frac{\beta + 1}{\beta} i_C \tag{8.5}$$

that is,

$$i_E = \frac{\beta + 1}{\beta} I_S e^{v_{BE}/V_T} \tag{8.6}$$

Alternatively, we can express Eq. (8.5) in the form

$$i_C = \alpha i_E \tag{8.7}$$

where the constant α is related to β by

$$\alpha = \frac{\beta}{\beta + 1} \tag{8.8}$$

Thus the emitter current in Eq. (8.6) can be written

$$i_E = (I_S/\alpha)e^{v_{BE}/V_T} \tag{8.9}$$

Finally, we can use Eq. (8.8) to express β in terms of α; that is,

$$\beta = \frac{\alpha}{1 - \alpha} \tag{8.10}$$

It can be seen from Eq. (8.8) that α is a constant (for the particular transistor) less than but very close to unity. For instance, if $\beta = 100$, then $\alpha \simeq 0.99$. Equation (8.10) reveals an important fact: small changes in α correspond to very large changes in β. This mathematical observation manifests itself physically, with the result that transistors of the same type may have widely different values of β. For reasons that will become apparent later, α is called the *common-base current gain*.

Recapitulation

We have presented a first-order model for the operation of the *npn* transistor in the active mode. Basically, the forward-bias voltage v_{BE} causes an exponentially related current i_C to flow in the collector terminal. The collector current i_C is independent of the value of the collector voltage as long as the collector–base junction remains reverse-biased; that is, $v_{CB} \geq 0$. Thus in the active mode the collector terminal behaves as an ideal constant-current source where the value of the current is determined by v_{BE}. The base current i_B is a factor $1/\beta$ of the collector current, and the emitter current is equal to the sum of the collector and base currents. Since i_B is much smaller than i_C (that is, $\beta \gg 1$), $i_E \simeq i_C$. More precisely, the collector current is a fraction α of the emitter current, with α smaller than, but close to, unity.

Equivalent Circuit Models

The first-order model of transistor operation described above can be represented by the equivalent circuit shown in Fig. 8.5a. Here diode D_E has a current scale factor equal to (I_S/α) and thus provides a current i_E related to v_{BE} according to Eq. (8.9). The current of the controlled source, which is equal to the collector current, is controlled by v_{BE} according to the exponential relationship indicated, a restatement of Eq. (8.1). This model is in essence a nonlinear voltage-controlled current source. It can be converted to the current-controlled current-source model shown in Fig. 8.5b by expressing the current of the controlled source as αi_E. Note that this model is also nonlinear because of the exponential relationship of the current i_E through diode D_E and the voltage v_{BE}. From this model we observe that if the transistor is used as a

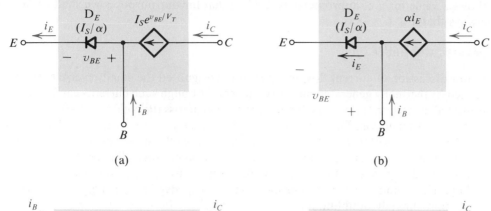

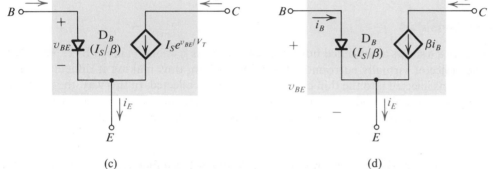

Fig. 8.5 Large-signal equivalent circuit models of the *npn* BJT operating in the active mode.

two-port network with the input port between E and B and the output port between C and B (that is, with B as a common terminal), then the current gain observed is equal to α. Thus α is called the common-base current gain.

Two other equivalent circuit models, shown in Fig. 8.5c and d, may be used to represent the large-signal operation of the BJT. The model of Fig. 8.5c is essentially a voltage-controlled current source. However, here diode D_B conducts the base current and thus its current scale factor is I_S/β, resulting in the i_B-v_{BE} relationship given in Eq. (8.3). By simply expressing the collector current as βi_B we obtain the current-controlled current-source model shown in Fig. 8.5d. From this latter model we observe that if the transistor is used as a two-port network with the input port between B and E and the output port between C and E (that is, with E as the common terminal), then the current gain observed is equal to β. Thus β is called the common-emitter current gain.

The Constant *n*

In the diode equation (Chapter 4) we used a constant n in the exponential and mentioned that its value is between 1 and 2. For modern bipolar junction transistors the constant n is close to unity except in special cases: (1) at high currents (that is, high relative to the normal current range of the particular transistor) the i_C-v_{BE} relationship exhibits a value for n that is close to 2, and (2) at low currents, the i_B-v_{BE} relationship shows a value for n of approximately 2. Note that for our purposes we shall assume always that $n = 1$.

The Collector–Base Reverse Current (I_{CBO})

In our discussion of current flow in transistors we ignored the small reverse currents carried by thermally generated minority carriers. Although such currents can be safely neglected in modern transistors, the reverse current across the collector–base junction deserves some mention. This current, denoted I_{CBO}, is the reverse current flowing from collector to base with the emitter open-circuited (hence the subscript O). This current is usually in the nanoampere range, a value that is many times higher than its theoretically predicted value. As with the diode reverse current, I_{CBO} stems mainly from leakage effects, and its value is dependent on v_{CB}. I_{CBO} depends strongly on temperature, approximately doubling for every 10°C rise.[2]

The Structure of Actual Transistors

Figure 8.6 shows a simplified but more realistic cross section of an *npn* BJT. Note that the collector virtually surrounds the emitter region, thus making it difficult for the electrons injected into the thin base to escape being collected. In this way, the resulting α is close to unity and β is large. Also, observe that the device is not symmetrical. For more details on the physical structure of actual devices the reader is referred to Appendix A.

Exercises

8.1 Consider an *npn* transistor with $v_{BE} = 0.7$ V at $i_C = 1$ mA. Find v_{BE} at $i_C = 0.1$ mA and 10 mA.

Ans. 0.64 V; 0.76 V

8.2 Transistors of a certain type are specified to have β values in the range 50 to 150. Find the range of their α values.

Ans. 0.980 to 0.993

8.3 Measurement of an *npn* BJT in a particular circuit shows the base current to be 14.46 μA, the emitter current to be 1.460 mA, and the base–emitter voltage to be 0.7 V. For these conditions calculate α, β, and I_S.

Ans   0.99; 100; 10^{-15} A

8.4 Calculate β for two transistors for which $\alpha = 0.99$ and 0.98. For collector currents of 10 mA, find the base current of each transistor.

Ans. 99; 49; 0.1 mA; 0.05 mA

8.5 Consider the BJT model shown in Fig. 8.5d. Find the value of the scale current of D_B given that $I_S = 10^{-14}$ A and $\beta = 100$. If this transistor is operated in the common-emitter configuration, with the base fed with a constant-current source supplying a 10-μA current and with the collector connected to a $+10$-V dc supply (E is at ground), find V_{BE} and I_C.

Ans. 10^{-16} A; 0.633 V; 1 mA

---

[2] The temperature coefficient of I_{CBO} is different from that of I_S because I_{CBO} is mostly a leakage current.

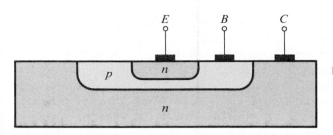

Fig. 8.6 Cross section of an *npn* BJT.

8.3 THE *pnp* TRANSISTOR

The *pnp* transistor operates in a manner similar to that of the *npn* device described in Section 8.2. Figure 8.7 shows a *pnp* transistor biased to operate in the active mode. Here the voltage V_{EB} causes the *p*-type emitter to be higher in potential than the *n*-type base, thus forward-biasing the base–emitter junction. The collector–base junction is reverse-biased by the voltage V_{BC}, which keeps the *n*-type base higher in potential than the *p*-type collector.

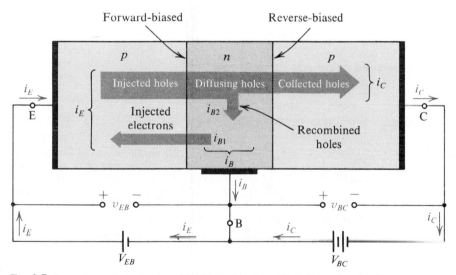

Fig. 8.7 Current flow in a *pnp* transistor biased to operate in the active mode.

Unlike the *npn* transistor, current in the *pnp* device is mainly conducted by holes injected from the emitter into the base as a result of the forward-bias voltage V_{EB}. Since the component of emitter current contributed by electrons injected from base to emitter is kept small by using a lightly doped base, most of the emitter current will be due to holes. The electrons injected from base to emitter give rise to the dominant component of base current, i_{B1}. Also, a number of the holes injected into the base will recombine with the majority carriers in the base (electrons) and will thus be lost. The disappearing base electrons will have to be replaced from the external circuit, giving rise to the second component of base current, i_{B2}. The holes that succeed in reaching the boundary of the depletion region of the collector–base junction will be attracted by the negative voltage on the collector. Thus these holes will be swept across the depletion region into the collector and appear as collector current.

It can easily be seen from the above description that the current–voltage relationships of the *pnp* transistor will be identical to those of the *npn* transistor except that v_{BE} has to be replaced by v_{EB}. Also, the large-signal operation of the *pnp* transistor can be modeled by any one of four possible equivalent circuit models that parallel those given for the *npn* transistor in Fig. 8.5. For illustration, two of the four *pnp* models are depicted in Fig. 8.8.

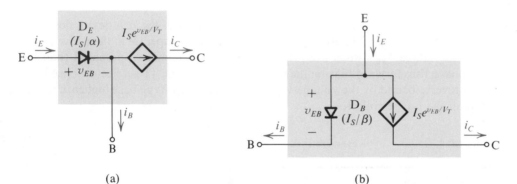

(a)   (b)

Fig. 8.8 Two large-signal models for the *pnp* transistor operating in the active mode.

Exercises

8.6 Consider the model in Fig. 8.8a applied in the case of a *pnp* transistor whose base is grounded, the emitter is fed by a constant-current source that supplies a 2-mA current into the emitter terminal, the collector is connected to a -10-V dc supply. Find the emitter voltage, the base current, and the collector current if for this transistor $\beta = 50$ and $I_S = 10^{-14}$ A.

Ans. 0.650 V; 39.2 μA; 1.96 mA

8.7 For a *pnp* transistor having $I_S = 10^{-11}$ A and $\beta = 100$, calculate v_{BE} for $i_C = 1.5$ A.

Ans. 0.643 V

8.4 CIRCUIT SYMBOLS AND CONVENTIONS

The physical structure used thus far to explain transistor operation is rather cumbersome to employ in drawing the schematic of a multitransistor circuit. Fortunately, a very descriptive and convenient circuit symbol exists for the BJT. Figure 8.9a shows the symbol for the *npn* transistor while the *pnp* symbol is given in Fig. 8.9b. In both symbols the emitter is distinguished by an arrowhead. This distinction is important because practical BJTs are not symmetric devices. That is, interchanging the emitter

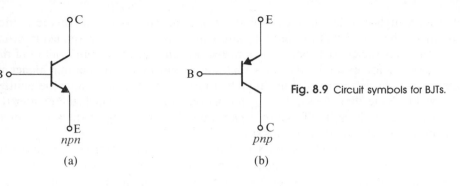

Fig. 8.9 Circuit symbols for BJTs.

and collector will result in a different, and much lower, value for α, a value called the *inverse or reverse α*.[3]

   The polarity of the device—*npn* or *pnp*—is indicated by the direction of the arrowhead on the emitter. This arrowhead points in the direction of normal current flow in the emitter. Since we have adopted a drawing convention by which currents flow from top to bottom, we will always draw *pnp* transistors in the manner shown in Fig. 8.9 (that is, with their emitters on top).

   Figure 8.10 shows *npn* and *pnp* transistors biased to operate in the active mode. It should be mentioned in passing that the biasing arrangement shown, utilizing two dc sources, is not a usual one and is used merely to illustrate operation. Practical biasing schemes will be presented in Section 8.8. Figure 8.10 also indicates the reference and actual directions of current flow throughout the transistor. Our convention will be to take the reference direction to coincide with the normal direction of current flow. Hence, normally, we should not encounter a negative value for i_E, i_B, or i_C.

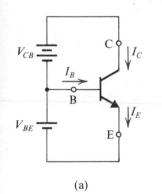
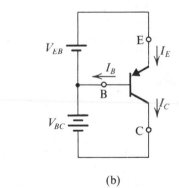

Fig. 8.10 Voltage polarities and current flow in transistors biased in the active mode.

   The convenience of the circuit drawing convention that we have adopted should be obvious from Fig. 8.10. Note that currents flow from top to bottom and that

---

[3] The inverse mode of operation of the BJT will be studied in Chapter 16.

voltages are higher at the top and lower at the bottom. The arrowhead on the emitter also implies the polarity of the emitter–base voltage that should be applied in order to forward-bias the emitter–base junction. Just a glance at the circuit symbol of the *pnp* transistor, for example, indicates that we should make the emitter higher in voltage than the base (by v_{EB}) in order to cause current to flow into the emitter (downward). Note that the symbol v_{EB} means the voltage by which the emitter (E) is higher than the base (B). Thus for a *pnp* transistor operating in the active mode v_{EB} is positive, while in an *npn* transistor v_{BE} is positive.

From the discussion of Section 8.3 it follows that an *npn* transistor whose EBJ is forward-biased will operate in the active mode *as long as the collector is higher in potential than the base.* Active-mode operation will be maintained even if the collector voltage falls to equal that of the base, since a silicon *pn* junction is essentially non-conducting when the voltage across it is zero. The collector voltage should not be allowed, however, to fall below that of the base if active-mode operation is required. If it does fall below the base voltage, the collector–base junction could become forward-biased, with the transistor entering a new mode of operation, saturation. We will discuss the saturation mode later.

In a parallel manner, the *pnp* transistor will operate in the active mode *if the potential of the collector is lower than (or equal to) that of the base.* The collector voltage should not be allowed to rise above that of the base if active-mode operation is to be maintained.

Exercises

8.8 In the circuit shown in Fig. E8.8 the voltage at the emitter was measured and found to be -0.7 V. If $\beta = 50$, find I_E, I_B, I_C, and V_C.

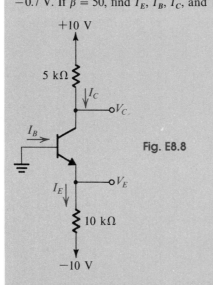

Fig. E8.8

Ans. 0.93 mA; 18.2 μA; 0.91 mA; $+5.44$ V

8.9 In the circuit shown in Fig. E8.9, measurement indicates V_B to be $+1.0$ V and V_E to be $+1.7$ V. What are α and β for this transistor? What voltage V_C do you expect at the collector?

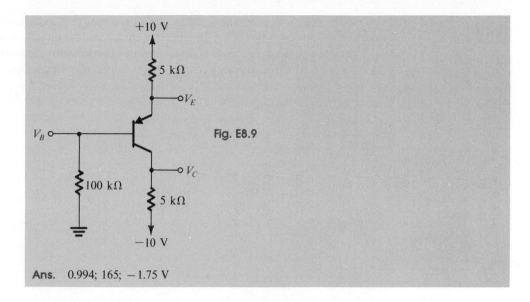

Fig. E8.9

Ans.   0.994; 165; -1.75 V

8.5 GRAPHICAL REPRESENTATION OF TRANSISTOR CHARACTERISTICS

It is sometimes useful to describe the transistor i-v characteristics graphically. Figure 8.11 shows the i_C-v_{BE} characteristic, which is the exponential relationship

$$i_C = I_S e^{v_{BE}/V_T}$$

which is identical (except for the value of constant n) to the diode i-v relationship. The i_E-v_{BE} and i_B-v_{BE} characteristics appear quite similar to that of the i_C-v_{BE} curve of Fig. 8.11. Since the constant of the exponential characteristic, $1/V_T$, is quite high ($\simeq 40$), the curve rises very sharply. For v_{BE} smaller than about 0.5 V the current is negligibly small. Also, over most of the normal current range v_{BE} lies in the range 0.6 to 0.8 V. In performing rapid first-order dc calculations we normally will assume that $V_{BE} \simeq 0.7$ V, which is similar to the approach used in the analysis of diode circuits (Chapter 4). For a *pnp* transistor the i_C-v_{EB} characteristic will look identical to that of Fig. 8.11.

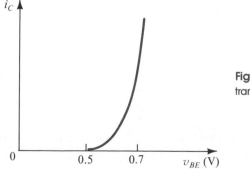

Fig. 8.11 The i_C-v_{BE} characteristic for an *npn* transistor.

As in silicon diodes, the voltage across the emitter–base junction decreases by about 2 mV for each rise of 1°C in temperature, provided that the junction is operating at a constant current. Figure 8.12 illustrates this temperature dependence by depicting i_C-v_{BE} curves at three different temperatures for an *npn* transistor.

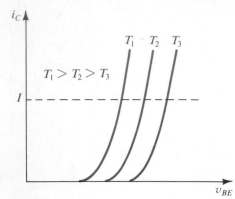

Fig. 8.12 Effect of temperature on the i_C-v_{BE} characteristic. At a constant emitter current (broken line) v_{BE} changes by -2 mV/°C.

Figure 8.13b shows the i_C versus v_{CB} characteristics of an *npn* transistor for various values of the emitter current i_E. These characteristics can be measured using the circuit shown in Fig. 8.13a. Only active-mode operation is shown, since only the portion of the characteristics for $v_{CB} \geq 0$ is drawn. As can be seen, the curves are horizontal straight lines, corroborating the fact that the collector behaves as a constant-current source. In this case the value of the collector current is controlled by that of the emitter current ($i_C = \alpha i_E$), and the transistor may be thought of as a current-controlled current source (see the model of Fig. 8.5b).

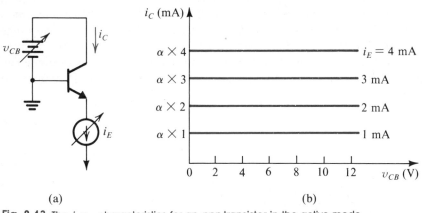

Fig. 8.13 The i_C-v_{CB} characteristics for an *npn* transistor in the active mode.

When operated in the active region, practical BJTs show some dependence of the collector current on the collector voltage, with the result that their i_C-v_{CB} characteristics are not perfectly horizontal straight lines. To see this dependence more clearly,

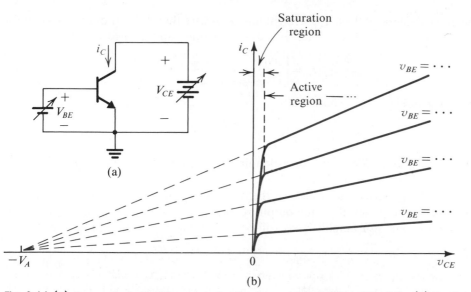

Fig. 8.14 **(a)** Conceptual circuit for measuring the i_C-v_{CE} characteristics of the BJT.   **(b)** The i_C-v_{CE} characteristics of a practical BJT.

consider the conceptual circuit shown in Fig. 8.14a. The transistor is connected in the common-emitter configuration and its V_{BE} can be set to any desired value by adjusting the dc source connected between base and emitter. At each value of V_{BE}, the corresponding i_C-v_{CE} characteristic curve can be measured point-by-point by varying the dc source connected between collector and emitter and measuring the corresponding collector current. The result is the family of i_C-v_{CE} characteristic curves shown in Fig. 8.14b.

At low values of v_{CE}, as the collector voltage drops below that of the base, the collector–base junction becomes forward-biased and the transistor leaves the active mode and enters the saturation mode. We shall study the saturation mode of operation in a later section. At this time we wish to examine the characteristic curves in the active region in detail. We observe that the characteristic curves, though still straight lines, have finite slope. In fact, when extrapolated, the characteristic lines meet at a point on the negative v_{CE} axis, at $v_{CE} = -V_A$. The voltage V_A, a positive number, is a parameter for the particular BJT, with typical values in the range of 50–100 V. It is called the Early voltage, after the scientist who first studied this phenomenon.

At a given value of v_{BE}, increasing v_{CE} increases the reverse-bias voltage on the collector–base junction and thus increases the width of the depletion region of this junction (refer to Fig. 8.3). This in turn results in a decrease in the *effective base width* W. Recalling that I_S is inversely proportional to W, we see that I_S will increase and i_C increases proportionally. This is the Early effect.

The linear dependence of i_C on v_{CE} can be accounted for by assuming that I_S remains constant and including the factor $(1 + v_{CE}/V_A)$ in the equation for i_C as follows:

$$i_C = I_S e^{v_{BE}/V_T}\left(1 + \frac{v_{CE}}{V_A}\right) \tag{8.11}$$

The nonzero slope of the i_C-v_{CE} straight lines indicates that the output resistance looking into the collector is not infinite. Rather it is finite and defined by

$$r_o = \left[\frac{\partial i_C}{\partial v_{CE}} \bigg|_{v_{BE} = \text{constant}} \right]^{-1} \tag{8.12}$$

Using Eq. (8.11) we can show that

$$r_o \simeq \frac{V_A}{I_C} \tag{8.13}$$

where I_C is the current level corresponding to the constant value of v_{BE}, near the boundary of the active region.

It is rarely necessary to include the dependence of i_C on v_{CE} in dc bias design and analysis. However, the finite output resistance r_o can have a significant effect on the gain of transistor amplifiers, as will be seen in later sections and chapters.

Exercises

8.10 Consider a *pnp* transistor with $v_{EB} = 0.7$ V at $i_E = 1$ mA. Let the base be grounded, the emitter be fed by a 2-mA constant-current source and the collector be connected to a -5-V supply through a 1-kΩ resistance. If the temperature increases by 30°C, find the changes in emitter and collector voltages.

Ans. -60 mV; 0 V

8.11 Find the output resistance of a BJT for which $V_A = 100$ V, at $I_C = 0.1$, 1, and 10 mA.

Ans. 1 MΩ; 100 kΩ; 10 kΩ

8.12 Consider the circuit in Fig. 8.14a. At $V_{CE} = 1$ V, V_{BE} is adjusted to yield a collector current of 1 mA. Then, while V_{BE} is kept constant, V_{CE} is raised to 11 V. Find the new value of I_C. For this transistor $V_A = 100$ V.

Ans. 1.1 mA

8.6 DC ANALYSIS OF TRANSISTOR CIRCUITS

We are now ready to consider the analysis of some simple transistor circuits to which only dc voltages are applied. In the following examples we will use the simple constant-V_{BE} model, which is similar to that discussed in Section 4.8 for the junction diode. Specifically we will assume that $V_{BE} = 0.7$ V irrespective of the exact value of current. If it is desired, this approximation can be refined using techniques similar to those employed in the diode case. The emphasis here, however, is on the essence of transistor circuit analysis.

EXAMPLE 8.1

Consider the circuit shown in Fig. 8.15a, which is redrawn in Fig. 8.15b to remind the reader of the convention employed throughout this book for indicating connections to dc sources. We wish to analyze this circuit to determine all node voltages and branch currents. We will assume that β is specified to be 100.

Solution
We do not know initially whether the transistor is in the active mode or not. A simple approach would be to assume that the device is in the active mode, proceed with the

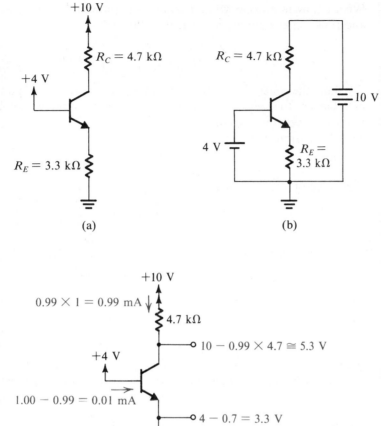

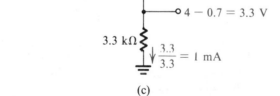

Fig. 8.15 Analysis of the circuit for Example 8.1.

solution, and finally check whether or not the transistor is in fact in the active mode. If we find that the conditions for active-mode operation are met, then our work is completed. Otherwise, the device is in another mode of operation and we have to solve the problem again. Obviously, at this stage we have learned only about the active mode of operation and thus will not be able to deal with circuits that we find are not in the active mode.

Glancing at the circuit in Fig. 8.15a, we note that the base is connected to +4 V and the emitter is connected to ground through a resistance R_E. It therefore is safe to conclude that the base–emitter junction will be forward-biased. Assuming that this is the case and assuming that V_{BE} is approximately 0.7 V, it follows that the emitter voltage will be

$$V_E = 4 - V_{BE} \simeq 4 - 0.7 = 3.3 \text{ V}$$

We are now in an opportune position; we know the voltages at the two ends of R_E and thus can determine the current I_E through it,

$$I_E = \frac{V_E - 0}{R_E} = \frac{3.3}{3.3} = 1 \text{ mA}$$

Since the collector is connected through R_C to the $+10$-V power supply, it appears possible that the collector voltage will be higher than the base voltage, which is essential for active-mode operation. Assuming that this is the case, we can evaluate the collector current from

$$I_C = \alpha I_E$$

The value of α is obtained from

$$\alpha = \frac{\beta}{\beta + 1} = \frac{100}{101} \simeq 0.99$$

Thus I_C will be given by

$$I_C = 0.99 \times 1 = 0.99 \text{ mA}$$

We are now in a position to use Ohm's law to determine the collector voltage V_C,

$$V_C = 10 - I_C R_C = 10 - 0.99 \times 4.7 \simeq +5.3 \text{ V}$$

Since the base is at $+4$ V, the collector–base junction is reverse-biased by 1.3 V, and the transistor is indeed in the active mode as assumed.

It remains only to determine the base current I_B, as follows:

$$I_B = \frac{I_E}{\beta + 1} = \frac{1}{101} \simeq 0.01 \text{ mA}$$

Before leaving this example we wish to emphasize strongly the value of carrying out the analysis directly on the circuit diagram. Only in this way will one be able to analyze complex circuits in a reasonable length of time. Figure 8.15c illustrates the above analysis on the circuit diagram.

EXAMPLE 8.2

We wish to analyze the circuit of Fig. 8.16a to determine the voltages at all nodes and the currents through all branches. Note that this circuit is identical to that of Fig. 8.15 except that the voltage at the base is now $+6$ V.

Solution

Assuming active-mode operation, we have

$$V_E = +6 - V_{BE} \simeq +6 - 0.7 = 5.3 \text{ V}$$

$$I_E = \frac{5.3}{3.3} = 1.6 \text{ mA}$$

$$V_C = +10 - 4.7 \times I_C \simeq 10 - 7.52 = 2.48 \text{ V}$$

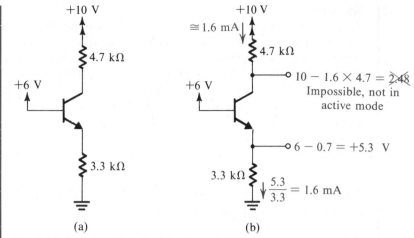

Fig. 8.16 Analysis of the circuit for Example 8.2.

Since the collector voltage calculated appears to be less than the base voltage by 3.52 V, it follows that our original assumption of active-mode operation is incorrect. In fact, the transistor has to be in the *saturation* mode. Since we have not yet studied the saturation mode of operation, we shall defer the analysis of this circuit to a later section.

The details of the analysis performed above are illustrated in Fig. 8.16b.

EXAMPLE 8.3

We wish to analyze the circuit in Fig. 8.17a to determine the voltages at all nodes and the currents through all branches. Note that this circuit is identical to that considered in Examples 8.1 and 8.2 except that now the base voltage is zero.

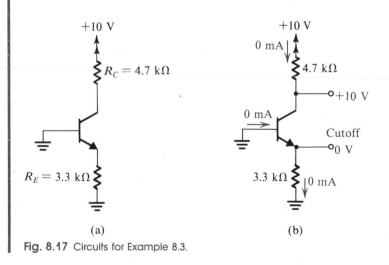

Fig. 8.17 Circuits for Example 8.3.

Solution

Since the base is at zero volts, the emitter–base junction cannot conduct and the emitter current is zero. Also, the collector–base junction cannot conduct, since the *n*-type collector is connected through R_C to the positive power supply while the *p*-type base is at ground. It follows that the collector current will be zero. The base current will also have to be zero, and the transistor is in the *cutoff* mode of operation.

The emitter voltage will obviously be zero, while the collector voltage will be equal to $+10$ V, since the voltage drop across R_C is zero. Figure 8.17 shows the analysis details.

EXAMPLE 8.4

We desire to analyze the circuit of Fig. 8.18a to determine the voltages at all nodes and the currents through all branches.

Solution

The base of this *pnp* transistor is grounded, while the emitter is connected to a positive supply ($V^+ = +10$ V) through R_E. It follows that the emitter–base junction will be forward-biased with

$$V_E = V_{EB} \simeq 0.7 \text{ V}$$

Thus the emitter current will be given by

$$I_E = \frac{V^+ - V_E}{R_E} = \frac{10 - 0.7}{2} = 4.65 \text{ mA}$$

Since the collector is connected to a negative supply (more negative than the base voltage) through R_C, is it *possible* that this transistor is operating in the active mode.

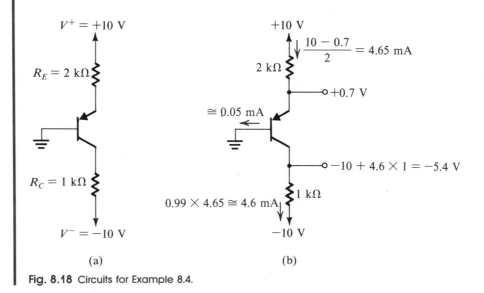

(a)   (b)

Fig. 8.18 Circuits for Example 8.4.

Assuming this to be the case, we obtain

$$I_C = \alpha I_E$$

Since no value for β has been given, we shall assume $\beta = 100$, which results in $\alpha = 0.99$. Since large variations in β result in small differences in α, this assumption will not be critical as far as determining the value of I_C is concerned. Thus

$$I_C = 0.99 \times 4.65 = 4.6 \text{ mA}$$

The collector voltage will be

$$V_C = V^- + I_C R_C$$
$$= -10 + 4.6 \times 1 = -5.4 \text{ V}$$

Thus the collector–base junction is reverse-biased by 5.4 V and the transistor is indeed in the active mode, which supports our original assumption.

It remains only to calculate the base current,

$$I_B = \frac{I_E}{\beta + 1} = \frac{4.65}{101} \simeq 0.0460 \text{ mA}$$

Obviously, the value of β critically affects the base current. Note, however, that in this circuit the value of β will have no effect on the mode of operation of the transistor. Since β is generally an ill-specified parameter, this circuit represents a good design. As a rule, one should strive to *design the circuit such that its performance is as insensitive to the value of β as possible.* The analysis details are illustrated in Fig. 8.18b.

EXAMPLE 8.5

We want to analyze the circuit in Fig. 8.19a to determine the voltages at all nodes and the currents in all branches. Assume $\beta = 100$.

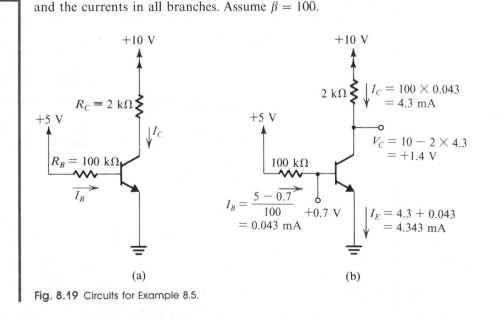

(a)   (b)

Fig. 8.19 Circuits for Example 8.5.

Solution

The base–emitter junction is clearly forward-biased. Thus

$$I_B = \frac{+5 - V_{BE}}{R_B} \simeq \frac{5 - 0.7}{100} = 0.043 \text{ mA}$$

Assume that the transistor is operating in the active mode. We now can write

$$I_C = \beta I_B = 100 \times 0.043 = 4.3 \text{ mA}$$

The collector voltage can now be determined as

$$V_C = +10 - I_C R_C = 10 - 4.3 \times 2 = +1.4 \text{ V}$$

Since the base voltage V_B is

$$V_B = V_{BE} \simeq +0.7 \text{ V}$$

it follows that the collector–base junction is reverse-biased by 0.7 V and the transistor is indeed in the active mode. The emitter current will be given by

$$I_E = (\beta + 1)I_B = 101 \times 0.043 \simeq 4.3 \text{ mA}$$

We note from this example that the collector and emitter currents depend critically on the value of β. In fact, if β were 10% higher, the transistor would leave the active mode and enter saturation. Therefore this clearly is a bad design. The analysis details are illustrated in Fig. 8.19b.

EXAMPLE 8.6

We want to analyze the circuit of Fig. 8.20a to determine the voltages at all nodes and the currents through all branches. Assume $\beta = 100$.

Solution

The first step in the analysis consists of simplifying the base circuit using Thévenin's theorem. The result is shown in Fig. 8.20b, where

$$V_{BB} = +15 \frac{R_{B2}}{R_{B1} + R_{B2}} = 15 \frac{50}{100 + 50} = +5 \text{ V}$$

$$R_{BB} = (R_{B1} \| R_{B2}) = (100 \| 50) = 33.3 \text{ k}\Omega$$

To evaluate the base or the emitter current we have to write a loop equation around the loop marked L in Fig. 8.20b. Note, though, that the current through R_{BB} is different from the current through R_E. The loop equation will be

$$V_{BB} = I_B R_{BB} + V_{BE} + I_E R_E$$

Substituting for I_B by

$$I_B = \frac{I_E}{\beta + 1}$$

and rearranging the equation gives

$$I_E = \frac{V_{BB} - V_{BE}}{R_E + [R_{BB}/(\beta + 1)]}$$

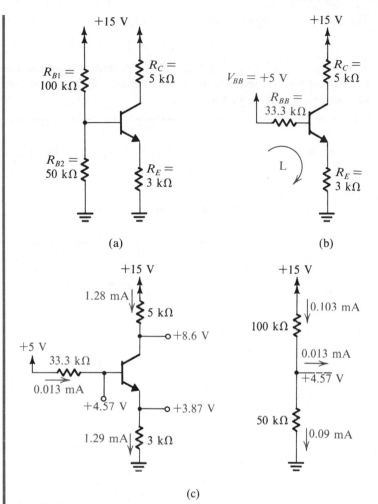

Fig. 8.20 Circuits for Example 8.6.

For the numerical values given we have

$$I_E = \frac{5 - 0.7}{3 + (33.3/100)} = 1.29 \text{ mA}$$

The base current will be

$$I_B = \frac{1.29}{101} = 0.0128 \text{ mA}$$

The base voltage is given by

$$V_B = V_{BE} + I_E R_E$$
$$= 0.7 + 1.29 \times 3 = 4.57 \text{ V}$$

Assume active-mode operation. We can evaluate the collector current as

$$I_C = \alpha I_E = 0.99 \times 1.29 = 1.28 \text{ mA}$$

The collector voltage can now be evaluated as

$$V_C = +15 - I_C R_C = 15 - 1.28 \times 5 = 8.6 \text{ V}$$

It follows that the collector is higher in potential than the base by 4.03 V, which means that the transistor is in the active mode, as had been assumed. The results of the analysis are given in Fig. 8.20c.

EXAMPLE 8.7

We wish to analyze the circuit in Fig. 8.21a to determine the voltages at all nodes and the currents through all branches.

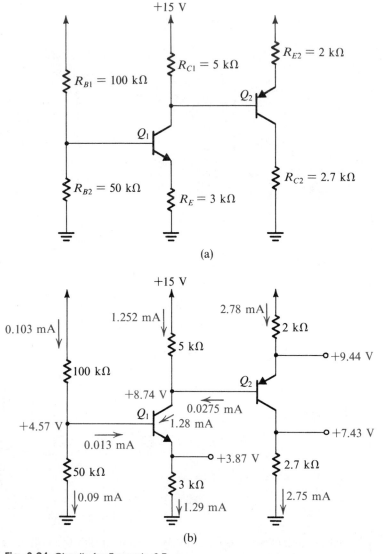

(a)

(b)

Fig. 8.21 Circuits for Example 8.7.

Solution

We first recognize that part of this circuit is identical to the circuit we analyzed in Example 8.6—namely, the circuit of Fig. 8.20a. The difference, of course, is that in the new circuit we have an additional transistor Q_2 together with its associated resistors R_{E2} and R_{C2}. Assume that Q_1 is still in the active mode. The following values will be identical to those obtained in the previous example.

$$V_{B1} = +4.57 \text{ V}$$

$$I_{B1} = 0.0128 \text{ mA}$$

$$I_{E1} = 1.29 \text{ mA}$$

$$I_{C1} = 1.28 \text{ mA}$$

However, the collector voltage will be different than previously calculated, since part of the collector current I_{C1} will flow in the base lead of Q_2 (I_{B2}). As a first approximation we may assume that I_{B2} is much smaller than I_{C1}; that is, we may assume that the current through R_{C1} is almost equal to I_{C1}. This will enable us to calculate V_{C1}:

$$V_{C1} \simeq +15 - I_{C1}R_{C1}$$
$$= 15 - 1.28 \times 5 = +8.6 \text{ V}$$

Thus Q_1 is in the active mode, as had been assumed.

As far as Q_2 is concerned, we note that its emitter is connected to $+15$ V through R_{E2}. It is therefore safe to assume that the emitter–base junction of Q_2 will be forward-biased. Thus the emitter of Q_2 will be at a voltage V_{E2} given by

$$V_{E2} = V_{C1} + V_{EB}|_{Q_2} \simeq 8.6 + 0.7 = +9.3 \text{ V}$$

The emitter current of Q_2 may now be calculated as

$$I_{E2} = \frac{+15 - V_{E2}}{R_{E2}} = \frac{15 - 9.3}{2} = 2.85 \text{ mA}$$

Since the collector of Q_2 is returned to ground via R_{C2}, it is possible that Q_2 is operating in the active mode. Assume this to be the case. We now find I_{C2} as

$$I_{C2} = \alpha_2 I_{E2}$$
$$= 0.99 \times 2.85 = 2.82 \text{ mA} \qquad \text{(assuming } \beta_2 = 100\text{)}$$

The collector voltage of Q_2 will be

$$V_{C2} = I_{C2}R_{C2} = 2.82 \times 2.7 = 7.62 \text{ V}$$

which is lower than V_{B2} by 0.98 V. Thus Q_2 is in the active mode, as assumed.

It is important at this stage to find the magnitude of the error incurred in our calculation by the assumption that I_{B2} is negligible. The value of I_{B2} is given by

$$I_{B2} = \frac{I_{E2}}{\beta_2 + 1} = \frac{2.85}{101} = 0.028 \text{ mA}$$

which is indeed much smaller than I_{C1} (1.28 mA). If desired, we can obtain more accurate results by iterating once more time, assuming I_{B2} to be 0.028 mA. The new values will be

$$\text{Current in } R_{C1} = I_{C1} - I_{B2} = 1.28 - 0.028 = 1.252 \text{ mA}$$

$$V_{C1} = 15 - 5 \times 1.252 = 8.74 \text{ V}$$

$$V_{E2} = 8.74 + 0.7 = 9.44 \text{ V}$$

$$I_{E2} = \frac{15 - 9.44}{2} = 2.78 \text{ mA}$$

$$I_{C2} = 0.99 \times 2.78 = 2.75 \text{ mA}$$

$$V_{C2} = 2.75 \times 2.7 = 7.43 \text{ V}$$

$$I_{B2} = \frac{2.78}{101} = 0.0275 \text{ mA}$$

Note that the new value of I_{B2} is very close to the value used in our iteration, and no further iterations are warranted. The final results are indicated in Fig. 8.21b.

The reader justifiably might be wondering about the necessity for using an iterative scheme in solving a linear (or linearized) problem. Indeed, we can obtain the exact solution (if we can call anything we are doing with a first-order model exact!) by writing appropriate equations. The reader is encouraged to find this solution and compare the results with those obtained above. It is important to emphasize, however, that in most such problems it is quite sufficient to obtain an approximate solution, provided that we can obtain it quickly and, of course, correctly.

Important Note

In Examples 8.1 through 8.7 we frequently used the exact value of α to calculate the collector current. Since $\alpha \simeq 1$, the error in such calculations will be very small if one assumes $\alpha = 1$ and $i_C = i_E$. Therefore, except in calculations that depend critically on the value of α (such as the calculation of base current), one usually assumes $\alpha \simeq 1$.

Exercise

8.13 For the circuit shown in Fig. E8.13, assume β to be very high and find I_E, V_E, and V_C.

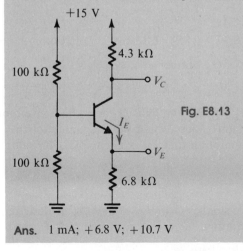

Fig. E8.13

Ans.   1 mA; $+6.8$ V; $+10.7$ V

8.7 THE TRANSISTOR AS AN AMPLIFIER

To operate as an amplifier a transistor must be biased in the active region. The biasing problem is that of establishing a constant dc current in the emitter (or the collector). This current should be predictable and insensitive to variations in temperature, value of β, and so on. While deferring the study of different biasing techniques to Section 8.8, we will demonstrate in the following the need to bias the transistor at a constant collector current. This requirement stems from the fact that the operation of the transistor as an amplifier is highly influenced by the value of the quiescent (or bias) current, as shown below.

DC Conditions

To understand how the transistor operates as an amplifier, consider the conceptual circuit shown in Fig. 8.22. Here the bias current I_C is assumed to be established by a dc voltage V_{BE} (battery), clearly an impractical arrangement. The reverse bias of the CBJ is established by connecting the collector to another power-supply voltage V_{CC} through a resistor R_C. Consider first the dc conditions; that is, assume that the signal v_{be} is zero. We have the following dc quantities.

$$I_C = I_S e^{V_{BE}/V_T} \tag{8.14}$$

$$I_E = \frac{I_C}{\alpha} \tag{8.15}$$

$$I_B = \frac{I_C}{\beta} \tag{8.16}$$

$$V_C = V_{CE} = V_{CC} - I_C R_C \tag{8.17}$$

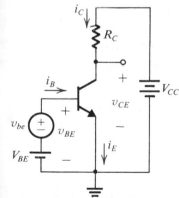

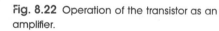

Fig. 8.22 Operation of the transistor as an amplifier.

Obviously, for active-mode operation, V_C should be greater than V_B by an amount that allows for a reasonable signal swing at the collector yet maintains the transistor in the active region at all times. We shall return to this point later.

Transconductance

If a signal v_{be} is applied as shown in Fig. 8.22, the total instantaneous base–emitter voltage v_{BE} becomes

$$v_{BE} = V_{BE} + v_{be} \tag{8.18}$$

Correspondingly, the collector current becomes

$$
\begin{aligned}
i_C &= I_S e^{v_{BE}/V_T} \\
&= I_S e^{(V_{BE}+v_{be})/V_T} \\
&= I_S e^{(V_{BE}/V_T)} e^{(v_{be}/V_T)} \tag{8.19}
\end{aligned}
$$

Use of Eq. (8.14) yields

$$i_C = I_C e^{v_{be}/V_T} \tag{8.20}$$

Now, if $v_{be} \ll V_T$, we may approximate Eq. (8.20) as

$$i_C \simeq I_C \left(1 + \frac{v_{be}}{V_T}\right) \tag{8.21}$$

Here we have expanded the exponential in Eq. (8.20) in a series and retained only the first two terms. This approximation, which is valid only for v_{be} less than about 10 mV, is referred to as the *small-signal approximation*. Under this approximation the total collector current is given by Eq. (8.21) and can be rewritten

$$i_C = I_C + \frac{I_C}{V_T} v_{be} \tag{8.22}$$

Thus the collector current is composed of the dc bias value I_C and a signal component i_c.

$$i_c = \frac{I_C}{V_T} v_{be} \tag{8.23}$$

This equation relates the signal current in the collector to the corresponding base–emitter signal voltage. It can be rewritten as

$$i_c = g_m v_{be} \tag{8.24}$$

where g_m is called the *transconductance*, and from Eq. (8.23) it is given by

$$g_m = \frac{I_C}{V_T} \tag{8.25}$$

We observe that the transconductance of the BJT is directly proportional to the collector bias current I_C. This should be contrasted with the case of the JFET [Eq. (6.40)] and the MOSFET [Eq. (7.30b)], where g_m is proportional to $\sqrt{I_D}$. Another important difference is that in the BJT the value of g_m is independent of the device dimensions; rather, it is completely determined by I_C and the thermal voltage V_T. Thus to obtain a constant, predictable value for g_m we need a constant, predictable

I_C. Finally, we note that BJTs have relatively high transconductance (as compared to FETs); for instance, at $I_C = 1$ mA, $g_m \simeq 40$ mA/V.

A graphical interpretation for g_m is given in Fig. 8.23, where it is shown that g_m is equal to the slope of the i_C-v_{BE} characteristic curve at $i_C = I_C$. Thus

$$g_m = \left. \frac{\partial i_C}{\partial v_{BE}} \right|_{i_C = I_C} \tag{8.26}$$

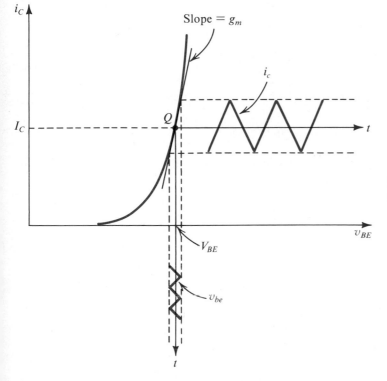

Fig. 8.23 Linear operation of the transistor under the small-signal condition.

The small-signal approximation implies keeping the signal amplitude sufficiently small so that operation is restricted to an almost linear segment of the i_C-v_{BE} exponential curve. Increasing the signal amplitude will result in the collector current having components nonlinearly related to v_{be}. Similar types of approximation were employed for diodes (Section 4.8), JFETs (Section 6.7) and MOSFETs (Section 7.4).

The analysis above suggests that for small signals ($v_{be} \ll V_T$), the transistor behaves as a voltage-controlled current source. The input port of this controlled source is between base and emitter, and the output port is between collector and emitter. The transconductance of the controlled source is g_m, and the output resistance is infinite. This latter ideal property is a result of our first-order model of transistor operation in which the collector voltage has no effect on the collector current in the active mode. As we have seen in Section 8.5, practical BJTs have finite output resistance. The effect of the output resistance on amplifier performance will be considered later.

The Input Resistance at the Base

To determine the resistance seen by v_{be}, we first evaluate the total base current i_B using Eq. (8.22), as follows:

$$i_B = \frac{i_C}{\beta} = \frac{I_C}{\beta} + \frac{1}{\beta}\frac{I_C}{V_T}v_{be}$$

Thus

$$i_B = I_B + i_b \tag{8.27}$$

where I_B is equal to I_C/β and the signal component i_b is given by

$$i_b = \frac{1}{\beta}\frac{I_C}{V_T}v_{be} \tag{8.28}$$

Substituting for I_C/V_T by g_m gives

$$i_b = \frac{g_m}{\beta}v_{be} \tag{8.29}$$

The small-signal input resistance between base and emitter, *looking into the base*, is denoted by r_π and is defined

$$r_\pi \equiv \frac{v_{be}}{i_b} \tag{8.30}$$

Using Eq. (8.29) gives

$$r_\pi = \frac{\beta}{g_m} \tag{8.31}$$

Thus r_π is directly dependent on β and is inversely proportional to the bias current I_C. Substituting for g_m in Eq. (8.31) from Eq. (8.25) and replacing I_C/β by I_B gives an alternate expression for r_π,

$$r_\pi = \frac{V_T}{I_B} \tag{8.32}$$

The Hybrid-π Model

For small signals the BJT can be represented by the equivalent circuit model shown in Fig.8.24a. Note that this equivalent circuit applies at a particular bias point, since the two parameters g_m and r_π depend on the value of I_C. In order to be consistent with the literature, we use v_π to denote the base–emitter signal voltage v_{be}.

Since the output current $g_m v_\pi$ can be written

$$g_m v_\pi = g_m r_\pi i_b = \beta i_b$$

the equivalent circuit model can be converted to the current-controlled current-source form depicted in Fig. 8.24b. It may be observed that the equivalent circuit of Fig. 8.24a is the incremental, or small-signal, version of the model of Fig. 8.5c. This can

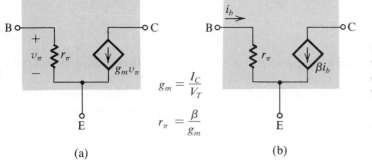

Fig. 8.24 A first-order small-signal model for the BJT in two equivalent forms. This model is called the hybrid-π model and is applicable to both *npn* and *pnp* transistors.

$$g_m = \frac{I_C}{V_T}$$

$$r_\pi = \frac{\beta}{g_m}$$

(a)   (b)

be easily verified by noting from Eq. (8.32) that r_π is the incremental resistance of diode D_B, which conducts a dc bias current I_B. Similarly, we can show that the equivalent circuit of Fig. 8.24b is the incremental, or small-signal, version of the model of Fig. 8.5d.

For reasons that will become apparent later, the equivalent circuit model of Fig. 8.24 is called the *simplified hybrid-π model*. It should be noted that this model applies to both *npn* and *pnp* transistors with *no change* in polarities required.

The Emitter Resistance r_e

The total emitter current i_E can be determined using Eq. (8.22) and

$$i_E = \frac{i_C}{\alpha}$$

Thus

$$i_E = I_E + i_e \tag{8.33}$$

where I_E is equal to I_C/α, and the signal current i_e is given by

$$i_e = \frac{I_C}{\alpha V_T} v_{be} = \frac{I_E}{V_T} v_{be} \tag{8.34}$$

This relationship can be expressed in the form

$$i_e = \frac{v_{be}}{r_e} \tag{8.35}$$

where r_e, called the *emitter resistance*, is defined as

$$r_e \equiv \frac{V_T}{I_E} \tag{8.36}$$

Comparison with Eq. (8.25) reveals that

$$r_e = \frac{\alpha}{g_m} \simeq \frac{1}{g_m} \tag{8.37}$$

An extremely useful interpretation for r_e is that it is the resistance between base and emitter *looking into the emitter*. This is to distinguish it from the resistance between base and emitter *looking into the base*, which is r_π. It can be shown that r_e and r_π are related by

$$r_\pi = (\beta + 1)r_e \tag{8.38}$$

which is intuitively obvious because $\beta + 1$ is the ratio of i_e (which is the current flowing through r_e) and i_b (which is the current flowing through r_π).

In the equivalent circuit models of Fig. 8.24, r_e is not explicitly shown. Nevertheless it is implicitly included, as can easily be demonstrated (Exercise 8.16). An alternative small-signal model that explicitly employs r_e is given in Fig. 8.25. It can be shown that this model is the incremental version of the model given in Fig. 8.5b, by noting that r_e is the incremental, or small-signal, resistance of diode D_E, which conducts a bias current I_E.

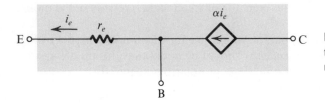

Fig. 8.25 An alternative small-signal model for the BJT, explicitly showing the emitter resistance r_e.

Voltage Gain

Up until now we have established only that the transistor senses the base–emitter signal v_{be} or v_π and causes a proportional current $g_m v_\pi$ to flow in the collector lead at a high (ideally infinite) impedance level. In this way the transistor is acting as a voltage-controlled current source. To obtain an output voltage signal we may force this current to flow through a resistor, as is done in Fig. 8.22. Then the total collector voltage v_C will be

$$\begin{aligned} v_C &= V_{CC} - i_C R_C \\ &= V_{CC} - (I_C + i_c)R_C \\ &= (V_{CC} - I_C R_C) - i_c R_C \\ &= V_C - i_c R_C \end{aligned} \tag{8.39}$$

Here the quantity V_C is the dc bias voltage at the collector, and the signal voltage is given by

$$\begin{aligned} v_c &= -i_c R_C \\ &= -g_m v_{be} R_C \\ &= (-g_m R_C)v_{be} \end{aligned} \tag{8.40}$$

Thus the voltage gain of this amplifier is as follows:

$$\text{Voltage gain} \equiv \frac{v_c}{v_{be}} = -g_m R_C \tag{8.41}$$

Here again we note that because g_m is directly proportional to the collector bias current, the gain will be as stable as the collector bias current is made.

Use of Small-Signal Equivalent Circuits

From the above analysis we conclude that under the small-signal approximation each of the currents and voltages in the circuit will be composed of a dc bias component and a signal component. We may therefore simplify matters and consider the two components separately. That is, we first establish a suitable dc operating point and subsequently ignore all dc quantities while we carry out a signal analysis. From a signal point of view the transistor may be replaced by one of the models in Figs. 8.24 and 8.25. All these models are, of course, equivalent.

Although equivalent circuit models are indispensable in the analysis of complicated circuits, it is far quicker and certainly more insightful to analyze simple circuits with the model *implicitly* assumed. This is especially true if one is interested only in obtaining approximate values using first-order models such as those developed above. At a later stage we will add to the hybrid-π model the components that represent second-order effects.

EXAMPLE 8.8

We wish to analyze the transistor amplifier shown in Fig. 8.26a. Assume $\beta = 100$.

Solution

The first step in the analysis consists of determining the quiescent operating point. For this purpose we assume that $v_i = 0$. The dc base current will be

$$I_B = \frac{V_{BB} - V_{BE}}{R_{BB}}$$

$$\simeq \frac{3 - 0.7}{100} = 0.023 \text{ mA}$$

The dc collector current will be

$$I_C = \beta I_B = 100 \times 0.023 = 2.3 \text{ mA}$$

The dc voltage at the collector will be

$$V_C = V_{CC} - I_C R_C$$
$$= +10 - 2.3 \times 3 = +3.1 \text{ V}$$

Since $V_B = +0.7$ V, it follows that in the quiescent condition the transistor will be operating in the active mode.

Having determined the operating point, we may now proceed to determine the small-signal model parameters:

$$r_e = \frac{V_T}{I_E} = \frac{25 \text{ mV}}{(2.3/0.99) \text{ mA}} = 10.8 \text{ } \Omega$$

$$g_m = \frac{I_C}{V_T} = \frac{2.3 \text{ mA}}{25 \text{ mV}} = 92 \text{ mA/V}$$

$$r_\pi = \frac{\beta}{g_m} = \frac{100}{92} = 1.09 \text{ k}\Omega$$

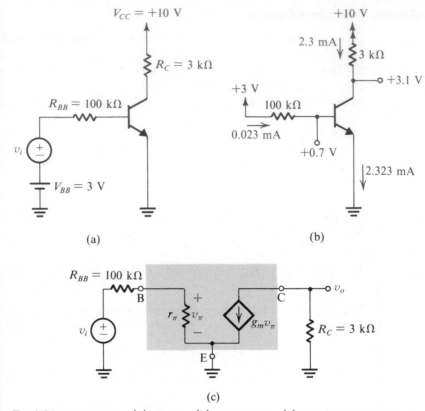

Fig. 8.26 Example 8.8: **(a)** circuit; **(b)** dc analysis; **(c)** small-signal model.

To carry out a small-signal analysis it is convenient to employ the first-order hybrid-π equivalent circuit model of Fig. 8.24a. The resulting small-signal equivalent circuit model of the amplifier circuit is given in Fig. 8.26c. Note that no dc quantities are included in this equivalent circuit. It is most important to note that the dc supply voltage V_{CC} has been replaced by a *short circuit* in the signal equivalent circuit because the circuit terminal connected to V_{CC} will always have a constant voltage. That is, the signal voltage at this terminal will be zero. In other words, *a circuit terminal connected to a constant dc source can always be considered as a signal ground.*

Analysis of the equivalent circuit in Fig. 8.26c proceeds as follows:

$$v_\pi = v_i \frac{r_\pi}{r_\pi + R_{BB}}$$

$$= v_i \frac{1.09}{101.09} = 0.011 v_i \tag{8.42}$$

The output voltage v_o is given by

$$v_o = -g_m v_\pi R_C$$

$$= -92 \times 0.011 v_i \times 3 = -3.04 v_i$$

Thus the voltage gain will be

$$\frac{v_o}{v_i} = -3.04 \tag{8.43}$$

where the minus sign indicates a phase reversal.

To gain more insight into the behavior of this circuit, let us consider the waveforms at various points. For this purpose we shall assume v_i to have a triangular waveform. The first question facing us will be, How high can the amplitude of v_i become? One constraint on signal amplitude is the small-signal approximation on which all of the above analysis is based. This constraint stipulates that v_{be} (that is, v_π) not exceed about 10 mV. If we take the triangular waveform v_π to be 20 mV peak-to-peak and work backward, Eq. (8.42) can be used to determine the maximum possible peak of v_i,

$$\hat{V}_i = \frac{\hat{V}_\pi}{0.011} = 0.91 \text{ V}$$

To check whether or not the transistor remains in the active mode with v_i having a peak value $\hat{V}_i = 0.91$ V, we have to evaluate the collector voltage. The voltage at the collector will consist of a triangular wave v_c superimposed on the dc value $V_C = 3.1$ V. The peak voltage of the triangular waveform will be

$$\hat{V}_c = \hat{V}_i \times \text{gain} = 0.91 \times 3.04 = 2.77 \text{ V}$$

It follows that when the output swings negative, the collector voltage reaches a minimum of $3.1 - 2.77 = 0.33$ V, which is lower than the base voltage $\simeq 0.7$ V. Thus the transistor will not remain in the active mode with v_i having a peak value of 0.91 V. We can easily determine, though, the maximum value of the peak of the input signal such that the transistor remains active at all times. This can be done by finding the value of $\hat{V}_i$ that corresponds to the minimum value of the collector voltage being equal to the base voltage, which is approximately 0.7 V. Thus

$$\hat{V}_i = \frac{3.1 - 0.7}{3.04} = 0.79 \text{ V}$$

Let us choose $\hat{V}_i$ to be approximately 0.8 V, as shown in Fig. 8.27a, and complete the analysis of this problem. The signal current in the base will be triangular, with a peak value $\hat{I}_b$ of

$$\hat{I}_b = \frac{\hat{V}_i}{R_{BB} + r_\pi} = \frac{0.8}{100 + 1.09} = 0.008 \text{ mA}$$

This triangular-wave current will be superimposed on the quiescent base current I_B, as shown in Fig. 8.27b. The base–emitter voltage will consist of a triangular-wave component superimposed on the dc V_{BE} that is approximately 0.7 V. The peak value of the triangular waveform will be

$$\hat{V}_\pi = \hat{V}_i \frac{r_\pi}{r_\pi + R_{BB}} = 0.8 \frac{1.09}{100 + 1.09} = 8.6 \text{ mV}$$

The total v_{BE} is sketched in Fig. 8.27c.

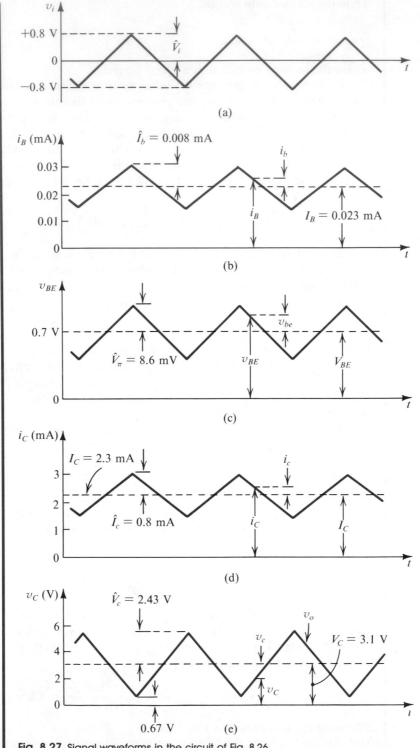

Fig. 8.27 Signal waveforms in the circuit of Fig. 8.26.

The signal current in the collector will be triangular in waveform, with a peak value $\hat{I}_c$ given by

$$\hat{I}_c = \beta \hat{I}_b = 100 \times 0.008 = 0.8 \text{ mA}$$

This current will be superimposed on the quiescent collector current I_C ($= 2.3$ mA), as shown in Fig. 8.27d.

Finally, the signal voltage at the collector can be obtained by multiplying v_i by the voltage gain; that is,

$$\hat{V}_c = 3.04 \times 0.8 = 2.43 \text{ V}$$

Figure 8.27e shows a sketch of the total collector voltage v_C versus time. Note the phase reversal between the input signal v_i and the output signal v_c. Also note that although the minimum collector voltage is slightly lower than the base voltage, the transistor will remain in the active mode. In fact, BJTs remain in the active mode even with their collector–base junctions forward-biased by as much as 0.3 or 0.4 V.

Before we leave this example, we wish to make an important observation. Although we have drawn the equivalent circuit in Fig. 8.26b and used it in the analysis, this step could have been avoided. In fact, for first-order calculations the equivalent circuit model is usually used implicitly, as will be illustrated in Example 8.9.

EXAMPLE 8.9

We need to analyze the circuit of Fig. 8.28a to determine the voltage gain and the signal waveforms at various points. The capacitor C is a coupling capacitor whose purpose is to couple the signal v_i to the emitter while blocking dc. In this way the dc bias established by V^+ and V^- together with R_E and R_C will not be disturbed when the signal v_i is connected. For the purpose of this example, C will be assumed to be infinite—that is, acting as a perfect short circuit at signal frequencies of interest.

Solution

We shall start by determining the dc operating point as follows:

$$I_E = \frac{+10 - V_E}{R_E} \simeq \frac{+10 - 0.7}{10} = 0.93 \text{ mA}$$

Assuming $\beta = 100$, then $\alpha = 0.99$ and

$$I_C = 0.99 I_E = 0.92 \text{ mA}$$

$$V_C = -10 + I_C R_C$$
$$= -10 + 0.92 \times 5 = -5.4 \text{ V}$$

Thus the transistor is in the active mode. Furthermore, the collector signal can swing from -5.4 V to zero (which is the base voltage) without the transistor going into saturation. However, a negative 5.4-V swing in the collector voltage will (theoretically) cause the minimum collector voltage to be -10.8 V, which is more negative than the power-supply voltage. It follows that if we attempt to apply an input that results in such an output signal, the transistor will cut off and the negative peaks of the output

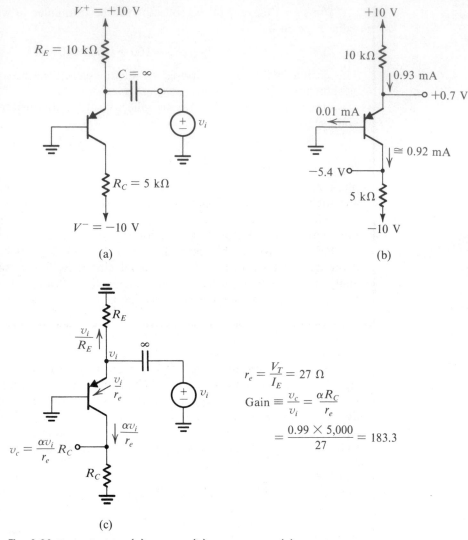

(a)

(b)

(c)

Fig. 8.28 Example 8.9: **(a)** circuit; **(b)** dc analysis; **(c)** signal analysis.

signal will be clipped off, as illustrated in Fig. 8.29. The waveform in Fig. 8.29, however, is shown to be linear (except for the clipped peak); that is, the effect of the nonlinear i_C-v_{BE} characteristic is not taken into account. This is not correct, since if we are driving the transistor into cutoff at the negative signal peaks, then we will surely be exceeding the small-signal limit, as will be shown later.

Let us now proceed to determine the small-signal voltage gain. Toward that end, we evaluate the emitter resistance r_e,

$$r_e = \frac{V_T}{I_E} = \frac{25}{0.93} \simeq 27\ \Omega$$

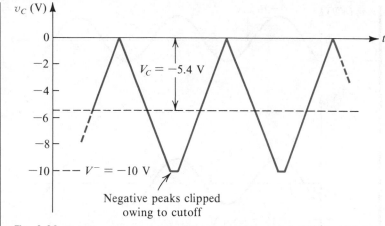

Fig. 8.29 Distortion in output signal due to transistor cutoff. Note that it is assumed that no distortion due to transistor nonlinear characteristics is occurring.

We also note that

$$v_{eb} = v_i$$

Thus for small-signal operation the peak value of v_i should be limited to less than 10 mV; otherwise nonlinear distortion will occur.

The signal current in the emitter can be easily obtained as

$$i_e = \frac{v_{eb}}{r_e} = \frac{v_i}{r_e}$$

The signal current in the collector will be

$$i_c = \alpha i_e = \frac{\alpha v_i}{r_e}$$

The signal component of the collector voltage can now be evaluated assuming the negative power supply (V^-) to be a signal ground. Thus

$$v_c = i_c R_C = \frac{\alpha v_i}{r_e} R_C$$

which results in the voltage gain

$$\frac{v_c}{v_i} = \frac{\alpha R_C}{r_e} = g_m R_C$$

$$= \frac{0.99 \times 5,000}{27} = 183.3$$

Note that the voltage gain is positive, indicating that the output is in phase with the input signal. This property is due to the fact that the input signal is applied to the emitter rather than to the base, as was done in Example 8.8. We should emphasize that the positive gain has noting to do with the fact that the transistor used in this example is of the *pnp* type.

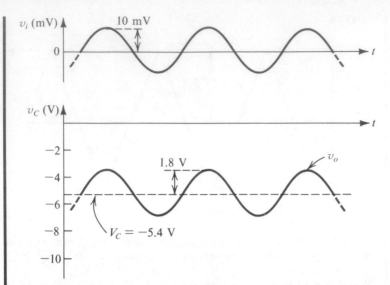

Fig. 8.30 Input and output waveforms for the circuit of Fig. 8.28.

Assuming an input sine-wave signal with a peak of 10 mV, as shown in Fig. 8.30, it follows that the signal component of the collector voltage will have a peak value

$$\hat{V}_c = 183.3 \times 0.01 = 1.833 \text{ V}$$

Thus, as expected, the output signal voltage is limited by the maximum signal that we can apply at the input without running the risk of nonlinear distortion.

The small-signal analysis above was carried out directly on the circuit, as illustrated in Fig. 8.28c, with the model of Fig. 8.25 implicitly employed. An alternative, more systematic, analysis method consists of explicitly replacing the BJT with its hybrid-π model as shown in Fig. 8.31. The reader is urged to show that analysis of this circuit yields results identical to those obtained above.

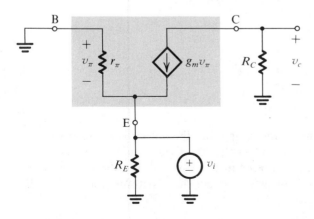

Fig. 8.31 The circuit of Fig. 8.28a with the transistor replaced with its hybrid-π equivalent circuit model.

Transistor Capacitances and Output Resistance

In our study of the *pn* junction in Chapter 4 we learned that two capacitive effects exist. The first such effect is the change in charge stored in the depletion region as the voltage across the junction changes, which is modeled by the space-charge or depletion-layer capacitance C_j. The second capacitive effect relates to the change in the excess minority carrier charge stored in the *p* and *n* materials as the voltage across the junction is changed, which is modeled by the diffusion capacitance C_d. The diffusion capacitance C_d is proportional to the current flowing across the junction and is zero for a reverse-biased junction.

Applying this to the BJT, we see that for a device operating in the active mode the forward-biased EBJ displays two parallel capacitances: a depletion-layer capacitance C_{je} and a diffusion capacitance C_{de}. On the other hand, because the CBJ is reverse-biased it displays only a depletion capacitance C_{jc}. Figure 8.32 shows the small-signal equivalent circuit model of a BJT operating in the active mode with the junction capacitances included. The EBJ capacitances have been added together to obtain the total capacitance C_π. Also, in order to conform with the literature, the CBJ depletion capacitance has been denoted by C_μ. The values of C_π and C_μ depend on the transistor type, junction areas, and, of course, the dc bias point. We will have more to say about these capacitances in Chapter 11, when we deal with amplifier frequency response. For the time being it should be mentioned that while C_π can range from few picofarads to a few tens of picofarads, C_μ is usually about 1 or 2 picofarads. Nevertheless, C_μ plays an important role in determining the effective input capacitance of a transistor amplifier through the Miller multiplication effect (see Section 2.5). It will be shown in Chapter 11 that C_π and C_μ determine the high-frequency response of transistor amplifiers. For gain calculations at low and medium frequencies, these capacitances can be ignored.

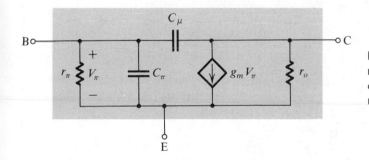

Fig. 8.32 The hybrid-π equivalent circuit model of the BJT, including the two junction capacitances C_π and C_μ and the output resistance r_o.

The equivalent circuit model of Fig. 8.32 also includes the output resistance r_o. We have studied the origin of r_o in Section 8.5 and found that its value is given by

$$r_o = \frac{V_A}{I_C}$$

where I_C is the dc collector bias current and V_A is the Early voltage of the given transistor, typically 50–100 V. If the transistor emitter is connected to ground and a load resistance is connected to the collector, then r_o appears across R_C, resulting in a reduction in gain. For instance, the gain expression in Eq. (8.41) becomes

$$\text{Voltage gain} \equiv \frac{v_c}{v_{be}} = -g_m(R_C \| r_o) \tag{8.44}$$

In applications where $R_C \ll r_o$, the effect of r_o on the value of voltage gain is negligible. However, we will encounter at a later point amplifier circuits in which R_C is made very large and consequently r_o plays an important role in determining the voltage gain.

Exercises

8.14 Use Eq. (8.26) to derive the expression for g_m in Eq. (8.25).

8.15 For a transistor with $\beta = 100$ operating at $I_C = 0.5$ mA, find g_m, r_e, and r_π.

Ans. 20 mA/V; 50 Ω; 5 kΩ

8.16 We wish to show that the resistance between emitter and base looking into the emitter is equal to r_e. Use the hybrid-π model of Fig. 8.24a. Ground the base and apply a test voltage v_x to the emitter. Find the current drawn from v_x and thus show that the input resistance looking into the emitter is equal to $1/(g_m + 1/r_\pi)$. Then show that this expression reduces to r_e.

8.17 In the circuit of Fig. 8.22, V_{BE} is adjusted to yield a dc collector current $I_C = 1$ mA. Let $V_{CC} = 15$ V, $R_C = 10$ kΩ, and $\beta = 100$. If $v_{be} = 0.005 \sin \omega t$ volts, find $v_C(t)$ and $i_B(t)$.

Ans. $(5 - 2 \sin \omega t)$ volts; $(10 + 2 \sin \omega t)$ μA

8.18 The transistor in the circuit of Fig. E8.18 is biased to operate in the active mode. (In fact, the bias of this circuit was analyzed in Exercise 8.13.) Find the value of r_e. Use the equivalent circuit of Fig. 8.25 to show that

$$\frac{v_{o1}}{v_i} = \frac{R_E}{R_E + r_e}$$

$$\frac{v_{o2}}{v_i} = \frac{-\alpha R_C}{R_E + r_e}$$

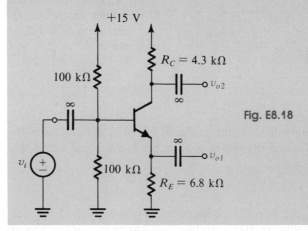

Fig. E8.18

If $\beta = 100$, find the values of these voltage gains. Then use the equivalent circuit of Fig. 8.24a to obtain an alternate derivation of the above expressions.

Ans.   25 Ω; 0.996 V/V; -0.62 V/V

8.19 The transistor in the circuit of Fig. E8.19 has $\beta = 100$.

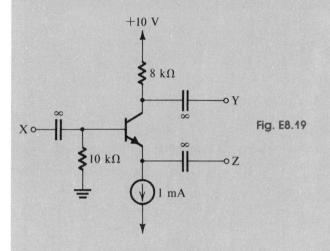

Fig. E8.19

(a) Find the dc voltages at the base, emitter, and collector.
(b) Find g_m and r_π.
(c) If terminal Z is connected to ground, X connected to signal source v_i, and Y to an 8-kΩ load resistance, use the BJT small-signal model of Fig. 8.24a to find the voltage gain v_y/v_I.

Ans.   **(a)** -0.1 V,   0.8 V, $+2$ V;   **(b)** 40 mA/V, 2.5 kΩ;   **(c)** -160 V/V

8.8 BIASING THE BJT FOR DISCRETE-CIRCUIT DESIGN

The biasing problem is that of establishing a constant dc current in the emitter of the BJT. This current has to be calculable, predictable, and insensitive to variations in temperature and to the large variations in the value of β encountered among transistors of the same type. In this section we shall deal with the classical approach to solving the bias problem in transistor circuits designed from discrete devices. Bias methods for integrated-circuit design are presented in Chapter 9.

Bias Arrangement Using a Single Power Supply

Figure 8.33a shows the arrangement most commonly used for biasing a transistor amplifier if only a single power supply is available. The technique, which is identical to that used with enhancement MOSFETs, consists of supplying the base of the transistor with a fraction of the supply voltage V_{CC} through the voltage divider R_1, R_2. In addition, a resistor R_E is connected to the emitter.

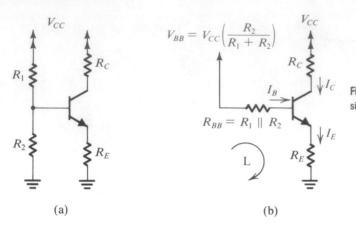

Fig. 8.33 Classical biasing for BJTs using a single power supply.

(a)   (b)

Figure 8.33b shows the same circuit with the voltage-divider network replaced by its Thévenin equivalent,

$$V_{BB} = \frac{R_2}{R_1 + R_2} V_{CC} \tag{8.45}$$

$$R_B = \frac{R_1 R_2}{R_1 + R_2} \tag{8.46}$$

The current I_E can be determined by writing a Kirchhoff loop equation for the base–emitter–ground loop and substituting $I_B = I_E/(\beta + 1)$:

$$I_E = \frac{V_{BB} - V_{BE}}{R_E + R_B/(\beta + 1)} \tag{8.47}$$

To make I_E insensitive to temperature and β variations, we design the circuit to satisfy the following two constraints.

$$V_{BB} \gg V_{BE} \tag{8.48}$$

$$R_E \gg \frac{R_B}{\beta + 1} \tag{8.49}$$

Condition (8.48) ensures that small variations in V_{BE} (around 0.7 V) will be swamped by the much larger V_{BB}. Typically, one designs for V_{BB} from 3 to 5 V; a rule of thumb states that V_{BB} should be chosen approximately equal to $V_{CC}/3$ and R_C chosen so that the voltage drop across it is $V_{CC}/3$. This leaves $V_{CC}/3$ for the signal swing at the collector.

Condition (8.49) makes I_E insensitive to variations in β and could be satisfied by selecting R_B small. This in turn is achieved by using low values for R_1 and R_2. Lower values for R_1 and R_2, however, will mean a higher current drain from the power supply and normally will result in a lowering of the input resistance of the amplifier, which is the trade-off involved in this design problem. It should be noted that Condition (8.49) means that we want to make the base voltage independent of the value of β and determined solely by the voltage divider. This will obviously be satisfied if the current in the divider is made much larger than the base current. Typically one selects R_1 and R_2 such that their current is in the range of I_E to $0.1I_E$.

EXAMPLE 8.10

We wish to design the bias network of the amplifier in Fig. 8.33 to establish a current $I_E = 1$ mA using a power supply $V_{CC} = +12$ V.

Solution
We shall follow the rule of thumb mentioned above and allocate one-third of the supply voltage to the voltage drop across R_2 and another third to the voltage drop across R_C, leaving one-third for possible signal swing at the collector. Thus

$$V_B = +4 \text{ V}$$

$$V_E = 4 - V_{BE} \simeq 3.3 \text{ V}$$

and R_E is determined from

$$R_E = \frac{V_E}{I_E} = 3.3 \text{ k}\Omega$$

From the discussion above we select a voltage-divider current of $0.1I_E$. Neglecting the base current, we find

$$R_1 + R_2 = \frac{12}{0.1I_E} = 120 \text{ k}\Omega$$

$$\frac{R_2}{R_1 + R_2} V_{CC} = 4 \text{ V}$$

Thus $R_2 = 40$ kΩ and $R_1 = 80$ kΩ.

At this point it is desirable to find a more accurate estimate for I_E, taking into account the nonzero base current. Using Eq. (8.47), and assuming that β is specified to be 100, we obtain

$$I_E = \frac{3.3}{3.3 + 0.267} = 0.93 \text{ mA}$$

We could, of course, have obtained a value much closer to the desired 1 mA by designing with exact equations. However, since our work is based on first-order models, it does not make sense to strive for accuracy better than 5% or 10%.

It should be noted that if we are willing to draw a higher current from the power supply and if we are prepared to accept a lower input resistance for the amplifier, then we may use a voltage-divider current equal, say, to I_E, resulting in $R_1 = 8$ kΩ and $R_2 = 4$ kΩ. The effect of this on the amplifier input resistance will be analyzed in Section 8.9. We shall refer to the circuit using these latter values as design 2, for which the actual value of I_E will be

$$I_E = \frac{3.3}{3.3 + 0.026} \simeq 1 \text{ mA}$$

The value of R_C can be determined from

$$R_C = \frac{12 - V_C}{I_C}$$

Thus for design 1 we have

$$R_C = \frac{12 - 8}{0.99 \times 0.93} = 4.34 \text{ k}\Omega$$

whereas for design 2 we have

$$R_C = \frac{12 - 8}{0.99 \times 1} = 4.04 \text{ k}\Omega$$

For simplicity we shall select $R_C = 4$ kΩ for both designs.

Exercise

8.20 For design 1 in Example 8.10 calculate the expected range of I_E if the transistor used has β in the range 50 to 150. Repeat for design 2.

Ans.   For design 1, 0.86 to 0.95 mA; for design 2, 0.98 to 0.995 mA

Biasing Using Two Power Supplies

A somewhat simpler bias arrangement is possible if two power supplies are available, as shown in Fig. 8.34. Writing a loop equation for the loop labeled L gives

$$I_E = \frac{V_{EE} - V_{BE}}{R_E + R_B/(\beta + 1)} \tag{8.50}$$

This equation is identical to Eq. (8.47) except for V_{EE} replacing V_{BB}. Thus the two constraints of Eqs. (8.48) and (8.49) apply here as well. Note that if the transistor is to be used with the base grounded (that is, in the common-base configuration dis-

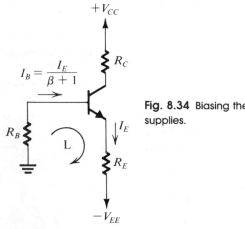

Fig. 8.34 Biasing the BJT using two power supplies.

cussed in the next section), then R_B can be eliminated altogether. On the other hand, if the input signal is to be coupled to the base, then R_B is needed.

Exercise

8.21 The bias arrangement of Fig. 8.34 is to be used for a common-base amplifier. Design the circuit to establish a dc emitter current of 1 mA and allowing for a maximum signal swing at the collector of ± 2 V. Use $+10$-V and -5-V power supplies.

Ans. $R_B = 0$; $R_E = 4.3$ kΩ; $R_C = 8$ kΩ

8.9 CLASSICAL SINGLE-STAGE TRANSISTOR AMPLIFIERS

In this section we study two of the three basic configurations of BJT amplifiers: the common-emitter and the common-base circuits, together with variants on their basic structure. The third basic configuration, the common-collector, or emitter-follower, circuit will be studied in the next section. Although the circuits will be presented here in capacitively coupled form, the three topologies and their variants are also used in direct-coupled form in the design of IC amplifiers, as will be shown in Chapters 9 and 13.

The Common-Emitter Amplifier

Figure 8.35 shows the complete circuit of a classical BJT amplifier in the *common-emitter (grounded-emitter) configuration*. The circuit uses the biasing arrangement described in the previous section. The common-emitter circuit is quite similar to the common-source JFET and MOSFET amplifiers studied in Chapters 6 and 7.

As indicated in Fig. 8.35, the signal source v_s has a resistance R_s and is coupled to the base of the transistor through capacitor C_{C1}. The value of C_{C1} should be chosen sufficiently large so that it acts almost as a short circuit over the frequency range of interest. The output signal at the collector is coupled to a load resistance R_L through another coupling capacitor C_{C2} that (just as C_{C1}) should be chosen sufficiently large. Finally, the effect of the emitter bias resistance R_E on the signal

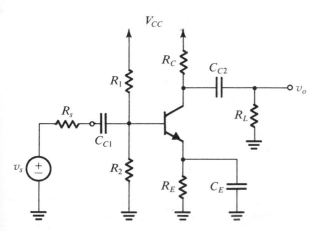

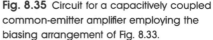

Fig. 8.35 Circuit for a capacitively coupled common-emitter amplifier employing the biasing arrangement of Fig. 8.33.

performance of the amplifier is eliminated by connecting a large capacitor C_E across R_E. This capacitance acts as a short circuit at signal frequencies of interest such that R_E is effectively eliminated as far as signals are concerned. An alternative way of visualizing the action of C_E is to note that it presents a low-impedance path between the emitter and ground. Thus while the dc emitter current will continue to flow through R_E, the signal current i_e will flow through C_E, bypassing R_E. For this reason C_E is called an *emitter bypass capacitor* and the circuit is called a grounded-emitter or common-emitter amplifier. (Recall that we have used bypass capacitors previously in the design of FET amplifiers.) The effect of leaving part of R_E unbypassed will be discussed shortly.

At this point it is important to note that the amplifier circuit in Fig. 8.35 is an ac amplifier whose gain deteriorates at low frequencies. This is due, of course, to the finite values of C_{C1}, C_{C2}, and C_E. The analysis of amplifier frequency response will be considered in Chapter 11; for the time being we shall assume that these capacitances are of infinite value. Doing this, replacing the transistor with its hybrid-π model, and replacing the dc supply with a short circuit, results in the amplifier equivalent circuit shown in Fig. 8.36. Analysis of this circuit proceeds as follows: The transmission from the signal source to the amplifier input is given by

$$\frac{v_\pi}{v_s} = \frac{R_{in}}{R_{in} + R_s} \tag{8.51}$$

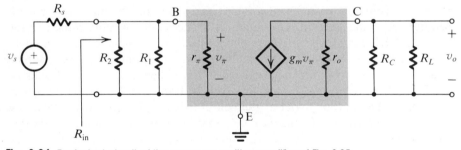

Fig. 8.36 Equivalent circuit of the common-emitter amplifier of Fig. 8.35.

where the amplifier input resistance R_{in} is given by

$$R_{in} = R_1 \| R_2 \| r_\pi$$
$$= R_B \| r_\pi \tag{8.52}$$

Here we note that to keep R_{in} high, R_B should be selected high, in conflict with the requirement for bias stability.

The gain from base to collector is given by,

$$\frac{v_o}{v_\pi} = -g_m[R_L \| R_C \| r_o] \tag{8.53}$$

In many cases $r_o \gg (R_L \| R_C)$ and may be neglected. The overall voltage gain, from source to load, can be obtained by combining Eqs. (8.51) and (8.53).

The common-emitter amplifier is capable of providing high voltage gain. Its input resistance, however, is rather low because r_π is usually only a few kilohms. The output resistance is $(R_C \| r_o)$ and is moderately high since R_C is in the kilohm range. The magnitude of the input signal is limited by the fact that v_π should not exceed about 10 mV; otherwise, nonlinear distortion sets in. A final, but very important, problem encountered with the common-emitter configuration is its limited high-frequency response. To see the origin of this limitation recall that the BJT collector–base junction capacitance C_μ appears between the two nodes whose voltages are related by the large gain given by Eq. (8.53). Thus C_μ gets multiplied by the Miller factor $(1 - v_o/v_\pi)$ and gives rise to a large capacitance between the transistor input terminals. This input capacitance, together with R_{in} and R_s, form a low-pass filter with a relatively low cutoff frequency. Detailed frequency response analysis of the CE configuration will be carried out in Chapter 11 where methods for extending the high-frequency response are also discussed.

Exercises

8.21 For the common-emitter amplifier whose bias design was carried out in Example 8.10, calculate R_{in}, v_b/v_s, and v_o/v_s for both design 1 and design 2. Let $\beta = 100$, $R_s = 4$ kΩ, and $R_L = 4$ kΩ. Neglect r_o.

Ans. Design 1: $R_{in} = 2.5$ kΩ, $v_b/v_s = 0.38$, $v_o/v_s = -28.0$ V/V; design 2: $R_{in} = 1.3$ kΩ, $v_b/v_s = 0.25$, $v_o/v_s = -19.8$ V/V

8.22 For the amplifier of Exercise 8.21 above, find the maximum allowed amplitude of v_s (for both design 1 and design 2) under the constraint that the amplitude of v_{be} is limited to 10 mV. Also find the output signal amplitude that would result in each case.

Ans. Design 1: $\hat{v}_s = 26.3$ mV, $\hat{v}_o = 0.74$ V; design 2: $\hat{v}_s = 40$ mV, $\hat{v}_o = 0.79$ V

The results of Exercise 8.21 indicate that the gain obtained with design 2 is considerably smaller than that obtained with design 1. Thus the price paid for the more stable operating point is a reduction in gain (in addition to an increase in current drain on the supply).

The Common-Emitter Circuit with Unbypassed Emitter Resistance

Leaving part of the emitter resistance unbypassed as shown in Fig. 8.37 provides us with an additional design parameter that can be used to improve some performance aspects of the CE circuit at the expense of a reduction in gain. We shall now analyze the circuit of Fig. 8.37 and point out the design trade-offs available. Analysis can be performed by replacing the BJT with its hybrid-π model, as we have done for the CE circuit above. For variety, and because it is in fact simpler, we shall carry out the analysis directly on the circuit diagram, as indicated in Fig. 8.37. To simplify matters we shall neglect the effect of r_o.

In order to find the fraction of the input signal that appears at the transistor base, v_b, we first need to evaluate the input resistance R_{in}. Specifically, the transmission from the source to the transistor base is given by

$$\frac{v_b}{v_s} = \frac{R_{in}}{R_s + R_{in}} \tag{8.54}$$

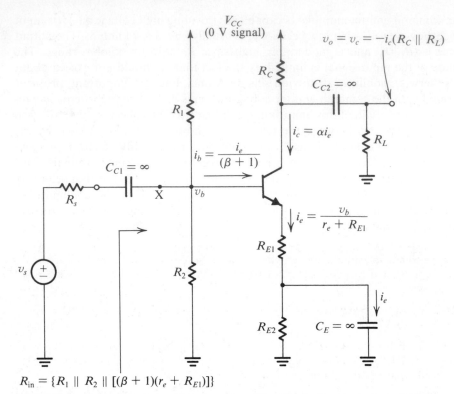

$$R_{in} = \{R_1 \parallel R_2 \parallel [(\beta + 1)(r_e + R_{E1})]\}$$

Fig. 8.37 Small-signal analysis of the common-emitter circuit with an unbypassed emitter resistance.

To evaluate R_{in} imagine yourself standing at point X (Fig. 8.37) and looking to the right. Between X and ground you will see R_2 in parallel with R_1 (note that a power supply is considered a signal ground) in parallel with the input resistance looking into the base of the transistor. The latter resistance is defined as

$$R_b \equiv \frac{v_b}{i_b} \qquad (8.55)$$

But we have

$$v_b = i_e r_e + i_e R_{E1}$$

and

$$i_b = \frac{i_e}{\beta + 1}$$

leading to

$$R_b = (\beta + 1)(r_e + R_{E1}) \qquad (8.56)$$

This is an important result: It says that *the input resistance looking into the base of a transistor is equal to the total resistance in its emitter multiplied by the factor $\beta + 1$.* This factor obviously arises because the base current is $\beta + 1$ times smaller than the emitter current. [Are you wondering where r_π has gone? It is still there since $(\beta + 1)r_e = r_\pi$.]

In summary, the input resistance is given by

$$R_{\text{in}} = \{R_1 \parallel R_2 \parallel [(\beta + 1)(r_e + R_{E1})]\} \tag{8.57}$$

This formula clearly shows the effect of the unbypassed emitter resistance R_{E1} on increasing the input resistance. This attribute of emitter resistance is exploited in the design of amplifiers with high input resistance, as will be illustrated in Chapter 9.

Figure 8.37 illustrates some of the details of calculating the voltage gain v_o/v_s. Starting from the signal at the transistor base v_b we find the signal current in the emitter i_e as

$$i_e = \frac{v_b}{r_e + R_{E1}}$$

Thus the collector signal current i_c will be

$$i_c = \alpha i_e$$

$$= \frac{\alpha v_b}{r_e + R_{E1}}$$

To obtain the output voltage v_o, we multiply i_c by the total resistance between collector and ground, which is seen to be $(R_C \parallel R_L)$; thus

$$v_o = v_c = -\alpha i_e (R_C \parallel R_L)$$

$$= \frac{-\alpha v_b}{r_e + R_{E1}} (R_C \parallel R_L)$$

Since $\alpha \simeq 1$, the voltage gain between base and collector is given by

$$\frac{v_o}{v_b} \simeq -\frac{(R_C \parallel R_L)}{r_e + R_{E1}} \tag{8.58}$$

which is simply *the ratio of the total resistance in the collector lead to the total resistance in the emitter lead.* This is a handy, easy-to-remember rule for calculating gain. The overall voltage gain v_o/v_s can be obtained by multiplying v_o/v_b [Eq. (8.58)] by the transmission from the source to the base, v_b/v_s [Eq. (8.54)].

Exercise

> **8.23** Repeat Exercise 8.21 for the case in which a resistance $R_{E1} = 425\ \Omega$ is left unbypassed.
>
> **Ans.**   Design 1: $R_{\text{in}} = 16.8\ \text{k}\Omega$, $v_b/v_s = 0.81$, $v_o/v_s = -3.6$ V/V; design 2: $R_{\text{in}} = 2.5\ \text{k}\Omega$, $v_b/v_s = 0.38$, $v_o/v_s = -1.7$ V/V

Comparison of the results of Exercise 8.23 with those of Exercise 8.21 indicates that leaving part of the emitter resistance unbypassed increases the input resistance and thus decreases the loss of signal strength in coupling the source to the amplifier input. The price paid for these improvements is a substantial reduction in *overall* voltage gain. Not apparent from these results, however, are two important advantages of including R_{E1}. The first is that the bandwidth of the amplifier is extended, as will be shown in Chapter 11. The second is that one is able to apply signals with larger amplitude without running the risk of inducing large amounts of nonlinear distortion. The reason for the increased signal-handling capability is that only a fraction of the input signal v_s appears as v_{be},

$$v_{be} = v_b \frac{r_e}{r_e + R_{E1}}$$

Thus the signal-handling capability is increased by the factor $(r_e + R_{E1})/r_e$.

The results of this exercise show that substantially larger signals can be applied to the amplifier with unbypassed emitter resistance than to the standard CE configuration. Note, however, that in all cases the resulting signal swing at the collector is about ± 0.8 V. This points out a serious shortcoming in our bias design in Example 8.10. Having allowed 4 V for the collector signal swing, we are forced by considerations of nonlinear distortion to operate with only a fraction of this. Obviously we can increase R_C and thus obtain higher gain and a larger output signal swing.

The Common-Base Configuration

Figure 8.38 shows the *common-base (CB)* or *grounded-base* BJT amplifier. The circuit uses the same biasing arrangement employed in the common-emitter configuration. Here the base is short-circuited to ground, at signal frequencies, using capacitor C_B. The signal source is coupled to the emitter via capacitor C_{C1} and the collector signal is coupled to the load R_L via capacitor C_{C2}.

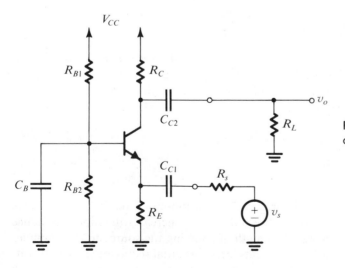

Fig. 8.38 The common-base amplifier configuration.

Assuming that C_B, C_{C1}, and C_{C2} are behaving as perfect short circuits, and replacing the BJT with its hybrid-π model we obtain the equivalent circuit shown in

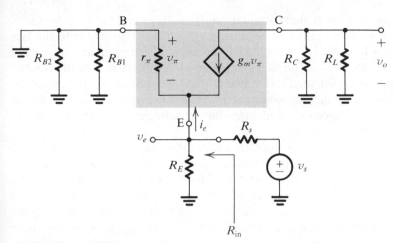

Fig. 8.39 Equivalent circuit of the common-base amplifier of Fig. 8.38 assuming that C_B, C_{C1}, and C_{C2} behave as signal short circuits.

Fig. 8.39. Note that for simplicity we have neglected the effect of r_o. To find the fraction of the source signal that appears at the emitter, v_e, we use the voltage divider rule

$$\frac{v_e}{v_s} = \frac{R_{in}}{R_{in} + R_s} \tag{8.59}$$

where R_{in}, the input resistance, consists of R_E in parallel with the resistance looking into the emitter. Since the base is grounded, the resistance looking into the emitter is r_e. This can be verified using the equivalent circuit of Fig. 8.39. Specifically note that

$$i_e = \frac{-v_\pi}{r_\pi} - g_m v_\pi = -v_\pi \left(\frac{1}{r_\pi} + g_m \right) = -\frac{v_\pi}{r_e}$$

But $v_\pi = -v_e$, leading to

$$\frac{v_e}{i_e} = \frac{-v_\pi}{-v_\pi / r_e}$$

Thus

$$R_{in} = R_E \| r_e \tag{8.60}$$

Since $r_e \ll R_E$,

$$R_{in} \simeq r_e \tag{8.61}$$

The output voltage v_o can be determined as

$$v_o = -g_m v_\pi (R_C \| R_L)$$
$$= g_m v_e (R_C \| R_L) \tag{8.62}$$

Combining Eqs. (8.59) and (8.62) gives the overall gain. We observe that the gain has a positive sign, indicating that the CB configuration does not provide signal inversion.

An important point to note is that although the gain from emitter to collector is large [Eq. (8.62)], the overall gain can be small if R_s is large. This is due to the fact that R_{in} is very small [Eq. (8.61)]. Thus the CB amplifier is seldom used as a voltage amplifier. Rather it finds application as a current amplifier where its low input resistance becomes an asset rather than a disadvantage. More specifically, the common-base configuration is usually employed as a *current buffer* or *current follower*, accepting input current at a low impedance and providing an almost equal current at the collector at a high impedance level.

The most important feature of the CB configuration is that it does not suffer from the Miller capacitance multiplication effect that limits the high frequency response of the common-emitter configuration. This is because each of C_μ and C_π has a grounded terminal. It follows that combining the common-emitter and the common-base circuits together in the configuration shown in Fig. 8.40, known as the *cascode configuration*, can result in an amplifier having the advantages of both separate configurations. Here transistor Q_1 operates in the common-emitter configuration, thus providing the amplifier with a relatively high input resistance. The collector current of Q_1 is fed to the emitter of transistor Q_2, which operates in the common-base configuration. The low input resistance of Q_2 appears as the load resistance of Q_1. Thus the voltage gain from the base to collector of Q_1 is low; in fact about -1. Hence the Miller effect on C_μ of Q_1 is considerably reduced and the high-frequency response extended. Meanwhile, Q_2 delivers a collector current, almost equal to the collector current of Q_1, to the combination of $(R_L \| R_C)$. Thus the overall voltage gain will be almost equal to that obtained with a common-emitter amplifier. The frequency response of the cascode circuit will be analyzed in Chapter 11. The cascode circuit finds application in the design of IC amplifiers, as will be shown in Chapters 9 and 13.

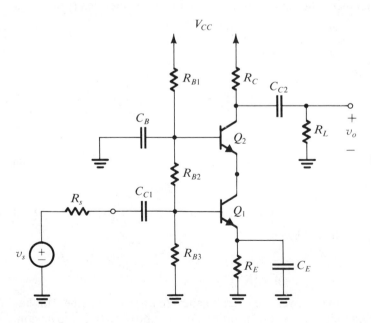

Fig. 8.40 The cascode configuration.

Exercises

8.10 THE EMITTER FOLLOWER

In this section we shall study an extremely useful single-transistor amplifier circuit: the *emitter follower*, or the *common-collector configuration*. Like the source follower, studied in Sections 6.9 and 7.5, the emitter follower is characterized by a high input resistance and a low output resistance. It therefore is useful as an isolation or buffer amplifier to connect a high-resistance source to a low-resistance load.

Figure 8.41 shows a capacitively coupled emitter follower utilizing a biasing arrangement similar to that discussed in Section 8.8. The emitter follower can be used as a direct-coupled amplifier stage, as will be illustrated later. The results derived in this section apply also to the direct-coupled emitter follower.

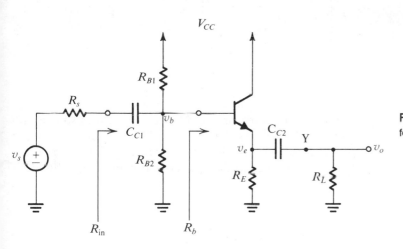

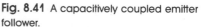

Fig. 8.41 A capacitively coupled emitter follower.

As shown in Fig. 8.41, the input signal is applied to the base and the output signal is taken from the emitter. Since the collector is connected to the dc supply, it will have zero signal voltage and thus act as a signal ground, giving rise to the name *grounded-collector* or *common-collector configuration*. The name *emitter follower* arises because the voltage at the emitter follows that at the input, as will become apparent shortly.

DC Analysis

Analysis of the emitter-follower circuit to determine dc bias quantities is straightforward and follows the techniques previously explained. It should be pointed out, however, that in designing an emitter follower one normally chooses large values for the bias resistances R_{B1} and R_{B2} in order to keep the input resistance high and in spite of the resulting increased dependence of I_E on the value of β. It will be shown that this dependence is not critical in emitter-follower design because the voltage gain is not highly dependent on the value of I_E.

Exercise

8.27 Find all dc voltages and currents in the circuit of Fig. 8.41 for $V_{CC} = +15$ V, $R_{B1} = R_{B2} = 100$ kΩ, $R_E = 2$ kΩ, and $\beta = 100$.

Ans. See Fig. E8.27

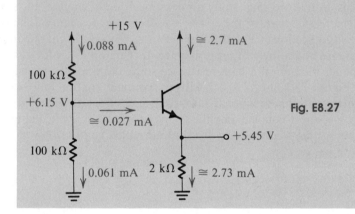

Fig. E8.27

Voltage Gain

We now wish to find the voltage gain of the emitter follower driven by a source with resistance R_s and connected to a load R_L, as shown in Fig. 8.41. To obtain the input resistance R_b between base and ground, we use the resistance reflection rule formulated in Section 8.9; that is, the total resistance in the emitter lead is multiplied by $\beta + 1$ to obtain R_b:

$$R_b = (\beta + 1)[r_e + (R_E \| R_L)]$$

Using this value for R_b we can obtain R_{in} from

$$R_{in} = (R_{B1} \| R_{B2} \| R_b) \tag{8.63}$$

Having determined R_{in}, we can now find the transmission between the source and the base as

$$\frac{v_b}{v_s} = \frac{R_{in}}{R_{in} + R_s} \tag{8.64}$$

The output voltage v_o is equal to the emitter voltage v_e and can be determined using the voltage-divider rule relating v_e and v_b:

$$\frac{v_e}{v_b} = \frac{(R_E \| R_L)}{(R_E \| R_L) + r_e} \tag{8.65}$$

Combining Eqs. (8.64) and (8.65) and replacing v_e by v_o gives the gain expression

$$\frac{v_o}{v_s} = \frac{R_{in}}{R_{in} + R_s} \frac{(R_E \| R_L)}{(R_E \| R_L) + r_e} \tag{8.66}$$

where R_{in} is given by Eq. (8.63). We observe that the voltage gain is less than unity. However, since R_{in} is usually large (because of the β multiplication effect) and since r_e is usually quite small, the gain is usually close to unity. The advantage of the emitter follower, of course, lies in its high input resistance, which enables us to couple a high-resistance source to a low-resistance load without loss of signal.

Exercise

8.28 Consider the emitter follower whose dc analysis was carried out in Exercise 8.27. If $R_s = 5 \text{ k}\Omega$ and $R_L = 1 \text{ k}\Omega$, find the input resistance R_{in} and the voltage gain.

Ans. 28.9 kΩ; 0.84 V/V

Use of the Equivalent Circuit Model

In the above we have performed signal analysis directly on the circuit—that is, with the equivalent circuit model *implicitly* assumed. It is worthwhile at this point to consider the explicit use of equivalent circuit models. Figure 8.42 shows the equivalent circuit model of the emitter follower where the transistor has been replaced by its simplified hybrid-π model (r_o has been neglected). The equivalent circuit of Fig. 8.42 can be used to find R_{in} and the voltage gain by performing straightforward circuit analysis. The reader is urged to carry out such an analysis and verify that the results are identical to those obtained above. Specifically, we urge the reader to derive the

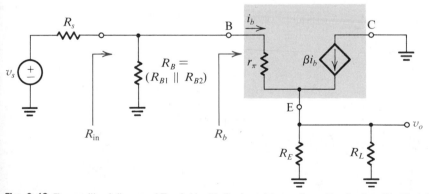

Fig. 8.42 The emitter follower of Fig. 8.41 with the transistor replaced by its simplified hybrid-π model.

resistance-reflection relationship. We note that if r_o were included, it would simply appear in parallel with R_E and R_L.

Output Resistance

An alternative way of describing the performance of the emitter follower is to specify its open-circuit voltage gain and its output resistance—that is, to specify the Thévenin equivalent circuit at the output of the follower. The open-circuit voltage gain A_{vo} can be obtained from the formula derived earlier with R_L set equal to ∞:

$$A_{vo} = \frac{R_i}{R_i + R_s} \frac{R_E}{R_E + r_e} \tag{8.67}$$

where

$$R_i = [R_{B1} \| R_{B2} \| (\beta + 1)(r_e + R_E)] \tag{8.68}$$

To obtain the output resistance R_o we will make use of the equivalent circuit of Fig. 8.42. In this way we shall develop a simple rule for reflecting resistances from the base side to the emitter side. As will be seen, this rule is the inverse of the one that we have been using to reflect resistances from the emitter side to the base side.

Figure 8.43 shows the equivalent circuit of the emitter follower with the voltage source v_s reduced to zero and the source resistance R_s left in the circuit. To obtain the output resistance R_o we have applied a voltage source v. Our task now is to find the current i drawn from this source, since by definition

$$R_o \equiv \frac{v}{i} \tag{8.69}$$

From the circuit in Fig. 8.43 we find

$$i = \frac{v}{R_E} - (\beta + 1)i_b \tag{8.70}$$

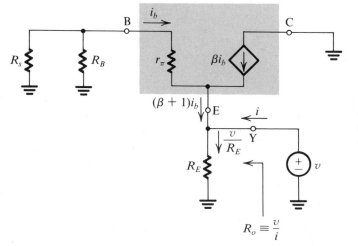

Fig. 8.43 Illustrating the method for evaluating the output resistance R_o of the emitter follower.

We now need a relationship between i_b and v. This is easily obtained by noting that v is the voltage across r_π in series with $(R_B \| R_s)$,

$$i_b = -\frac{v}{r_\pi + (R_B \| R_s)} \tag{8.71}$$

where the minus sign comes about because of the reference directions of v and i_b. Substituting for i_b from Eq. (8.71) into Eq. (8.70) gives

$$i = \frac{v}{R_E} + \frac{(\beta + 1)v}{r_\pi + (R_B \| R_s)} \tag{8.72}$$

from which we obtain

$$\frac{i}{v} = \frac{1}{R_E} + \frac{1}{r_\pi/(\beta + 1) + (R_B \| R_s)/(\beta + 1)}$$

Thus R_o is given by

$$R_o = R_E \| \left[r_\pi/(\beta + 1) + (R_B \| R_s)/(\beta + 1) \right]$$

Since $r_e = r_\pi/(\beta + 1)$, we may write R_o as

$$R_o = R_E \| \left[r_e + (R_B \| R_s)/(\beta + 1) \right] \tag{8.73}$$

which is the final result.

Clearly the above analysis is lengthy, and we have to find a simple rule that enables us to write the final result directly. The rule used is based on the fact that the emitter current is $\beta + 1$ times the base current. Therefore *all resistances on the base side may be reflected to the emitter side after dividing their values by $\beta + 1$.* To apply this rule to the evaluation of the output resistance of the emitter follower, "grab hold" of the point marked Y in Fig. 8.41 and look to the left. Between Y and ground we see the resistance R_E in parallel with a path through the emitter of the transistor. This latter path consists of r_e in series with the resistance between the base and ground reflected to the emitter side. Between the base and ground we see three parallel resistances: R_{B1}, R_{B2}, and R_s. Thus the second path between Y and ground has a resistance $r_e + (R_B \| R_s)/(\beta + 1)$, and the output resistance of the emitter follower will be given by Eq. (8.73).

Examination of Eq. (8.73) reveals that the output resistance of the emitter follower is usually quite small. The reason for the low output resistance is that the source resistance appears divided by $\beta + 1$.

Exercise

8.29 Calculate the open-circuit voltage gain and the output resistance of the emitter follower considered in Exercises 8.27 and 8.28.

Ans.   0.885 V/V; 52.7 Ω

The open-circuit voltage gain A_{vo} and output resistance R_o can be used to evaluate the gain for any load resistance R_L using

$$\text{Voltage gain} = A_{vo} \frac{R_L}{R_L + R_o} \tag{8.74}$$

The result obtained this way should be identical to that found from direct analysis [Eq. (8.66)].

Signal Swing

Finally, we consider the problem of the maximum allowed input signal swing in the emitter-follower circuit. Since only a small fraction of the input signal appears across the base–emitter junction, the emitter follower exhibits linear performance for a large range of input signal amplitude. In fact, the upper limit on the value of the input signal amplitude is usually imposed by transistor cutoff. To see how this comes about, consider as an example an input sine-wave signal (refer to Fig. 8.41). As the input goes negative, the output v_o will also go negative, and the current in R_L will be flowing from ground into the emitter terminal. The total voltage at the emitter will remain positive but will decrease, since $v_E = V_E + v_e$ and since we are considering the time at which v_e is negative. Thus the current in R_E will still be flowing from emitter to ground, but its value will be reduced.

Now, by writing a node equation at the emitter we can find the value of v_e at which the emitter current will be reduced to zero (that is, the transistor cuts off). Denote the peak value of v_e by $\hat{V}_e$. We may write

$$\frac{V_E - \hat{V}_e}{R_E} = \frac{\hat{V}_e}{R_L} \tag{8.75}$$

from which $\hat{V}_e$ can be obtained as

$$\hat{V}_e = \frac{V_E}{1 + R_E/R_L} \tag{8.76}$$

The corresponding value of the amplitude of v_s can be obtained from

$$\hat{V}_s = \frac{\hat{V}_e}{\text{gain}} \tag{8.77}$$

Increasing the amplitude of v_s above this value results in the transistor becoming cut off, and the negative peaks of the output signal waveform will be clipped off.

Exercise

8.30 For the emitter follower considered in Exercises 8.27, 8.28, and 8.29, find the input signal amplitude that causes the transistor to cut off.

Ans. 2.1 V

8.11 THE TRANSISTOR AS A SWITCH—CUTOFF AND SATURATION

Having studied the active mode of operation in detail, we are now ready to complete the picture by considering what happens when the transistor leaves the active region. At one extreme the transistor will enter the cutoff region, while at the other extreme the transistor enters the saturation region. These two extreme modes of operation are very useful if the transistor is to be used as a switch, such as in digital logic circuits. In the present section we shall study the cutoff and saturation modes of operation. The application of the BJT in switching circuits is presented in Chapter 16, which also includes a detailed study of the saturation mode of operation.

Cutoff Region

To help introduce cutoff and saturation, we consider the simple circuit shown in Fig. 8.44, which is fed with a voltage source v_I. We wish to analyze this circuit for different values of v_I.

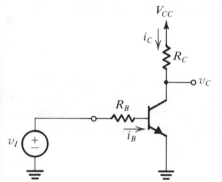

Fig. 8.44 A simple circuit used to illustrate the different modes of operation of the BJT.

If v_I is smaller than about 0.5 V, the EBJ will conduct negligible current. In fact, the EBJ could be considered "reverse"-biased, and the device will be in the cutoff mode. It follows that

$$i_B = 0 \qquad i_E = 0 \qquad i_C = 0 \qquad v_C = V_{CC}$$

Note that the CBJ is, of course, reverse-biased.

Active Region

To turn the transistor on, we have to increase v_I above 0.5 V. In fact, for appreciable currents to flow v_{BE} should be about 0.7 V and v_I should be higher. For $v_I > 0.7$ V we have

$$i_B = \frac{v_I - V_{BE}}{R_B} \tag{8.78}$$

which may be approximated by

$$i_B \simeq \frac{v_I - 0.7}{R_B} \tag{8.79}$$

provided that $v_I \gg 0.7$ V (for example, ≥ 2 V) and that the resulting collector current is in the normal range for this particular transistor. The collector current is given by

$$i_C = \beta i_B \tag{8.80}$$

which applies only if the device is in the active mode. How do we know that the device is in the active mode? We do not know; therefore we assume that it is in the active mode, calculate i_C using Eq. (8.80) and v_C from

$$v_C = V_{CC} - R_C i_C \tag{8.81}$$

and then check whether $v_{CB} \geq 0$ or not. In our case we merely check whether $v_C \geq 0.7$ V or not. If $v_C \geq 0.7$ V, then our original assumption is correct and we have completed the analysis for the particular value of v_I. On the other hand, if v_C is found to be less than 0.7 V, then the device has left the active region and entered the saturation region.

Obviously as v_I is increased, i_B will increase [Eq. (8.79)], i_C will correspondingly increase [Eq. (8.80)], and v_C will decrease [Eq. (8.81)]. Eventually, v_C will become less than v_B (0.7 V) and the device will enter the saturation region.

Saturation Region

Saturation occurs when we attempt to force a current in the collector higher than the collector circuit can support while maintaining active-mode operation. For the circuit in Fig. 8.44 the maximum current that the collector "can take" without the transistor leaving the active mode can be evaluated by setting $v_{CB} = 0$, which results in

$$\hat{I}_C = \frac{V_{CC} - V_B}{R_C} \simeq \frac{V_{CC} - 0.7}{R_C} \tag{8.82}$$

This collector current is obtained by forcing a base current $\hat{I}_B$, given by

$$\hat{I}_B = \frac{\hat{I}_C}{\beta} \tag{8.83}$$

and the corresponding required value of v_I can be obtained from Eq. (8.79). Now, if we increase i_B above $\hat{I}_B$, the collector current will increase and the collector voltage will fall below that of the base. This will continue until the CBJ becomes forward-biased with a forward-bias voltage of about 0.4 to 0.5 V. Note that the forward voltage drop of the collector–base junction is small because the CBJ has a relatively large area (see Fig. 8.6). This situation is referred to as saturation, since any further increase in the base current will result in a very small increase in the collector current and a corresponding small decrease in the collector voltage. This means that in saturation the *incremental* β ($\Delta i_C / \Delta i_B$) is negligibly small. Any "extra" current that we force into the base terminal mostly will flow through the emitter terminal. Thus the ratio of the collector current to the base current of a saturated transistor is *not* equal to β and can be set to any desired value—smaller than β—simply by pushing more current into the base.

Let us now return to the circuit of Fig. 8.44, which we have redrawn in Fig. 8.45 with the assumption that the transistor is in saturation. The value of V_{BE} of a saturated transistor is usually slightly higher than that of the device operating in the active mode. The increase in V_{BE} is due to the increased base current producing a sizable ohmic (IR) voltage drop across the bulk resistance of the base region. In other words, part of V_{BE} will appear across the base semiconductor material as an IR drop, and the remainder will appear across the EBJ. For simplicity, we shall assume V_{BE} to remain around 0.7 V even if the device is in saturation.

Since in saturation the base voltage is higher than the collector voltage by about 0.4 or 0.5 V, it follows that the collector voltage will be higher than the emitter voltage by 0.3 or 0.2 V. This latter voltage is referred to as V_{CEsat}, and we will normally assume

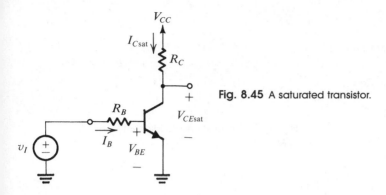

Fig. 8.45 A saturated transistor.

that $V_{CEsat} \simeq 0.3$ V. Note, however, that if we push more current into the base, we drive the transistor "deeper" into saturation and the CBJ forward-bias increases, which means that V_{CEsat} will decrease.

The value of the collector current in saturation will be almost constant. We denote this value by $I_{C_{sat}}$. It follows that for the circuit in Fig. 8.45 we have

$$I_{Csat} = \frac{V_{CC} - V_{CEsat}}{R_C} \tag{8.84}$$

In order to ensure that the transistor is driven into saturation, we have to force a base current of at least

$$I_{Bsat} = \frac{I_{Csat}}{\beta} \tag{8.85}$$

Normally one designs the circuit such that I_B is higher than I_{Bsat} by a factor of 2 to 10 (called the *overdrive factor*). The ratio of I_{Csat} to I_B is called the *forced* β (β_{forced}), since its value can be set at will,

$$\beta_{forced} = \frac{I_{Csat}}{I_B} \tag{8.86}$$

Transistor Inverter

Figure 8.46 shows the transfer characteristic of the circuit of Fig. 8.44. The shape of this curve should be obvious to the reader who has followed the above discussion. The three regions of operation—cutoff, active, and saturation—are indicated in Fig. 8.46. For the transistor to be operated as an amplifier, it should be biased somewhere in the active region, such as at the point marked X. The voltage gain of the amplifier is equal to the slope of the transfer characteristic at this point.

For switching applications the transistor is usually operated in cutoff and saturation.[4] That is, one state of the switch will correspond to the transistor being cut off, and the other state corresponds to the transistor in saturation. There are a number

---

[4] An exception to this is found in emitter-coupled logic (ECL), which is studied in detail in Chapter 16.

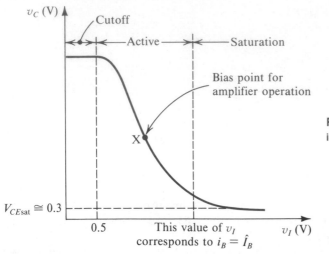

Fig. 8.46 Transfer characteristic for the circuit in Fig. 8.44.

of reasons for choosing these two extreme modes of operation. One reason is that in both cutoff and saturation the currents and voltages in the transistor are well defined and do not depend on such ill-specified parameters as β. Another reason is that in both cutoff and saturation the power dissipated in the transistor is minimal. Although this is obvious for the cutoff mode, it is also the case in the saturation mode, since the voltage V_{CEsat} is small. Finally we note that the circuit we have been using as an example is in fact the basic transistor logic inverter. More will be said about BJT logic circuits in Chapter 16.

Model for the Saturated BJT

From the discussion above we obtain a simple model for transistor operation in the saturation mode, as shown in Fig. 8.47. Normally, we use such a model implicitly in the analysis of a given circuit.

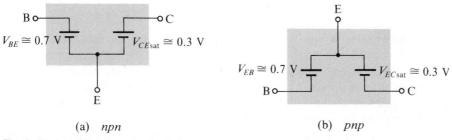

(a) *npn* (b) *pnp*

Fig. 8.47 Model for the saturated BJT.

For quick approximate calculations one may consider V_{BE} and V_{CEsat} to be zero and use the three-terminal short circuit shown in Fig. 8.48 to model a saturated transistor.

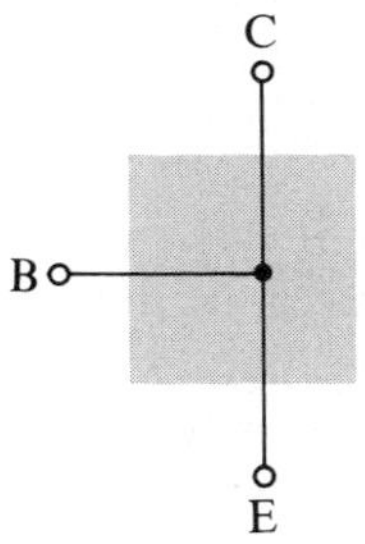

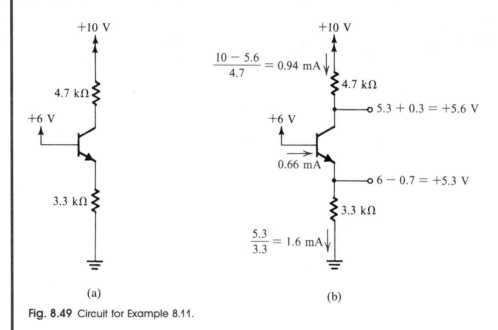

Fig. 8.48 An approximate model for the saturated BJT.

EXAMPLE 8.11

We wish to analyze the circuit in Fig. 8.49a to determine the voltages at all nodes and the currents in all branches. Assume the transistor β is specified to be *at least* 50.

(a)

(b)

Fig. 8.49 Circuit for Example 8.11.

Solution

We have already considered this circuit in Example 8.2 and discovered that the transistor has to be in saturation. Assuming this to be the case, we have

$$V_E = +6 - 0.7 = +5.3 \text{ V}$$

$$V_C = V_E + V_{CEsat} \simeq +5.3 + 0.3 = +5.6 \text{ V}$$

$$I_E = \frac{V_E}{3.3} = \frac{5.3}{3.3} = 1.6 \text{ mA}$$

$$I_C = \frac{+10 - 5.6}{4.7} = 0.94 \text{ mA}$$

$$I_B = I_E - I_C = 1.6 - 0.94 = 0.66 \text{ mA}$$

Thus the transistor is operating at a forced β of

$$\beta_{\text{forced}} = \frac{I_C}{I_B} = \frac{0.94}{0.66} = 1.4$$

Since β_{forced} is less than the *minimum* specified value of β, the transistor is indeed saturated. We should emphasize here that in testing for saturation the minimum value of β should be used. By the same token, if we are designing a circuit in which a transistor is to be saturated, the design should be based on the minimum specified β. Obviously, if a transistor with this minimum β is saturated, then transistors with higher values of β will also be saturated. The details of the analysis are shown in Fig. 8.49b.

EXAMPLE 8.12

The transistor in Fig. 8.50 is specified to have β in the range 50 to 150. Find the value of R_B that results in saturation with an overdrive factor of at least 10.

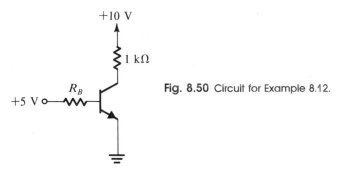

Fig. 8.50 Circuit for Example 8.12.

Solution

When the transistor is saturated the collector voltage will be

$$V_C = V_{CE\text{sat}} \simeq 0.3 \text{ V}$$

Thus the collector current is given by

$$I_{C\text{sat}} = \frac{+10 - 0.3}{1} = 9.7 \text{ mA}$$

To saturate the transistor with the lowest β we need to provide a base current of at least

$$I_{B\text{sat}} = \frac{I_{C\text{sat}}}{\beta_{\text{min}}} = \frac{9.7}{50} = 0.194 \text{ mA}$$

For an overdrive factor of 10, base current should be

$$I_B = 10 \times 0.194 = 1.94 \text{ mA}$$

Thus we require a value of R_B such that

$$\frac{+5 - 0.7}{R_B} = 1.94$$

$$R_B = \frac{4.3}{1.94} = 2.2 \text{ k}\Omega$$

EXAMPLE 8.13

We want to analyze the circuit of Fig. 8.51 to determine the voltages at all nodes and the currents through all branches. The minimum value of β is specified to be 30.

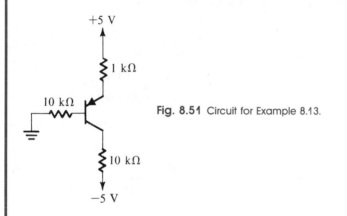

Fig. 8.51 Circuit for Example 8.13.

Solution
A quick glance at this circuit reveals that the transistor will be either active or saturated. Assuming active-mode operation and neglecting the base current, we see that the base voltage will be approximately zero volts, the emitter voltage will be approximately $+0.7$ V, and the emitter current will be approximately 4.3 mA. Since the maximum current that the collector can support while the transistor remains in the active mode is approximately 0.5 mA, it follows that the transistor is definitely saturated.

Assuming that the transistor is saturated and denoting the voltage at the base by V_B, it follows that

$$V_E = V_B + V_{EB} \simeq V_B + 0.7$$

$$V_C = V_E - V_{EC\text{sat}} \simeq V_B + 0.7 - 0.3 = V_B + 0.4$$

$$I_E = \frac{+5 - V_E}{1} = \frac{5 - V_B - 0.7}{1} = 4.3 - V_B \qquad \text{mA}$$

$$I_B = \frac{V_B}{10} = 0.1 V_B \qquad \text{mA}$$

$$I_C = \frac{V_C - (-5)}{10} = \frac{V_B + 0.4 + 5}{10} = 0.1 V_B + 0.54 \qquad \text{mA}$$

Using the relationship

$$I_E = I_B + I_C$$

we obtain

$$4.3 - V_B = 0.1V_B + 0.1V_B + 0.54$$

which results in

$$V_B = \frac{3.76}{1.2} \simeq 3.13 \text{ V}$$

Substituting in the equations above, we obtain

$$V_E = 3.83 \text{ V}$$

$$V_C = 3.53 \text{ V}$$

$$I_E = 1.17 \text{ mA}$$

$$I_C = 0.853 \text{ mA}$$

$$I_B = 0.313 \text{ mA}$$

(note that I_E does not exactly equal $I_B + I_C$ because the value of V_B is approximate). It is clear that the transistor is saturated, since the value of forced β is

$$\beta_{\text{forced}} = \frac{0.853}{0.313} \simeq 2.7$$

which is much smaller than the specified minimum β.

EXAMPLE 8.14

We desire to evaluate the voltages at all nodes and the currents through all branches in the circuit of Fig. 8.52a. Assume $\beta = 100$.

Solution

By examining the circuit we conclude that the two transistors Q_1 and Q_2 cannot be simultaneously conducting. Thus if Q_1 is on, Q_2 will be off, and vice versa. Assume that Q_2 is on. It follows that current will flow from ground through the 1-kΩ load resistor into the emitter of Q_2. Thus the base of Q_2 will be at a negative voltage, and base current will be flowing out of the base through the 10-kΩ resistor and into the +5-V supply. This is impossible, since if the base is negative, current in the 10-kΩ resistor will have to flow into the base. Thus we conclude that our original assumption—that Q_2 is on—is incorrect. It follows that Q_2 will be off and Q_1 will be on.

The question now is whether Q_1 is active or saturated. The answer in this case is obvious. Since the base is fed with a +5-V supply and since base current flows into the base of Q_1, it follows that the base of Q_1 will be at a voltage lower than +5 V. Thus the collector–base junction of Q_1 is reverse-biased and Q_1 is in the active mode. It remains only to determine the currents and voltages using techniques already described in detail. The results are given in Fig. 8.52b.

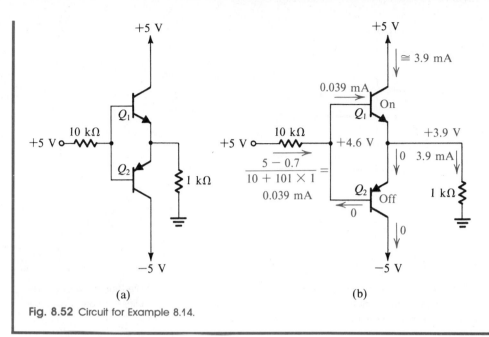

Fig. 8.52 Circuit for Example 8.14.

Exercises

8.31 Consider the circuit of Fig. 8.45 with the input connected to $+5$ V. Let $V_{CC} = 5$ V, $R_C = 1$ kΩ, $R_B = 10$ kΩ, and $\beta = 50$. What is the value of forced β? Find the value of v_I required to establish $\beta_{forced} = \beta/2$.

Ans. 10.9; 2.6 V

8.32 Repeat Example 8.12 with an overdrive factor of 5.

Ans. $R_B = 4.4$ kΩ

8.33 Solve the problem in Example 8.14 with the voltage feeding the bases changed to $+10$ V. Assume that $\beta_{min} = 30$, and find V_E, V_B, I_{C1}, and I_{C2}.

Ans. $+4.7$ V; $+5.4$ V; 4.24 mA; 0

8.12 COMPLETE STATIC CHARACTERISTICS AND GRAPHICAL ANALYSIS

We conclude this chapter with a discussion of the complete static characteristics of the BJT as they appear on its data sheets. We also study a number of secondary characteristics that limit the operation of the BJT in the more advanced circuits presented in subsequent chapters. Graphical analysis of transistor circuits, which can yield additional insight into circuit operation, is also presented.

Common-Base Characteristics

Figure 8.53 shows the complete set of i_C-v_{CB} characteristic curves for an *npn* transistor. As mentioned in Section 8.5, the i_C-v_{CB} characteristics are measured at constant values

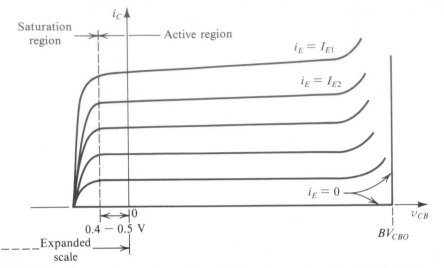

Fig. 8.53 The i_C-v_{CB} or common-base characteristics of an *npn* transistor. Note that in the active region there is a slight dependence of i_C on the value of v_{CB}. The result is a finite output resistance that decreases as the current level in the device is increased.

of emitter current i_E (see Fig. 8.13a). Since in such an arrangement the base is connected to a constant voltage, the i_C-v_{CB} curves are called the *common-base characteristics*.

The curves in Fig. 8.53 differ from those presented in Fig. 8.13 in three aspects. First, the avalanche breakdown of the CBJ at large voltages is indicated and will be briefly explained at a later stage. Second, the saturation-region characteristics are included. As indicated, as v_{CB} goes negative the CBJ becomes forward-biased and the collector current decreases. Since for each curve i_E is held constant, the decrease in i_C results in an equal increase in i_B. The large effect that v_{CB} has on the collector current in saturation is evident from Fig. 8.53 and is consistent with our earlier description of the saturation mode of operation.

The third difference between the characteristic curves of Fig. 8.53 and the curves presented earlier is that in the active region the characteristic curves are shown to have a very small slope. This slope indicates that in the common-base configuration the collector current depends to a small extent on the collector–base voltage, which is a manifestation of the Early effect discussed previously. It should be noted, however, that the slope of the i_C-v_{CB} curves measured with a constant i_E is much smaller than the slope of the i_C-v_{CE} curves measured with a constant v_{BE}. In other words, the output resistance of the common-base configuration is much greater than that of the common-emitter circuit with a constant v_{BE} (that is, r_o). Another important point to note here is that since each i_C-v_{CB} curve is measured at a constant i_E, the increase in i_C with v_{CB} implies a corresponding decrease in i_B. The dependence of i_B on v_{CB} can be modeled by the addition of a resistor r_μ between collector and base in the hybrid-π model, resulting in the augmented model shown in Fig. 8.54. The resistance r_μ is very large, greater than βr_o.

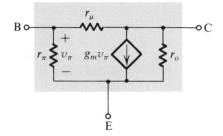

Fig. 8.54 The hybrid-π model, including the resistance r_μ, which models the effect of v_C on i_b.

The augmented hybrid-π model of Fig. 8.54 can be used to find the output resistance of the common-base configuration, which is the inverse of the slope of the i_C-v_{CB} characteristic lines in Fig. 8.53. To do that, simply ground the base, leave the emitter open-circuited (because i_E is constant), apply a test voltage between collector and ground, and find the current drawn from the test voltage. The result (Problem 8.56) is that the output resistance is approximately equal to the parallel equivalent of r_μ and βr_o, and thus is very large.

Common-Emitter Characteristics

An alternative way of graphically displaying the transistor characteristics is shown in Fig. 8.55, where i_C is plotted versus v_{CE} for various values of the base current i_B. These characteristics are measured in a different way from those of Fig. 8.14. While in the latter case v_{BE} is held constant for every curve, here i_B is held constant. As a result the slope in the active region is different from $1/r_o$; in fact, the slope here is greater. It can be shown using the hybrid-π model of Fig. 8.54 that the output resistance of the common-emitter configuration with i_B held constant is approximately equal to $[r_o \,\|\, (r_\mu/\beta)]$; see Problem 8.55.

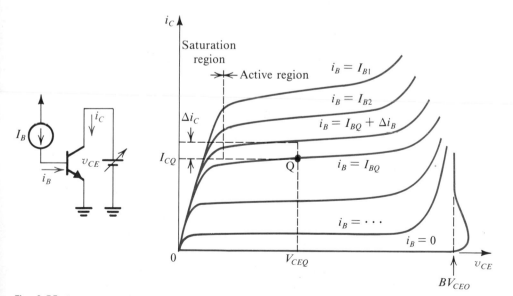

Fig. 8.55 Common-emitter characteristics.

The saturation region is evident also on the common-emitter characteristics of Fig. 8.55. We note that whereas the transistor in the active region acts as a current source with a high (but finite) output resistance, in the saturation region it behaves as a "closed switch" with a small "closure resistance" R_{CEsat}. Since the characteristic curves are all "bunched together" in saturation, we show an expanded view of the saturation portion of the characteristics in Fig. 8.56. Note that the curves do not

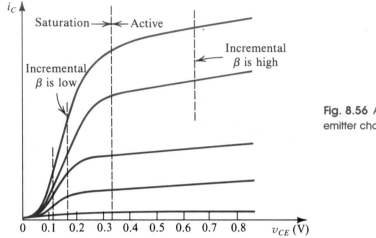

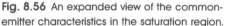

Fig. 8.56 An expanded view of the common-emitter characteristics in the saturation region.

extend directly to the origin. In fact, for a given value of i_B the i_C-v_{CE} characteristic in saturation can be approximated by a straight line intersecting the v_{CE} axis at a point V_{CEoff}, as illustrated in Fig. 8.57. The voltage V_{CEoff} is called the *offset voltage* of the transistor switch. Field-effect transistors (Chapters 6 and 7) do not exhibit such offset voltages and hence make superior switches. FETs, however, display higher values of closure resistance.

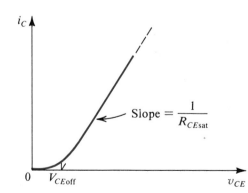

Fig. 8.57 The i_C-v_{CE} characteristic in the saturation region. Note that the characteristic can be modeled by an offset voltage V_{CEoff} and a small resistance R_{CEsat}.

The Transistor β

Earlier we defined β as the ratio of the total current in the collector to the total current in the base when the transistor is operating in the active mode. Let us be more specific.

Assume that the transistor is operating at a base current I_{BQ}, a collector current I_{CQ}, and a collector-to-emitter voltage V_{CEQ}. These quantities define the operating or bias point Q in Fig. 8.55. The ratio of I_{CQ} to I_{BQ} is called the dc β or h_{FE} (the reason for the latter name will be explained in Chapter 11),

$$h_{FE} = \beta_{dc} = \frac{I_{CQ}}{I_{BQ}} \qquad (8.87)$$

When the transistor is used as an amplifier it is first biased at a point such as Q. Applied signals then cause incremental changes in i_B, i_C, and v_{CE} around the bias point. We may therefore define an incremental or ac β as follows: Let the collector-to-emitter voltage be maintained constant at V_{CEQ} (in order to eliminate the Early effect), and change the base current by an increment Δi_B. If the collector current changes by an increment Δi_C (see Fig. 8.53), then β_{ac} (or h_{fe}, as it is usually called) at the operating point Q is defined as

$$h_{fe} \equiv \beta_{ac} \equiv \frac{\Delta i_C}{\Delta i_B}\bigg|_{V_{CE}=\text{constant}} \qquad (8.88)$$

The fact that V_{CE} is held constant implies that the incremental voltage v_{ce} is zero; therefore h_{fe} is called the *short-circuit current gain*.

When we perform small-signal analysis, the β should be the ac β (h_{fe}). On the other hand, if we are analyzing or designing a switching circuit, β_{dc} (h_{FE}) is the appropriate β. The difference in value between β_{dc} and β_{ac} is usually small, and we will not normally distinguish between the two. A point worth mentioning, however, is that the value of β depends on the current level in the device, and the relationship takes the form shown in Fig. 8.58.

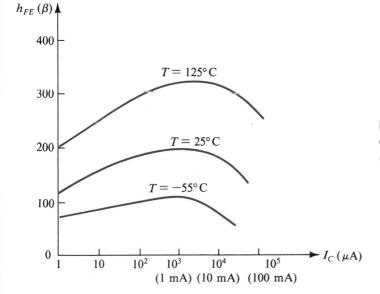

Fig. 8.58 Typical dependence of β on i_C and on temperature in a modern integrated-circuit *npn* silicon transistor.

Transistor Breakdown

The maximum voltages that can be applied to a BJT are limited by the EBJ and CBJ breakdown effects that follow the avalanche multiplication mechanism described in Section 4.12. Consider first the common-base configuration. The i_C-v_{CB} characteristics in Fig. 8.53 indicate that for $i_E = 0$ (that is, with the emitter open-circuited) the collector–base junction breaks down at a voltage denoted by BV_{CBO}. For $i_E > 0$ breakdown occurs at voltages smaller than BV_{CBO}. Typically BV_{CBO} is greater than 50 V.

Next consider the common-emitter characteristics of Fig. 8.55, which show breakdown occurring at a voltage BV_{CEO}. Here, although breakdown is still of the avalanche type, the effects on the characteristics are more complex than in the common-base configuration. We will not explain these details; it is sufficient to point out that typically BV_{CEO} is about half BV_{CBO}. On the transistor data sheets BV_{CEO} is sometimes referred to as the *sustaining voltage* LV_{CEO}.

Breakdown of the CBJ in either the common-base or common-emitter configuration is not destructive as long as the power dissipation in the device is kept within safe limits. This, however, is not the case with the breakdown of the emitter–base junction. The EBJ breaks down in an avalanche manner at a voltage BV_{EBO} much smaller than BV_{CBO}. Typically BV_{EBO} is in the range 6 to 8 V, and the breakdown is destructive in the sense that the β of the transistor is permanently reduced. This does not prevent use of the EBJ as a zener diode to generate reference voltages in IC design. In such applications, however, one is not concerned with the β-degradation effect. A circuit arrangement to prevent EBJ breakdown in IC amplifiers will be discussed in Chapter 13. Transistor breakdown and maximum allowable power dissipation are important parameters in the design of power amplifiers (Chapter 10).

Graphical Analysis of Transistor Circuits

Although it is of little practical value in the analysis and design of most transistor circuits, it is illustrative to portray with graphical techniques the operation of a simple transistor circuit. Consider the simple circuit shown in Fig. 8.59, which we have already analyzed in a number of examples. A graphical analysis of the circuit may

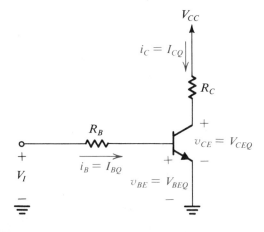

Fig. 8.59 Circuit whose operation is analyzed graphically.

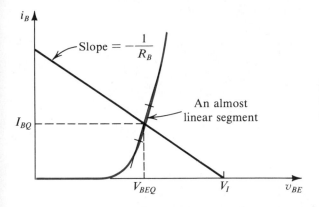

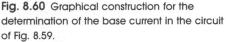

Fig. 8.60 Graphical construction for the determination of the base current in the circuit of Fig. 8.59.

be performed as follows: First, we have to determine the base bias current I_{BQ} using the technique illustrated in Fig. 8.60 (we have employed this technique in the analysis of diode circuits in Chapter 4). We next move to the i_C-v_{CE} characteristics, shown in Fig. 8.61. We know that the operating point will lie on the curve corresponding to the value of base current (I_{BQ}) that we have just determined. Where it lies on the curve will be determined by the collector circuit. Specifically, the collector circuit imposes the constraint

$$v_{CE} = V_{CC} - i_C R_C$$

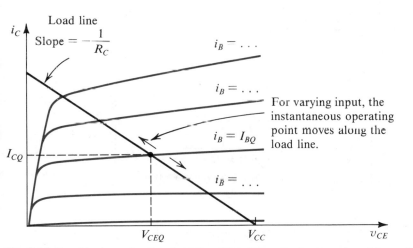

Fig. 8.61 Graphical construction for determining the collector current and the collector-to-emitter voltage in the circuit of Fig. 8.59.

which can be rewritten

$$i_C = \frac{V_{CC}}{R_C} - \frac{1}{R_C} v_{CE} \qquad (8.89)$$

and which represents a linear relationship between v_{CE} and i_C. This relationship can be represented by a straight line, as shown in Fig. 8.61. The operating point Q will be

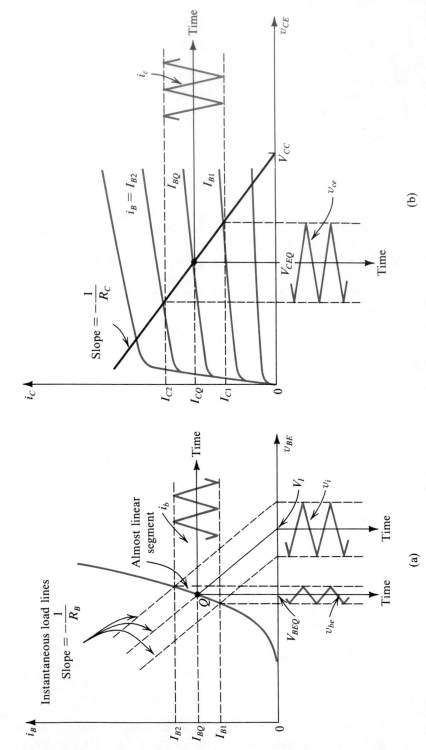

Fig. 8.62 Graphical determination of the signal components v_{be}, i_b, i_c, and v_{ce} when a signal component v_i is superimposed on the dc voltage v_I (see Fig. 8.59).

at the intersection of this straight line, called the load line, and the characteristic curve corresponding to the base current I_{BQ}.

If the load line intersects the characteristic curve corresponding to I_{BQ} at a point in the active region, the transistor will be operating in the active mode. This is the case illustrated in Fig. 8.61. If an input signal v_i is superimposed on V_I, there will be a corresponding base current signal i_b and a base–emitter voltage signal v_{be}. As shown in Fig. 8.62a, if v_i is "small enough," the instantaneous operating point will move along a linear segment of the i_B-v_{BE} exponential curve. On the i_C-v_{CE} characteristics the instantaneous operating point will move along the load line, as illustrated in Fig. 8.62b.

Finally, Fig. 8.63 shows a load line that intersects the i_C-v_{CE} characteristics at a point in the saturation region. Note that in this case changes in the base current result in very small changes in i_C and v_{CE} and that in saturation the incremental β (β_{ac}) is negligibly small.

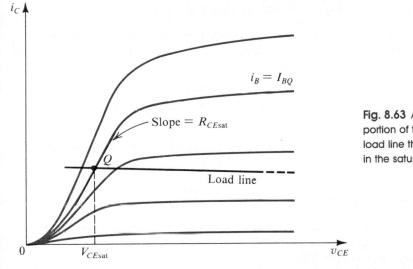

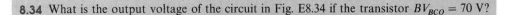

Fig. 8.63 An expanded view of the saturation portion of the characteristics together with a load line that results in operation at a point Q in the saturation region.

Exercises

8.34 What is the output voltage of the circuit in Fig. E8.34 if the transistor $BV_{BCO} = 70$ V?

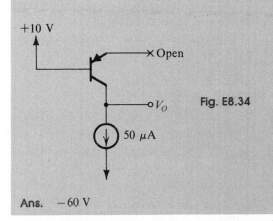

Fig. E8.34

Ans. -60 V

8.35 Measurements made on a BJT switch having a constant base-current dirve at low values of v_{CE} provide the following data: at $i_C = 5$ mA, $v_{CE} = 170$ mV; at $i_C = 2$ mA, $v_{CE} = 110$ mV. What are the values of the offset voltage and saturation resistance of this switch?

Ans. 70 mV; 20 Ω

8.13 SUMMARY

- Depending on the bias conditions on its two junctions, the BJT can operate in one of three possible modes: cutoff (both junctions reverse-biased), active (the EBJ forward-biased and the CBJ reverse-biased), and saturation (both junctions forward-biased).

- For amplifier applications the BJT is operated in the active mode. Switching applications make use of the cutoff and saturation modes.

- A BJT operating in the active mode provides a collector current $i_C = I_S e^{|v_{BE}|/V_T}$. The base current $i_B = i_C/\beta$, and the emitter current $i_E = i_C + i_B$. Also, $i_C = \alpha i_E$ and thus $\beta = \alpha/(1 - \alpha)$ and $\alpha = \beta/(\beta + 1)$.

- To ensure operation in the active mode, the collector voltage of an *npn* transistor must be kept greater than the base voltage. For a *pnp* transistor the collector voltage must be lower than the base voltage.

- The dc analysis of transistor circuits is greatly simplified by assuming that $|V_{BE}| \simeq 0.7$ V.

- To operate as a linear amplifier the BJT is biased in the active region and the signal v_{be} is kept small ($v_{be} \ll V_T$).

- For small signals, the BJT functions as a linear voltage-controlled current source with a transconductance $g_m = I_C/V_T$. The input resistance between base and emitter, looking into the base, is $r_\pi = \beta/g_m$.

- Bias design seeks to establish a dc collector current that is as independent of the value of β as possible.

- In the common-emitter configuration, the emitter is at signal ground, the input signal is applied to the base, and the output is taken at the collector. A high voltage gain and a reasonably high input resistance are obtained, but the high-frequency response is limited.

- The input resistance of the common-emitter amplifier can be increased by including an unbypassed resistance in the emitter lead.

- In the common-base configuration, the base is at signal ground, the input signal is applied to the emitter, and the output is taken at the collector. A high voltage gain and an excellent high-frequency response are obtained, but the input resistance is low.

- The cascode configuration combines the advantages of the common-emitter and the common-base circuits.

● In the emitter follower the collector is at signal ground, the input signal is applied to the base, and the output taken at the emitter. Although the voltage gain is less than unity, the input resistance is very high and the output resistance is very low. The circuit is useful as a buffer amplifier.

● In a saturated transistor, $|V_{CE\text{sat}}| \simeq 0.3$ V and $I_{C\text{sat}} = (V_{CC} - V_{CE\text{sat}})/R_C$. The ratio of $I_{C\text{sat}}$ to the base current is the forced β, which is lower than β. The collector-to-emitter resistance, $R_{CE\text{sat}}$, is small (few tens of ohms).

● With the emitter open-circuited ($i_E = 0$), the CBJ breaks down at a reverse voltage BV_{CBO} that is typically > 50 V. For $i_E > 0$, the breakdown voltage is less than BV_{CBO}. In the common-emitter configuration the breakdown voltage specified is BV_{CEO}, which is about half BV_{CBO}. The emitter base junction breaks down at a reverse bias of 6–8 V. This breakdown usually has an adverse effect on β.

BIBLIOGRAPHY

E. J. Angelo, Jr., *Electronics: BJTs, FETs, and Microcircuits*, New York: McGraw-Hill, 1969.

Circuit Properties of Transistors, Vol. 3 of the SEEC Series, New York: Wiley, 1964.

I. Getreu, *Modeling the Bipolar Transistor*, Beaverton, Ore.: Tektronix, Inc., 1976.

P. R. Gray and R. G. Meyer, *Analysis and Design of Analog Integrated Circuits*, 2nd ed. New York: Wiley, 1984.

P. E. Gray and C. L. Searle, *Electronic Principles*, New York: Wiley, 1971.

C. L. Searle, A. R. Boothroyd, E. J. Angelo, Jr., P. E. Gray, and D. O. Pederson, *Elementary*

PROBLEMS

8.1 Find the base–emitter voltage of an *npn* transistor conducting a current $i_C = 10$ mA and having $I_S = 10^{-14}$ A.

8.2 A transistor similar to that described in Problem 8.1 in all respects, except for having a base–emitter junction area ten times as large, has a base–emitter voltage of 0.70 V. What is the collector current it must be conducting?

8.3 A BJT having a common-emitter current gain $\beta = 100$ operates with a collector current of 10 mA. What base current corresponds? If the available base current is reduced to 1 μA, what must the collector current become?

8.4 For a particular transistor having a very thin base, a base current of 15 μA and a corresponding collector current of 10 mA are measured. What is β for this device?

8.5 For a BJT having a base current of 50 μA and a collector current of 5 mA, what is the emitter current? What is β for this transistor? Using your computed value

of i_E and the given value of i_C, find the value of the common-base current gain α. Verify that $\alpha = \beta/(\beta + 1)$ and that $\beta = \alpha/(1 - \alpha)$.

8.6 The leakage current I_{CBO} of a small transistor is measured to be 10 nA at 25°C. If the temperature of the device is raised to 125°C, what do you expect the leakage current to become?

8.7 Using the *npn* transistor model of Fig. 8.5b, consider the case of a transistor whose base is connected to ground, the collector is connected to a $+5$-V dc supply, and a 10-mA current source is connected to the emitter with the polarity so that current is drawn out of the emitter terminal. If $\beta = 100$ and $I_S = 10^{-14}$ A calculate the emitter voltage, collector current, and base current.

8.8 Figure 8.8 shows two large-signal models for the *pnp* transistor operating in the active mode. Sketch two additional models that parallel those given for the *npn* transistor in Fig. 8.5b and d.

8.9 Consider the *pnp* large-signal model of Fig. 8.8b applied for a transistor having $I_S = 10^{-13}$ A and $\beta = 50$. If the emitter is connected to ground, the base is connected to a current source that pulls out of the base terminal a current of 10 μA, and the collector is connected to a −5-V supply, find the base voltage and the collector current.

8.10 A medium-power *pnp* transistor operates at $i_C = 1$ A with $v_{EB} = 0.7$ V. What do you expect v_{EB} to become at $i_C = 10$ A? At 1 mA?

8.11 From Fig. 8.6 we note that the transistor is not a symmetrical device; that is, interchanging the collector and emitter terminals will result in a device with different values of α and β, called the inverse or reverse values and denoted α_R and β_R. An *npn* transistor is accidentally connected with collector and emitter leads interchanged. The resulting emitter and base currents are 5 mA and 1 mA, respectively. What are the values of α_R and β_R?

8.12 For the circuits in Fig. P8.12 find the labeled currents and voltages. Let $\beta = 100$ and $|V_{BE}| = 0.7$ V.

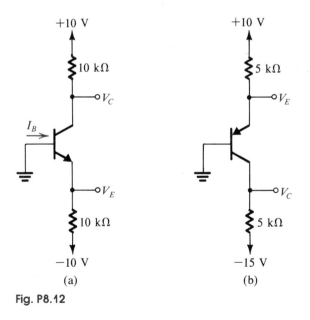

Fig. P8.12

8.13 The transistors in the circuits shown in Fig. P8.13 have very large values of β, so we may assume the base currents to be negligibly small. If, in addition, it is determined by measurement that $|V_{BE}| = 0.7$ V, find the values of the labeled voltages.

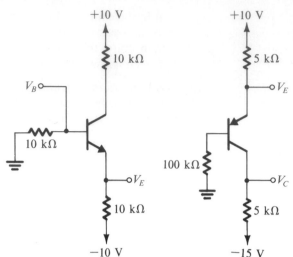

Fig. P8.13

8.14 A single measurement indicates the emitter voltage of the transistor in the circuit of Fig. P8.14 to be 1.0 V. Under the assumption that $V_{BE} = 0.7$ V, what are V_B, I_B, I_E, I_C, β, and α? (Isn't it surprising what a little measurement can lead to?)

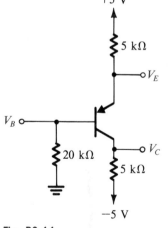

Fig. P8.14

8.15 An *npn* transistor has its base connected to −5 V, its collector connected to ground via a 1-kΩ resistor, and its emitter connected to a 2-mA constant-current source that pulls current out of the emitter terminal. If the base voltage is raised by 0.4 V, what voltage changes are measured at the emitter and at the collector?

*8.16 A *pnp* BJT operated at a constant emitter current is found to have an emitter–base voltage of 0.715 V when an instrument cooling fan is operating. However, when the fan is turned off, the emitter–base voltage drops to 0.700 V in 1 minute and to 0.685 V after 5 minutes or more. What is the eventual temperature rise of the transistor emitter–base junction? Assuming an exponential temperature change with time, what is the thermal time constant of the transistor and its packaging?

8.17 Consider the circuit of Fig. 8.14a. Let V_{BE} be adjusted to yield a current of $I_C = 1$ mA at $V_{CE} = 1$ V. Then, while keeping V_{BE} constant, V_{CE} is raised to $+11$ V and I_C is measured to be 1.2 mA. Find r_o and V_A for this BJT.

8.18 Identify whether the circuits in Fig. P8.18 operate in the active or saturation mode. What is the emitter voltage in each case? If active, what is the collector voltage? $|V_{BE}| = 0.7$ V, $\beta = 100$.

8.19 Identify whether the circuits in Fig. P8.19 operate in the active or saturation mode. What is the base voltage in each case? If active, what is the collector voltage? $|V_{BE}| = 0.7$ V. Consider the cases $\beta = 50$ and $\beta = 100$.

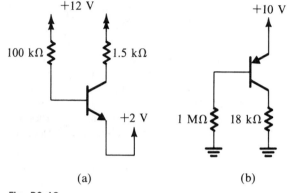

(a)   (b)

Fig. P8.19

8.20 Reconsider Example 8.6 for the case in which $\beta = \infty$. What are the node voltages and branch currents?

8.21 Find the voltages at all nodes and the currents through all branches in the circuit of Fig. P8.21. $V_{BE} = 0.7$ V, $\beta = 50$.

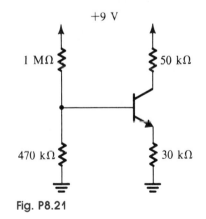

Fig. P8.21

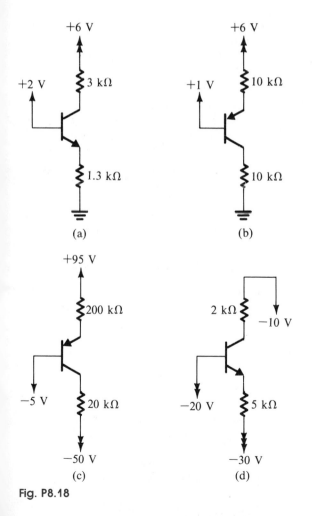

Fig. P8.18

8.22 Consider the circuit of Fig. 8.20 with component values changed as follows: $R_{B1} = 47$ kΩ, $R_{B2} = 22$ kΩ, $R_E = 1$ kΩ, and $R_C = 1.5$ kΩ. For $V_{BE} = 0.7$ V and $\beta = 30$ calculate all node voltages.

8.23 Reconsider the circuit of Fig. 8.21 with the supply reduced to 10 V, initially for $\beta = \infty$, then for $\beta = 100$.

*8.24 In obtaining Eq. (8.21), higher-order terms in the exponential expression have been ignored. What is their total value for $v_{be} = 10$ mV. For what value of v_{be} are they **(a)** 10% of (v_{be}/V_T) and **(b)** 1% of (v_{be}/V_T)?

8.25 A transistor is biased at $I_C = 1$ mA and its base current is measured to be 0.01 mA. Find g_m and r_π at this bias point.

8.26 Using the relationships for r_π and r_e in Eqs. (8.32) and (8.36), show that $r_\pi = (\beta + 1)r_e$.

8.27 Using the BJT model of Fig. 8.25, find **(a)** the resistance looking into the emitter with the base grounded and **(b)** the resistance looking into the base with the emitter grounded. Repeat using the model of Fig. 8.24a, thus showing that identical results can be obtained using either model. (Note that each model has a situation for which it is best suited, and one for which it is adequate, but less convenient.)

8.28 In the circuit of Fig. 8.22 let the BJT be biased so that $g_m = 40$ mA/V and let $R_C = 10$ kΩ. What is the magnitude of the voltage gain? If the bias current is reduced by a factor of 2, what does the gain become?

8.29 Reconsider the analysis of Example 8.8 for the case in which $\beta = 50$. What is the overall voltage gain? What is the peak input signal that is usable? For what value of β does linear operation become impossible?

8.30 Reconsider Example 8.9, particularly the effect of β on the design. What is the voltage gain for $\beta = \infty$? For $\beta = 10$?

8.31 In the design of BJT integrated-circuit amplifiers the arrangement shown in Fig. P8.31 is often used. Here the transistor is biased with a constant-current source feeding the collector. The external circuit (not shown) is arranged so that a stable dc voltage develops at the collector. Using the hybrid-π equivalent circuit model, including r_o, show that the small-signal voltage gain obtained from base to collector is equal to $-(V_A/V_T)$. For $V_A = 100$ V find the value of the gain.

8.32 A redesign of the circuit of Fig. E8.18 results in doubling all resistor values. Calculate the voltage gains v_{o1}/v_i and v_{o2}/v_i that can be expected. If each output is coupled to a load of 10 kΩ, what gains result?

8.33 Consider the circuit of Fig. E8.19.
(a) If X is connected to ground, Z to an input signal source v_i, and Y to a load resistance of 8 kΩ, find the voltage gain v_y/v_i.
(b) If Y is connected to ground, X to an input signal source v_i, and Z to a load resistance of 1 kΩ, find the voltage gain v_z/v_i.

8.34 Consider the bias arrangement for a BJT shown in Fig. 8.33. Provide a design employing a 15-V supply which for $\beta = \infty$ provides $I_E = 10$ mA utilizing $V_{BB} = V_{CC}/3$ and a bias network current of $I_E/10$. For $R_C = R_E$ what are I_E and V_{CE} for $\beta = 100$ and for $\beta = 10$?

8.35 Repeat the design requested in Problem 8.34 but using a bias network current equal to the emitter current for $\beta = \infty$. What are the values of R_1, R_2, and R_E? For $R_C = R_E$ what are I_E and V_{CE} for $\beta = 100$ and for $\beta = 10$?

8.36 Design a bias circuit of the type shown in Fig. 8.33 to establish a bias current of 2 mA using a 9-V supply and the "one-third" rule. Choose the bias network current so that I_E reduces by no more than 10% when β falls from ∞ to 100. What are R_E, R_2, R_1, and R_C for your design?

*8.37 An amplifier having the topology shown in Fig. 8.35 is to be designed using a 12-V power supply to drive a 5-kΩ load resistance R_L. For this design $R_C = R_L$, $V_{BB} = V_{CC}/3$, $V_{CE} = V_{CC}/6$, and $R_2 = 10R_E$. What are the required values of R_C, R_E, R_2, R_1, and I_E for $\beta = \infty$. For $\beta = 100$ what does I_E become? With $C_E = \infty$ what is the input resistance? What is the voltage gain from the base to the load? From the input with $R_s = 10$ kΩ to the load?

*8.38 In the circuit of Fig. 8.37, $R_s = 10$ kΩ, $R_1 = 50$ kΩ, $R_2 = 27$ kΩ, $R_{E2} = 2.7$ kΩ, $R_{E1} = 270$ Ω, $R_C = 4.7$ kΩ, $V_{CC} = 12$ V, and $R_L = 5$ kΩ. Neglecting the effect of finite β, find the dc bias current in the emitter. For $\beta = 150$ find the input resistance and the voltage gain from source to load. What is the source voltage and corresponding load voltage for which v_{be} is limited to 10 mV?

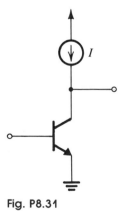

Fig. P8.31

8.39 If in Problem 8.38, R_1 and R_2 are reduced to 5 kΩ and 2.7 kΩ to ensure bias stability at low β, what does the voltage gain from source to load become with $\beta = 150$?

*__8.40__ For a BJT amplifier having a collector resistance $R_C = 5$ kΩ, $R_L = \infty$, and $g_m = 40$ mA/V, what is the peak output voltage for which the peak v_{be} is 10 mV? For an unbypassed emitter resistor of 175 Ω what peak signal voltage is allowed at the base? To maintain operation in the range where the transistor just leaves the active region ($v_{CB} = 0$) with this signal, what is the minimum value of the collector–base dc voltage? If one-third of V_{CC} is used between base and ground (to obtain dc bias stability), find the minimum required value of V_{CC}.

**_8.41_ Under the conditions that an input signal source can accept a small amount of dc current, the circuit of Fig. P8.41 provides a convenient way to bias an amplifier in the common-base configuration. Assuming that the source signal has a zero dc component and that $\beta = 100$, find the collector bias current. Calculate the value of the input resistance and the overall voltage gain v_o/v_s. What limits the size of input signal for linear operation: the 10-mV limitation on v_{be} or the transistor leaving the active region?

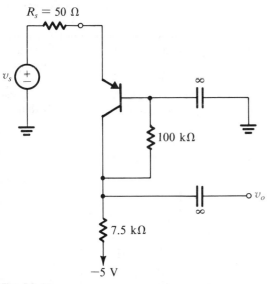

Fig. P8.41

**_8.42_ The bias arrangement of Fig. P8.41 suffers from a severe limitation on signal size, especially for high-β transistors. A modified arrangement is shown in Fig.

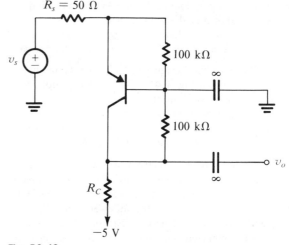

Fig. P8.42

P8.42. Find the value of R_C that will establish a dc collector current of 0.5 mA. Compare the output signal swing possible with this circuit with that obtained in the circuit of Fig. P8.41.

8.43 For the cascode circuit in Fig. 8.40 let $V_{CC} = 12$ V, $R_{B1} = 6$ kΩ, $R_{B2} = 2$ kΩ, $R_{B3} = 4$ kΩ, $R_E = 3.3$ kΩ, $R_C = R_L = 4$ kΩ, $\beta = 100$, $R_s = 1$ kΩ, and $C_{C1} = C_{C2} = C_B = C_E = \infty$. Find the voltage gain v_o/v_s.

8.44 The emitter follower shown in Fig. P8.44 uses a BJT having a β of 100.

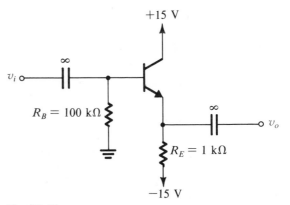

Fig. P8.44

(a) Find the dc emitter current.
(b) For the unloaded circuit find the input resistance, the voltage gain v_o/v_i, and the output resistance.

(c) For the circuit loaded by a 1-kΩ resistor find the input resistance and the voltage gain.

(d) For the loaded circuit driven by a source v_s having a source resistance of 10 kΩ, find the overall voltage gain.

*8.45 Consider the design of a follower using the topology shown in Fig. 8.41. Use a 10-V supply with $V_{BB} = 5$ V and $I_E = 10$ mA (for $\beta = \infty$). Select $R_{B1} = R_{B2}$ so that I_E reduces by at most 10% for $\beta = 50$. What are the values of R_{B1}, R_{B2}, and R_E? Find the input resistance of the follower with a load of 1 kΩ and $\beta = 100$.

*8.46 The follower in the circuit of Fig. P8.46 connects the signal source (which has a zero dc component) directly to the transistor base. The transistor has $\beta = 100$. What is the output resistance of the follower? Find the gain v_o/v_s with no load and with a load of 1 kΩ. With the 1-kΩ load connected, find the largest possible negative output signal. What is the largest possible positive output signal if operation is satisfactory up to the point that the base–collector junction is forward-biased by 0.2 V?

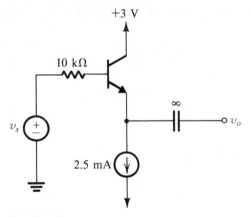

Fig. P8.46

*8.47 An *npn* emitter follower loaded by 1 kΩ through a capacitor utilizes an emitter resistance $R_E = 3$ kΩ across which a 4-V bias exists for no signal. What is the largest available output signal swing in the direction that cuts off the transistor? If, for no signal, $V_{CE} = 3$ V and operation can proceed until $V_{BC} = 0.2$ V, what is the largest available signal of the other polarity?

8.48 A very simple approach to biasing a follower for small-signal operation is shown in Fig. P8.48. The difficulty is that for high-β transistors, operation is very near the edge of the active region. Find V_O, I_E, v_o/v_i, and R_i for $\beta = \infty$ and for $\beta = 100$.

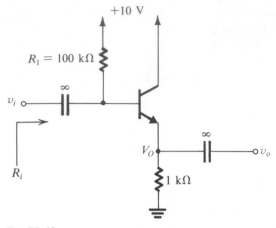

Fig. P8.48

**8.49 An improvement can be made in the circuit of Fig. P8.48 by shunting the emitter–base junction with a resistor R_2 to provide current to R_1 when β is large. Repeat Problem 8.48 for $R_2 = 21$ kΩ, $V_{BE} = 0.7$ V. (*Hint:* To find R_i apply Miller's theorem to R_2.)

8.50 An approach to raising the input resistance of a follower while maintaining bias stability is to use the idea called *bootstrapping*. It is particularly convenient for use with a grounded load using a variation of the simple circuit introduced in Problem 8.44, as shown in Fig. P8.50. **(a) Find I_E, v_o/v_i, and R_{in} for $\beta = \infty$ and for $\beta = 100$. Note that the connection of R_B to the output terminal utilizes the Miller effect to raise the component of the input resistance due to R_B.

(b) Use the method of Fig. 8.43 to find the output resistance of the follower in Fig. P8.50 excluding R_L (but in-

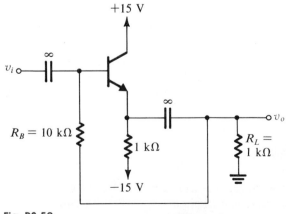

Fig. P8.50

cluding the connection of R_B) when the source resistance is (i) zero and (ii) 100 kΩ.

8.51 Consider the circuit of Fig. 8.44. For $V_{CC} = 5$ V, $R_C = 500$ Ω, $R_B = 5$ kΩ, $V_{BE} = 0.7$ V, and $\beta = 100$, what is the voltage v_I that causes the transistor to reach the boundary between active mode and saturation ($v_{CB} = 0$)? At this point find the small-signal gain v_o/v_i.

*8.52 Using the three-terminal-short model for a saturated transistor find the collector voltages in the circuits of Fig. P8.52. Also calculate the forced β for each of the transistors.

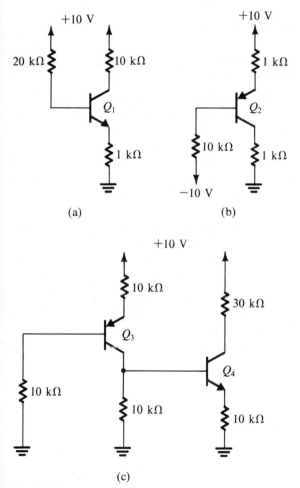

(a)

(b)

(c)

Fig. P8.52

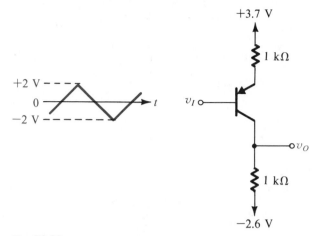

Fig. P8.53

0.3 V. Sketch and label the output waveform corresponding to the 4-V peak-to-peak input shown.

8.54 The transistors in the circuit in Fig. P8.54 are matched, with very high β, $|V_{BE}| = 0.7$ V, and $|V_{CE_{sat}}| = 0.3$ V. Find the voltages at nodes B, C, D, and E when node A is at: **(a) 0 V, **(b)** +1 V, **(c)** +2 V, **(d)** −4 V, and **(e)** −5 V.

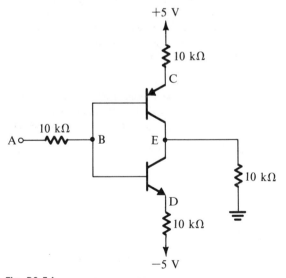

Fig. P8.54

*8.53 The transistor in the circuit of Fig. P8.53 has a very high β in the active region, $V_{EB} = 0.7$ V, and $V_{EC_{sat}} =$

*8.55 Using the equivalent circuit of Fig. 8.54 show that the output resistance when the emitter is grounded and

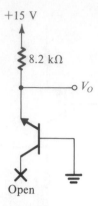

+15 V

8.2 kΩ

V_O

Open

Fig. P8.58

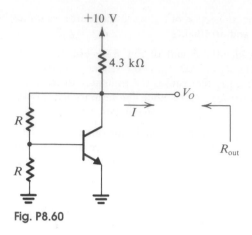

+10 V

4.3 kΩ

V_O

I

R

R

R_{out}

Fig. P8.60

the base is open-circuited is approximately equal to $[r_o \| (r_\mu/\beta)]$.

**8.56 Use the equivalent circuit of Fig. 8.54 to model the common-base measurement circuit of Fig. 8.13a. Note that, from a small-signal point of view, the emitter is open-circuited and the base is grounded. The change in v_{CB} can be represented by a signal source v_{cb} connected between C and B. Find the current drawn from this signal source. Note that this current is equal to i_c and $-i_b$. Show that the output resistance of the CB configuration (defined as v_{cb}/i_c) is approximately equal to $[r_\mu \| (\beta + 1)r_o]$. If for this transistor it is known that $r_\mu \simeq 10\beta r_o$, that at $I_E = 10$ mA and $V_{CB} = 1$ V, I_C is measured to be 9.900, and that when V_{CB} is raised to 11 V, I_C increases to 9.908, find β, r_o, and r_μ.

8.57 A BJT operating at a collector current of 9.0 mA and a base current of 100 μA with a collector-to-emitter voltage held at 4 V, when subjected to a 10-μA increase

in base current, conducts 10.0 mA in the collector. Find the dc and the ac β values.

8.58 What is the value of V_O in the circuit of Fig. P8.58? The transistor has an emitter–base breakdown voltage with the collector open, V_{EBO}, of 6.8 V.

8.59 On the output characteristics shown in Fig. 8.61 sketch load lines corresponding to **(a)** a constant-current load at I_{CQ} and **(b)** a constant-voltage load at V_{CEQ}. Note the large transresistance ($\partial v_{ce}/\partial i_b$) available with a very-high-resistance load.

***8.60 The transistor in the circuit of Fig. P8.60 has $v_{BE} = 0.7$ V at $I_E = 1$ mA and has $n = 1$.
(a) With $I = 0$ calculate V_O and the small-signal output resistance R_{out} for the cases $R = 700$ Ω and $R = 7$ kΩ and for $\beta = \infty$ and $\beta = 50$.
(b) With $R = 7$ kΩ and $\beta = 50$ what does the output V_O become when I is varied from 0 to 1 mA?
Note that this circuit can be used as a voltage regulator.

Chapter

Differential and Multistage Amplifiers

INTRODUCTION

The differential amplifier is the most widely used circuit building block in analog integrated circuits. For instance, the input stage of every op amp is a differential amplifier. Also, the BJT differential amplifier is the basis of a very-high-speed logic circuit family, called emitter-coupled logic (ECL), which we shall study in Chapter 16.

In this chapter we shall study differential amplifiers implemented with BJTs, JFETs, and MOSFETs. Also presented are the biasing techniques employed in the design of bipolar and MOS integrated circuits. The chapter concludes with a discussion of the structure of multistage amplifiers. The analysis of such amplifiers is illustrated by a detailed example. □

9.1 THE BJT DIFFERENTIAL PAIR

Qualitative Description of Operation

Figure 9.1 shows the basic BJT differential-pair configuration. It consists of two matched transistors, Q_1 and Q_2, whose emitters are joined together and biased by a constant-current source I. The latter is usually implemented by a transistor circuit

485

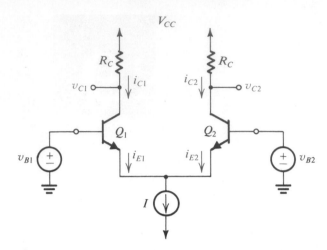

Fig. 9.1 The basic BJT differential-pair configuration.

of the type described in Section 9.4. Although each collector is connected to the positive supply voltage V_{CC} through a resistance R_C, this connection is not essential to the operation of the differential pair—that is, in some applications the two collectors may be connected to other transistors rather than to resistive loads. It is essential, though, that the collector circuits be such that Q_1 and Q_2 never enter saturation.

To see how the differential pair works, consider first the case where the two bases are joined together and connected to a voltage v_{CM}, called the *common-mode voltage*. That is, as shown in Fig. 9.2a, $v_{B1} = v_{B2} = v_{CM}$. Since Q_1 and Q_2 are matched, it follows from symmetry that the current I will divide equally between the two devices. Thus $i_{E1} = i_{E2} = I/2$, and the voltage at the emitters will be $v_{CM} - V_{BE}$, where V_{BE} is the voltage corresponding to an emitter current of $I/2$. The voltage at each collector will be $V_{CC} - \frac{1}{2}\alpha I R_C$, and the difference in voltage between the two collectors will be zero.

Now let us vary the value of the common-mode input signal v_{CM}. Obviously, as long as Q_1 and Q_2 remain in the active region the current I will still divide equally between Q_1 and Q_2, and the voltages at the collectors will not change. Thus the differential pair does not respond to (*rejects*) common-mode input signals.

As another experiment, let the voltage v_{B2} be set to a constant value, say, zero (by grounding B_2), and let $v_{B1} = +1$ V (see Fig. 9.2b). With a bit of reasoning it can be seen that Q_1 will be on and conducting all of the current I and that Q_2 will be off. For Q_1 to be on, the emitter has to be at approximately $+0.3$ V, which keeps the EBJ of Q_2 reverse-biased. The collector voltages will be $v_{C1} = V_{CC} - \alpha I R_C$ and $v_{C2} = V_{CC}$.

Let us now change v_{B1} to -1 V (Fig. 9.2c). Again with some reasoning it can be seen that Q_1 will turn off, and Q_2 will carry all the current I. The common emitter will be at -0.7 V, which means that the EBJ of Q_1 will be reverse-biased by 0.3 V. The collector voltages will be $v_{C1} = V_{CC}$ and $v_{C2} = V_{CC} - \alpha I R_C$.

From the above we see that the differential pair certainly responds to *difference-mode or differential signals*. In fact, with relatively small difference voltages we are able to steer the entire bias current from one side of the pair to the other. This

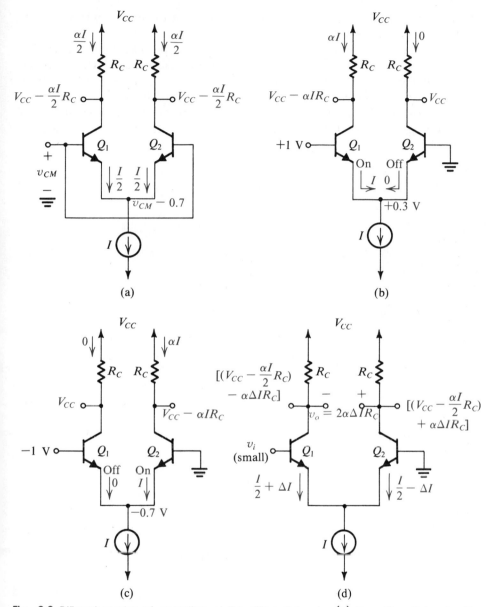

Fig. 9.2 Different modes of operation of the differential pair. **(a)** The differential pair with a common-mode input signal v_{CM}. **(b)** The differential pair with a "large" differential input signal. **(c)** The differential pair with a large differential input signal of polarity opposite to that in (b). **(d)** The differential pair with a small differential input signal v_i.

current-steering property of the differential pair allows it to be used in logic circuits, as will be demonstrated in Chapter 16.

To use the differential pair as a linear amplifier we apply a very small differential signal (a few millivolts), which will result in one of the transistors conducting a current of $I/2 + \Delta I$; the current in the other transistor will be $I/2 - \Delta I$, with ΔI being proportional to the difference input voltage (see Fig. 9.2d). The output voltage taken between the two collectors will be $2\alpha\,\Delta I R_C$, which is proportional to the differential input signal v_i. The small-signal operation of the differential pair will be studied in Section 9.2.

9.1 Find v_E, v_{C1}, and v_{C2} in the circuit of Fig. E9.1.

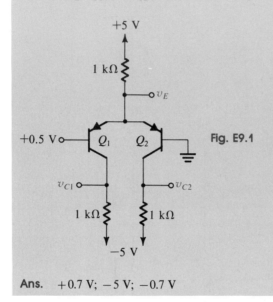

Fig. E9.1

Ans. $+0.7$ V; -5 V; -0.7 V

Large-Signal Operation of the BJT Differential Pair

We now present a general analysis of the BJT differential pair of Fig. 9.1. If we denote the voltage at the common emitter by v_E, the exponential relationship applied to each of the two transistors may be written

$$i_{E1} = \frac{I_S}{\alpha}\,e^{(v_{B1} - v_E)/V_T} \tag{9.1}$$

$$i_{E2} = \frac{I_S}{\alpha}\,e^{(v_{B2} - v_E)/V_T} \tag{9.2}$$

These two equations can be combined to obtain

$$\frac{i_{E1}}{i_{E2}} = e^{(v_{B1} - v_{B2})/V_T} \tag{9.3}$$

which can be manipulated to yield

$$\frac{i_{E1}}{i_{E1} + i_{E2}} = \frac{1}{1 + e^{(v_{B2} - v_{B1})/V_T}} \tag{9.4}$$

$$\frac{i_{E2}}{i_{E1} + i_{E2}} = \frac{1}{1 + e^{(v_{B1} - v_{B2})/V_T}} \tag{9.5}$$

The circuit imposes the additional constraint

$$i_{E1} + i_{E2} = I \tag{9.6}$$

Using Eq. (9.6) together with Eqs. (9.4) and (9.5) gives

$$i_{E1} = \frac{I}{1 + e^{(v_{B2} - v_{B1})/V_T}} \tag{9.7}$$

$$i_{E2} = \frac{I}{1 + e^{(v_{B1} - v_{B2})/V_T}} \tag{9.8}$$

The collector currents i_{C1} and i_{C2} can be obtained simply by multiplying the emitter currents in Eqs. (9.7) and (9.8) by α, which is normally very close to unity.

The fundamental operation of the differential amplifier is illustrated by Eqs. (9.7) and (9.8). First, note that the amplifier responds only to the difference voltage $v_{B1} - v_{B2}$. That is, if $v_{B1} = v_{B2} = v_{CM}$, the current I divides equally between the two transistors irrespective of the value of the common-mode voltage v_{CM}. This is the essence of differential-amplifier operation, which also gives rise to its name.

Another important observation is that a relatively small difference voltage $v_{B1} - v_{B2}$ will cause the current I to flow almost entirely in one of the two transistors. Figure 9.3 shows a plot of the two collector currents (assuming $\alpha \simeq 1$) as a function

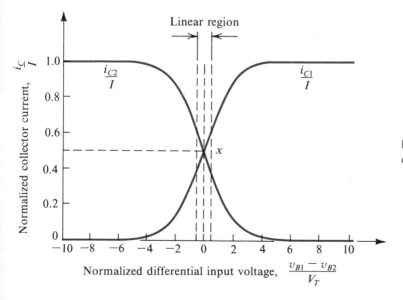

Fig. 9.3 Transfer characteristics of the BJT differential pair of Fig. 9.1.

of the difference signal. This is a normalized plot that can be used universally. Note that a difference voltage of about $4V_T$ ($\simeq 100$ mV) is sufficient to switch the current almost entirely to one side of the pair.

The nonlinear transfer characteristics of the differential pair, shown in Fig. 9.3, will not be utilized any further in this chapter. In the following we shall be interested specifically in the application of the differential pair as a small-signal amplifier. For this purpose the difference input signal is limited to less than about $V_T/2$ in order that we may operate on a linear segment of the characteristics around the midpoint x.

Exercise

9.2 Find the value of input differential signal sufficient to cause $i_{E1} = 0.99I$.

Ans.   115 mV

9.2 SMALL-SIGNAL OPERATION OF THE BJT DIFFERENTIAL AMPLIFIER

In this section we shall study the application of the BJT differential pair in small-signal amplification. Figure 9.4 shows the differential pair with a difference voltage signal v_d applied between the two bases. Implied is that the dc level at the input—that is, the common-mode input signal—has been somehow established. For instance, one of the two input terminals can be grounded and v_d applied to the other input terminal. Alternatively, the differential amplifier may be fed from the output of another differential amplifier. In this case the voltage at one of the input terminals will be $v_{CM} + v_d/2$ while that at other input terminals will be $v_{CM} - v_d/2$. We will consider common-mode operation at a later stage.

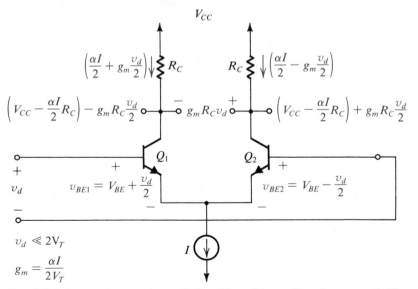

Fig. 9.4 The currents and voltages in the differential amplifier when a small difference signal v_d is applied.

The Collector Currents When v_d Is Applied

Return now to the circuit of Fig. 9.4. We may use Eqs. (9.7) and (9.8) to find the total currents i_{C1} and i_{C2} as functions of the differential signal v_d by substituting $v_{B1} - v_{B2} = v_d$.

$$i_{C1} = \frac{\alpha I}{1 + e^{-v_d/V_T}} \tag{9.9}$$

$$i_{C2} = \frac{\alpha I}{1 + e^{v_d/V_T}} \tag{9.10}$$

Multiplying the numerator and the denominator of the right-hand side of Eq. (9.9) by $e^{(v_d/2V_T)}$ gives

$$i_{C1} = \frac{\alpha I e^{(v_d/2V_T)}}{e^{(v_d/2V_T)} + e^{(-v_d/2V_T)}} \tag{9.11}$$

Assume that $v_d \ll 2V_T$. We may thus expand the exponential $e^{(\pm v_d/2V_T)}$ in a series and retain only the first two terms:

$$i_{C1} \simeq \frac{\alpha I(1 + v_d/2V_T)}{1 + v_d/2V_T + 1 - v_d/2V_T}$$

Thus

$$i_{C1} = \frac{\alpha I}{2} + \frac{\alpha I}{2V_T}\frac{v_d}{2} \tag{9.12}$$

Similar manipulations can be applied to Eq. (9.10) to obtain

$$i_{C2} = \frac{\alpha I}{2} - \frac{\alpha I}{2V_T}\frac{v_d}{2} \tag{9.13}$$

Equations (9.12) and (9.13) tell us that when $v_d = 0$, the bias current I divides equally between the two transistors of the pair. Thus each transistor is biased at an emitter current of $I/2$. When a "small-signal" v_d is applied differentially (that is, between the two bases), the collector current of Q_1 increases by an increment i_c and that of Q_2 decreases by an equal amount. This ensures that the sum of the total currents in Q_1 and Q_2 remains constant, as constrained by the current-source bias. The incremental or signal current component i_c is given by

$$i_c = \frac{\alpha I}{2V_T}\frac{v_d}{2} \tag{9.14}$$

Equation (9.14) has an easy interpretation. First, note from the symmetry of the circuit (Fig. 9.4) that the differential signal v_d should divide equally between the base-emitter junctions of the two transistors. Thus the total base-emitter voltages will be

$$v_{BE}|_{Q_1} = V_{BE} + \frac{v_d}{2} \tag{9.15}$$

$$v_{BE}|_{Q_2} = V_{BE} - \frac{v_d}{2} \tag{9.16}$$

where V_{BE} is the dc BE voltage corresponding to an emitter current of $I/2$. Therefore the collector current of Q_1 will increase by $g_m v_d/2$ and the collector current of Q_2 will decrease by $g_m v_d/2$. Here g_m denotes the transconductance of Q_1 and of Q_2, which are equal and given by

$$g_m = \frac{I_C}{V_T} = \frac{\alpha I/2}{V_T} \tag{9.17}$$

Thus Eq. (9.14) simply states that $i_c = g_m v_d/2$.

An Alternative Viewpoint

There is an extremely useful alternative interpretation of the above results. Assume the current source I to be ideal. Its incremental resistance then will be infinite. Thus the voltage v_d appears across a total resistance of $2r_e$, where

$$r_e = \frac{V_T}{I_E} = \frac{V_T}{I/2} \tag{9.18}$$

Correspondingly there will be a signal current i_e, as illustrated in Fig. 9.5, given by

$$i_e = \frac{v_d}{2r_e} \tag{9.19}$$

Thus the collector of Q_1 will exhibit a current increment i_c and the collector of Q_2 will exhibit a current decrement i_c:

$$i_c = \alpha i_e = \frac{\alpha v_d}{2r_e} = g_m \frac{v_d}{2} \tag{9.20}$$

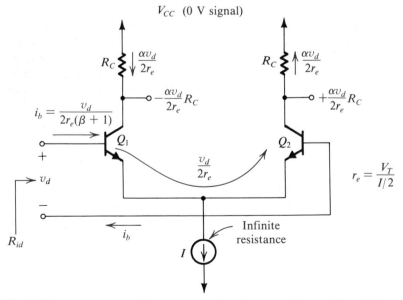

Fig. 9.5 Simple technique for determining the signal currents in a differential amplifier excited by a differential voltage signal v_d; dc quantities are not shown.

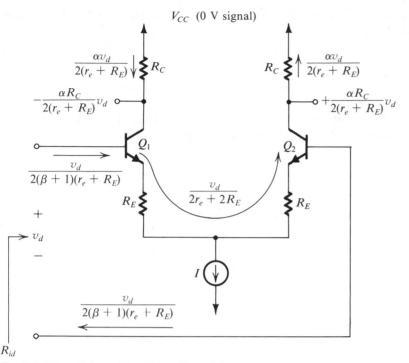

Fig. 9.6 Differential amplifier with emitter resistances.

Note that in Fig. 9.5 we have shown signal quantities only. It is implied, of course, that each transistor is biased at an emitter current of $I/2$.

This method of analysis is particularly useful when resistances are included in the emitters, as shown in Fig. 9.6. For this latter circuit we have

$$i_e = \frac{v_d}{2r_e + 2R_E} \tag{9.21}$$

Input Differential Resistance

The input differential resistance is the resistance seen between the two bases; that is, it is the resistance seen by the differential input signal v_d. For the differential amplifier in Figs. 9.4 and 9.5 it can be seen that the base current of Q_1 shows an increment i_b and the base current of Q_2 shows an equal decrement,

$$i_b = \frac{i_e}{\beta + 1} = \frac{v_d/2r_e}{\beta + 1} \tag{9.22}$$

Thus the differential input resistance R_{id} is given by

$$R_{id} \equiv \frac{v_d}{i_b} = (\beta + 1)2r_e = 2r_\pi \tag{9.23}$$

This result is just a restatement of the familiar resistance-reflection rule; namely, *the resistance seen between the two bases is equal to the total resistance in the emitter circuit multiplied by* $\beta + 1$. We can employ this rule to find the input differential resistance for the circuit in Fig. 9.6 as

$$R_{id} = (\beta + 1)(2r_e + 2R_E) \tag{9.24}$$

Differential Voltage Gain

We have established that for small difference input voltages ($v_d \ll 2V_T$; that is, v_d smaller than about 20 mV) the collector currents are given by

$$i_{C1} = I_C + g_m \frac{v_d}{2} \tag{9.25}$$

$$i_{C2} = I_C - g_m \frac{v_d}{2} \tag{9.26}$$

where

$$I_C = \frac{\alpha I}{2} \tag{9.27}$$

Thus the total voltages at the collectors will be

$$v_{C1} = (V_{CC} - I_C R_C) - g_m R_C \frac{v_d}{2} \tag{9.28}$$

$$v_{C2} = (V_{CC} - I_C R_C) + g_m R_C \frac{v_d}{2} \tag{9.29}$$

The quantities in parentheses are simply the dc voltages at each of the two collectors.

The output voltage signal of a differential amplifier can be taken either *differentially* (that is, between the two collectors) or *single-ended* (that is, between one collector and ground). If the output is taken differentially, then the differential gain (as opposed to the common-mode gain) of the differential amplifier will be

$$A_d = \frac{v_{c1} - v_{c2}}{v_d} = -g_m R_C \tag{9.30}$$

On the other hand, if we take the output single ended (say, between the collector of Q_1 and ground), then the differential gain will be given by

$$A_d = \frac{v_{c1}}{v_d} = -\tfrac{1}{2}g_m R_C \tag{9.31}$$

For the differential amplifier with resistances in the emitter leads (Fig. 9.6) the differential gain when the output is taken differentially is given by

$$A_d = -\frac{\alpha(2R_C)}{2r_e + 2R_E} \simeq -\frac{R_C}{r_e + R_E} \tag{9.32}$$

This equation is a familiar one: it states that *the voltage gain is equal to the ratio of the total resistance in the collector circuit* ($2R_C$) *to the total resistance in the emitter circuit* ($2r_e + 2R_E$).

Equivalence of the Differential Amplifier to a Common-Emitter Amplifier

The above analysis and results are quite similar to those obtained in the case of a common-emitter amplifier stage. That the differential amplifier is in fact equivalent to a common-emitter amplifier is illustrated in Fig. 9.7. Figure 9.7a shows a differential amplifier fed by a differential signal v_d with the differential signal applied in a *complementary* (*push-pull* or *balanced*) manner. That is, while the base of Q_1 is raised by $v_d/2$, the base of Q_2 is lowered by $v_d/2$. We have also included the output resistance R of the bias current source. From symmetry it follows that the signal voltage at the common emitter will be zero. Thus the circuit is equivalent to the two common-emitter amplifiers shown in Fig. 9.7b, where each of the two transistors is biased at an emitter current of $I/2$. Note that the finite output resistance R of the current source will have no effect on the operation. The equivalent circuit in Fig. 9.7b is valid for differential operation only.

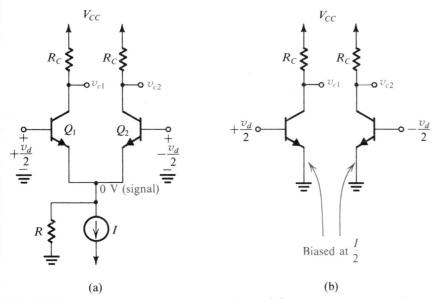

(a)   (b)

Fig. 9.7 Equivalence of the differential amplifier in (a) to the two common-emitter amplifiers in (b). This equivalence applies only for differential input signals. Either of the two common-emitter amplifiers in (b) can be used to evaluate the differential gain, input differential resistance, frequency response, and so on, of the differential amplifier.

In many applications the differential amplifier is not fed in a complementary fashion; rather, the input signal may be applied to one of the input terminals while the other terminal is grounded, as shown in Fig. 9.8. In this case the signal voltage at the emitters will not be zero, and thus the resistance R will have an effect on the operation. Nevertheless, if R is large ($R \gg r_e$), as is usually the case, then v_d will still divide equally (approximately) between the two junctions, as shown in Fig. 9.8. Thus the operation of the differential amplifier in this case will be almost identical to that in the case of symmetric feed, and the common-emitter equivalence can still be employed.

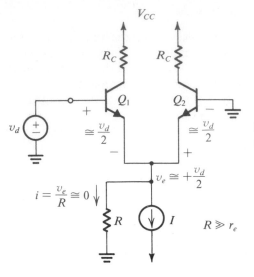

Fig. 9.8 The differential amplifier fed in a single-ended manner.

Since in Fig. 9.7 $v_{c2} = -v_{c1}$, the two common-emitter transistors in Fig. 9.7b yield similar results about the performance of the differential amplifier. Thus only one is needed to analyze the differential small-signal operation of the differential amplifier, and is known as the *differential half-circuit*. If we take the common-emitter transistor fed with $+v_d/2$ as the differential half-circuit and replace the transistor with its low-frequency equivalent circuit model, the circuit in Fig. 9.9 results. In evaluating the values of the model parameters r_π, g_m, and r_o we must recall that the half-circuit is biased at $I/2$. The voltage gain of the differential amplifier (with the output taken differentially) is equal to the voltage gain of the half-circuit—that is, $v_{c1}/(v_d/2)$. Here we note that including r_o will modify the gain expression in Eq. (9.30) to

$$A_d = -g_m(R_C \| r_o) \tag{9.33}$$

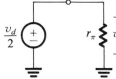 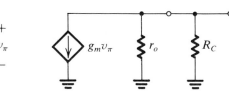

Fig. 9.9 Equivalent circuit model of the differential half-circuit.

The input differential resistance of the differential amplifier is twice that of the half-circuit—that is, $2r_\pi$. Finally, we note that the differential half-circuit of the amplifier of Fig. 9.6 is a common-emitter transistor with a resistance R_E in the emitter lead.

Common-Mode Gain

Figure 9.10a shows a differential amplifier fed by a common-mode voltage signal v_{CM}. The resistance R is the incremental output resistance of the bias current source. From symmetry it can be seen that the circuit is equivalent to that shown in Fig. 9.10b,

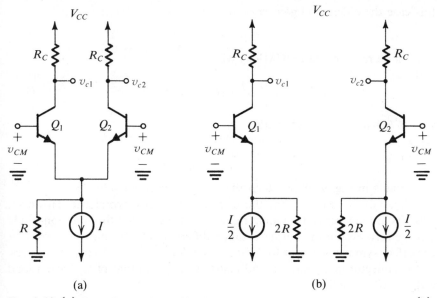

Fig. 9.10 (a) The differential amplifier fed by a common-mode voltage signal. **(b)** Equivalent "half-circuits" for common-mode calculations.

where each of the two transistors Q_1 and Q_2 is biased at an emitter current $I/2$ and has a resistance $2R$ in its emitter lead. Thus the common-mode output voltage v_{c1} will be

$$v_{c1} = -v_{CM}\frac{\alpha R_C}{2R + r_e} \simeq -v_{CM}\frac{\alpha R_C}{2R} \tag{9.34}$$

At the other collector we have an equal common-mode signal v_{c2},

$$v_{c2} \simeq -v_{CM}\frac{\alpha R_C}{2R} \tag{9.35}$$

Now, if the output is taken differentially, then the output common-mode voltage $v_{c1} - v_{c2}$ will be zero and the common-mode gain also will be zero. On the other hand, if the output is taken single-ended, the common mode gain A_{cm} will be finite and given by[1]

$$A_{cm} = -\frac{\alpha R_C}{2R} \tag{9.36}$$

---

[1] The expressions in Eqs. (9.34) and (9.35) are obtained by neglecting r_o and r_μ. A detailed derivation using the complete hybrid-π equivalent circuit model shows that v_{c1}/v_{CM} and v_{c2}/v_{CM} are approximately

$$\frac{-\alpha R_C}{2R}\left[1 - 2R\left(\frac{1}{\beta r_o} + \frac{1}{\alpha r_\mu}\right)\right]$$

This expression reduces to those in Eqs. (9.34) and (9.35) when $2R \ll \beta r_o$ and $2R \ll \alpha r_\mu$.

Since in this case the differential gain is

$$A_d = \tfrac{1}{2} g_m R_C \tag{9.37}$$

the common-mode rejection ratio (CMRR) will be

$$\text{CMRR} = \left| \frac{A_d}{A_{cm}} \right| \simeq g_m R \qquad \alpha \simeq 1 \tag{9.38}$$

Normally the CMRR is expressed in dB:

$$\text{CMRR} = 20 \log \left| \frac{A_d}{A_{cm}} \right| \tag{9.39}$$

Each of the circuits in Fig. 9.10b is called the *common-mode half-circuit*.

The above analysis assumes that the circuit is perfectly symmetric. However, practical circuits are not perfectly symmetric, with the result that the common-mode gain will not be zero even if the output is taken differentially. To illustrate, consider the case of perfect symmetry except for a mismatch ΔR_C in the collector resistances. That is, let the collector of Q_1 have a load resistance R_C and that of Q_2 have a load resistance $R_C + \Delta R_C$. It follows that

$$v_{c1} = -v_{CM} \frac{\alpha R_C}{2R + r_e}$$

$$v_{c2} = -v_{CM} \frac{\alpha (R_C + \Delta R_C)}{2R + r_e}$$

Thus the common-mode signal at the output will be

$$v_o = v_{c1} - v_{c2} = v_{CM} \frac{\alpha \, \Delta R_C}{2R + r_e}$$

and the common-mode gain will be

$$A_{cm} = \frac{\alpha \, \Delta R_C}{2R + r_e} \simeq \frac{\Delta R_C}{2R}$$

This expression can be rewritten

$$A_{cm} = \frac{R_C}{2R} \frac{\Delta R_C}{R_C} \tag{9.40}$$

Compare the common-mode gain in Eq. (9.40) with that for the case of single-ended output in Eq. (9.36). We see that the common-mode gain is much smaller in the case of differential output. Therefore the input differential stage of an op amp, for example, is usually a balanced one, with the output taken differentially. This ensures that the op amp will have a low common-mode gain or, equivalently, a high CMRR.

The input signals v_1 and v_2 to a differential amplifier usually contain a common-mode component, v_{CM},

$$v_{CM} \equiv \frac{v_1 + v_2}{2} \tag{9.41}$$

and a differential component v_d,

$$v_d \equiv v_1 - v_2 \qquad (9.42)$$

Thus the output signal will be given by

$$v_o = A_d(v_1 - v_2) + A_{cm}\left(\frac{v_1 + v_2}{2}\right) \qquad (9.43)$$

Input Common-Mode Resistance

The definition of the *common-mode input resistance* R_{icm} is illustrated in Fig. 9.11a. Figure 9.11b shows the equivalent common-mode half-circuit; its input resistance is $2R_{icm}$.

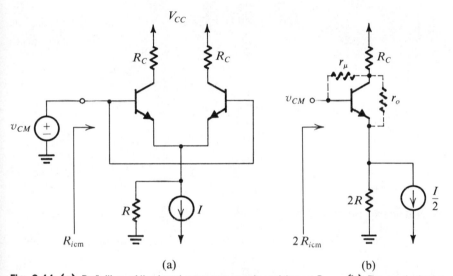

Fig. 9.11 (a) Definition of the input common-mode resistance R_{icm}. **(b)** The equivalent common-mode half-circuit.

Since the input common-mode resistance is usually very large, its value will be affected by the transistor resistances r_o and r_μ. These resistances are indicated on the equivalent common-mode half-circuit in Fig. 9.11b. Now, since the common-mode gain is usually small, the signal at the collector will be very small. We can simplify matters considerably by assuming that the signal at the collector is 0 V; that is, the collector is at signal ground. Under this assumption the input resistance can be found by inspection, as follows:

$$2R_{icm} = r_\mu \| [(\beta + 1)(2R) \| [(\beta + 1)r_o]$$

Thus

$$R_{icm} = \left(\frac{r_\mu}{2}\right) \Big\| [(\beta + 1)R] \Big\| \left[(\beta + 1)\frac{r_o}{2}\right] \qquad (9.44)$$

Exercise

9.3 The differential amplifier in Fig. E9.3 uses transistors with $\beta = 100$. Evaluate the following:

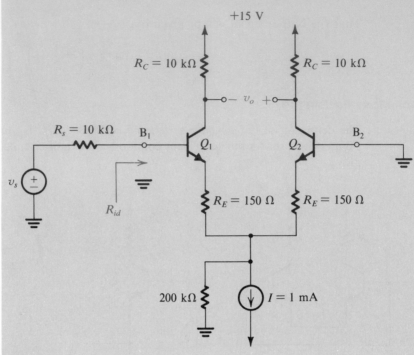

Fig. E9.3

(a) The input differential resistance R_{id}.
(b) The overall voltage gain v_o/v_s (neglect the effect of r_o).
(c) The worst-case common-mode gain if the two collector resistances are accurate to within $\pm 1\%$.
(d) The CMRR, in dB.
(e) The input common-mode resistance (assuming that the Early voltage $V_A = 100$ V and that $r_\mu = 10 \, \beta r_o$).
Ans. (a) 40 kΩ; (b) 40 V/V; (c) 5×10^{-4}; (d) 98 dB; (e) 6.3 MΩ

9.3 OTHER NONIDEAL CHARACTERISTICS OF THE DIFFERENTIAL AMPLIFIER

Input Offset Voltage

Consider the basic BJT differential amplifier with both inputs grounded, as shown in Fig. 9.12a. If the two sides of the differential pair were perfectly matched (that is, Q_1 and Q_2 identical and $R_{C1} = R_{C2} = R_C$), then current I would split equally between Q_1 and Q_2 and V_O would be zero. Practical circuits exhibit mismatches that result in a dc output voltage V_O even with both inputs grounded. We call V_O the output dc offset voltage. More commonly, we divide V_O by the differential gain of the amplifier,

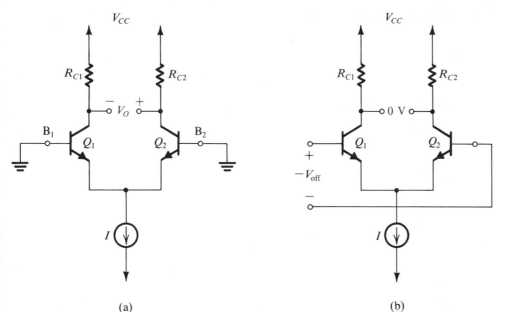

(a)

(b)

Fig. 9.12 (a) The BJT differential pair with both inputs grounded. Due to device mismatches, a finite dc output V_O results. **(b)** Application of the input offset voltage $V_{off} \equiv V_O/A_d$ to the input terminals with opposite polarity reduces V_O to zero.

A_d, to obtain a quantity known as the input offset voltage, V_{off},

$$V_{off} = V_O/A_d \tag{9.45}$$

Obviously, if we apply a voltage $-V_{off}$ between the input terminals of the differential amplifier, then the output voltage will be reduced to zero (see Fig. 9.12b). This observation gives rise to the usual definition of the input offset voltage. It should be noted, however, that since the offset voltage is a result of device mismatches, its polarity is not known a priori.

The offset voltage results from mismatches in the load resistances R_{C1} and R_{C2} and from mismatches in Q_1 and Q_2. Consider first the effect of the load mismatch. Let

$$R_{C1} = R_C + \frac{\Delta R_C}{2} \tag{9.46}$$

$$R_{C2} = R_C - \frac{\Delta R_C}{2} \tag{9.47}$$

and assume that Q_1 and Q_2 are perfectly matched. It follows that current I will divide equally between Q_1 and Q_2, and thus

$$V_{C1} = V_{CC} - \left(\frac{\alpha I}{2}\right)\left(R_C + \frac{\Delta R_C}{2}\right)$$

$$V_{C2} = V_{CC} - \left(\frac{\alpha I}{2}\right)\left(R_C - \frac{\Delta R_C}{2}\right)$$

Thus the output voltage will be

$$V_O = V_{C2} - V_{C1}$$

$$= \alpha \left(\frac{I}{2} \right) (\Delta R_C)$$

and the input offset voltage will be

$$V_{\text{off}} = \frac{\alpha(I/2)(\Delta R_C)}{A_d} \tag{9.48}$$

Substituting $A_d = -g_m R_C$ and

$$g_m = \frac{\alpha I/2}{V_T}$$

gives

$$|V_{\text{off}}| = V_T \left(\frac{\Delta R_C}{R_C} \right) \tag{9.49}$$

As an example consider the situation where the collector resistors are accurate to within $\pm 1\%$. Then the worst case mismatch will be

$$\frac{\Delta R_C}{R_C} = 0.02$$

and the resulting input offset voltage will be

$$|V_{\text{off}}| = 25 \times 0.02 = 0.5 \text{ mV}$$

Next consider the effect of mismatches in transistors Q_1 and Q_2. In particular, let the transistors have a mismatch in their emitter–base junction areas. Such an area mismatch gives rise to a proportional mismatch in the scale currents I_S,

$$I_{S1} = I_S + \left(\frac{\Delta I_S}{2} \right) \tag{9.50}$$

$$I_{S2} = I_S - \left(\frac{\Delta I_S}{2} \right) \tag{9.51}$$

Refer to Fig. 9.12a and note that $V_{BE1} = V_{BE2}$. Thus, the current I will split between Q_1 and Q_2 as

$$I_{E1} = \frac{I}{2} \left(1 + \frac{\Delta I_S}{2I_S} \right) \tag{9.52}$$

$$I_{E2} = \frac{I}{2} \left(1 - \frac{\Delta I_S}{2I_S} \right) \tag{9.53}$$

It follows that the output offset voltage will be

$$V_O = \alpha \left(\frac{I}{2} \right) \left(\frac{\Delta I_S}{I_S} \right) R_C$$

and the corresponding input offset voltage will be

$$|V_{\text{off}}| = V_T \left(\frac{\Delta I_S}{I_S} \right) \tag{9.54}$$

As an example, an area mismatch of 4% gives rise to $\Delta I_S/I_S = 0.04$ and an input offset voltage of 1 mV.

Finally, we note that since the two contributions to the input offset voltage are not correlated, an estimate of the total input offset voltage can be found as

$$V_{\text{off}} = \sqrt{ \left(V_T \frac{\Delta R_C}{R_C} \right)^2 + \left(V_T \frac{\Delta I_S}{I_S} \right)^2 }$$

$$= V_T \sqrt{ \left(\frac{\Delta R_C}{R_C} \right)^2 + \left(\frac{\Delta I_S}{I_S} \right)^2 } \tag{9.55}$$

Input Bias and Offset Currents

In a perfectly symmetric differential pair the two input terminals carry equal dc currents; that is,

$$I_{B1} = I_{B2} = \frac{I/2}{\beta + 1} \tag{9.56}$$

This is the input bias current of the differential amplifier.

Mismatches in the amplifier circuit and most importantly a mismatch in β make the two input dc currents unequal. The resulting difference is the input offset current, I_{off}, given as

$$I_{\text{off}} = |I_{B1} - I_{B2}| \tag{9.57}$$

Let

$$\beta_1 = \beta + \left(\frac{\Delta \beta}{2} \right) \tag{9.58}$$

$$\beta_2 = \beta - \left(\frac{\Delta \beta}{2} \right) \tag{9.59}$$

then

$$I_{B1} = \left(\frac{I}{2} \right) \frac{1}{\beta + 1 + \Delta\beta/2} \simeq \frac{I}{2} \frac{1}{\beta + 1} \left(1 - \frac{\Delta \beta}{2\beta} \right) \tag{9.60}$$

$$I_{B2} = \left(\frac{I}{2} \right) \frac{1}{\beta + 1 - \Delta\beta/2} \simeq \frac{I}{2} \frac{1}{\beta + 1} \left(1 + \frac{\Delta \beta}{2\beta} \right) \tag{9.61}$$

Thus

$$I_{\text{off}} = \frac{I}{2(\beta + 1)} \left(\frac{\Delta \beta}{\beta} \right) \tag{9.62}$$

Formally, the input bias current I_B is defined as follows:

$$I_B \equiv \frac{I_{B1} + I_{B2}}{2}$$

$$= \frac{I}{2(\beta + 1)} \tag{9.63}$$

Thus

$$I_{\text{off}} = I_B \left(\frac{\Delta\beta}{\beta} \right) \tag{9.64}$$

As an example, a 10% β mismatch results in an offset current one-tenth the value of the input bias current.

Input Common-Mode Range

The input common-mode range of a differential amplifier is the range of the input voltage v_{CM} over which the differential pair behaves as a linear amplifier for differential input signals. The upper limit of the common-mode range is determined by Q_1 and Q_2 leaving the active mode and entering the saturation mode of operation. Thus, the upper limit is approximately equal to the dc collector voltage of Q_1 and Q_2. The lower limit is determined by the transistor that supplies the biasing current I leaving its active region of operation and thus no longer functioning as a constant-current source. Current-source circuits are studied in the next section.

We conclude this section by noting that the definitions presented above are identical to those presented in Chapter 3 for op amps. In fact, as will be seen in Chapter 13, it is the input differential stage in an op-amp circuit that primarily determines the op-amp dc offset voltage, input bias and offset currents, and input common-mode range.

Exercise

9.4 For a BJT differential amplifier utilizing transistors having $\beta = 100$, matched to 10% or better, and areas that are matched to 10% or better and collector resistors that are matched to 2% or better, find V_{off}, I_B, and I_{off}. The dc bias current is 100 μA.

Ans. 2.5 mV; 0.5 μA; 50 nA

9.4 BIASING IN BJT INTEGRATED CIRCUITS

The BJT biasing techniques discussed in Chapter 8 are not suitable for the design of IC amplifiers. This shortcoming stems from the need for a large number of resistors (three per amplifier stage) as well as large coupling and bypass capacitors. With present IC technology it is almost impossible to fabricate large capacitors, and it is uneconomical to manufacture large resistances. On the other hand, IC technology provides the designer with the possibility of using many transistors, which can be produced cheaply. Furthermore, it is easy to make transistors with matched characteristics that track with changes in environmental conditions. The limitations of, and opportunities available in, IC technology dictate a biasing philosophy that is quite different from that employed in discrete BJT amplifiers.

Basically, biasing in integrated-circuit design is based on the use of constant-current sources. We have already seen that the differential pair utilizes constant-current-source bias. On an IC chip with a number of amplifier stages a constant dc current is generated at one location and is then reproduced at various other locations for biasing the various amplifier stages. This approach has the advantage that the bias currents of the various stages track each other in case of changes in power-supply voltage or in temperature.

In this section we shall study a variety of current-source and current-steering circuits. Although these circuits can be used in discrete-circuit design, they are primarily intended for application in IC design.

The Diode-Connected Transistor

Shorting the base and collector of a BJT together results in a two-terminal device having an i-v characteristic identical to the i_E-v_{BE} characteristic of the BJT. Figure 9.13 shows two *diode-connected transistors*, one *npn* and the other *pnp*. Observe that since the BJT is still operating in the active mode ($v_{CB} = 0$ results in active-mode operation) the current i divides between base and collector according to the value of the BJT β, as indicated in Fig. 9.13. Thus, internally the BJT still operates as a transistor in the active mode. This is the reason the i-v characteristic of the resulting diode is identical to the i_E-v_{BE} relationship of the BJT.

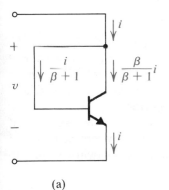

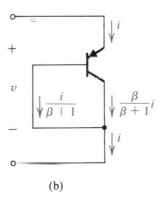

(a)   (b)

Fig. 9.13 Diode-connected BJTs.

It can be shown (Exercise 9.5) that the incremental resistance of the diode-connected transistor is approximately equal to r_e. In the following we shall make extensive use of the diode-connected BJT.

Exercise

9.5 Replace the BJT in the diode-connected transistor of Fig. 9.13a with its complete low-frequency hybrid-π model. Thus show that the incremental resistance of the two-terminal device is $[r_\pi \| (1/g_m) \| r_o] \simeq r_e$. Evaluate the incremental resistance for $i = 0.5$ mA.

Ans. 50 Ω

The Current Mirror

The *current mirror*, shown in its simplest form in Fig. 9.14, is the most basic building block in the design of IC current sources and current-steering circuits. (MOS current mirrors were studied in Chapter 7). The current mirror consists of two matched transistors Q_1 and Q_2 with their bases and emitters connected together, and which thus have the same v_{BE}. In addition, Q_1 is connected as a diode by shorting its collector to its base.

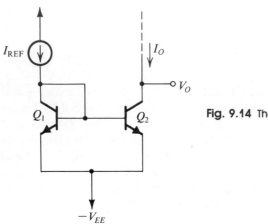

Fig. 9.14 The basic BJT current mirror.

The current mirror is shown fed with a constant-current source I_{REF} and the output current is taken from the collector of Q_2. The circuit fed by the collector of Q_2 should ensure active-mode operation for Q_2 (by keeping its collector voltage higher than that of the base) at all times. Assume that the BJTs have high β, and thus their base currents are negligibly small. The input current I_{REF} flows through the diode-connected transistor Q_1 and thus establishes a voltage across Q_1 that corresponds to the value of I_{REF}. This voltage in turn appears between the base and emitter of Q_2. Since Q_2 is identical to Q_1, the emitter current of Q_2 will be equal to I_{REF}. It follows that as long as Q_2 is maintained in the active region, its collector current I_O will be approximately equal to I_{REF}. Note that the mirror operation is independent of the value of the voltage $-V_{EE}$ as long as Q_2 remains active.

Next we consider the effect of finite transistor β on the operation of the current mirror. The analysis proceeds as follows: Since Q_1 and Q_2 are matched and since they have equal v_{BE}, their emitter currents will be equal. This is the key point. The rest of the analysis is straightforward and is illustrated in Fig. 9.15. It follows that

$$I_O = \frac{\beta}{\beta + 1} I_E$$

$$I_{REF} = \frac{\beta + 2}{\beta + 1} I_E$$

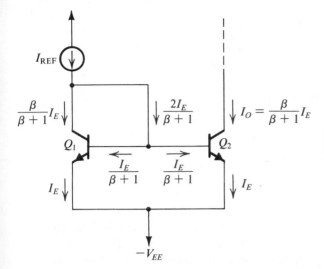

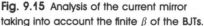

Fig. 9.15 Analysis of the current mirror taking into account the finite β of the BJTs.

Thus the current gain of the mirror is given by

$$\frac{I_O}{I_{REF}} = \frac{\beta}{\beta + 2} = \frac{1}{1 + 2/\beta} \tag{9.65}$$

which approaches unity for $\beta \gg 1$. Note, however, that the deviation of current gain from unity can be relatively high; $\beta = 100$ results in a 2% error.

Another factor that makes I_O unequal to I_{REF} is the linear dependence of the collector current of Q_2, which is I_O, on the collector voltage of Q_2. In fact, even if we ignore the effect of finite β and assume that Q_1 and Q_2 are perfectly matched, the current I_O will be equal to I_{REF} only when the voltage at the collector of Q_2 is equal to the base voltage. As the collector voltage is increased, I_O increases. Since Q_2 is operated at a constant v_{BE} (as determined by I_{REF}) the dependence of I_O on V_O is determined by r_o of Q_2. In other words, the output resistance of the current mirror of Fig. 9.14 is equal to r_o of Q_2.

Exercise

9.6 The current-mirror circuit of Fig. 9.14 utilizes BJTs having $\beta = 100$ and an Early voltage $V_A = 100$ V. Find the output resistance and the output current for $I_{REF} = 1$ mA, $V_{EE} = 5$ V, and $V_O = +5$ V. Assume that $V_{BE} \simeq 0.7$ V.

Ans. 100 kΩ; 1.093 mA

A Simple Current Source

Figure 9.16 shows a simple BJT constant-current-source circuit. It utilizes a pair of matched transistors in a current-mirror configuration, with the input reference current to the mirror, I_{REF}, determined by a resistor R connected to the positive power supply V_{CC}. The current I_{REF} is given by

$$I_{REF} = \frac{V_{CC} - V_{BE}}{R} \tag{9.66}$$

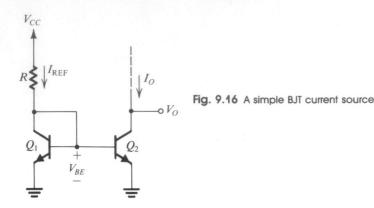

Fig. 9.16 A simple BJT current source.

where V_{BE} is the base–emitter voltage corresponding to an emitter current I_{REF}. Neglecting the effect of finite β and the dependence of I_O on V_O, the output current I_O will be equal to I_{REF}. The circuit will operate as a constant-current source as long as Q_2 remains in the active region—that is, for $V_O \geq V_{BE}$. The output resistance of this current source is r_o of Q_2.

Exercise

> **9.7** For the circuit in Fig. 9.16 find the value of R that results in $I_O = 1$ mA with $V_{CC} = 5$ V. Assume that $V_{BE} \simeq 0.7$ V and neglect the effect of r_o.
>
> **Ans.** 4.3 kΩ

Current-Steering Circuits

As mentioned earlier, in an IC a dc reference current is generated in one location and is then reproduced at other locations for the purpose of biasing the various amplifier stages on the IC. As an example consider the circuit shown in Fig. 9.17. This circuit utilizes two power supplies, V_{CC} and $-V_{EE}$. The dc reference current I_{REF} is generated in the branch that consists of the diode-connected transistor Q_1, resistor R, and diode-connected transistor Q_2. If we assume that all transistors have high β and thus the base currents are negligibly small, then

$$I_{REF} = \frac{V_{CC} + V_{EE} - V_{EB1} - V_{BE2}}{R} \qquad (9.67)$$

Diode-connected transistor Q_1 forms a current mirror with Q_3. Thus Q_3 will supply a constant current I_1 equal to I_{REF}. Transistor Q_3 can supply this current to any load as long as the voltage that develops at the collector does not exceed that at the base ($V_{CC} - V_{EB3}$).

To generate a dc current twice the value of I_{REF} two transistors Q_5 and Q_6 are connected in parallel, and the combination forms a mirror with Q_1. Thus $I_3 = 2I_{REF}$. Note that the parallel combination of Q_5 and Q_6 is equivalent to a transistor whose EBJ area is double that of Q_1, which is precisely what would be done if this circuit were to be fabricated in IC form. Current mirrors are indeed used to provide multiples

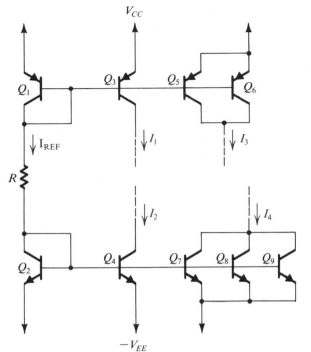

Fig. 9.17 Generation of a number of constant currents.

of the reference current by simply designing the transistors to have an area ratio equal to the desired multiple.

Transistor Q_4 forms a mirror with Q_2 and thus Q_4 provides a constant current I_2 equal to I_{REF}. Note an important difference between Q_3 and Q_4: Although both supply equal currents, Q_3 *sources* its current to parts of the circuit whose voltage should not exceed $V_{CC} - V_{EB3}$. On the other hand, Q_4 *sinks* its current from parts of the circuit whose voltage should not decrease below $-V_{EE} + V_{BE4}$. Finally, to generate a current three times the reference, three transistors Q_7, Q_8, and Q_9 are paralleled and the combination placed in a mirror configuration with Q_2. Again, in an IC implementation, Q_7, Q_8, and Q_9 would be replaced with a transistor having a junction area three times that of Q_2.

The above description ignored the effects of the finite transistor β. We have analyzed this effect in the case of a mirror having a single output. The effect of finite β becomes more severe as the number of outputs of the mirror is increased. This is not surprising since the addition of more transistors means that their base currents have to be supplied by the reference current source.

Exercise

9.8 Figure E9.8 shows an N-output current mirror. Assuming all transistors to be matched and have finite β and ignoring the effect of finite output resistances, show that

$$I_1 = I_2 = \cdots = I_N = \frac{1}{1 + (N + 1)/\beta}$$

For $\beta = 100$ find the maximum number of outputs for an error not exceeding 10%.

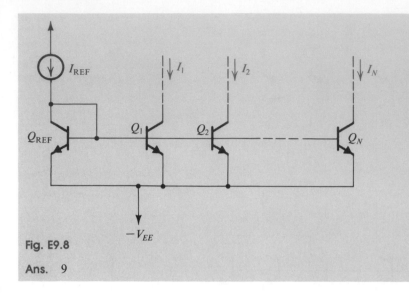

Fig. E9.8

Ans.   9

Improved Current-Source Circuits

There are two performance parameters of the BJT current source that need improvement. The first is the dependence of I_O on β, which is a result of the error in the mirror current gain introduced by the finite BJT β. The second is the output resistance of the current source, which was found to be equal to the BJT r_o and thus limited to the order of 100 kΩ. The need to increase the current-source output resistance can be seen if we recall that the common-mode gain of the differential amplifier is directly determined by the output resistance of its biasing current source. Also, it will be seen at a later stage that the BJT current sources are usually used in place

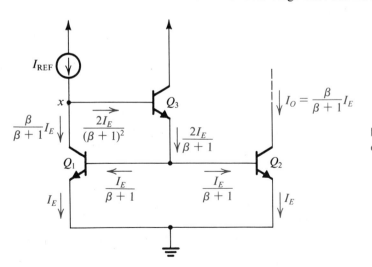

Fig. 9.18 Current mirror with base-current compensation.

of the load resistances R_C of the differential amplifier. Thus, to obtain high voltage gain, a large output resistance is required.

We shall now discuss several circuit techniques that result in reduced dependence on β and/or increased output resistance. The first circuit, shown in Fig. 9.18, includes a transistor Q_3 whose emitter supplies the base currents of Q_1 and Q_2. The sum of the base currents is then divided by $(\beta + 1)$ of Q_3, resulting in a much smaller current that has to be supplied by I_{REF}. Detailed analysis, shown on the circuit diagram, is based on the assumption that Q_1 and Q_2 are matched and thus have equal emitter currents, I_E. A node equation at the node labeled x gives

$$I_{REF} = \left[\frac{\beta}{\beta + 1} + \frac{2}{(\beta + 1)^2}\right] I_E$$

Since

$$I_O = \frac{\beta}{\beta + 1} I_E$$

it follows that the current gain of this mirror is given by

$$\frac{I_O}{I_{REF}} = \frac{1}{1 + 2/(\beta^2 + \beta)} \tag{9.68a}$$

$$\cong \frac{1}{1 + 2/\beta^2} \tag{9.68b}$$

which means that the error due to finite β has been reduced from $2/\beta$ to $2/\beta^2$, a tremendous improvement. Finally, note that for simplicity the circuit is shown fed with a current I_{REF}. To use the circuit as a current source we connect a resistance R between node x and the positive supply V_{CC}, then

$$I_{REF} = \frac{V_{CC} - V_{BE1} - V_{BE3}}{R} \tag{9.69}$$

An alternative mirror circuit that achieves both base-current compensation and increased output resistance is the Wilson mirror shown in Fig. 9.19. Analysis of this circuit taking into account the finite β results in a current-gain expression identical to that in Eq. (9.68).

Exercise

> **9.9** For the Wilson current mirror in Fig. 9.19 assume all BJTs are matched and have finite β. Denoting the currents in the emitters of Q_1 and Q_2 by I_E, find I_{REF} and I_O in terms of I_E and hence show that I_O/I_{REF} is given by Eq. (9.68).

The Wilson current source features an output resistance approximately equal to $\beta r_o/2$, a factor of $\beta/2$ greater than that of the simple current source of Fig. 9.16 (see Problem 9.29).

Increased output resistance can be also obtained using a cascode current mirror (Problem 9.30).

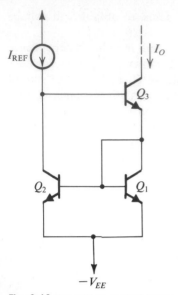

Fig. 9.19 The Wilson current mirror.

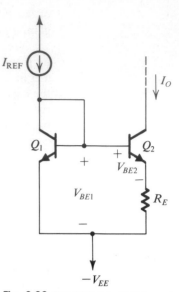

Fig. 9.20 The Widlar current source.

Our final current-source circuit, known as the Widlar current source, is shown in Fig. 9.20. It differs from the basic current mirror circuit in an important way: A resistor R_E is included in the emitter lead of Q_2. Neglecting base currents we can write:

$$V_{BE1} = V_T \ln \left(\frac{I_{\text{REF}}}{I_S} \right) \tag{9.70}$$

and

$$V_{BE2} = V_T \ln \left(\frac{I_O}{I_S} \right) \tag{9.71}$$

where we have assumed that Q_1 and Q_2 are matched devices. Combining Eqs. (9.70) and (9.71) gives

$$V_{BE1} - V_{BE2} = V_T \ln \left(\frac{I_{\text{REF}}}{I_O} \right) \tag{9.72}$$

But from the circuit we see that

$$V_{BE1} = V_{BE2} + I_O R_E \tag{9.73}$$

Thus

$$I_O R_E = V_T \ln \left(\frac{I_{\text{REF}}}{I_O} \right) \tag{9.74}$$

EXAMPLE 9.1

Figure 9.21 shows two circuits for generating a constant current $I_O = 10\ \mu A$. Determine the values of the required resistors assuming that V_{BE} is 0.7 V at a current of 1 mA and neglecting the effect of finite β.

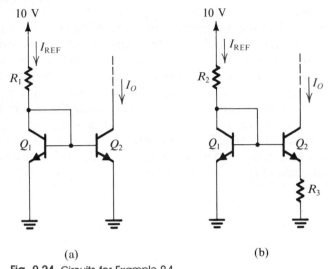

(a)   (b)

Fig. 9.21 Circuits for Example 9.1.

Solution

For the basic current source circuit in Fig. 9.21a we choose a value for R_1 to result in $I_{REF} = 10\ \mu A$. At this current, the voltage drop across Q_1 will be

$$V_{BE1} = 0.7 + V_T \ln \left(\frac{10\ \mu A}{1\ mA} \right) = 0.58\ V$$

Thus

$$R_1 = \frac{10 - 0.58}{0.01} = 942\ k\Omega$$

For the Widlar circuit in Fig. 9.21b we must first decide on a suitable value for I_{REF}. Selecting $I_{REF} = 1$ mA, then $V_{BE1} = 0.7$ V and R_2 is given by

$$R_2 = \frac{10 - 0.7}{1} = 9.3\ k\Omega$$

The value of R_3 can be determined using Eq. (9.74), as follows:

$$10 \times 10^{-6} R_3 = 0.025 \ln \left(\frac{1\ mA}{10\ \mu A} \right)$$

$$R_3 = 11.5\ k\Omega$$

From the above example we observe that using the Widlar circuit allows the generation of a small constant current using relatively small resistors. This is an important advantage that results in considerable savings in chip area. In fact the circuit of Fig. 9.21a, requiring a 942-kΩ resistance, is totally impractical for implementation in IC form.

Another important characteristic of the Widlar current source is that its output resistance is high. The increase in the output resistance, above that achieved in the basic current source of Fig. 9.16, is due to the emitter resistance R_E. To determine the output resistance of Q_2, we replace the BJT with its low-frequency hybrid-π model and apply a test voltage v_x to the collector, as shown in Fig. 9.22a. Note that the base of Q_2 is shown grounded, which is not quite the case in the original circuit in Fig. 9.20. Indeed, the base of Q_2 is connected to $-V_{EE}$ (signal ground) via the diode-connected transistor Q_1. The latter, however, has a small incremental resistance (r_e), and thus to simplify matters we shall assume that this resistance is small enough to place the base of Q_2 at signal ground.

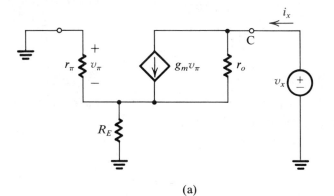

(a)

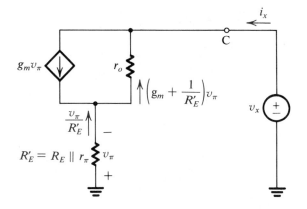

(b)

Fig. 9.22 Determination of the output resistance of the Widlar current source of Fig. 9.20.

The circuit of Fig. 9.22a is simplified by combining R_E and r_π in parallel to form R'_E, as illustrated in Fig. 9.22b, which shows some of the analysis details. A loop equation yields

$$v_x = -v_\pi - \left(g_m + \frac{1}{R'_E}\right)v_\pi r_o \tag{9.75}$$

and a node equation at C provides

$$i_x = g_m v_\pi - \left(g_m + \frac{1}{R'_E}\right)v_\pi \tag{9.76}$$

Dividing Eq. (9.75) by Eq. (9.76) gives the output resistance

$$R_o \equiv \frac{v_x}{i_x} = \frac{1 + \left(g_m + \dfrac{1}{R'_E}\right)r_o}{1/R'_E}$$

which can be rearranged into the form

$$R_o = R'_E + (1 + g_m R'_E)r_o \tag{9.77}$$

$$\cong (1 + g_m R'_E)r_o \tag{9.78}$$

Thus the output resistance is increased by the factor $1 + g_m R'_E = 1 + g_m(R_E \parallel r_\pi)$. Finally note that in the analysis above we have neglected r_μ. However, it can be easily taken into account since it appears in parallel with the resistance given by Eq. (9.78).

Exercise

9.10 Find the output resistance of each of the two current sources designed in Example 9.1. Let $V_A = 100$ V and $\beta = 100$.

Ans. (a) 10 MΩ; (b) 54 MΩ

EXAMPLE 9.2

Figure 9.23 shows the circuit of a simple operational amplifier. Terminals 1 and 2, shown connected to ground, are the op amp's input terminals, and terminal 3 is the output terminal.

(a) Perform an approximate dc analysis (assuming $\beta \gg 1$) to calculate the dc currents and voltages everywhere in the circuit. Note that Q_6 has four times the area of each of Q_9 and Q_3.

(b) Calculate the quiescent power dissipation in this circuit.

(c) If transistors Q_1 and Q_2 have $\beta = 100$, calculate the input bias current of the op amp.

(d) What is the common-mode range of this op amp?

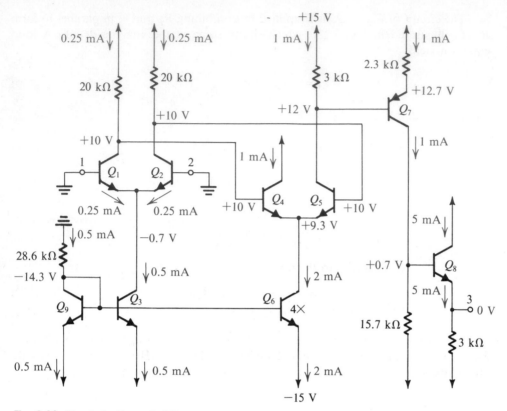

Fig. 9.23 Circuits for Example 9.2.

Solution

(a) The values of all dc currents and voltages are indicated on the circuit diagram. These values were calculated by ignoring the base current of every transistor—that is, by assuming β to be very high. The analysis starts by determining the current through the diode-connected transistor Q_9 to be 0.5 mA. Then we see that transistor Q_3 conducts 0.5 mA and transister Q_6 conducts 2 mA. The current-source transistor Q_3 feeds the differential pair (Q_1, Q_2) with 0.5 mA. Thus each of Q_1 and Q_2 will be biased at 0.25 mA. The collectors of Q_1 and Q_2 will be at $[+15 - 0.25 \times 20] = +10$ V.

Proceeding to the second differential stage formed by Q_4 and Q_5, we find the voltage at their emitters to be $[+10 - 0.7] = 9.3$ V. This differential pair is biased by the current-source transistor Q_6, which supplies a current of 2 mA; thus Q_4 and Q_5 will each be biased at 1 mA. We can now calculate the voltage at the collector of Q_5 as $+15 - 1 \times 3 = +12$ V. This will cause the voltage at the emitter of the *pnp* transistor Q_7 to be $+12.7$ V, and the emitter current of Q_7 will be $(+15 - 12.7)/2.3 = 1$ mA.

The collector current of Q_7, 1 mA, causes the voltage at the collector to be $-15 + 1 \times 15.7 = +0.7$ V. The emitter of Q_8 will be 0.7 V below the base; thus output terminal 3 will be at 0 V. Finally, the emitter current of Q_8 can be calculated to be $[0 - (-15)]/3 = 5$ mA.

(b) To calculate the power dissipated in the circuit in the quiescent state (that is, with zero input signal) we simply evaluate the dc current that the circuit draws from each of the two power supplies. From the $+15$-V supply the dc current is $I^+ = 0.25 + 0.25 + 1 + 1 + 1 + 5 = 8.5$ mA. Thus the power supplied by the positive power supply is $P^+ = 15 \times 8.5 = 127.5$ mW. The -15-V supply provides a current I^- given by $I^- = 0.5 + 0.5 + 2 + 1 + 5 = 9$ mA. Thus the power provided by the negative supply is $P^- = 15 \times 9 = 135$ mW. Adding P^+ and P^- provides the total power dissipated in the circuit P_D: $P_D = P^+ + P^- = 262.5$ mW.

(c) The input bias current of the op amp is the average of the dc currents that flow in the two input terminals (that is, in the bases of Q_1 and Q_2). These two currents are equal (because we have assumed matched devices); thus the bias current is given by

$$I_B = \frac{I_{E1}}{\beta + 1} \simeq 2.5\ \mu A$$

(d) The upper limit on the input common-mode voltage is determined by the voltage at which Q_1 and Q_2 leave the active mode and enter saturation. This will happen if the input voltage equals or exceeds the collector voltage, which is $+10$ V. Thus the upper limit of the common-mode range is $+10$ V.

  The lower limit of the input common-mode range is determined by the voltage at which Q_3 leaves the active mode and thus ceases to act as a constant-current source. This will happen if the collector voltage of Q_3 goes below the voltage at its base, which is -14.3 V. It follows that the input common-mode voltage should not go lower than $-14.3 + 0.7 = -13.6$ V. Thus the common-mode range is   13.6 to $+10$ V.

9.5 THE BJT DIFFERENTIAL AMPLIFIER WITH ACTIVE LOADS

Active devices (transistors) occupy much less silicon area than medium- and large-sized resistors. For this reason we studied in Chapter 7 a variety of MOSFET amplifier circuits that utilize MOSFETs as load devices. Similarly, many practical BJT integrated-circuit amplifiers use BJT loads in place of the resistive loads, R_C. In such circuits the BJT load transistor is usually connected as a constant-current source and thus presents the amplifier transistor with a very-high-resistance load (the output resistance of the current source). Thus amplifiers that utilize *active loads* can achieve higher voltage gains than those with passive (resistive) loads. In this section we study a circuit configuration that has become very popular in the design of BJT ICs.

  The active-load differential amplifier circuit is shown in Fig. 9.24. Transistors Q_1 and Q_2 form the differential pair biased with constant current I. The load circuit consists of transistors Q_3 and Q_4 connected in a current mirror configuration. The output is taken single-endedly from the collector of Q_2.

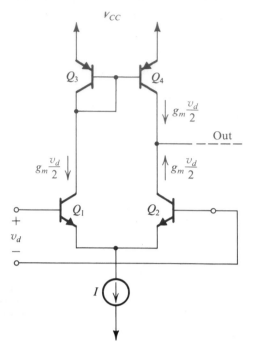

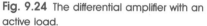

Fig. 9.24 The differential amplifier with an active load.

Consider first the case when no input signal is applied (that is, the two input terminals are grounded.) The current I splits equally between Q_1 and Q_2. Thus Q_1 draws a current approximately $I/2$ from the diode-connected transistor Q_3. Assuming $\beta \gg 1$, the mirror supplies an equal current $I/2$ through the collector of Q_4. Since this current is equal to that through the collector of Q_2, no output current flows through the output terminal. It should be noted, however, that in practical circuits the dc quiescent voltage at the output terminal is determined by the subsequent amplifier stage. An example of that will be seen in Chapter 13 where the internal circuit of the 741-type op amp is studied in detail.

Next consider the situation when a differential signal v_d is applied at the input. Current signals $g_m(v_d/2)$ will result in the collectors of Q_1 and Q_2 with the polarities indicated in Fig. 9.24. The current mirror reproduces the current signal $g_m(v_d/2)$ through the collector of Q_4. Thus, at the output node we have two current signals that add together to produce a total current signal of $(g_m v_d)$. Now if the resistance presented by the subsequent amplifier stage is very large, the voltage signal at the output terminal will be determined by the total signal current $(g_m v_d)$ and the total resistance between the output terminal and ground, R_o; that is,

$$v_o = g_m v_d R_o \tag{9.79}$$

The output resistance R_o is the parallel equivalent of the output resistance of Q_2 and the output resistance of Q_4. Since Q_2 is in effect operating in the common-emitter configuration, its output resistance will be equal to r_{o2}. Also, from our study of the basic current mirror circuit in the previous section we know that its output resistance is equal to r_O of Q_4—that is, r_{o4}. Thus,

$$R_o = r_{o2} \| r_{o4} \tag{9.80}$$

For the case $r_{o2} = r_{o4} = r_o$,

$$R_o = r_o/2 \tag{9.81}$$

and the output voltage will be

$$v_o = g_m v_d(r_o/2) \tag{9.82}$$

leading to a voltage gain

$$\frac{v_o}{v_d} = \frac{g_m r_o}{2} \tag{9.83}$$

Substituting $g_m = I_C/V_T$ and $r_o = V_A/I_C$, where $I_C = I/2$, we obtain

$$g_m r_o = \frac{V_A}{V_T} \tag{9.84}$$

which is a constant for a given transistor. Typically, $V_A = 100$ V, leading to $g_m r_o = 4{,}000$ and a stage voltage gain of about 2,000.

In some cases the input resistance of the subsequent amplifier stage may be of the same order as R_o and thus must be taken into account in determining voltage gain. In such situations it is convenient to represent the amplifier of Fig. 9.24 by the transconductance amplifier model shown in Fig. 9.25. Here R_i is the differential input resistance, for our case $R_i = 2r_\pi$. The amplifier transconductance G_m is the short-circuit transconductance and for our case

$$G_m = g_m = \frac{I/2}{V_T}$$

Finally, R_o is the output resistance given by Eq. (9.81).

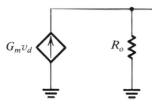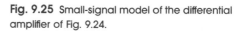

Fig. 9.25 Small-signal model of the differential amplifier of Fig. 9.24.

In order to obtain yet higher voltage gains, more elaborate current-mirror circuits can be utilized. Also, the basic differential amplifier configuration can be modified to increase the output resistance of Q_2. For instance, the cascode configuration studied in Chapter 8 can be used.

As a final note, observe the role that the current-mirror circuit in Fig. 9.24 plays: It inverts the current signal $g_m(v_d/2)$ supplied by the collector of Q_1 and provides an equal current at the collector of Q_4 with such a polarity that it adds to the current signal in the collector of Q_2. Without the current mirror (that is, using only a simple current source) the voltage gain would be half the value found above.

9.11 The finite β (which is usually low for the standard IC process) of the *pnp* transistors that form the current mirror in the circuit of Fig. 9.24 results in a dc offset voltage.
(a) With both inputs grounded show that there will be an output current directed into the amplifier of approximately I/β_p, where β_p is the β of the *pnp* transistors. (Assume that β of the *npn* transistors is high.)
(b) Find the differential input voltage that would reduce the output current found in (a) to zero. This is the input dc offset voltage due to the finite β_p.
(c) For $\beta_p = 25$, find V_{off}.

Ans. (b) $2V_T/\beta_p$; (c) 2 mV

9.6 THE JFET DIFFERENTIAL PAIR

Because of their high input resistance, FETs are popular in the design of the differential input stage of op amps. FET-input amplifiers are available commercially and exhibit very small input bias currents (in the picoampere range). In this section we shall study the operation of the JFET differential amplifier.

Figure 9.26 shows the basic JFET differential amplifier fed by two voltage sources v_{G1} and v_{G2} and biased by a constant-current source I. We wish to derive expressions for the currents i_{D1} and i_{D2} in terms of v_{G1} and v_{G2}. To do this we will assume that the JFETs are operating in the pinch-off region. Therefore we may use the square-law i_D-v_{GS} relationship

$$i_D = I_{DSS}\left(1 - \frac{v_{GS}}{V_P}\right)^2$$

where for *n*-channel devices V_P is negative. Furthermore, we will assume that the two FETs are perfectly matched. For transistor Q_1 we have

$$i_{D1} = I_{DSS}\left(1 - \frac{v_{G1} - v_S}{V_P}\right)^2$$

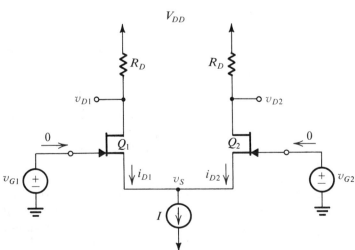

Fig. 9.26 The JFET differential pair.

which can be rewritten

$$\sqrt{i_{D1}} = \sqrt{I_{DSS}}\left(1 - \frac{v_{G1}}{V_P} + \frac{v_S}{V_P}\right)$$ (9.85)

Similarly for Q_2 we have

$$i_{D2} = I_{DSS}\left(1 - \frac{v_{G2} - v_S}{V_P}\right)^2$$

which can be rewritten

$$\sqrt{i_{D2}} = \sqrt{I_{DSS}}\left(1 - \frac{v_{G2}}{V_P} + \frac{v_S}{V_P}\right)$$ (9.86)

Subtracting Eq. (9.86) from Eq. (9.85) gives

$$\sqrt{i_{D1}} - \sqrt{i_{D2}} = \sqrt{I_{DSS}}\,\frac{v_{G1} - v_{G2}}{-V_P}$$

Substituting $v_{G1} - v_{G2} = v_{id}$, where v_{id} is the differential input voltage, gives

$$\sqrt{i_{D2}} = \sqrt{i_{D1}} - \sqrt{I_{DSS}}\,\frac{v_{id}}{-V_P}$$ (9.87)

The current-source bias imposes the following constraint:

$$i_{D1} + i_{D2} = I$$ (9.88)

Substituting for i_{D2} from Eq. (9.87) into Eq. (9.88) results in a quadratic equation that can be solved to yield ultimately

$$i_{D1} = \frac{I}{2} + v_{id}\frac{I}{-2V_P}\sqrt{2\,\frac{I_{DSS}}{I} - \left(\frac{v_{id}}{V_P}\right)^2\left(\frac{I_{DSS}}{I}\right)^2}$$ (9.89)

$$i_{D2} = \frac{I}{2} - v_{id}\frac{I}{-2V_P}\sqrt{2\,\frac{I_{DSS}}{I} - \left(\frac{v_{id}}{V_P}\right)^2\left(\frac{I_{DSS}}{I}\right)^2}$$ (9.90)

Equations (9.89) and (9.90) indicate that the differential pair responds to a difference signal v_{id} by changing the proportion in which the bias current I divides between the two transistors. For $v_{id} = 0$, $i_{D1} = i_{D2} = I/2$, as should be expected. If v_{id} is positive, the current i_{D1} increases by the amount given by the second term on the right-hand side of Eq. (9.89), and i_{D2} decreases by an equal amount. The maximum value of the increment in i_{D1} (the decrement in i_{D2}) is limited to $I/2$ and is obtained when v_{id} is of such a value that

$$v_{id}\frac{I}{-2V_P}\sqrt{2\,\frac{I_{DSS}}{I} - \left(\frac{v_{id}}{V_P}\right)^2\left(\frac{I_{DSS}}{I}\right)^2} = \frac{I}{2}$$

which results in

$$\left|\frac{v_{id}}{V_P}\right| = \sqrt{\frac{I}{I_{DSS}}}$$ (9.91)

That is, the value of v_{id} given by Eq. (9.91) results in the current I being entirely carried by one of the two transistors (Q_1 for positive v_{id} and Q_2 for negative v_{id}). At this point we should note that the bias current I should be smaller than I_{DSS}; otherwise one of the two FETs would carry a current greater than I_{DSS}, which would result in its gate-channel junction becoming forward-biased.

As in the case of the BJT differential pair, the input common-mode range is determined at the low end by the current-source device (which can be a FET or a BJT) leaving the active region, and at the high end by Q_1 and Q_2 leaving the active (pinch-off) region. This upper limit on v_{CM} will therefore be $|V_P|$ volts below the quiescent voltage at the drain, $V_{DD} - (I/2)R_D$.

Small-Signal Operation

Small-signal analysis of the JFET differential amplifier can be carried out in a manner identical to that given in detail for the BJT circuit. For instance, to evaluate the small-signal differential gain we use Eqs. (9.89) and (9.90) and assume that the terms involving v_{id}^2 are negligibly small; that is

$$\left|\frac{v_{id}}{V_P}\right| \ll \sqrt{\frac{2I}{I_{DSS}}} \tag{9.92}$$

Note, however, that unlike the BJT case, where v_{id} is limited to about 20 mV, here the small-signal condition limits v_{id} to a volt or so. Under this condition Eqs. (9.89) and (9.90) can be approximated to

$$i_{D1} \simeq \frac{I}{2} + i_d \tag{9.93}$$

$$i_{D2} \simeq \frac{I}{2} - i_d \tag{9.94}$$

where the current signal i_d is given by

$$i_d = \frac{v_{id}}{2}\left(\frac{2I_{DSS}}{-V_P}\sqrt{\frac{I/2}{I_{DSS}}}\right) \tag{9.95}$$

We immediately recognize the quantity in parentheses as g_m of Q_1 and of Q_2. Thus Eq. (9.95) simply reaffirms our expectations that the input difference voltage v_{id} divides equally between the two transistors,

$$v_{gs1} = v_{sg2} = \frac{v_{id}}{2} \tag{9.96}$$

and thus the signal current in Q_1 is $g_m v_{id}/2$ and that in Q_2 is $-g_m v_{id}/2$. Alternatively, we can think of the voltage v_{id} as appearing across a total source resistance of $2/g_m$; thus a current

$$i_d = \frac{v_{id}}{2/g_m}$$

flows, as illustrated in Fig. 9.27.

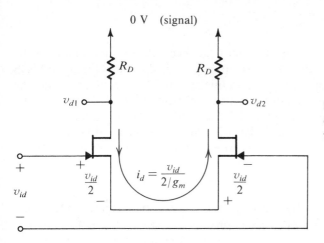

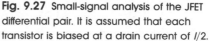

Fig. 9.27 Small-signal analysis of the JFET differential pair. It is assumed that each transistor is biased at a drain current of $I/2$.

At the drain of Q_1 the voltage signal v_{d1} will be

$$v_{d1} = -i_d R_D = -g_m \frac{v_{id}}{2} R_D \tag{9.97}$$

and at the drain of Q_2 we have

$$v_{d2} = +i_D R_D = +g_m \frac{v_{id}}{2} R_D \tag{9.98}$$

If the output is taken differentially (that is, between the two drains), then

$$v_o = v_{d1} - v_{d2} = -g_m R_D v_{id} \tag{9.99}$$

and the differential gain will be

$$\frac{v_o}{v_{id}} = -g_m R_D \tag{9.100}$$

If the JFET output resistance r_o is taken into account, the expression for the gain in Eq. (9.100) is modified to

$$\frac{v_o}{v_{id}} = -g_m (R_D \| r_o) \tag{9.101}$$

Finally, it should be pointed out that the current source supplying the bias current I will inevitably have a finite output resistance, which causes the differential amplifier to have a nonzero common-mode gain. The common-mode gain can be evaluated in a manner identical to that used for the BJT case.

Exercise

9.12 Consider the circuit in Fig. 9.26 with $V_{DD} = +15\,\text{V}$, $I = 1\,\text{mA}$, $R_D = 10\,\text{k}\Omega$, $I_{DSS} = 2\,\text{mA}$, and $V_P = -2\,\text{V}$. Find the value of v_{id} required to switch the current entirely to Q_1. Also calculate the small-signal voltage gain if the output is taken differentially.

Ans. 1.4 V; 10 V/V

9.7 MOS DIFFERENTIAL AMPLIFIERS

During the past few years, the MOS transistor has become very prominent in the design of analog integrated circuits. We have already studied some of the basic MOS IC amplifier circuits in Chapter 7. Building on this material, this section presents the MOS differential pair, which is the most important building block in MOS ICs. We shall also discuss MOS current mirrors, which are used for biasing and as loads for the differential pair. The section concludes with an active-load MOS differential amplifier.

The MOS Differential Pair

Figure 9.28 shows the basic MOS differential pair. It consists of two matched enhancement MOSFETs, Q_1 and Q_2, biased with a constant-current source I. The latter is usually implemented using a current-mirror configuration, much like the case in BJT circuits. Note that the differential amplifier loads are not shown. At this point our purpose is to relate the drain currents to the input voltage. It is, of course, assumed that the load circuit is such that the two MOSFETs in the pair operate in pinch-off (the active region).

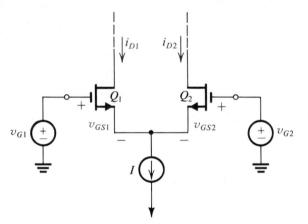

Fig. 9.28 The MOSFET differential pair.

Assuming that the two devices are identical and neglecting the output resistance and body effect, we can express the drain currents as

$$i_{D1} = K(v_{GS1} - V_t)^2 \tag{9.102}$$

$$i_{D2} = K(v_{GS2} - V_t)^2 \tag{9.103}$$

where

$$K = \tfrac{1}{2}\mu_n C_{ox}(W/L) \tag{9.104}$$

Equations (9.102) and (9.103) can be rewritten

$$\sqrt{i_{D1}} = \sqrt{K}(v_{GS1} - V_t) \tag{9.105}$$

$$\sqrt{i_{D2}} = \sqrt{K}(v_{GS2} - V_t) \tag{9.106}$$

Subtracting Eq. (9.106) from Eq. (9.105) and substituting

$$v_{GS1} - v_{GS2} = v_{id}$$

where v_{id} is the differential input voltage, gives

$$\sqrt{i_{D1}} - \sqrt{i_{D2}} = \sqrt{K}\,v_{id} \tag{9.107}$$

The current-source bias imposes the constraint

$$i_{D1} + i_{D2} = I \tag{9.108}$$

Equations (9.107) and (9.108) are two equations in the two unknowns i_{D1} and i_{D2}. They can be solved together to yield

$$i_{D1} = \frac{I}{2} + \sqrt{2KI}\left(\frac{v_{id}}{2}\right)\sqrt{1 - \frac{(v_{id}/2)^2}{(I/2K)}} \tag{9.109}$$

$$i_{D2} = \frac{I}{2} - \sqrt{2KI}\left(\frac{v_{id}}{2}\right)\sqrt{1 - \frac{(v_{id}/2)^2}{(I/2K)}} \tag{9.110}$$

At the bias (quiescent) point, $v_{id} = 0$, leading to

$$i_{D1} = i_{D2} = \frac{I}{2} \tag{9.111}$$

Correspondingly,

$$v_{GS1} = v_{GS2} = V_{GS}$$

where

$$\frac{I}{2} = K(V_{GS} - V_t)^2 \tag{9.112}$$

This relationship can be used to rewrite Eqs. (9.109) and (9.110) in the form

$$i_{D1} = \frac{I}{2} + \left(\frac{I}{V_{GS} - V_t}\right)\left(\frac{v_{id}}{2}\right)\sqrt{1 - \left(\frac{v_{id}/2}{V_{GS} - V_t}\right)^2} \tag{9.113}$$

$$i_{D2} = \frac{I}{2} - \left(\frac{I}{V_{GS} - V_t}\right)\left(\frac{v_{id}}{2}\right)\sqrt{1 - \left(\frac{v_{id}/2}{V_{GS} - V_t}\right)^2} \tag{9.114}$$

For $v_{id}/2 \ll V_{GS} - V_t$ (small-signal approximation),

$$i_{D1} \cong \frac{I}{2} + \left(\frac{I}{V_{GS} - V_t}\right)\left(\frac{v_{id}}{2}\right) \tag{9.115}$$

$$i_{D2} \cong \frac{I}{2} - \left(\frac{I}{V_{GS} - V_t}\right)\left(\frac{v_{id}}{2}\right) \tag{9.116}$$

From Chapter 7 we recall that a MOSFET biased at a drain current I_D has $g_m = 2I_D/(V_{GS} - V_t)$. Thus we see that, for each transistor in the differential pair,

$$g_m = \frac{2(I/2)}{V_{GS} - V_t} = \frac{I}{V_{GS} - V_t} \tag{9.117}$$

and Eqs. (9.115) and (9.116) simply state that for small differential input signals, $v_{id} \ll 2(V_{GS} - V_t)$, the current in Q_1 increases by i_d and that in Q_2 decreases by i_d, where

$$i_d = g_m(v_{id}/2) \tag{9.118}$$

Returning to Eqs. (9.113) and (9.114), we can find the value of v_{id} at which full switching occurs (that is, $i_{D1} = I$ and $i_{D2} = 0$, or vice versa for negative v_{id}) by equating the second term in Eq. (9.113) to $I/2$. The result is

$$v_{id}|_{max} = \sqrt{2}(V_{GS} - V_t)$$

Figure 9.29 shows plots of the normalized currents i_{D1}/I and i_{D2}/I versus the normalized differential input voltage $v_{id}/(V_{GS} - V_t)$.

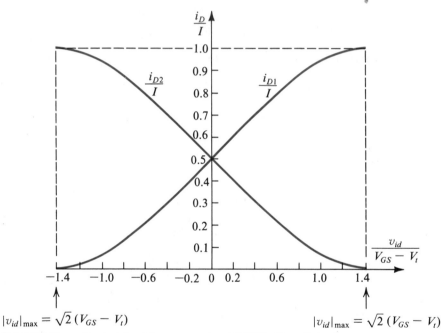

Fig. 9.29 Normalized plots of the currents in a MOSFET differential pair.

Finally, we note that for differential input signals, each MOSFET in the pair operates as a common-source amplifier and thus has an output resistance equal to r_o.

Exercise

9.13 An MOS differential amplifier utilizes a bias current $I = 25\ \mu A$. The devices have $V_t = 1$ V, $W = 120\ \mu m$, $L = 6\ \mu m$, and $(\mu_n C_{ox})$ for this technology is $20\ \mu A/V^2$. Find V_{GS}, g_m, and the value of v_{id} for full current switching.

Ans.   1.25 V; 0.1 mA/V; 0.35 V

Offset Voltage

Three factors contribute to the dc offset voltage of the MOS differential pair: mismatch in load resistances, mismatch in K, and mismatch in V_t. We shall consider the three contributing factors one at a time.

Consider the differential pair shown in Fig. 9.30, in which resistive loads are used in order to simplify the analysis. Since both inputs are grounded, the output voltage V_O is the dc output offset voltage. Consider first the case where Q_1 and Q_2 are perfectly matched but R_{D1} and R_{D2} show a mismatch ΔR_D; that is,

$$R_{D1} = R_D + \left(\frac{\Delta R_D}{2}\right) \tag{9.119}$$

$$R_{D2} = R_D - \left(\frac{\Delta R_D}{2}\right) \tag{9.120}$$

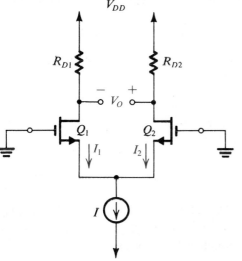

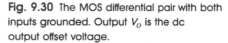

Fig. 9.30 The MOS differential pair with both inputs grounded. Output V_O is the dc output offset voltage.

The current I will split equally between Q_1 and Q_2. Nevertheless, because of the mismatch in load resistances, an output voltage V_O develops,

$$V_O = \left(\frac{I}{2}\right)\Delta R_D \tag{9.121}$$

The corresponding input offset voltage is obtained by dividing V_O by the gain $g_m R_D$ and substituting for g_m from Eq. (9.117). The result is

$$V_{\text{off}} = \left(\frac{V_{GS} - V_t}{2}\right)\left(\frac{\Delta R_D}{R_D}\right) \tag{9.122}$$

Next consider the effect of a mismatch in the W/L ratios of Q_1 and Q_2,

$$\left(\frac{W}{L}\right)_1 = \frac{W}{L} + \frac{1}{2}\Delta\left(\frac{W}{L}\right) \tag{9.123}$$

$$\left(\frac{W}{L}\right)_2 = \frac{W}{L} - \frac{1}{2}\Delta\left(\frac{W}{L}\right) \tag{9.124}$$

Such a mismatch gives rise to a proportional mismatch in the conductivity parameter $K = \frac{1}{2}(\mu_n C_{OX})(W/L)$; that is,

$$K_1 = K + \frac{\Delta K}{2} \tag{9.125}$$

$$K_2 = K - \frac{\Delta K}{2} \tag{9.126}$$

The currents I_1 and I_2 will no longer be equal; rather, it can be easily shown that

$$I_1 = \frac{I}{2} + \frac{I}{2}\left(\frac{\Delta K}{2K}\right) \tag{9.127}$$

$$I_2 = \frac{I}{2} - \frac{I}{2}\left(\frac{\Delta K}{2K}\right) \tag{9.128}$$

Dividing the current increment

$$\frac{I}{2}\left(\frac{\Delta K}{2K}\right)$$

by g_m gives half the input offset voltage (due to the mismatch in K values). Thus

$$V_{\text{off}} = \left(\frac{V_{GS} - V_t}{2}\right)\left(\frac{\Delta K}{K}\right) \tag{9.129}$$

Finally, we consider the effect of a mismatch ΔV_t between the two threshold voltages,

$$V_{t1} = V_t + \frac{\Delta V_t}{2} \tag{9.130}$$

$$V_{t2} = V_t - \frac{\Delta V_t}{2} \tag{9.131}$$

The current I_1 will be given by

$$I_1 = K\left(V_{GS} - V_t - \frac{\Delta V_t}{2}\right)^2$$

$$= K(V_{GS} - V_t)^2\left[1 - \frac{\Delta V_t}{2(V_{GS} - V_t)}\right]^2$$

which, for $\Delta V_t \ll 2(V_{GS} - V_t)$, can be approximated as

$$I_1 \simeq K(V_{GS} - V_t)^2 \left(1 - \frac{\Delta V_t}{V_{GS} - V_t}\right)$$

Similarly,

$$I_2 \simeq K(V_{GS} - V_t)^2 \left(1 + \frac{\Delta V_t}{V_{GS} - V_t}\right)$$

It follows that

$$K(V_{GS} - V_t)^2 = \frac{I}{2}$$

and the current increment (decrement) in Q_2 (Q_1) is

$$\Delta I = \frac{I}{2} \frac{\Delta V_t}{V_{GS} - V_t}$$

Dividing ΔI by g_m gives half the input offset voltage (due to ΔV_t). Thus,

$$V_{\text{off}} = \Delta V_t \qquad (9.132)$$

For modern silicon-gate MOS technology ΔV_t can be easily as high as 2 mV. We note that ΔV_t has no counterpart in BJT differential amplifiers. Also, comparison of V_{off} for the MOS differential pair in Eqs. (9.122) and (9.129) to V_{off} of the BJT differential pair in Eqs. (9.49) and (9.54) shows that the offset voltage is larger in the MOS pair because $(V_{GS} - V_t)/2$ is usually much greater than V_T. Finally, we observe from Eqs. (9.122) and (9.129) that to keep V_{off} small, one attempts to operate Q_1 and Q_2 at low values of $V_{GS} - V_t$.

Exercise

> **9.14** For the MOS differential pair specified in Exercise 9.13, find the three components of input offset voltage. Let $\Delta R_D/R_D = 2\%$, $\Delta K/K = 2\%$, and $\Delta V_t = 2$ mV.
>
> **Ans.** 2.5 mV; 2.5 mV; 2 mV

Current Mirrors

As in BJT integrated circuits, current mirrors are used in the design of current sources for biasing as well as to operate as active loads. The basic MOS current-mirror circuit, shown in Fig. 9.31a, was studied in Section 7.10. The inaccuracy in current transfer ratio due to the finite β of the BJT has no counterpart in MOS mirrors. Thus the only performance parameter of interest here is the output resistance. For the simple mirror in Fig. 9.31a, the output resistance is approximately equal to r_{o2}.

The output resistance can be increased by using either the cascode mirror of Fig. 9.31b or the Wilson mirror of Fig. 9.31c. To determine the output resistance of the cascode mirror we use the equivalent circuit shown in Fig. 9.32a. Note that since the incremental resistance of each of the diode-connected transistors Q_1 and Q_2 is equal to $1/g_m$ and thus is relatively small, we have assumed that the signal voltages

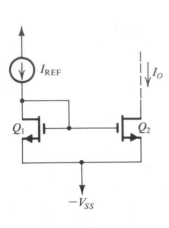

(a)

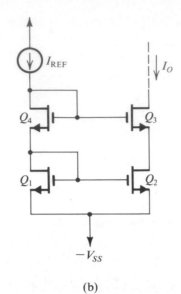

(b)

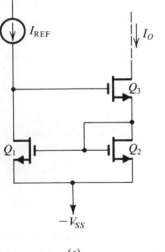

(c)

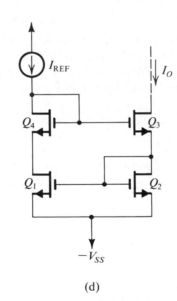

(d)

Fig. 9.31 MOS current mirrors: **(a)** basic, **(b)** cascode, **(c)** Wilson, **(d)** modified Wilson.

at the gates of Q_2 and Q_3 are approximately zero. Replacing Q_2 by its output resistance r_{o2} and replacing Q_3 by its equivalent circuit model leads to the circuit in Fig. 9.32b. Analysis of the latter circuit is straightforward and leads to

$$R_o \equiv \frac{v_x}{i_x} = r_{o3} + r_{o2} + g_{m3}r_{o3}r_{o2} \tag{9.133}$$

For $r_{o2} = r_{o3} = r_o$ we have

$$R_o = r_o(2 + g_m r_o) \tag{9.134}$$

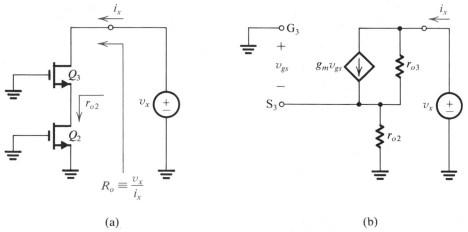

Fig. 9.32 Determining the output resistance of the cascode mirror of Fig. 9.31b.

Thus the cascode configuration increases the output resistance by a factor approximately equal to $g_m r_o$. Similar results are obtained with the Wilson circuit in Fig. 9.31c. The Wilson circuit, however, suffers from the fact that the drain voltages of Q_1 and Q_2 are not equal, and thus their currents will be unequal. This problem can be solved by including the diode-connected transistor Q_4, as shown in Fig. 9.31d.

Exercises

9.15 Consider the current mirror circuit of Fig. 9.31a with $V_{SS} = -5$ V and $I_{REF} = 10$ μA. Let Q_1 and Q_2 be identical, with $V_t = 1$ V, $\mu_n C_{OX} = 20$ μA/V^2, $L = 10$ μm, $W = 40$ μm, and $V_A = 20$ V. Find the output resistance, V_{GS}, and the lowest allowable output voltage.

Ans.   2 MΩ; 1.5 V; -4.5 V

9.16 Repeat Exercise 9.15 for the cascode mirror of Fig. 9.31b, assuming all devices to be identical.

Ans.   164 MΩ; 1.5 V; -3 V

An Active-Loaded CMOS Amplifier

We conclude this section with a discussion of a popular configuration for a differential amplifier in CMOS technology. The circuit, shown in Fig. 9.33, consists of the differential pair Q_1 and Q_2 loaded by the current mirror formed by Q_3 and Q_4. The dc bias voltage at the output is normally set by the subsequent amplifier stage, as will be shown in Chapter 13.

The circuit is analogous to the BJT version in Fig. 9.24. The signal current i is given by

$$i = g_m(v_{id}/2)$$

where

$$g_m = \frac{I}{V_{GS} - V_t}$$

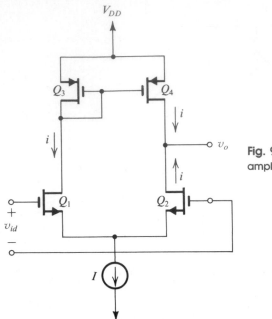

Fig. 9.33 An active-loaded differential amplifier in CMOS technology.

The output signal voltage is given by

$$v_o = 2i(r_{o2} \parallel r_{o4}) \tag{9.135}$$

For

$$r_{o2} = r_{o4} = r_o = \frac{V_A}{I/2}$$

the voltage gain becomes

$$A_v \equiv \frac{v_o}{v_{id}} = g_m \frac{r_o}{2} \tag{9.136}$$

$$= \frac{V_A}{V_{GS} - V_t} \tag{9.137}$$

To obtain higher voltage gains, a cascode current mirror and a cascode differential stage can be used. This, however, reduces the allowable output signal swing. We will have more to say about CMOS differential amplifiers in Chapter 13.

Exercise

9.17 Find the voltage gain of the differential amplifier circuit of Fig. 9.33 under the condition that $I = 25\ \mu\text{A}$, $V_t = 1\ \text{V}$, $W_1 = W_2 = 120\ \mu\text{m}$, $L_1 = L_2 = 6\ \mu\text{m}$, $\mu_n C_{OX} = 20\ \mu\text{A/V}^2$, $V_A = 20\ \text{V}$.

Ans. 80 V/V

9.8 MULTISTAGE AMPLIFIERS

Practical transistor amplifiers usually consist of a number of stages connected in cascade. In addition to providing gain, the first (or input) stage is usually required to provide a high input resistance in order to avoid loss of signal level when the amplifier is fed with a high-resistance source. In a differential amplifier the input stage must also provide large common-mode rejection. The function of the middle stages of an amplifier cascade is to provide the bulk of the voltage gain. In addition, the middle stages provide such other functions as the conversion of the signal from differential mode to single-ended mode and the shifting of the dc level of the signal. These two functions and others will be illustrated later in this section and in greater detail in Chapter 13.

Finally, the main function of the last (or output) stage of an amplifier is to provide a low output resistance in order to avoid loss of gain when a low-valued load resistance is connected to the amplifier. Also, the output stage should be able to supply the current required by the load in an efficient manner—that is, without dissipating an unduly large amount of power in the output transistors. We have already studied one type of amplifier configuration suitable for implementing output stages, namely, the source follower and the emitter follower. It will be shown in Chapter 10 that the source and emitter followers are not optimum from the point of view of power efficiency and that other, more appropriate circuit configurations exist for output stages required to supply large amounts of output power.

To illustrate the structure and method of analysis of multistage amplifiers, we will conclude this chapter with a detailed example. The amplifier circuit to be analyzed is shown in Fig. 9.34. The dc analysis of this simple op-amp circuit was presented in Example 9.2, which we urge the reader to review before studying the following material.

The op-amp circuit in Fig. 9.34 consists of four stages. The input stage is *differential-in, differential-out* and consists of transistors Q_1 and Q_2, which are biased by current source Q_3. The second stage is also a differential-input amplifier, but its output is taken single-ended at the collector of Q_5. This stage is formed by Q_4 and Q_5, which are biased by the current source Q_6. Note that the conversion from differential to single-ended as performed by the second stage results in a loss of gain of a factor of 2. A more elaborate method for accomplishing this conversion was studied in Sections 9.5 and 9.7; it involves using the current mirror as an active load for the first stage.

In addition to providing some voltage gain, the third stage, consisting of the *pnp* transistor Q_7, provides the essential function of *shifting the dc level* of the signal. Thus while the signal at the collector of Q_5 is not allowed to swing below the voltage at the base of Q_5 ($+10$ V), the signal at the collector of Q_7 can swing negative (and positive, of course). From our study of op amps in Chapter 3 we know that the output terminal of the op amp should be capable of positive and negative voltage swings. Therefore every op-amp circuit includes a *level-shifting* arrangement. Although the use of the complementary *pnp* transistor provides a simple solution to the level-shifting problem, other forms of level shifters exist, one of which will be discussed in Chapter 13.

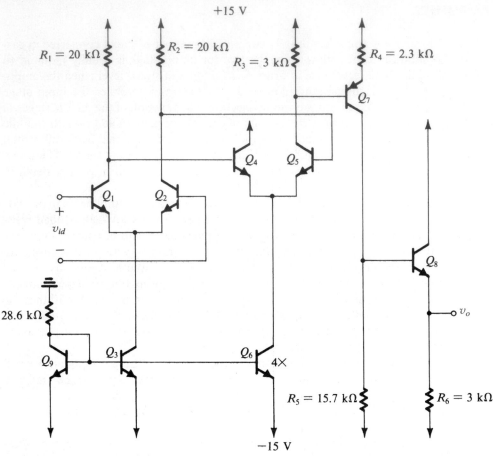

Fig. 9.34 A multistage amplifier circuit (Example 9.3).

Finally, we note that the output stage consists of emitter follower Q_8 and that ideally the dc level at the output is zero volts (as was calculated in Example 9.2).

EXAMPLE 9.3

Use the dc bias quantities evaluated in Example 9.2 and analyze the circuit in Fig. 9.34 to determine the input resistance, the voltage gain, and the output resistance.

Solution

The input differential resistance R_{id} is given by

$$R_{id} = r_{\pi 1} + r_{\pi 2}$$

Since Q_1 and Q_2 are each operating at an emitter current of 0.25 mA, it follows that

$$r_{e1} = r_{e2} = \frac{25}{0.25} = 100 \ \Omega$$

Assume $\beta = 100$; then

$$r_{\pi 1} = r_{\pi 2} = 101 \times 100 = 10.1 \text{ k}\Omega$$

Thus $R_{id} = 20.2 \text{ k}\Omega$.

To evaluate the gain of the first stage we first find the input resistance of the second stage, R_{i2},

$$R_{i2} = r_{\pi 4} + r_{\pi 5}$$

Q_4 and Q_5 are each operating at an emitter current of 1 mA; thus

$$r_{e4} = r_{e5} = 25 \ \Omega$$

$$r_{\pi 4} = r_{\pi 5} = 101 \times 25 = 2.525 \text{ k}\Omega$$

Thus $R_{i2} = 5.05 \text{ k}\Omega$. This resistance appears between the collectors of Q_1 and Q_2, as shown in Fig. 9.35. Thus the gain of the first stage will be

$$A_1 \equiv \frac{v_{o1}}{v_{id}} \simeq \frac{\text{Total resistance in collector circuit}}{\text{Total resistance in emitter circuit}}$$

$$= \frac{[R_{i2} \| (R_1 + R_2)]}{r_{e1} + r_{e2}}$$

$$= \frac{(5.05 \text{ k}\Omega \| 40 \text{ k}\Omega)}{200 \ \Omega} = 22.4 \text{ V/V}$$

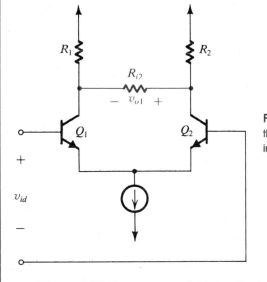

Fig. 9.35 Equivalent circuit for calculating the gain of the input stage of the amplifier in Fig. 9.34.

Figure 9.36 shows an equivalent circuit for calculating the gain of the second stage. As indicated, the input voltage to the second stage is the output voltage of the first stage, v_{o1}. Also shown is the resistance R_{i3}, which is the input resistance of the third stage formed by Q_7. The value of R_{i3} can be found by multiplying the total

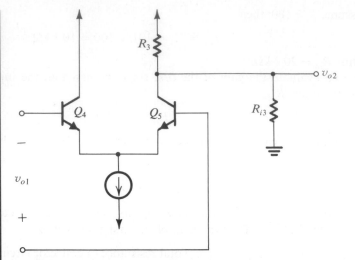

Fig. 9.36 Equivalent circuit for calculating the gain of the second stage of the amplifier in Fig. 9.34.

resistance in the emitter of Q_7 by $\beta + 1$:

$$R_{i3} = (\beta + 1)(R_4 + r_{e7})$$

Since Q_7 is operating at an emitter current of 1 mA,

$$r_{e7} = \frac{25}{1} = 25 \ \Omega$$

$$R_{i3} = 101 \times 2.325 = 234.8 \ \text{k}\Omega$$

We can now find the gain A_2 of the second stage as the ratio of the total resistance in the collector circuit to the total resistance in the emitter circuit:

$$A_2 \equiv \frac{v_{o2}}{v_{o1}} \simeq -\frac{(R_3 \parallel R_{i3})}{r_{e4} + r_{e5}}$$

$$= -\frac{(3 \ \text{k}\Omega \parallel 234.8 \ \text{k}\Omega)}{50 \ \Omega} = -59.2 \ \text{V/V}$$

To obtain the gain of the third stage we refer to the equivalent circuit shown in Fig. 9.37, where R_{i4} is the input resistance of the output stage formed by Q_8. Using the resistance-reflection rule we calculate the value of R_{i4} as

$$R_{i4} = (\beta + 1)(r_{e8} + R_6)$$

where

$$r_{e8} = \frac{25}{5} = 5 \ \Omega$$

$$R_{i4} = 101(5 + 3,000) = 303.5 \ \text{k}\Omega$$

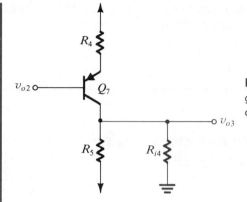

Fig. 9.37 Equivalent circuit for evaluating the gain of the third stage in the amplifier circuit of Fig. 9.34.

The gain of the third stage is given by

$$A_3 \equiv \frac{v_{o3}}{v_{o2}} \simeq \frac{(R_5 \parallel R_{i4})}{r_{e7} + R_4}$$

$$= -\frac{(15.7 \text{ k}\Omega \parallel 303.5 \text{ k}\Omega)}{2.325 \text{ k}\Omega} = -6.42 \text{ V/V}$$

Finally, to obtain the gain A_4 of the output stage we refer to the equivalent circuit in Fig. 9.38 and write

$$A_4 \equiv \frac{v_o}{v_{o3}} = \frac{R_6}{R_6 + r_{e8}}$$

$$= \frac{3,000}{3,000 + 5} = 0.998 \simeq 1$$

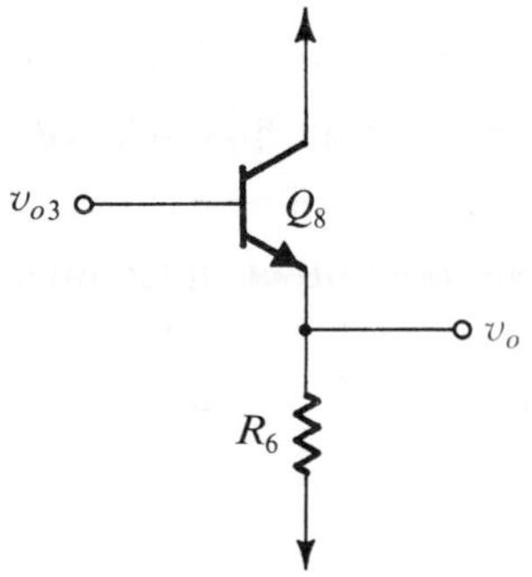

Fig. 9.38 The output stage of the amplifier circuit of Fig. 9.34.

The overall voltage gain of the amplifier can then be obtained as follows:

$$\frac{v_o}{v_{id}} = A_1 A_2 A_3 A_4 = 8,513 \text{ V/V}$$

or 78.6 dB.

To obtain the output resistance R_o we "grab hold" of the output terminal in Fig. 9.34 and look into the circuit. By inspection we find

$$R_o = \{R_6 \parallel [r_{e8} + R_5/(\beta + 1)]\}$$

which gives $R_o = 152\ \Omega$.

Exercise

9.18 Use the results of Example 9.3 to calculate the overall voltage gain of the amplifier in Fig. 9.34 when it is connected to a source having a resistance of 10 kΩ and a load of 1 kΩ.

Ans. 4,943 V/V

9.9 SUMMARY

● The differential pair is the most important building block in analog IC design. The input stage of every op amp is a differential amplifier.

● Differential amplifiers are implemented using BJTs, JFETs, or MOSFETs. FETs offer the advantage of very small input bias current and very high input resistance. BJTs provide smaller input offset voltages and wider bandwidth.

● The differential amplifier has high differential gain and low common-mode gain; the ratio of the two is the CMRR and is usually expressed in decibels.

● In the BJT differential pair, a differential input voltage of about 100 mV is sufficient to steer the bias current to one side of the pair.

● For differential input signals, operation of the differential amplifier can be analyzed using the differential half-circuit. Common-mode gain and input resistance can be obtained from the common-mode half-circuit.

● To obtain low common-mode gain, and thus high CMRR, the bias current source must be designed to have a high output resistance.

● High CMRR is achieved by taking the output differentially and ensuring a high degree of matching between the two sides of the differential amplifier.

● Mismatches between the two sides of the differential amplifier give rise to an input offset voltage, and, in BJT amplifiers, an input offset current.

● Biasing in analog ICs is based on using constant-current sources. The basic BJT current-mirror circuit suffers from the dependence on transistor β and has a relatively low output resistance. The β-dependence can be minimized and the output resistance increased by using more elaborate mirror circuits. No β-dependence problems exist in MOS mirrors.

● The differential amplifier utilizing a current-mirror active load is a popular circuit for both bipolar and MOS analog IC design.

● A multistage amplifier usually consists of an input stage having high input resistance and, if differential, high CMRR, one or more intermediate stages to realize the bulk of the gain, and an output stage having low output resistance. In analyzing a multistage amplifier, the loading effect of each stage on the one that precedes it must be taken into account.

BIBLIOGRAPHY

J. N. Giles, *Linear Integrated Circuits Applications Handbook*, Mountain View, Calif.: Fairchild Semiconductors, 1967.

P. R. Gray and R. G. Meyer, *Analysis and Design of Analog Integrated Circuits*, 2nd ed., New York: Wiley, 1984.

A. B. Grebene, *Bipolar and MOS Analog Integrated Circuit Design*, New York: Wiley, 1984.

D. J. Hamilton and W. G. Howard, *Basic Integrated Circuit Engineering*, New York: McGraw-Hill, 1975.

J. K. Roberge, *Operational Amplifiers: Theory and Practice*, New York: Wiley, 1975.

PROBLEMS

9.1 Consider the differential amplifier of Fig. 9.2a. For $I = 0.8$ mA, $V_{CC} = 5$ V, $v_{CM} = -2$V, $R_C = 1.25$ kΩ, and $\alpha = 0.98$, find the output voltages.

9.2 For the circuit of Fig. 9.2b with input of 1 V and other conditions as specified in Problem 9.1, find the output voltages at the collectors of Q_1 and Q_2.

9.3 Repeat Exercise 9.1 for an input of zero volts.

9.4 For the BJT differential amplifier of Fig. 9.1 find the value of the input differential signal, $(v_{B1} - v_{B2})$, sufficient to cause $i_{E1} = 0.90I$.

9.5 To provide insight into the possibility of nonlinear distortion resulting from large differential-input signals applied to the differential amplifier of Fig. 9.1, evaluate the ratio i_{E1}/I for differential input signals of 5, 10, 20, 30, and 40 mV.

9.6 In the series expansion of the exponential referred to in the development of Eq. (9.12), what is the ratio of the first term that is dropped to the last term that is included? What is their ratio when $v_d = 2V_T$? $v_d = V_T$? For what value of v_d is their ratio 1 to 10?

9.7 A BJT differential amplifier biased from a 4-mA constant-current source includes a 100-Ω resistor in each emitter. What is the output signal voltage produced across one collector resistor of value 2 kΩ for a differen-

tial voltage of 0.1 V applied between the bases of the differential pair? Assume $\alpha = 1$.

9.8 A BJT differential amplifier, biased so that for each transistor $g_m = 10$ mA/V, utilizes two 10-kΩ collector resistors. What is the voltage gain of the amplifier for a differential input if the output is taken differentially?

9.9 Find the voltage gain and input resistance of the amplifier in Fig. P9.9. Assume $\beta = 100$.

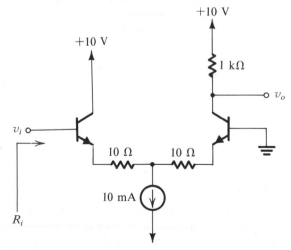

Fig. P9.9

9.10 A BJT differential amplifier having the configuration of Fig. 9.1 is found to have an input resistance of 10 kΩ and a voltage gain of 100 with the output taken differentially between collector resistors of 5 kΩ. What is the bias current for the amplifier and β for the transistors used?

9.11 Find the voltage gain and input resistance of the amplifier shown in Fig. P9.11. Assume β = 100.

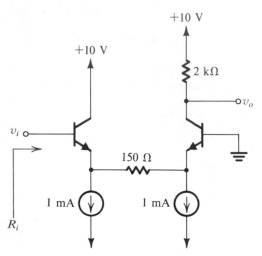

Fig. P9.11

9.12 Consider a BJT differential amplifier with collector load resistors of 10 kΩ biased with a constant-current source supplying 0.5 mA and having a 2-MΩ output resistance. Find the differential gain, the common-mode gain, and the CMRR for (a) differential output and (b) single-ended output. Assume β = ∞.

9.13 A BJT differential amplifier with 10-kΩ collector loads is found to have a common-mode gain for single-ended outputs of −40 dB. Find the value of the output resistance of the biasing current source.

9.14 A BJT differential amplifier is found to have a common-mode gain for single-ended outputs of −40 dB and for differential outputs of −80 dB. If this is due entirely to a mismatch in the load resistors, what must the percentage mismatch be?

***9.15** In a particular BJT differential amplifier a production error results in one of the transistors having an emitter–base junction area that is twice that of the other.

How will the emitter bias current split in this case between the two transistors? If the current-source output resistance is 1 MΩ, the collector resistors are 10 kΩ, and the output is taken differentially, what common-mode gain is expected?

9.16 A BJT differential amplifier utilizes a 0.1-mA bias current from a source for which the output resistance is 4 MΩ. The amplifier transistors have a β of 200, $V_A = 100$ V and $r_\mu = 10\beta r_o$. What is the common-mode input resistance? What is the differential-input resistance?

***9.17** The BJT differential amplifier described in Problem 9.16 utilizes 50-kΩ collector resistors and its output is taken differentially. The amplifier is driven by two sources each having a 100-kΩ resistance. What is the common-mode gain of this amplifier, for signals common to both inputs, if the BJTs have β's that differ by 20%? Neglect the effect of r_o and r_μ. (*Hint*: Realize that the β difference affects the split of bias current as well as the split of the signal current caused by the finite output resistance of the biasing current source.)

9.18 Reconsider Exercise 9.3 for a corresponding circuit in which the bias current is reduced to half its value and all resistors (including R) are doubled.

***9.19** In a particular BJT differential amplifier a production error results in one of the transistors having an emitter–base junction area twice that of the other. With both inputs grounded, find the current in each of the two transistors and hence the dc offset voltage at the output assuming that the collector resistances are equal. Use small-signal analysis to find the input voltage that would restore current balance to the differential pair. Repeat using large-signal analysis and compare results. Also find the input bias and offset currents assuming I = 0.1 mA and $\beta_1 = \beta_2 = 100$.

9.20 The circuit in Fig. P9.20 provides a constant current I_O as long as the circuit to which the collector is connected maintains the BJT in the active mode. Show that

$$I_O \equiv \alpha \frac{V_{CC}[R_2/(R_1 + R_2)] - V_{BE}}{R_E + (R_1 \parallel R_2)/(\beta + 1)}$$

*9.21** For the circuit in Fig. P9.21, assuming all transistors to be identical with β ≫ 1, show that by selecting

$$R_1 = R_2 = R_E\left(1 - \frac{2V_{BE}}{V_{CC}}\right)$$

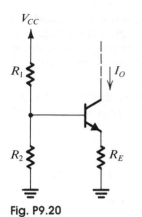

Fig. P9.20

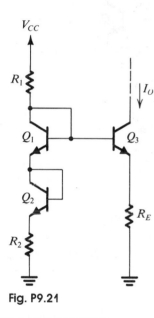

Fig. P9.21

the collector current of Q_3 will be

$$I_O = \frac{\alpha V_{CC}}{2R_E}$$

which is independent of V_{BE}. For $V_{CC} = 15$ V, and assuming $\alpha \simeq 1$ and $V_{BE} = 0.7$ V, design the circuit to obtain an output current of 1 mA. What is the lowest voltage that can be applied to the collector of Q_3?

9.22 In a diode-connected transistor with $\beta = 100$ find the percentage of total current that flows through the base.

9.23 Neglecting r_o, find the incremental resistance of a diode-connected transistor with a resistance R in the base lead.

*9.24 For the circuit in Fig. P9.24, find the value of R such that v_{CE} is constant, for small variations in I around 1 mA.

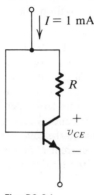

Fig. P9.24

9.25 For the circuit in Fig. P9.25 find the value of R that will result in $I_O \simeq 1$ mA. What is the largest voltage that can be applied to the collector? Assume $|V_{BE}| = 0.7$ V.

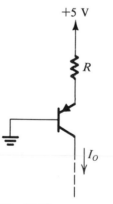

Fig. P9.25

9.26 Consider the current mirror of Fig. 9.14. For what value of β is the current transfer ratio $I_O/I_{REF} \geq 90\%$? $\geq 99\%$?

*9.27 Using a matched pair of *pnp* transistors connected in the basic current-mirror configuration, ± 5-V dc supplies, and resistor R, design a current source that provides 1 mA to a load whose voltage lies in the range -5 V to +4.3 V. If the BJTs have an Early voltage of 100 V, find the change in the output current of your current source corresponding to the range of voltage over which it is required to operate. Assume $|V_{BE}| = 0.7$ V.

*9.28 Repeat Problem 9.27 using the Wilson current mirror (with *pnp* devices). Note that in this case the voltage range can only be -5 V to $+3.6$ V. Assume $\beta = 100$. (Recall that the output resistance of the Wilson mirror is $\beta r_o/2$.)

**9.29 For the Wilson current mirror of Fig. 9.19, replace the diode-connected transistor Q_1 with its incremental resistance r_e, and replace Q_2 and Q_3 with their low-frequency hybrid-π models including r_o but excluding r_μ (for simplicity). Note that all three transistors operate at equal dc currents and thus have identical model parameters. Apply a test voltage v_x between the collector of Q_3 and ground and determine the current i_x drawn from the source v_x. Hence, show that the output resistance $R_o \equiv v_x/i_x \cong \beta r_o/2$.

***9.30 For the cascode mirror shown in Fig. P9.30 show that

$$\frac{I_O}{I_{REF}} \cong \frac{1}{1 + (4/\beta)}$$

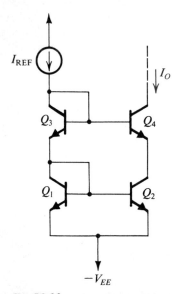

Fig. P9.30

Replace Q_1 and Q_3 with their incremental resistance r_e, and Q_2 and Q_4 with their hybrid-π model including r_o but excluding r_μ (for simplicity). Note that all transistors

operate at approximately the same bias current and thus have equal model parameters. Apply a test voltage v_x to the collector of Q_4 and determine the current i_x drawn from v_x. Hence, show that the output resistance $R_o \equiv v_x/i_x \cong \beta r_o/2$.

9.31 For the circuit in Fig. P9.31, find I_{O1} and I_{O2} in terms of I_{REF}. Assume all transistors to be matched with current gain β.

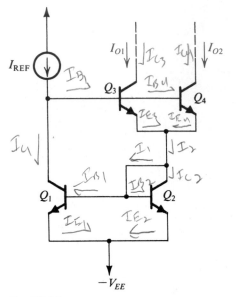

Fig. P9.31

9.32 For the circuit in Fig. P9.32, find I_O in terms of I_{REF} and β.

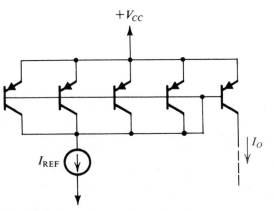

Fig. P9.32

*9.33 Using the *pnp* version of the Widlar circuit in Fig. 9.20 with power supplies of ± 5 V and I_{REF} generated using a resistance connected to the negative power supply, design a current source that delivers 20 μA of current. Design for as wide a range of dc voltages at the output terminal of the current source as possible. The largest

resistance to be used is not to exceed 10 kΩ. Assume the BJTs to have 0.7-V V_{EB} drop at a current of 1 mA, $V_A = 100$ V, and high β. Calculate the output resistance of your current source. Hence find the change in current as the voltage at the output terminal changes from -5 V to the highest allowable positive voltage.

*9.34 The BJT in Fig. P9.34 has $V_{BE} = 0.7$ V, $\beta = 100$, $V_A = 100$ V, and $r_\mu = 10\beta r_o$. Find the output resistance.

9.35 Find the voltages at all nodes and the currents through all branches in the circuit of Fig. P9.35. Assume $|V_{BE}| = 0.7$ V and $\beta = \infty$.

9.36 Consider the current mirror of Fig. 9.14 with Q_2 having an emitter–base junction five times the area of that of Q_1. Find the current gain I_O/I_{REF} assuming that the BJTs have finite β.

9.37 As a means of increasing the positive limit of the input common-mode range, modify the values of the four resistors connected to $+15$ V in Fig. 9.23 so that the

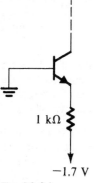

1 kΩ

-1.7 V

Fig. P9.34

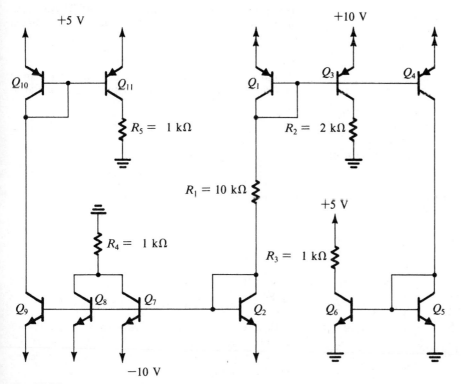

Fig. P9.35

collectors of Q_1, Q_2, and Q_5 operate at $+12$ V, $+12$ V, and $+13$ V, respectively. Determine the new resistor values and the new input common-mode range.

9.38 For the circuit in Fig. P9.38 let $|V_{BE}| = 0.7$ V and $\beta = \infty$. Find I, V_1, V_2, V_3, V_4 and V_5 for **(a)** $R = 10$ kΩ and **(b)** $R = 100$ kΩ.

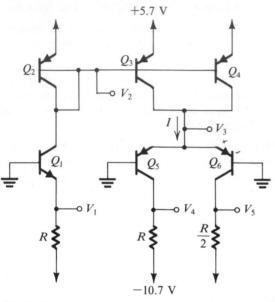

+5.7 V

−10.7 V

Fig. P9.38

*9.39 For the circuit in Fig. P9.39, find i as a function of v and R for the case $v > 0$. What restriction applies to the voltage on the collector of Q_2? What effect on this restriction does taking the feedback from a tap on R at (say) 0.1 R have?

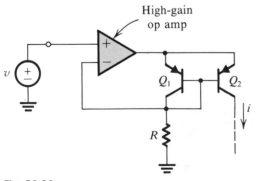

High-gain op amp

Fig. P9.39

**9.40 Repeat Exercise 9.11 with the current mirror replaced by the Wilson circuit.

***9.41 Taking into account the finite β of the pnp transistors, β_p, show that the common-mode rejection ratio of the differential amplifier in Fig. 9.24 is $\beta_p g_m R$, where $g_m = (I/2)/V_T$ and R is the output resistance of the biasing current source.

**9.42 Replace the current mirror in the circuit of Fig. 9.24 with the Wilson mirror. What is the open-circuit voltage gain of the resulting circuit? (*Hint:* The output resistance of the Wilson circuit is $\beta r_o/2$.)

9.43 A JFET differential amplifier, utilizing JFETS for which $I_{DSS} = 2$ mA and $|V_P| = 2$ V, is operated with a current bias of 0.5 mA. What is the value of differential-input voltage that just cuts off one of the devices?

9.44 A differential amplifier, utilizing JFETs for which $I_{DSS} = 2$ mA and $|V_P| = 2$ V, is biased at a constant current of 2 mA. For drain resistors of 10 kΩ, what is the gain of the amplifier for differential output? If the drain resistors have $\pm1\%$ tolerance and the current source has an output resistance of 100 kΩ, what is the worst-case common-mode gain and CMRR? Also find the worst-case input offset voltage due to the mismatch in R_D.

9.45 The JFET circuit shown in Fig. P9.45 can be used to implement the current-source bias in a differential amplifier. If $I_{DSS} = 2$ mA, $V_P = -2$ V, and $V_A = 100$ V, find the output resistance.

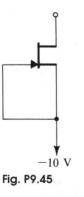

−10 V

Fig. P9.45

*9.46 Calculate the output resistance of the JFET current source shown in Fig. P9.46. For the JFET, $I_{DSS} = 8$ mA, $V_P = -2$ V, $V_A = 100$ V, and R is chosen so that the output current is 2 mA.

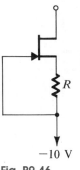

−10 V

Fig. P9.46

*9.47 For a JFET with $I_{DSS} = 8$ mA, $V_P = -2$ V, and $V_A = 100$ V, find R_o of the circuit in Fig. P9.47 for $I_1 = 2$ mA and (a) the generator I_1 ideal, (b) the generator I_1, having an output resistance of $100/I_1$. Use the result of Problem 9.46.

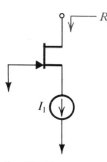

Fig. P9.47

*9.48 (a) For an MOS differential pair with $W/L = 20$ and $\mu_n C_{OX} = 20$ μA/V², find the bias current I that will result in device operation at $V_{GS} - V_t = 0.2$ V.

(b) Find g_m for each of the two transistors.

(c) Find the input differential voltage that results in full current switching.

(d) Find r_o of each device if $V_A = 24$ V.

(e) Find the input offset voltage due to a mismatch of 4% in the W/L ratios.

(f) If the differential pair is loaded by a current mirror, as in Fig. 9.33, find the voltage gain.

*9.49 Derive the relationship in Eq. (9.133).

*9.50 Find the output resistance of the current mirror in Fig. P9.50. To simplify matters, assume that the incremental voltage at the gates of Q_1, Q_2, and Q_3 is zero. [Hint: Use the relationship in Eq. (9.133)].

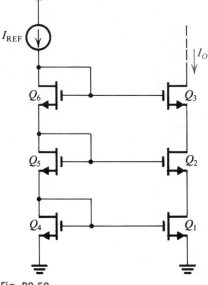

Fig. P9.50

***9.51 The two MOSFETs in Fig. P9.51 are supposed to be matched, with $K_1 = K_2 = K$ and $V_{t1} = V_{t2} = V_t$. Under these ideal conditions they conduct equal currents $I_{D1} = I_{D2} = I_D - K(V_{GS} - V_t)^2$. Show that if the two devices have a K mismatch (caused by a W/L mismatch), ΔK, and a V_t mismatch, ΔV_t, then their currents will differ by ΔI_D,

$$\frac{\Delta I_D}{I_D} = \frac{\Delta K}{K} - \frac{2\,\Delta V_t}{V_{GS} - V_t}$$

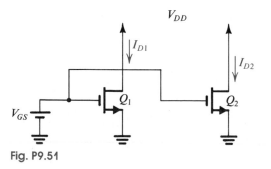

Fig. P9.51

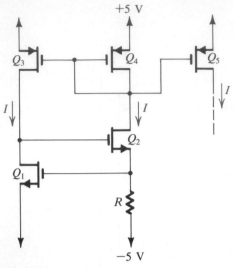

Fig. P9.52

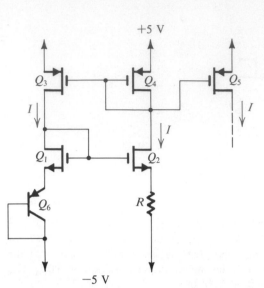

Fig. P9.53

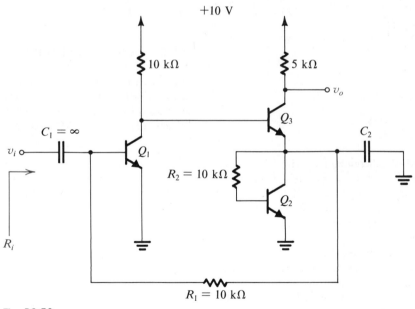

Fig. P9.58

*9.52 The circuit in Fig. P9.52 represents a technique for generating a current I that is determined almost entirely by V_t and R. Show that

$$IR = V_t + \sqrt{\frac{2I}{\mu_n C_{OX}(W/L)_1}}$$

Assume Q_3, Q_4, and Q_5 to be matched. If $\mu_n C_{OX} = 20\ \mu\text{A/V}^2$, $(W/L)_1 = 25$ and $V_t = 1$ V, find the value of R that results in $I = 10\ \mu\text{A}$.

**9.53 If the pnp transistor in the circuit of Fig. P9.53 is characterized by its exponential relationship with a scale current I_S, show that the dc current I is given by

$$IR = V_T \ln(I/I_S)$$

Assume Q_1 and Q_2 to be matched and Q_3, Q_4, and Q_5 to be matched. Find the value of R that yields a current $I = 10\ \mu\text{A}$. For the BJT, $V_{EB} = 0.7$ V at $I_E = 1$ mA.

9.54 A BJT differential amplifier, biased to have $r_e = 10\ \Omega$ and utilizing two 40-Ω emitter resistors and 5-kΩ loads, drives a second differential stage biased to have $r_e = 2\ \Omega$ and $\beta = 100$. What is the effective gain of the first stage?

9.55 In the multistage amplifier of Fig. 9.34, emitter resistors are to be introduced—100 Ω in stage 1 and 25 Ω in stage 2. What is the effect on input resistance, the voltage gain of the first stage, and the overall voltage gain? Use the bias values found in Example 9.3.

9.56 If, in the multistage amplifier of Fig. 9.34, the resistor R_5 is replaced by a constant-current source $\simeq 1$ mA, such that the bias situation is essentially unaffected, what does the overall voltage gain of the amplifier become? Assume that the output resistance of the current source is very high. Use the results of Example 9.3.

*9.57 With the modification suggested in the previous problem to the multistage amplifier, what is the effect of the change on output resistance? What is the overall gain of the amplifier when loaded by 100 Ω to ground? The original amplifier (before modification) has an output resistance of 152 Ω and a voltage gain of 8,513 V/V. What is its gain when loaded by 100 Ω? Use $\beta = 100$.

***9.58 The circuit shown in Fig. P9.58 uses a modified Wilson current mirror to bias a two-stage BJT amplifier. R_1 is introduced to allow a signal to be applied to Q_1. R_2 is introduced to maintain bias mirror balance. Calculate the voltage gain v_o/v_i and the input resistance R_i for (a) $C_2 = \infty$, (b) $C_2 = 0$. Use $\beta = 100$ and $V_{BE} = 0.7$ V.

Chapter

10

Output Stages and Power Amplifiers

INTRODUCTION

An important function of the output stage is to provide the amplifier with a low output resistance so that it can deliver the output signal to the load without loss of gain. Since the output stage is the final stage of the amplifier, it usually deals with relatively large signals. Thus the small-signal approximations and models are either not applicable or must be used with care. Nevertheless, linearity remains a very important requirement. In fact, a measure of goodness of the design of the output stage is the *total harmonic distortion* (THD) it introduces. This is the rms value of the harmonic components of the output signal, excluding the fundamental, expressed as a percentage of the rms of the fundamental. A high-fidelity audio power

amplifier features a THD of the order of a fraction of a percent.

The most challenging requirement in the design of the output stage is that it deliver the required amount of power to the load in an *efficient* manner. This implies that the power *dissipated* in the output-stage transistors must be as low as possible. This requirement stems mainly from the fact that the power dissipated in a transistor raises its internal *junction temperature*, and there is a maximum temperature (in the range of 150°C to 200°C for silicon devices) above which the transistor is destroyed. Other reasons for requiring a high power-conversion efficiency are to prolong the life of batteries employed in battery-powered circuits, to permit a smaller, lower-cost power supply, or to obviate the need for cooling fans.

We begin this chapter with a study of the various output-stage configurations employed in amplifiers that handle both low and high power. In this context, "high power" generally means greater than 1 W. We then consider the specific requirements of BJTs employed in the design of high-power output stages, called *power transistors*. Special at-tention will be paid to the thermal properties of the transistor.

A power amplifier is simply an amplifier with a high-power output stage. Examples of discrete- and integrated-circuit power amplifiers will be presented. The chapter concludes with a brief discussion of MOSFET structures that are currently finding application in power-circuit design. □

10.1 CLASSIFICATION OF OUTPUT STAGES

Output stages are classified according to the collector current waveform that results when an input signal is applied. Figure 10.1 illustrates the classification for the case of a sinusoidal input signal. The class A stage, whose associated waveform is shown in Fig. 10.1a, is biased at a current I_C greater than the amplitude of the signal current, $\hat{I}_c$. Thus the transistor in a class A stage conducts for the entire cycle of the input signal; that is, the conduction angle is 360°. In contrast, the class B stage, whose associated waveform is shown in Fig. 10.1b, is biased at zero dc current. Thus a transistor in a class B stage conducts for only half of the cycle of the input sine wave, resulting in a conduction angle of 180°. As will be seen later, the negative halves of the sinusoid will be supplied by another transistor that also operates in the class B mode and conducts during the alternate half cycles.

An intermediate class between A and B, appropriately named class AB, involves biasing the transistor at a nonzero dc current much smaller than the peak current of the sine-wave signal. As a result, the transistor conducts for an interval slightly greater than half a cycle, as illustrated in Fig. 10.1c. The resulting conduction angle is greater than 180° but much less than 360°. The class AB stage has another transistor that conducts for an interval slightly greater than that of the negative half-cycle, and the currents from the two transistors are combined in the load. It follows that, during the intervals near the zero crossings of the input sinusoid, both transistors conduct.

Figure 10.1d shows the collector-current waveform for a transistor operated as a class C amplifier. Observe that the transistor conducts for an interval shorter than that of a half-cycle; that is, the conduction angle is less than 180°. The result is the periodically pulsating current waveform shown. To obtain a sinusoidal output voltage, this current is passed through a parallel LC circuit, tuned to the frequency of the input sinusoid. The tuned circuit acts as a bandpass filter and provides an output voltage proportional to the amplitude of the fundamental component in the Fourier series representation of the current waveform.

Class A, AB, and B amplifiers are studied in this chapter. They are employed as output stages of op amps and audio power amplifiers. In the latter application, class AB is the preferred choice, for reasons that will be explained in the following sections. Class C amplifiers are usually employed for radio-frequency (RF) power amplification

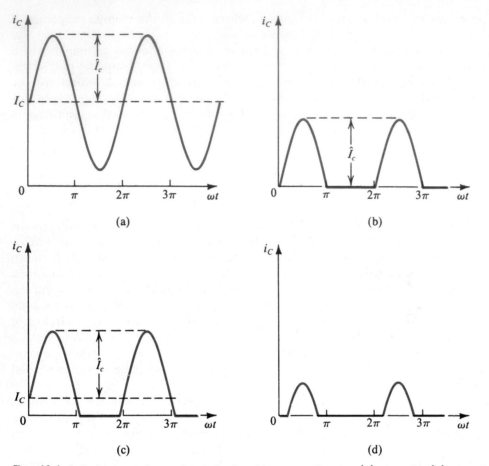

Fig. 10.1 Collector current waveforms for transistors operating in **(a)** class A, **(b)** class B, **(c)** class AB, and **(d)** class C stages.

(required, for example, in radio and TV transmitters). The design of class C amplifiers is a rather specialized topic and is not included in this book.

Although the BJT has been used to illustrate the definition of the various output-stage classes, the same classification applies to output stages implemented with MOSFETs. Furthermore, the above classification extends to amplifier stages other than those used at the output. In this regard, all the common-emitter, common-base, and common-collector amplifiers (and their FET counterparts) studied in previous chapters fall into the class A category.

10.2 CLASS A OUTPUT STAGE

Because of its low output resistance, the emitter follower is the most popular class A output stage. We have already studied the emitter follower in Chapters 8 and 9; in the following we consider its large-signal operation.

Transfer Characteristic

Figure 10.2 shows an emitter follower Q_1 biased with a constant current I supplied by transistor Q_2. Since the emitter current $i_{E1} = I + i_L$, the bias current I must be greater than the largest negative load current; otherwise, Q_1 cuts off and class A operation will no longer be maintained.

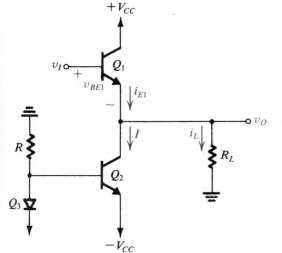

Fig. 10.2 An emitter follower biased with a constant current source.

The transfer characteristic of the emitter follower of Fig. 10.2 is described by

$$v_O = v_I - v_{BE1} \tag{10.1}$$

where v_{BE1} depends on the emitter current i_{E1} and thus on the load current i_L. If we neglect the relatively small changes in v_{BE1} (60 mV for every factor of 10 change in emitter current), the linear transfer curve shown in Fig. 10.3 results. As indicated, the positive limit of the linear region is determined by the saturation of Q_1; thus

$$v_{O\text{max}} = V_{CC} - V_{CE1\text{sat}} \tag{10.2}$$

In the negative direction, the limit of the linear region is determined either by Q_1 turning off,

$$v_{O\text{min}} = -IR_L \tag{10.3}$$

or by Q_2 saturating,

$$v_{O\text{min}} = -V_{CC} + V_{CE2\text{sat}} \tag{10.4}$$

depending on the values of I and R_L. The absolute minimum output voltage is that given by Eq. (10.4) and is achieved provided that the bias current I is greater than the magnitude of the corresponding load current,

$$I \geq \frac{|-V_{CC} + V_{CE2\text{sat}}|}{R_L} \tag{10.5}$$

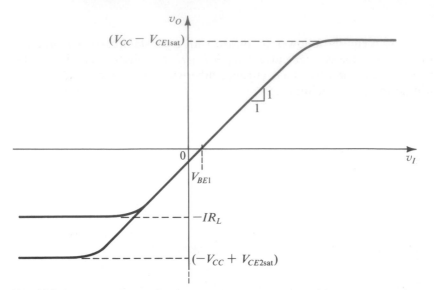

Fig. 10.3 Transfer characteristic of the emitter follower in Fig. 10.2.

Exercises

10.1 For the emitter follower in Fig. 10.2 let $V_{CC} = 15$ V, $V_{CEsat} = 0.2$ V, $V_{BE} = 0.7$ V and constant, β is very high. Find the value of R that will establish a bias current sufficiently large to allow the largest possible output signal swing for $R_L = 1$ kΩ. Determine the resulting output signal swing and the minimum and maximum emitter currents.

Ans. 0.97 kΩ; -14.8 V to $+14.8$ V; 0 to 29.6 mA

10.2 For the emitter follower of Exercise 10.1, in which $I = 14.8$ mA, consider the case in which v_O is limited to the range -10 V to $+10$ V. Let Q_1 have $v_{BE} = 0.6$ V at $i_C = 1$ mA and assume $\alpha \simeq 1$. Find v_I corresponding to $v_O = -10$ V, 0 and $+10$ V. At each of these points, use small-signal analysis to determine the voltage gain v_o/v_i. Note that the incremental voltage gain gives the slope of the v_O versus v_I characteristic.

Ans. -9.36 V, 0.67 V, and 10.68 V; 0.995 V/V, 0.998 V/V, and 0.999 V/V

Signal Waveforms

Consider the operation of the emitter-follower circuit of Fig. 10.2 for sine-wave input. Neglecting V_{CEsat}, we see that the output voltage can swing from $-V_{CC}$ to $+V_{CC}$ with the quiescent value being zero, as shown in Fig. 10.4a. Figure 10.4b shows the corresponding waveform of $v_{CE1} = V_{CC} - v_O$. Now, assuming that the bias current I is selected to allow a maximum negative load current of V_{CC}/R_L, the collector current of Q_1 will have the waveform shown in Fig. 10.4c. Finally, Fig. 10.4d shows the waveform of the *instantaneous power dissipation* in Q_1,

$$p_{D1} = v_{CE1}i_{C1} \tag{10.6}$$

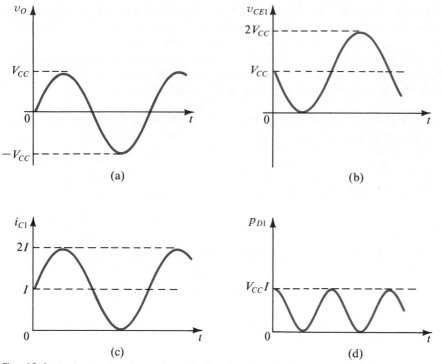

Fig. 10.4 Maximum signal waveforms in the class A output stage of Fig. 10.2 under the condition $R_L = V_{CC}/I$.

Power Dissipation

Figure 10.4d indicates that the maximum instantaneous power dissipation in Q_1 is $V_{CC}I$. This is equal to the quiescent power dissipation in Q_1. Thus the emitter-follower transistor dissipates the largest amount of power when $v_O = 0$. Since this condition (no input signal) can easily prevail for prolonged periods of time, transistor Q_1 must be able to withstand a continuous power dissipation of $V_{CC}I$.

The power dissipation in Q_1 depends on the value of R_L. Consider the extreme case of an output open circuit, that is $R_L = \infty$. In this case $i_{C1} = I$ is constant and the instantaneous power dissipation in Q_1 will depend on the instantaneous value of v_O. The maximum power dissipation will occur when $v_O = -V_{CC}$, for in this case v_{CE1} is a maximum of $2V_{CC}$ and $p_{D1} = 2 V_{CC}I$. This condition, however, would not normally persist for a prolonged interval, so the design need not be that conservative. Observe that with an open-circuit load the average power dissipation in Q_1 is $V_{CC}I$. A far more dangerous situation occurs at the other extreme of R_L—specifically, $R_L = 0$. In the event of an output short circuit, a positive input voltage would theoretically result in an infinite load current. In practice, a very large current may flow through Q_1, and if the short-circuit condition persists, the resulting large power dissipation in Q_1 can raise its junction temperature beyond the specified maximum, causing Q_1 to burn up. To guard against such a situation, output stages are usually equipped with *short-circuit protection*, as will be explained later.

The power dissipation in Q_2 also must be taken into account in designing an emitter-follower output stage. Since Q_2 conducts a constant current I, and the maximum value of v_{CE2} is $2V_{CC}$, the maximum instantaneous power dissipation in Q_2 is $2V_{CC}I$. This maximum, however, occurs when $v_O = V_{CC}$, a condition that would not normally prevail for a prolonged period of time. A more significant quantity for design purposes is the average power dissipation in Q_2, which is $V_{CC}I$.

Exercise

10.3 Consider the emitter follower in Fig. 10.2 with $V_{CC} = 10$ V, $I = 100$ mA, and $R_L = 100 \, \Omega$. Find the power dissipated in Q_1 and Q_2 under quiescent conditions ($v_O = 0$). For a sinusoidal output voltage of maximum possible amplitude (neglecting V_{CEsat}) find the average power dissipation in Q_1 and Q_2. Also find the load power.

Ans. 1 W, 1 W; 0.5 W, 1 W; 0.5 W

Power Conversion Efficiency

The power conversion efficiency of an output stage is defined as

$$\eta \equiv \frac{\text{load power } (P_L)}{\text{supply power } (P_S)} \tag{10.7}$$

For the emitter follower of Fig. 10.2, assuming that the output voltage is a sinusoid with the peak value $\hat{V}_o$, the average load power will be

$$P_L = \frac{1}{2} \frac{\hat{V}_o^2}{R_L} \tag{10.8}$$

Since the current in Q_2 is constant (I), the power drawn from the negative supply is $V_{CC}I$. The *average* current in Q_1 is equal to I, and thus the average power drawn from the positive supply is $V_{CC}I$. Thus the total average supply power is

$$P_S = 2V_{CC}I \tag{10.9}$$

Equations (10.8) and (10.9) can be combined to yield

$$\eta = \frac{1}{4} \frac{\hat{V}_o^2}{IR_L V_{CC}}$$
$$= \frac{1}{4} \left(\frac{\hat{V}_o}{IR_L} \right) \left(\frac{\hat{V}_o}{V_{CC}} \right) \tag{10.10}$$

Since $\hat{V}_o \leq V_{CC}$ and $\hat{V}_o \leq IR_L$, maximum efficiency is obtained when

$$\hat{V}_o = V_{CC} = IR_L \tag{10.11}$$

The maximum efficiency attainable is 25%. Because this is a rather low figure, the class A output stage is rarely used in large-power ($>$1-W) applications. Note also that in practice the output voltage is limited to lower values in order to avoid transistor saturation and associated nonlinear distortion. Thus the efficiency achieved is usually in the 10% to 20% range.

Exercise

10.3 CLASS B OUTPUT STAGE

Figure 10.5 shows a class B output stage. It consists of a complementary pair of transistors (that is, an *npn* and a *pnp*) connected in such a way that both cannot conduct simultaneously.

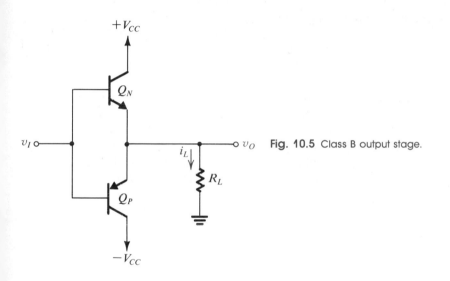

Fig. 10.5 Class B output stage.

Circuit Operation

When the input voltage v_I is zero, both transistors are cut off and the output voltage v_O is zero. As v_I goes positive and exceeds about 0.5 V, Q_N conducts and operates as an emitter follower. In this case v_O follows v_I (that is, $v_O = v_I - v_{BEN}$) and Q_N supplies the load current. Meanwhile, the emitter–base junction of Q_P will be reverse-biased by the V_{BE} of Q_N, which is approximately 0.7 V. Thus Q_P will be cut off.

If the input goes negative by more than about 0.5 V, Q_P turns on and acts as an emitter follower. Again v_O follows v_I (that is, $v_O = v_I + v_{EBP}$), but in this case Q_P supplies the load current and Q_N will be cut off.

We conclude that the transistors in the class B stage of Fig. 10.5 are biased at zero current and conduct only when the input signal is present. The circuit operates in a *push-pull* fashion: Q_N *pushes* (sources) current into the load when v_I is positive and Q_P *pulls* (sinks) current from the load when v_I is negative.

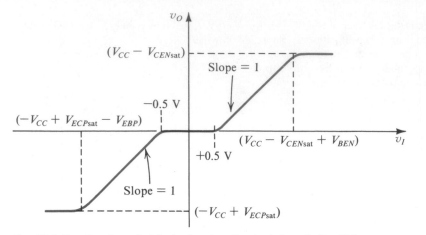

Fig. 10.6 Transfer characteristic for the class B output stage in Fig. 10.5.

Transfer Characteristic

A sketch of the transfer characteristic of the class B stage is shown in Fig. 10.6. Note that there exists a range of v_I centered around zero where both transistors are cut off and v_O is zero. This *dead band* results in the *crossover distortion* illustrated in Fig. 10.7 for the case of an input sine wave. The effect of crossover distortion will be most pronounced when the amplitude of the input signal is small. Crossover distortion in audio power amplifiers gives rise to unpleasant sounds.

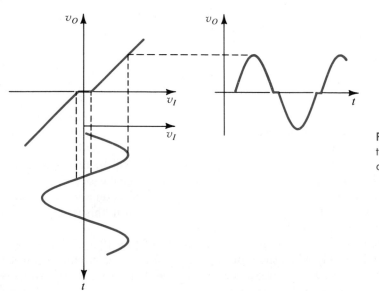

Fig. 10.7 Illustrating how the dead band in the class B transfer characteristic results in crossover distortion.

Power Conversion Efficiency

To calculate the power conversion efficiency, η, of the class B stage, we neglect the crossover distortion and consider the case of an output sinusoid of peak amplitude $\hat{V}_o$. The average load power will be

$$P_L = \frac{1}{2} \frac{\hat{V}_o^2}{R_L} \tag{10.12}$$

The current drawn from each supply will consist of half sine waves of peak amplitude $(\hat{V}_o/R_L)$. Thus the average current drawn from each of the two power supplies will be $\hat{V}_o/\pi R_L$. It follows that the average power drawn from each of the two power supplies will be the same,

$$P_{S+} = P_{S-} = \frac{1}{\pi} \frac{\hat{V}_o}{R_L} V_{CC} \tag{10.13}$$

and the total supply power will be

$$P_S = \frac{2}{\pi} \frac{\hat{V}_o}{R_L} V_{CC} \tag{10.14}$$

Thus the efficiency will be given by

$$\eta = \frac{\pi}{4} \frac{\hat{V}_o}{V_{CC}} \tag{10.15}$$

It follows that the maximum efficiency is obtained when $\hat{V}_o$ is at its maximum. This maximum is limited by the saturation of Q_N and Q_P to $V_{CC} - V_{CEsat} \simeq V_{CC}$. At this value of peak output voltage, the power conversion efficiency is

$$\eta_{max} = \frac{\pi}{4} = 78.5\% \tag{10.16}$$

This value is much larger than that obtained in the class A stage (25%). Finally, we note that the maximum average power available from a class B output stage is obtained by substituting $\hat{V}_o = V_{CC}$ in Eq. (10.12),

$$P_{Lmax} = \frac{1}{2} \frac{V_{CC}^2}{R_L} \tag{10.17}$$

Power Dissipation

Unlike the class A stage, which dissipates maximum power under quiescent conditions ($v_O = 0$), the quiescent power dissipation of the class B stage is zero. When an input signal is applied, the *average* power dissipated in the class B stage is given by

$$P_D = P_S - P_L \tag{10.18}$$

Substituting for P_S from Eq. (10.14) and for P_L from Eq. (10.12) results in

$$P_D = \frac{2}{\pi} \frac{\hat{V}_o}{R_L} V_{CC} - \frac{1}{2} \frac{\hat{V}_o^2}{R_L} \tag{10.19}$$

From symmetry we see that half of P_D is dissipated in Q_N and the other half in Q_P. Thus Q_N and Q_P must be capable of safely dissipating $\frac{1}{2}P_D$ watts. Since P_D depends on $\hat{V}_o$, we must find the worst-case power dissipation, $P_{D\text{max}}$. Differentiating Eq. (10.19) with respect to $\hat{V}_o$ and equating the derivative to zero gives the value of $\hat{V}_o$ that results in maximum average power dissipation as

$$\hat{V}_o|_{P_{D\text{max}}} = \frac{2}{\pi} V_{CC} \tag{10.20}$$

Substituting this value in (10.19) gives

$$P_{D\text{max}} = \frac{2V_{CC}^2}{\pi^2 R_L} \tag{10.21}$$

Thus

$$P_{DN\text{max}} = P_{DP\text{max}} = \frac{V_{CC}^2}{\pi^2 R_L} \tag{10.22}$$

At the point of maximum power dissipation the efficiency can be evaluated by substituting for $\hat{V}_o$ from Eq. (10.20) into Eq. (10.15); hence

$$\eta = 50\%$$

Figure 10.8 shows a sketch of P_D [Eq. (10.19)] versus the peak output voltage $\hat{V}_o$. Curves such as this are usually given on the data sheets of IC power amplifiers. (Usually, however, P_D is plotted versus

$$P_L = \frac{1}{2} \frac{\hat{V}_o^2}{R_L}$$

rather than $\hat{V}_o$.) An interesting observation follows from Fig. 10.8: Increasing $\hat{V}_o$ beyond $2V_{CC}/\pi$ *decreases* the power dissipated in the class B stage while increasing

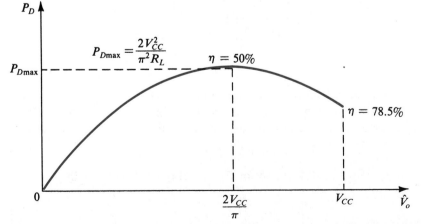

Fig. 10.8 Power dissipation of the class B output stage versus amplitude of the output sinusoid.

the load power. The price paid is an increase in nonlinear distortion as a result of approaching the saturation region of operation of Q_N and Q_P. Transistor saturation flattens the peaks of the output sine waveform. Unfortunately, this type of distortion cannot be significantly reduced by the application of negative feedback (see Chapter 12), and thus transistor saturation should be avoided in applications requiring low THD.

EXAMPLE 10.1

It is required to design a class B output stage to deliver an average power of 20 W to an 8-Ω load. The power supply is to be selected such that V_{CC} is about 5 V greater than the peak output voltage. This avoids transistor saturation and the associated nonlinear distortion, and allows for including short-circuit protection circuitry. (The latter will be discussed in Section 10.7.) Determine the supply voltage required, the peak current drawn from each supply, the total supply power, and the power-conversion efficiency. Also determine the maximum power that each transistor must be able to dissipate safely.

Solution
Since

$$P_L = \frac{1}{2}\frac{\hat{V}_o^2}{R_L}$$

then

$$\hat{V}_o = \sqrt{2P_L R_L}$$
$$= \sqrt{2 \times 20 \times 8} = 17.9 \text{ V}$$

Therefore we select $V_{CC} - 23$ V.

The peak current drawn from each supply is

$$\hat{I}_o = \frac{\hat{V}_o}{R_L} = \frac{17.9}{8} = 2.24 \text{ A}$$

The average power drawn from each supply is

$$P_{S+} = P_{S-} = \frac{1}{\pi} \times 2.24 \times 23 = 16.4 \text{ W}$$

for a total supply power of 32.8 W. The power conversion efficiency is

$$\eta = \frac{P_L}{P_S} = \frac{20}{32.8} \times 100 = 61\%$$

The maximum power dissipated in each transistor is given by Eq. (10.22); thus

$$P_{DN\text{max}} = P_{DP\text{max}} = \frac{V_{CC}^2}{\pi^2 R_L}$$

$$= \frac{(23)^2}{\pi^2 \times 8} = 6.7 \text{ W}$$

Reducing Crossover Distortion

The crossover distortion of a class B output stage can be reduced substantially by employing a high-gain op amp and overall negative feedback, as shown in Fig. 10.9. The ± 0.7-V dead band is reduced to $\pm 0.7/A_0$ volts, where A_0 is the dc gain of the op amp. Nevertheless, the slew-rate limitation of the op amp will cause the alternate turning on and off of the output transistors to be noticeable, especially at high frequencies. A more practical method for reducing and almost eliminating crossover distortion is found in the class AB stage, which will be studied in the next section.

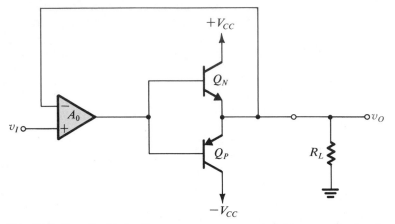

Fig. 10.9 Class B circuit with an op amp connected in a negative-feedback loop to reduce crossover distortion.

Single-Supply Operation

The class B stage can be operated from a single power supply, in which case the load is capacitively coupled, as shown in Fig. 10.10. Note that in order to make the formulas derived above directly applicable, the single power supply is denoted $2V_{CC}$.

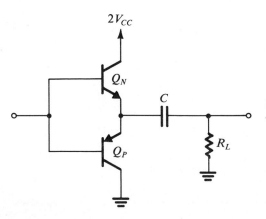

Fig. 10.10 Class B output stage operated with a single power supply.

Exercise

10.5 For the class B output stage of Fig. 10.5 let $V_{CC} = 6$ V and $R_L = 4\ \Omega$. If the output is a sinusoid with 4.5-V peak amplitude, find
(a) the output power;
(b) the average power drawn from each supply;
(c) the power efficiency obtained at this output voltage;
(d) the peak currents supplied by v_I, assuming that $\beta_N = \beta_P = 50$;
(e) the maximum power that each transistor must be capable of dissipating safely.

Ans. **(a)** 2.53 W;   **(b)** 2.15 W;   **(c)** 59%;   **(d)** 22.1 mA;   **(e)** 0.91 W

10.4 CLASS AB OUTPUT STAGE

Crossover distortion can be virtually eliminated by biasing the complementary output transistors at a small, nonzero current. The result is the class AB output stage shown in Fig. 10.11. A bias voltage V_{BB} is applied between the bases of Q_N and Q_P. For $v_I = 0$, $v_O = 0$ and a voltage $V_{BB}/2$ appears across the base–emitter junction of each of Q_N and Q_P. Assuming matched devices,

$$i_N = i_P = I_Q = I_S e^{V_{BB}/2V_T} \tag{10.23}$$

The value of V_{BB} is selected so as to yield the required quiescent current I_Q.

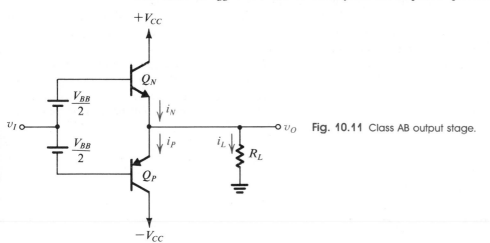

Fig. 10.11 Class AB output stage.

Circuit Operation

When v_I goes positive by a certain amount, the voltage at the base of Q_N increases by the same amount and the output becomes positive at an almost equal value,

$$v_O = v_I + \frac{V_{BB}}{2} - v_{BEN} \tag{10.24}$$

The positive v_O causes a current i_L to flow through R_L, and thus i_N must increase; that is,

$$i_N - i_P + i_L \tag{10.25}$$

The increase in i_N will be accompanied by a corresponding increase in v_{BEN} (above the quiescent value of $V_{BB}/2$). However, since the voltage between the two bases remains constant at V_{BB}, the increase in v_{BEN} will result in an equal decrease in v_{EBP} and hence in i_P. The relationship between i_N and i_P can be derived as follows:

$$v_{BEN} + v_{EBP} = V_{BB}$$

$$V_T \ln\left(\frac{i_N}{I_S}\right) + V_T \ln\left(\frac{i_P}{I_S}\right) = 2V_T \ln\left(\frac{I_Q}{I_S}\right)$$

$$i_N i_P = I_Q^2 \tag{10.26}$$

Thus as i_N increases, i_P decreases by the same ratio while the product remains constant. Equations (10.25) and (10.26) can be combined to yield i_N for a given i_L as the solution to the quadratic equation

$$i_N^2 - i_L i_N - I_Q^2 = 0 \tag{10.27}$$

From the above, we can see that for positive output voltages, the load current is supplied by Q_N, which acts as the output emitter follower. Meanwhile, Q_P will be conducting a current that decreases as v_O increases; for large v_O the current in Q_P can be ignored altogether.

For negative input voltages the opposite occurs; the load current will be supplied by Q_P, which acts as the output emitter follower, while Q_N conducts a current that gets smaller as v_I becomes more negative. Equation (10.26) relating i_N and i_P holds for negative inputs as well.

We conclude that the class AB stage operates in much the same manner as the class B circuit, with one important exception: For small v_I, both transistors conduct, and as v_I is increased or decreased, one of the two transistors takes over the operation. Since the transition is a smooth one, crossover distortion will be almost totally eliminated. Figure 10.12 shows the transfer characteristic of the class AB stage.

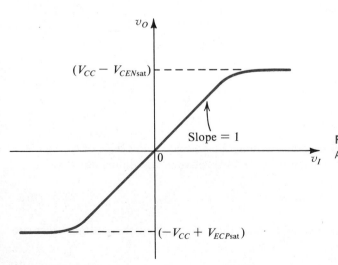

Fig. 10.12 Transfer characteristic of the class AB stage in Fig. 10.11.

The power relationships in the class AB stage are almost identical to those derived for the class B circuit in the previous section. The only difference is the fact that under quiescent conditions the class AB circuit dissipates a power of $V_{CC}I_Q$ per transistor. Since I_Q is usually much smaller than the peak load current, the quiescent power dissipation is usually small. Nevertheless, it can be taken into account easily. Specifically, we can simply add the quiescent dissipation per transistor to its maximum power dissipation with an input signal applied, to obtain the total power dissipation that the transistor must be able to handle safely.

Output Resistance

If we assume that the source supplying v_I is ideal, then the output resistance of the class AB stage can be determined from the circuit in Fig. 10.13 as

$$R_{\text{out}} = r_{eN} \| r_{eP} \tag{10.28}$$

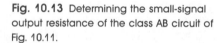

Fig. 10.13 Determining the small-signal output resistance of the class AB circuit of Fig. 10.11.

where r_{eN} and r_{eP} are the small-signal emitter resistances of Q_N and Q_P, respectively. At a given input voltage, the currents i_N and i_P can be determined, and r_{eN} and r_{eP} are given by

$$r_{eN} = \frac{V_T}{i_N} \tag{10.29}$$

$$r_{eP} = \frac{V_T}{i_P} \tag{10.30}$$

Thus

$$R_{\text{out}} = \frac{V_T}{i_N} \left\| \frac{V_T}{i_P} = \frac{V_T}{i_P + i_N} \right. \tag{10.31}$$

Since as i_N increases, i_P decreases, and vice versa, the output resistance remains approximately constant in the region around $v_I = 0$. This is in effect the reason for the virtual absence of crossover distortion. At larger load currents either i_N or i_P will be significant, and R_{out} decreases as the load current increases.

Exercise

10.6 Consider a class AB circuit with $V_{CC} = 15$ V, $I_Q = 2$ mA, and $R_L = 100$ Ω. Determine V_{BB}. Construct a table giving i_L, i_N, i_P, v_{BEN}, v_{EBP}, v_I, v_O/v_I, R_{out}, and v_o/v_i versus v_O for $v_O = 0, 0.1, 0.2, 0.5, 1, 5, 10, -0.1, -0.2, -0.5, -1, -5$ and -10 V. Note that v_O/v_I is the large-signal voltage gain and v_o/v_i is the incremental gain obtained as $R_L/(R_L + R_{out})$. The incremental gain is equal to the slope of the transfer curve. Assume Q_N and Q_P to be matched, with $I_S = 10^{-13}$ A.

Ans. $V_{BB} = 1.186$ V

v_O (V)	i_L (mA)	i_N (mA)	i_P (mA)	v_{BEN} (V)	v_{EBP} (V)	v_I (V)	v_O/v_I	R_{out} (Ω)	v_o/v_i
0	0	2	2	0.593	0.593	0	—	6.25	0.94
+0.1	1	2.56	1.56	0.599	0.587	0.106	0.94	6.07	0.94
+0.2	2	3.24	1.24	0.605	0.581	0.212	0.94	5.58	0.95
+0.5	5	5.70	0.70	0.619	0.567	0.526	0.95	4.03	0.96
+1.0	10	10.39	0.39	0.634	0.552	1.041	0.96	2.32	0.98
+5.0	50	50.08	0.08	0.673	0.513	5.08	0.98	0.50	1.00
+10.0	100	100.04	0.04	0.691	0.495	10.1	0.99	0.25	1.00
−0.1	−1	1.56	2.56	0.587	0.599	−0.106	0.94	6.07	0.94
−0.2	−2	1.24	3.24	0.581	0.605	−0.212	0.94	5.58	0.95
−0.5	−5	0.70	5.70	0.567	0.619	−0.526	0.95	4.03	0.96
−1.0	−10	0.39	10.39	0.552	0.634	−1.041	0.96	2.32	0.98
−5.0	−50	0.08	50.08	0.513	0.673	−5.08	0.98	0.50	1.00
−10	−100	0.04	100.04	0.495	0.691	−10.1	0.99	0.25	1.00

10.5 BIASING THE CLASS AB CIRCUIT

In this section we discuss two approaches for generating the voltage V_{BB} required for biasing the class AB output stage.

Biasing Using Diodes

Figure 10.14 shows a class AB circuit in which the bias voltage V_{BB} is generated by passing a constant current I_{bias} through a pair of diodes, or diode-connected transistors, D_1 and D_2. In circuits that supply large amounts of power the output transistors are large-geometry devices. The biasing diodes, however, need not be large devices, and thus the quiescent current I_Q established in Q_N and Q_P will be

$$I_Q = nI_{bias}$$

where n is the ratio of the emitter junction area of the output devices to the junction area of the biasing diodes. In other words, the saturation (or scale) current I_S of the output transistors is n times that of the biasing diodes. Area ratioing is simple to implement in integrated circuits but difficult to realize in discrete-circuit designs.

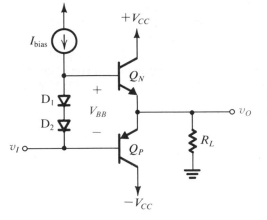

Fig. 10.14 Class AB output stage utilizing diodes for biasing.

When the output stage of Fig. 10.14 is sourcing current to the load, the base current of Q_N increases from I_Q/β_N (which is usually small) to approximately i_L/β_N. This base current drive must be supplied by the current source I_{bias}. It follows that I_{bias} must be greater than the maximum anticipated base drive for Q_N. This sets a lower limit on the value of I_{bias}. Now, since $I_Q = nI_{\text{bias}}$ and since I_Q is usually much smaller than the peak load current (less than 10%), we see that we cannot make n a large number. In other words, we cannot make the diodes much smaller than the output devices. This is a disadvantage of the diode biasing scheme.

From the above discussion we see that the current through the biasing diodes will decrease when the output stage is sourcing current to the load. Thus the bias voltage V_{BB} will also decrease, and the analysis of the previous section must be modified to take this effect into account.

The diode biasing arrangement has an important advantage: It can provide thermal stabilization of the quiescent current in the output stage. To appreciate this point recall that the class AB output stage dissipates power under quiescent conditions. Power dissipation raises the internal temperature of the BJTs. From Chapter 8 we know that a rise in transistor temperature results in a decrease in its V_{BE} (approximately $-2\text{mV}/°\text{C}$) if the collector current is held constant. Alternatively, if V_{BE} is held constant and the temperature increases, the collector current increases. The increase in collector current increases the power dissipation, which in turn increases the collector current. Thus a positive feedback mechanism exists that can result in a phenomenon called *thermal runaway*. Unless checked, thermal runaway can lead to the ultimate destruction of the BJT. Diode biasing can be arranged to provide a compensating effect that can protect the output transistors against thermal runaway under quiescent conditions. Specifically, if the diodes are in close thermal contact with the output transistors, their temperature will increase by the same amount as that of Q_N and Q_P. Thus V_{BB} will decrease at the same rate as $V_{BEN} + V_{EBP}$, with the result that I_Q remains constant. Close thermal contact is easily achieved in IC fabrication. It is obtained in discrete circuits by mounting the bias diodes on the metal case of Q_N or Q_P.

EXAMPLE 10.2

Consider the class AB output stage under the conditions that $V_{CC} = 15$ V, $R_L = 100$ Ω, and the output is sinusoidal with a maximum amplitude of 10 V. Let Q_N and Q_P be matched with $I_S = 10^{-13}$ A and $\beta = 50$. Assume that the biasing diodes have one-third the junction area of the output devices. Find the value of I_{bias} that guarantees a minimum of 1 mA through the diodes at all times. Determine the quiescent current and the quiescent power dissipation in the output transistors (i.e., at $v_O = 0$). Also find V_{BB} for $v_O = 0$, $+10$ V, and -10 V.

Solution

The maximum current through Q_N is approximately equal to $i_{Lmax} = 10$ V/0.1 kΩ = 100 mA. Thus the maximum base current in Q_N is approximately 2 mA. To maintain a minimum of 1 mA through the diodes, we select $I_{bias} = 3$ mA. The area ratio of 3 yields a quiescent current of 9 mA through Q_N and Q_P. The quiescent power dissipation is

$$P_{DQ} = 2 \times 15 \times 9 = 270 \text{ mW}$$

For $v_O = 0$, the base current of Q_N is 9/51 $\simeq$ 0.18 mA, leaving a current of $3 - 0.18 = 2.82$ mA to flow through the diodes. Since the diodes have $I_S = \frac{1}{3} \times 10^{-13}$ A, the voltage V_{BB} will be

$$V_{BB} = 2V_T \ln \frac{2.82 \text{ mA}}{I_S} = 1.26 \text{ V}$$

At $v_O = +10$ V, the current through the diodes will decrease to 1 mA resulting in $V_{BB} \simeq 1.21$ V. At the other extreme of $v_O = -10$ V, Q_N will be conducting a very small current; thus its base current will be negligibly small and all of I_{bias} (3 mA) flows through the diodes, resulting in $V_{BB} \simeq 1.26$ V.

Exercises

10.7 For the circuit of Example 10.2 find i_N and i_P for $v_O = +10$ V and $v_O = -10$ V.

Ans.   100.1 mA, 0.1 mA; 0.8 mA, 100.8 mA

10.8 If the collector current of a transistor is held constant, its v_{BE} decreases by 2 mV for every °C rise in temperature. Alternatively, if v_{BE} is held constant, then i_C increases by approximately $g_m \times 2$ mV for every °C rise in temperature. For a device operating at $I_C = 10$ mA find the change in collector current resulting from an increase in temperature of 5°C.

Ans.   4 mA

Biasing Using the V_{BE} Multiplier

An alternative biasing arrangement that provides the designer with considerably more flexibility in both discrete and integrated designs is shown in Fig. 10.15. The bias circuit consists of transistor Q_1 with a resistor R_1 connected between base and emitter and a feedback resistor R_2 connected between collector and base. The resulting two-terminal network is fed with a constant-current source I_{bias}. If we neglect the base current of Q_1, then R_1 and R_2 will carry the same current I_R, given by

$$I_R = \frac{V_{BE1}}{R_1} \tag{10.32}$$

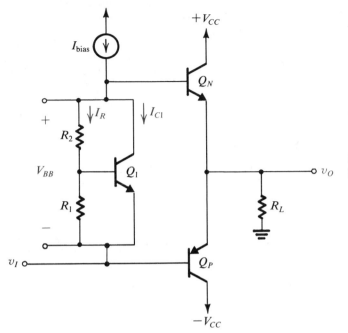

Fig. 10.15 Class AB output stage utilizing a V_{BE} multiplier for biasing.

and the voltage V_{BB} across the bias network will be

$$V_{BB} = I_R(R_1 + R_2)$$

$$= V_{BE1}\left(1 + \frac{R_2}{R_1}\right) \tag{10.33}$$

Thus the circuit simply multiplies V_{BE1} by the factor $(1 + R_2/R_1)$, and is known as the "V_{BE} multiplier." The multiplication factor is obviously under the designer's control and can be used to establish the value of V_{BB} required to yield a desired quiescent current I_Q. In IC design it is relatively easy to accurately control the ratio of two resistances. In discrete-circuit design, a potentiometer can be used, as shown in Fig. 10.16, and is manually set to produce the desired value of I_Q.

The value of V_{BE1} in Eq. (10.33) is determined by the portion of I_{bias} that flows through the collector of Q_1; that is,

$$I_{C1} = I_{\text{bias}} - I_R \tag{10.34}$$

$$V_{BE1} = V_T \ln\left(\frac{I_{C1}}{I_{S1}}\right) \tag{10.35}$$

where we have neglected the base current of Q_N, which is normally small both under quiescent conditions, and when the output voltage is swinging negative. However, for positive v_O, especially at and near its peak value, the base current of Q_N can become sizable and reduces the current available for the V_{BE} multiplier. Nevertheless, since large changes in I_{C1} correspond to only small changes in V_{BE1}, the decrease in current will be mostly absorbed by Q_1, leaving I_R, and hence V_{BB}, almost constant.

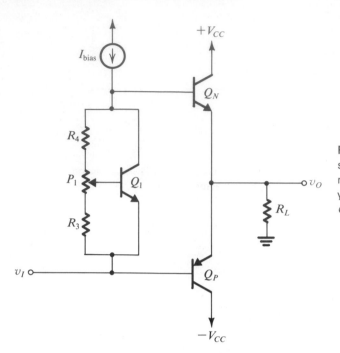

Fig. 10.16 A discrete-circuit class AB output stage with a potentiometer used in the V_{BE} multiplier. The potentiometer is adjusted to yield the desired value of quiescent current in Q_N and Q_P.

Exercise

10.9 Consider a V_{BE} multiplier with $R_1 = R_2 = 1.2$ kΩ, utilizing a transistor that has $V_{BE} = 0.6$ V at $I_C = 1$ mA, and a very high β.

(a) Find the value of the current I that should be supplied to the multiplier to obtain a terminal voltage of 1.2 V.

(b) Find the value of I that will result in the terminal voltage changing (from the 1.2-V value) by $+50$ mV, $+100$ mV, $+200$ mV, -50 mV, -100 mV, -200 mV.

Ans. **(a)** 1.5 mA; **(b)** 3.24 mA, 7.93 mA, 55.18 mA, 0.85 mA, 0.59 mA, 0.43 mA

Like the diode biasing network, the V_{BE}-multiplier circuit can provide thermal stabilization of I_Q. This is especially true if $R_1 = R_2$, and if Q_1 is in close thermal contact with the output transistors.

EXAMPLE 10.3

It is required to redesign the output stage of Example 10.2 utilizing a V_{BE} multiplier for biasing. Use a small-geometry transistor for Q_1 with $I_S = 10^{-14}$ A and design for a quiescent current $I_Q = 2$ mA.

Solution

Since the peak positive current is 100 mA, the base current of Q_N can be as high as 2 mA. We shall therefore select $I_{bias} = 3$ mA, thus providing the multiplier with a minimum current of 1 mA.

Under quiescent conditions ($v_O = 0$ and $i_L = 0$) the base current of Q_N can be neglected and all of I_{bias} flows through the multiplier. We now must decide on how this current (3 mA) is to be divided between I_{C1} and I_R. If we select I_R greater than

1 mA, the transistor will be almost cut off at the positive peak of v_O. Therefore, we shall select $I_R = 0.5$ mA, leaving 2.5 mA for I_{C1}.

To obtain a quiescent current of 2 mA in the output transistors, V_{BB} should be

$$V_{BB} = 2V_T \ln \frac{2 \times 10^{-3}}{10^{-13}} = 1.19 \text{ V}$$

We can now determine $R_1 + R_2$ as follows:

$$R_1 + R_2 = \frac{V_{BB}}{I_R} = \frac{1.19}{0.5} = 2.38 \text{ k}\Omega$$

At a collector current of 2.5 mA, Q_1 has

$$V_{BE1} = V_T \ln \frac{2.5 \times 10^{-3}}{10^{-14}} = 0.66 \text{ V}$$

The value of R_1 can now be determined as

$$R_1 = \frac{0.66}{0.5} = 1.32 \text{ k}\Omega$$

and R_2 as

$$R_2 = 2.38 - 1.32 = 1.06 \text{ k}\Omega$$

10.6 POWER BJTs

Transistors that are required to conduct currents in the ampere range and withstand power dissipation in the watts and tens-of-watts range differ in their physical structure, packaging and specification from the small-signal transistors considered in previous chapters. In this section we consider some of the important properties of power transistors, especially those aspects that pertain to the design of circuits of the type discussed in the previous sections. There are, of course, other important applications of power transistors, such as their use as switching elements in power inverters and motor-control circuits. Such applications are not studied in this book.

Junction Temperature

Power transistors dissipate large amounts of power in their collector–base junctions. The dissipated power is converted into heat, which raises the junction temperature. However, the junction temperature, T_J, must not be allowed to exceed a specified maximum, $T_{J\max}$, otherwise the transistor could suffer permanent damage. For silicon devices $T_{J\max}$ is in the range from 150°C to 200°C.

Thermal Resistance

Consider first the case of a transistor operating in free air—that is, with no special arrangements for cooling. The heat dissipated in the transistor junction will be conducted away from the junction to the transistor case, and from the case to the surrounding environment. In a steady state in which the transistor is dissipating P_D watts, the temperature rise of the junction relative to the surrounding ambience can be expressed as

$$T_J - T_A = \theta_{JA} P_D \tag{10.36}$$

where θ_{JA} is the *thermal resistance* between junction and ambience, having the units of °C per watt. Note that θ_{JA} simply gives the rise in junction temperature over ambient temperature for each watt of dissipated power. Since we wish to be able to dissipate large amounts of power without raising the junction temperature above $T_{J\max}$, it is desirable to have as small a value for the thermal resistance θ_{JA} as possible. For operation in free air, θ_{JA} depends primarily on the type of case in which the transistor is packaged. The value of θ_{JA} is usually specified on the transistor data sheet.

Equation (10.36), which describes the thermal-conduction process, is analogous to Ohm's law, which describes the electrical-conduction process. In this analogy, power dissipation corresponds to current, temperature difference corresponds to voltage difference, and thermal resistance corresponds to electrical resistance. Thus, we may represent the thermal-conduction process by the electric circuit shown in Fig. 10.17.

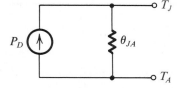

Fig. 10.17 Electrical equivalent circuit of the thermal conduction process; $T_J - T_A = P_D \theta_{JA}$.

Power Dissipation versus Temperature

The transistor manufacturer usually specifies $T_{J\max}$, the maximum power dissipation at a particular ambient temperature T_{A0} (usually, 25°C), and the thermal resistance θ_{JA}. In addition, a graph such as that shown in Fig. 10.18 is usually provided. The graph simply states that for operation at ambient temperatures below T_{A0}, the device can safely dissipate the rated value of P_{D0} watts. However, if the device is to be operated at higher ambient temperatures, the maximum allowable power dissipation must be *derated* according to the straight line shown in Fig. 10.18. The power-derating curve is a graphical representation of Eq. (10.36). Specifically, note that if the ambient temperature is T_{A0} and the power dissipation is at the maximum allowed (P_{D0}), then the junction temperature will be $T_{J\max}$. Substituting these quantities in Eq. (10.36) results in

$$\theta_{JA} = \frac{T_{J\max} - T_{A0}}{P_{D0}}, \tag{10.37}$$

which is the inverse of the slope of the power-derating straight line. At an ambient temperature T_A, higher than T_{A0}, the maximum allowable power dissipation $P_{D\max}$

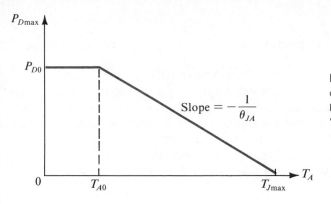

Fig. 10.18 Maximum allowable power dissipation versus ambient temperature for a BJT operated in free air. This is known as a "power-derating" curve.

can be obtained from Eq. (10.36) by substituting $T_J = T_{J\text{max}}$, thus

$$P_{D\text{max}} = \frac{T_{J\text{max}} - T_A}{\theta_{JA}} \quad (10.38)$$

Observe that as T_A approaches $T_{J\text{max}}$, the allowable power dissipation decreases; the lower thermal gradient limits the amount of heat that can be removed from the junction. In the extreme case of $T_A = T_{J\text{max}}$, no power can be dissipated because no heat can be removed from the junction.

EXAMPLE 10.4

A BJT is specified to have a maximum power dissipation P_{D0} of 2 watts at an ambient temperature T_{A0} of 25°C, and a maximum junction temperature $T_{J\text{max}}$ of 150°C. Find the following.

(a) The thermal resistance θ_{JA}.

(b) The maximum power that can be safely dissipated at an ambient temperature of 50°C.

(c) The junction temperature if the device is operating at $T_A = 25$°C and is dissipating 1 W.

Solution

(a) $\theta_{JA} = \dfrac{T_{J\text{max}} - T_{A0}}{P_{D0}}$

$= \dfrac{150 - 25}{2} = 62.5$°C/W

(b) $P_{D\text{max}} = \dfrac{T_{J\text{max}} - T_A}{\theta_{JA}}$

$= \dfrac{150 - 50}{62.5} = 1.6$ W

(c) $T_J = T_A + \theta_{JA} P_D$

$= 25 + 62.5 \times 1 = 87.5$°C

Transistor Case and Heat Sink

The thermal resistance between junction and ambience, θ_{JA}, can be expressed as

$$\theta_{JA} = \theta_{JC} + \theta_{CA} \tag{10.39}$$

where θ_{JC} is the thermal resistance between junction and transistor case (package) and θ_{CA} is the thermal resistance between case and ambience. For a given transistor, θ_{JC} is fixed by the device design and packaging. The device manufacturer can reduce θ_{JC} by encapsulating the device in a relatively large metal case and placing the collector (where most of the heat is dissipated) in direct contact with the case. Most high-power transistors are packaged in this fashion. Fig. 10.19 shows a sketch of a typical package.

Fig. 10.19 A popular package for power transistors. The case is metal with a diameter of about 2.2 cm while the outside dimension of the "seating plane" is about 4 cm. The seating plane has two holes for screws to bolt it to a heat sink. The collector is electrically connected to the case.

Although the circuit designer has no control over θ_{JC} (once a particular transistor is selected), the designer can considerably reduce θ_{CA} below its free-air value (specified by the manufacturer as part of θ_{JA}). Reduction of θ_{CA} can be effected by providing means to facilitate heat transfer from case to ambience. A popular approach is to bolt the transistor to the chassis or to an extended metal surface. Such a metal surface then functions as a *heat sink*. Heat is easily conducted from the transistor case to the heat sink; that is, the thermal resistance θ_{CS} is usually very small. Also, heat is efficiently transferred (by convection) from the heat sink to the ambience, resulting in a low thermal resistance θ_{SA}. Thus, if a heat sink is utilized, the case-to-ambience thermal resistance will be given by

$$\theta_{CA} = \theta_{CS} + \theta_{SA} \tag{10.40}$$

and can be small because its two components can be made small by the choice of an appropriate heat sink.[1] For example, in very high-power applications the heat sink is usually equipped with fins that further facilitate cooling by radiation.

The electrical analog of the thermal-conduction process when a heat sink is employed is shown in Fig. 10.20, from which we can write

$$T_J - T_A = P_D(\theta_{JC} + \theta_{CS} + \theta_{SA}) \tag{10.41}$$

---

[1] As mentioned earlier, the metal case of a power transistor is electrically connected to the collector. Thus an electrically insulating material such as mica is usually placed between the metal case and the metal heat sink. Also, insulating bushes are generally used in bolting the transistor to the heat sink.

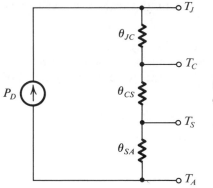

Fig. 10.20 Electrical analog of the thermal conduction process when a heat sink is utilized.

As well as specifying θ_{JC}, the device manufacturer usually supplies a derating curve for $P_{D\text{max}}$ versus the case temperature, T_C. Such a curve is shown in Fig. 10.21. Note that the slope of the power-derating straight line is $-1/\theta_{JC}$. For a given transistor, the maximum power dissipation at a *case temperature* T_{C0} (usually 25°C) is much greater than that at an *ambient temperature* T_{A0} (usually 25°C). If the device can be maintained at a case temperature T_C, $T_{C0} \leq T_C \leq T_{J\text{max}}$ then the maximum safe power dissipation is obtained when $T_J = T_{J\text{max}}$,

$$P_{D\text{max}} = \frac{T_{J\text{max}} - T_C}{\theta_{JC}} \qquad (10.42)$$

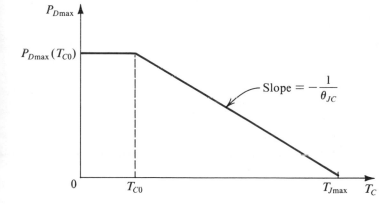

Fig. 10.21 Maximum allowable power dissipation versus the transistor case temperature.

EXAMPLE 10.5   A BJT is specified to have $T_{J\text{max}} = 150°C$ and to be capable of dissipating the following maximum power:

    40 W at $T_C = 25°C$
    2 W at $T_A = 25°C$

Above 25°C, the maximum power dissipation is to be derated linearly with $\theta_{JC} = 3.12°C/W$ and $\theta_{JA} = 62.5°C/W$. Find the following.

(a) The maximum power that can be dissipated safely by this transistor when operated in free air at $T_A = 50°C$.

(b) The maximum power that can be dissipated safely by this transistor when operated at an ambient temperature of 50°C, but with a heat sink for which $\theta_{CS} = 0.5°C/W$ and $\theta_{SA} = 4°C/W$. In this case find the temperature of the case and of the heat sink.

(c) The maximum power that can be dissipated safely if an *infinite heat sink* is used and $T_A = 50°C$.

Solution

(a) $$P_{D\text{max}} = \frac{T_{J\text{max}} - T_A}{\theta_{JA}}$$

$$= \frac{150 - 50}{62.5} = 1.6 \text{ W}$$

(b) With a heat sink, θ_{JA} becomes

$$\theta_{JA} = \theta_{JC} + \theta_{CS} + \theta_{SA}$$
$$= 3.12 + 0.5 + 4 = 7.62°C/W$$

Thus

$$P_{D\text{max}} = \frac{150 - 50}{7.62} = 13.1 \text{ W}$$

Figure 10.22 shows the thermal equivalent circuit with the various temperatures indicated.

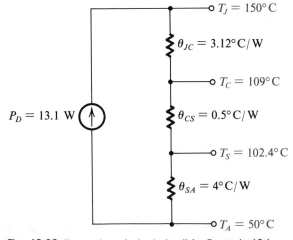

Fig. 10.22 Thermal equivalent circuit for Example 10.6.

(c) An infinite heat sink, if it existed, would cause the case temperature T_C to equal the ambient temperature T_A. The infinite heat sink has $\theta_{CA} = 0$. Obviously, one cannot buy an infinite heat sink; nevertheless, this terminology is used by some manufacturers to describe the power-derating curve of Fig. 10.21. The abcissa is then labeled T_A and the curve is called "power dissipation versus ambient temperature with infinite heat sink." For our example, with infinite heat sink,

$$P_{D\max} = \frac{T_{J\max} - T_A}{\theta_{JC}}$$

$$= \frac{150 - 50}{3.12} = 32 \text{ W}$$

The advantage of using a heat sink is clearly evident from the above example: With a heat sink, the maximum allowable power dissipation increases from 1.6 W to 13.1 W. Also note that although the transistor considered can be called a "40-W transistor," this level of power dissipation cannot be achieved in practice; it would require an infinite heat sink and an ambient temperature $T_A \leq 25°C$.

Exercise

10.10 The 2N6306 power transistor is specified to have $T_{J\max} = 200°C$ and $P_{D\max} = 125$ W for $T_C \leq 25°C$. For $T_C \geq 25°C$, $\theta_{JC} = 1.4°C/W$. If in a particular application this device is to dissipate 50 W and operate at an ambient temperature of 25°C, find the maximum thermal resistance of the heat sink that must be used (i.e., θ_{SA}). Assume $\theta_{CS} = 0.6°C/W$. What is the case temperature, T_C?

Ans. 1.5°C/W; 130°C

The BJT Safe Operating Area

In addition to specifying the maximum power dissipation at different case temperatures, power transistor manufacturers usually provide a plot of the boundary of the safe operating area (SOA) in the i_C-v_{CE} plane. The SOA specification takes the form illustrated by the sketch in Fig. 10.23. The paragraph numbers below correspond to the boundaries on the sketch.

1. The maximum allowable current $I_{C\max}$. Exceeding this current on a continuous basis can result in melting the wires that bond the device to the package terminals.

2. The maximum power dissipation hyperbola. This is the locus of the points for which $v_{CE}i_C = P_{D\max}$ (T_{C0}). For temperatures $T_C > T_{C0}$, the power-derating curves described above should be used to obtain the applicable $P_{D\max}$ and thus a correspondingly lower hyperbola. Although the operating point can be allowed to move temporarily above the hyperbola, the *average* power dissipation should not be allowed to exceed $P_{D\max}$.

3. The *second-breakdown* limit. Second breakdown is a phenomenon that results because current flow across the emitter–base junction is not uniform. Rather,

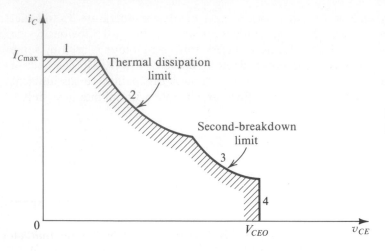

Fig. 10.23 Safe operating area (SOA) of a BJT.

the current density is greatest near the periphery of the junction. This "current crowding" gives rise to increased localized power dissipation and hence temperature rise (at locations called hot spots). Since a temperature rise causes an increase in current, a form of thermal runaway can occur, leading to junction destruction.

4. The collector-to-emitter breakdown voltage, BV_{CEO}. The instantaneous value of v_{CE} should never be allowed to exceed BV_{CEO}; otherwise, avalanche breakdown of the collector–base junction will occur (see Section 8.12).

Finally, it should be mentioned that logarithmic scales are usually used for i_C and v_{CE}, leading to an SOA boundary that consists of straight lines.

Parameter Values of Power Transistors

Owing to their large geometry and high operating currents, power transistors display typical parameter values that can be quite different from those of small-signal transistors. The important differences are as follows:

1. At high currents, the exponential i_C-v_{BE} relationship exhibits a constant $n = 2$; that is, $i_C = I_S e^{v_{BE}/2V_T}$.

2. β is low, typically 30–80, but it can be as low as 5. Here it is important to note that β has a positive temperature coefficient.

3. At high currents, r_π becomes very small and r_x (a few ohms) becomes important. (r_x is defined and explained in Chapter 11.)

4. f_T is low (a few MHz), C_μ is large (hundreds of pF) and C_π is even larger. (These parameters are defined and explained in Chapter 12.)

5. I_{CBO} is large (a few tens of μA) and, as usual, doubles for every 10°C rise in temperature.

6. BV_{CEO} is typically 50–100 V, but it can be as high as 500 V.

7. I_{Cmax} is typically in the ampere range, but it can be as high as 100 A.

10.7 VARIATIONS ON THE CLASS AB CONFIGURATION

In this section we discuss a number of circuit improvements and protection techniques for the class AB output stage.

Use of Input Emitter Followers

Figure 10.24 shows a class AB circuit biased using transistors Q_1 and Q_2 which also function as emitter followers, thus providing the circuit with a high input resistance. In effect, the circuit functions as a unity-gain buffer amplifier. Since all four transistors are usually matched, the quiescent current ($v_I = 0$, $R_L = \infty$) in Q_3 and Q_4 is equal to that in Q_1 and Q_2. Resistors R_3 and R_4 are usually very small and are included to compensate for possible mismatches between Q_3 and Q_4 and to guard against the possibility of thermal runaway due to temperature differences between the input- and output-stage transistors. The latter point can be appreciated by noting that an increase in the current of, say, Q_3 causes an increase in the voltage drop across R_3 and a corresponding decrease in V_{BE3}. Thus R_3 provides negative feedback that helps stabilize the current through Q_3.

Because the circuit of Fig. 10.24 requires high-quality *pnp* transistors, it is not suitable for implementation in monolithic IC technology. However, excellent results have been obtained with this circuit implemented in hybrid thick-film technology (Wong & Sherwin, 1979). This technology permits component trimming, for instance,

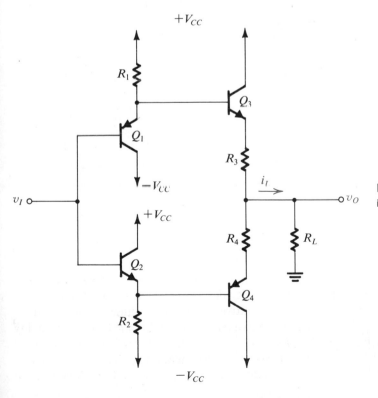

Fig. 10.24 Class AB output stage with an input buffer.

to minimize the output offset voltage. The circuit can be used alone or together with an op amp to provide increased output driving capability. The latter application will be discussed in the next section.

Exercise

> **10.11** (*Note:* Although very instructive, this exercise is rather long.) Consider the circuit of Fig. 10.24 with $R_1 = R_2 = 5 \text{ k}\Omega$, $R_3 = R_4 = 0 \Omega$, and $V_{CC} = 15$ V. Let the transistors be matched with $I_S = 3.3 \times 10^{-14}$ A, $n = 1$, and $\beta = 200$. (These are the values used in the LH002 manufactured by National Semiconductor, except that $R_3 = R_4 = 2 \Omega$.)
> (a) For $v_I = 0$ and $R_L = \infty$ find the quiescent current in each of the four transistors and v_O.
> (b) For $R_L = \infty$, find i_{C1}, i_{C2}, i_{C3}, i_{C4} and v_O for $v_I = +10$ V and -10 V.
> (c) Repeat (b) for $R_L = 100 \Omega$.
>
> **Ans.** (a) 2.87 mA; 0 V; (b) for $v_I = +10$ V: 0.88 mA, 4.87 mA, 1.95 mA, 1.95 mA, $+9.98$ V and for $v_I = -10$ V: 4.87 mA, 0.88 mA, 1.95 mA, 1.95 mA, -9.98 V; (c) for $v_I = +10$ V: 0.38 mA, 4.87 mA, 100 mA, 0.02 mA, $+9.86$ V and for $v_I = -10$ V: 4.87 mA, 0.38 mA, 0.02 mA, 100 mA, -9.86 V

Use of Compound Devices

In order to increase the current gain of the output-stage transistors, and thus reduce the required base current drive, the Darlington configuration shown in Fig. 10.25 is frequently used to replace the *npn* transistor of the class AB stage. The Darlington configuration is equivalent to a single *npn* transistor having $\beta \simeq \beta_1 \beta_2$, but almost twice the value of V_{BE}.

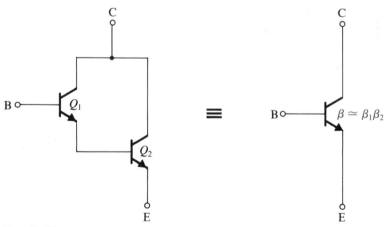

Fig. 10.25 The Darlington configuration.

The Darlington configuration can be also used for *pnp* transistors, and this is indeed done in discrete-circuit design. In IC design, the lack of good-quality *pnp* transistors prompted the use of the alternative compound configuration shown in Fig. 10.26. This compound device is equivalent to a single *pnp* transistor having $\beta \simeq \beta_1 \beta_2$. When fabricated with the standard IC technology, Q_1 is usually a lateral *pnp* having a low β ($\beta = 5$–10) and poor high-frequency response ($f_T \simeq 5$ MHz); see Appendix A. The compound device, although it has a relatively high equivalent

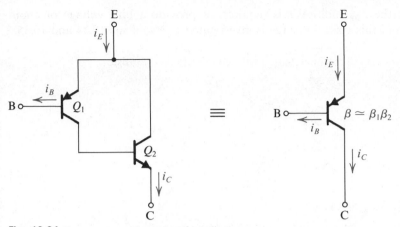

Fig. 10.26 The compound-*pnp* configuration.

β, still suffers from a poor high-frequency response. It also suffers from another problem: The feedback loop formed by Q_1 and Q_2 is prone to high-frequency oscillations (with frequency near f_T of the *pnp* device; that is, about 5 MHz). Methods exist for preventing such oscillations. The subject of feedback amplifier stability will be studied in Chapter 12.

To illustrate the application of the Darlington configuration and the compound *pnp*, we show in Fig. 10.27 an output stage utilizing both. The class AB biasing is achieved using a V_{BE} multiplier. Note that the Darlington *npn* adds one more V_{BE}

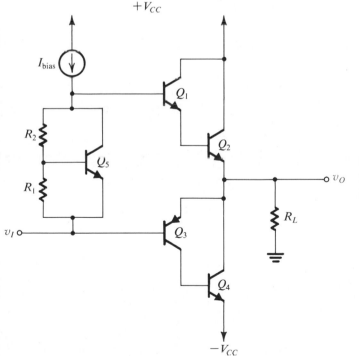

Fig. 10.27 A class AB output stage utilizing a Darlington *npn* and a compound *pnp*. Biasing is obtained using a V_{BE} multiplier.

drop, and thus the V_{BE} multiplier is required to provide a bias voltage of about 2 V. The design of this class AB stage is investigated in Problems 10.34 and 10.35.

Exercise

10.12 (a) Refer to Fig. 10.26. Show that, for the composite *pnp* transistor,

$$i_B \simeq \frac{i_C}{\beta_N \beta_P}$$

and

$$i_E \simeq i_C$$

Hence show that

$$i_C \simeq \beta_N I_{SP} e^{v_{EB}/V_T}$$

and thus the transistor has an effective scale current

$$I_S = \beta_N I_{SP}$$

(b) For $\beta_P = 20$, $\beta_N = 50$, $I_{SP} = 10^{-14}$ A, find the effective current gain of the compound device and its v_{EB} when $i_C = 100$ mA. Let $n = 1$.

Ans. (b) 1,000; 0.651 V

Short-Circuit Protection

Figure 10.28 shows a class AB output stage equipped with protection against the effect of short-circuiting the output while the stage is sourcing current. The large current that flows through Q_1 in the event of a short circuit will develop a voltage

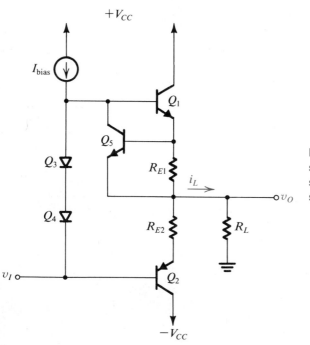

Fig. 10.28 A class AB output stage with short-circuit protection. The protection circuit shown operates in the event of an output short circuit while v_O is positive.

drop across R_{E1} of sufficient value to turn Q_5 on. The collector of Q_5 will then conduct most of the current I_{bias}, robbing Q_1 of its base drive. The current through Q_1 will thus be reduced to a safe operating level.

This method of short-circuit protection is effective in ensuring device safety, but it has the disadvantage that under normal operation about 0.5 V drop might appear across each R_E. This means that the voltage swing at the output will be reduced by that much, in each direction. On the other hand, the inclusion of emitter resistors provides the additional benefit of protecting the output transistors against thermal runaway.

Exercise

> **10.13** In the circuit of Fig. 10.28 let $I_{\text{bias}} = 2$ mA. Find the value of R_{E1} that causes Q_5 to turn on and absorb all 2 mA when the output current being sourced reaches 150 mA. For Q_5, $I_S = 10^{-14}$ A and $n = 1$. If the normal peak output current is 100 mA, find the voltage drop across R_{E1} and the collector current of Q_5.
>
> **Ans.**   4.3 Ω; 430 mV; 0.3 μA

Thermal Shutdown

In addition to short-circuit protection, most IC power amplifiers are usually equipped with a circuit that senses the temperature of the chip and turns on a transistor in the event that the temperature exceeds a safe preset value. The turned-on transistor is connected in such a way that it absorbs the bias current of the amplifier, thus virtually shutting down its operation.

Figure 10.29 shows a thermal shutdown circuit. Here transistor Q_2 is normally off. As the chip temperature rises, the combination of the positive temperature coefficient of zener diode Z_1 and the negative temperature coefficient of V_{BE1} causes the

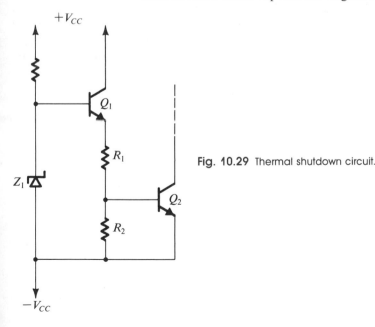

Fig. 10.29 Thermal shutdown circuit.

voltage at the emitter of Q_1 to rise. This in turn raises the voltage at the base of Q_2 to the point at which Q_2 turns on.

10.8 IC POWER AMPLIFIERS

A variety of IC power amplifiers are available. Most consist of a high-gain small-signal amplifier followed by a class AB output stage. Some have overall negative feedback already applied, resulting in a fixed closed-loop voltage gain. Others do not have on-chip feedback and are, in effect, op amps with large output power capability. In fact, the output current driving capability of any general-purpose op amp can be increased by cascading it with a class B or class AB output stage and applying overall negative feedback. The additional output stage can be either a discrete circuit or a hybrid IC such as the buffer discussed in the previous section. In the following we discuss some power amplifier examples.

A Fixed-Gain IC Power Amplifier

Our first example is the LM380 (a product of National Semiconductor Corporation), which is a fixed-gain monolithic power amplifier. A simplified version of the internal circuit of the amplifier[2] is shown in Fig. 10.30. The circuit consists of an input differential amplifier utilizing Q_1 and Q_2 as emitter followers for input buffering, and Q_3 and Q_4 as a differential pair with an emitter resistor R_3. The two resistors R_4 and R_5 provide dc paths to ground for the base currents of Q_1 and Q_2, thus enabling the input signal source to be capacitively coupled to either of the two input terminals.

The differential amplifier transistors Q_3 and Q_4 are biased by two separate direct currents: Q_3 is biased by a current from the dc supply V_S through the diode–connected transistor Q_{10}, and resistor R_1; Q_4 is biased by a dc current from the output terminal through R_2. Under quiescent conditions (that is, with no input signal applied) the two bias currents will be equal and the current through, and the voltage across, R_3 will be zero. For Q_3 we can write

$$I_3 \simeq \frac{V_S - V_{EB10} - V_{EB3} - V_{EB1}}{R_1}$$

where we have neglected the small dc voltage drop across R_4. Assuming, for simplicity, all V_{EB} to be equal,

$$I_3 \simeq \frac{V_S - 3V_{EB}}{R_1} \tag{10.43}$$

---

[2] The main objective of showing this circuit is to point out some interesting design features. The circuit is *not* a detailed schematic diagram of what is actually on the chip.

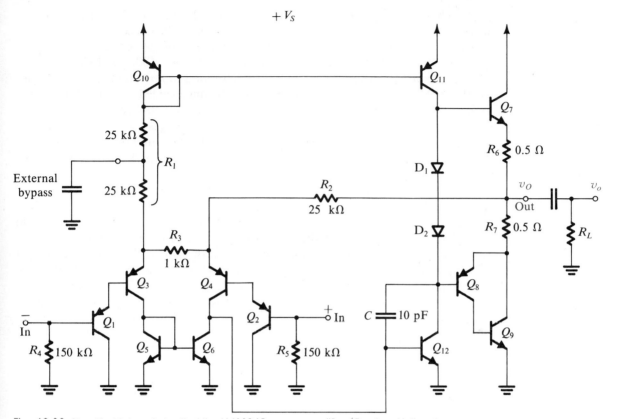

Fig. 10.30 Simplified internal circuit of the LM380 IC power amplifier. (Courtesy National Semiconductor Corporation.)

For Q_4 we have

$$I_4 = \frac{V_O - V_{EB4} - V_{EB2}}{R_2}$$

$$\simeq \frac{V_O - 2V_{EB}}{R_2} \tag{10.44}$$

where V_O is the dc voltage at the output. Equating I_3 and I_4 and using the fact that $R_1 = 2R_2$ results in

$$V_O = \tfrac{1}{2}V_S + \tfrac{1}{2}V_{EB} \tag{10.45}$$

Thus the output is biased at approximately half the power supply voltage, as desired. An important feature is the dc feedback from the output to the emitter of Q_4, through R_2. This dc feedback acts to stabilize the output dc bias voltage at the value in Eq. (10.45). Qualitatively, the dc feedback functions as follows: If for some reason V_O increases, a corresponding current increment will flow through R_2 and into the emitter of Q_4. Thus the collector current of Q_4 increases, resulting in a positive increment in the voltage at the base of Q_{12}. This, in turn, causes the collector current of Q_{12} to increase, thus bringing down the voltage at the base of Q_7 and hence V_O.

Continuing with the description of the circuit in Fig. 10.30, we observe that the differential amplifier (Q_3, Q_4) has a current-mirror load composed of Q_5 and Q_6 (refer to Section 9.5 for a discussion of active loads). The single-ended output voltage signal of the first stage appears at the collector of Q_6 and thus is applied to the base of the second-stage common-emitter amplifier Q_{12}. Transistor Q_{12} is biased by the constant-current source Q_{11}, which also acts as its active load. In actual operation, however, the load of Q_{12} will be dominated by the reflected resistance due to R_L. Capacitor C provides frequency compensation (see Chapter 12).

The output stage is class AB, utilizing a compound *pnp* transistor (Q_8 and Q_9). Negative feedback is applied from the output to the emitter of Q_4 via resistor R_2. To find the closed-loop gain consider the small-signal equivalent circuit shown in Fig. 10.31. We have replaced the second-stage common-emitter amplifier and the output stage with an inverting amplifier block with gain A. We shall assume that the amplifier A has high gain and high input resistance, and thus the input signal current into A is negligibly small. Under this assumption, Fig. 10.31 shows the analysis details with an input signal v_i applied to the inverting input terminal. Note that since the input differential amplifier has a relatively large resistance, R_3, in the emitter

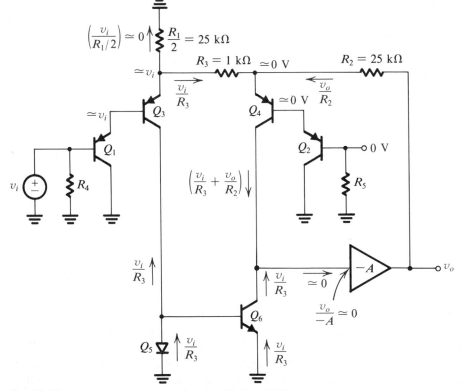

Fig. 10.31 Small-signal analysis of the circuit in Fig. 10.30.

circuit, most of the applied input voltage appears across R_3. In other words, the signal voltages across the emitter–base junctions of Q_1, Q_2, Q_3, and Q_4 are small compared to the voltage across R_3. The voltage gain can be found by writing a node equation at the collector of Q_6:

$$\frac{v_i}{R_3} + \frac{v_o}{R_2} + \frac{v_i}{R_3} = 0$$

which yields

$$\frac{v_o}{v_i} = -\frac{2R_2}{R_3} \simeq -50 \text{ V/V}$$

Exercise

10.14 Denoting the total resistance between the collector of Q_6 and ground R, show, using Fig. 10.31, that

$$\frac{v_o}{v_i} = \frac{-2R_2/R_3}{1 + (R_2/AR)}$$

which reduces to $(-2R_2/R_3)$ under the condition that $AR \gg R_2$.

As will be demonstrated in Chapter 12, one of the advantages of negative feedback is the reduction of nonlinear distortion. This is the case in the circuit of the LM380.

The LM380 is designed to operate from a single supply V_S in the range 12–22 V. The selection of supply voltage depends on the value of R_L and the required output power P_L. The manufacturer supplies curves for the device power dissipation versus output power for a given load resistance and different supply voltages. One such set of curves for $R_L = 8 \ \Omega$ is shown in Fig. 10.32. Note the similarity with the class B power dissipation curve of Fig. 10.8. In fact, the reader can easily verify that the

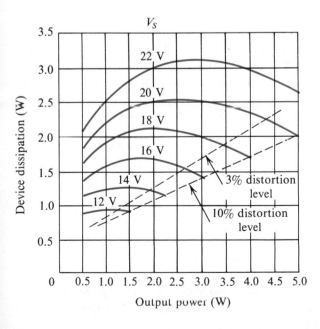

Fig. 10.32 Power dissipation (P_D) versus output power (P_L) for the LM380 with $R_L = 8 \ \Omega$. (Courtesy National Semiconductor Corporation.)

location and value of the peaks of the curves in Fig. 10.32 are accurately predicted by Eqs. (10.20) and (10.21), respectively. The line labeled "3% distortion" in Fig. 10.32 is the locus of the points on the various curves at which the distortion (THD) reaches 3%. A THD of 3% represents the onset of peak clipping due to output transistor saturation.

The manufacturer also supplies curves for maximum power dissipation versus temperature (derating curves) similar to those discussed in Section 10.6 for discrete power transistors.

Exercises

10.15 The manufacturer specifies that for ambient temperatures below 25°C the LM380 can dissipate a maximum of 3.6 W. This is obtained under the condition that the dual-in-line package is soldered onto a printed circuit board in thermal contact with 6 square inches of 2-ounce copper foil. Above $T_A = 25°C$ the thermal resistance is $\theta_{JA} = 35°C/W$. $T_{J\max}$ is specified to be 150°C. Find the maximum power dissipation possible if the ambient temperature is to be 50°C.

Ans. 2.9 W

10.16 It is required to use the LM380 to drive an 8-Ω loudspeaker. Use the curves of Fig. 10.32 to determine the maximum power supply possible while limiting the maximum power dissipation to the 2.9 W determined in Exercise 10.15. If for this application a 3% THD is allowed, find P_L and the peak-to-peak output voltage.

Ans. 20 V; 4.2 W; 16.4 V

Power Op Amps

Figure 10.33 shows the general structure of a power op amp. It consists of an op amp followed by a class AB buffer similar to that discussed in Section 10.7. The buffer consists of transistors Q_1, Q_2, Q_3, and Q_4, with bias resistors R_1 and R_2 and emitter degeneration resistors R_5 and R_6. The buffer supplies the required load current until the current increases to the point that the voltage drop across R_3 (in the current sourcing mode) becomes sufficiently large to turn Q_5 on. Transistor Q_5 then supplies the additional load current required. In the current sinking mode, Q_4 supplies the load current until sufficient voltage develops across R_4 to turn Q_6 on. Then, Q_6 sinks the additional load current. Thus the stage formed by Q_5 and Q_6 acts as a *current booster*. The power op amp is intended to be used with negative feedback in the usual closed-loop configurations. A circuit based on the structure of Fig. 10.33 is commercially available from National Semiconductor as LH0101. This op amp is capable of providing a continuous output current of 2 A, and with appropriate heat sinking can provide 40 W of output power (Wong & Johnson, 1981). The LH0101 is fabricated using hybrid thick-film technology.

The Bridge Amplifier

We conclude this section with a discussion of a circuit configuration that is popular in high-power applications. This is the bridge amplifier configuration shown in Fig. 10.34 utilizing two power op amps, A_1 and A_2. While A_1 is connected in the noninverting configuration with a gain $K = 1 + (R_2/R_1)$, A_2 is connected as an inverting

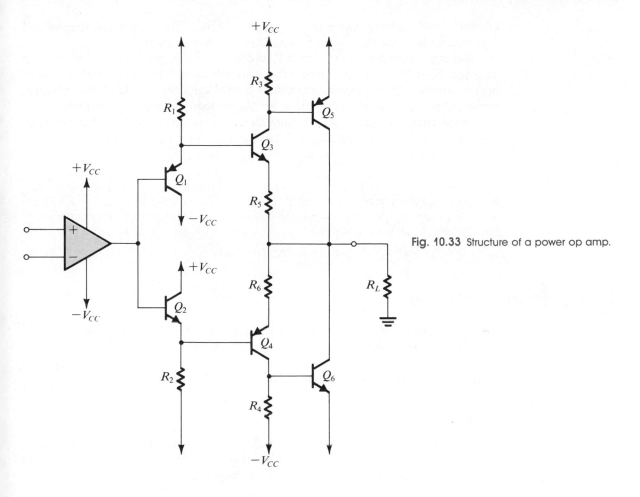

Fig. 10.33 Structure of a power op amp.

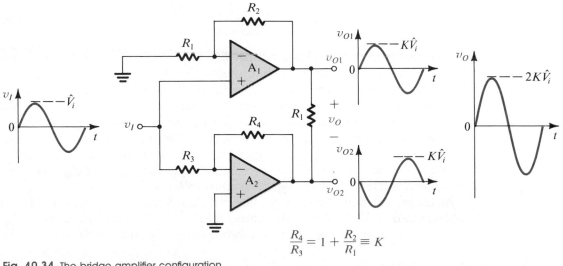

$$\frac{R_4}{R_3} = 1 + \frac{R_2}{R_1} \equiv K$$

Fig. 10.34 The bridge amplifier configuration.

amplifier with a gain of equal magnitude $K = R_4/R_3$. The load R_L is floating and is connected between the output terminals of the two op amps.

If v_I is a sinusoid with amplitude $\hat{V}_i$, the voltage swing at the output of each op amp will be $\pm K\hat{V}_i$, and that across the load will be $\pm 2K\hat{V}_i$. Thus with op amps operated from ± 15-V supplies and capable of providing, say, a ± 12-V output swing, an output swing of ± 24 V is obtained across the load of the bridge amplifier.

In designing bridge amplifiers, note should be taken of the fact that the peak current drawn from each op amp is $2K\hat{V}_i/R_L$. This effect can be taken into account by considering the load seen by each op amp to be $R_L/2$.

Exercise

10.17 Consider the circuit of Fig. 10.34 with $R_1 = R_3 = 10$ kΩ, $R_2 = 5$ kΩ, $R_4 = 15$ kΩ, and $R_L = 8$ Ω. Find the voltage gain and input resistance. The power supply used is ± 18 V. If v_I is a 20-V peak-to-peak sine wave, what is the peak-to-peak output voltage? What is the peak load current? What is the load power?

Ans. 3 V/V; 10 kΩ; 60 V; 3.75 A; 56.25 W

10.9 MOS POWER TRANSISTORS

Thus far in this chapter we have dealt exclusively with BJT circuits. However, recent technological developments have resulted in MOS power transistors with specifications that are quite competitive with those of BJTs. In this section we consider the structure, characteristics, and application of power MOSFETs.

Structure of the Power MOSFET

The enhancement-MOSFET structure studied in Chapter 7 (Fig. 7.9) is not suitable for high-power applications. To appreciate this fact, recall that the drain current of an n-channel MOSFET operating in the pinch-off region is given by

$$i_D = \frac{1}{2} \mu_n C_{ox} \left(\frac{W}{L}\right) (v_{GS} - V_t)^2 \tag{10.46}$$

It follows that to increase the current capability of the MOSFET its width W should be made large and its channel length L should be made as short as possible. Unfortunately, however, reducing the channel length of the standard MOSFET structure results in a drastic reduction in its breakdown voltage. Specifically, the depletion region of the reverse-biased body-to-drain junction spreads into the short channel, resulting in breakdown at a relatively low voltage. Thus the resulting device would not be capable of handling the high voltages typical of power transistor applications. For this reason, new structures had to be found for fabricating short-channel (1- to 2-micrometer) MOSFETs with high breakdown voltages.

At the present time the most popular structure for a power MOSFET is the double-diffused or DMOS transistor shown in Fig. 10.35. As indicated, the device is fabricated on a lightly doped n-type substrate with a heavily doped region at the

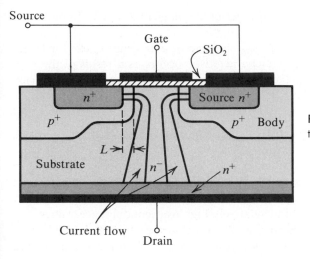

Fig. 10.35 Double-diffused vertical MOS transistor (DMOS).

bottom for the drain contact. Two diffusions[3] are employed, one to form the *p*-type body region and another to form the *n*-type source region.

The DMOS device operates as follows. Application of a positive gate voltage, v_{GS}, greater than the threshold voltage V_t, induces a lateral *n* channel in the *p*-type body region underneath the gate oxide. The resulting channel is short and its length is denoted L in Fig. 10.35. Current is then conducted by electrons from the source moving through the resulting short channel to the substrate and then vertically down the substrate to the drain. This should be contrasted with the lateral current flow in the standard small-signal MOSFET structure (Chapter 7).

Despite the fact the DMOS transistor has a short channel, its breakdown voltage can be very high (as high as 600 V). This is because the depletion region between the substrate and the body extends mostly in the lightly doped substrate and does not spread into the channel. The result is a MOS transistor that simultaneously has a high current capability (50 A is possible) as well as the high breakdown voltage just mentioned. Finally, we note that the vertical structure of the device provides efficient utilization of the silicon area.

An earlier structure used for power MOS transistors deserves mention. This is the V-groove MOS device (see Severns, 1984). Although still in use, the V-groove MOSFET has lost application ground to the vertical DMOS structure of Fig. 10.35, except possibly for high-frequency applications. Because of space limitations, we shall not describe the V-groove MOSFET.

Characteristics of Power MOSFETs

In spite of their radically different structure, power MOSFETs exhibit characteristics that are quite similar to those of the small-signal MOSFETs studied in Chapter 7. Important differences exist, however, and these are discussed in the following.

---

[3] See Appendix A for a description of the IC fabrication process.

Power MOSFETs have threshold voltages in the range of 2 to 4 V. In pinch-off, the drain current is related to v_{GS} by the square-law characteristic of Eq. (10.46). However, as shown in Fig. 10.36, the i_D-v_{GS} characteristic becomes linear for larger values of v_{GS}. The linear portion of the characteristic occurs as a result of the high electric field along the short channel, causing the velocity of charge carriers to reach an upper limit, a phenomenon known as *velocity saturation*. The drain current is then given by

$$i_D = \tfrac{1}{2}C_{\mathrm{OX}}\,WU_{\mathrm{sat}}(v_{GS} - V_t) \tag{10.47}$$

where U_{sat} is the saturated velocity value (5×10^6 cm/s for electrons in silicon). The linear i_D-v_{GS} relationship implies a constant g_m in the velocity saturation region. It is interesting to note that g_m is proportional to W which is usually large for power devices; thus power MOSFETs exhibit relatively high transconductance values.

The i_D-v_{GS} characteristic shown in Fig. 10.36 includes a segment labeled "subthreshold." Though of little significance for power devices, the subthreshold region of operation is of interest in very-low-power applications.

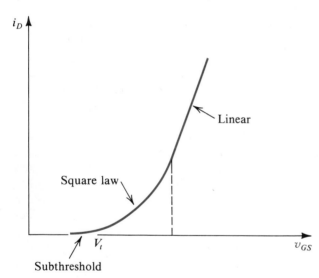

Fig. 10.36 Typical i_D-V_{GS} characteristic for a power MOSFET.

Temperature Effects

Of most interest in the design of MOS power circuits is the variation of MOSFET characteristics with temperature, illustrated in Fig. 10.37. Observe that there is a value of v_{GS} (in the range of 4–6 V for most power MOSFETs) at which the temperature coefficient of i_D is zero. At higher values of v_{GS}, i_D exhibits a negative temperature coefficient. This is a significant property; it implies that a MOSFET operating beyond the zero-temperature-coefficient point does not suffer from the possibility of thermal

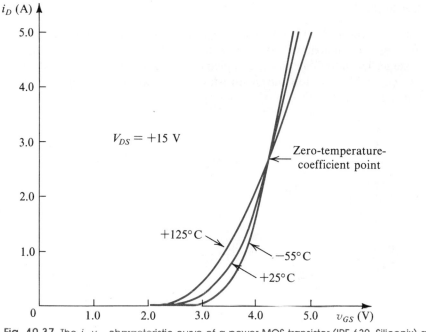

Fig. 10.37 The i_D-v_{GS} characteristic curve of a power MOS transistor (IRF 630, Siliconix) at a case temperature of $-55°C$, $+25°C$, and $+125°C$. (Courtesy Siliconix Inc.)

runaway. This is *not* the case, however, at low currents (that is, lower than the zero-temperature-coefficient point). In the (relatively) low-current region, the temperature coefficient of i_D is positive, and the power MOSFET can easily suffer thermal runaway (with unhappy consequences). Since class AB output stages are biased at low currents, means must be provided to guard against thermal runaway.

The reason for the positive temperature coefficient of i_D at low currents is that $v_{GS} - V_t$ is relatively low, and the temperature dependence is dominated by the negative temperature coefficient of V_t (in the range of -3 to -6 mV/°C).

Comparison with BJTs

The power MOSFET does not suffer from second breakdown, which limits the safe operating area of BJTs. Also, power MOSFETs do not require the large base-drive currents of power BJTs. Note, however, that the driver stage in a MOS power amplifer should be capable of supplying sufficient current to charge and discharge the MOSFET large and nonlinear input capacitance in the time allotted. Finally, the power MOSFET features, in general, a higher speed of operation than the power BJT. This makes MOS power transistors especially suited to switching applications—for instance, in motor-control circuits.

A Class AB Output Stage Utilizing MOSFETs

As an application of power MOSFETs, we show in Fig. 10.38 a class AB output stage utilizing a pair of complementary MOSFETs and employing BJTs for biasing and in the driver stage. The latter consists of complementary Darlington emitter followers formed by Q_1–Q_4 and has the low output-resistance necessary for driving the output MOSFETs at high speeds.

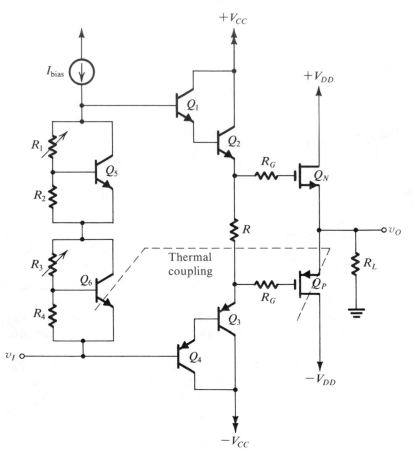

Fig. 10.38 A class AB amplifier with MOS output transistors and BJT drivers. Resistor R_3 is adjusted to provide temperature compensation while R_1 is adjusted to yield the desired value of quiescent current in the output transistors.

Of special interest in the circuit of Fig. 10.38 is the bias circuit utilizing two V_{BE} multipliers formed by Q_5 and Q_6 and their associated resistors. Transistor Q_6 is placed in direct thermal contact with the output transistors, which is achieved by simply mounting Q_6 on their common heat sink. Thus, by the appropriate choice of the V_{BE} multiplication factor of Q_6, the bias voltage V_{GG} (between the gates of the output transistors) can be made to decrease with temperature at the same rate as the sum of the threshold voltages, $(V_{tN} + |V_{tP}|)$, of the output MOSFETs. In this way the quiescent current of the output transistors can be stabilized against temperature variations.

Analytically, V_{GG} is given by

$$V_{GG} = \left(1 + \frac{R_3}{R_4}\right)V_{BE6} + \left(1 + \frac{R_1}{R_2}\right)V_{BE5} - 4V_{BE} \tag{10.48}$$

Since V_{BE6} is thermally coupled to the output devices while the other BJTs remain at constant temperature, we have

$$\frac{\partial V_{GG}}{\partial T} = \left(1 + \frac{R_3}{R_4}\right)\frac{\partial V_{BE6}}{\partial T} \tag{10.49}$$

which is the relationship needed to determine R_3/R_4. The other V_{BE} multiplier is then adjusted to yield the required value of V_{GG} and hence the desired quiescent current in Q_N and Q_P.

Exercises

10.18 For the circuit in Fig. 10.38, find the ratio R_3/R_4 that provides temperature stabilization of the quiescent current in Q_N and Q_P. Assume that $|V_t|$ changes at $-3\,\text{mV/°C}$ and that $\partial V_{BE}/\partial T = -2\,\text{mV/°C}$.

Ans. 2

10.19 For the circuit in Fig. 10.38 assume that the BJTs have a nominal V_{BE} of 0.7 V and that the MOSFETs have $V_t = 3$ V and $K \equiv \frac{1}{2}\mu_n C_{ox}(W/L) = 1\,\text{A/V}^2$. It is required to establish a quiescent current of 100 mA in the output stage and 20 mA in the driver stage. Find $|V_{GS}|$, V_{GG}, R, and R_1/R_2. Use the value of R_3/R_4 found in Exercise 10.18.

Ans. 3.32 V; 6.64 V; 332 Ω; 9.5

10.10 SUMMARY

● Output stages are classified according to the transistor conduction angle: class A (360°); class AB (slightly greater than 180°); class B (180°); and class C (smaller than 180°).

● The most common class A output stage is the emitter follower. It is biased at a current greater than the peak load current.

● The class A output stage dissipates its maximum power under quiescent conditions ($v_O = 0$). It achieves a maximum power conversion efficiency of 25%.

● The class B stage is biased at zero current and thus dissipates no power in quiescence.

● The class B stage can achieve a power conversion efficiency as high as 78.5%. It dissipates its maximum power for $\hat{V}_o = (2/\pi)V_{CC}$.

● The class B stage suffers from crossover distortion.

● The class AB output stage is biased at a small current; thus both transistors conduct for small input signals and crossover distortion is virtually eliminated.

● Except for an additional small quiescent power dissipation, the power relationships of the class AB stage are similar to those in class B.

● To guard against the possibility of thermal runaway, the bias voltage of the class AB circuit is made to vary with temperature in the same manner as does V_{BE} of the output transistors.

● To facilitate the removal of heat from the silicon chip, power devices are usually mounted on heat sinks. The maximum power that can be safely dissipated in the device is given by

$$P_{D\max} = \frac{T_{J\max} - T_A}{\theta_{JC} + \theta_{CS} + \theta_{SA}}$$

where $T_{J\max}$ and θ_{JC} are specified by the manufacturer, while θ_{CS} and θ_{SA} depend on the heat sink design.

● Use of the Darlington configuration in the class AB output stage reduces the base-current drive requirement. In integrated circuits, the compound *pnp* configuration is commonly used.

● Output stages are usually equipped with circuitry that, in the event of a short circuit, can turn on and limit the base-current drive, and hence the emitter current, of the output transistors.

● IC power amplifiers consist of a small-signal voltage amplifier cascaded with a high-power output stage. Overall feedback is applied either on chip or externally.

● The bridge-amplifier configuration provides, across a floating load, a peak-to-peak output voltage twice that possible from a single amplifier with a grounded load.

● The DMOS transistor is a short-channel power device capable of both high-current and high-voltage operation.

● At low currents, the drain current of a power MOSFET exhibits a positive temperature coefficient and thus the device can suffer thermal runaway. At high currents the temperature coefficient of i_D is negative.

BIBLIOGRAPHY

C. A. Holt, *Electronic Circuits*, New York: Wiley, 1978.

National Semiconductor Corporation, *Audio/Radio Handbook*, Santa Clara, Calif.: National Semiconductor Corporation, 1980.

D. L. Schilling and C. Belove, *Electronic Circuits*, 2nd ed., New York: McGraw-Hill, 1979.

R. Severns, (Ed.), *MOSPOWER Applications Handbook*, Santa Clara, Calif.: Siliconix, 1984.

S. Soclof, *Applications of Analog Integrated Circuits*, Englewood Cliffs, N.J.: Prentice-Hall, 1985.

Texas Instruments, Inc., *Power-Transistor and TTL Integrated-Circuit Applications*, New York: McGraw-Hill, 1977.

J. Wong and R. Johnson, "Low-distortion wideband power op amp," *Application Note 261*, National Semiconductor Corporation, July 1981.

J. Wong and J. Sherwin, "Applications of wide-band buffer amplifiers," *Application Note 227*, National Semiconductor Corporation, October 1979.

PROBLEMS

10.1 Consider the follower circuit shown in Fig. 10.2, for the condition in which all transistors are identical, $R = R_L$, $V_{CC} = 10$ V, $V_{CEsat} = 0.2$ V, $V_{BE} = 0.7$ V, and β is very high. What is the most negative output voltage available?

10.2 Consider an emitter follower having the topology depicted in Fig. 10.2, but for which the load R_L is connected to -5 V rather than to ground. For $V_{CC} = 15$ V and $R_L = 1$ kΩ, what value of R is required to ensure that an output level of -10 V can be reached? If an additional requirement exists that the follower small-signal gain exceed 0.98, what value of R should be chosen? In this case what is the current in Q_1 for an output voltage of $+10$ V?

*10.3 Consider the operation of the follower circuit of Fig. 10.2 for which $R_L = V_{CC}/I$, when driven by a square wave such that the output ranges from $+V_{CC}$ to $-V_{CC}$ (ignoring V_{CEsat}). For this situation sketch the equivalent of Fig. 10.4 for v_O, i_{C1}, v_{CE1} and p_{D1}. Repeat for a square-wave output that has peak levels of $+V_{CC}/2$. What is the average power dissipation in Q_1 in each case? Compare these results for sine waves of peak amplitude V_{CC} and $V_{CC}/2$, respectively.

10.4 Consider the situation described in Problem 10.3. For a square-wave output having a peak-to-peak value of $2V_{CC}$ and V_{CC}, and for sine waves of the same peak-to-peak values, find the average power loss in the current-source transistor Q_2.

10.5 Reconsider the situation described in Exercise 10.4 for variation in V_{CC} and specifically for $V_{CC} = 16$, 12, 10, and 8 V. For the latter case assume V_{CEsat} is nearly zero. What is the power conversion efficiency in each case?

**10.6 An estimate of the crossover distortion produced by a class B output stage can be found by considering the distorted sine-wave output of V volts amplitude to consist of a sine-wave output of amplitude $(V + \Delta V)$ from which a square wave of amplitude ΔV

has been subtracted. By using Eq. (1.1) to resolve the square wave into fundamental and harmonic components, show that the percent total harmonic distortion of the class B stage output is given by

$$\%\text{THD} = \frac{61.6\,\Delta V}{V - 0.27\,\Delta V}$$

Evaluate the %THD for outputs of measured amplitudes of 1 and 10 volts, and $\Delta V = 0.5$ V. Note the importance of using large input (and output) signals for low distortion.

10.7 A class B output stage operating from regulated power supplies of ± 5 V supplies a relatively undistorted sine wave of 5-V peak amplitude to a 10-Ω load. Determine the output power, the supply power, and the power loss in *each* of the output transistors. Find the power conversion efficiency. Repeat for an output peak amplitude of 1 V (ignore the crossover distortion, which is more pronounced in this case).

10.8 Repeat Problem 10.7 for square-wave outputs.

10.9 Reconsider the situation in Example 10.1 if the amplifier designed is subject to the cumulative effect of 10% increase in supply voltage due to manufacturing variation and 10% due to power-line variation. Also, the load resistance is known to vary by as much as $\pm 10\%$. For the worst case combination of these conditions, find the maximum power dissipation in each output transistor.

***10.10 A class B output stage operates from power supplies whose open circuit voltage is ± 10 V and whose source resistance (for each) is 10 Ω. Assuming a transistor saturation voltage that is nearly zero, what is the maximum square-wave peak voltage that can be presented to a 10-Ω load with the transistor reaching the onset of saturation? Now if two very large capacitors (infinite, for simplicity) are used to reduce the ac impedance of the two supplies, what square-wave peak output is possible? Estimate the value of the capacitors required to achieve a square-wave amplitude of 90% of that obtained with infinite capacitors for input frequencies as low as 100 Hz.

10.11 Consider the design of the class AB output stage of Fig. 10.11 for a quiescent current bias of 1 mA. What value of V_{BB} is required for complementary transistor pairs for which **(a)** $|V_{BE}| = 0.7$ V at 1 mA, **(b)** $|V_{BE}| = 0.7$ V at 10 mA, and **(c)** $I_S = 10^{-14}$ A.

*10.12 Reconsider Exercise 10.6 for the case in which $R_L = 10 \Omega$ for $v_O = 0$, ± 0.1 V and ± 1.0 V. Construct a table similar to that in Exercise 10.6. Compare the results in both cases.

*10.13 A class AB output stage, resembling that in Fig. 10.11 but utilizing a single supply of $+10$ V and biased at $V_I = 6$ V, is capacitively coupled to a 100-Ω load. For transistors for which $|V_{BE}| = 0.7$ V at 1 mA and for a bias voltage $V_{BB} = 1.4$ V, what quiescent current results? For a step change in output from 0 to -1 V, what input signal v_I is required? Assuming transistor saturation voltages of zero, find the largest possible positive-going and negative-going steps at the output.

*10.14 A class AB output stage using a two-diode bias network as shown in Fig. 10.14 utilizes diodes having the same junction area as the output transistors. For $V_{CC} = 10$ V, $I_{bias} = 0.5$ mA, $R_L = 100 \Omega$, $\beta_N = 50$, and $|V_{CEsat}| = 0$ V, what is the quiescent current? What is the largest possible positive and negative output signal levels? To achieve a positive peak output level equal to the negative peak level, what value of β_N is needed if I_{bias} is not changed? What value of I_{bias} is needed if β_N is held at 50? For this value, what does I_Q become?

*10.15 A class AB output circuit of the type shown in Fig. 10.14, utilizing diodes having half the junction area of the output transistors and $I_{bias} = 1$ mA, is operated with the diodes and transistors thermally isolated from one another. When all devices are at 25°C, what quiescent current flows? Following a burst of output signal activity the junction temperatures of Q_N and Q_P rise to 125°C. What does the quiescent current now become? For power supplies of ± 20 V, what is the increase in quiescent power dissipation that now results? Assume that I_S changes by $+14\%/°C$ and recall that $V_T = kT/q$, where T is the temperature in kelvins $= 273 +$ temperature in °C. Also, at 25°C, $|V_{BE}| = 700$ mV.

*10.16 Repeat Problem 10.18 for the case in which the diodes are thermally coupled to the transistors, but where coupling is not perfect and the diode junction temperature rise above 25°C is half that of the transistor

junctions. Assume that the diode voltage drop decreases by 2 mV/C° of temperature rise.

*10.17 Consider the enhancement-MOSFET class AB output stage shown in Fig. P10.17. All transistors have $|V_t| = 1$ V and $K_1 = K_2 = nK_3 = nK_4$. Also $K_3 = 1$ mA/V².

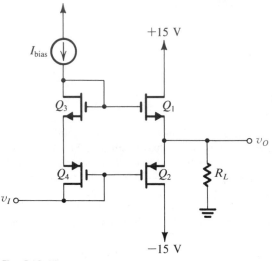

Fig. P10.17

(a) With $v_O = O$ show that $I_Q = nI_{bias}$.
(b) For $I_{bias} = 100 \mu A$ and $R_L = 1$ kΩ find the value of n that results in a small-signal gain of 0.99 for outputs around zero.

*10.18 The diode-biased class AB output stage in Fig. 10.14 operates at $v_O = 0$ with $I_{bias} = 1$ mA, $I_Q = 5$ mA, and $v_I = -0.7$ V. For $v_O = 5$ V and $R_L = 100 \Omega$, what minimum β is necessary? For β 10% higher, what input voltage is required? For β twice a large as necessary, what input voltage is required?

10.19 A V_{BE} multiplier with $R_1 = R_2 = 1.2$ kΩ utilizing a transistor for which $V_{BE} = 0.6$ V at $I_C = 1$ mA is supplied with a current of 1.5 mA. What is the collector–emitter terminal voltage for **(a)** $\beta = \infty$? **(b)** $\beta = 100$? and **(c)** $\beta = 10$? For simplicity, assume that V_{BE} remains constant.

10.20 A power transistor operating at an embient temperature of 50°C and an average emitter current of 3 A dissipates 30 W. If the thermal resistance of the transistor is known to be less than 3°C/W, what is the greatest junction temperature you would expect? If the transistor

V_{BE} measured using a pulsed emitter current of 3 A at a junction temperature of 25°C is 0.80 V, what average V_{BE} would you expect under normal operation? (Use a temperature coefficient of -2 mV/°C.)

10.21 A transistor having a thermal resistance, junction to ambient, of 10°C/W is operated at a power level of 13 W. If damage to its structure is known to occur after long-term operation at junction temperatures in excess of 175°C, what is the maximum ambient temperature that can be tolerated?

10.22 An improved manufacturing technique allows the maximum junction temperature of the transistor in Example 10.4 to be raised to 200°C. What are the results of this change if θ_{JA} remains unchanged?

10.23 For a particular application of the transistor specified in Example 10.4, extreme reliability is essential. To improve reliability the maximum junction temperature is to be limited to 100°C. What are the consequences of this decision for the conditions specified?

10.24 A low-power transistor is said to have a maximum power dissipation of 100 mW at 25°C and a thermal resistance of 1.6°C/mW. What is the maximum safe junction temperature that the transistor can tolerate?

10.25 A power transistor for which $T_{Jmax} = 180$°C can dissipate 50 W at a case temperature of 50°C. If it is connected to a heat sink using an insulating washer for which the thermal resistance is 0.6°C/W, what heat-sink temperature is necessary to ensure safe operation at 30 W? For an ambient temperature of 39°C, what heat-sink thermal resistance is required? If, for a particular extruded-aluminum-finned heat sink, the thermal resistance in still air is 4.5°C/W per cm of length, how long a heat sink is needed?

10.26 If, in the situation described in Exercise 10.10, the calculated heat sink is doubled in size and θ_{CS} is reduced to 0.4°C/W using "heat sink grease," what is the reduction in junction temperature for the same power level? What additional power can be dissipated if the limiting junction temperature is not exceeded?

10.27 Use the results given in the answer to Exercise 10.11 to determine the input current of the circuit in Fig. 10.24 for $v_I = 0$ and ± 10 V with infinite and 100-Ω loads.

*__10.28__ Reconsider Exercise 10.11 with R_1 and R_2 replaced by constant-current sources of 2.87 mA each. Find the device currents and output voltages under the conditions of inputs at 0 and ± 10 V with no load and

100-Ω load. Compare your results to those given in Exercise 10.11.

10.29 Consider the Darlington configuration of Fig. 10.25. Find the exact expressions for i_C and i_E in terms of i_B, β_1, and β_2. Hence, find the effective β of the Darlington configuration in terms of β_1 and β_2. Also find v_{BE} in terms of i_E, β_1, β_2, I_{S1}, and I_{S2}. For $I_{S1} = 3.3 \times 10^{-15}$ A, $I_{S2} = 3.3 \times 10^{-14}$ A, and $\beta_1 = \beta_2 = 100$, find V_{BE}, I_B, and I_C for $I_E = 10$ mA.

*__**10.30__ Consider the class AB output stage of Fig. 10.27 under the conditions that $V_{CC} = 15$ V, $R_L = 100$ Ω, and the output is sinusoidal with a maximum amplitude of 10 V. For Q_2 and Q_4, $I_S = 10^{-13}$ A and $\beta = 50$. For Q_1 and Q_5, $I_S = 10^{-14}$ A and $\beta = 100$. For Q_3, $I_S = 10^{-14}$ A and $\beta = 10$. Design the circuit using a value of I_{bias} that is twice that required to supply the maximum base current of Q_1, and a multiplier for which the transistor current variation is at most 5 to 1. The quiescent current in Q_2 and Q_4 is to be 10 mA. What input voltages are required for outputs at 0 and ± 10 V?

*__*10.31__ In the thermal shutdown circuit of Fig. 10.29, at 25°C, $V_{Z1} = 6.8$ V with a temperature coefficient of 400 parts per million (ppm)/°C (that is, for every °C rise in temperature the voltage is multiplied by the factor 1.000400) at the bias current chosen. For Q_1 and Q_2, $I_S = 10^{-15}$ A and β is very large. Design for $I_{C1} = 100$ μA at 25°C and $I_{C2} - 1$ mA with all junctions at 150°C. Assume that V_{BE} of Q_1 changes by -2 mV/°C (even though its current will not remain constant). Also assume that I_S has a temperature coefficient of $+14\%$/°C. What is I_{C2} at 25°C? (Recall that $V_T = kT/q$.)

10.32 Consider the front end of the circuit shown in Fig. 10.30. For $V_S = 20$ V, calculate approximate values for the bias currents in Q_1 through Q_6. Assume $\beta_{npn} = 100$, $\beta_{pnp} = 20$, and $|V_{BE}| = 0.7$ V. Also find the dc voltage at the output.

*__10.33__ For the circuit in Fig. 10.30, consider the case in which $R_L = 0$. Find an approximate value of the transconductance measured between the two input terminals and the base of Q_{12}. What is the differential input resistance?

*__10.34__ Consider the current-boosted output circuit shown in Fig. 10.33 with $R_5 = R_6 = 0$. Let $V_{CC} = 15$ V and the total quiescent current = 10 mA, 90% of which flows in Q_3 and Q_4. Q_1 and Q_2 are matched, with $I_S = 10^{-14}$ A and high β. Q_3 and Q_4 are matched, with $I_S = 10^{-13}$ A and $\beta = 100$. Q_5 and Q_6 are matched, with $I_S = 10^{-12}$ A and $\beta = 20$. Arrange that Q_5 and Q_3 conduct

equal currents when the load current reaches 100 mA. Find the values of $R_1 = R_2$ and $R_3 = R_4$.

10.35 For the bridge amplifier of Fig. 10.34, let $R_2 = R_4 = 100$ kΩ. Find R_1 and R_3 to obtain an overall gain of 10.

*10.36 An alternate bridge amplifier configuration with high input resistance is shown in Fig. P10.36. What is the gain v_O/v_I? For op amps using ± 15-V supplies that limit at 1 V from the supplies, what is the peak-to-peak voltage of the largest sine-wave output available? Design the circuit for a gain of 20 using $R_1 = 10$ kΩ.

10.37 For a DMOS transistor that exhibits velocity saturation at a gate-to-source voltage 4 V above V_t, use Eqs. (10.46) and (10.47) to find the value of mobility μ_n (in cm²/V·s). The channel length is 2 μm and $U_{\text{sat}} = 5 \times 10^6$ cm/s.

10.38 Find g_m for a DMOS transistor operating in the velocity saturation region [see Eq. (10.47)]. Compare it with g_m of the same device operating in the conventional square-law region.

10.39 Consider the design of the class AB amplifier of Fig. 10.38 under the following conditions: $V_t = 2$ V, $K = 100$ mA/V², $|V_{BE}| = 0.7$ V, β is high, $I_{QN} = I_{QP} = I_R = 10$ mA, $I_{\text{bias}} = 100$ μA, $I_{Q5} = I_{Q6} = I_{\text{bias}}/2$, $R_2 = R_4$, temperature coefficient of $V_{BE} = -2$ mV/°C, and temperature coefficient of $V_t = -3$ mV/°C in the low-current region. Find the values of R, R_1, R_2, R_3, and R_4. Assume Q_6, Q_P, and Q_N to be thermally coupled.

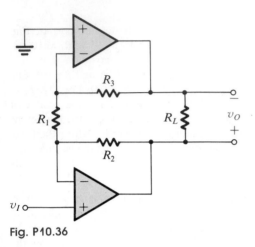

Fig. P10.36

Chapter 11

Frequency Response

INTRODUCTION

In Chapter 2 we introduced the topic of amplifier frequency response and briefly mentioned the various frequency-response shapes encountered. In addition, the frequency, step, and pulse responses of single-time-constant (STC) networks were studied in detail. This material was found to be directly applicable to the evaluation of the frequency response of op-amp circuits (Chapter 3). Apart from this and the occasional mention of limitations imposed on amplifier frequency response, the detailed study of this important topic has been deferred to the present chapter. This was done for a number of reasons, the most important of which is the need for circuit theoretic concepts and methods that the reader might not have been exposed to at the beginning of the book. We refer specifically to the complex frequency variable s and associated concepts such as poles and zeros. These topics are normally found in introductory texts on

circuit analysis [see, for example, Van Valkenburg (1974) and Bobrow (1987)]. In the following it will be assumed that the reader is familiar with this material.

After a brief review of s-domain analysis (Section 11.1) and a study of amplifier transfer functions (Section 11.2), the frequency response of the classical common-source amplifier is presented (Section 11.3). The BJT hybrid-π model is studied in Section 11.4 and used to find the frequency response of the common-emitter amplifier (Section 11.5), the common-base amplifier (Section 11.6), and the common-collector amplifier (Section 11.7). A number of two-stage amplifier configurations having the advantage of extended bandwidth are also studied in this chapter. Also, the differential amplifier frequency response is considered in detail.

The single-stage and two-stage amplifiers analyzed in this chapter can be used either singly or as building blocks for more complex multistage amplifiers. Frequency-response analysis of multistage amplifiers uses the methods studied in this chapter as will be illustrated further in Chapter 13 in the context of analyzing op-amp circuits.

Although the emphasis in this chapter is on analysis, the material is readily applicable to design. This is achieved by keeping the analysis relatively simple and thus focusing attention on the mechanisms that limit frequency response and on methods for extending amplifier bandwidth. Of course, once an initial design is obtained, its exact frequency response can be found using computer-aided analysis (see Appendix C). The results obtained this way can be used to improve the design further. □

11.1 *s*-DOMAIN ANALYSIS

Most of our work in this chapter will be concerned with finding amplifier voltage gain as a transfer function of the complex frequency *s*. In this *s*-domain analysis a capacitance C is replaced by an admittance sC, or equivalently an impedance $1/sC$, and an inductance L is replaced by an impedance sL. Then, using usual circuit-analysis techniques, one derives the voltage transfer function $T(s) \equiv V_o(s)/V_i(s)$.

Exercise

11.1 Find the voltage transfer function $T(s) \equiv V_o(s)/V_i(s)$ for the STC network shown in Fig. E11.1.

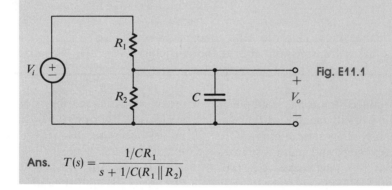

Fig. E11.1

Ans. $T(s) = \dfrac{1/CR_1}{s + 1/C(R_1 \| R_2)}$

Once the transfer function $T(s)$ is obtained, it can be evaluated for *physical frequencies* by replacing s by $j\omega$. The resulting transfer function $T(j\omega)$ is in general a complex quantity whose magnitude gives the magnitude response (or transmission) and whose angle gives the phase response of the amplifier.

In many cases it will not be necessary to substitute $s = j\omega$ and evaluate $T(j\omega)$; rather, the form of $T(s)$ will reveal many useful facts about the circuit performance. In general, for all the circuits dealt with in this chapter, $T(s)$ can be expressed in the form

$$T(s) = \frac{a_m s^m + a_{m-1} s^{m-1} + \cdots + a_0}{s^n + b_{n-1} s^{n-1} + \cdots + b_0} \tag{11.1}$$

where the coefficients a and b are real numbers and the order m of the numerator is smaller than or equal to the order n of the denominator; the latter is called the *order of the network*. Furthermore, for a *stable* circuit—that is, one that does not generate signals on its own—the denominator coefficients should be such that *the roots of the denominator polynomial all have negative real parts*. We shall study the problem of amplifier stability in Chapter 12.

Poles and Zeros

An alternative form for expressing $T(s)$ is

$$T(s) = a_m \frac{(s - Z_1)(s - Z_2) \cdots (s - Z_m)}{(s - P_1)(s - P_2) \cdots (s - P_n)} \tag{11.2}$$

where a_m is a multiplicative constant (the coefficient of s^m in the numerator), Z_1, $Z_2, \ldots, Z_m$ are the roots of the numerator polynomial, and $P_1, P_2, \ldots, P_n$ are the roots of the denominator polynomial. $Z_1, Z_2, \ldots, Z_m$ are called the *transfer-function zeros* or *transmission zeros*, and $P_1, P_2, \ldots, P_n$ are *the transfer-function poles* or *the natural modes* of the network. A transfer function is completely specified in terms of its poles and zeros together with the value of the multiplicative constant.

The poles and zeros can be either real or complex numbers. However, since the a and b coefficients are real numbers, the complex poles (or zeros) must occur in *conjugate pairs*. That is, if $5 + j3$ is a zero, then $5 - j3$ also must be a zero. A zero that is pure imaginary ($\pm j\omega_z$) causes the transfer function $T(j\omega)$ to be exactly zero at $\omega = \omega_z$. Thus the "trap" one places at the input of a television set is a circuit that has a transmission zero at the particular interfering frequency. Real zeros, on the other hand, do not produce transmission nulls (why not?). Finally, note that for values of s much greater than all the poles and zeros, the transfer function in Eq. (11.1) becomes $T(s) \simeq a_m/s^{n-m}$. Thus the transfer function has $(n \quad m)$ zeros at $s = \infty$.

First-Order Functions

All the transfer functions encountered in this chapter have real poles and zeros and can therefore be written as the product of first-order transfer functions of the general form

$$T(s) = \frac{a_1 s + a_0}{s + \omega_0} \tag{11.3}$$

where $-\omega_0$ is the location of the real pole. The quantity ω_0 is called the *pole frequency* and is equal to the inverse of the time constant of this single-time-constant (STC) network (see Chapter 2). The constants a_0 and a_1 determine the type of STC network. Specifically, we studied in Chapter 2 two types of STC networks, low pass and high pass. For the low-pass first-order network we have

$$T(s) = \frac{a_0}{s + \omega_0} \tag{11.4}$$

In this case the dc gain is a_0/ω_0, and ω_0 is the corner or 3-dB frequency. Note that this transfer function has one zero at $s = \infty$. On the other hand, the first-order high-pass transfer function has a zero at dc and can be written

$$T(s) = \frac{a_1 s}{s + \omega_0} \tag{11.5}$$

At this point the reader is strongly urged to review the material on STC networks and their frequency and pulse responses in Chapter 2. Of specific interest are the plots of the magnitude and phase responses of the two special kinds of STC networks. Such plots can be employed to generate the magnitude and phase plots of a high-order transfer function, as explained below.

Bode Plots

A simple technique exists for obtaining an approximate plot of the magnitude and phase of a transfer function given its poles and zeros. The technique is particularly useful in the case of real poles and zeros. The method was developed by H. Bode, and the resulting diagrams are called *Bode plots*.

A transfer function of the form depicted in Eq. (11.2) consists of a product of factors of the form $s + a$, where such a factor appears on top if it corresponds to a zero and on the bottom if it corresponds to a pole. It follows that the magnitude response in decibels of the network can be obtained by summing together terms of the form $20 \log_{10} \sqrt{a^2 + \omega^2}$, and the phase response can be obtained by summing terms of the form $\tan^{-1}(\omega/a)$. In both cases the terms corresponding to poles are summed with negative signs. For convenience we can extract the constant a and write the typical magnitude term in the form $20 \log \sqrt{1 + (\omega/a)^2}$. On a plot of deci-

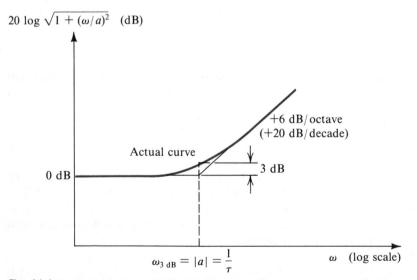

Fig. 11.1 Bode plot for the typical magnitude term. The curve shown applies for the case of a zero. For a pole the high-frequency asymptote should be drawn with a -6-dB/octave slope.

bels versus log frequency this term gives rise to the curve and straight-line asymptotes shown in Fig. 11.1. Here the low-frequency asymptote is a horizontal straight line at 0 dB level and the high-frequency asymptote is a straight line with a slope of 6 dB/octave or, equivalently, 20 dB/decade. The two asymptotes meet at the frequency $\omega = |a|$, which is called *the corner frequency*. As indicated, the actual magnitude plot differs slightly from the value given by the asymptotes; the maximum difference is 3 dB and occurs at the corner frequency.

For $a = 0$—that is, a pole or a zero at $s = 0$—the plot is simply a straight line of 6dB/octave slope intersecting the 0-dB line at $\omega = 1$.

In summary, to obtain the Bode plot for the magnitude of a transfer function, the asymptotic plot for each pole and zero is first drawn. The slope of the high-frequency asymptote of the curve corresponding to a zero is $+20$ dB/decade, while that for a pole is -20 dB/decade. The various plots are then added together, and the overall curve is shifted vertically by an amount determined by the multiplicative constant of the transfer function.

EXAMPLE 11.1

An amplifier has the voltage transfer function

$$T(s) = \frac{10s}{(1 + s/10^2)(1 + s/10^5)}$$

Find the poles and zeros and sketch the magnitude of the gain versus frequency. Find approximate values for the gain at $\omega = 10$, 10^3, and 10^6 rad/s.

Solution
The zeros are as follows: one at $s = 0$ and one at $s = \infty$. The poles are as follows: one at $s = -10^2$ rad/s and one at $s = -10^5$ rad/s.

Figure 11.2 shows the asymptotic Bode plots of the different factors of the transfer function. Curve 1, which is a straight line with $+20$ dB/decade slope, corresponds to the s term (that is, the zero at $s = 0$) in the numerator. The pole at $s = -10^2$ results in curve 2, which consists of two asymptotes intersecting at $\omega = 10^2$. Similarly, the pole at $s = -10^5$ is represented by curve 3, where the intersection of the asymptotes is at $\omega = 10^5$. Finally, curve 4 represents the multiplicative constant of value 10.

Adding the four curves results in the asymptotic Bode diagram of the amplifier gain (curve 5). Note that since the two poles are widely separated, the gain will be very close to 10^3 (60 dB) over the frequency range 10^2 to 10^5 rad/s. At the two corner frequencies (10^2 and 10^5 rad/s) the gain will be approximately 3 dB below the maximum of 60 dB. At the three specific frequencies the values of the gain as obtained from the Bode plot and from exact evaluation of the transfer function are as follows:

ω	Approximate Gain	Exact Gain
10	40 dB	39.96 dB
10^3	60 dB	59.96 dB
10^6	40 dB	39.96 dB

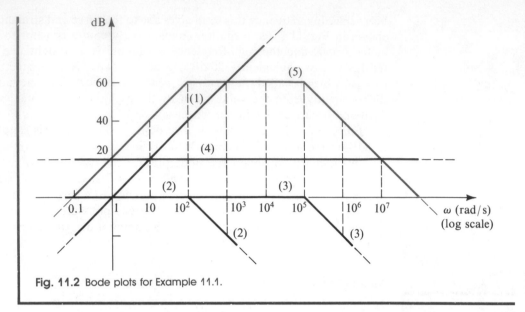

Fig. 11.2 Bode plots for Example 11.1.

We next consider the phase plot. Figure 11.3 shows a plot of the typical phase term $\tan^{-1}(\omega/a)$, assuming that a is negative. Also shown is an asymptotic straight-line approximation of the arctan function. The asymptotic plot consists of three straight lines. The first is horizontal at $\phi = 0$ and extends up to $\omega = 0.1|a|$. The second line has a slope of $-45°$/decade and extends from $\omega = 0.1|a|$ to $\omega = 10|a|$. The third line has a zero slope and a level of $\phi = -90°$. The complete phase response can be obtained by summing the asymptotic Bode plots of the phase of all poles and zeros.

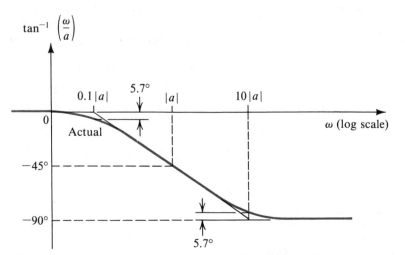

Fig. 11.3 Bode plot of the typical phase term $\tan^{-1}(\omega/a)$ when a is negative.

EXAMPLE 11.2

Find the Bode plot for the phase of the transfer function of the amplifier considered in Example 11.1.

Solution

The zero at $s = 0$ gives rise to a constant $+90°$ phase function represented by curve 1 in Fig. 11.4. The pole at $s = -10^2$ gives rise to the phase function

$$\phi_1 = -\tan^{-1}\frac{\omega}{10^2}$$

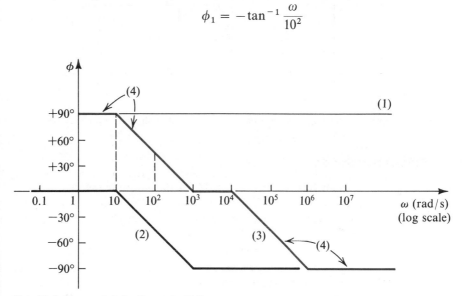

Fig. 11.4 Phase plots for Example 11.2.

(the leading minus sign is due to the fact that this singularity is a pole). The asymptotic plot for this function is given by curve 2 in Fig. 11.4. Similarly, the pole at $s = -10^5$ gives rise to the phase function

$$\phi_2 = -\tan^{-1}\frac{\omega}{10^5}$$

whose asymptotic plot is given by curve 3. The overall phase response (curve 4) is obtained by direct summation of the three plots.

11.2 THE AMPLIFIER TRANSFER FUNCTION

The amplifiers considered in this chapter have voltage-gain functions of either of the two forms shown in Fig. 11.5. Figure 11.5a applies for direct-coupled or dc amplifiers and Fig. 11.5b for capacitively coupled or ac amplifiers. The only difference between the two types is that the gain of the ac amplifier falls off at low frequencies. In the following we shall study the more general response shown in Fig. 11.5b. The response of the dc amplifier follows as a special case.

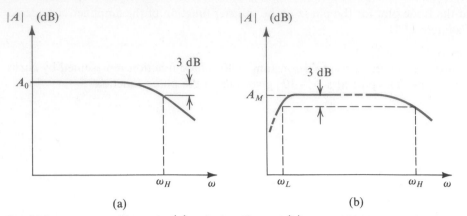

Fig. 11.5 Frequency response for **(a)** a dc amplifier and **(b)** a capacitively coupled amplifier.

The Three Frequency Bands

As can be seen from Fig. 11.5b the amplifier gain is almost constant over a wide frequency range called the midband. In this frequency range all capacitances (coupling, bypass, and transistor internal capacitances) have negligible effects and can be ignored in gain calculations. At the high-frequency end of the spectrum the gain drops owing to the effect of the internal capacitances of the device. On the other hand, at the low-frequency end of the spectrum the coupling and bypass capacitances no longer act as perfect short circuits and thus cause the gain to drop. The extent of the midband is usually defined by the two frequencies ω_L and ω_H. These are the frequencies at which the gain drops by 3 dB below the value at midband. The amplifier bandwidth is usually defined as

$$BW = \omega_H - \omega_L \tag{11.6}$$

and, since $\omega_L \ll \omega_H$,

$$BW \simeq \omega_H \tag{11.7}$$

A figure of merit for the amplifier is its *gain-bandwidth product*, defined as

$$GB \equiv A_M \omega_H \tag{11.8}$$

where A_M is the magnitude of midband gain in volts per volt. As will be shown in later sections, it is generally possible to trade off gain for bandwidth.

The Gain Function A(s)

The amplifier gain as a function of the complex frequency s can be expressed in the general form

$$A(s) = A_M F_L(s) F_H(s) \tag{11.9}$$

where $F_L(s)$ and $F_H(s)$ are functions that account for the dependence of gain on frequency in the low-frequency band and in the high-frequency band, respectively. For frequencies ω much greater than ω_L the function $F_L(s)$ approaches unity. Similarly, for frequencies ω much smaller than ω_H the function $F_H(s)$ approaches unity. Thus for $\omega_L \ll \omega \ll \omega_H$,

$$A(s) \simeq A_M$$

as should have been expected. It also follows that the gain of the amplifier in the low-frequency band, $A_L(s)$, can be expressed as

$$A_L(s) \simeq A_M F_L(s) \tag{11.10}$$

and the gain in the high-frequency band can be expressed as

$$A_H(s) \simeq A_M F_H(s) \tag{11.11}$$

The midband gain is determined by analyzing the amplifier equivalent circuit with the assumption that the coupling and bypass capacitors are acting as perfect short circuits, and the internal capacitors of the transistor model are acting as perfect open circuits. The low-frequency transfer function, $A_L(s)$, is determined from analysis of the amplifier equivalent circuit including the coupling and bypass capacitors but assuming that the transistor-model capacitances behave as perfect open circuits. On the other hand, the high-frequency transfer function, $A_H(s)$, is determined from analysis of the amplifier equivalent circuit including the transistor-model capacitors but assuming that the coupling and bypass capacitors behave as perfect short circuits.

The Low-Frequency Response

The function $F_L(s)$, which characterizes the low-frequency response of the amplifier, takes the general form

$$F_L(s) = \frac{(s + \omega_{Z1})(s + \omega_{Z2}) \cdots (s + \omega_{Zn_L})}{(s + \omega_{P1})(s + \omega_{P2}) \cdots (s + \omega_{Pn_L})} \tag{11.12}$$

where $\omega_{P1}, \omega_{P2}, \ldots, \omega_{Pn_L}$ are positive numbers representing the frequencies of the n_L low-frequency poles and $\omega_{Z1}, \omega_{Z2}, \ldots, \omega_{Zn_L}$ are positive, negative, or zero numbers representing the n_L zeros. It should be noted from Eq. (11.12) that as s approaches infinity (in fact, as $s = j\omega$ approaches midband frequencies), $F_L(s)$ approaches unity.

In many cases the zeros are at such low frequencies (much smaller than ω_L) as to be of little importance in determining the lower 3-dB frequency ω_L. Also, usually one of the poles—say, ω_{P1}—has a much higher frequency than all other poles. It follows that for frequencies ω close to the midband, $F_L(s)$ can be approximated by

$$F_L(s) \simeq \frac{s}{s + \omega_{P1}} \tag{11.13}$$

which is the transfer function of a first-order high-pass network. In this case the low-frequency response of the amplifier is *dominated* by the pole at $s = -\omega_{P1}$ and the lower 3-dB frequency is approximately equal to ω_{P1},

$$\omega_L \simeq \omega_{P1} \tag{11.14}$$

If this *dominant-pole approximation* holds, it becomes a simple matter to determine ω_L. Otherwise one has to find the complete Bode plot for $|F_L(j\omega)|$ and thus determine ω_L. As a rule of thumb the dominant-pole approximation can be made if the highest-frequency pole is separated from the nearest pole or zero by at least two octaves (that is, a factor of four).

If a dominant low-frequency pole does not exist, an approximate formula can be derived for ω_L in terms of the poles and zeros. For simplicity consider the case of a circuit having two poles and two zeros at the low-frequency end; that is,

$$F_L(s) = \frac{(s + \omega_{Z1})(s + \omega_{Z2})}{(s + \omega_{P1})(s + \omega_{P2})} \tag{11.15}$$

Substituting $s = j\omega$ and taking the squared magnitude gives

$$|F_L(j\omega)|^2 = \frac{(\omega^2 + \omega_{Z1}^2)(\omega^2 + \omega_{Z2}^2)}{(\omega^2 + \omega_{P1}^2)(\omega^2 + \omega_{P2}^2)} \tag{11.16}$$

By definition, at $\omega = \omega_L$, $|F_L|^2 = \frac{1}{2}$, and thus

$$\frac{1}{2} = \frac{(\omega_L^2 + \omega_{Z1}^2)(\omega_L^2 + \omega_{Z2}^2)}{(\omega_L^2 + \omega_{P1}^2)(\omega_L^2 + \omega_{P2}^2)}$$

$$= \frac{1 + (1/\omega_L^2)(\omega_{Z1}^2 + \omega_{Z2}^2) + (1/\omega_L^4)(\omega_{Z1}^2\omega_{Z2}^2)}{1 + (1/\omega_L^2)(\omega_{P1}^2 + \omega_{P2}^2) + (1/\omega_L^4)(\omega_{P1}^2\omega_{P2}^2)} \tag{11.17}$$

Since ω_L is greater than the frequencies of all the poles and zeros, we may neglect the terms containing $(1/\omega_L^4)$ and solve for ω_L to obtain

$$\omega_L \simeq \sqrt{\omega_{P1}^2 + \omega_{P2}^2 - 2\omega_{Z1}^2 - 2\omega_{Z2}^2} \tag{11.18}$$

This relationship can be extended to any number of poles and zeros. Note that if one of the poles, say P_1, is dominant, then $\omega_{P1} \gg \omega_{P2}, \omega_{Z1}, \omega_{Z2}$, and Eq. (11.18) reduces to Eq. (11.14).

EXAMPLE 11.3

The low-frequency response of an amplifier is characterized by the transfer function

$$F_L(s) = \frac{s(s + 10)}{(s + 100)(s + 25)}$$

Determine its 3-dB frequency, approximately and exactly.

Solution
Noting that the highest-frequency pole at 100 rad/s is two octaves higher than the second pole and a decade higher than the zero, we find that a dominant-pole situation almost exists and $\omega_L \simeq 100$ rad/s. A better esitmate of ω_L can be obtained using Eq. (11.18), as follows:

$$\omega_L = \sqrt{100^2 + 25^2 - 2 \times 10^2} = 102 \text{ rad/s}$$

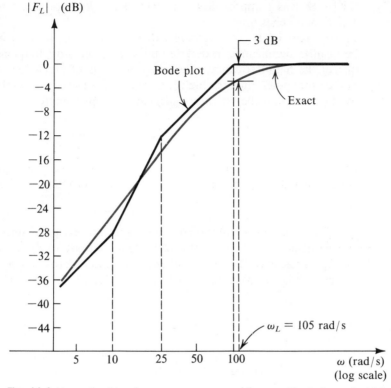

Fig. 11.6 Normalized low-frequency response of the amplifier in Example 11.3.

The exact value of ω_L can be determined from the given transfer function as 105 rad/s. Finally, we show in Fig. 11.6 a Bode plot and an exact plot for the magnitude of the given transfer function. Note that this is a plot of the low-frequency response of the amplifier normalized relative to the midband gain. That is, if the midband gain is 100 dB, then the entire plot should be shifted upward by 100 dB.

The High-Frequency Response

Consider next the high-frequency end. The function $F_H(s)$ can be expressed in the general form

$$F_H(s) = \frac{(1 + s/\omega_{Z1})(1 + s/\omega_{Z2}) \cdots (1 + s/\omega_{Zn_H})}{(1 + s/\omega_{P1})(1 + s/\omega_{P2}) \cdots (1 + s/\omega_{Pn_H})} \tag{11.19}$$

where $\omega_{P1}, \omega_{P2}, \ldots, \omega_{Pn_H}$ are positive numbers representing the frequencies of the n_H high-frequency real poles and $\omega_{Z1}, \omega_{Z2}, \ldots, \omega_{Zn_H}$ are positive, negative, or infinite numbers representing the frequencies of the n_H high-frequency zeros. Note from Eq.

(11.19) that as s approaches 0 (in fact as $s = j\omega$ approaches midband frequencies), $F_H(s)$ approaches unity.

In many cases the zeros are either at infinity or at such high frequencies as to be of little significance in determining the upper 3-dB frequency ω_H. If in addition one of the high-frequency poles—say, ω_{P1}—is of much lower frequency than any of the other poles, then the high-frequency response of the amplifier will be *dominated* by this pole, and the function $F_H(s)$ can be approximated by

$$F_H(s) \simeq \frac{1}{1 + s/\omega_{P1}} \tag{11.20}$$

which is the transfer function of a first-order low-pass network. It follows that if a dominant high-frequency pole exists, then the determination of ω_H is greatly simplified:

$$\omega_H \simeq \omega_{P1} \tag{11.21}$$

If a dominant high-frequency pole does not exist, the upper 3-dB frequency ω_H can be determined from a plot of $|F_H(j\omega)|$. Alternatively, an approximate formula for ω_H in terms of the high-frequency poles and zeros can be derived in a manner similar to that used above in deriving Eq. (11.18). The formula for ω_H is

$$\omega_H \simeq 1 \bigg/ \sqrt{\frac{1}{\omega_{P1}^2} + \frac{1}{\omega_{P2}^2} + \cdots - \frac{2}{\omega_{Z1}^2} - \frac{2}{\omega_{Z2}^2} \cdots} \tag{11.22}$$

Note that if one of the poles, say P_1, is dominant then $\omega_{P1} \ll \omega_{P2}, \omega_{P3}, \ldots, \omega_{Z1}, \omega_{Z2}, \ldots$ and Eq. (11.22) reduces to Eq. (11.21).

EXAMPLE 11.4

The high-frequency response of an amplifier is characterized by the transfer function

$$F_H(s) = \frac{1 - s/10^5}{(1 + s/10^4)(1 + s/4 \times 10^4)}$$

Determine the 3-dB frequency approximately and exactly.

Solution
Noting that the lowest-frequency pole at 10^4 rad/s is two octaves lower than the second pole and a decade lower than the zero, we find that a dominant-pole situation almost exists and $\omega_H \simeq 10^4$ rad/s. A better estimate of ω_H can be obtained using Eq. (11.22), as follows:

$$\omega_H = 1 \bigg/ \sqrt{\frac{1}{10^8} + \frac{1}{16 \times 10^8} - \frac{2}{10^{10}}}$$

$$\doteq 9,800 \text{ rad/s}$$

The exact value of ω_H can be determined from the given transfer function as 9,537 rad/s. Finally, we show in Fig. 11.7 a Bode plot and an exact plot for the given transfer function. Note that this is a plot of the high-frequency response of the amplifier normalized relative to its midband gain. That is, if the midband gain is 100 dB, then the entire plot should be shifted upward by 100 dB.

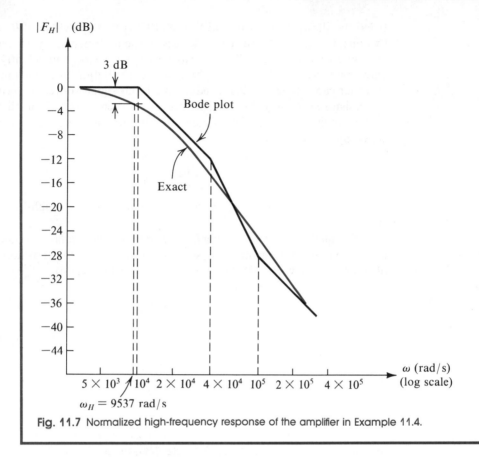

Fig. 11.7 Normalized high-frequency response of the amplifier in Example 11.4.

Using Short-Circuit and Open-Circuit Time Constants
for the Approximate Determination of ω_L and ω_H

If the poles and zeros of the amplifier transfer function can be determined easily, then we can determine ω_L and ω_H using the techniques described above. In many cases, however, it is not a simple matter to determine the poles and zeros. In such cases, approximate values of ω_L and ω_H can be obtained using the following method.

Consider first the high-frequency response. The function $F_H(s)$ of Eq. (11.19) can be expressed in the alternate form

$$F_H(s) = \frac{1 + a_1 s + a_2 s^2 + \cdots + a_{n_H} s^{n_H}}{1 + b_1 s + b_2 s^2 + \cdots + b_{n_H} s^{n_H}} \tag{11.23}$$

where the a and b coefficients are related to the zero and pole frequencies, respectively. Specifically, the coefficient b_1 is given by

$$b_1 = \frac{1}{\omega_{P_1}} + \frac{1}{\omega_{P_2}} + \cdots + \frac{1}{\omega_{Pn_H}} \tag{11.24}$$

It can be shown [see Gray and Searle (1969)] that the value of b_1 can be obtained by considering the various capacitances in the high-frequency equivalent circuit one at a time while reducing all other capacitors to zero (or, equivalently, replacing them with open circuits). That is, to obtain the contribution of capacitance C_i we reduce all other capacitances to zero, reduce the input signal source to zero, and determine the resistance R_{io} seen by C_i. This process is then repeated for all other capacitors in the circuit. The value of b_1 is computed by summing the individual time constants, called *open-circuit time constants*,

$$b_1 = \sum_{i=1}^{n_H} C_i R_{io} \tag{11.25}$$

where we have assumed that there are n_H capacitors in the high-frequency equivalent circuit.

This method for determining b_1 is *exact*; the approximation comes about in using the value of b_1 to determine ω_H. Specifically, if the zeros are not dominant and if one of the poles—say, P_1—is dominant, then from Eq. (11.24)

$$b_1 \simeq \frac{1}{\omega_{P_1}} \tag{11.26}$$

and the upper 3-dB frequency will be approximately equal to ω_{P1}, leading to

$$\omega_H \simeq \frac{1}{\left[\sum_i C_i R_{io} \right]} \tag{11.27}$$

Here it should be pointed out that in complex circuits we usually do not know whether or not a dominant pole exists. Nevertheless, using Eq. (11.27) to determine ω_H normally yields remarkably good results[1] even if a dominant pole does not exist. The method will be illustrated by an example.

EXAMPLE 11.5

Figure 11.8a shows the high-frequency equivalent circuit of a common-source FET amplifier. The amplifier is fed with a signal generator having a resistance R. Resistance R_{in} is due to the biasing network. Resistance R'_L is the parallel equivalent of the load resistance R_L, the drain bias resistance R_D, and the FET output resistance r_o. For $R = 100$ kΩ, $R_{in} = 420$ kΩ, $C_{gs} = C_{gd} = 1$ pF, $g_m = 4$ mA/V, and $R'_L = 3.33$ kΩ, find the midband voltage gain, $A_M = V_o/V_i$, and the upper 3-dB frequency, f_H.

Solution

The midband voltage gain is determined by assuming that the capacitors in the FET model are perfect open circuits. This results in the midband equivalent circuit shown

---

[1] The method of open-circuit time constants yields good results only when all the poles are real, as is the case in this chapter.

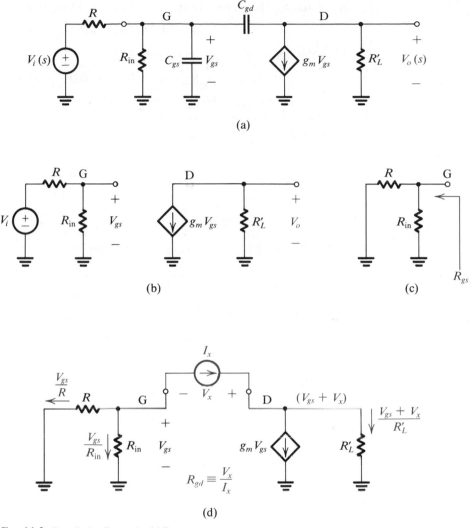

Fig. 11.8 Circuits for Example 11.5.

in Fig. 11.8b, from which we find

$$A_M \equiv \frac{V_o}{V_i} = -\frac{R_{in}}{R_{in} + R}(g_m R'_L)$$

$$= -\frac{420}{420 + 100} \times 4 \times 3.33 = -10.8 \text{ V/V}$$

We shall determine ω_H using the method of open-circuit time constants. The resistance R_{gs} seen by C_{gs} is found by setting $C_{gd} = 0$ and short-circuiting the signal generator V_i. This results in the circuit of Fig. 11.8c, from which we find that

$$R_{gs} = R_{in} \| R = 420 \text{ k}\Omega \| 100 \text{ k}\Omega = 80.8 \text{ k}\Omega$$

Thus the open-circuit time constant of C_{gs} is

$$\tau_{gs} \equiv C_{gs}R_{gs} = 1 \times 10^{-12} \times 80.8 \times 10^3 = 80.8 \text{ ns}$$

The resistance R_{gd} seen by C_{gd} is found by setting $C_{gs} = 0$ and short-circuiting V_i. The result is the circuit in Fig. 11.8d, to which we apply a test current I_x. Writing a node equation at G gives

$$I_x = -\frac{V_{gs}}{R_{in}} - \frac{V_{gs}}{R}$$

Thus,

$$V_{gs} = -I_x R' \qquad (11.28)$$

where $R' = R_{in} \| R$. A node equation at D provides

$$I_x = g_m V_{gs} + \frac{V_{gs} + V_x}{R_L'}$$

Substituting for V_{gs} from Eq. (11.28) and rearranging terms yields

$$R_{gd} \equiv \frac{V_x}{I_x} = R' + R_L' + g_m R_L' R' = 1.16 \text{ M}\Omega$$

Thus, the open-circuit time constant of C_{gd} is

$$\tau_{gd} \equiv C_{gd}R_{gd}$$
$$= 1 \times 10^{-12} \times 1.16 \times 10^6 = 1{,}160 \text{ ns}$$

The upper 3-dB frequency ω_H can now be determined from

$$\omega_H \simeq \frac{1}{\tau_{gs} + \tau_{gd}}$$

$$= \frac{1}{(80.8 + 1{,}160) \times 10^{-9}} = 806 \text{ krad/s}$$

Thus

$$f_H = \frac{\omega_H}{2\pi} = 128.3 \text{ kHz}$$

The method of open-circuit time constants has an important advantage in that it tells the circuit designer which of the various capacitances is significant in determining the amplifier frequency response. Specifically, the relative contribution of the various capacitances to the effective time constant b_1 is immediately obvious. For instance, in the above example we see that C_{gd} is the dominant capacitance in determining f_H. We also note that to effectively increase f_H either we use an FET with smaller C_{gd} or, for a given FET, we use a smaller R' or R'_L, thus reducing R_{gd}. If R' is fixed, then for a given FET the only way to increase bandwidth is by reducing the load resistance. Unfortunately, this also decreases the midband gain.

Next we outline the use of short-circuit time constants to determine the lower 3-dB frequency, ω_L. The function $F_L(s)$ of Eq. (11.12) can be expressed in the alternate form

$$F_L(s) = \frac{s^{n_L} + d_1 s^{n_L - 1} + \cdots}{s^{n_L} + e_1 s^{n_L - 1} + \cdots} \tag{11.29}$$

where the coefficients d and e are related to the zero and pole frequencies, respectively. Specifically, the coefficient e_1 is given by

$$e_1 = \omega_{P1} + \omega_{P2} + \cdots + \omega_{P n_L} \tag{11.30}$$

As shown in Gray and Searle (1969), the exact value of e_1 can be obtained by analyzing the amplifier low-frequency equivalent circuit, considering the various capacitors one at a time while setting all other capacitors to ∞ (or, equivalently, replacing them with short circuits). Thus if capacitor C_i is under consideration, we replace all other capacitors with short circuits, and also reduce the input signal to zero, and determine the resistance R_{is} seen by C_i. The process is then repeated for all other capacitors, and the value of e_1 is computed from

$$e_1 = \sum_{i=1}^{n_L} \frac{1}{C_i R_{is}} \tag{11.31}$$

where it is assumed that there are n_L capacitors in the low-frequency equivalent circuit.

The value of e_1 can be used to obtain an approximate value of the 3-dB frequency ω_L provided that none of the zeros is dominant and that a dominant pole exists. This condition is satisfied if one of the poles, say P_1, has a frequency ω_{P1} much higher than (at least four times) that of all the other poles and zeros. If this is the case then $\omega_L \simeq \omega_{P1}$ and from Eq. (11.30) we see that $e_1 \simeq \omega_{P1}$, leading to

$$\omega_L \simeq \sum_i \frac{1}{C_i R_{is}} \tag{11.32}$$

Of course, in a complex circuit it is usually difficult to ascertain whether a dominant low-frequency pole exists or not. Nevertheless, the method of short-circuit time constants usually provides a reasonable estimate of ω_L. Such an estimate is quite sufficient for an initial paper-and-pencil design. The method also allows the designer considerable insight into which of the various capacitors most severely limits the low-frequency response. These points will be further illustrated in subsequent sections.

Exercises

11.2 A first-order circuit, having a gain of 10 at dc and a gain of 1 at infinite frequency, has its pole at 10 kHz. Find its transfer function.

Ans. $\dfrac{s + 2\pi \times 10^5}{s + 2\pi \times 10^4}$

11.3 A direct-coupled amplifier has a gain of 1,000 and an upper 3-dB frequency of 100 kHz. What is its gain–bandwidth product in hertz?

Ans. 10^8 Hz

11.4 The high-frequency response of an amplifier is characterized by two zeros at $s = \infty$ and two poles at ω_{P1} and ω_{P2}. Expressing $\omega_{P2} = k\omega_{P1}$, find the value of k that results in the exact value of ω_H being $0.9\omega_{P1}$. Repeat for $\omega_H = 0.99\omega_{P1}$.

Ans.   2.78; 9.88

11.5 For the amplifier described in Exercise 11.4, find the exact and approximate values [using Eq. (11.22)] of ω_H (as a function of ω_{P1}) for the cases $k = 1, 2, 4$.

Ans.   0.64, 0.71; 0.84, 0.89; 0.95, 0.97

11.6 For the amplifier in Example 11.5, find the gain–bandwidth product in megahertz. Find the value of R'_L that will result in $f_H = 180$ kHz. Find the new values of the midband gain and of the gain–bandwidth product.

Ans.   1.39 MHz; 2.23 kΩ; -7.2 V/V; 1.30 MHz

11.3 FREQUENCY RESPONSE OF THE COMMON-SOURCE AMPLIFIER

In this section the frequency response of the classical capacitor-coupled common-source amplifier stage, shown in Fig. 11.9, will be analyzed. The analysis given applies equally well to MOSFETs. Furthermore, the limitations on the high-frequency response of direct-coupled common-source amplifiers are identical to those of the capacitively coupled circuit. Fig. 11.10 shows the equivalent circuit model that we shall use for the FET at high frequencies. Typically, C_{gs} and C_{gd} are in the range of 1 to 3 pF.

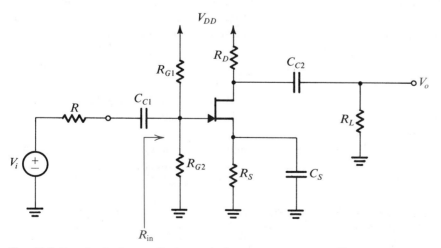

Fig. 11.9 The classical capacitively coupled common-source amplifier.

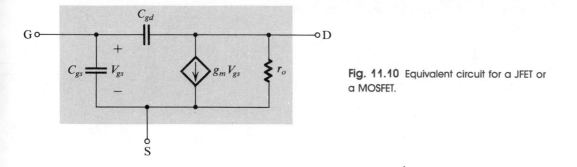

Fig. 11.10 Equivalent circuit for a JFET or a MOSFET.

Analysis in the Low-Frequency Band

To find the low-frequency gain of the circuit in Fig. 11.9, we shall start at the signal source and proceed toward the load in a step-by-step manner. Using the voltage-divider rule at the input side, we can find the voltage V_g (between gate and ground) as

$$V_g(s) = V_i(s) \frac{R_{\text{in}}}{R_{\text{in}} + R + 1/sC_{C1}}$$

where $R_{\text{in}} = (R_{G1} \parallel R_{G2})$. Thus the transfer function from the input to the gate is given by

$$\frac{V_g(s)}{V_i(s)} = \frac{R_{\text{in}}}{R_{\text{in}} + R} \frac{s}{s + 1/C_{C1}(R_{\text{in}} + R)} \tag{11.33}$$

which is a high-pass function indicating that C_{C1} introduces a zero at zero frequency (dc) and a real pole with a frequency ω_{P1},

$$\omega_{P1} = \frac{1}{C_{C1}(R_{\text{in}} + R)} \tag{11.34}$$

Note that we could have arrived at this result by inspection of the input circuit of the amplifier using the techniques of STC network analysis of Chapter 2. Specifically, the input circuit is a high-pass STC network with a time constant equal to C_{C1} multiplied by the total resistance seen by C_{C1}; ω_{P1} is simply the inverse of this time constant.

The next step in the analysis is to find the drain current $I_d(s)$:

$$I_d(s) = I_s(s) = \frac{V_g(s)}{1/g_m + Z_S} \tag{11.35}$$

where we have made use of the fact that the equivalent resistance between gate and source is equal to $1/g_m$; thus the total impedance between gate and ground, on the source side, is $1/g_m$ in series with Z_S, which denotes the parallel equivalent of R_S and

C_S. Equation (11.35) can be rewritten

$$I_d(s) = g_m V_g(s) \frac{Y_S}{g_m + Y_S}$$

where

$$Y_S = \frac{1}{Z_S} = \frac{1}{R_S} + sC_S$$

Thus

$$I_d(s) = g_m V_g(s) \frac{1/R_S + sC_S}{g_m + 1/R_S + sC_S}$$

that is,

$$I_d(s) = g_m V_g(s) \frac{s + 1/C_S R_S}{s + (g_m + 1/R_S)/C_S} \tag{11.36}$$

which indicates that the bypass capacitor C_S introduces a real zero and a real pole. The real zero has a frequency ω_Z,

$$\omega_Z = \frac{1}{C_S R_S} \tag{11.37}$$

while the frequency of the real pole is given by

$$\omega_{P2} = \frac{g_m + 1/R_S}{C_S} = \frac{1}{C_S(R_S \parallel 1/g_m)} \tag{11.38}$$

It thus can be seen that ω_Z will always be lower in value than ω_{P2}.

It is instructive to interpret physically the above results regarding the effects of C_S. C_S introduces a zero at the value of s that makes Z_S infinite, which makes physical sense because an infinite Z_S will cause I_d, and hence V_o, to be zero. The pole frequency is the inverse of the time constant formed by multiplying C_S by the resistance seen by the capacitor. To evaluate the latter resistance we ground the signal source (note that the network poles, or natural modes, are independent of the excitation) and grab hold of the terminals of C_S. The resistance seen by C_S will be R_S in parallel with the resistance between source and gate, which is $1/g_m$.

Having determined $I_d(s)$, we can now obtain the output voltage using the output equivalent circuit in Fig. 11.11a. There is a slight approximation in this equivalent circuit: The resistance r_o is shown connected between drain and ground rather than between drain and source in spite of the fact that the source terminal is no longer at ground potential because C_S is not acting as a perfect bypass. However, since the effect of r_o is small anyway, this approximation is valid.

Figure 11.11b shows the equivalent output circuit after application of Thévenin's theorem. From this figure we obtain

$$V_o(s) = -I_d(s)(R_D \parallel r_o \parallel R_L) \frac{s}{s + 1/C_{C2}[R_L + (R_D \parallel r_o)]} \tag{11.39}$$

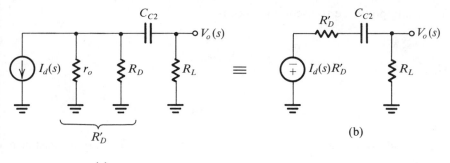

Fig. 11.11 The output equivalent circuit (at low frequency) for the amplifier in Fig. 11.9.

Thus C_{C2} introduces a zero at zero frequency (dc) and a real pole with a frequency ω_{P3},

$$\omega_{P3} = \frac{1}{C_{C2}[R_L + (R_D \| r_0)]} \tag{11.40}$$

Again the frequency of this pole could have been found by inspection: it equals the inverse of the time constant found by multiplying C_{C2} by the total resistance seen by that capacitor.

The low-frequency amplifier gain $A_L(s)$ can be found by combining Eqs. (11.33), (11.36), and (11.39):

$$A_L(s) = \frac{V_o(s)}{V_i(s)} = A_M \frac{s}{s + \omega_{P1}} \frac{s + \omega_Z}{s + \omega_{P2}} \frac{s}{s + \omega_{P3}} \tag{11.41}$$

where the midband gain A_M is given by

$$A_M = -\frac{R_{in}}{R_{in} + R} g_m(R_D \| r_0 \| R_L) \tag{11.42}$$

and where ω_{P1}, ω_Z, ω_{P2}, and ω_{P3} are given by Eqs. (11.34), (11.37), (11.38), and (11.40), respectively. Note from Eq. (11.41) that as the frequency $s = j\omega$ becomes much larger in magnitude than ω_{P1}, ω_{P2}, ω_{P3}, and ω_Z, the gain approaches the midband value A_M.

Having determined the low-frequency poles and zeros, we can employ the techniques of Section 11.2 to find the lower 3-dB frequency ω_L.

EXAMPLE 11.6

We wish to select appropriate values for the coupling capacitors C_{C1} and C_{C2} and the bypass capacitor C_S of the amplifier in Fig. 11.9 so that the low-frequency response will be dominated by a pole at 100 Hz and that the nearest pole or zero will be at least a decade away. Let $V_{DD} = 20$ V, $R = 100$ kΩ, $R_{G1} = 1.4$ MΩ, $R_{G2} = 0.6$ MΩ, $R_S = 3.5$ kΩ, $R_D = 5$ kΩ, $r_0 = \infty$, $R_L = 10$ kΩ, $V_P = -2$ V, and $I_{DSS} = 8$ mA. Also, determine the midband gain.

Solution
With the methods of Chapter 6, the following dc operating point is determined.

$$I_D = 2 \text{ mA} \qquad V_{GS} = -1 \text{ V} \qquad V_D = +10 \text{ V}$$

At this operating point the transconductance is

$$g_m = \frac{2I_{DSS}}{-V_P} \sqrt{\frac{I_D}{I_{DSS}}}$$

Thus

$$g_m = \frac{2 \times 8}{2} \sqrt{\frac{2}{8}} = 4 \text{ mA/V}$$

The midband voltage gain can be determined as follows: The input resistance R_{in} is given by

$$R_{in} = \frac{R_{G1}R_{G2}}{R_{G1} + R_{G2}} = \frac{1.4 \times 0.6}{2} = 420 \text{ k}\Omega$$

and the midband voltage gain A_M can be written

$$A_M = \frac{R_{in}}{R_{in} + R} \times -g_m(R_D \| R_L)$$

$$= -\frac{420}{520} \times 4 \times \frac{5 \times 10}{5 + 10} = -10.8 \text{ V/V}$$

Thus the amplifier has a midband gain of 20.7 dB.

To find which of the three capacitors, C_{C1}, C_S, and C_{C2} should be made to cause the dominant low-frequency pole at 100 Hz, we first determine the resistance associated with each, as follows:

$$R_{C_{C1}} = R + R_{in} = 520 \text{ k}\Omega$$

$$R_{C_S} = R_S \| (1/g_m) = 0.233 \text{ k}\Omega$$

$$R_{C_{C2}} = R_D + R_L = 15 \text{ k}\Omega$$

We select the smallest resistance, R_{C_S}, as the one to form the highest-frequency (and thus, dominant) pole. Thus,

$$C_S = \frac{1}{2\pi f_L R_{C_S}}$$

$$= \frac{1}{2\pi \times 100 \times 0.233 \times 10^3} = 6.83 \ \mu\text{F}$$

The zero due to C_S can be found, using Eq. (11.37), as

$$f_Z = \frac{1}{2\pi C_S R_S} = \frac{1}{2\pi \times 6.83 \times 10^{-6} \times 3.5 \times 10^3} = 6.7 \text{ Hz}$$

To place the two other poles, due to C_{C1} and C_{C2}, at least a decade away from f_L (that is, equal to or lower than 10 Hz) we select the capacitors as follows:

$$C_{C1} \geq \frac{1}{2\pi \times 10 \times 520 \times 10^3} = 0.03 \ \mu\text{F}$$

and

$$C_{C2} \geq \frac{1}{2\pi \times 10 \times 15 \times 10^3} = 1.06 \ \mu F$$

Note that selecting the smallest resistance to cause the highest-frequency pole results in reasonably small values for the two other capacitors associated with the non-dominant poles.

Exercises

11.7 A common-source FET amplifier utilizes a resistance R_S of 1 kΩ bypassed by a capacitor C_S. It is found that the pole and zero due to C_S are at 100 rad/s and 10 rad/s, respectively. Find the values of C_S and g_m.

Ans.   100 μF; 9 mA/V

11.8 If, in the common-source FET amplifier, R_S is replaced with an ideal constant-current source, find the frequencies ω_Z and ω_P caused by C_S.

Ans. $\omega_Z = 0$; $\omega_P = g_m/C_S$

Analysis in the High-Frequency Band

The high-frequency equivalent circuit of the common-source amplifier is shown in Fig. 11.12a, where R_{in} denotes $(R_{G1} \| R_{G2})$. A simplified version of the equivalent circuit, as shown in Fig. 11.12b, is obtained by applying Thévenin's theorem at the input side and by combining the three resistances r_o, R_D, and R_L.

There are a number of ways to analyze the high-frequency equivalent circuit of Fig. 11.12b to determine the upper 3-dB frequency ω_H. One approach is to use the method of open-circuit time constants. We have in fact already done this in Example 11.5. Another approximate method that yields considerable insight into high-frequency limitations involves the application of Miller's theorem (Section 2.5) to replace C_{gd} by an equivalent input capacitance between the gate and ground. This method is based on the observation that since C_{gd} is small, the current through it will be much smaller than that of the controlled source $g_m V_{gs}$. Thus, neglecting the current through C_{gd} in determining the output voltage V_o, we can write

$$V_o \simeq -g_m V_{gs} R'_L \tag{11.43}$$

Using the ratio of the voltages at the two sides of C_{gd} enables us to replace C_{gd} at the input (gate) side with the equivalent Miller capacitance

$$C_{eq} = C_{gd}(1 + g_m R'_L) \tag{11.44}$$

as shown in Fig. 11.12c. We recognize the circuit at the input side as that of a first-order low-pass filter whose time constant is determined by the total input capacitance

$$C_T = C_{gs} + C_{gd}(1 + g_m R'_L) \tag{11.45}$$

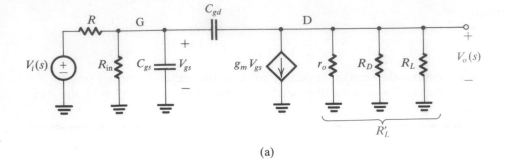

(a)

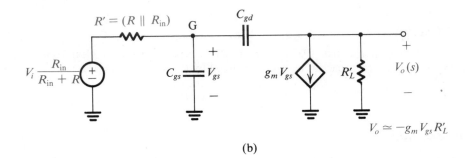

(b)

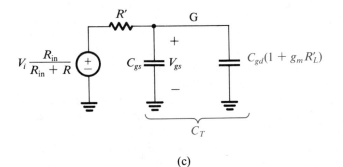

(c)

Fig. 11.12 Equivalent circuits for evaluating the high-frequency response of the amplifier of Fig. 11.9.

and the effective generator resistance

$$R' = R \parallel R_{\text{in}} \tag{11.46}$$

This first-order circuit determines the high-frequency response of the common-source amplifier, introducing a dominant high-frequency pole. Thus the upper 3-dB frequency will be

$$\omega_H = \frac{1}{C_T R'} \tag{11.47}$$

We may thus express the high-frequency gain as

$$A_H(s) = A_M \frac{1}{1 + s/\omega_H} \qquad (11.48)$$

where, as before, A_M denotes the midband gain given in Eq. (11.42).

From the above, we note the important role played by the small feedback capacitance C_{gd} in determining the high frequency response of the common-source amplifier. Because the voltages at the two sides of C_{gd} are in the ratio of $-g_m R'_L$, which is a large number approximately equal to the midband gain, C_{gd} gives rise to a large capacitance, $C_{gd}(1 + g_m R'_L)$, across the input terminals of the amplifier. This is the Miller effect, discussed in Section 2.5. It follows that to increase the upper 3-dB or cutoff frequency of the amplifier, one has to reduce either $g_m R'_L$, which reduces the midband gain, or reduce the source resistance, which might not always be possible. Alternatively, one can use circuit configurations that do not suffer from the Miller effect, such as the cascode circuit introduced in Section 7.5 for MOSFETs and in Section 8.9 for BJTs. This and other special configurations for *wideband amplifiers* will be studied in later sections.

EXAMPLE 11.7

Use the approximate method based on the Miller effect to find the upper 3-dB frequency of the common-source amplifier whose component values are specified in Example 11.6. Let $C_{gs} = C_{gd} = 1$ pF. Compare the result with that obtained, for the same amplifier, in Example 11.5 using the method of open-circuit time constants.

Solution

The total input capacitance is obtained from Eq. (11.45) as follows:

$$C_T = 1 + 1 \times (1 + 4 \times 3.33) = 15.3 \text{ pF}$$

The effective generator resistance is obtained from Eq. (11.46):

$$R' = 100 \text{ k}\Omega \parallel 420 \text{ k}\Omega = 80.8 \text{ k}\Omega$$

Thus, using Eq. (11.47), we obtain f_H as follows:

$$f_H = \frac{\omega_H}{2\pi} = \frac{1}{2\pi \times 15.3 \times 10^{-12} \times 80.8 \times 10^3} = 128.7 \text{ kHz}$$

which is very close to the value (128.3 kHz) obtained in Example 11.5.

The approximation involved in the above method for determining ω_H is equivalent to assuming that a dominant high-frequency pole exists. To verify that this indeed is the case we shall derive the exact high-frequency transfer function of the common-source amplifier. Converting the input-signal generator to the Norton's form results in the circuit shown in Fig. 11.13. Writing a node equation at G yields

$$\frac{V_i(s)}{R} = \frac{V_{gs}}{R'} + sC_{gs}V_{gs} + sC_{gd}(V_{gs} - V_o) \qquad (11.49)$$

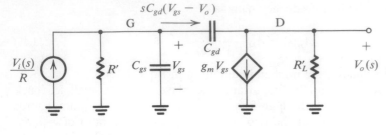

Fig. 11.13 Determination of the exact high-frequency transfer function of the common-source amplifier.

Writing a node equation at D gives

$$sC_{gd}(V_{gs} - V_o) = g_m V_{gs} + \frac{V_o(s)}{R'_L} \tag{11.50}$$

Eliminating V_{gs} from Eqs. (11.49) and (11.50) gives the transfer function

$$\frac{V_o(s)}{V_i(s)} = -A_M \frac{1 - \dfrac{s}{(g_m/C_{gd})}}{1 + s[C_{gs} + C_{gd}(1 + g_m R'_L) + C_{gd}(R'_L/R')]R' + s^2 C_{gs} C_{gd} R' R'_L} \tag{11.51}$$

Thus the amplifier has a zero with frequency $\omega_Z = g_m/C_{gd}$ and two poles whose frequencies can be determined from the denominator polynomial. Note that the coefficient of the s term in the denominator is, as expected, equal to the value derived in Example (11.5) using open-circuit time constants. Apart from this observation, the denominator polynomial is unfortunately too complex to draw useful information from. We can, however, substitute numerical values and obtain the frequencies of the poles, as is requested in the following exercise.

Exercise

11.9 For the common-source amplifier specified in Examples 11.6 and 11.7 use Eq. (11.51) to determine the frequencies of its finite zero and two poles.

Ans. $f_Z = 637$ MHz; $f_{P1} = 128.2$ kHz; $f_{P2} = 734$ MHz

The answers to Exercise 11.9 show that the zero and second pole are indeed at much higher frequencies than the dominant pole. The fact that the two poles are so widely separated enables us to factor the denominator of Eq. (11.51) as shown below. The denominator polynomial $D(s)$ can be written as

$$D(s) = \left(1 + \frac{s}{\omega_{P1}}\right)\left(1 + \frac{s}{\omega_{P2}}\right)$$

$$= 1 + s\left(\frac{1}{\omega_{P1}} + \frac{1}{\omega_{P2}}\right) + \frac{s^2}{\omega_{P1}\omega_{P2}}$$

$$\simeq 1 + \frac{s}{\omega_{P1}} + \frac{s^2}{\omega_{P1}\omega_{P2}} \tag{11.52}$$

Equating the coefficients of the s terms in Eqs. (11.51) and (11.52) gives

$$\omega_{P1} = \frac{1}{[C_{gs} + C_{gd}(1 + g_m R'_L) + C_{gd}(R'_L/R')]R'} \tag{11.53}$$

which is slightly different from the value obtained using the Miller effect but identical to the value of ω_H obtained using open-circuit time constants. Equating the coefficients of s^2 in Eqs. (11.51) and (11.52) and using (11.53) gives the frequency of the second pole

$$\omega_{P2} = \frac{C_{gs} + C_{gd}(1 + g_m R'_L) + C_{gd}(R'_L/R')}{C_{gs} C_{gd} R'_L} \tag{11.54}$$

For $g_m R'_L \gg 1$ and $R'_L < R'$, this expression can be approximated as

$$\omega_{P2} \simeq \frac{g_m}{C_{gs}} \tag{11.55}$$

which shows that ω_{P2} will usually be very high.

EXAMPLE 11.8

If the amplifier analyzed in Examples 11.5 to 11.7 is excited by a pulse of 0.2-V height and 0.2-ms width, find the height, the rise and fall times, and the sag of the output pulse.

Solution

The output pulse will have a height V,

$$V = A_M \times 0.2 = 10.8 \times 0.2 = 2.16 \text{ V}$$

The rise and fall times of the output pulse will be determined by the dominant high-frequency pole (see Chapter 2),

$$t_r = t_f \simeq 2.2\tau_H$$

where

$$\tau_H = \frac{1}{\omega_H} = \frac{1}{2\pi f_H} = \frac{1}{2\pi \times 128.7 \times 10^3} \simeq 1.2 \text{ }\mu s$$

Thus

$$t_r = t_f = 2.64 \text{ }\mu s$$

On the other hand, the sag or decay in the amplitude of the output pulse will be determined by the dominant low-frequency pole (see Chapter 2). Specifically, if the loss in pulse height is denoted by ΔV, then

$$\text{sag} = \frac{\Delta V}{V} = \frac{t_p}{\tau_L} = \omega_L t_p = 2\pi f_L t_p$$

Thus

$$\text{sag} = 2\pi \times 100 \times 0.2 \times 10^{-3} = 0.126 = 12.6\%$$

11.4 THE HYBRID-π EQUIVALENT CIRCUIT MODEL

Before considering the frequency response analysis of BJT amplifiers, we shall take a closer look at the hybrid-π model and give methods for the determination of the model parameters from terminal measurements or from data-sheet specifications

The Low-Frequency Model

Figure 11.14 shows the complete low-frequency hybrid-π equivalent circuit model. In addition to the intrinsic model parameter r_π and g_m, this model includes three resistances r_o, r_μ, and r_x.

Resistance r_o models the slight effect of the collector voltage on the collector current in the active region of operation. Typically r_o is in the range of tens to hundreds of kΩ and its value is inversely proportional to the dc bias current ($r_o = V_A/I_C$). Since g_m is directly proportional to the bias current, the product $g_m r_o$, denoted μ, where $\mu = V_A/V_T$, is a constant for a given transistor, with a value of a few thousand.

The resistance r_μ models the effect of the collector voltage on the base current, and its value is usually much larger than r_o. It can be shown from physical considerations of device operation that r_μ is at least equal to $\beta_0 r_o$ (β_0 denotes the value of β at low frequencies). For modern IC transistors r_μ is closer to $10\beta_0 r_o$. Because of its extremely large value and because including r_μ in the equivalent circuit model destroys its *unilateral* character and thus complicates the analysis, one usually ignores r_μ. There are very special situations, however, where one has to include r_μ. We have already encountered one such case (in evaluating the common-mode input resistance of a differential amplifier).

The resistance r_x models the resistance of the silicon material of the base region between the base terminal B and a fictitious internal, or intrinsic, base terminal B'. The latter node represents the base side of the emitter–base junction. Typically, r_x is a few tens of ohms, and its value depends on the current level in a rather complicated manner. Since r_x is much smaller than r_π, the effect of r_x is negligible at low frequencies. Its presence is felt, however, at high frequencies, where the input impedance of the transistor becomes highly capacitive. This point will become apparent later in this section. In conclusion, r_x can usually be neglected in low-frequency applications.

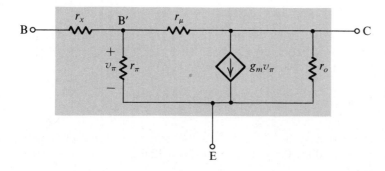

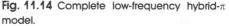

Fig. 11.14 Complete low-frequency hybrid-π model.

Exercise

11.10 Each of the common-emitter i_C-v_{CE} curves is measured with the value of base current held constant. This is equivalent to assuming that the signal current $i_b = 0$; that is, from a signal point of view, the base is open-circuited. Use the equivalent circuit in Fig. E11.10 to determine the slope of the i_C-v_{CE} curves in the active region.

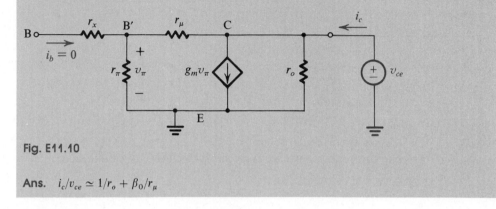

Fig. E11.10

Ans. $i_c/v_{ce} \simeq 1/r_o + \beta_0/r_\mu$

Determination of the Low-Frequency Model Parameters

The transistor is a three-terminal device that can be converted into a two-port network by grounding one of its terminals. It therefore can be characterized by one of the various two-port parameter sets (see Appendix B). For the BJT at low frequencies, the *h* parameters have been found to be the most convenient. In the following we briefly discuss methods for measuring the *h* parameters for a transistor biased to operate in the active region. We also derive formulas relating the hybrid-π model parameters to the measured *h* parameters. These formulas allow us to determine the hybrid-π parameters. It should be emphasized that because the hybrid-π model is closely related to the physical operation of the transistor, its use provides the circuit designer with considerable insight into circuit operation. Thus our interest in *h* parameters is solely for the purpose of determining the hybrid-π component values.

If the emitter of a transistor biased in the active mode is grounded, port 1 is defined to be between base and emitter, and port 2 is defined to be between collector and emitter, then for small signals around the given bias point we can write

$$v_b = h_{ie}i_b + h_{re}v_c \tag{11.56}$$

$$i_c = h_{fe}i_b + h_{oe}v_c \tag{11.57}$$

These are the defining equations of the common-emitter *h* parameters where, rather than using the notation h_{11}, h_{12}, and so on, we have assigned more descriptive subscripts to the *h* parameters: *i* means input, *r* means reverse, *f* means forward, *o* means output, and the added *e* denotes a common emitter.

For measurement of h_{ie} and h_{fe}, the circuit shown in Fig. 11.15 can be used. Here R_B is a large resistance that together with V_{BB} determines I_B. The resistance R_C is used to establish the desired dc voltage at the collector, and a very small resistance R_L is used to enable measuring the signal current in the collector. Since R_L is small,

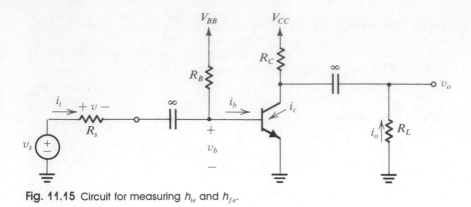

Fig. 11.15 Circuit for measuring h_{ie} and h_{fe}.

the collector is effectively short-circuited to ground and

$$i_c \simeq i_o = -\frac{v_o}{R_L}$$

The input signal current i_i is determined by measuring the voltage v across a known resistance R_s. If R_B is large, then

$$i_b \simeq i_i = \frac{v}{R_s}$$

The input signal voltage v_b can be measured directly at the base. Using these measured values one can compute h_{ie} and h_{fe}:

$$h_{ie} = \frac{v_b}{i_b} \qquad \text{and} \qquad h_{fe} = \frac{i_c}{i_b}$$

To measure h_{re} we use the circuit shown in Fig. 11.16. Here again R_B should be large (much larger than r_π), and the voltmeter used to measure v_b should have a high input resistance to ensure that the base is effectively open-circuited. The value of h_{re}

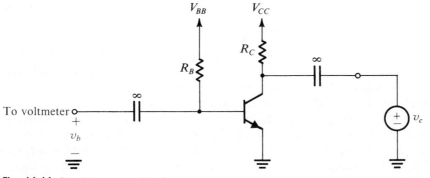

Fig. 11.16 Circuit for measuring h_{re}.

can then be determined, as follows:

$$h_{re} = \frac{v_b}{v_c}$$

Finally, from Eq. (11.57) we note that h_{oe} is the output conductance with the base open-circuited, which by definition is the slope of the i_C-v_{CE} characteristic curves. Thus h_{oe} can be most conveniently determined from the static common-emitter characteristics.

Expressions for h_{ie} and h_{fe} in terms of the hybrid-π model parameters can be derived by analyzing the equivalent circuit in Fig. 11.17. This analysis yields

$$h_{ie} = r_x + (r_\pi \| r_\mu) \tag{11.58}$$

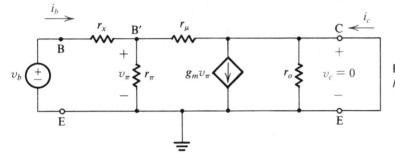

Fig. 11.17 Circuit for deriving expressions for h_{ie} and h_{fe}.

which can be approximated as

$$h_{ie} \simeq r_x + r_\pi \tag{11.59}$$

and

$$h_{fe} = g_m r_\pi \tag{11.60}$$

Here we should note that, by definition, h_{fe} is identical to the ac β, or β_{ac}, and that the value given by Eq. (11.60) indeed corresponds to the formula we used earlier for the low-frequency β, or β_0.

To derive expressions for h_{re} and h_{oe} we use the equivalent circuit model in Fig. 11.18 and obtain

$$h_{re} = \frac{r_\pi}{r_\pi + r_\mu}$$

which can be approximated as

$$h_{re} \simeq \frac{r_\pi}{r_\mu} \tag{11.61}$$

$$h_{oe} \simeq \frac{1}{r_o} + \frac{\beta_0}{r_\mu} \tag{11.62}$$

This last expression is identical to that given for the slope of the i_C-v_{CE} characteristics with the base open-circuited (Exercise 11.10).

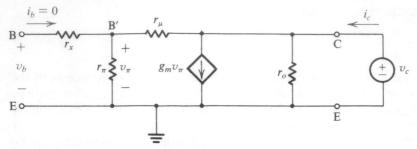

Fig. 11.18 Circuit for deriving expressions for h_{re} and h_{oe}.

The above expressions can be used to determine the values of the hybrid-π model parameters from the measured h parameters as follows:

$$g_m = \frac{I_C}{V_T} \tag{11.63}$$

$$r_\pi = \frac{h_{fe}}{g_m} \tag{11.64}$$

$$r_x = h_{ie} - r_\pi \tag{11.65}$$

$$r_\mu = \frac{r_\pi}{h_{re}} \tag{11.66}$$

$$r_o = \left(h_{oe} - \frac{h_{fe}}{r_\mu} \right)^{-1} \tag{11.67}$$

$$= \frac{V_A}{I_C} \tag{11.68}$$

It should be noted, however, that since normally $r_x \ll r_\pi$, Eq. (11.65) does not provide an accurate determination of r_x. In fact, there is no accurate way to determine r_x at low frequencies, which should come as no surprise since r_x plays a minor role at low frequencies.

Exercise

11.11 The following parameters were measured on a transistor biased at $I_C = 1$ mA: $h_{ie} = 2.6$ kΩ, $h_{fe} = 100$, $h_{re} = 0.5 \times 10^{-4}$, $h_{oe} = 1.2 \times 10^{-5}$ A/V. Determine the values of g_m, r_π, r_x, r_μ, and r_o, and V_A.

Ans. 40 mA/V; 2.5 kΩ; 100 Ω; 50 MΩ; 100 kΩ; 100 V

The High-Frequency Model

With the exception of the extremely large resistance r_μ, the high-frequency hybrid-π model shown in Fig. 11.19 includes all the resistances of the low-frequency model as well as two capacitances: the emitter–base capacitance C_π and the collector–base capacitance C_μ. The resistance r_μ is omitted because, even at moderate frequencies, the reactance of C_μ is much smaller than r_μ. As mentioned in Section 8.7, the

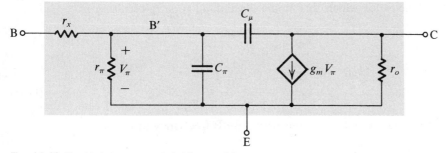

Fig. 11.19 The high-frequency hybrid-π model.

emitter–base capacitance C_π is composed of two parts: a diffusion capacitance, which is proportional to the dc bias current, and a depletion-layer capacitance, which depends on the value of V_{BE}. The collector–base capacitance C_μ is entirely a depletion capacitance, and its value depends on V_{CB}. Typically C_π is in the range of a few picofarads to a few tens of picofarads, and C_μ is in the range of a fraction of a picofarad to a few picofarads.

The Cutoff Frequency

The transistor data sheets do not usually specify the value of C_π. Rather, the behavior of h_{fe} versus frequency is normally given. In order to determine C_π and C_μ we shall derive an expression for h_{fe} as a function of frequency in terms of the hybrid-π components. For this purpose consider the circuit shown in Fig. 11.20, in which the collector is shorted to the emitter. The short-circuit collector current I_c is

$$I_c = (g_m - sC_\mu)V_\pi \qquad (11.69)$$

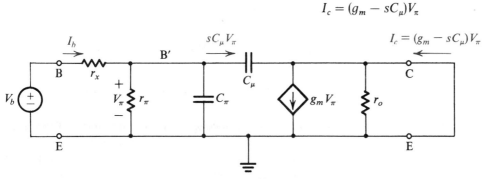

Fig. 11.20 Circuit for deriving an expression for $h_{fe}(s) \equiv I_c/I_b$.

A relationship between V_π and I_b can be established by multiplying I_b by the impedance seen between B′ and E:

$$V_\pi = I_b(r_\pi \parallel C_\pi \parallel C_\mu) \qquad (11.70)$$

Thus h_{fe} can be obtained by combining Eqs. (11.69) and (11.70):

$$h_{fe} \equiv \frac{I_c}{I_b} = \frac{g_m - sC_\mu}{1/r_\pi + s(C_\pi + C_\mu)}$$

At the frequencies for which this model is valid, $g_m \gg \omega C_\mu$, resulting in

$$h_{fe} \simeq \frac{g_m r_\pi}{1 + s(C_\pi + C_\mu)r_\pi}$$

Thus

$$h_{fe} = \frac{\beta_0}{1 + s(C_\pi + C_\mu)r_\pi} \tag{11.71}$$

Thus h_{fe} has a single-pole response with a 3-dB frequency at $\omega = \omega_\beta$,

$$\omega_\beta = \frac{1}{(C_\pi + C_\mu)r_\pi} \tag{11.72}$$

Figure 11.21 shows a Bode plot for $|h_{fe}|$. From the -6 dB/octave slope it follows that the frequency at which $|h_{fe}|$ drops to unity, which is called the *unity-gain bandwidth* ω_T, is given by

$$\omega_T = \beta_0 \omega_\beta \tag{11.73}$$

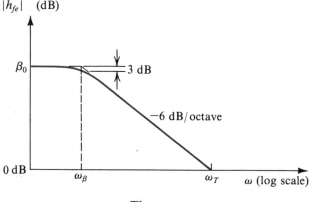

Fig. 11.21 Bode plot for $|h_{fe}|$.

Thus

$$\omega_T = \frac{g_m}{C_\pi + C_\mu} \tag{11.74}$$

The unity-gain bandwidth f_T is usually specified on the data sheets of the transistor. In some cases f_T is given as a function of I_C and V_{CE}. To see how f_T changes with I_C, recall that g_m is directly proportional to I_C, but only part of C_π (the diffusion capacitance) is directly proportional to I_C. It follows that f_T decreases at low currents, as shown in Fig. 11.22. However, the decrease in f_T at high currents, also shown in Fig. 11.22, cannot be explained by this argument; rather it is due to the same phenomenon that causes β_0 to decrease at high currents. In the region where f_T is almost constant, C_π is dominated by the diffusion part.

Typically, f_T is in the range of 100 MHz to a few GHz, with 400 MHz being a common figure for IC *npn* transistors operating at normal current levels. The value of f_T can be used in Eq. (11.74) to determine $C_\pi + C_\mu$. The capacitance C_μ is usually determined separately by measuring the capacitance between base and collector at the desired reverse-bias voltage V_{CB}.

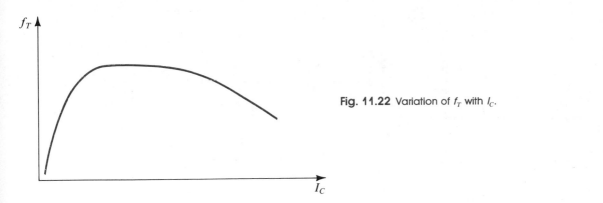

Fig. 11.22 Variation of f_T with I_C.

Before leaving this section we should mention that the hybrid-π model of Fig. 11.19 characterizes transistor operation fairly accurately up to a frequency of about $0.2\omega_T$. At higher frequencies one has to add other parasitic elements to the model as well as refine the model to account for the fact that the transistor is in fact a distributed-parameter network that we are trying to model with a lumped-component circuit. One such refinement consists of splitting r_x into a number of parts and replacing C_μ by a number of capacitors each connected between the collector and one of the taps of r_x. This topic is beyond the scope of this book.

An important observation to make from the high-frequency model of Fig. 11.19 is that at frequencies above $5\omega_\beta$ or $10\omega_\beta$ one may ignore the resistance r_π. It can be seen then that r_x becomes the only resistive part of the input impedance at high frequencies. Thus r_x plays an important role in determining the frequency response of transistors at high frequencies. It follows that an accurate determination of r_x should be made from a high-frequency measurement.

Exercises

11.12 For the same transistor considered in Exercise 11.11 and at the same bias point, determine f_T and C_π if $C_\mu = 2$ pF and $|h_{fe}| = 10$ at 50 MHz.

Ans.   500 MHz; 10.7 pF

11.13 If C_π of the BJT in Exercise 11.12 includes a relatively constant depletion-layer capacitance of 2 pF, find f_T of the BJT when operated at $I_C = 0.1$ mA.

Ans.   130.7 MHz

11.5 FREQUENCY RESPONSE OF THE COMMON-EMITTER AMPLIFIER

Analysis of the frequency response of the classical common-emitter amplifier stage in Fig. 11.23 follows a procedure identical to that used for the FET common-source amplifier in Section 11.3. Thus the overall gain can be written in the form

$$A(s) = A_M \frac{1}{1 + s/\omega_H} \frac{s^2(s + \omega_Z)}{(s + \omega_{P1})(s + \omega_{P2})(s + \omega_{P3})} \tag{11.75}$$

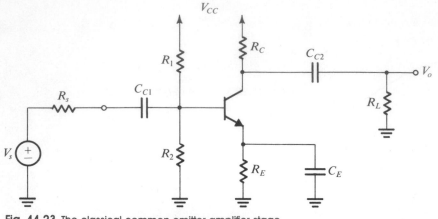

Fig. 11.23 The classical common-emitter amplifier stage.

where the midband gain A_M is evaluated by ignoring all capacitive effects, and where ω_H is the frequency of the dominant high-frequency pole obtained in a manner identical to that used in the FET case, and ω_Z, ω_{P1}, ω_{P2}, and ω_{P3} are the zero and three poles introduced in the low-frequency band by the coupling and the bypass capacitors. Because of the finite input resistance of the BJT, determination of the low-frequency singularities is more complicated than it is for the FET case. This can be seen from the low-frequency equivalent circuit shown in Fig. 11.24. Although we can certainly analyze this circuit and determine its transfer function and hence the poles and zeros, the expressions derived will be too complicated to yield useful insights. Rather, we will make use of the method of short-circuit time constants, described in Section 11.2, to obtain an estimate of the lower 3-dB frequency ω_L.

The determination of ω_L proceeds as follows: First, we set C_E and C_{C2} to infinity and find the resistance R_{C1} seen by C_{C1}. From the equivalent circuit in Fig. 11.24, with C_E set to ∞, we find

$$R_{C1} = R_s + [R_B \| (r_x + r_\pi)] \tag{11.76}$$

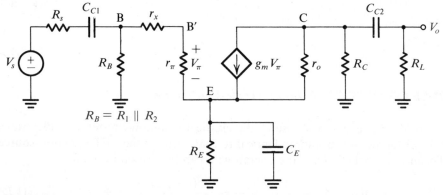

Fig. 11.24 Equivalent circuit for the amplifier of Fig. 11.23 in the low-frequency band.

Next, we set C_{C1} and C_{C2} to ∞ and determine the resistance R'_E seen by C_E. Again from the equivalent circuit in Fig. 11.24, or simply using the rule for reflecting resistances from the base to the emitter circuit, we obtain

$$R'_E = R_E \left\| \frac{r_\pi + r_x + (R_B \| R_s)}{\beta_0 + 1} \right. \tag{11.77}$$

Finally, we set both C_{C1} and C_E to infinity and obtain the resistance seen by C_{C2}:

$$R_{C2} = R_L + (R_C \| r_o) \tag{11.78}$$

An approximate value for the lower 3-dB frequency can now be determined from

$$\omega_L \simeq \frac{1}{C_{C1}R_{C1}} + \frac{1}{C_E R'_E} + \frac{1}{C_{C2}R_{C2}} \tag{11.79}$$

At this point we should note that the zero introduced by C_E is at the value of s that makes $Z_E = 1/(1/R_E + sC_E)$ infinite,

$$s_Z = -\frac{1}{C_E R_E} \tag{11.80}$$

The frequency of the zero is usually much lower than ω_L, justifying the approximation involved in using the method of short-circuit time constants.

Given a desired value for ω_L, Eq. (11.79) can be used in design as follows: Since R'_E is usually the smallest of the three resistances R_{C1}, R'_E, and R_{C2}, we select a value for C_E so that $(1/C_E R'_E)$ is the dominant term on the right-hand side of Eq. (11.79), say $1/C_E R'_E = 0.8\,\omega_L$. This is equivalent to making C_E form the dominant low-frequency pole; in other words, at $\omega = \omega_L$ the two other capacitors will have small reactances and will thus be playing a minor role. The remaining 20% of ω_L is then split equally between the two other terms in Eq. (11.79). Finally, practical values for the three capacitors are used so that the realized ω_L is equal to or smaller than the specified value.

All the equations derived in the previous section for the high-frequency response of the FET amplifier apply equally well to the BJT amplifier by simply changing symbols (that is, C_{gs} is replaced with C_π, C_{gd} with C_μ, and so on). Thus the comments made, and the conclusions reached, in the previous section apply here as well. Finally, if we assume that the common-emitter amplifier is indeed properly characterized by a dominant low-frequency pole, Eq. (11.75) can be approximated by

$$A(s) \simeq A_M \frac{1}{1 + s/\omega_H} \frac{s}{s + \omega_L} \tag{11.81}$$

Exercises

The following exercises pertain to the common-emitter amplifier in Fig. 11.23 with $R_s = 4$ kΩ, $R_1 = 8$ kΩ, $R_2 = 4$ kΩ, $R_E = 3.3$ kΩ, $R_C = 6$ kΩ, $R_L = 4$ kΩ, and $V_{CC} = 12$ V. The dc emitter current can be shown to be $I_E \simeq 1$ mA. At this current the transistor has $\beta_0 = 100$, $C_\pi = 13.9$ pF, $C_\mu = 2$ pF, $r_o = 100$ kΩ, and $r_x = 50$ Ω.

11.14 Find the midband gain.

Ans. $A_M = -22.5$ V/V

11.15 Find R_{C1}, R'_E, and R_{C2}, and hence f_L, for the case $C_{C1} = C_{C2} = 1\ \mu F$ and $C_E = 10\ \mu F$. Also find the frequency of the zero.

Ans.   5.3 kΩ; 40.5 Ω; 9.66 kΩ; 439.5 Hz; 4.8 Hz

11.16 It is required to change the values of C_{C1}, C_{C2}, and C_E so that f_L becomes 100 Hz. Use the design procedure described in the text to obtain the new capacitor values.

Ans. $C_{C1} = 3\ \mu F$; $C_E = 49.1\ \mu F$; $C_{C2} = 1.65\ \mu F$

11.17 Use the Miller-effect method to determine the total input capacitance and hence the dominant high-frequency pole.

Ans.   203.1 pF; 788 kHz.

11.18 Use Eq. (11.53), with the symbols replaced with those for the BJT, to determine a better estimate of the dominant pole. Also, use Eq. (11.54) to determine the second pole.

Ans.   770 kHz; 508.5 MHz

11.19 Consider the BJT amplifier with midband gain of -22.5 V/V, a lower 3-dB frequency of 439.5 Hz and an upper 3-dB frequency of 770 kHz. If a negative input pulse of 10-mV height and 10-μs width is applied at the input, find the rise time, height, and sag of the output pulse waveform.

Ans.   0.45 μs; 225 mV; 6.3 mV

11.6 THE COMMON-BASE AND CASCODE CONFIGURATIONS

In the previous sections it was shown that the high-frequency response of the common-source amplifier and the common-emitter amplifier is limited by the Miller effect introduced by the feedback capacitance (C_{gd} in the FET and C_μ in the BJT). It follows that to extend the upper frequency limit of a transistor amplifier stage one has to reduce or eliminate the Miller capacitance multiplication. In the following we shall show that this can be achieved in the common-base amplifier configuration. An almost identical analysis can be applied to the common-gate configuration.

We shall also analyze the frequency response of the cascode configuration, and show that it combines the advantages of the common-emitter and the common-base circuits (the common-source and the common-gate circuits in the FET case). To be general and to show the parallels with the common-emitter amplifier, we shall present the circuits in their capacitively coupled form. The high-frequency analysis, however, applies directly to direct-coupled circuits.

Analysis of the Common-Base Amplifier

Figure 11.25 shows a common-base amplifier stage of the capacitively coupled type. Recall that such a configuration was analyzed at midband frequencies (that is, with all capacitive effects neglected) in Chapter 8. In the following we shall be interested specifically in the high-frequency analysis; the low-frequency response can be determined using techniques similar to those of the previous sections.

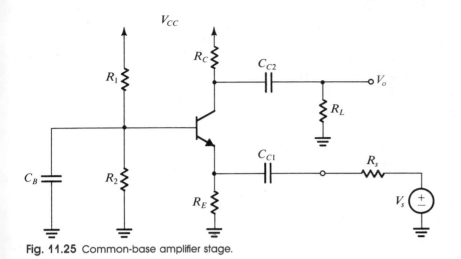

Fig. 11.25 Common-base amplifier stage.

The high-frequency equivalent circuit of the common-base amplifier is shown in Fig. 11.26a. To simplify matters and to focus attention on the special features of the common-base circuit, r_o and r_x have been omitted.

In the circuit of Fig. 11.26a we observe that the voltage at the emitter terminal V_e is equal to $-V_\pi$. We can write a node equation at the emitter terminal that enables us to express the emitter current I_e as

$$I_e = -V_\pi\left(\frac{1}{r_\pi} + sC_\pi\right) - g_m V_\pi = V_e\left(\frac{1}{r_\pi} + g_m + sC_\pi\right)$$

Thus the input admittance looking into the emitter is

$$\frac{I_e}{V_e} = \frac{1}{r_\pi} + g_m + sC_\pi = \frac{1}{r_e} + sC_\pi \tag{11.82}$$

Therefore at the input of the circuit we may replace the transistor by this input admittance, as shown in Fig. 11.26b.

At the output side (Fig. 11.26a) we see that V_o is determined by the current source $g_m V_\pi$ feeding $(R_C \parallel R_L \parallel C_\mu)$. This observation is used in drawing the output part in the simplified equivalent circuit shown in Fig. 11.26b.

The simplified equivalent circuit of Fig. 11.26b clearly shows the most important feature of the common-base configuration: the absence of an internal feedback capacitance. Unlike the common-emitter circuit, here C_μ has one terminal grounded, and no Miller effect is present. We therefore expect that the upper cutoff frequency will be much higher than that of the common-emitter configuration.

The high-frequency poles can be directly determined from the equivalent circuit of Fig. 11.26b. At the input side we have a pole whose frequency ω_{P1} can be written by inspection as

$$\omega_{P1} = \frac{1}{C_\pi(r_e \parallel R_E \parallel R_s)} \tag{11.83}$$

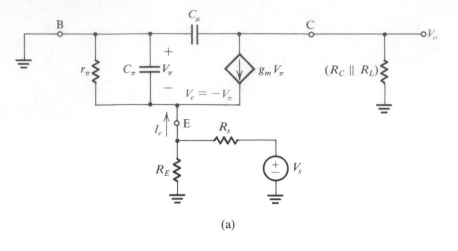

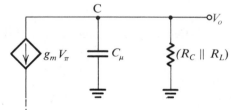

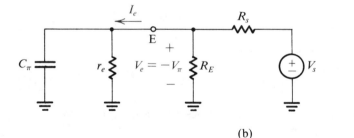

(b)

Fig. 11.26 (a) Equivalent circuit of the common-base amplifier in Fig. 11.25; **(b)** simplified version of the circuit in (a).

Since r_e is usually very small, the frequency ω_{P1} will be quite high. At the output side there is a pole with frequency ω_{P2} given by

$$\omega_{P2} = \frac{1}{C_\mu(R_C \parallel R_L)} \tag{11.84}$$

Since C_μ is quite small, ω_{P2} also will be quite high.

The question now arises as to the accuracy of the above analysis. Since we are dealing with poles at very high frequencies, we should take into account effects normally thought to be negligible. For instance, the parasitic capacitance usually present between collector and substrate (ground) in an IC transistor will obviously have a considerable effect on the value of ω_{P2}. Also, it is not clear that r_x could be neglected. It follows that for the accurate determination of the high-frequency response of a common-base amplifier a more elaborate transistor model should be used, and one normally employs a computer circuit-analysis program (Appendix C). Nevertheless, the point that we wish to emphasize here is that the common-base amplifier has a much higher upper cutoff frequency than that of the common-emitter circuit.

The Cascode Configuration

As discussed in Section 8.8 (and in Section 7.5 for MOSFETs), the cascode configuration combines the advantages of the common-emitter and the common-base circuits. Figure 11.27 shows a capacitively coupled cascode amplifier designed using bipolar transistors. The following analysis applies equally well to FET cascode circuits and to mixed circuits (that is, a cascode in which Q_1 is an FET and Q_2 is a BJT).

In the cascode circuit, Q_1 is connected in the common-emitter configuration and therefore presents a relatively high input resistance to the signal source. The collector signal current of Q_1 is fed to the emitter of Q_2, which is connected in the common-base configuration. Thus the load resistance seen by Q_1 is simply the input resistance r_e of Q_2. This low load resistance of Q_1 considerably reduces the Miller multiplier effect of $C_{\mu 1}$ and thus extends the upper cutoff frequency. This is achieved without reducing the midband gain, since the collector of Q_2 carries a current almost equal to the collector current of Q_1. Furthermore, since it is in the common-base configuration, Q_2 does not suffer from the Miller effect and hence does not limit the high-frequency response. Transistor Q_2 acts essentially as a current buffer or an impedance

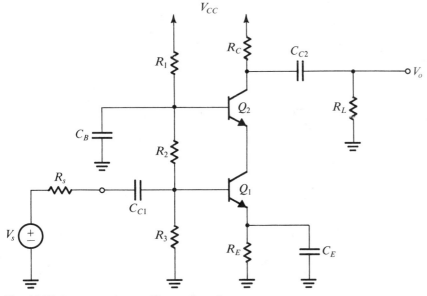

Fig. 11.27 The cascode amplifier configuration.

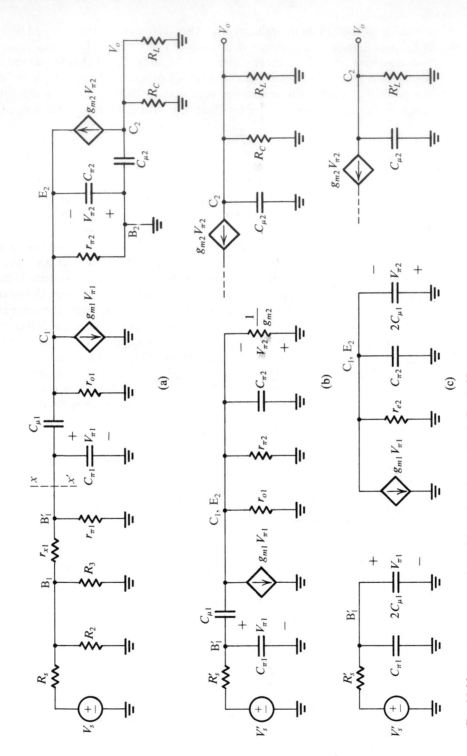

Fig. 11.28 High-frequency analysis of the cascode amplifier in Fig. 11.27.

transformer, faithfully passing on the signal current to the load while presenting a low load resistance to the amplifying device Q_1.

A detailed analysis of the cascode amplifier of Fig. 11.27 will now be presented: Figure 11.28a shows the high-frequency equivalent circuit. To simplify matters, r_{x2} and r_{o2} have been omitted. Although the two transistors are operating at equal bias currents and therefore their corresponding parameters are equal, we have for clarity kept the identity of the two sets of parameters separate.

Application of Thévenin's theorem enables us to reduce the circuit to the left of line xx' (Fig. 11.28a) to a source V_s' and a resistance R_s', as shown in Fig. 11.28b, where

$$V_s' = V_s \frac{(R_2 \parallel R_3)}{R_s + (R_2 \parallel R_3)} \frac{r_{\pi 1}}{r_{\pi 1} + r_{x1} + (R_2 \parallel R_3 \parallel R_s)} \tag{11.85}$$

$$R_s' = \{ r_{\pi 1} \parallel [r_{x1} + (R_3 \parallel R_2 \parallel R_s)] \} \tag{11.86}$$

Another important simplification included in the circuit of Fig. 11.28b is the replacement of the current source $g_{m2} V_{\pi 2}$ by a resistance $1/g_{m2}$ (see the source absorption theorem in Section 2.5). This resistance is then combined with the parallel resistance $r_{\pi 2}$ to obtain r_{e2}. Since $r_{e2} \ll r_{o1}$, we see that between the collector of Q_1 and ground the total resistance is approximately r_{e2}. Capacitance $C_{\pi 2}$ together with resistance r_{e2} produces a transfer-function pole with a frequency

$$\omega_2 = \frac{1}{C_{\pi 2} r_{e2}} \simeq \omega_T \tag{11.87}$$

which is much higher than the frequency of the pole that arises due to the interaction of R_s' and the input capacitance of Q_1. It follows that in the frequency range of interest $C_{\pi 2}$ can be ignored in calculating the voltage at the collector of Q_1; that is,

$$V_{c1} \simeq -g_{m1} V_{\pi 1} r_{e2} \simeq -V_{\pi 1}$$

Thus the gain between B_1' and C_1 (Fig. 11.28b) is approximately -1, and we can employ Miller's theorem to replace the bridging capacitance $C_{\mu 1}$ by a capacitance $2C_{\mu 1}$ between B_1' and ground and a capacitance $2C_{\mu 1}$ between C_1 and ground. The resulting equivalent circuit is shown in Fig. 11.28c, from which we can now evaluate the frequency of the pole due to the RC low-pass circuit at the input as

$$\omega_1 = \frac{1}{R_s'(C_{\pi 1} + 2C_{\mu 1})} \tag{11.88}$$

This frequency will normally be much lower than both ω_2 [Eq. (11.87)] and the frequency of the pole produced by the output part of the circuit,

$$\omega_3 = \frac{1}{C_{\mu 2} R_L'} \tag{11.89}$$

That is, the input circuit produces a dominant high-frequency pole, and the upper 3-dB frequency ω_H is given by

$$\omega_H \simeq \omega_1 \tag{11.90}$$

A slightly better estimate of ω_H can be obtained by combining the frequencies of the three poles using Eq. (11.22). Also, it should be pointed out that the method of open-circuit time constants could have been applied directly to the circuit in Fig. 11.28b.

The midband gain can be easily evaluated by ignoring the capacitances in the equivalent circuit of Fig. 11.28c together with substituting for V'_s and R'_s from Eqs. (11.85) and (11.86):

$$A_M = \frac{V_o}{V_i} = -g_m(R_L \parallel R_C) \frac{(R_2 \parallel R_3)}{(R_2 \parallel R_3) + R_s} \frac{r_\pi}{r_\pi + r_x + (R_2 \parallel R_3 \parallel R_s)} \qquad (11.91)$$

This expression is identical in form to the expression for the gain of a common-emitter circuit.

Exercise

> **11.20** Consider the cascode circuit in Fig. 11.27 with the following component values: $R_s = 4$ kΩ, $R_1 = 18$ kΩ, $R_2 = 4$ kΩ, $R_3 = 8$ kΩ, $R_E = 3.3$ kΩ, $R_C = 6$ kΩ, $R_L = 4$ kΩ, $C_{C1} = 1$ μF, $C_{C2} = 1$ μF, $C_B = 10$ μF, $C_E = 10$μF, and $V_{CC} = +15$ V. Show that each transistor is operating at $I_E \simeq 1$ mA. Note that this design is identical to that of the common-emitter amplifier in Exercises 11.13–11.18. If we thus assume that the transistors are of the same type, we are able to compare results and draw conclusions. Calculate A_M, f_1, f_2, and f_3. Then, use the sum-of-squares formula in Eq. (11.22) to find f_H.
>
> **Ans.** -23.1; 8.95 MHz; 456 MHz; 33 MHz; 8.64 MHz

11.7 FREQUENCY RESPONSE OF THE EMITTER FOLLOWER

The emitter-follower or common-collector configuration was studied in Section 8.10. In the following we consider the high-frequency response of this important circuit configuration. The results apply, with simple modifications, to the FET source follower.

Consider the direct-coupled emitter-follower circuit shown in Fig. 11.29a, where R_s represents the source resistance and R_E represents the combination of emitter-biasing resistance and load resistance. The high-frequency equivalent circuit is shown in Fig. 11.29b and is redrawn in a slightly different form in Fig. 11.29c. Analysis of this circuit results in the emitter-follower transfer function $V_o(s)/V_s(s)$, which can be shown to have two poles and one real zero:

$$\frac{V_o(s)}{V_s(s)} = A_M \frac{1 + s/\omega_Z}{(1 + s/\omega_{P1})(1 + s/\omega_{P2})} \qquad (11.92)$$

where A_M denotes the value of the gain at low and medium frequencies. Unfortunately, though, symbolic analysis of the circuit will not reveal whether or not one of the poles is dominant. To gain more insight we shall take an alternative route. Writing a node equation at the emitter (Fig. 11.29b) results in

$$V_o = (g_m + y_\pi)V_\pi R_E \qquad (11.93)$$

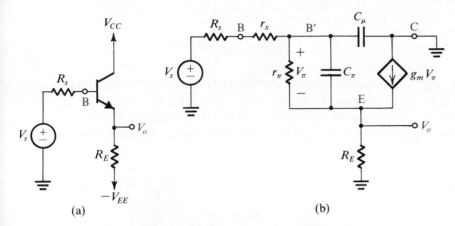

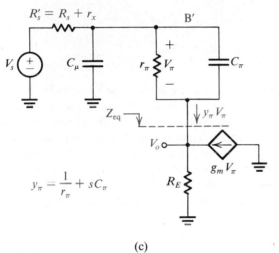

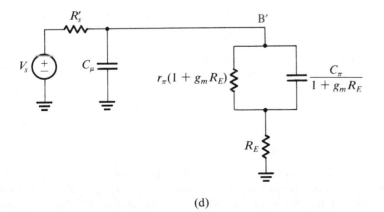

Fig. 11.29 High-frequency analysis of the emitter follower.

where

$$y_\pi = \frac{1}{r_\pi} + sC_\pi$$

Thus V_o will be zero at the value of s that makes $V_\pi = 0$ and at the value of s that makes $g_m + y_\pi = 0$. In turn, V_π will be zero at the value of s that makes $z_\pi = 0$ or equivalently $y_\pi = \infty$, namely, $s = \infty$. The fact that a transmission zero exists at $s = \infty$ correlates with Eq. (11.92). The other transmission zero is obtained from

$$g_m + y_\pi = 0$$

that is,

$$g_m + \frac{1}{r_\pi} + s_Z C_\pi = 0$$

which yields

$$s_Z = -\frac{g_m + 1/r_\pi}{C_\pi} = -\frac{1}{C_\pi r_e} \simeq -\omega_T \tag{11.94}$$

Since the frequency of this zero is quite high, it will normally play a minor role in determining the high-frequency response of the emitter follower.

Next we consider the poles. Whether or not one of the two poles is dominant will depend on the particular application—specifically, on the values of R_s and R_E. In most applications R_s is large, and it together with the input capacitance provides a dominant pole. To see this more clearly, consider the equivalent circuit of Fig. 11.29c. By invoking the source-absorption theorem (Section 2.5), we can replace the circuit below the broken line by its equivalent impedance Z_{eq},

$$Z_{eq} \equiv \frac{V_o}{y_\pi V_\pi}$$

Thus

$$Z_{eq} = \frac{(g_m + y_\pi)R_E}{y_\pi} \tag{11.95}$$

Note that Z_{eq} is simply R_E reflected to the base side through use of a generalized form of the reflection rule: R_E is multiplied by $(h_{fe} + 1)$. The total impedance between B' and ground is

$$Z_{b'} = \frac{1}{y_\pi} + Z_{eq} = \frac{1 + g_m R_E}{y_\pi} + R_E$$

As shown in Fig. 11.29d, this impedance can be represented by a resistance R_E in series with an RC network consisting of a resistance $(1 + g_m R_E)r_\pi$ in parallel with a capacitance $C_\pi/(1 + g_m R_E)$. Since the impedance of the parallel RC circuit is usually much larger than R_E, we may neglect the latter impedance and obtain a simple STC low-pass network. From this STC circuit it follows that a pole exists at

$$\omega_P = \left[\left(C_\mu + \frac{C_\pi}{1 + g_m R_E}\right)\left[R'_s \| (1 + g_m R_E) r_\pi\right]\right]^{-1} \tag{11.96}$$

Even though this pole is usually dominant, its frequency is normally quite high, giving the emitter follower a wide bandwidth.[2]

An alternative approach for finding an approximate value of the 3-dB frequency ω_H is to use the open-circuit time-constants method on the equivalent circuit in Fig. 11.29d.

Exercises

11.21 For an emitter follower biased at $I_C = 1$ mA and having $R_s = R_E = 1$ kΩ, and using a transistor specified to have $f_T = 400$ MHz, $C_\mu = 2$ pF, $r_x = 100$ Ω, and $\beta_0 = 100$, evaluate the midband gain A_M and the frequency of the dominant high-frequency pole.

Ans. 0.97 V/V; 62.5 MHz

11.22 For the emitter follower specified in Exercise 11.21 use the method of open-circuit time constants to estimate the upper 3-dB frequency f_H.

Ans. 55.3 MHz

11.8 THE COMMON-COLLECTOR COMMON-EMITTER CASCADE

The excellent high-frequency response of the emitter follower is due to the absence of the Miller capacitance-multiplication effect. The problem with it, though, is that it does not provide voltage gain. It appears possible that we can obtain both gain and wide bandwidth by using a cascade of a common-collector and a common-emitter stage, as shown in Fig. 11.30. Here the emitter-follower transistor Q_1 is shown biased by a current source I, as is usually the case in integrated-circuit design. Because the collector of Q_1 is at a signal ground, $C_{\mu 1}$ does not get multiplied by the stage gain, as is the case in the common-emitter amplifier. Thus the pole caused by the interaction of the source resistance and the input capacitance will be at a high frequency.

The voltage gain is provided by the common-emitter transistor Q_2. This transistor suffers from the Miller effect; that is, the total effective capacitance between its base and ground will be large. Nevertheless, this will not be detrimental; the resistance seen by that capacitance will be small because of the low output resistance of the emitter follower Q_1.

Before considering a numerical example, we wish to draw the reader's attention to the similarities between the circuit of Fig. 11.30 and the cascode amplifier studied

---

[2] Although we have not considered the emitter follower with a capacitive load, this case deserves a comment: A capacitive load for the emitter follower leads to a *negative* input conductance; and if the source impedance is inductive, the circuit may oscillate. Stability and oscillations are studied in Chapters 12 and 14.

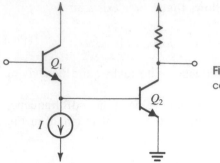

Fig. 11.30 The common-collector common-emitter cascade amplifier.

in Section 11.6. Both circuits employ a common-emitter amplifier to obtain voltage gain. Both circuits achieve wider bandwidth (than that obtained in a common-emitter amplifier) through minimizing the effect of the Miller multiplier. In the cascode circuit this is achieved by isolating the load resistance from the collector of the common-emitter stage by a low-input-resistance common-base stage. In the present circuit, although Miller multiplication occurs, the resulting large capacitance is isolated from the source resistance by an emitter follower.

EXAMPLE 11.9

Figure 11.31 shows a capacitively coupled amplifier designed as the cascade of a common-collector stage and a common-emitter stage. Assume that the transistors used have $f_T = 400$ MHz and $C_\mu = 2$ pF and neglect r_x and r_o. We wish to evaluate the midband gain and the high-frequency response of this circuit. Note that the load and source resistances and the transistor parameters are identical to those used in the common-emitter case (Exercises 11.14 to 11.19) and in the cascode case (Exercise 11.20); hence comparisons can be made ($\beta = 100$).

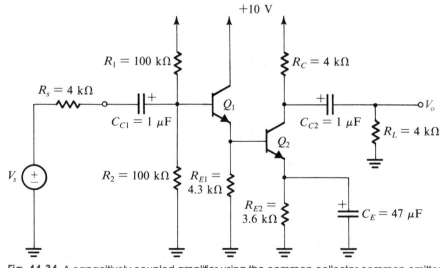

Fig. 11.31 A capacitively coupled amplifier using the common-collector common-emitter cascade configuration.

Solution

We first determine the dc bias currents as follows:

$$V_{B1} \simeq 5 \text{ V} \qquad V_{E1} \simeq 4.3 \text{ V} \qquad I_{E1} = \frac{4.3 \text{ V}}{4.3 \text{ k}\Omega} = 1 \text{ mA}$$

$$V_{E2} \simeq 3.6 \text{ V} \qquad I_{E2} = \frac{3.6 \text{ V}}{3.6 \text{ k}\Omega} = 1 \text{ mA}$$

Thus both transistors are operating at emitter currents of approximately 1 mA, and both are in the active mode. At this operating point the equivalent circuit components are

$$g_m \simeq 40 \text{ mA/V} \qquad r_e \simeq 25 \text{ }\Omega \qquad r_\pi \simeq 2.5 \text{ k}\Omega$$

$$C_\pi + C_\mu = \frac{g_m}{\omega_T} = 15.9 \text{ pF} \qquad C_\mu = 2 \text{ pF} \qquad C_\pi = 15.9 - 2 = 13.9 \text{ pF}$$

To evaluate the midband gain we shall first determine the value of the input resistance R_{in}. Toward that end note that the input resistance between the base of Q_2 and ground is equal to $r_{\pi 2}$. Thus in the emitter circuit of Q_1 we have R_{E1} in parallel with $r_{\pi 2}$. The input resistance R_{in} will therefore be given by

$$R_{\text{in}} = R_1 \| R_2 \| \{(\beta_1 + 1)[r_{e1} + (R_{E1} \| r_{\pi 2})]\}$$

which leads to $R_{\text{in}} \simeq 38 \text{ k}\Omega$. The transmission from the input to the base of Q_1 is

$$\frac{V_{b1}}{V_s} = \frac{R_{\text{in}}}{R_{\text{in}} + R_s} = 0.9 \tag{11.97}$$

Next, the gain of the emitter follower Q_1 can be obtained as

$$\frac{V_{e1}}{V_{b1}} = \frac{(R_{E1} \| r_{\pi 2})}{(R_{E1} \| r_{\pi 2}) + r_{e1}} = 0.98 \tag{11.98}$$

Finally, the gain of the common-emitter amplifier Q_2 can be evaluated as

$$\frac{V_o}{V_{e1}} = -g_{m2}(R_C \| R_L) = -80 \tag{11.99}$$

The overall voltage gain can be obtained by combining Eqs. (11.97) through (11.99):

$$\frac{V_o}{V_s} = -0.9 \times 0.98 \times 80 = -70.6 \text{ V/V} \tag{11.100}$$

The high-frequency response can be determined from the equivalent circuit shown in Fig. 11.32a. We apply Miller's theorem to the second stage[3] and perform a number

---

[3] Assuming that $g_{m2}R'_L \gg 1$, application of Miller's theorem results in a capacitance approximately equal to $C_{\mu 2}$ in parallel with R'_L. Note that this is *not* the output capacitance of the amplifier (see Section 2.5). Nevertheless, it can be used in the computation of f_H using the method of open-circuit time constants.

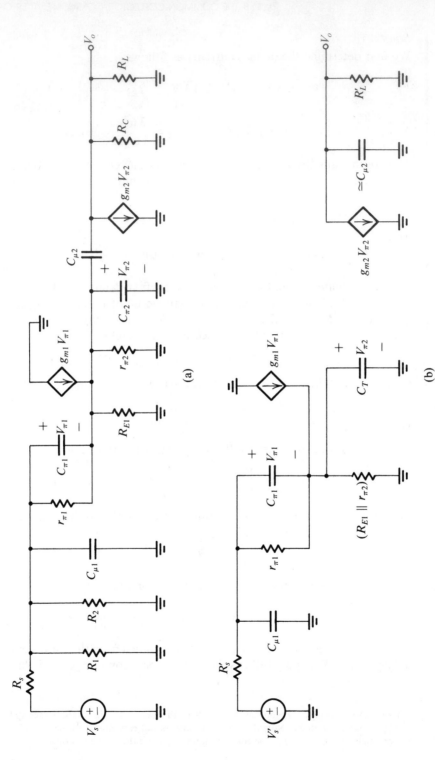

Fig. 11.32 Equivalent circuits for the determination of the high-frequency response of the amplifier in Fig. 11.31.

of other simplifications, and we obtain the circuit in Fig. 11.32b, where

$$R'_s = (R_s \parallel R_1 \parallel R_2)$$

$$V_s = V_s \frac{(R_1 \parallel R_2)}{(R_1 \parallel R_2) + R_s}$$

$$C_T = C_{\pi 2} + C_{\mu 2}(1 + g_{m2}R'_L)$$

$$R'_L = (R_L \parallel R_C)$$

This circuit is still quite complex, and an exact pencil-and-paper analysis would be quite tedious. Alternatively, we employ the technique for open-circuit time constants discussed in Section 11.2 to determine f_H as follows: Capacitor $C_{\mu 1}$ sees a resistance $R_{\mu 1}$ given by

$$R_{\mu 1} = (R_s \parallel R_{\text{in}}) = (4 \parallel 38) = 3.62 \text{ k}\Omega$$

It can be shown that $C_{\pi 1}$ sees a resistance $R_{\pi 1}$ given by

$$R_{\pi 1} = \left(r_{\pi 1} \parallel \frac{R'_s + R'_{E1}}{1 + g_{m1}R'_{E1}} \right)$$

where $R'_{E1} = (R_{E1} \parallel r_{\pi 2})$. Thus $R_{\pi 1} = 80 \ \Omega$.

Capacitance C_T, which is equal to 175.9 pF, sees a resistance R_T given by

$$R_T = \left(R'_{E1} \parallel \frac{r_{\pi 1} + R'_s}{\beta_1 + 1} \right) = 59 \ \Omega$$

Capacitance $C_{\mu 2}$ sees a resistance R'_L given by

$$R'_L = (R_L \parallel R_C) = 2 \text{ k}\Omega$$

Thus the effective time constant is given by

$$\tau = C_{\mu 1}R_{\mu 1} + C_{\pi 1}R_{\pi 1} + C_T R_T + C_{\mu 2}R'_L = 22.7 \text{ ns}$$

which corresponds to an upper 3-dB frequency of

$$f_H \simeq \frac{1}{2\pi\tau} = 7 \text{ MHz}$$

Although the upper 3-dB frequency is not as high as the value obtained for the cascode amplifier (8.95 MHz), the midband gain here, 70.6, is higher than that found for the cascode (23.1). A figure of merit for an amplifier is its gain-bandwidth product.

Exercises

11.23 We wish to use the method of short-circuit time constants to determine an approximate value of the lower 3-dB frequency f_L of the amplifier circuit in Fig. 11.32. Also find the frequency of the zero introduced by C_E.

Ans. 156.8 Hz; 0.94 Hz

11.24 The common-collector common-emitter configuration studied in this section is a modified version of the composite device obtained by connecting two transistors in the form shown in Fig. E11.24a. This configuration, known as the *Darlington configuration*, is equivalent to a single transistor with $\beta \simeq \beta_1 \beta_2$. It can therefore be used as a high-performance follower, as illustrated in Fig. E11.24b. For the latter circuit assume that Q_2 is biased at $I_E = 5$ mA and let $R_s = 100$ kΩ, $R_E = 1$ kΩ, and $\beta_1 = \beta_2 = 100$. Find R_{in}, V_o/V_s, and R_{out}.

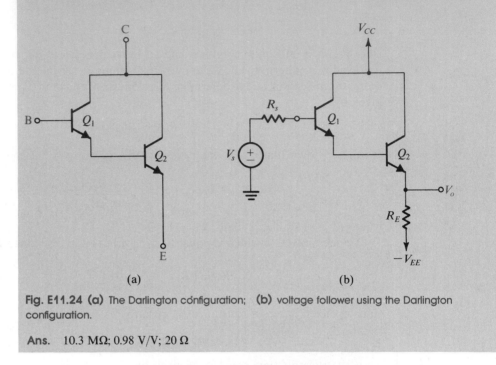

(a)   (b)

Fig. E11.24 (a) The Darlington configuration;   **(b)** voltage follower using the Darlington configuration.

Ans.   10.3 MΩ; 0.98 V/V; 20 Ω

11.9 FREQUENCY RESPONSE OF THE DIFFERENTIAL AMPLIFIER

The differential pair studied in Chapter 9 is the most important building block in analog integrated-circuit design. In this section we analyze its frequency response. Although only the BJT differential pair is considered, the method applies equally well to the FET pair.

Frequency Response in the Case of Symmetric Excitation

Consider the differential amplifier shown in Fig. 11.33a. The input signal V_s is applied in a complementary (push-pull) fashion, and the source resistance R_s is equally distributed between the two sides of the pair. This situation arises, for instance, if the differential amplifier is fed from the output of another differential stage.

Since the circuit is symmetric and is fed in a complementary fashion, its frequency response will be identical to that of the equivalent common-emitter circuit shown in Fig. 11.33b. We have analyzed the common-emitter circuit in detail in Section 11.5. Since the differential pair is a direct-coupled amplifier, its gain will extend down to

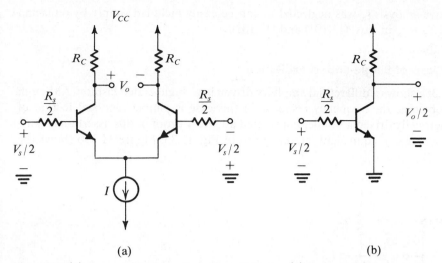

(a)   (b)

Fig. 11.33 (a) A symmetrically excited differential pair; **(b)** its equivalent half-circuit.

zero frequency with a low-frequency value of

$$\frac{V_o}{V_s} = -\frac{r_\pi}{r_\pi + R_s/2} g_m R_C \tag{11.101}$$

The high-frequency response will be dominated by a real pole with a frequency ω_P,

$$\omega_P = \frac{1}{[(R_s/2) \| r_\pi][C_\pi + C_\mu(1 + g_m R_C)]} \tag{11.102}$$

Denote the low-frequency gain given in Eq. (11.101) by A_0. The transfer function of the differential pair will be given by

$$\frac{V_o}{V_s} = \frac{A_0}{1 + s/\omega_P} \tag{11.103}$$

Thus the plot of gain versus frequency will have the standard single-pole shape shown in Fig. 11.34. The 3-dB frequency ω_H is equal to the pole frequency,

$$\omega_H - \omega_P$$

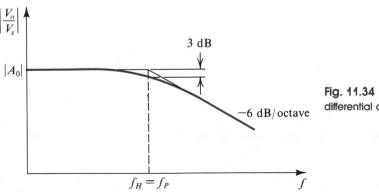

Fig. 11.34 Frequency response of the differential amplifier.

In the above analysis r_x was neglected. It can be easily included simply by replacing $R_s/2$ by $R_s/2 + r_x$ in Eqs. (11.101) and (11.102).

Frequency Response in the Case of Single-Ended Excitation

Figure 11.35a shows a differential amplifier driven in a single-ended fashion. Although it is not obvious, the frequency response in this case is almost identical to that of the symmetrically driven amplifier considered above. A proof of this assertion is illustrated by a series of equivalent circuits given in Fig. 11.35. Figure 11.35b shows the

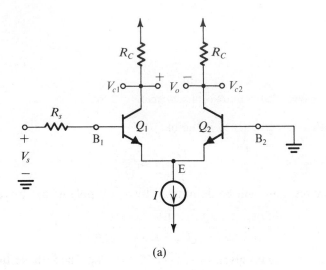

(a)

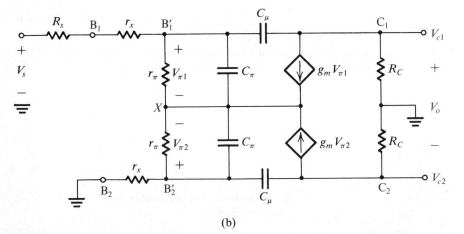

(b)

Fig. 11.35 (a) The differential amplifier excited in a single-ended fashion. **(b)** Equivalent circuit of the amplifier in (a).

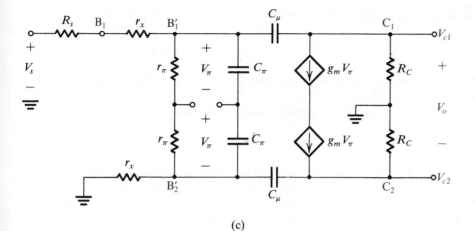

(c)

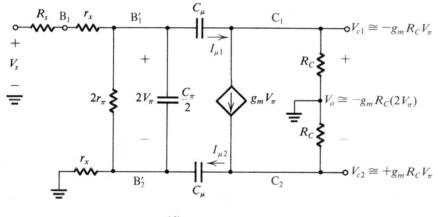

(d)

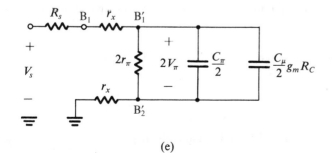

(e)

Fig. 11.35 (c) Simplified equivalent circuit using the fact that $V_{\pi 1} = -V_{\pi 2}$. **(d)** A further simplification of the equivalent circuit. **(e)** The input equivalent circuit after use of Miller's theorem.

complete equivalent circuit. We can write a node equation at node X and obtain

$$V_{\pi 1}\left(\frac{1}{r_\pi} + sC_\pi + g_m\right) + V_{\pi 2}\left(\frac{1}{r_\pi} + sC_\pi + g_m\right) = 0$$

Thus $V_{\pi 1} = -V_{\pi 2}$. This leads to the equivalent circuit in Fig. 11.35c, which is further simplified in Fig. 11.35d.

Consider Fig. 11.35d. If we neglect $I_{\mu 1}$ and $I_{\mu 2}$ in comparison with $g_m V_\pi$, it follows that

$$V_{c1} \simeq -g_m R_C V_\pi \qquad V_{c2} \simeq +g_m R_C V_\pi$$

Thus $V_o \simeq -g_m R_C(2V_\pi)$. If we further assume that r_x is small and that $g_m R_C \gg 2$, then Miller's theorem now can be applied to obtain the simplified input equivalent circuit shown in Fig. 11.35e, from which we see that the high-frequency response is dominated by a pole at $s = -\omega_P$, where ω_P is given by

$$\omega_P = \frac{1}{[2r_\pi \,\|\, (R_s + 2r_x)][C_\pi/2 + (C_\mu/2)(g_m R_C)]} \tag{11.104}$$

The low-frequency gain A_0 is given by

$$A_0 = \frac{V_o}{V_s} = -g_m R_C \frac{2r_\pi}{2r_\pi + R_s + 2r_x} \tag{11.105}$$

These results are almost identical to those obtained in the case of symmetric excitation.

Effect of Emitter Resistance on the Frequency Response

The bandwidth of the differential amplifier can be widened (that is, ω_H can be increased) by including two equal resistances R_E in the emitters. This is achieved at the expense of a reduction in the low-frequency gain. To evaluate the effect of the emitter resistances on frequency response, consider the equivalent half-circuit shown in Fig. 11.36a. The low-frequency gain is given by

$$A_0 \equiv \frac{V_o}{V_s} = \frac{-(\beta + 1)(r_e + R_E)}{R_s/2 + r_x + (\beta + 1)(r_e + R_E)} \frac{\alpha R_C}{R_E + r_e} \tag{11.106}$$

The high-frequency equivalent circuit is shown in Fig. 11.36b. Since it is no longer convenient to apply Miller's theorem, we shall use the technique of open-circuit time constants explained in Section 11.2. The method proceeds as follows: We first eliminate C_μ and determine the resistance seen by C_π, which we shall call R_π. Figure 11.36c shows the circuit for finding R_π,

$$R_\pi = \left(r_\pi \,\Big\|\, \frac{R'_s + R_E}{1 + g_m R_E}\right) \tag{11.107}$$

Next, the resistance R_μ seen by C_μ can be determined from the circuit in Fig. 11.36d,

$$R_\mu = R_C + \frac{1 + R_E/r_e + g_m R_C}{1/r_\pi + (1/R'_s)(1 + R_E/r_e)} \tag{11.108}$$

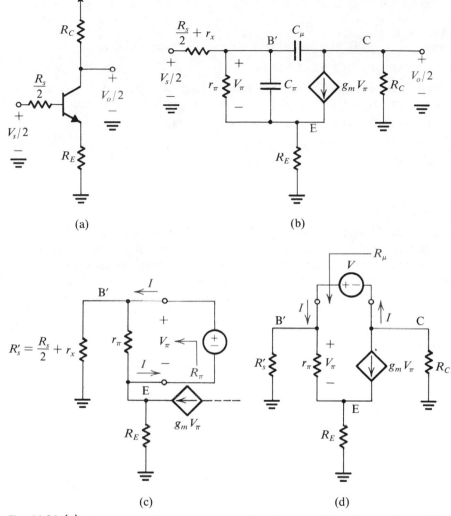

Fig. 11.36 (a) Equivalent half-circuit of the differential amplifier with emitter resistance R_E. (b) Equivalent circuit of the half-circuit in (a). (c) Circuit for determining the resistance R_π seen by C_π. (d) Circuit for determining the resistance R_μ seen by C_μ.

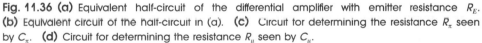

The overall effective time constant will be given by

$$\tau = C_\pi R_\pi + C_\mu R_\mu \tag{11.109}$$

and the 3-dB frequency ω_H will be

$$\omega_H \simeq \frac{1}{\tau} \tag{11.110}$$

Exercises

11.25 Consider a differential amplifier biased with a current source $I = 1$ mA and having $R_C = 10$ kΩ. Let the amplifier be fed with a source having $R_s = 10$ kΩ. Also let the transistors be specified to have $\beta_0 = 100$, $C_\pi = 6$ pF, $C_\mu = 2$ pF, and $r_x = 50$ Ω. Find the dc differential gain A_0, the 3-dB frequency f_H, and the gain–bandwidth product.

Ans.   100 V/V (40 dB); 156 kHz; 15.6 MHz

11.26 Consider the differential amplifier of Exercise 11.25 but with a 150-Ω resistance included in each emitter lead. Find A_0, R_π, R_μ, f_H, and the gain–bandwidth product.

Ans.   40 V/V (32 dB); 1.03 kΩ; 215.6 kΩ; 364 kHz; 14.6 MHz

(*Note:* The difference between the gain-bandwidth products calculated in Exercises 11.25 and 11.26 is due mainly to the different approximations made.)

Variation of the CMRR with Frequency

The common-mode rejection ratio (CMRR) of a differential amplifier falls off at high frequencies because of a number of factors, the most important of which is the increase of the common-mode gain with frequency. To see how this comes about, consider the common-mode equivalent half-circuit shown in Fig. 11.37. Here the resistance R is the output resistance and the capacitance C is the output capacitance of the bias current source. From our study of the frequency response of the common-emitter amplifier in Section 11.5 we know that the components $2R$ and $C/2$ will introduce a zero in the common-mode gain function. This zero will be at a frequency f_Z,

$$f_Z = \frac{1}{2\pi(2R)(C/2)} = \frac{1}{2\pi RC} \tag{11.111}$$

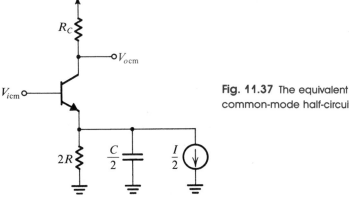

Fig. 11.37 The equivalent common-mode half-circuit.

Since R is usually very large, even a very small output capacitance C will result in f_Z having a relatively low value. The result is that the common-mode gain will start increasing, with a slope of $+6$ dB/octave at a relatively low frequency as shown in Fig. 11.38a. The common-mode gain falls off at higher frequencies because of the internal capacitances C_π and C_μ and because of the pole created by $C/2$.

The behavior of the common-mode gain shown in Fig. 11.38a, together with the high-frequency rolloff of the differential gain (Fig. 11.38b), results in the CMRR having the frequency response shown in Fig. 11.38c.

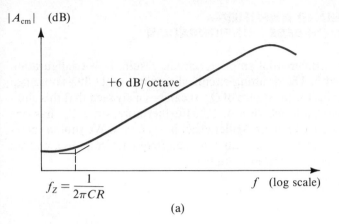

$|A_{cm}|$ (dB)

+6 dB/octave

$$f_Z = \frac{1}{2\pi CR}$$

f (log scale)

(a)

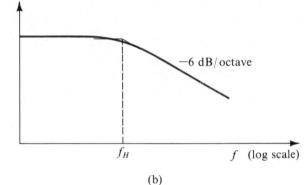

$|A_d|$ (dB)

−6 dB/octave

f_H

f (log scale)

(b)

Fig. 11.38 Variation of **(a)** common-mode gain, **(b)** differential gain, and **(c)** common-mode rejection ratio with frequency.

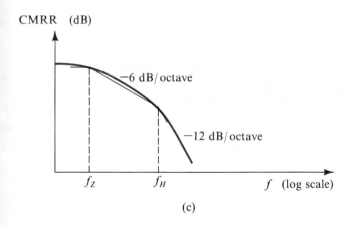

CMRR (dB)

−6 dB/octave

−12 dB/octave

f_Z f_H

f (log scale)

(c)

Exercise

11.27 A differential amplifier is biased by a constant-current source having an output resistance of 30 MΩ and an output capacitance of 2 pF. Find the frequency at which the CMRR decreases by 3 dB.

Ans. 2.65 kHz

11.10 THE DIFFERENTIAL PAIR AS A WIDEBAND AMPLIFIER—
THE COMMON-COLLECTOR COMMON-BASE CONFIGURATION

A slight modification of the differential-amplifier circuit results in a configuration with a much higher bandwidth. The resulting circuit, shown in Fig. 11.39, is obtained by simply eliminating the collector resistance of Q_1. It can be easily seen that this eliminates the Miller capacitance multiplication of $C_{\mu 1}$. Furthermore, since $C_{\mu 2}$ has one of its terminals grounded, there will be no Miller effect in Q_2 either. We should therefore expect the circuit of Fig. 11.39 to have an extended frequency response, and we may add it to our repertoire of wideband amplifiers.

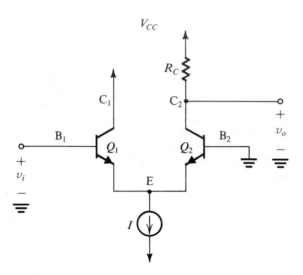

Fig. 11.39 The differential amplifier modified for wideband amplification. Eliminating the collector resistance of Q_1 eliminates the Miller capacitance multiplication.

Low-Frequency Gain

An alternative way of looking at the circuit of Fig. 11.39 is to consider it as a common-collector stage (Q_1) followed by a common-base stage (Q_2). The common-collector transistor Q_1 has in its emitter circuit the emitter resistance r_e of Q_2. Thus the signal current in the emitter of Q_1 is given by

$$i_e = \frac{v_i}{r_{e1} + r_{e2}} = \frac{v_i}{2r_e} \tag{11.112}$$

because

$$r_{e1} = r_{e2} = r_e = \frac{V_T}{I/2} \tag{11.113}$$

This signal current flows in the emitter of Q_2 and appears in its collector multiplied by α,

$$i_{c2} = \alpha i_e = \frac{\alpha v_i}{2r_e} \tag{11.114}$$

Thus the output signal voltage at the collector of Q_2 will be

$$v_o = i_{c2}R_C = \frac{\alpha v_i}{2r_e} R_C \tag{11.115}$$

and the voltage gain will be

$$\frac{v_o}{v_i} = \frac{\alpha R_C}{2r_e} \tag{11.116}$$

Frequency Response

To simplify matters we shall neglect the effect of r_x and thus obtain the equivalent circuit shown in Fig. 11.40a. It is assumed that the circuit is fed with a signal source having a voltage V_s and a resistance R_s. A node equation at node E reveals that

$$V_{\pi 1} = -V_{\pi 2}$$

This allows us to simplify the circuit to that shown in Fig. 11.40b, where $V_\pi = V_{\pi 1} = -V_{\pi 2}$. It can be seen that there are two real poles, one at the input with a frequency f_{P1},

$$f_{P1} = \frac{1}{2\pi(R_s \parallel 2r_\pi)(C_\pi/2 + C_\mu)} \tag{11.117}$$

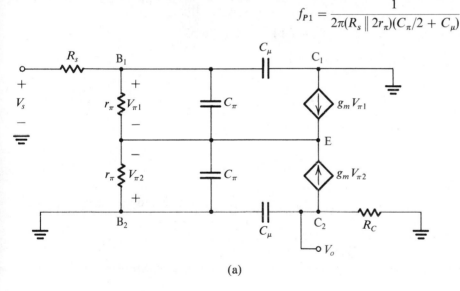

(a)

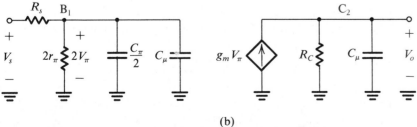

(b)

Fig. 11.40 (a) Equivalent circuit for the modified differential amplifier in Fig. 11.39. **(b)** Simplified equivalent circuit.

and one at the output with a frequency f_{P2},

$$f_{P2} = \frac{1}{2\pi R_C C_\mu} \qquad (11.118)$$

Whether one of these poles is dominant or not will depend on the particular application. If no dominant pole exists, then one has to use the overall transfer function to evaluate the 3-dB frequency f_H. This is demonstrated in Example 11.10.

EXAMPLE 11.10

Consider the modified differential amplifier discussed above and let $I = 1$ mA, $R_C = 10$ kΩ, $R_s = 10$ kΩ, $f_T = 400$ MHz, and $C_\mu = 2$ pF. Evaluate the low-frequency gain and the 3-dB frequency f_H. ($\beta = 100$)

Solution

Each transistor is biased at an emitter current of 0.5 mA. Thus

$$r_e \simeq 50\ \Omega \qquad g_m \simeq 20\ \text{mA/V} \qquad r_\pi \simeq 5\ \text{k}\Omega$$

$$C_\pi + C_\mu = \frac{g_m}{\omega_T} = \frac{20 \times 10^{-3}}{2\pi \times 400 \times 10^6} \simeq 8\ \text{pF}$$

$$C_\mu = 2\ \text{pF} \qquad C_\pi = 6\ \text{pF}$$

The low-frequency gain is given by

$$A_0 = \frac{2r_\pi}{R_s + 2r_\pi} \frac{\alpha R_C}{2r_e} \simeq \frac{10}{10 + 10} \frac{10}{0.1}$$

Thus $A_0 = 50$, or 34 dB. The pole at the input has a frequency ω_{P1},

$$\omega_{P1} = \frac{1}{(R_s \parallel 2r_\pi)(C_\pi/2 + C_\mu)}$$

$$= \frac{1}{5 \times 10^3 (3 + 2) \times 10^{-12}} = 40\ \text{Mrad/s}$$

The pole at the output has a frequency ω_{P2}, given by

$$\omega_{P2} = \frac{1}{C_\mu R_C} = \frac{1}{2 \times 10^{-12} \times 10 \times 10^3} = 50\ \text{Mrad/s}$$

Since the two poles are very close to each other, we have to use the overall transfer function to evaluate the 3-dB frequency f_H. The amplifier transfer function is given by

$$A(s) \equiv \frac{V_o}{V_s} = \frac{A_0}{(1 + s/\omega_{P1})(1 + s/\omega_{P2})}$$

Thus

$$|A(j\omega)| = \frac{A_0}{\sqrt{(1 + \omega^2/\omega_{P1}^2)(1 + \omega^2/\omega_{P2}^2)}}$$

At $\omega = \omega_H$, $|A(j\omega_H)| = A_0/\sqrt{2}$. Thus

$$2 = \left(1 + \frac{\omega_H^2}{\omega_{P1}^2}\right)\left(1 + \frac{\omega_H^2}{\omega_{P2}^2}\right)$$

which leads to $\omega_H = 28.53$ Mrad/s. Thus the upper 3-dB frequency f_H is $f_H = \omega_H/2\pi = 4.54$ MHz.

Before leaving this section we should point out a disadvantage of the modified differential-amplifier circuit. Since the output is taken single-ended, the CMRR will be much lower than that of the balanced differential amplifier. For this reason the first stage of an operational amplifier is usually a balanced one. The modified circuit can be used in subsequent stages or in wideband amplifiers, where the requirement of a high CMRR is not important.

11.11 SUMMARY

● Bode plots provide a convenient means for sketching the magnitude and phase of amplifier gain versus frequency. They are particularly suited for amplifiers with real poles and zeros.

● The amplifier gain remains almost constant over the midfrequency band. It falls off at high frequencies where transistor-model capacitors no longer have very high reactances. For ac amplifiers the gain falls off at low frequencies as well because the coupling and bypass capacitors no longer have very low reactances.

● The amplifier bandwidth is the frequency range over which the gain remains within 3 dB of the value at midband. The limits of the bandwidth are the frequencies f_L and f_H (for a dc amplifier only f_H is meaningful.)

● If the amplifier poles and zeros are known (or easy to find), then f_L and f_H can be determined exactly either graphically or analytically. Alternatively, simple, approximate formulas exist for obtaining reasonably good estimates of f_L and f_H.

● If the poles and zeros are not easy to find, an approximate value for f_L (or f_H) can be determined by evaluating the short-circuit (or open-circuit) time constants of the circuit. The accuracy of this method improves if the zeros are not important (that is, if the zeros in the low-frequency band are at very low frequencies and if those in the high-frequency band are at very high frequencies) and if one pole is dominant. The method pinpoints the capacitor(s) that is (are) most significant in determining the frequency response, and is thus useful in design.

● The amplifier is said to have a dominant pole in the low-frequency band if the nearest pole or zero is at least two octaves lower in frequency. A dominant pole in the high-frequency band exists if the nearest pole or zero is at least two octaves higher in frequency.

● If a dominant low-frequency pole exists, then the low-frequency response is essentially that of an STC high-pass network, and f_L is equal to the pole frequency. A dominant high-frequency pole makes the high-frequency response like that of an STC low-pass network with f_H equal to the pole frequency.

● The high-frequency response of the common-source (common-emitter) amplifier is limited by the Miller effect, which multiples the feedback capacitance C_{gd} (C_μ) and forms a dominant pole at the input. The bandwidth can be extended by reducing the generator resistance and/or the load resistance. The latter action reduces the gain.

● The components of the low-frequency hybrid-π model of the BJT can be determined from measurement of the common-emitter h-parameters.

● The common-emitter short-circuit current gain h_{fe} is equal to β at low frequencies. Its magnitude begins to fall off around f_β with -6 dB/octave slope, reaching unity at f_T.

● The common-base (common-gate) configuration does not suffer from the Miller effect and thus exhibits wide bandwidth. However, it has a low input resistance.

● The cascode configuration combines the wide bandwidth of the common-base (common-gate) circuit with the high input resistance of the common-emitter (common-source) circuit.

● The emitter follower (source follower) features a wide bandwidth that approaches the transistor f_T. [For the FET we define f_T as $g_m/2\pi(C_{gs} + C_{gd})$].

● The common-collector common-emitter cascade provides a wideband amplifier comparable to the cascode.

● The frequency response of the differential pair is obtained by analyzing its equivalent differential half-circuit, which is a common-emitter (common-source) configuration.

● The bandwidth of the differential amplifier can be extended (at the expense of gain reduction) by adding resistors in the emitter (source) leads.

● The CMRR of the differential pair decreases with frequency as a result of the zero introduced in the common-mode gain by the output capacitance of the biasing current source.

● Wideband amplification can be obtained by eliminating the collector resistor of the input transistor in the differential pair, thus converting the circuit into a common-collector common-base cascade.

BIBLIOGRAPHY

L. S. Bobrow, *Elementary Linear Circuit Analysis*, 2nd ed., New York: Holt, Rinehart and Winston, 1987.

P. E. Gray and C. L. Searle, *Electronic Principles*, New York: Wiley, 1969.

P. R. Gray and R. G. Meyer, *Analysis and Design of Analog Integrated Circuits*, 2nd ed. New York: Wiley, 1984.

M. E. Van Valkenburg, *Network Analysis*, 3rd ed., Englewood Cliffs, N.J.: Prentice-Hall, 1974.

PROBLEMS

11.1 Find the voltage transfer function $T(s) \equiv V_o(s)/V_i(s)$ for the STC networks shown in Fig. P11.1.

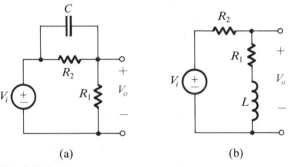

(a)   (b)

Fig. P11.1

11.2 Sketch and label the Bode magnitude plot for the circuit of Exercise 11.2. What is the gain magnitude at 5 kHz, 10 kHz, 50 kHz, 100 kHz, and 200 kHz? Also find the phase angle at each of these frequencies.

11.3 An amplifier with midband gain of 100 has zeros at zero and infinite frequencies and poles at 10 and 1,000 rad/s. Sketch its Bode magnitude and phase plots. What are the values of gain and phase at 1, 10, 10^2, 10^3, and 10^4 rad/s?

*11.4 An amplifier has one zero at 0, one zero at 10 rad/s (on the negative real axis), two zeros at infinity, poles at 1, 10^2, 10^4, and 10^5 rad/s, and a gain of 10^3 at 10^3 rad/s. Sketch its Bode magnitude and phase plots.

*11.5 An amplifier has zeros at 0, 10^6 rad/s (on the negative real axis) and ∞, poles at 10^4, 10^7, and 10^8 rad/s and a gain of 10^3 at 10^5 rad/s. Sketch Bode magnitude and phase plots. What is the peak gain of the amplifier and its phase shift at the corresponding frequency?

11.6 A direct-coupled differential amplifier has a differential gain of 100 with poles at 10^6 and 10^8 rad/s, and a common-mode gain of 10^{-3} with a zero at 10^4 rad/s and a pole at 10^8 rad/s. Sketch the Bode magnitude plots for the differential gain, the common-mode gain, and the CMRR. What is the CMRR at 10^7 rad/s? (*Hint:* Division of magnitudes corresponds to subtraction of logarithms.)

11.7 Consider the Bode phase plot shown in Fig. 11.3. At what fraction of the corner frequency a does the phase shift become 1°?

11.8 Consider a direct-coupled amplifier with n coincident poles at ω_0. At what fraction of ω_0 is **(a)** the gain reduced by 3 dB and **(b)** the phase shift 45°, for $n = 1$, 2, 3, and 4?

11.9 Write the transfer function for an amplifier having a gain of -100 at midband and a low-frequency response characterized by zeros at 1 and 10 rad/s (on the negative real axis) and poles at 5 and 100 rad/s. What is the dc gain of this amplifier? What is its 3-dB frequency?

11.10 If in Problem 11.9 the pole at 5 rad/s is shifted to 10 rad/s, what is the effect on the transfer function? This process is referred to as pole-zero cancellation. What is the dc gain of the resulting amplifier?

11.11 The low-frequency response of an amplifier is characterized by two zeros at 0, one zero at 10 rad/s, and poles at 100, 50, and 5 rad/s. Find an approximate value for its 3-dB frequency.

11.12 The high-frequency response of a direct-coupled amplifier having a dc gain of -100 incorporates zeros at ∞ and 10^6 rad/s and poles at 10^5 and 10^7 rad/s. Write an expression for the amplifier transfer function. Find ω_H: **(a)** using the dominant-pole approximation, **(b)** using the sum-of-squares approximation and **(c)** exactly. If a means is found to lower the frequency of the finite zero to 10^5 rad/s, what does the transfer function become? What is the upper 3-dB frequency of the resulting amplifier?

***11.13 (a) Show that the transfer function of the circuit in Fig. P11.13 is given by

$$\frac{V_o(s)}{V_i(s)} = \frac{s^2}{s^2 + s\left(\dfrac{1}{C_1 R_1} + \dfrac{1}{C_1 R_2} + \dfrac{1}{C_2 R_2}\right) + \dfrac{1}{C_1 R_1 C_2 R_2}}$$

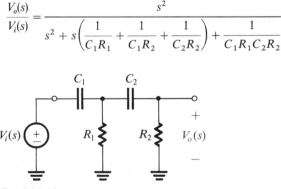

Fig. P11.13

(*Hint:* Start the analysis at the output and work your way back to the input.)

(b) Use the transfer function derived in (a) to find the low-frequency poles of the circuit for component value as follows:

	C_1 (μF)	C_2 (μF)	R_1 (kΩ)	R_2 (kΩ)
(i)	1	1	10	10
(ii)	1	0.1	10	100
(iii)	0.1	1	100	10

(c) For each of the cases in (b), find ω_L exactly.
(d) For each of the cases in (b), find ω_L using the dominant pole approximation.
(e) For each of the cases in (b), find ω_L using the sum-of-squares approximation.
(f) For each of the cases in (b), find ω_L using short-circuit constants.

11.14 Consider the amplifier in Fig. 11.9. For $R_{G2} = 2.2$ MΩ, $R_{G1} = 10$ MΩ, and $R = 100$ kΩ, find the value of C_{C1} for which the input pole is at 100 Hz. If only a limited range of capacitors is available, having capacitance 1×10^{-n} F, what capacitor would you choose for a conservative design? What is the pole frequency which results?

*11.15 The amplifier shown in Fig. 11.9 is augmented by a resistor R_{S1} in series with C_S. Using the ideas on physical interpretation presented following Eq. (11.38), find the effect of this change on the frequency of the zero and pole due to C_S.

11.16 For the amplifier in Fig. 11.9, $R_D = 10$ kΩ, $r_o = 100$ kΩ, and $R_L = 10$ kΩ. Find the value of C_{C2} that produces a pole at 100 Hz.

11.17 For the depletion MOSFET used in the amplifier of Fig. P11.17, $I_{DSS} = 1$ mA and $|V_P| = 2$ V. Sketch $|V_o/V_i|$ as a function of frequency.

11.18 A grounded-source amplifier for which $R_{in} = 1$ MΩ, $R_D = 10$ kΩ, and $g_m = 4$ mA/V is connected between a 100-kΩ source and a 10-kΩ load. If $C_{gs} = 3$ pF and $C_{gd} = 2$ pF, find the midband gain and the frequency of the dominant high-frequency pole. Also, find the frequency of the zero and of the second high-frequency pole.

11.19 A JFET amplifier has a dominant low-frequency pole at 100 Hz, a dominant high-frequency pole at 20 kHz, and a midband gain of -20. For a negative input pulse of 100-mV amplitude and 100-μs duration, what rise time, fall time, and sag would you expect in the output pulse?

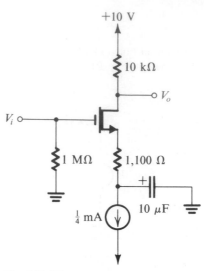

Fig. P11.17

*11.20 It is required to analyze the high-frequency response of the CMOS amplifier shown in Fig. P11.20. The dc bias current is 10 μA. For Q_1, $\mu_n C_{OX} = 20$ μA/V^2, $V_A = 50$ V, $W/L = 64$, $C_{gs} = C_{gd} = 1$ pF. For Q_2, $C_{gd} = 1$ pF and $V_A = 50$ V. Also, there is a 1-pF stray capacitance between the common drain connection and ground. Assume that the resistance of the input signal generator is negligibly small. Also, for simplicity assume that the signal voltage at the gate of Q_2 is zero. Find the frequency of the pole and zero.

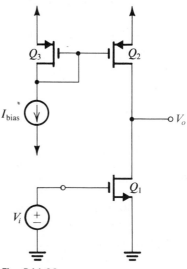

Fig. P11.20

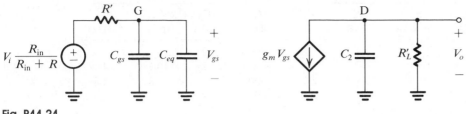

Fig. P11.21

**11.21 Figure P11.21 shows a more exact equivalent circuit for the common-source amplifier whose equivalent circuit is shown in Fig. 11.12b. This equivalent circuit can be obtained by applying Miller's theorem to the can be obtained by applying Miller's theorem to the circuit in Fig. 11.12b but without making the approximation $V_o \simeq -g_m R'_L V_{gs}$. Alternatively, we can find the values of C_{eq} and C_2 using the expressions for ω_{P1} and ω_{P2} in Eqs. (11.53) and (11.54), respectively. Find expressions for C_{eq} and C_2 and show that C_{eq} differs slightly from the approximate value given by Eq. (11.44). Also find numerical values for C_{eq} and C_2 using the component values given in Examples 11.5 and 11.7 and Exercise 11.9.

*11.22 A BJT biased by a 10-mA emitter-current source, and having its emitter bypassed to ground by a large capacitor, is fed a small-signal current through an external 1-kΩ series base resistor. Its collector is connected to a suitable supply via a 100-Ω resistor. Careful ac measurements taken with respect to ground provide the following voltages.
(a) On the emitter: 0.6 mV.
(b) On the base: 3.9 mV.
(c) On the collector: 120 mV.
(d) On the input resistor remote from the base: 13.9 mV.
Assuming the small voltage on the emitter is due to incomplete bypassing, find g_m, r_e, r_π, h_{fe}, and r_x.

11.23 For a BJT operating with emitter grounded and an external base-circuit resistance equal in value to $r_\pi - r_x$, what is the resistance seen looking into the collector? Use the equivalent circuit of Fig. 11.14 in your analysis.

*11.24 In the circuit of Fig. 11.15, $R_s = 10$ kΩ, $R_B = 100$ kΩ, $R_C = 10$ kΩ, and $R_L = 100$ Ω. AC measurements are taken as follows: $v_s = 15$ mV, $v_b = 5$ mV, and $v_o = 20$ mV.
(a) Calculate h_{ie} and h_{fe} both approximately and exactly (that is, taking the currents through R_B and R_C into account).

(b) If r_x is determined by a separate high-frequency measurement to be 50 Ω, calculate r_π, r_e, g_m, and I_E.
(c) The circuit is then converted to that in Fig. 11.16 and a signal $v_c = 5$ V is applied. Measurement indicates v_b to be 1 mV. What is the value of h_{re}? Note that this measurement must be done at very low frequencies. Why?
(d) Using the results in (b) and (c) calculate r_μ. At what frequency does the magnitude of the impedance of a 1-pF capacitance equal this value of r_μ?
(e) Consider the circuit of Fig. 11.16 augmented with a resistor of 100 Ω in series with the emitter to ground and a large capacitor connected from the emitter to a voltmeter whose reading is v_e. With $v_c = 5$ V, v_e is measured to be 8 mV. For this situation estimate h_{oe} and hence r_o.

11.25 The following parameters were measured on a transistor biased at $I_C = 2$ mA: $h_{ie} = 1.35$ kΩ, $h_{fe} = 100$, $h_{re} = 5 \times 10^{-5}$, $h_{oe} = 2.4 \times 10^{-5}$ A/V. Determine the values of g_m, r_π, r_e, r_x, r_o, and V_A.

11.26 A particular BJT operating at $I_C = 1$ mA has $C_\mu = 1$ pF, $C_\pi = 10$ pF, and $\beta = 150$. What are ω_T and ω_β for this situation?

11.27 For the transistor described in Problem 11.26, C_π includes a relatively constant depletion-layer capacitance of 2 pF. If the device is operated at $I_C = 0.1$ mA, what does its ω_T become?

11.28 A particular small-geometry BJT has f_T of 3 GHz and $C_\mu = 0.1$ pF when operated at $I_C = 0.2$ mA. What is C_π in this situation?

11.29 For a BJT whose unity-gain bandwidth is 1 GHz and $\beta_0 = 200$, at what frequency does the magnitude of h_{fe} become 10? What is ω_β?

*11.30 For a sufficiently high frequency, measurement of the complex input impedance of a BJT having (ac) grounded emitter and collector yields a real part approximating r_x. For what frequency, defined in terms of ω_β, is such an estimate of r_x good to within 10% under the condition that $r_x \le r_\pi/10$?

Transistor	I_E (mA)	r_e (Ω)	g_m (mA/V)	r_π (kΩ)	β_0	f_T (MHz)	C_μ (pF)	C_π (pF)	f_β (MHz)
(a)	1				100	400	2		
(b)		25					2	10.7	4
(c)			2.525			400		13.8	4
(d)	10				100	400	2		
(e)	0.1				100	100	2		
(f)	1				10	400	2		
(g)						800	1	9	80

*11.31 Complete the table entries above for transistors (a) through (g) under the conditions indicated.

*11.32 Consider the circuit of Fig. 11.23 augmented by a resistor r_E either **(a)** in series with the emitter (with R_E reduced accordingly) or **(b)** in series with the capacitor C_E. For each case, using the equivalent circuit of Fig. 11.24, appropriately augmented, find the resistance R'_E seen by C_E and thus find the effect of including r_E on ω_L. Also, find the frequency of the zero caused by C_E.

*11.33 For the circuit in Fig. P11.33 find R_{in} and $A_v = V_o/V_i$ for **(a)** $C = 0$ and **(b)** $C = \infty$. Also, find the transfer function $V_o(s)/V_i(s)$, evaluate its singularities (poles and zeros), and sketch a Bode plot for its magnitude for the case $C = 1$ μF. Assume $\beta = \infty$, $V_{BE} = 0.7$ V, and $r_o = \infty$, and neglect all BJT internal capacitances.

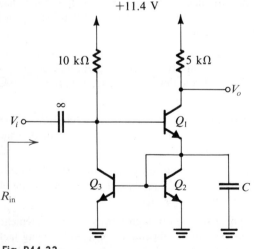

Fig. P11.33

*11.34 Refer to Fig. P11.34. Utilizing the BJT high-frequency hybrid-π model with $r_x = 0$ and $r_o = \infty$, derive

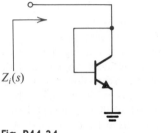

Fig. P11.34

an expression for $Z_i(s)$ as a function of r_e, C_π, and C_μ. Find the frequency at which the impedance has a phase angle of $45°$ for the case in which the BJT has $f_T = 400$ MHz and $\beta_0 = 100$.

*11.35 For the current mirror in Fig. P11.35 derive an expression for the current transfer function $I_o(s)/I_i(s)$ taking into account the BJT internal capacitances and neglecting r_x and r_o. Assume the BJTs to be identical. Observe that a signal ground appears on the collector of Q_2. If the mirror is biased at 1 mA and the BJTs at this operating point are characterized by $f_T = 400$ MHz, $C_\mu = 2$ pF, and $\beta_0 = \infty$, find the frequencies of the pole and zero of the transfer function.

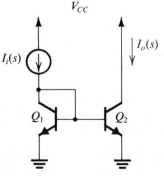

Fig. P11.35

11.36 Consider the circuit of Fig. 11.23. For $R_s = 10$ kΩ, $R_B \equiv R_1 \parallel R_2 = 10$ kΩ, $r_x = 100$ Ω, $r_\pi = 1$ kΩ, $\beta_0 = 100$, and $R_E = 1$ kΩ, what is the ratio C_E/C_{c1} that ensures that their contributions to the determination of ω_L are equal?

11.37 A common-base amplifier having the topology of Fig. 11.25, biased at 10 mA by means of an emitter resistor R_E of 200 Ω, is driven from a 50-Ω source. For a transistor having $f_T = 100$ MHz and $C_\mu = 5$ pF, find the frequency of the input-circuit pole. If $R_C = R_L = 500$ Ω, find the output-circuit pole.

***11.38** Consider the cascode amplifier of Fig. 11.27. For $V_{CC} = 9$ V, and $V_{BE} = 0.7$ V provide a bias design for operation at $I_{E1} \simeq I_{E2} = 0.1$ mA (for high-β devices) that distributes the power supply equally across R_E, R_C, and the two transistors in series, each of the latter having equal collector-to-emitter drops. Use a base-network bias current of one-tenth the emitter currents. What is the input resistance of your circuit in the midband for **(a)** β very large and **(b)** $\beta = 100$? If $\beta = \infty$, what is the midband voltage gain of the amplifier for **(c)** no load and **(d)** a load of 10 kΩ? If the BJTs have $f_T = 100$ MHz, $C_\mu = 1$ pF, and $\beta = 50$, and if the amplifier is driven with a source having $R_s = 10$ kΩ and loaded with 10 kΩ, find **(e)** the overall midband voltage gain, **(f)** the frequencies of the three high-frequency poles, and **(g)** the upper 3-dB frequency f_H.

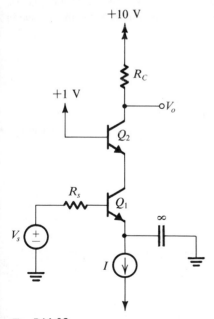

Fig. P11.39

11.39 For the circuit in Fig. P11.39 let $R_C = 5$ kΩ, $I = 1$ mA, $\beta = 100$, $C_\pi = 5$ pF, and $C_\mu = 1$ pF. Find the frequencies of the three high-frequency poles and estimate f_H for two cases: **(a)** $R_s = 1$ kΩ and **(b)** $R_s = 10$ kΩ.

11.40 The BJTs in the cascode circuit in Fig. P11.40 have $\beta = 100$ and $V_{BE} = 0.7$ V. Calculate the following.
(a) The dc voltages at nodes A, B, and C. (Ignore base currents.)
(b) The midband gain V_o/V_s.
(c) The low cutoff frequency f_L, using short-circuit time constants.

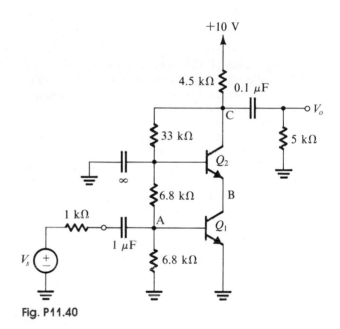

Fig. P11.40

11.41 Consider a JFET source follower characterized by a driving-signal generator with resistance R, a resistance in the source lead of R_S, a transconductance g_m, and internal transistor capacitances C_{gs} and C_{gd}. Repeat the analysis of Section 11.7 for this situation. Evaluate the midband gain and the frequency of the dominant high-frequency pole for $R = 100$ kΩ, $R_S = 10$ kΩ, $g_m = 2$ mA/V, $C_{gs} = 2$ pF, and $C_{gd} = 1$ pF.

***11.42** For the emitter follower shown in Fig. P11.42, find the midband gain, the frequency of the high-frequency dominant pole, an estimate of f_H using the method of open-circuit time constants, and the frequency of the transfer function zero for the cases **(a)** $R_s = 1$ kΩ, **(b)** $R_s = 10$ kΩ, and **(c)** $R_s = 100$ kΩ. Let $R_L = 1$ kΩ, $\beta_o = 100$, $f_T = 400$ MHz, and $C_\mu = 2$ pF.

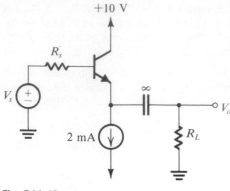

Fig. P11.42

***11.43** The CC-CE cascade circuit shown in Fig. P11.43 offers the advantage of operation with low values of supply voltage and of elimination of the large emitter bypass capacitor.

(a) For $V_{CC} = 5.5$ V, $R_C = 4$ kΩ, $R_B = 100$ kΩ, and $R_E = 7$ kΩ find the dc operating point of each transistor. Let $V_{BE} = 0.7$ V and $\beta = 100$.

(b) Find the midband gain from the base of Q_1 to the collector of Q_2 for $R_L = 4$ kΩ. Using Miller's theorem to replace R_B with two resistors, find R_{in} and R_{out}. Then, for $R_s = 4$ kΩ find the overall voltage gain at midband.

(c) For $C_{C1} = C_{C2} = 1$ μF, find the input and output low-frequency poles. Use the sum-of-squares formula to obtain an estimate of the lower 3-dB frequency, f_L.

(d) Use the method of open-circuit time constants to determine an approximate value for the upper 3-dB frequency, f_H. Let the BJTs have $f_T = 400$ MHz at $I_E = 1$ mA, $C_{je} = 1$ pF, and $C_\mu = 2$ pF.

11.44 Consider the symmetrically excited differential pair shown in Fig. 11.33 for which $I = 10$ mA, $R_C = 5$ kΩ, $R_s = 10$ kΩ, $r_x = 50$ Ω, $\beta = 50$, $f_T = 1$ GHz, and $C_\mu = 1$ pF. Find the midband gain and the dominant high-frequency pole.

11.45 The differential amplifier described in Problem 11.44 is augmented with emitter resistors of 100 Ω. What are its midband gain and upper 3-dB frequency? Compare the gain-bandwidth products of the basic and augmented designs.

*11.46 Consider a common-emitter amplifier having an external emitter resistor R_E for which $R_C = 100$ Ω, $R_s = 100$ Ω, $r_x = 10$ Ω, $g_m = 4$ A/V, $\beta = 100$, $C_\mu = 100$ pF, and $f_T = 100$ MHz, for the cases $R_E = 0$ and 2 Ω. For each case find the midband gain, upper 3-dB frequency, and gain-bandwidth product.

*11.47 A circuit improvement in the constant-current bias of a differential BJT amplifier results in an increase by a factor of 10 in the output resistance of the current source at the expense of a 50% increase in its output capacitance. Sketch a Bode plot for the CMRR before and after the circuit modification. What is the change in

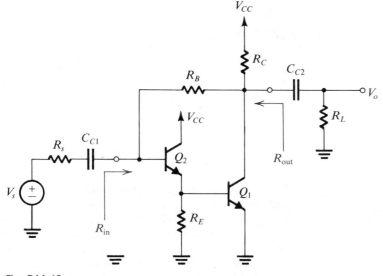

Fig. P11.43

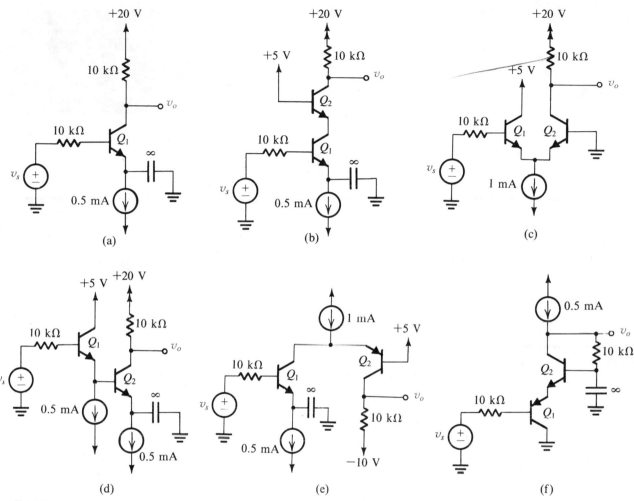

Fig. P11.49

the CMRR (in decibels) at very low frequencies? What is the change at the frequency of the dominant pole of the differential gain?

11.48 A wideband amplifier using the CC-CB configuration shown in Fig. 11.39 is biased with $I = 0.2$ mA and has a 10-kΩ collector resistor. $\beta = 50$. What is the volt-

age gain of the amplifier when fed with a 10-kΩ source? If the BJTs have $f_T = 10^9$ Hz, and $C_\mu = 0.5$ pF, find the frequencies of the poles and the 3-dB frequency.

**11.49 In each of the six circuits in Fig. P11.49, let $\beta = 100$, $C_\mu = 2$ pF, and $f_T = 400$ MHz. Calculate the midband gain and the upper 3-dB frequency.

Chapter 12

Feedback

INTRODUCTION

Most physical systems embody some form of feedback. It is interesting to note, though, that the theory of negative feedback has been developed by electronics engineers. In his search for methods for the design of amplifiers with stable gain for use in telephone repeaters, Harold Black, an electronics engineer with the Western Electric Company, invented the feedback amplifier in 1928. Since then the technique has been so widely used that it is almost impossible to think of electronic circuits without some form of feedback, either implicit or explicit. Furthermore, the concept of feedback and its associated theory is currently used in areas other than engineering, such as in the modeling of biological systems.

Feedback can be either *negative* (degenerative) or *positive* (regenerative). In amplifier design negative feedback is applied to effect one or more of the following properties:

1. *Desensitize the gain;* that is, make the value of the gain less sensitive to variations in the value of circuit components, such as variations that might be caused by changes in temperature.

2. *Reduce nonlinear distortion;* that is, make the output proportional to the input (in other words, make the gain constant independent of signal level).

3. *Reduce the effect of noise;* that is, minimize the contribution to the output of unwanted electric signals generated by the circuit components and extraneous interference.

4. *Control the input and output impedances;* that is, raise or lower input and output impedances by selection of appropriate feedback topology.

5. *Extend the bandwidth* of the amplifier.

All of the above desirable properties are obtained at the expense of a reduction in gain. It will be shown that the gain-reduction factor, called the *amount of feedback*, is the factor by which the circuit is desensitized, by which the input impedance of a voltage amplifier is increased, by which the bandwidth is extended, and so on. In short, the basic idea of negative feedback is to trade off gain for other desirable properties. This chapter is devoted to the study of negative-feedback amplifiers: their analysis, design, and characteristics.

Under certain conditions the negative feedback in an amplifier can become positive and of such a magnitude as to cause oscillations. The reader will recall the use of positive feedback in the design of oscillators and bistable circuits (Chapter 5). Furthermore, in Chapter 14 we shall employ positive feedback in the design of sinusoidal oscillators. In this chapter, however, we are interested in the design of stable amplifiers. We shall therefore study the stability problems of negative-feedback amplifiers.

It should not be implied, however, that positive feedback always leads to instability. Positive feedback is useful in a number of applications, such as the design of active filters, which are studied in Chapter 14.

Before we begin our study of negative feedback we wish to remind the reader that we have already encountered negative feedback in a number of applications. Almost all op-amp circuits employ negative feedback. Another popular application of negative feedback is the use of the emitter resistance R_E to stabilize the bias point of bipolar transistors and to increase the input resistance and bandwidth of a BJT differential amplifier. In addition, the emitter follower and the source follower employ a large amount of negative feedback. The question then arises as to the need for a formal study of negative feedback. As will be appreciated by the end of this chapter, the formal study of feedback provides an invaluable tool for the analysis and design of electronic circuits. Also, the insight gained by thinking in terms of feedback is extremely profitable. □

12.1 THE GENERAL FEEDBACK STRUCTURE

Figure 12.1 shows the basic structure of a feedback amplifier. Rather than showing voltages and currents, Fig. 12.1 is a *signal-flow* diagram, where each x can represent either a voltage or a current signal. The *open-loop* amplifier has a gain A; thus its output x_o is related to the input x_i by

$$x_o = Ax_i \tag{12.1}$$

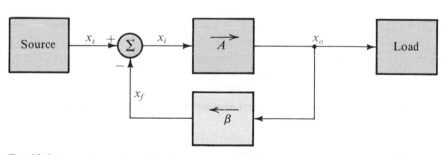

Fig. 12.1 General structure of the feedback amplifier. This is a signal-flow diagram, and the quantities x represent either voltage or current signals.

The output x_o is fed to the load as well as to a feedback network, which produces a sample of the output. This sample x_f is related to x_o by the feedback factor β,

$$x_f = \beta x_o \tag{12.2}$$

The feedback signal x_f is *subtracted* from the source signal x_s, which is the input to the complete feedback amplifier, to produce the signal x_i, which is the input to the basic amplifier,

$$x_i = x_s - x_f \tag{12.3}$$

Here we note that it is this subtraction that makes the feedback negative. In essence, negative feedback reduces the signal that appears at the input of the basic amplifier.

Implicit in the above description is that the source, the load, and the feedback network *do not* load the basic amplifier. That is, the gain A does not depend on any of these three networks. In practice this will not be the case, and we shall have to find a method for casting a real circuit into the ideal structure depicted in Fig. 12.1. Also implicit in Fig. 12.1 is that the forward transmission occurs entirely through the basic amplifier and the reverse transmission occurs entirely through the feedback network.

The gain of the feedback amplifier can be obtained by combining Eqs. (12.1) through (12.3):

$$A_f \equiv \frac{x_o}{x_s} = \frac{A}{1 + A\beta} \tag{12.4}$$

The quantity $A\beta$ is called the *loop gain*, a name that follows from Fig. 12.1. For the feedback to be negative, the loop gain $A\beta$ should be positive; that is, the feedback signal x_f should have the same sign as x_s, thus resulting in a smaller difference signal x_i. Equation (12.4) indicates that for positive $A\beta$ the gain with feedback will be smaller than the open-loop gain A by the quantity $1 + A\beta$, which is called the *amount of feedback*.

If, as is the case in many circuits, the loop gain $A\beta$ is large,

$$A\beta \gg 1$$

then from Eq. (12.4) it follows that

$$A_f \simeq \frac{1}{\beta}$$

which is a very interesting result: *the gain of the feedback amplifier is almost entirely determined by the feedback network*. Since the feedback network usually consists of passive components, which can be chosen to be as accurate as one wishes, the advantage of negative feedback in obtaining accurate, predictable, and stable gain should be apparent. In other words the overall gain will have very little dependence on the gain of the basic amplifier, A, a desirable property because the gain A is usually a function of many parameters, some of which might have wide tolerances. We have seen a dramatic illustration of all of these results in op-amp circuits, where the *closed-loop gain* (which is another name for the gain with feedback) is almost entirely determined by the feedback elements.

Equations (12.1) through (12.3) can be combined to obtain the following expression for the feedback signal x_f:

$$x_f = \frac{A\beta}{1 + A\beta} x_s$$

Thus for $A\beta \gg 1$ we see that

$$x_f \simeq x_s$$

which implies that the signal x_i at the input of the basic amplifier is reduced to almost zero. Thus if a large amount of negative feedback is employed, the feedback signal x_f becomes an almost identical replica of the input signal x_s. An outcome of this property is the tracking of the two input terminals of an op amp. The difference between x_s and x_f, which is x_i, is sometimes referred to as the "error signal". Accordingly the input differencing circuit is often also called a *comparator*. (It is also known as a *mixer*.)

Exercise

12.1 The noninverting op-amp configuration shown in Fig. E12.1 provides a direct implementation of the feedback loop of Fig. 12.1.
(a) Assume that the op amp has infinite input resistance and zero output resistance. Find an expression for the feedback factor β.
(b) If the open-loop voltage gain $A = 10^4$, find R_2/R_1 to obtain a closed-loop voltage gain A_f of 10.
(c) What is the amount of feedback in decibels?
(d) If $V_s = 1$ V, find V_o, V_f, and V_i.
(e) If A decreases by 20%, what is the corresponding decrease in A_f?

Ans. **(a)** $\beta = R_1/(R_1 + R_2)$; **(b)** 9.01; **(c)** 60 dB; **(d)** 10 V, 0.999 V, 0.001 V;
(e) 0.02%

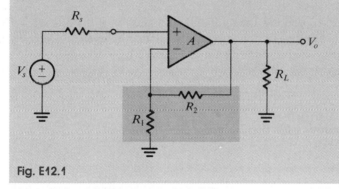

Fig. E12.1

12.2 SOME PROPERTIES OF NEGATIVE FEEDBACK

The properties of negative feedback were mentioned in the Introduction. In the following we shall consider some of these properties in more detail.

Gain Desensitivity

The effect of negative feedback on desensitizing the closed-loop gain was demonstrated in Exercise 12.1, where we saw that a 20% reduction in the gain of the basic amplifier gave rise to only a 0.02% reduction in the gain of the closed-loop amplifier. This sensitivity-reduction property can be analytically established as follows:

Assume that β is constant. Taking differentials of both sides of Eq. (12.4) results in

$$dA_f = \frac{dA}{(1 + A\beta)^2} \tag{12.5}$$

Dividing Eq. (12.5) by Eq. (12.4) yields

$$\frac{dA_f}{A_f} = \frac{1}{(1 + A\beta)} \frac{dA}{A} \tag{12.6}$$

which says that the percentage change in A_f (due to variations in some circuit parameter) is smaller than the change in A by the amount of feedback. For this reason the amount of feedback, $1 + A\beta$, is also known as the *desensitivity factor*.

Bandwidth Extension

Consider an amplifier whose high-frequency response is characterized by a single pole. Its gain can be expressed as

$$A(s) = \frac{A_M}{1 + s/\omega_H} \tag{12.7}$$

where A_M denotes the midband gain and ω_H is the uppper 3-dB frequency. Application of negative feedback, with a frequency-independent factor β, around this amplifier results in a closed-loop gain $A_f(s)$ given by

$$A_f(s) = \frac{A(s)}{1 + \beta A(s)}$$

Substituting for $A(s)$ from Eq. (12.7) results in

$$A_f(s) = \frac{A_M/(1 + A_M\beta)}{1 + s/\omega_H(1 + A_M\beta)}$$

Thus the feedback amplifier will have a midband gain of $A_M/(1 + A_M\beta)$ and an upper 3-dB frequency ω_{Hf} given by

$$\omega_{Hf} = \omega_H(1 + A_M\beta) \tag{12.8}$$

It follows that the upper 3-dB frequency is increased by a factor equal to the amount of feedback.

Similarly, it can be shown that if the open-loop gain is characterized by a dominant low-frequency pole giving rise to a lower 3-dB frequency ω_L, then the feedback amplifier will have a lower 3-dB frequency ω_{Lf},

$$\omega_{Lf} = \frac{\omega_L}{1 + A_M\beta} \tag{12.9}$$

Exercise

12.2 Consider the noninverting op-amp circuit of Exercise 12.1. Let the open-loop gain A have a low-frequency value of 10^4 and a uniform -6 dB/octave rolloff at high frequencies with a 3-dB frequency of 100 Hz. Find the low-frequency gain and the upper 3-dB frequency of a closed-loop amplifier with $R_1 = 1$ kΩ and $R_2 = 9$ kΩ.

Ans.   9.99 V/V; 100.1 kHz

Noise Reduction

Negative feedback can be employed to reduce the noise or interference in an amplifier or, more precisely, to increase the ratio of signal to noise. However, as we shall now explain, this noise-reduction process is possible only under certain conditions. Consider the situation illustrated in Fig. 12.2. Figure 12.2a shows an amplifier with gain A_1, an input signal V_s, and noise or interference V_n. It is assumed that for some reason this amplifier suffers from noise and that the noise can be assumed to be introduced at the input of the amplifier. The *signal-to-noise ratio* for this amplifier is

$$S/N = \frac{V_s}{V_n}$$

Consider next the circuit in Fig. 12.2b. Here we assume that it is possible to build another amplifier stage with gain A_2 that does not suffer from the noise problem. If

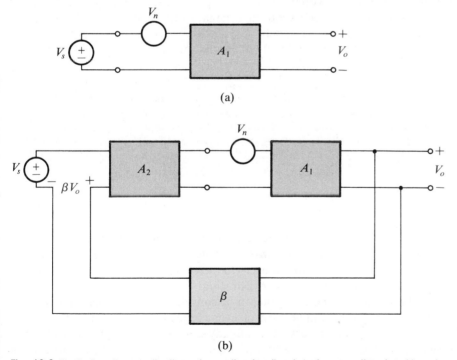

(a)

(b)

Fig. 12.2 Illustrating the application of negative feedback to improve the signal-to-noise ratio in amplifiers.

this is the case, then we may precede our original amplifier A_1 by the *clean* amplifier A_2 and apply negative feedback around the overall cascade of such an amount as to keep the overall gain constant. The output voltage of the circuit in Fig. 12.2b can be found by superposition:

$$V_o = V_s \frac{A_1 A_2}{1 + A_1 A_2 \beta} + V_n \frac{A_1}{1 + A_1 A_2 \beta}$$

Thus the signal-to-noise ratio at the output becomes

$$S/N = \frac{V_s}{V_n} A_2$$

which is A_2 times higher than in the original case.

We should emphasize once more that the improvement in signal-to-noise ratio by the application of feedback is possible only if one can precede the noisy stage by a (relatively) noise-free stage. This situation, however, is not uncommon in practice. The best example is found in the output power-amplifier stage of an audio amplifier. Such a stage usually suffers from a problem known as *power-supply hum*. The problem arises because of the large currents that this stage draws from the power supply and the difficulty in providing adequate power-supply filtering inexpensively.

The power-output stage is required to provide large power gain but little or no voltage gain. We may therefore precede the power-output stage by a small-signal amplifier that provides large voltage gain and apply a large amount of negative feedback, thus restoring the voltage gain to its original value. Since the small-signal amplifier can be fed from another, less hefty (and hence better regulated) power supply, it will not suffer from the hum problem. The hum at the output will then be reduced by the amount of the voltage gain of this added *preamplifier*.

Exercise

> **12.3** Consider a power-output stage with voltage gain $A_1 =$, an input signal $V_s = 1$ V, and a hum V_n of 1 V. Assume that this power stage is preceded by a small-signal stage with gain $A_2 = 100$ V/V and that overall feedback with $\beta = 1$ is applied. If V_s and V_n remain unchanged, find the signal and noise voltages at the output and hence the improvement in S/N.
>
> **Ans.** $\simeq 1$ V; $\simeq 0.01$ V; 100 (40 dB)

Reduction in Nonlinear Distortion

Curve (a) in Fig. 12.3 shows the transfer characteristic of an amplifier. As indicated, the characteristic is piecewise linear, with the voltage gain changing from 1,000 to 100 and then to 0. This nonlinear transfer characteristic will result in this amplifier generating a large amount of nonlinear distortion.

The amplifier transfer characteristic can be considerably *linearized* (that is, made less nonlinear) through the application of negative feedback. That this is possible should not be too suprising, since we have already seen that negative feedback reduces the dependence of the overall closed-loop amplifier gain on the open-loop gain of the basic amplifier. Thus large changes in open-loop gain (1,000 to 100 in this case) give rise to much smaller corresponding changes in the closed-loop gain.

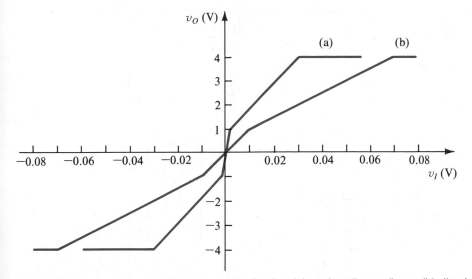

Fig. 12.3 Illustrating the application of negative feedback to reduce the nonlinear distortion in amplifiers. Curve (a) shows the amplifier transfer characteristic without feedback. Curve (b) shows the characteristic with negative feedback ($\beta = 0.01$) applied.

To illustrate, let us apply negative feedback with $\beta = 0.01$ to the amplifier whose open-loop voltage transfer characteristic is depicted in Fig. 12.3. The resulting transfer characteristic of the closed-loop amplifier is shown in Fig. 12.3 as curve (b). Here the slope of the steepest segment is given by

$$A_{f1} = \frac{1,000}{1 + 1,000 \times 0.01} = 90.9$$

and the slope of the next segment is given by

$$A_{f2} = \frac{100}{1 + 100 \times 0.01} = 50$$

Thus the order-of-magnitude change in slope has been considerably reduced. The price paid, of course, is a reduction in voltage gain. Thus if the overall gain has to be restored, then a preamplifier should be added. This preamplifier should not present a severe nonlinear distortion problem, since it will be dealing with smaller signals.

Finally, it should be noted that negative feedback does nothing about amplifier saturation, since in saturation the gain is very small (almost zero) and hence the amount of feedback is also very small (almost zero).

12.3 THE FOUR BASIC FEEDBACK TOPOLOGIES

Based on the quantity to be amplified (voltage or current) and on the desired form of output (voltage or current), amplifiers can be classified into four categories. These categories were discussed in Chapter 2. In the following we shall review this amplifier classification and point out the feedback topology appropriate in each case.

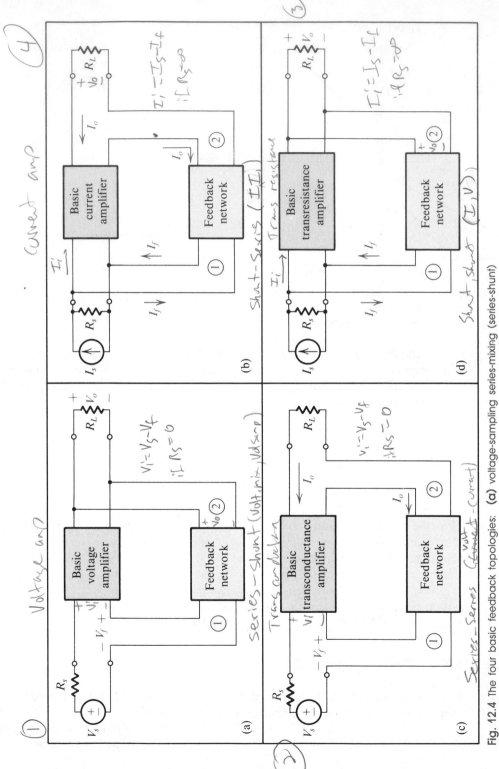

Fig. 12.4 The four basic feedback topologies: **(a)** voltage-sampling series-mixing (series-shunt) topology; **(b)** current-sampling shunt-mixing (shunt-series) topology; **(c)** current-sampling series-mixing (series-series) topology; **(d)** voltage-sampling shunt-mixing (shunt-shunt) topology.

Voltage Amplifiers

Voltage amplifiers are intended to amplify an input voltage signal and provide an output voltage signal. The voltage amplifier is essentially a voltage-controlled voltage source. The input impedance is required to be high, and the output impedance is required to be low. Since the signal source is essentially a voltage source, it is convenient to represent it in terms of a Thévenin equivalent circuit. In a voltage amplifier the output quantity of interest is the output voltage. It follows that the feedback network should *sample* the output *voltage*. Also, because of the Thévenin representation of the source, the feedback signal x_f should be a voltage that can be *mixed* with the source voltage in *series*.

Λ suitable feedback topology for the voltage amplifier is the *voltage-sampling series-mixing* one shown in Fig 12.4a. As will be shown, this topology not only stabilizes the voltage gain but also results in a higher input resistance and a lower output resistance, which are desirable properties for a voltage amplifier. The noninverting op-amp configuration of Fig. E12.1 is an example of this feedback topology. Finally, it should be mentioned that this feedback topology is also known as *series-shunt feedback*, where series refers to the connection at the input and shunt refers to the connection at the output.

Current Amplifiers

Here the input signal is essentially a current, and thus the signal source is most conveniently represented by its Norton equivalent. The output quantity of interest is current; hence the feedback network should *sample* the output *current*. The feedback signal should be in current form so that it may be *mixed* in *shunt* with the source current. Thus the feedback topology suitable for a current amplifier is the *current-sampling shunt-mixing* topology, illustrated in Fig. 12.4b. As will be shown, this topology not only stabilizes the current gain but also results in a lower input resistance and a higher output resistance, desirable properties for a current amplifier.

An example of the current-sampling shunt-mixing feedback topology is given in Fig. 12.5. Note that the bias details are not shown. Also note that the current being sampled is not the output current but the almost equal emitter current of Q_2. This is done for circuit design convenience and is quite usual in circuits involving current sampling.

The reference direction indicated in Fig. 12.5 for the feedback current I_f is such that it subtracts from I_s. This reference notation will be followed in all circuits in this chapter, since it is consistent with the notation used in the general feedback structure of Fig. 12.1. Therefore, in all circuits, for the feedback to be negative the loop gain $A\beta$ should be positive. The reader is urged to verify that in the circuit of Fig. 12.5 A is negative and β is negative.

It is of utmost importance to be able to qualitatively ascertain the feedback polarity (positive or negative). This can be done by "following the signal around the loop." For instance, let the current I_s in Fig. 12.5 increase. We see that the base current of Q_1 will increase, and thus its collector current will also increase. This will cause the collector voltage of Q_1 to decrease, and thus the collector current of Q_2, I_o, will

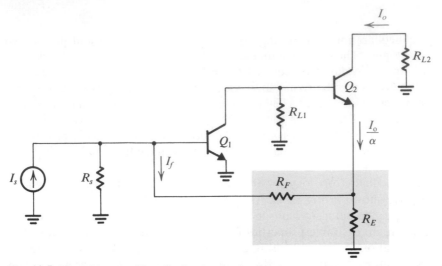

Fig. 12.5 A transistor amplifier with shunt-series feedback.

decrease. Thus the emitter current of Q_2, I_o/α (where α is the common-base current gain of the BJT), decreases. From the feedback network we see that if I_o/α decreases, then I_f (in the direction shown) will increase. The increase in I_f will subtract from I_s, causing a smaller increment to be seen by the amplifier. Hence the feedback is negative.

Finally, we should mention that this feedback topology is also known as *shunt-series* feedback.

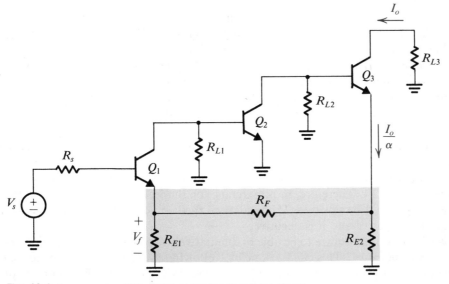

Fig. 12.6 An example of the series-series feedback topology.

Transconductance Amplifiers

Here the input signal is a voltage and the output signal is a current. It follows that the appropriate feedback topology is the *current-sampling series-mixing* topology, illustrated in Fig. 12.4c.

An example of this feedback topology is given in Fig. 12.6. Here note that as in the circuit of Fig. 12.5 the current sampled is not the output current but the almost equal emitter current of Q_3. In addition, the mixing loop is not a conventional one; it is not a simple series connection, since the feedback signal developed across R_{E1} is in the emitter circuit of Q_1, while the source is in the base circuit of Q_1. These two approximations are done for convenience of circuit design.

Finally, it should be mentioned that the current-sampling series-mixing feedback topology is also known as the *series-series* feedback configuration.

Transresistance Amplifiers

Here the input signal is current and the output signal is voltage. It follows that the appropriate feedback topology is of the *voltage-sampling current-mixing* type, shown in Fig. 12.4d.

An example of this feedback topology is found in the inverting op-amp configuration of Fig. 12.7a. The circuit is redrawn in Fig. 12.7b with the source converted to Norton's form.

This feedback topology is also known as *shunt-shunt* feedback.

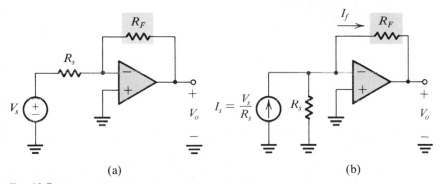

(a)   (b)

Fig. 12.7 The inverting op-amp configuration as an example of shunt-shunt feedback.

12.4 ANALYSIS OF THE SERIES-SHUNT FEEDBACK AMPLIFIER

The Ideal Situation

The ideal structure of the series-shunt feedback amplifier is shown in Fig. 12.8a. It consists of a *unilateral* open-loop amplifier (the A circuit) and an ideal voltage-sampling series-mixing feedback network (the β circuit). The A circuit has an input resistance R_i, a voltage gain A, and an output resistance R_o. It is assumed that the

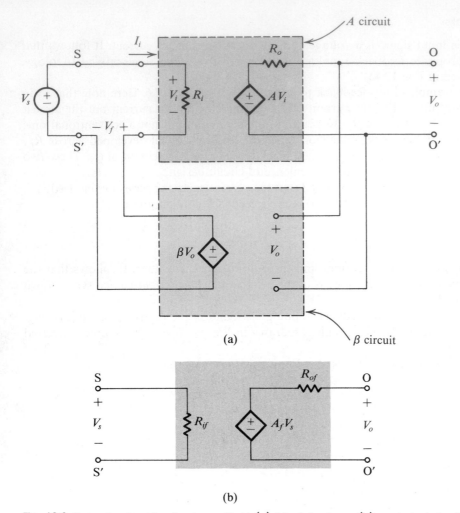

Fig. 12.8 The series-shunt feedback amplifier: **(a)** ideal structure; **(b)** equivalent circuit.

source and load resistances have been included inside the A circuit. Furthermore, note that the β circuit does *not* load the A circuit; that is, connecting the β circuit does not change the value of A (defined $A \equiv V_o/V_i$).

The circuit of Fig. 12.8a exactly follows the ideal feedback model of Fig. 12.1. Therefore the closed-loop voltage gain A_f is given by

$$A_f \equiv \frac{V_o}{V_s} = \frac{A}{1 + A\beta}$$

Note that A and β have reciprocal units. This in fact is always the case, resulting in a dimensionless loop gain $A\beta$.

The equivalent circuit model of the series-shunt feedback amplifier is shown in Fig. 12.8b. Here R_{if} and R_{of} denote the input and output resistances with feedback.

The relationship between R_{if} and R_i can be established by considering the circuit in Fig. 12.8a:

$$R_{if} \equiv \frac{V_s}{I_i} = \frac{V_s}{V_i/R_i}$$

$$= R_i \frac{V_s}{V_i}$$

$$= R_i \frac{V_i + \beta A V_i}{V_i}$$

Thus

$$R_{if} = R_i(1 + A\beta) \tag{12.10}$$

That is, the negative feedback in this case increases the input resistance by a factor equal to the amount of feedback. Since the above derivation does not depend on the method of sampling (shunt or series), it follows that the relationship between R_{if} and R_i is a function only of the method of mixing. We shall discuss this point further in later sections.

Note, however, that this result is not surprising and is physically intuitive: Since the feedback voltage V_f subtracts from V_s, the voltage that appears across R_i — that is, V_i—becomes quite small. Thus the input current I_i becomes correspondingly small and the resistance seen by V_s becomes large. Finally, it should be pointed out that Eq. (12.10) can be generalized to the form

$$Z_{if}(s) = Z_i(s)[1 + A(s)\beta(s)] \tag{12.11}$$

To find the output resistance, R_{of}, of the feedback amplifier in Fig. 12.8a we reduce V_s to zero and apply a test voltage V_t at the output, as shown in Fig. 12.9,

$$R_{of} \equiv \frac{V_t}{I}$$

From Fig. 12.9 we can write

$$I = \frac{V_t - A V_i}{R_o}$$

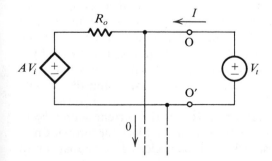

AV_i

Fig. 12.9 Measuring the output resistance of the feedback amplifier of Fig. 12.8a.

and since $V_s = 0$ it follows from Fig. 12.8a that

$$V_i = -V_f = -\beta V_o = -\beta V_t$$

Thus

$$I = \frac{V_t + A\beta V_t}{R_o}$$

leading to

$$R_{of} = \frac{R_o}{1 + A\beta} \qquad (12.12)$$

That is, the negative feedback in this case reduces the output resistance by a factor equal to the amount of feedback. With a little thought one can see that the derivation of Eq. (12.12) does not depend on the method of mixing. Thus the relationship between R_{of} and R_o depends only on the method of sampling. Again this result is not surprising and is physically intuitive: With reference to Fig. 12.9 we see that since the feedback samples V_t and is negative, the controlled voltage source will be $-A\beta V_t$, independent of the method of mixing. This large negative voltage causes the current I to be large, indicating that the effective output resistance is small. Finally, we note that Eq. (12.12) can be generalized to

$$Z_{of}(s) = \frac{Z_o(s)}{1 + A(s)\beta(s)} \qquad (12.13)$$

The Practical Situation

In a practical series-shunt feedback amplifier the feedback network will not be an ideal voltage-controlled voltage source. Rather, the feedback network is usually passive and hence will load the basic amplifier and thus affect the values of A, R_i, and R_o. In addition, the source and load resistances will affect these three parameters. Thus the problem we have is as follows: given a series-shunt feedback amplifier represented by the block diagram of Fig. 12.10a, find the A circuit and the β circuit.

Our problem essentially involves representing the amplifier of Fig. 12.10a by the ideal structure of Fig. 12.8a. As a first step toward that end we observe that the source and load resistances should be lumped with the basic amplifier. This, together with representing the two-port feedback network in terms of its h parameters (see Appendix B), is illustrated in Fig. 12.10b. The choice of h parameters is based on the fact that this is the only parameter set that represents the feedback network by a series network at port 1 and a parallel network at port 2. Such a representation is obviously convenient in view of the series connection at the input and the parallel connection at the output.

Examination of the circuit in Fig. 12.10b reveals that the current source $h_{21}I_1$ represents the forward transmission of the feedback network. Since the feedback network is usually passive, its forward transmission can be neglected in comparison to

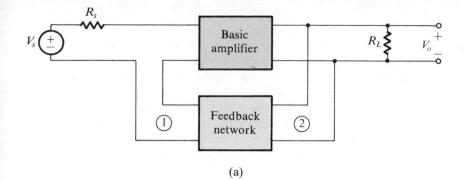

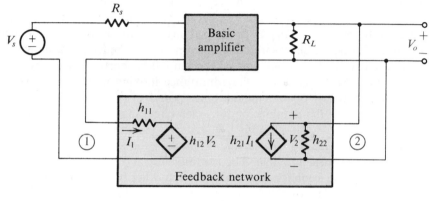

(b)

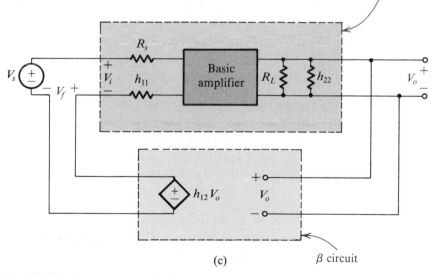

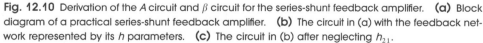

Fig. 12.10 Derivation of the A circuit and β circuit for the series-shunt feedback amplifier. **(a)** Block diagram of a practical series-shunt feedback amplifier. **(b)** The circuit in (a) with the feedback network represented by its h parameters. **(c)** The circuit in (b) after neglecting h_{21}.

the much larger forward transmission of the basic amplifier. We will therefore assume that h_{21} of the feedback network is approximately zero.

Compare the circuit of Fig. 12.10b (after eliminating the current source $h_{21}I_1$) to the ideal circuit of Fig. 12.8a. We see that by including h_{11} and h_{22} with the basic amplifier we obtain the circuit shown in Fig. 12.10c, which is very similar to the ideal circuit. Now, if the basic amplifier is unilateral (or almost unilateral), then the circuit of Fig. 12.10c is equivalent (or approximately equivalent) to the ideal circuit. It follows then that the A circuit is obtained by augmenting the basic amplifier at the input with the source impedance R_s and the impedance h_{11} of the feedback network, and augmenting it at the output with the load impedance R_L and the admittance h_{22} of the feedback network.

We conclude that the loading effect of the feedback network on the basic amplifier is represented by the components h_{11} and h_{22}. From the definitions of the h parameters in Appendix B we see that h_{11} is the impedance looking into port 1 of the feedback network with port 2 short-circuited. Since port 2 of the feedback network is connected in *shunt* with the output port of the amplifier, short-circuiting port 2 destroys the feedback. Similarly, h_{22} is the admittance looking into port 2 of the feedback network with port 1 open-circuited. Since port 1 of the feedback network is connected in *series* with the amplifier input, open-circuiting port 1 destroys the feedback.

These observations suggest a simple rule for finding the loading effects of the feedback network on the basic amplifier: The loading effect is found by looking into the appropriate port of the feedback network while the other port is open-circuited or short-circuited so as to destroy the feedback. If the connection is a shunt one, we short-circuit the port; if it is a series one, we open-circuit it. In Sections 12.5 and 12.6 it will be seen that this simple rule applies also to the other three feedback topologies.[1]

We next consider the determination of β. From Fig. 12.10c we see that β is equal to h_{12} of the feedback network,

$$\beta = h_{12} \equiv \left.\frac{V_1}{V_2}\right|_{I_1=0}$$

Thus to measure β one applies a voltage to port 2 and measures the voltage that appears at port 1 while the latter port is open-circuited. This result is intuitively appealing because the object of the feedback network is to sample the output voltage ($V_2 = V_o$) and provide a voltage signal ($V_1 = V_f$) that is mixed in series with the input source. The series connection at the input suggests that (as in the case of finding the loading effects of the feedback network) β should be found with port 1 open-circuited.

Summary

A summary of the rules for finding the A circuit and β for a given series-shunt feedback amplifier of the form in Fig. 12.10a is given in Fig. 12.11.

---

[1] A simple rule to remember is: If the connection is *shunt*, *sh*ort it; if *se*ries, *se*ver it.

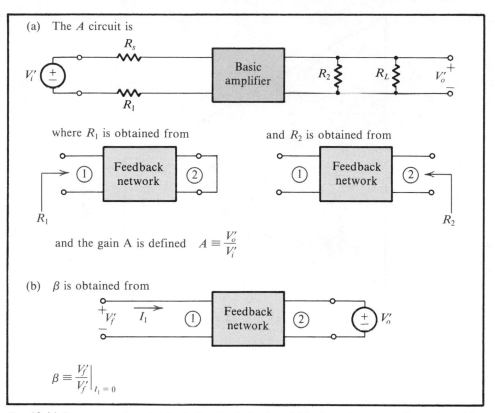

(a)   The A circuit is

where R_1 is obtained from   and R_2 is obtained from

and the gain A is defined $A \equiv \dfrac{V'_o}{V'_i}$

(b) β is obtained from

$\beta \equiv \dfrac{V'_f}{V'_f}\Bigg|_{I_1 = 0}$

Fig. 12.11 Summary of the rules for finding the A circuit and β for the voltage-sampling series-mixing case of Fig. 12.10a.

EXAMPLE 12.1

Figure 12.12a shows an op amp connected in the noninverting configuration. The op amp has an open-loop gain μ, a differential input resistance R_{id}, a common-mode input resistance R_{icm}, and an output resistance r_o. Find expressions for A, β, the closed-loop gain V_o/V_s, the input resistance R'_{if} (see Fig. 12.12a) and the output resistance R'_{of}. Also find numerical values, given $\mu = 10^4$, $R_{id} = 100 \ \mathrm{k\Omega}$, $R_{icm} = 10 \ \mathrm{M\Omega}$, $r_o = 1 \ \mathrm{k\Omega}$, $R_L = 2 \ \mathrm{k\Omega}$, $R_1 = 1 \ \mathrm{k\Omega}$, $R_2 = 1 \ \mathrm{M\Omega}$, and $R_s = 10 \ \mathrm{k\Omega}$.

Solution
We first observe that the existence of the two resistances labeled $2R_{icm}$ between the op-amp inputs and ground will complicate the analysis.[2] This problem, however, can be easily overcome using Thevenin's theorem, as shown in Fig. 12.12b. We now observe that the feedback network consists of R_2 and R'_1. This network samples the

---

[2] Refer to Fig. 12.10c and let the lower terminal of V_s be grounded. We can see that none of the input terminals of the A circuit can be grounded. Thus the resistance R_i is simply the resistance between the two input terminals of the A circuit. If, however, there exist inside the A circuit resistances with grounded terminals, the determination of R_i becomes quite lengthy.

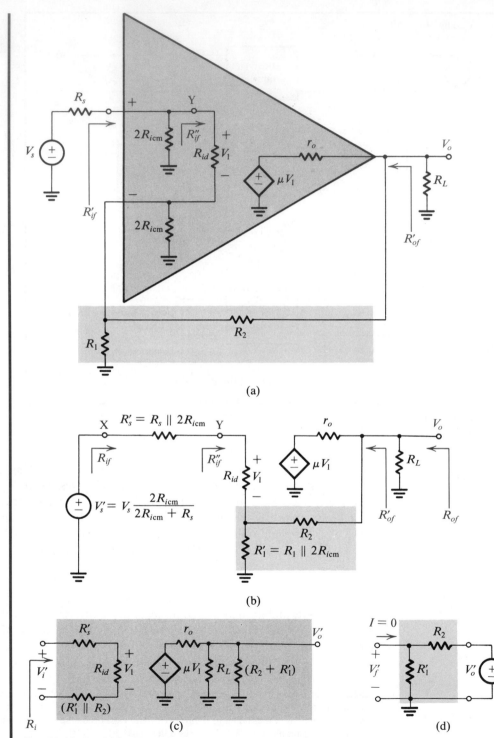

Fig. 12.12 Circuits for Example 12.1.

output voltage V_o and provides a voltage signal (across R'_1) that is mixed in series with the input source V'_s.

The A circuit is easily obtained with the rules of Fig. 12.11 and is shown in Fig. 12.12c. For this circuit we can write by inspection

$$A \equiv \frac{V'_o}{V'_i} = \mu \frac{[R_L \| (R'_1 + R_2)]}{[R_L \| (R'_1 + R_2)] + r_o} \frac{R_{id}}{R_{id} + R'_s + (R'_1 \| R_2)}$$

For the values given we find that $R'_1 \simeq R_1 = 1 \text{ k}\Omega$ and $R'_s \simeq R_s = 10 \text{ k}\Omega$, leading to $A \simeq 6{,}000$.

The circuit for obtaining β is shown in Fig. 12.12d, from which we obtain

$$\beta \equiv \frac{V'_f}{V'_o} = \frac{R'_1}{R'_1 + R_2} \simeq 10^{-3}$$

The voltage gain with feedback is now obtained as

$$A_f \equiv \frac{V_o}{V'_s} = \frac{A}{1 + A\beta} = \frac{6{,}000}{7} = 857 \text{ V/V}$$

This can be used to obtain V_o/V_s:

$$\frac{V_o}{V_s} = \frac{V_o}{V'_s} \frac{2R_{icm}}{2R_{icm} + R_s} \simeq \frac{V_o}{V'_s} = 857 \text{ V/V}$$

The input resistance R_{if} given by the feedback equations is the resistance seen by the external source (in the case V'_s); that is, it is the resistance between node X and ground in Fig. 12.12b. This resistance is given by

$$R_{if} = R_i(1 + A\beta)$$

where R_i is the input resistance of the A circuit in Fig. 12.12c:

$$R_i = R'_s + R_{id} + (R'_1 \| R_2)$$

For the values given, $R_i \simeq 111 \text{ k}\Omega$, resulting in

$$R_{if} = 111 \times 7 = 777 \text{ k}\Omega$$

This, however, is not the resistance asked for. What is required is R'_{if}, indicated in Fig. 12.12a. To obtain R'_{if} we first subtract R'_s from R_{if} and find R''_{if}, indicated in Fig. 12.12b and in Fig. 12.12a. From the latter figure we see that R'_{if} can be obtained by including $2R_{icm}$ in parallel with R''_{if}:

$$R'_{if} = [2R_{icm} \| (R_{if} - R'_s)]$$

For the values given, $R'_{if} = 739 \text{ k}\Omega$. The resistance R_{of} given by the feedback equations is the output resistance of the feedback amplifier, including the load resistance R_L, as indicated in Fig. 12.12b. R_{of} is given by

$$R_{of} = \frac{R_o}{1 + A\beta}$$

where R_o is the output resistance of the A circuit. R_o can be obtained by inspection of Fig. 12.12c as

$$R_o = [r_o \parallel R_L \parallel (R_2 + R_1')]$$

For the values given, $R_o \simeq 667 \ \Omega$ and

$$R_{of} = \frac{667}{7} = 95.3 \ \Omega$$

The resistance asked for, R_{of}', is the output resistance of the feedback amplifier excluding R_L. From Fig. 12.12b we see that

$$R_{of} = (R_{of}' \parallel R_L)$$

Thus

$$R_{of}' \simeq 100 \ \Omega$$

12.4 If the op amp of Example 12.1 has a uniform -6 dB/octave high-frequency rolloff with $f_{3dB} = 1$ kHz, find the 3-dB frequency of the closed-loop gain V_o/V_s.

Ans. 7 kHz

12.5 The circuit shown in Fig. E12.5 consists of a differential stage followed by an emitter follower, with series-shunt feedback supplied by the resistors R_1 and R_2. Assuming that the dc component of V_s is zero, find the dc operating current of each of the three transistors and show that the dc voltage at the output is approximately zero. Then find the values of A, β, $A_f \equiv V_o/V_s$, R_{if}', and R_{of}'. Assume that the transistors have $\beta = 100$.

Ans. 85.7 V/V; 0.1 V/V; 8.96 V/V; 191 kΩ; 19.1 Ω

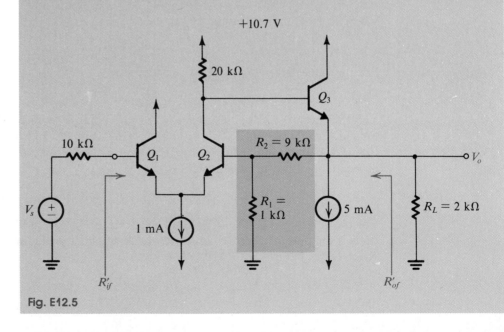

Fig. E12.5

12.5 ANALYSIS OF THE SERIES-SERIES FEEDBACK AMPLIFIER

The Ideal Case

As mentioned in Section 12.3, the series-series feedback topology stabilizes I_o/V_s and is therefore best suited for transconductance amplifiers. Figure 12.13a shows the ideal structure for the series-series feedback amplifier. It consists of a unilateral open-loop amplifier (the A circuit) and an ideal feedback network. Note that in this case A is a transconductance,

$$A \equiv \frac{I_o}{V_i} \tag{12.14}$$

while β is a transresistance. Thus the loop gain $A\beta$ remains a dimensionless quantity, as it should always be.

In the ideal structure of Fig. 12.13a, the load and source resistances have been absorbed inside the A circuit, and the β circuit does not load the A circuit. Thus the circuit follows the ideal feedback model of Fig. 12.1, and we can write

$$A_f \equiv \frac{I_o}{V_s} = \frac{A}{1 + A\beta} \tag{12.15}$$

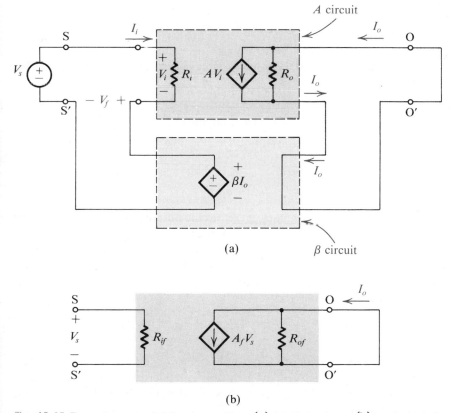

(a)

(b)

Fig. 12.13 The series-series feedback amplifier. (a) Ideal structure; (b) equivalent circuit.

This transconductance-with-feedback is included in the equivalent circuit model of the feedback amplifier, shown in Fig. 12.13b. In this model R_{if} is the input resistance with feedback. Using an analysis similar to that in Section 12.4 we can show that

$$R_{if} = R_i(1 + A\beta) \tag{12.16}$$

This relationship is identical to that obtained in the case of series-shunt feedback. This confirms our earlier observation that the relationship between R_{if} and R_i is a function only of the method of mixing. Series mixing therefore always increases the input resistance.

To find the output resistance R_{of} of the series-series feedback amplifier of Fig. 12.13a we reduce V_s to zero and break the output circuit to apply a test current I_t, as shown in Fig. 12.14:

$$R_{of} \equiv \frac{V}{I_t} \tag{12.17}$$

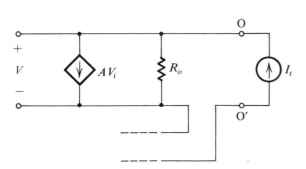

Fig. 12.14 Measuring the output resistance R_{of} of the series-series feedback amplifier.

In this case, $V_i = -V_f = -\beta I_o = -\beta I_t$. Thus for the circuit in Fig. 12.14 we obtain

$$V = (I_t - AV_i)R_o = (I_t + A\beta I_t)R_o$$

Hence

$$R_{of} = (1 + A\beta)R_o \tag{12.18}$$

That is, in this case the negative feedback increases the output resistance. This should have been expected, since the negative feedback tries to make I_o constant in spite of changes in the output voltage, which means increased output resistance. This result also confirms our earlier observation; the relationship betwen R_{of} and R_o is a function only of the method of sampling. While voltage (shunt) sampling reduces the output resistance, current (series) sampling increases it.

The Practical Case

Figure 12.15a shows a block diagram for a practical series-series feedback amplifier. To be able to apply the feedback equations to this amplifier we have to represent it by the ideal structure of Fig. 12.13a. Our objective therefore is to devise a simple method for finding A and β.

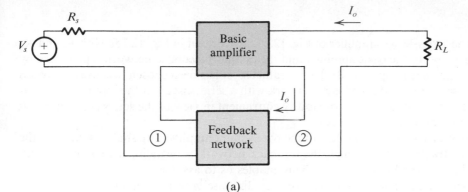

(a)

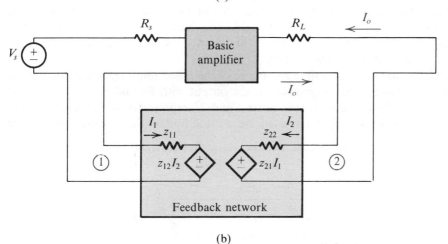

(b)

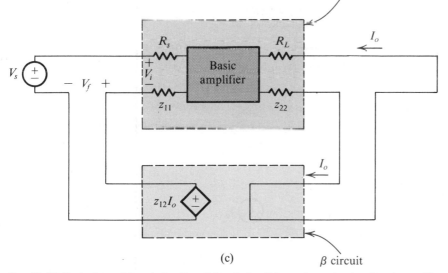

(c)

Fig. 12.15 Derivation of the A circuit and the β circuit for series-series feedback amplifiers. **(a)** A series-series feedback amplifier. **(b)** The circuit of (a) with the feedback network represented by its z parameters. **(c)** A redrawing of the circuit in (b) after neglecting z_{21}.

The series-series amplifier of Fig. 12.15a is redrawn in Fig. 12.15b with R_s and R_L shown closer to the basic amplifier and the two-port feedback network represented by its z parameters (Appendix B). This parameter set has been chosen because it provides a representation of the feedback network with a series circuit at the input and a series circuit at the output. This is obviously convenient in view of the series connections at input and output.

As we have done in the case of the series-shunt amplifier, we shall assume that the forward transmission through the feedback network is negligible as compared with that through the basic amplifier. This enables us to assume that z_{21} of the feedback network is almost zero, and thus we may dispense with the voltage source $z_{21}I_1$ in Fig. 12.15b. Doing this, and redrawing the circuit to include z_{11} and z_{22} with the basic amplifier, results in the circuit in Fig. 12.15c. Now if the basic amplifier is unilateral (or almost unilateral), we see that this circuit is equivalent (or almost equivalent) to the ideal circuit of Fig. 12.13a.

It follows that the A circuit is composed of the basic amplifier augmented at the input with R_s and z_{11} and augmented at the output with R_L and z_{22}. Since z_{11} and z_{22} are the impedances looking into ports 1 and 2, respectively, of the feedback network with the other port open-circuited, we see that finding the loading effects of the

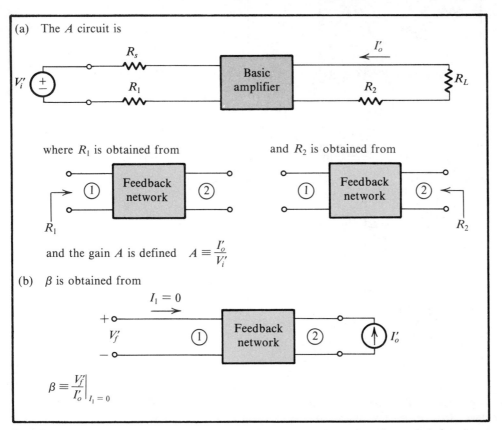

Fig. 12.16 Finding the A circuit and β for the current-sampling series-mixing (series-series) case.

feedback network on the basic amplifier follows the rule formulated in Section 12.4. That is, we look into one port of the feedback network while the other port is open-circuited or short-circuited so as to destroy the feedback (open if series and short if shunt).

From Fig. 12.15c we see that β is equal to z_{12} of the feedback network,

$$\beta = z_{12} \equiv \left.\frac{V_1}{I_2}\right|_{I_1 = 0} \tag{12.19}$$

This result is intuitively appealing. Recall that in this case the feedback network samples the output current $[I_2 = I_o]$ and provides a voltage $[V_f = V_1]$ that is mixed in series with the input source. Again, the series connection at the input suggests that β is measured with port 1 open.

Summary

For future reference we present in Fig. 12.16 a summary of the rules for finding A and β for a given series-series feedback amplifier of the type shown in Fig. 12.15a.

EXAMPLE 12.2

Figure 12.17a shows a voltage-controlled current-source circuit. It embodies feedback of the current-sampling series-mixing type. However, for practical reasons the current being sampled is *not* the output current in the collector but rather the almost-equal emitter current. We wish to carry out small-signal analysis to determine I_o/V_s, R'_{if}, and R'_{of} (see Figure 12.17a).

Solution

The small-signal equivalent circuit is shown in Fig. 12.17b, from which we see that series-series feedback is provided by resistance R_E. The rules of Fig. 12.16 can be directly applied to obtain the A circuit shown in Fig. 12.17c. For this circuit we can write by inspection

$$A \equiv \frac{I'_o}{V'_i} = \frac{g_m V_\pi r_o}{r_o + R_L + R_E} \frac{I}{V'_i}$$

$$= \frac{g_m r_o}{r_o + R_L + R_E} \frac{r_\pi}{r_\pi + R_s + R_E} \tag{12.20}$$

The circuit for determining β is shown in Fig. 12.17d, from which we obtain

$$\beta \equiv \frac{V'_f}{I'_o} = R_E \tag{12.21}$$

The overall transconductance A_f is then obtained using

$$A_f = \frac{A}{1 + A\beta}$$

Before substituting for A and β we shall make the approximation

$$r_o \gg R_L + R_E$$

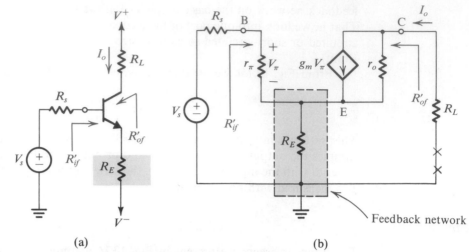

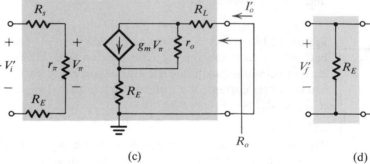

Fig. 12.17 Circuits for Example 12.2.

which reduces A to

$$A \simeq \frac{g_m r_\pi}{r_\pi + R_s + R_E} \tag{12.22}$$

Correspondingly we have for A_f

$$A_f \equiv \frac{I_o}{V_s} \simeq \frac{g_m r_\pi}{R_E(1 + g_m r_\pi) + r_\pi + R_s} \tag{12.23}$$

Substituting $g_m r_\pi = h_{fe}$ and

$$\frac{r_\pi}{h_{fe} + 1} = r_e$$

we obtain

$$\frac{I_o}{V_s} = \frac{h_{fe}}{R_E(1 + h_{fe}) + r_\pi + R_s} \tag{12.24}$$

a result that we could have written by inspection from our knowledge of transistor circuit analysis. Thus in simple circuits it is usually expedient to do the analysis directly. Nevertheless, the feedback method provides a quicker way for finding the output resistance, as will be demonstrated shortly.

The input resistance of the A circuit is obtained from Fig. 12.17c as

$$R_i = R_s + r_\pi + R_E$$

The input resistance with feedback, R_{if}, can be obtained as

$$R_{if} = (1 + A\beta)R_i = R_s + r_\pi + R_E + (g_m r_\pi)R_E \tag{12.25}$$

This is the input resistance of the feedback amplifier including R_s. To obtain the resistance required, R'_{if}, we simply subtract R_s from R_{if}:

$$R'_{if} = r_\pi + R_E(1 + g_m r_\pi) = r_\pi + R_E(1 + h_{fe})$$

an expression that we could have written by inspecting the circuit of Fig. 12.17a.

Finally, the output resistance of the A circuit can be obtained from Fig. 12.17c (after reducing the input V'_i to zero) as

$$R_o = R_L + r_o + R_E$$

The output resistance of the feedback circuit, R_{of}, is given by

$$R_{of} = (1 + A\beta)R_o$$

Use of the exact expression for A and substituting $\beta = R_E$ gives

$$R_{of} = R_L + R_E + r_o + \frac{g_m r_\pi R_E}{r_\pi + R_s + R_E} r_o \tag{12.26}$$

This resistance includes R_L. To obtain the output resistance R'_{of} excluding R_L we simply subtract R_L from R_{of}:

$$R'_{of} = R_E + r_o + \frac{g_m r_\pi R_E}{r_\pi + R_s + R_E} r_o$$

$$\simeq r_o\left(1 + \frac{h_{fe}R_E}{r_\pi + R_s + R_E}\right) \tag{12.27}$$

Exercise

12.6 Because negative feedback extends the amplifier bandwidth, it is commonly used in the design of broadband amplifiers. One such amplifier is the MC1553. Part of the circuit of the MC1553 is shown in Fig. E12.6. The circuit shown (called a *feedback triple*) is composed of three gain stages with series-series feedback provided by the network composed of R_{E1}, R_F, and R_{E2}. Assume that the bias circuit, which is not shown, causes $I_{C1} = 0.6$ mA, $I_{C2} = 1$ mA, and $I_{C3} = 4$ mA. Using these values and assuming that $h_{fe} = 100$ and $r_o = \infty$, find the open-loop gain A, the feedback factor β, the closed-loop gain $A_f \equiv I_o/V_s$, the voltage gain V_o/V_s, and the input resistance R_{if}. (*Hint:* In the A circuit you will find resistances in the emitters of Q_1 and Q_3. Do not use the feedback method to deal with these resistances; rather, it is simpler to use direct analysis.)

Ans. $A = 20.8$ A/V; $\beta = 12$ Ω; $A_f = 0.083$ A/V; $V_o/V_s = -49.8$; $R_{if} = 3.28$ MΩ

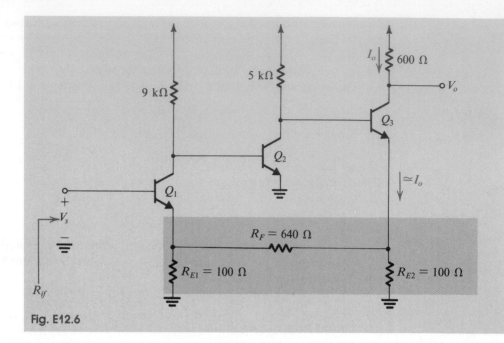

Fig. E12.6

12.6 ANALYSIS OF THE SHUNT-SHUNT AND THE SHUNT-SERIES FEEDBACK AMPLIFIERS

In this section we shall extend—without proof—the method of Sections 12.4 and 12.5 to the two remaining feedback topologies.

The Shunt-Shunt Configuration

Figure 12.18 shows the ideal structure for a shunt-shunt feedback amplifier. Here the A circuit has an input resistance R_i, a transresistance A, and an output resistance R_o. The β circuit is a voltage-controlled current source, and β is a transconductance. The closed-loop gain A_f is defined

$$A_f \equiv \frac{V_o}{I_s} \tag{12.28}$$

and is given by

$$A_f = \frac{A}{1 + A\beta}$$

The input resistance with feedback is given by

$$R_{if} = \frac{R_i}{1 + A\beta} \tag{12.29}$$

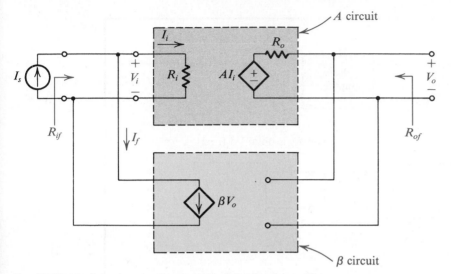

Fig. 12.18 Ideal structure for the shunt-shunt feedback amplifier.

where we note that the shunt connection at the input results in a reduced input resistance. Also note that the resistance R_{if} is the resistance seen by the source I_s, and it includes any source resistance.

The output resistance with feedback is given by

$$R_{of} = \frac{R_o}{1 + A\beta} \tag{12.30}$$

where we note that the shunt connection at the output results in a reduced output resistance. This resistance includes any load resistance.

Given a practical shunt-shunt feedback amplifier having the block diagram of Fig. 12.19, we use the method given in Fig. 12.20 to obtain the A circuit and the circuit for determining β. As in Sections 12.4 and 12.5, the method of Fig. 12.20 assumes that the basic amplifier is almost unilateral and that the forward transmission through the feedback network is negligibly small.

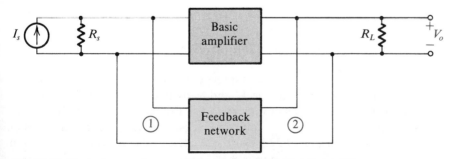

Fig. 12.19 Block diagram for a practical shunt-shunt feedback amplifier.

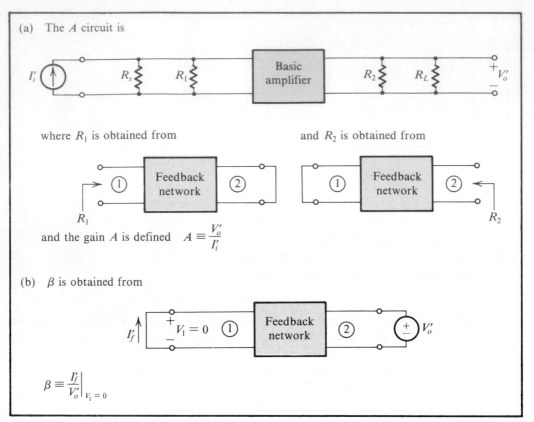

(a) The A circuit is

where R_1 is obtained from

and R_2 is obtained from

R_1

and the gain A is defined $A \equiv \dfrac{V'_o}{I'_i}$

(b) β is obtained from

$$\beta \equiv \left.\dfrac{I'_f}{V'_o}\right|_{V_1=0}$$

Fig. 12.20 Finding the A circuit and β for the voltage-sampling current-mixing (shunt-shunt) case.

EXAMPLE 12.3

We want to analyze the circuit of Fig. 12.21a to determine the small-signal voltage gain V_o/V_s, the input resistance R'_{if}, and the output resistance R_{of}. The transistor has $\beta = 100$.

Solution

First we determine the transistor dc operating point. The dc analysis is illustrated in Fig. 12.21b, from which we can write

$$V_C = 0.7 + (I_B + 0.07)47 = 3.99 + 47I_B \qquad \text{and} \qquad \frac{12 - V_C}{4.7} = (\beta + 1)I_B + 0.07$$

These two equations can be solved to obtain $I_B \simeq 0.015$ mA, $I_C \simeq 1.5$ mA, and $V_C = 4.7$ V.

To carry out small-signal analysis we first recognize that the feedback is provided by R_f, which samples the output voltage V_o and feeds back a current that is mixed with the source current. Thus it is convenient to use the Norton source representation, as shown in Fig. 12.21c. The A circuit can be easily obtained using the rules of Fig. 12.20, and it is shown in Fig. 12.22d. For the A circuit we can write by inspection

$$V_\pi = I'_i(R_s \parallel R_f \parallel r_\pi)$$

$$V'_o = -g_m V_\pi(R_f \parallel R_C)$$

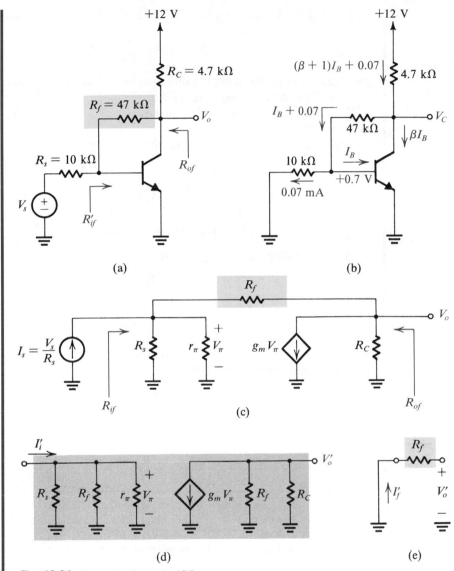

Fig. 12.21 Circuits for Example 12.3.

Thus

$$A = \frac{V'_o}{I'_i} = -g_m(R_f \parallel R_C)(R_s \parallel R_f \parallel r_\pi)$$

$$= -358.7 \text{ k}\Omega$$

The input and output resistances of the A circuit can be obtained from Fig. 12.21d as

$$R_i = (R_s \parallel R_f \parallel r_\pi) = 1.4 \text{ k}\Omega$$

$$R_o = (R_C \parallel R_f) = 4.27 \text{ k}\Omega$$

The circuit for determining β is shown in Fig. 12.21e, from which we obtain

$$\beta \equiv \frac{I'_f}{V'_o} = -\frac{1}{R_f} = -\frac{1}{47 \text{ k}\Omega}$$

Note that as usual the reference direction for I_f has been selected so that I_f subtracts from I_s. The resulting negative sign of β should cause no concern, since A is also negative, keeping the loop gain $A\beta$ positive, as it should be for the feedback to be negative.

We can now obtain A_f (for the circuit in Fig. 12.21c) as

$$A_f \equiv \frac{V_o}{I_s} = \frac{A}{1 + A\beta}$$

$$\frac{V_o}{I_s} = \frac{-358.7}{1 + 358.7/47} = \frac{-358.7}{8.63} = -41.6 \text{ k}\Omega$$

To find the voltage gain V_o/V_s we note that

$$V_s = I_s R_s$$

Thus

$$\frac{V_o}{V_s} = \frac{V_o}{I_s R_s} = \frac{-41.6}{10} \simeq -4.16 \text{ V/V}$$

The input resistance with feedback is given by

$$R_{if} = \frac{R_i}{1 + A\beta}$$

Thus

$$R_{if} = \frac{1.4}{8.63} = 162.2 \ \Omega$$

This is the resistance seen by the current source I_s in Fig. 12.21c. To obtain the input resistance of the feedback amplifier excluding R_s (that is, the required resistance R'_{if}) we subtract $1/R_s$ from $1/R_{if}$ and invert the result; thus $R'_{if} = 165 \ \Omega$. Finally, the amplifier output resistance R_{of} is evaluated using

$$R_{of} = \frac{R_o}{1 + A\beta} = \frac{4.27}{8.63} = 495 \ \Omega$$

The Shunt-Series Configuration

Figure 12.22 shows the ideal structure of the shunt-series feedback amplifier. It is a current amplifier whose gain with feedback is defined as

$$A_f \equiv \frac{I_o}{I_s} = \frac{A}{1 + A\beta} \tag{12.31}$$

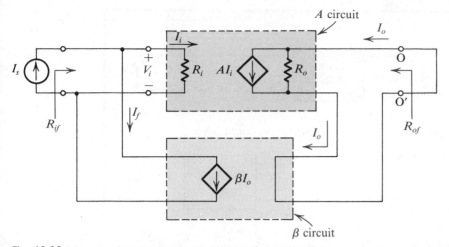

Fig. 12.22 Ideal structure for the shunt-series feedback amplifier.

The input resistance with feedback is the resistance seen by the current source I_s and is given by

$$R_{if} = \frac{R_i}{1 + A\beta} \tag{12.32}$$

Again we note that the shunt connection at the input reduces the input resistance. The output resistance with feedback is the resistance seen by breaking the output circuit, such as between O and O', and looking between the two terminals thus generated (that is, between O and O'). This resistance R_{of} is given by

$$R_{of} = R_o(1 + A\beta) \tag{12.33}$$

where we note that the increase in output resistance is due to the current (series) sampling.

Given a practical shunt-series feedback amplifier, such as that represented by the block diagram of Fig. 12.23, we follow the method given in Fig. 12.24 in order to obtain A and β.

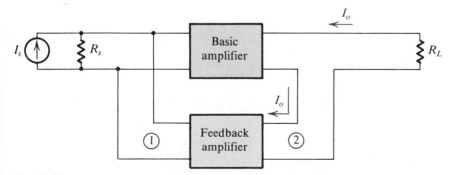

Fig. 12.23 Block diagram for a practical shunt-series feedback amplifier.

(a) The A circuit is

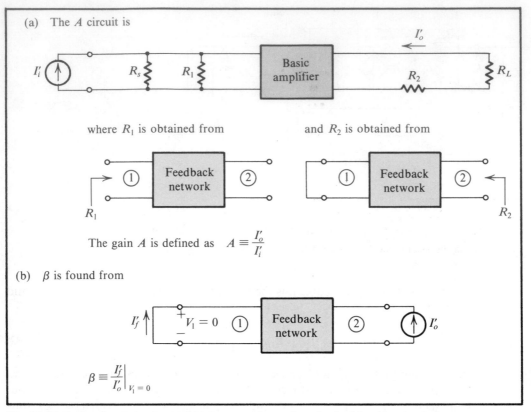

where R_1 is obtained from   and R_2 is obtained from

The gain A is defined as $A \equiv \dfrac{I'_o}{I'_i}$

(b) β is found from

$$\beta \equiv \left.\dfrac{I'_f}{I'_o}\right|_{V_1 = 0}$$

Fig. 12.24 Finding the A circuit and β for the current-sampling shunt-mixing (shunt-series) case.

EXAMPLE 12.4

Figure 12.25 shows a feedback circuit of the shunt-series type. Find $I_{\text{out}}/I_{\text{in}}$, R_{in}, and R_{out}. Assume the transistors to have $\beta = 100$ and $r_o = 100 \text{ k}\Omega$.

Solution

We begin by determining the dc operating points. In this regard we note that the feedback signal is capacitively coupled; thus the feedback has no effect on dc. The dc analysis proceeds as follows:

$$V_{B1} \simeq 12 \, \frac{15}{100 + 15} = 1.57 \text{ V}$$

$$V_{E1} \simeq 1.57 - 0.7 = 0.87 \text{ V}$$

$$I_{E1} = \frac{0.87}{0.87} = 1 \text{ mA}$$

$$V_{C1} \simeq 12 - 10 \times 1 = 2 \text{ V}$$

$$V_{E2} \simeq 2 - 0.7 = 1.3 \text{ V}$$

$$I_{E2} \simeq \frac{1.3}{1.3} \simeq 1 \text{ mA}$$

$$V_{C2} \simeq 12 - 1 \times 8 = 4 \text{ V}$$

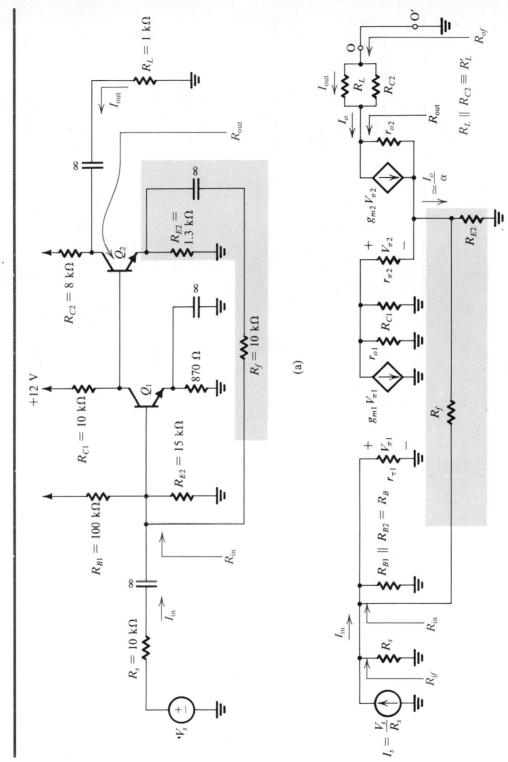

Fig. 12.25 Circuits for Example 12.4

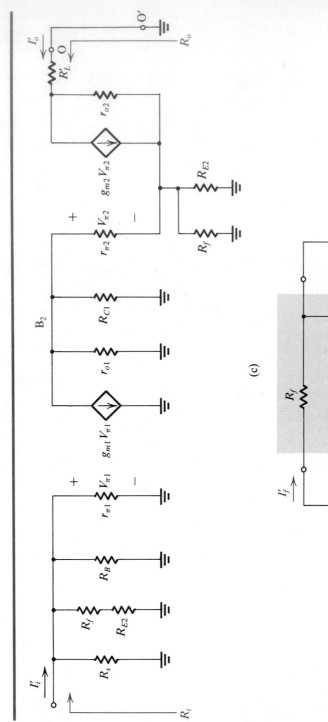

(c)

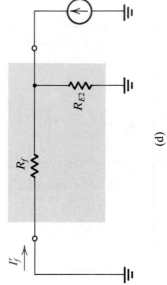

(d)

Fig. 12.25 (Continued).

The amplifier equivalent circuit is shown in Fig. 12.25b, from which we note that the feedback network is composed of R_{E2} and R_f. The feedback network samples the emitter current of Q_2, which is approximately equal to the collector current I_o. Also note that the required current gain, $I_{\text{out}}/I_{\text{in}}$, will be slightly different than the closed-loop current gain $A_f \equiv I_o/I_s$.

The A circuit is shown in Fig. 12.25c, where we have obtained the loading effects of the feedback network using the rules of Fig. 12.24. For the A circuit we can write

$$V_{\pi 1} = I_i'[R_s \,\|\, (R_{E2} + R_f) \,\|\, R_B \,\|\, r_{\pi 1}]$$

$$V_{b2} = -g_{m1}V_{\pi 1}\{r_{o1} \,\|\, R_{C1} \,\|\, [r_{\pi 2} + (\beta + 1)(R_{E2} \,\|\, R_f)]\}$$

$$V_{\pi 2} = V_{b2}\,\frac{r_{\pi 2}}{r_{\pi 2} + (\beta + 1)(R_{E2} \,\|\, R_f)}$$

$$I_o' \simeq g_{m2}V_{\pi 2}$$

These equations can be combined to obtain the open-loop current gain A,

$$A \equiv \frac{I_o'}{I_i'} \simeq 434.6$$

The input resistance R_i is given by

$$R_i = \lfloor R_s \,\|\, (R_{E2} + R_f) \,\|\, R_B \,\|\, r_{\pi 1}\rfloor = 1.5 \text{ k}\Omega$$

The output resistance R_o is that found by looking into the output loop of the A circuit (see Fig. 12.25c) with the input excitation I_i' set to zero. It can be shown that

$$R_o = R_L' + r_{o2} + \left[1 + (g_{m2}r_{o2})\,\frac{r_{\pi 2}}{r_{\pi 2} \,|\, (R_{C1} \,\|\, r_{o1})}\right][R_{E2} \,\|\, R_f \,\|\, (r_{\pi 2} + R_{C1} \,\|\, r_{o1})]$$

$$= 1.08 \text{ M}\Omega$$

The circuit for determining β is shown in Fig. 12.25d, from which we find

$$\beta \equiv \frac{I_f'}{I_o'} = -\frac{R_{E2}}{R_{E2} + R_f} = -\frac{1.3}{11.3} = -0.115$$

Thus

$$1 + A\beta = 51$$

The input resistance R_{if} is given by

$$R_{if} = \frac{R_i}{1 + A\beta} = 29.4 \ \Omega$$

The required input resistance R_{in} is given by

$$R_{\text{in}} = \frac{1}{1/R_{if} - 1/R_s} = 29.5 \ \Omega$$

Since $R_{in} \simeq R_{if}$, it follows from Fig. 12.25b that $I_{in} \simeq I_s$. The current gain A_f is given by

$$A_f \equiv \frac{I_o}{I_s} = \frac{A}{1 + A\beta} = -8.52$$

Note that because $A\beta \gg 1$ the closed-loop gain is approximately equal to $1/\beta$.

Now the required current gain is given by

$$\frac{I_{out}}{I_{in}} \simeq \frac{I_{out}}{I_s} = \frac{R_{C2}}{R_L + R_{C2}} \frac{I_o}{I_s}$$

Thus $I_{out}/I_{in} = -7.57$.

Finally, the output resistance R_{of} is given by

$$R_{of} = R_o(1 + A\beta) \simeq 55 \text{ M}\Omega$$

The required output resistance R_{out} can be obtained by subtracting R'_L from R_{of}:

$$R_{out} \simeq R_{of} = 55 \text{ M}\Omega$$

Exercise

12.7 Use the feedback method to find the voltage gain V_o/V_s, the input resistance R'_{if}, and the output resistance R'_{of} of the inverting op-amp configuration of Fig. E12.7. Let the op amp have open-loop gain $\mu = 10^4$, $R_{id} = 100$ kΩ, $R_{icm} = 10$ MΩ, and $r_o = 1$ kΩ. (*Hint:* The feedback is of the shunt-shunt type.)

Ans. -870 V/V; 150 Ω; 92 Ω

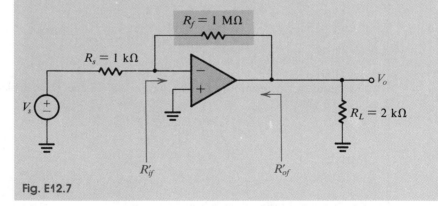

Fig. E12.7

12.7 DETERMINING THE LOOP GAIN

We have already seen that the loop gain $A\beta$ is a very important quantity that characterizes a feedback loop. Furthermore, in the following sections it will be shown that $A\beta$ determines whether the feedback amplifier is stable (as opposed to oscillatory). In this section we shall describe an alternative approach to the determination of loop gain.

Consider first the general feedback amplifier shown in Fig. 12.1. Let the external source x_s be set to zero. Open the feedback loop by breaking the connection of x_o to the feedback network and apply a test signal x_t. We see that the signal at the output of the feedback network is $x_f = \beta x_t$, that at the input of the basic amplifier is $x_i = -\beta x_t$ and the signal at the output of the amplifier, where the loop was broken, will be $x_o = -A\beta x_t$. It follows that the loop gain $A\beta$ is given by the negative of the ratio of the *returned* signal to the applied test signal; that is, $A\beta = -x_o/x_t$. It should also be obvious that this applies regardless of where the loop is broken.

In breaking the feedback loop of a practical amplifier circuit, we must ensure that the conditions that existed prior to breaking the loop do not change. This is achieved by terminating the loop where it is opened with an impedance equal to that seen (by the driving circuit) before the loop was broken. To be specific, consider the conceptual feedback loop shown in Fig. 12.26a. If we break the loop at XX', and apply a test voltage V_t to the terminals thus created to the left of XX', the terminals at the right of XX' should be loaded with an impedance Z_t as shown in Fig. 12.26b. The impedance Z_t is equal to that previously seen looking to the left of XX'. The loop gain $A\beta$ is then determined from

$$A\beta = -\frac{V_r}{V_t}$$

Finally, it should be noted that in some cases it may be convenient to determine $A\beta$ by applying a test current I_t and finding the returned current signal I_r. In this case, $A\beta = -I_r/I_t$.

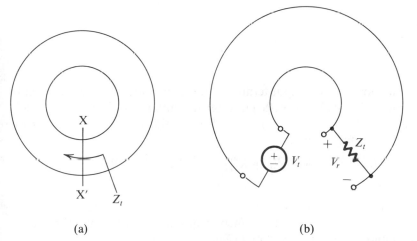

(a)   (b)

Fig. 12.26 A conceptual feedback loop is broken at XX' and a test voltage V_t is applied. The impedance Z_t is equal to that previously seen looking to the left of XX'. The loop gain $A\beta = -V_r/V_t$.

To illustrate the above procedure we consider the feedback loop shown in Fig. 12.27a. This feedback loop represents both the inverting and the noninverting op-amp configurations. Using a simple equivalent circuit model for the op amp we obtain the circuit of Fig. 12.27b. Examination of this latter circuit reveals that a convenient

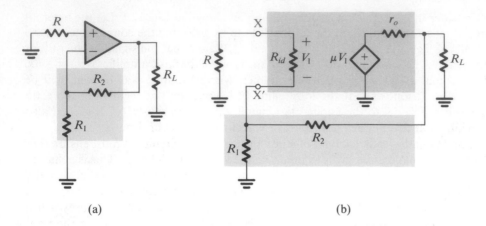

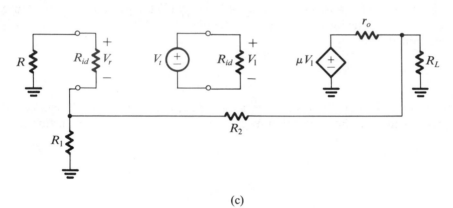

(a)   (b)

(c)

Fig. 12.27 Determination of the loop gain of the feedback loop in (a).

place to break the loop is at the input terminals of the op amp. The loop, broken in this manner, is shown in Fig. 12.27c with a test signal V_t applied to the right-hand-side terminals and a resistance R_{id} terminating the left-hand-side terminals. The returned voltage V_r is found by inspection as

$$V_r = -\mu V_1 \frac{\{R_L \| [R_2 + R_1 \| (R_{id} + R)]\}}{\{R_L \| [R_2 + R_1 \| (R_{id} + R)]\} + r_o} \frac{[R_1 \| (R_{id} + R)]}{[R_1 \| (R_{id} + R)] + R_2} \frac{R_{id}}{R_{id} + R}$$

This equation can be used directly to find the loop gain $L = A\beta = -V_r/V_t = -V_r/V_1$.

Since the loop gain L is generally a function of frequency, it is usual to call it *loop transmission* and denote it by $L(s)$ or $L(j\omega)$.

Equivalence of Circuits from a Feedback-Loop Point of View

From the study of circuit theory we know that the poles of a circuit are independent of the external excitation. In fact the poles, or the natural modes (which is a more appropriate name), are determined by setting the external excitation to zero. It follows that the poles of a feedback amplifier depend only on the feedback loop. This will be confirmed in a later section where we show that the characteristic equation (whose roots are the poles) is completely determined by the loop gain. Thus a given feedback loop may be used to generate a number of circuits having the same poles but different transmission zeros. The closed-loop gain and the transmission zeros depend on how and where the input signal is injected into the loop.

As an example consider the feedback loop of Fig. 12.27a. This loop can be used to generate the noninverting op-amp circuit by feeding the input voltage signal to the terminal of R that is connected to ground; that is, we lift this terminal off ground and connect it to V_s. The same feedback loop can be used to generate the inverting op-amp circuit by feeding the input voltage signal to the terminal of R_1 that is connected to ground.

Recognition of the fact that two or more circuits are equivalent from a feedback-loop point of view is very useful because (as will be shown in Section 12.8) stability is a function of the loop. Thus one needs to perform the stability analysis only once for a given loop.

In Chapter 14 we shall employ the concept of loop equivalence in the synthesis of active filters.

Exercises

12.8 For the circuit in Fig. 12.17b, determine the loop gain by breaking the feedback loop to the left of R_E. Note that Z_t will be $(r_\pi + R_s)$. Compare the result with that obtained by multiplying A and β obtained in Example 12.2. Note that small differences in results should be expected because of the approximations introduced in the basic feedback analysis.

Ans. $A\beta = \dfrac{g_m r_\pi}{\left(1 + \dfrac{R_s + r_\pi}{R_E}\right)\left(1 + \dfrac{R_L}{r_o}\right) + \dfrac{R_s + r_\pi}{r_o}}$

12.9 Find the numerical value of $A\beta$ for the amplifier in Exercise 12.7.

Ans. 3314 V/V

12.8 THE STABILITY PROBLEM

In a feedback amplifier such as that represented by the general structure of Fig. 12.1, the open-loop gain A is generally a function of frequency, and it should therefore be more accurately called the *open-loop transfer function, A(s)*. Also, we have been assuming for the most part that the feedback network is resistive and hence that the feedback factor β is constant, but this need not be always the case. We shall therefore assume that in the general case the *feedback transfer function is β(s)*. It follows that the *closed-loop transfer function $A_f(s)$* is given by

$$A_f(s) = \frac{A(s)}{1 + A(s)\beta(s)} \tag{12.34}$$

To focus attention on the points central to our discussion in this section, we shall assume that the amplifier is direct-coupled with constant dc gain A_0 and with poles and zeros occurring in the high-frequency band. Also, for the time being let us assume that at low frequencies $\beta(s)$ reduces to a constant value. Thus at low frequencies the loop gain $A(s)\beta(s)$ becomes a constant, which should be a positive number; otherwise the feedback would not be negative. The question then arises as to what happens at higher frequencies.

For physical frequencies $s = j\omega$, Eq. (12.34) becomes

$$A_f(j\omega) = \frac{A(j\omega)}{1 + A(j\omega)\beta(j\omega)} \tag{12.35}$$

Thus the loop gain $A(j\omega)\beta(j\omega)$ is a complex number that can be represented by its magnitude and phase,

$$
\begin{aligned}
L(j\omega) &\equiv A(j\omega)\beta(j\omega) \\
&= |A(j\omega)\beta(j\omega)| e^{j\phi(\omega)}
\end{aligned} \tag{12.36}
$$

It is the manner in which the loop-gain varies with frequency that determines the stability or instability of the feedback amplifier. To appreciate this fact, consider the frequency at which the phase angle $\phi(\omega)$ becomes $180°$. At this frequency, ω_{180}, the loop gain $A(j\omega)\beta(j\omega)$ will be a real number with a negative sign. Thus at this frequency the feedback will become positive. If at $\omega = \omega_{180}$ the magnitude of the loop-gain is less than unity, then from Eq. (12.35) we see that the closed-loop gain $A_f(j\omega)$ will be greater than the open-loop gain $A(j\omega)$, since the denominator of Eq. (12.35) will be smaller than unity. Nevertheless, the feedback amplifier will be stable.

On the other hand, if at the frequency ω_{180} the magnitude of the loop-gain is equal to unity, it follows from Eq. (12.35) that $A_f(j\omega)$ will be infinite. This means that the amplifier will have an output for zero input, which is by definition an *oscillator*. To visualize how this feedback loop may oscillate, consider the general loop of Fig. 12.1 with the external input x_s set to zero. Any disturbance in the circuit, such as the closure of the power-supply switch, will generate a signal $x_i(t)$ at the input to the amplifier. Such a noise signal usually contains a wide range of frequencies, and we shall now concentrate on the component with frequency $\omega = \omega_{180}$, that is, the signal $X_i \sin(\omega_{180}t)$. This input signal will result in a feedback signal given by

$$X_f = A(j\omega_{180})\beta(j\omega_{180})X_i = -X_i$$

Since X_f is further multiplied by -1 in the summer block at the input, we see that the feedback causes the signal X_i at the amplifier input to be *sustained*. That is, from this point on, there will be sinusoidal signals at the amplifier input and output of frequency ω_{180}. Thus the amplifier is said to oscillate at the frequency ω_{180}.

The question now is:.What happens if at ω_{180} the magnitude of the loop gain is greater than unity? We shall answer this question, not in general, but for the restricted yet very important class of circuits in which we are interested here. The answer, which is not obvious from Eq. (12.35), is that the circuit will oscillate, and the oscillations will grow in amplitude until some nonlinearity (which is always present is some form) reduces the magnitude of the loop gain to exactly unity, at which point sustained oscillations will be obtained. This mechanism for starting oscillations by using positive feedback with a loop gain greater than unity, and then using a nonlinearity to reduce the loop gain to unity at the desired amplitude, will be exploited in the design of sinusoidal oscillators in Chapter 14. Our objective here is just the opposite: now that we know how oscillations could occur in a negative-feedback amplifier, we wish to find methods to prevent their occurrence.

The Nyquist Plot

The Nyquist plot is a formalized approach for testing for stability based on the above discussion. It is simply a polar plot of loop gain with frequency used as a parameter. Figure 12.28 shows such a plot. Note that the radial distance is $|A\beta|$ and the angle is the phase angle ϕ. The solid-line plot is for positive frequencies. Since the loop gain—and for that matter any gain function of a physical network—has a magnitude that is an even function of frequency and a phase that is an odd fucntion of frequency, the $A\beta$ plot for negative frequencies can be drawn as a mirror image through the Re axis and is shown in Fig. 12.28 as a broken line.

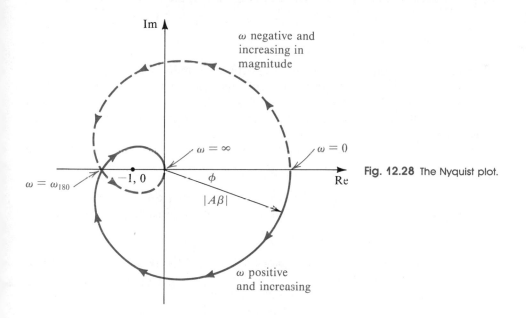

Fig. 12.28 The Nyquist plot.

The Nyquist plot intersects the negative real axis at the frequency ω_{180}. Thus if this intersection occurs to the left of the point $(-1, 0)$, we know that the magnitude of loop gain at this frequency is greater than unity and the amplifier will be unstable. On the other hand, if the intersection occurs to the right of the point $(-1, 0)$ the amplifier will be stable. It follows that if the Nyquist plot *encircles* the point $(-1, 0)$ then the amplifier will be unstable. It should be mentioned, however, that this statement is a simplified version of the *Nyquist criterion*; nevertheless, it applies to all the circuits in which we are interested. For the full theory behind the Nyquist method and for details on its application, consult Haykin (1970).

Exercise

12.10 Consider a feedback amplifier for which the open-loop transfer function $A(s)$ is given by

$$A(s) = \left(\frac{10}{1 + s/10^4}\right)^3$$

Let the feedback factor β be a constant independent of frequency. Find the frequency ω_{180} at which the phase shift is 180°. Then, show that the feedback amplifier will be stable if the feedback factor β is less than a critical value β_{cr} and unstable if $\beta \geq \beta_{cr}$, and find the value of β_{cr}.

Ans. $\omega_{180} = \sqrt{3} \times 10^4$ rad/s; $\beta_{cr} = 0.008$

12.9 EFFECT OF FEEDBACK ON THE AMPLIFIER POLES

The amplifier frequency response and stability are determined directly by its poles. We shall therefore investigate the effect of feedback on the poles of the amplifier.

Stability and Pole Location

We shall begin by considering the relation between stability and pole location. For an amplifier or any other system to be stable, its poles should lie in the left half of the *s* plane. A pair of complex conjugate poles on the $j\omega$ axis gives rise to sustained sinusoidal oscillations. Poles in the right half of the *s* plane give rise to growing oscillations.

To verify the above statement, consider an amplifier with a pole pair at

$$s = \sigma_0 \pm j\omega_n$$

If this amplifier is subjected to a disturbance, such as that caused by closure of the power-supply switch, its transient response will contain terms of the form

$$v(t) = e^{\sigma_0 t}[e^{+j\omega_n t} + e^{-j\omega_n t}] = 2e^{\sigma_0 t}\cos(\omega_n t)$$

This is a sinusoidal signal with an envelope $e^{\sigma_0 t}$. Now if the poles are in the left half of the *s* plane, then σ_0 will be negative and the oscillations will decay exponentially toward zero, as shown in Fig. 12.29a, indicating that the system is stable. If, on the other hand, the poles are in the right half-plane, then σ_0 will be positive and the oscillations will grow exponentially (until some nonlinearity limits their growth), as shown in Fig. 12.29b. Finally, if the poles are on the $j\omega$ axis, then σ_0 will be zero and the oscillations will be sustained, as shown in Fig. 12.29c.

Although the above discussion is in terms of complex conjugate poles, it can be shown that the existence of any right-half-plane poles results in instability. Here it is interesting to note that instability does not necessarily lead to sinusoidal oscillations. For instance, the bistable circuit studied in Chapter 5 has an unstable state in which it has right-half-plane poles. Nevertheless, it does not show explicit oscillations. It can, however, be thought of as oscillating at zero frequency.

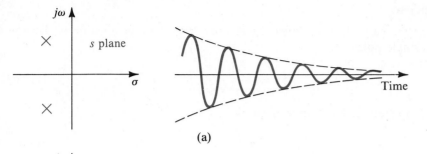

(a)

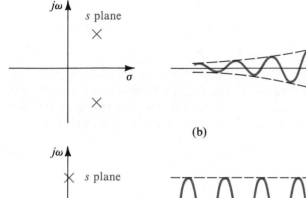

(b)

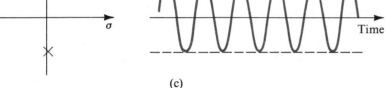

(c)

Fig. 12.29 Relationship between pole location and transient response.

Poles of the Feedback Amplifier

From the closed-loop transfer function in Eq. (12.34) we see that the poles of the feedback amplifier are the zeros of $1 + A(s)\beta(s)$. That is, the feedback amplifier poles are obtained by solving the equation

$$1 + A(s)\beta(s) = 0 \qquad (12.37)$$

which is called the *characteristic equation* of the feedback loop. It should therefore be apparent that applying feedback to an amplifier changes its poles.

In the following we shall consider how feedback affects the amplifier poles. For this purpose we shall assume that the open-loop amplifier has real poles and no finite zeros (that is, all the zeros are at $s = \infty$). This will simplify the analysis and enable us to focus our attention on the fundamental concepts involved. We shall also assume that the feedback factor β is independent of frequency.

Amplifier With Single-Pole Response

Consider first the case of an amplifier whose open-loop transfer function is characterized by a single pole.

$$A(s) = \frac{A_0}{1 + s/\omega_P} \tag{12.38}$$

The closed-loop transfer function is given by

$$A_f(s) = \frac{A_0/(1 + A_0\beta)}{1 + s/\omega_P(1 + A_0\beta)} \tag{12.39}$$

Thus the feedback moves the pole along the negative real axis to a frequency ω_{Pf},

$$\omega_{Pf} = \omega_P(1 + A_0\beta) \tag{12.40}$$

This process is illustrated in Fig. 12.30a. Figure 12.30b shows Bode plots for $|A|$ and $|A_f|$. Note that while at low frequencies the difference between the two plots is $20 \log(1 + A_0\beta)$, the two curves coincide at high frequencies. One can show that this indeed is the case by approximating Eq. (12.39) for frequencies $\omega \gg \omega_P(1 + A_0\beta)$:

$$A_f(s) \simeq \frac{A_0\omega_P}{s} \simeq A(s) \tag{12.41}$$

Physically speaking, at such high frequencies the loop gain is much smaller than unity and the feedback is ineffective.

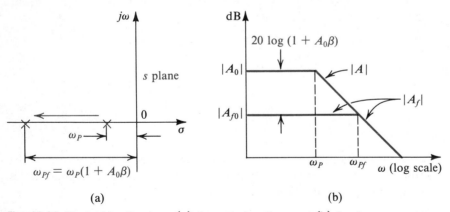

Fig. 12.30 Effect of feedback on **(a)** the pole location, and **(b)** the frequency response of an amplifier having a single-pole open-loop response.

Figure 12.30b clearly illustrates the fact that applying negative feedback to an amplifier results in extending its bandwidth at the expense of a reduction in gain. Since the pole of the closed-loop amplifier never enters the right half of the s plane, the single-pole amplifier is stable for any value of β. Thus this amplifier is said to be *unconditionally stable*. This result, however, is hardly surprising, since the phase lag associated with a single-pole response can never be greater than 90°. Thus the loop gain never achieves the 180° phase shift required for the feedback to become positive.

Exercise

12.11 An op amp having a single-pole rolloff at 100 Hz and a low-frequency gain of 10^5 is operated in a feedback loop with $\beta = 0.01$. What is the factor by which feedback shifts the pole? To what frequency? For a closed-loop gain of 1, to what frequency does the pole shift?

Ans.   1001; 100.1 kHz; 10 MHz

Amplifier with Two-Pole Response

Consider next an amplifier whose open-loop transfer function is characterized by two real-axis poles:

$$A(s) = \frac{A_0}{(1 + s/\omega_{P1})(1 + s/\omega_{P2})} \tag{12.42}$$

In this case the closed-loop poles are obtained from

$$1 + A(s)\beta = 0$$

which leads to

$$s^2 + s(\omega_{P1} + \omega_{P2}) + (1 + A_0\beta)\omega_{P1}\omega_{P2} = 0 \tag{12.43}$$

Thus the closed-loop poles are given by

$$s = -\tfrac{1}{2}(\omega_{P1} + \omega_{P2}) \pm \tfrac{1}{2}\sqrt{(\omega_{P1} + \omega_{P2})^2 - 4(1 + A_0\beta)\omega_{P1}\omega_{P2}} \tag{12.44}$$

From this equation we see that as the loop gain $A_0\beta$ is increased from zero, the poles are brought closer together. Then a value of loop gain is reached at which the poles become coincident. If the loop gain is further increased, the poles become complex conjugate and move along a vertical line. Figure 12.31 shows the locus of the poles

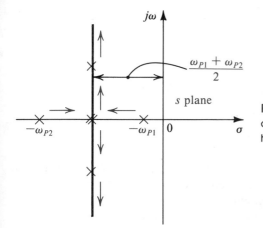

Fig. 12.31 Root locus diagram for a feedback amplifier whose open-loop transfer function has two real poles.

for increasing loop gain. This plot is called a *root-locus diagram*, where "root" refers to the fact that the poles are the roots of the characteristic equation.

From the root-locus diagram of Fig. 12.31 we see that this feedback amplifier also is unconditionally stable. Again, this result should come as no surprise; the maximum phase shift of $A(s)$ in this case is 180° (90° per pole), but this value is reached at $\omega = \infty$. Thus there is no finite frequency at which the phase shift reaches 180°.

Another observation to make on the root-locus diagram of Fig. 12.31 is that the open-loop amplifier might have a dominant pole, but this is not necessarily the case for the closed-loop amplifier. The response of the closed-loop amplifier can, of course, always be plotted once the poles are found from Eq. (12.44). As is the case with second-order responses, the closed-loop response can show a peak (see Chapter 14). To be more specific, the characteristic equation of a second-order network can be written in the standard form

$$s^2 + s\frac{\omega_0}{Q} + \omega_0^2 = 0 \qquad (12.45)$$

where ω_0 is called the *pole frequency* and Q is called *pole Q factor*. The poles are complex if Q is greater than 0.5. A geometric interpretation for ω_0 and Q of a pair of complex conjugate poles is given in Fig. 12.32, from which we note that ω_0 is the radial distance of the poles and that Q indicates the distance of the poles from the $j\omega$ axis. Poles on the $j\omega$ axis have $Q = \infty$.

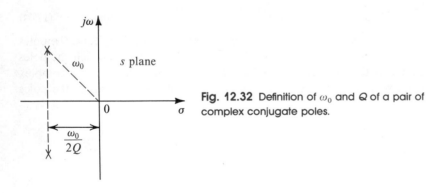

Fig. 12.32 Definition of ω_0 and Q of a pair of complex conjugate poles.

By comparing Eqs. (12.43) and (12.45) we obtain the Q factor for the poles of the feedback amplifier as

$$Q = \frac{\sqrt{(1 + A_0\beta)\omega_{P1}\omega_{P2}}}{(\omega_{P1} + \omega_{P2})} \qquad (12.46)$$

From the study of second-order network responses in Chapter 14 it will be seen that the response of the feedback amplifier under consideration shows no peaking for $Q \leq 0.707$. The boundary case corresponding to $Q = 0.707$ results in the *maximally flat* response. Figure 12.33 shows a number of possible responses obtained for various values of Q (or, correspondingly, various values of $A_0\beta$).

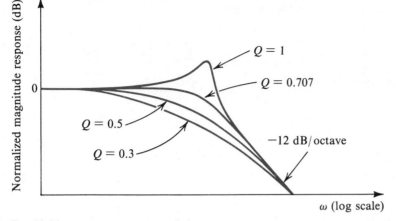

Fig. 12.33 Normalized magnitude response of a two-pole feedback amplifier for various values of Q

Exercise

12.12 An amplifier with a low-frequency gain of 100 and poles at 10^4 and 10^6 rad/s is incorporated in a negative-feedback loop with feedback factor β. For what value of β do the poles of the closed-loop amplifier coincide? What is the corresponding Q of the resulting second-order system? For what value of β is a maximally flat response achieved? What is the low-frequency closed loop gain in the maximally flat case?

Ans.   0.245; 0.5; 0.5; 1.96 V/V

EXAMPLE 12.5

As an illustration of some of the ideas discussed above we consider the positive feedback circuit shown in Fig. 12.34a. Find the loop transmission $L(s)$ and the characteristic equation. Sketch a root-locus diagram for varying K and find the value of K that results in a maximally flat response and the value of K that makes the circuit oscillate. Assume that the amplifier has infinite input impedance and zero output impedance.

Solution

To obtain the loop transmission we short-circuit the signal source and break the loop at the amplifier input. We then apply a test voltage V_t and find the returned voltage V_r, as indicated in Fig. 12.34b. The loop transmission $L(s) \equiv A(s)\beta(s)$ is given by

$$L(s) = -\frac{V_r}{V_t} = -KT(s) \tag{12.47}$$

where $T(s)$ is the transfer function of the two-port RC network shown inside the broken-line box in Fig. 12.34b:

$$T(s) \equiv \frac{V_r}{V_1} = \frac{s(1/CR)}{s^2 + s(3/CR) + (1/CR)^2} \tag{12.48}$$

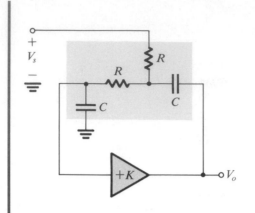

(a)

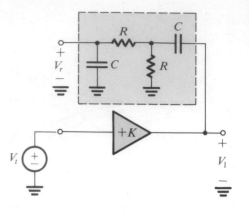

(b)

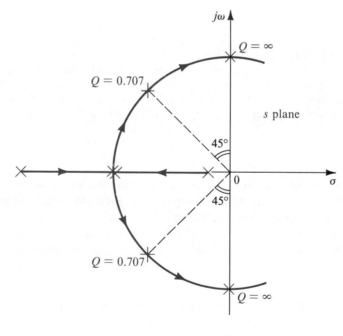

(c)

Fig. 12.34 Circuits and plot for Example 12.5.

Thus

$$L(s) = \frac{-s(K/CR)}{s^2 + s(3/CR) + (1/CR)^2}$$ (12.49)

The characteristic equation is

$$1 + L(s) = 0$$ (12.50)

that is,

$$s^2 + s\frac{3}{CR} + \left(\frac{1}{CR}\right)^2 - s\frac{K}{CR} = 0$$

$$s^2 + s\frac{3-K}{CR} + \left(\frac{1}{CR}\right)^2 = 0 \tag{12.51}$$

By comparing this equation to the standard form of the second-order characteristic equation [Eq. (12.45)] we see that the pole frequency ω_0 is given by

$$\omega_0 = \frac{1}{CR} \tag{12.52}$$

and the Q factor is

$$Q = \frac{1}{3-K} \tag{12.53}$$

Thus for $K = 0$ the poles have $Q = \frac{1}{3}$ and are therefore located on the negative real axis. As K is increased the poles are brought closer together and eventually coincide ($Q = 0.5$, $K = 1$). Increasing K further results in the poles becoming complex and conjugate. The root locus is then a circle because the radial distance ω_0 remains constant independent of the value of K.

The maximally flat response is obtained when $Q = 0.707$, which results when $K = 1.586$. In this case the poles are at $45°$ angles, as indicated in Fig. 12.34c. The poles cross the $j\omega$ axis into the right half of the s plane at the value of K that results in $Q = \infty$, that is, $K = 3$. Thus for $K \geq 3$ this circuit becomes unstable. This might appear to contradict our earlier conclusion that the feedback amplifier with a second-order response is unconditionally stable. Note, however, that the circuit in this example is quite different from the negative-feedback amplifier that we have been studying. Here we have an amplifier with a positive gain K and a feedback network whose transfer function $T(s)$ is frequency dependent. This feedback is in fact *positive*, and the circuit will oscillate at the frequency for which the phase of $T(j\omega)$ is zero.

Example 12.5 illustrates the use of feedback (positive feedback in this case) to move the poles of an RC network from their negative real-axis locations to complex conjugate locations. One can accomplish the same task using negative feedback, as the root-locus diagram of Fig. 12.31 demonstrates. The process of pole control is the essence of *active-filter design*, as will be discussed in Chapter 14.

Amplifiers with Three or More Poles

Figure 12.35 shows the root-locus diagram for a feedback amplifier whose open-loop response is characterized by three poles. As indicated, increasing the loop gain from zero moves the highest-frequency pole outward while the two other poles are brought closer together. As $A_0\beta$ is increased further, the two poles become coincident

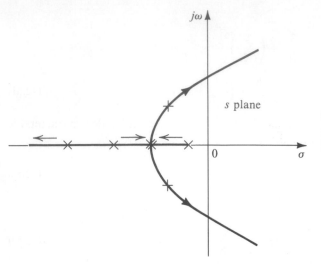

Fig. 12.35 Root-locus diagram for an amplifier with three poles. The arrows indicate the pole movement as $A_0\beta$ is increased.

and then become complex and conjugate. A value of $A_0\beta$ exists at which this pair of complex-conjugate poles enters the right half of the s plane, thus causing the amplifier to become unstable.

This result is not entirely unexpected, since an amplifier with three poles has a phase shift that reaches $-270°$ as ω approaches ∞. Thus there exists a finite frequency ω_{180}, at which the loop gain has $180°$ phase shift.

From the root-locus diagram of Fig. 12.35 we observe that one can always maintain amplifier stability by keeping the loop gain $A_0\beta$ smaller than the value corresponding to the poles entering the right half-plane. In terms of the Nyquist diagram, the critical value of $A_0\beta$ is that for which the diagram passes through the $(-1, 0)$ point. Reducing $A_0\beta$ below this value causes the Nyquist plot to shrink, and thus intersect the negative real axis to the right of the $(-1, 0)$ point, indicating stable amplifier performance. On the other hand, increasing $A_0\beta$ above the critical value causes the Nyquist plot to expand, thus encircling the $(-1, 0)$ point and indicating unstable performance.

For a given open-loop gain the above conclusions can be stated in terms of the feedback factor β. That is, there exists a *maximum value* for β above which the feedback amplifier becomes unstable. Alternatively, we can state that there exists a *minimum value* for the closed-loop gain A_{f0} below which the amplifier becomes unstable. To obtain lower values of closed-loop gain one needs therefore to alter the loop transfer function $L(s)$. This is the process known as *frequency compensation*. We shall study the theory and techniques of frequency compensation in Section 12.11.

Before leaving this section we should point out that construction of the root locus diagram for amplifiers having three or more poles as well as finite zeros is an involved process for which a systematic procedure exists. However, such a procedure will not be presented here, and the interested reader should consult Haykin (1970). Although the root locus diagram provides the amplifier designer with considerable insight, other, simpler techniques based on Bode plots can be effectively employed, as will be explained in Section 12.10

Exercise

12.13 Consider a feedback amplifier for which the open-loop transfer function $A(s)$ is given by

$$A(s) = \left(\frac{10}{1 + s/10^4} \right)^3$$

Let the feedback factor β be frequency-independent. Find the closed-loop poles as functions of β and hence show that the root locus is that of Fig. E12.13. Also find the value of β at which the amplifier becomes unstable. (*Note:* This is the same amplifier that was considered in Exercise 12.10.)

Ans.   See Fig. E12.13; $\beta_{\text{critical}} = 0.008$

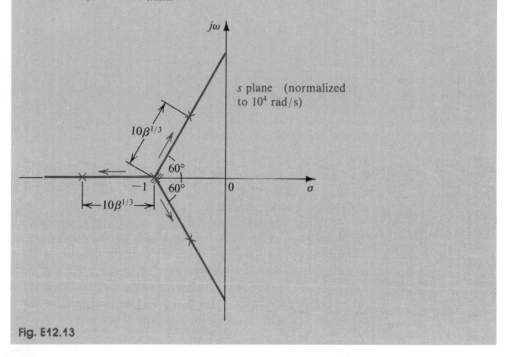

Fig. E12.13

12.10 STABILITY STUDY USING BODE PLOTS

Gain and Phase Margins

From Sections 12.8 and 12.9 we know that one can determine whether the feedback amplifier is stable or not by examining the loop gain $A\beta$ as a function of frequency. One of the simplest and most effective means for doing this is through the use of a Bode plot for $A\beta$, such as the one shown in Fig. 12.36. (Note that because the phase approaches $-360°$, the network examined is a fourth-order one.) The feedback amplifier whose loop gain is plotted in Fig. 12.36 will be stable, since at the frequency of 180° phase shift, ω_{180}, the magnitude of loop gain is less than unity (negative dB). The difference between the value of $|A\beta|$ at ω_{180} and unity is called the *gain margin* and is usually expressed in dB. The gain margin represents the amount by which the

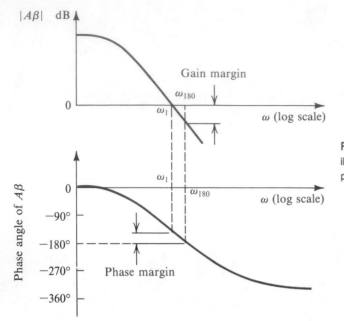

Fig. 12.36 Bode plot for the loop gain $A\beta$ illustrating the definitions of the gain and phase margins.

loop gain can be increased while stability is maintained. Feedback amplifiers are usually designed to have sufficient gain margin to allow for the inevitable changes in loop gain with temperature, time, and so on.

Another way to investigate the stability and to express its degree is to examine the Bode plot at the frequency for which $|A\beta| = 1$, which is the point at which the magnitude plot crosses the 0-dB line. If at this frequency the phase angle is less (in magnitude) than $180°$, then the amplifier is stable. This is the situation illustrated in Fig. 12.36. The difference between the phase angle at this frequency and $180°$ is termed the *phase margin*. On the other hand, if at the frequency of unity loop-gain magnitude the phase lag is in excess of $180°$, the amplifier will be unstable.

Exercise

12.14 Consider an op amp having a single-pole open-loop response with $A_0 = 10^5$ and $f_P = 10$ Hz. Let the op amp be ideal otherwise (infinite input impedance, zero output impedance, and so on). If this amplifier is connected in the noninverting configuration with a nominal low-frequency closed-loop gain of 100, find the frequency at which $|A\beta| = 1$. Also, find the phase margin.

Ans. 10^4 Hz; $90°$

Effect of Phase Margin on Closed-Loop Response

Feedback amplifiers are normally designed with a phase margin of at least $45°$. The amount of phase margin has a profound effect on the shape of the closed-loop magnitude response. To see this relationship, consider a feedback amplifier with a large low-frequency loop gain, $A_0\beta \gg 1$. It follows that the closed-loop gain at low frequencies is approximately $1/\beta$. Denoting the frequency at which the magnitude of loop gain is unity by ω_1 we have

$$A(j\omega_1)\beta = 1 \times e^{-j\theta} \tag{12.54}$$

where

$$\theta = 180° - \text{phase margin} \tag{12.55}$$

At ω_1 the closed-loop gain is

$$A_f(j\omega_1) = \frac{A(j\omega_1)}{1 + A(j\omega_1)\beta}$$

Substituting from Eq. (12.54) gives

$$A_f(j\omega_1) = \frac{(1/\beta)e^{-j\theta}}{1 + e^{-j\theta}}$$

Thus the magnitude of the gain at ω_1 is

$$|A_f(j\omega_1)| = \frac{1/\beta}{|1 + e^{-j\theta}|} \tag{12.56}$$

For a phase margin of $45°$, $\theta = 135°$ and we obtain

$$|A_f(j\omega_1)| = 1.3 \frac{1}{\beta}$$

That is, the gain peaks by a factor of 1.3 above the low-frequency value of $1/\beta$. This peaking increases as the phase margin is reduced, eventually reaching ∞ when the phase margin is zero. Zero phase margin, of course, implies that the amplifier can sustain oscillations [poles on the $j\omega$ axis; Nyquist plot passes through $(-1, 0)$].

Exercise

12.15 Find the closed-loop gain at ω_1 relative to the low-frequency gain when the phase margin is $30°$, $60°$, and $90°$.

Ans. 1.93; 1; 0.707

An Alternative Approach

Investigating stability by constructing Bode plots for the loop gain $A\beta$ can be a tedious and time-consuming process, especially if we have to investigate the stability of a given amplifier for a variety of feedback networks. An alternative approach, which is much simpler, is to construct a Bode plot for the open-loop gain $A(j\omega)$ only. Assuming for the time being that β is independent of frequency, we can plot $20 \log(1/\beta)$ as a horizontal straight line on the same plane used for $20 \log|A|$. The difference between the two curves will be

$$20 \log|A(j\omega)| - 20 \log \frac{1}{\beta} = 20 \log|A\beta| \tag{12.57}$$

which is the loop gain expressed in dB. We may therefore study stability by examining the difference between the two plots. If we wish to evaluate stability for a different feedback factor we simply draw another horizontal straight line at the level $20 \log(1/\beta)$.

To illustrate, consider an amplifier whose open-loop transfer function is characterized by three poles. For simplicity let the three poles be widely separated—say, at 0.1, 1, and 10 MHz, as shown in Fig. 12.37. Note that because the poles are widely separated, the phase is approximately $-45°$ at the first pole frequency, $-135°$ at the second, and $-225°$ at the third. The frequency at which the phase of $A(j\omega)$ is $-180°$ lies on the -40 dB/decade segment, as indicated in Fig. 12.37.

Consider next the horizontal straight line labeled (a). This line represents a feedback factor for which $20 \log(1/\beta) = 85$ dB, which corresponds to a closed-loop gain of approximately 85 dB. The loop gain is the difference between the $|A|$ curve and the

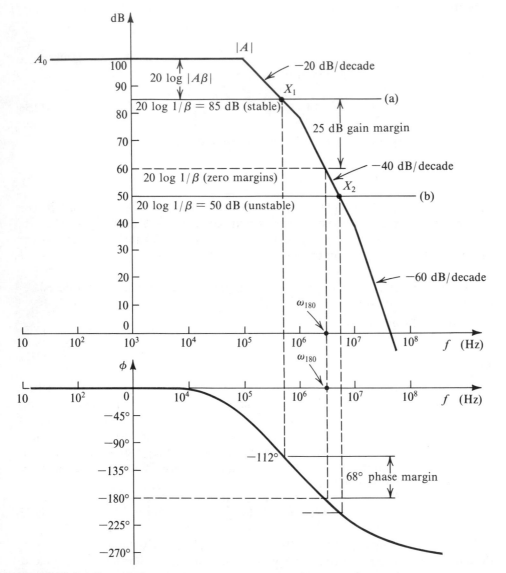

Fig. 12.37 Stability analysis.

$1/\beta$ line. The point of intersection X_1 corresponds to the frequency at which $|A\beta| = 1$. At this frequency we see from the phase plot that the phase is $-112°$. Thus the closed-loop amplifier, for which $20 \log(1/\beta) = 85$ dB, will be stable with a phase margin of $68°$. The gain margin can be easily obtained from Fig. 12.37; it is 25 dB.

Next, suppose that we wish to use this amplifier to obtain a closed-loop gain of 50-dB nominal value. Since $A_0 = 100$ dB, we see that $A_0\beta \gg 1$ and $20 \log(A_0\beta) \simeq 50$ dB, resulting in $20 \log(1/\beta) \simeq 50$ dB. To see whether this closed-loop amplifier is stable or not, we draw line (b) in Fig. 12.37 with a height of 50 dB. This line intersects the open-loop gain curve at point X_2, where the corresponding phase is greater than $180°$. Thus the closed-loop amplifier with 50 dB gain will be unstable.

In fact, it can easily be seen from Fig. 12.37 that the *minimum* value of $20 \log(1/\beta)$ that can be used, with the resulting amplifier being stable, is 60 dB. In other words, the minimum value of stable closed-loop gain obtained with this amplifier is approximately 60 dB. At this value of gain, however, the amplifier may still oscillate, since no margin is left to allow for possible changes in gain.

Since the $180°$-phase point always occurs on the -40-dB/decade segment of the Bode plot for $|A|$, a rule of thumb to guarantee stability is as follows: The closed-loop amplifier will be stable if the $20 \log(1/\beta)$ line intersects the $20 \log|A|$ curve at a point on the -20-dB/decade segment. Following this rule ensures that a phase margin of at least $45°$ is obtained. For the example of Fig. 12.37, the rule implies that the maximum value of β is 10^{-4}, which corresponds to a closed-loop gain of approximately 80 dB.

The above rule of thumb can be generalized for the case in which β is a function of frequency. The general rule states that at the intersection of $20 \log[1/|\beta(j\omega)|]$ and $20 \log|A(j\omega)|$ the difference of slopes (called the *rate of closure*) should not exceed 20 dB/decade.

Exercise

> **12.16** Consider an op amp whose open-loop gain is identical to that of Fig. 12.37. Assume that the op amp is ideal otherwise. Let the op amp be connected as a differentiator. Use the above rule of thumb to show that for stable performance the differentiator time constant should be greater than 159 ms. (*Hint:* Recall that for a differentiator, the Bode plot for $1/|\beta(j\omega)|$ has a slope of $+20$ dB/decade and intersects the 0-dB line at $1/\tau$, where τ is the differentiator time constant.)

12.11 FREQUENCY COMPENSATION

In this section we shall discuss methods for modifying the open-loop transfer function $A(s)$ of an amplifier having three or more poles so that the closed-loop amplifier is stable for any desired value of closed-loop gain.

Theory

The simplest method of frequency compensation consists of introducing a new pole in the function $A(s)$ at a sufficiently low frequency, f_D, such that the modified open-loop gain, $A'(s)$, intersects the $20 \log(1/|\beta|)$ curve with a slope difference of 20 dB/decade. As an example, let it be required to compensate the amplifier whose $A(s)$ is shown in

Fig. 12.38 such that closed-loop amplifiers with β as high as 10^{-2} (that is, closed-loop gains as low as approximately 40 dB) will be stable. First, we draw a horizontal straight line at the 40-dB level to represent $20 \log(1/\beta)$, as shown in Fig. 12.38. We then locate point Y on this line at the frequency of the first pole, f_{P1}. From Y we draw a line with -20 dB/decade slope and determine the point at which this line intersects the dc gain line, point Y'. This latter point gives the frequency f_D of the new pole that has to be introduced in the open-loop transfer function.

The compensated open-loop response $A'(s)$ is indicated in Fig. 12.38. It has four poles: at f_D, f_{P1}, f_{P2}, and f_{P3}. Thus $|A'|$ begins to roll off with a slope of -20 dB/decade at f_D. At f_{P1} the slope changes to -40 dB/decade, at f_{P2} it changes to -60 dB/decade, and so on. Since the $20 \log(1/\beta)$ line intersects the $20 \log|A'|$ curve at point Y on the -20-dB/decade segment, the closed-loop amplifier with this β value (or lower values) will be stable.

A serious disadvantage of this compensation method is that at most frequencies the open-loop gain has been drastically reduced. This means that at most frequencies the amount of feedback available will be small. Since all the advantages of negative feedback are directly proportional to the amount of feedback, the performance of the compensated amplifier has been impaired.

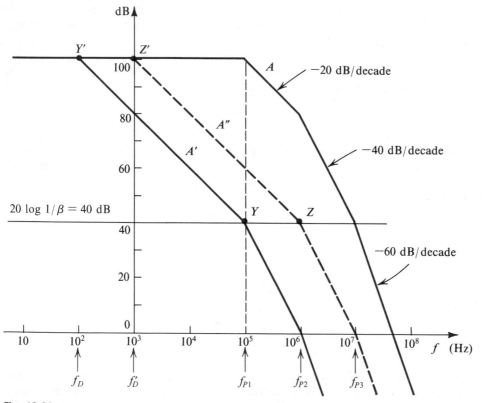

Fig. 12.38 Frequency compensation for $\beta = 10^{-2}$. The response labeled A' is obtained by introducing an additional pole at f_D. The A'' response is obtained by moving the low-frequency pole to f_D'.

Careful examination of Fig. 12.38 shows that the reason the gain $A'(s)$ is low is the pole at f_{P1}. If we can somehow eliminate this pole, then—rather than locating point Y, drawing YY', and so on—we can now start from point Z (at the frequency of the second pole) and draw the line ZZ'. This would result in the open-loop curve $A''(s)$, which shows considerably higher gain than $A'(s)$.

Although it is not possible to eliminate the pole at f_{P1}, it is usually possible to shift that pole from $f = f_{P1}$ to $f = f'_D$. This makes the pole dominant and eliminates the need for introducing an additional lower-frequency pole, as will be explained next.

Implementation

We shall now address the question of implementing the frequency-compensation scheme discussed above. The amplifier circuit normally consists of a number of cascaded gain stages, with each stage responsible for one or more of the transfer-function poles. Through manual and/or computer analysis of the circuit, one identifies which stage introduces each of the important poles f_{P1}, f_{P2}, and so on. For the purpose of our discussion, assume that the first pole f_{P1} is introduced at the interface between the two cascaded differential stages shown in Fig. 12.39a. In Fig. 12.39b we show a simple small-signal model of the circuit at this interface. Current source I_x represents the

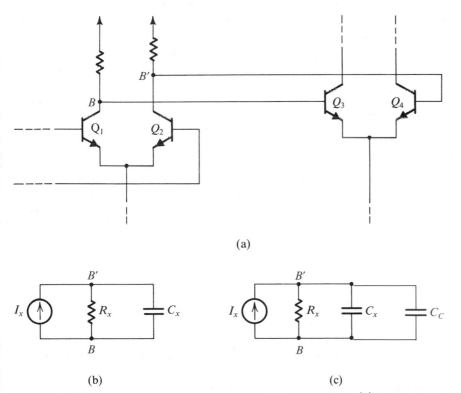

(a)

(b)   (c)

Fig. 12.39 (a) Two cascaded gain stages of a multistage amplifier. **(b)** Equivalent circuit for the interface between the two stages in (a). **(c)** Same circuit as in (b) but with a compensating capacitor C_C added.

output signal current of the $Q_1 - Q_2$ stage. Resistance R_x and capacitance C_x represent the total resistance and capacitance between the two nodes B and B′. It follows that the pole f_{P1} is given by

$$f_{P1} = \frac{1}{2\pi C_x R_x} \qquad (12.58)$$

Let us now connect the compensating capacitor C_C between nodes B and B′. This will result in the modified equivalent circuit shown in Fig. 12.39c, from which we see that the pole introduced will no longer be at f_{P1}; rather the pole can be at any desired lower frequency f'_D

$$f'_D = \frac{1}{2\pi(C_x + C_C)R_x} \qquad (12.59)$$

We thus conclude that one can select an appropriate value for C_C so as to shift the pole frequency from f_{P1} to the value f'_D determined by point Z' in Fig. 12.38.

At this juncture it should be pointed out that adding the capacitor C_C will usually result in changes in the location of the other poles (those at f_{P2} and f_{P3}). One might therefore need to calculate the new location of f_{P2} and perform a few iterations to arrive at the required value for C_C.

A disadvantage of this implementation method is that the required value of C_C is usually quite large. Thus if the amplifier to be compensated is an IC op amp, it will be difficult, and probably impossible, to include this compensating capacitor on the IC chip. (As pointed out in Chapter 13 and in Appendix A, the maximum practical size of a monolithic capacitor is about 100 pF.) An elegant solution to this problem is to connect the compensating capacitor in the feedback path of an amplifier stage. Because of the Miller effect, the compensating capacitance will be multiplied by the stage gain, resulting in a much larger effective capacitance. Furthermore, as explained below, another unexpected benefit accrues.

Miller Compensation and Pole Splitting

Figure 12.40a shows one gain stage in a multistage amplifier. For simplicity, the stage is shown as a common-emitter amplifier, but in practice it can be a more elaborate circuit. In the feedback path of this common-emitter stage we have placed a compensating capacitor C_f.

Figure 12.40b shows a simplified equivalent circuit of the gain stage of Fig. 12.40a. Here R_1 and C_1 represent the total resistance and total capacitance between node B and ground. Similarly, R_2 and C_2 represent the total resistance and total capacitance between node C and ground. Furthermore, it is assumed that C_1 and C_2 include the Miller components due to capacitance C_μ, and C_2 includes the input capacitance of the succeeding amplifier stage. Finally, I_i represents the output signal current of the preceding stage.

In the absence of the compensating capacitor C_f, we can see from Fig. 12.40b that there are two poles—one at the input and one at the output. Let us assume that these two poles are f_{P1} and f_{P2} of Fig. 12.38; thus

$$f_{P1} = \frac{1}{2\pi C_1 R_1} \qquad f_{P2} = \frac{1}{2\pi C_2 R_2} \qquad (12.60)$$

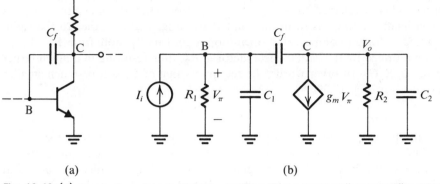

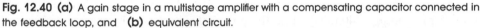

(a)   (b)

Fig. 12.40 (a) A gain stage in a multistage amplifier with a compensating capacitor connected in the feedback loop, and **(b)** equivalent circuit.

With C_f present, analysis of the circuit yields the transfer function

$$\frac{V_o}{I_i} = \frac{(sC_f - g_m)R_1R_2}{1 + s[C_1R_1 + C_2R_2 + C_f(g_mR_1R_2 + R_1 + R_2)] + s^2[C_1C_2 + C_f(C_1 + C_2)]R_1R_2}$$

(12.61)

The zero is usually at a much higher frequency than the dominant pole, and we shall neglect its effect. The denominator polynomial $D(s)$ can be written in the form

$$D(s) = \left(1 + \frac{s}{\omega'_{P1}}\right)\left(1 + \frac{s}{\omega'_{P2}}\right) = 1 + s\left(\frac{1}{\omega'_{P1}} + \frac{1}{\omega'_{P2}}\right) + \frac{s^2}{\omega'_{P1}\omega'_{P2}}$$

(12.62)

where ω'_{P1} and ω'_{P2} are the new frequencies of the two poles. Normally one of the poles will be dominant; $\omega'_{P1} \ll \omega'_{P2}$. Thus

$$D(s) \simeq 1 + \frac{s}{\omega'_{P1}} + \frac{s^2}{\omega'_{P1}\omega'_{P2}}$$

(12.63)

Equating the coefficients of s in the denominator of Eq. (12.61) and in Eq. (12.63) results in

$$\omega'_{P1} = \frac{1}{C_1R_1 + C_2R_2 + C_f(g_mR_1R_2 + R_1 + R_2)}$$

which can be approximated by

$$\omega'_{P1} \simeq \frac{1}{g_mR_2C_fR_1}$$

(12.64)

To obtain ω'_{P2} we equate the coefficients of s^2 in the denominator of Eq. (12.61) and in Eq. (12.63) and use Eq. (12.64):

$$\omega'_{P2} \simeq \frac{g_mC_f}{C_1C_2 + C_f(C_1 + C_2)}$$

(12.65)

From Eqs. (12.64) and (12.65) we see that as C_f is increased, ω'_{P1} is reduced and ω'_{P2} is increased. This is referred to as *pole splitting*. Note that the increase in ω'_{P2} is highly beneficial; it allows us to move point Z (see Fig. 12.38) further to the right, thus resulting in higher compensated open-loop gain. Finally, note from Eq. (12.64) that C_f is multiplied by the Miller-effect factor $g_m R_2$, thus resulting in a much larger capacitance, $g_m R_2 C_f$. In other words, the required value of C_f will be much smaller than that of C_C in Fig. 12.39.

EXAMPLE 12.6

Consider an op amp whose open-loop transfer function is identical to that shown in Fig. 12.37. We wish to compensate this op amp so that the closed-loop amplifier with resistive feedback is stable for any gain (that is, for β up to unity). Assume that the op-amp circuit includes a stage such as that of Fig. 12.40 with $C_1 = 100$ pF, $C_2 = 5$ pF, and $g_m = 40$ mA/V, that the pole at f_{P1} is caused by the input circuit of that stage, and that the pole at f_{P2} is introduced by the output circuit. Find the value of the compensating capacitor if it is connected either between the input node B and ground or in the feedback path of the transistor.

Solution

First we determine R_1 and R_2 from

$$f_{P1} = 0.1 \text{ MHz} = \frac{1}{2\pi C_1 R_1}$$

Thus

$$R_1 = \frac{10^5}{2\pi} \ \Omega$$

$$f_{P2} = 1 \text{ MHz} = \frac{1}{2\pi C_2 R_2}$$

Thus

$$R_2 = \frac{10^5}{\pi} \ \Omega$$

If a compensating capacitor C_C is connected across the input terminals of the transistor stage, then the frequency of the first pole changes from f_{P1} to f'_D:

$$f'_D = \frac{1}{2\pi(C_1 + C_C)R_1}$$

The second pole remains unchanged. The required value for f'_D is determined by drawing a -20 dB/decade line from the 1-MHz frequency point on the $20 \log(1/\beta) = 20 \log 1 = 0$ dB line. This line will intersect the 100-dB dc gain line at 10 Hz. Thus

$$f'_D = 10 \text{ Hz} = \frac{1}{2\pi(C_1 + C_C)R_1}$$

which results in $C_C \simeq 1\ \mu\text{F}$, which is quite large and which certainly cannot be included on the IC chip.

Next, if a compensating capacitor C_f is connected in the feedback path of the transistor, then both poles change location to the values given by Eqs. (12.64) and (12.65):

$$f'_{P1} \simeq \frac{1}{2\pi g_m R_2 C_f R_1}$$

$$f'_{P2} \simeq \frac{g_m C_f}{2\pi[C_1 C_2 + C_f(C_1 + C_2)]} \tag{12.66}$$

To determine where we should locate the first pole we need to know the value of f'_{P2}. As an approximation let us assume that $C_f \gg C_2$, which enables us to obtain

$$f'_{P2} \simeq \frac{g_m}{2\pi(C_1 + C_2)} = 60.6\ \text{MHz}$$

Thus it appears that this pole will move to a frequency higher than f_{P3} (which is 10 MHz). Let us therefore assume that the second pole will be at f_{P3}. This requires that the first pole be located at 100 Hz:

$$f'_{P1} = 100\ \text{Hz} = \frac{1}{2\pi g_m R_2 C_f R_1}$$

which results in $C_f = 78.5$ pF. Although this value is indeed much greater than C_2, we can determine the location of the pole f'_{P2} from Eq. (12.66) which yields $f'_{P2} = 57.2$ MHz, confirming the fact that this pole has indeed been moved past f_{P3}.

We conclude that using Miller compensation not only results in a much smaller compensating capacitor but, owing to pole splitting, also enables us to place the dominant pole a decade higher in frequency. This results in a wider bandwidth for the compensated op amp.

Exercises

12.17 A multipole amplifier having a first pole at 1 MHz and an open-loop gain of 100 dB is to be compensated for closed-loop gains as low as 20 dB by the introduction of a new dominant pole. At what frequency must the new pole be placed?

Ans. 100 Hz

12.18 For the amplifier described in Exercise 12.17, rather than introducing a new dominant pole we can use additional capacitance at the circuit node at which the first pole is formed to reduce the frequency of the first pole. If the frequency of the second pole is 10 MHz and if it remains unchanged while additional capacitance is introduced as mentioned, find the frequency to which the first pole must be lowered so that the resulting amplifier is stable for closed-loop gains as low as 20 dB. By what factor is the capacitance at the controlling node increased?

Ans. 1,000 Hz; 1,000

Table 12.1 SUMMARY OF RELATIONSHIPS FOR THE FOUR FEEDBACK-AMPLIFIER TOPOLOGIES

Feedback Amplifier	x_i	x_o	x_f	x_s	A	β	A_f	Source Form	Loading of feedback network is obtained		To find β, apply to port 2 of feedback network	Z_{if}	Z_{of}	Refer to Figs.
									At input	At output				
Series-shunt (voltage amplifier)	V_i	V_o	V_f	V_s	$\dfrac{V_o}{V_i}$	$\dfrac{V_f}{V_o}$	$\dfrac{V_o}{V_s}$	Thévenin	By short-circuiting port 2 of feedback network	By open-circuiting port 1 of feedback network	A voltage and find the open-circuit voltage at port 1	$Z_i(1 + A\beta)$	$\dfrac{Z_o}{1 + A\beta}$	12.4a 12.8 12.10 12.11
Shunt-series (current amplifier)	I_i	I_o	I_f	I_s	$\dfrac{I_o}{I_i}$	$\dfrac{I_f}{I_o}$	$\dfrac{I_o}{I_s}$	Norton	By open-circuiting port 2 of feedback network	By short-circuiting port 1 of feedback network	A current and find the short-circuit current at port 1	$\dfrac{Z_i}{1 + A\beta}$	$Z_o(1 + A\beta)$	12.4b 12.22 12.23 12.24
Series-series (transconductance amplifier)	V_i	I_o	V_f	V_s	$\dfrac{I_o}{V_i}$	$\dfrac{V_f}{I_o}$	$\dfrac{I_o}{V_s}$	Thévenin	By open-circuiting port 2 of feedback network	By open-circuiting port 1 of feedback network	A current and find the open-circuit voltage at port 1	$Z_i(1 + A\beta)$	$Z_o(1 + A\beta)$	12.4c 12.13 12.15 12.16
Shunt-shunt (transresistance amplifier)	I_i	V_o	I_f	I_s	$\dfrac{V_o}{I_i}$	$\dfrac{I_f}{V_o}$	$\dfrac{V_o}{I_s}$	Norton	By short-circuiting port 2 of feedback network	By short-circuiting port 1 of feedback network	A voltage and find the short-circuit current at port 1	$\dfrac{Z_i}{1 + A\beta}$	$\dfrac{Z_o}{1 + A\beta}$	12.4d 12.18 12.19 12.20

12.12 SUMMARY

● Negative feedback is employed to make the amplifier gain less sensitive to component variations, in order to control input and output impedances, to extend bandwidth, to reduce nonlinear distortion, and to enhance signal-to-noise (and signal-to-interference) ratio.

● The above advantages are obtained at the expense of a reduction in gain and at the risk of the amplifier becoming unstable (that is, oscillating). The latter problem is solved by careful design.

● For each of the four basic types of amplifiers, there is an appropriate feedback topology. The four topologies, together with their analysis procedure and their effects on input and output impedances, are summarized in Table 12.1

● The key feedback parameters are the loop gain ($A\beta$), which for negative feedback must be a positive dimensionless number, and the amount of feedback ($1 + A\beta$). The latter directly determines gain reduction, gain desensitivity, bandwidth extension, and changes in Z_i and Z_o.

● Since A and β are in general frequency-dependent, the poles of the feedback amplifier are obtained by solving the characteristic equation $1 + A(s)\beta(s) = 0$.

● For the feedback amplifier to be stable, its poles must all be in the left half of the s plane.

● Stability is guaranteed if at the frequency for which the phase angle of $A\beta$ is 180°, (that is, ω_{180}), $|A\beta|$ is less than unity; the amount by which it is less than unity, expressed in decibels, is the gain margin. Alternatively, the amplifier is stable if, at the frequency at which $|A\beta| = 1$, the phase angle is less than 180°; the difference is the phase margin.

● The stability of a feedback amplifier can be analyzed by constructing Bode plots for $|A|$ and $1/|\beta|$. Stability is guaranteed if the two plots intersect with a difference in slope no greater than 6 dB/octave.

● To make a given amplifier stable for a given feedback factor β, the open-loop frequency response can be suitably modified by a process known as frequency compensation.

● A popular method for frequency compensation involves connecting a feedback capacitor to an inverting stage in the amplifier. This causes the pole formed at the input of the amplifier stage to shift to a lower frequency and thus become dominant, while the pole formed at the output of the amplifier stage is moved to a very high frequency and thus becomes unimportant. This process is known as pole splitting.

BIBLIOGRAPHY

P. E. Gray and C. L. Searle, *Electronic Principles*, New York: Wiley, 1969.

P. R. Gray and R. G. Meyer, *Analysis and Design of Analog Integrated Circuits*, 2nd ed., New York: Wiley, 1984.

S. S. Haykin, *Active Network Theory*, Reading, Mass.: Addison-Wesley, 1970.

E. S. Kuh and R. A. Rohrer, *Theory of Linear Active Networks*, San Francisco: Holden-Day, Inc., 1967. (This is an advanced-level text.)

Linear Integrated Circuits, Harrison, N.J.: RCA, 1967.

G. S. Moschytz, *Linear Integrated Networks: Design*, New York: Van Nostrand Reinhold, 1974.

E. Renschler, *The MC1539 Operational Amplifier and Its Applications*, Application Note AN-439, Phoenix, Ariz.: Motorola Semiconductor Products.

J. K. Roberge, *Operational Amplifiers: Theory and Practice*, New York: Wiley, 1975.

PROBLEMS

12.1 A feedback amplifier intended to have a closed-loop gain of 100 utilizes a feedback network for which $\beta = 0.0100$. For what values of open-loop gain A will the actual closed-loop gain A_f be at least **(a)** 99 and **(b)** 99.9?

12.2 The noninverting buffer op-amp configuration shown in Fig. P12.2 provides a direct implementation of the feedback loop of Fig. 12.1. Assuming that the op amp has infinite input resistance and zero output resistance, what is β? If $A = 10$, what is the closed-loop voltage gain? What is the amount of feedback in dB? For $V_s = 1$ V, find V_o and V_i. If A decreases by 10%, what is the corresponding decrease in A_f?

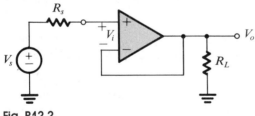

Fig. P12.2

12.3 Repeat Problem 12.2 for $A = 100$.

12.4 Repeat Exercise 12.1 for $A = 10^3$ and $A_f = 100$. For part (d), use $V_s = 0.1$ V.

12.5 In a feedback amplifier for which A is 10^4 and $A_f = 10^3$, what is the gain-desensitivity factor? Find A_f exactly, and approximately using Eq. (12.6) in the two cases: **(a)** A drops by 10%; and **(b)** A drops by 50%.

12.6 An amplifier having a midband gain of 10^3 and a single-pole high-frequency rolloff at 10^3 Hz is connected in a negative-feedback circuit with $\beta = 0.1$. What is the 3-dB frequency of the closed-loop amplifier?

12.7 An amplifier having a midband gain of 10^3 and a single low-frequency pole at 10^3 Hz is connected in a negative-feedback circuit that reduces its gain at 10^4 Hz to 10. What is the lower 3-dB frequency of the closed-loop amplifier?

*12.8 Consider an amplifier whose high-frequency response is characterized by two poles, one at ω_H and one at $100\omega_H$, and whose midband gain is $A_M = 10^3$, connected in a closed-loop for which $\beta = 10^{-2}$. What are the gain and poles of the resulting closed-loop amplifier?

12.9 Repeat Exercise 12.2 for an op amp with an open-loop gain of 10^5 and an upper rolloff at 10 Hz.

12.10 Repeat Exercise 12.2 for an op amp with an open-loop gain of 10^3 and an upper rolloff at 10^3 Hz.

*12.11 The complementary BJT follower shown in Fig. P12.11a has the approximate transfer characteristic shown in Fig. P12.11b. Observe that for -0.7 V $\le v_1 \le +0.7$ V, the output is zero. This "dead band" leads to crossover distortion (see Section 10.3). Consider this follower driven by the output of a differential amplifier of gain 100 whose positive input terminal is connected to the input signal source v_S and whose negative input terminal is connected to the emitters of the follower. Sketch the transfer characteristic v_O versus v_S of the resulting feedback amplifier. What are the limits of the dead band and what are the gains outside the dead band?

12.12 An op amp having a differential input resistance of 10 kΩ is connected in the noninverting configuration utilizing two resistors of 10 kΩ and 100 kΩ in order to have a nominal closed-loop gain of 11. It is found to have a closed-loop gain of 10. Find the input resistance of the closed-loop amplifier.

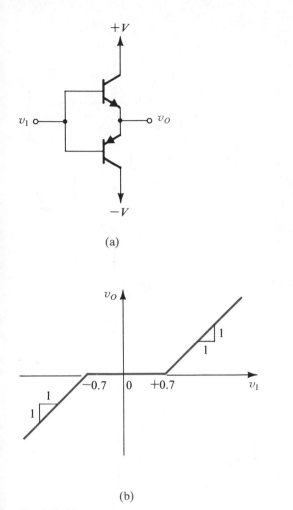

(a)

(b)

Fig. P12.11

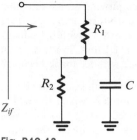

Z_{if}

Fig. P12.13

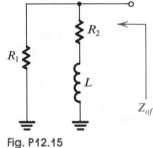

Z_{of}

Fig. P12.15

*12.13 A series-shunt feedback amplifier employs an op amp having a differential input resistance of 10 kΩ and a dc open-loop gain of 10^4 with a 3-dB rolloff at 100 Hz. The feedback network consists of two resistors of 1 kΩ and 9 kΩ (for a β of 0.1). Show that the input impedance of the closed-loop amplifier can be represented by the circuit in Fig. P12.13. Find the values of R_1, R_2 and C.

12.14 A feedback amplifier utilizing shunt sampling and employing a basic amplifier with a gain of 100 and an output resistance of 1,000 ohms has a closed-loop output resistance of 100 Ω. What is the closed-loop gain? If the basic amplifier is used to implement a unity-gain buffer, what output resistance do you expect to achieve?

*12.15 A feedback amplifier utilizing shunt sampling employs an op amp having an output resistance of 1000 Ω and a dc open-loop gain of 10^4 with a 3-dB rolloff at 100 Hz. For feedback β = 0.1, independent of frequency, and ignoring the loading effect of the feedback network, show that the closed-loop output impedance can be represented by the circuit shown in Fig. P12.15. Find the values of R_1, R_2 and L.

12.16 The circuit of Fig. E12.5 is modified by replacing R_2 by a short circuit. Find the values of A, $β$, A_f, R'_{if}, and R'_{of}.

12.17 The circuit of Fig. E12.5 is modified by incorporating a capacitance of 1 μF in series with R_1. Find A, $β$, A_f, R'_{if} and R'_{of} at **(a)** relatively high frequencies, where the capacitor can be considered a short circuit (while the transistor dynamics can still be ignored) and **(b)** relatively low frequencies, where the capacitor can be considered an open circuit. By examining the circuit, convince yourself that the frequency response exhibits a zero at $s = -1/C(R_1 + R_2)$. (You may do this by considering the amplifier to approximate an ideal op amp.) Using the calculated frequency of the zero, give a sketch of the amplifier frequency response and calculate the pole frequency.

12.18 A feedback amplifier employing series sampling utilizes a basic amplifier with output resistance of 10 kΩ and a transconductance of 100 mA/V, with a feedback network whose transresistance is 1 V/mA. For this circuit what does the output resistance with feedback become?

*12.19 Repeat the analysis outlined in Example 12.2 for a voltage-controlled current source in which the BJT is replaced by a FET characterized by g_m and r_o. Label the generator resistance R and the resistance in the source lead R_S. Perform the analysis two ways:
(a) From first principles utilizing the procedure in Fig. 12.16.
(b) By adapting the relationships developed in the BJT example. In this process try first to interpret the final results (e.g., for R'_{of}). If (as is likely) you have difficulty, revert to earlier and earlier stages of the process for adaptation.

12.20 Consider the circuit of Fig. E12.6 with $R_F = 0$. Assuming, as in Exercise 12.6, that $I_{C1} = 0.6$ mA, $I_{C2} = 1.0$ mA, and $I_{C3} = 4.0$ mA and that $h_{fe} = 100$ and $r_o = \infty$, find A, β, A_f, V_o/V_s, and R_{if}.

12.21 If the basic amplifier in Exercise 12.6 (the feedback triple) has a dominant single-pole rolloff at 1 MHz, what is the 3-dB frequency of the closed-loop circuit shown?

12.22 Consider the inverting op-amp configuration with a resistance R_1 from source to inverting op-amp input of 1 kΩ. The op amp has open-loop gain of 1,000 and $R_{id} = 10$ kΩ. The closed-loop gain was measured to be 10. Find the input resistance of the closed-loop amplifier excluding R_1.

12.23 Negative feedback is to be used to modify the characteristics of a particular amplifier for various purposes. Identify the feedback topology to be used if:
(a) Input resistance is to be lowered and output resistance raised.
(b) Both input and output resistances are to be raised.
(c) Both input and output resistances are to be lowered.

*12.24 For $h_{fe} = 100$, find the gain (V_o/V_s) and input and output resistances of the circuit in Fig. P12.24 using feedback analysis. Verify by direct analysis.

*12.25 For $V_t = 2$ V and $K = 0.25$ mA/V^2, find the voltage gain (V_o/V_s) and input and output resistances of the circuit in Fig. P12.25 using feedback analysis. Verify by direct analysis.

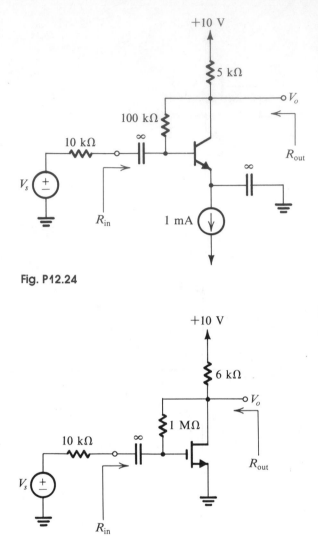

Fig. P12.24

Fig. P12.25

*12.26 Repeat Example 12.4 with the emitter resistor of Q_1 left unbypassed. What do the input and output resistances and closed-loop gain become?

*12.27 Reconsider Example 12.4 and Problem 12.26. Ignoring all other time constants, for what value of C, the capacitor across the emitter resistor of Q_1, does the lower 3-dB frequency of the amplifier without feedback (the A circuit) become 1 kHz? With this capacitor in place, what is the corresponding 3-dB frequency when feedback is applied?

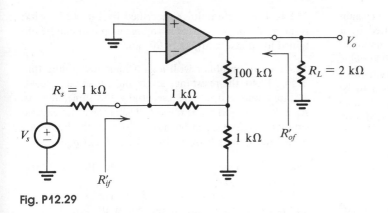

Fig. P12.29

**12.28 If in Exercise 12.7, the op amp has a single-pole frequency response with a unity-gain frequency of 1 MHz, find the input and output impedances, Z'_{if} and Z'_{of}.

*12.29 For the circuit in Fig. P12.29, use the feedback method to find the voltage gain V_o/V_s, the input resistance R'_{if} and the output resistance R'_{of}. The op amp has open-loop gain $\mu = 10^4$ V/V, $R_{id} = 100$ kΩ, $R_{icm} = \infty$, and $r_o = 1$ kΩ.

*12.30 Consider the amplifier of Fig. 12.25 to have its output at the emitter of the rightmost transistor Q_2. Use the technique for a shunt-shunt feedback amplifier to

calculate (V_{out}/I_{in}) and R_{in}. Using this result, calculate I_{out}/I_{in}. Compare this with the results obtained in Example 12.4.

*12.31 Neglecting r_{o2} (for simplicity), find the numerical value of the loop gain ($A\beta$) of the circuit in Fig. 12.25b. (*Hint:* Open the loop at the input to transistor Q_1; that is, $V_t = V_{\pi 1}$.)

*12.32 Assuming ideal op amps, find the transfer functions of the two circuits in Fig. P12.32 and thus show that they have the same poles. Then verify using the techniques discussed in Section 12.7 that the two circuits indeed have the same feedback loop.

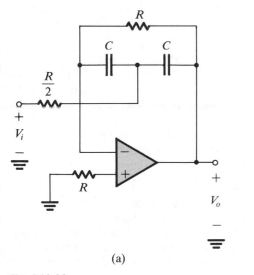

(a)

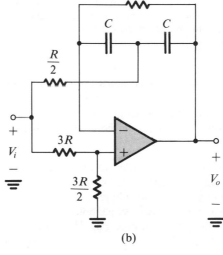

(b)

Fig. P12.32

12.33 An op amp designed to have a low-frequency gain of 10^5 and a high-frequency response dominated by a single pole at 100 rad/s acquires, through a manufacturing error, a pair of additional poles at 10,000 rad/s. At what frequency does the total phase shift reach 180°? At this frequency, for what value of β, assumed to be frequency-independent, does the loop gain reach a value of unity? What is the corresponding value of closed-loop gain at low frequencies?

*12.34 For the situation described in Problem 12.33, sketch a Nyquist plot for $\beta = 1.0$ and 10^{-3}. (Plot for $\omega = 0$, 100, 10^3, 10^4, and ∞ rad/s.)

12.35 An op amp having a low-frequency gain of 10^3 and a single-pole rolloff at 10^4 rad/s is connected in a negative-feedback loop via a feedback network having a transmission k and a two-pole rolloff at 10^4 rad/s. Find the value of k above which the closed-loop amplifier becomes unstable.

12.36 Consider a feedback amplifier for which the open-loop gain $A(s)$ is given by

$$A(s) = \frac{1,000}{(1 + s/10^4)(1 + s/10^5)^2}$$

If the feedback factor β is independent of frequency, find the frequency at which the phase shift is 180°, and the critical value of β at which oscillation will commence.

12.37 Consider an op amp having a two-pole response with voltage gain $A_0 = 10^5$, $f_{P1} = 10$ Hz, and $f_{P2} = 10^4$ Hz. If the amplifier is connected in a noninverting configuration with a nominal low-frequency closed-loop voltage gain of 100, find the frequency at which $|A\beta| = 1$ and the corresponding phase margin.

12.38 An op amp with open-loop gain of 80 dB and poles at 10^5, 10^6, and 2×10^6 Hz is connected as a differentiator. On the basis of the "rate-of-closure" rule, what is the smallest time constant that should be used for stable operation?

12.39 An op amp, when connected in a noninverting configuration with nominal low-frequency voltage gain of 100, is found to have unity gain with 90° phase shift at 10^4 Hz. If the op amp is assumed to have a single-pole open-loop response and low-frequency voltage gain of 10^5, what is the location of the pole?

*12.40 For the amplifier described by Fig. 12.37 what is the minimum closed-loop voltage gain that can be obtained for phase margins of 90° and 45°?

12.41 In an amplifier with a gain stage resembling that in Fig. 12.40, the resistance levels and capacitance levels at input and output are essentially the same; that is, $R_1 \simeq R_2 = R$ and $C_1 \simeq C_2 = C$. Using the exact expression for ω'_{P1} find C_f in terms of C so that the frequency of the first pole is lowered by a factor l. Let $g_m = \mu/R$.

*12.42 Reconsider Example 12.6 under the conditions that (separately)
(a) f_{P2} and f_{P3} are coincident at 10^6 Hz.
(b) f_{P1}, f_{P2}, and f_{P3} are coincident at 10^6 Hz.
(c) $C_1 = 50$ pF (a result of using a BJT with higher f_T).

12.43 Contemplate the effects of pole splitting by considering Eqs. (12.64) and (12.65) under the conditions that $R_1 \simeq R_2 = R, C_2 \simeq C_1/10 = C, C_f \gg C$, and $g_m = 100/R$, by calculating ω_{P1}, ω_{P2}, and $\omega'_{P1}, \omega'_{P2}$.

12.44 An op amp with open-loop voltage gain of 10^4 and poles at 10^5, 10^6, and 10^7 Hz is to be compensated by the addition of a fourth dominant pole to operate stably with unity feedback ($\beta = 1$). What is the frequency of the required dominant pole? The compensation network is to consist of an RC low-pass network placed in the negative-feedback path of the op amp. The dc bias conditions are such that a 1-MΩ resistor can be tolerated in series with each of the negative and positive input terminals. What capacitor is required between the negative input and ground to implement the required fourth pole?

*12.45 An op amp with an open-loop voltage gain of 80 dB and poles at 10^5, 10^6, and 2×10^6 Hz is to be compensated to be stable for unity β. Assume that the op amp incorporates an amplifier equivalent to that in Fig. 12.40, with $C_1 = 150$ pF, $C_2 = 5$ pF, and $g_m = 40$ mA/V, and that f_{P1} is caused by the input circuit and f_{P2} by the output circuit of this amplifier. Find the required value of the compensating Miller capacitance and the new frequency of the output pole.

12.46 Consider the general feedback structure shown in Fig. 12.1. What kind of feedback results **(a)** If A is negative and β is positive? **(b)** If both A and β are negative? **(c)** If A is positive and β is negative? How is the magnitude of A_f related to that of A if $A\beta$ is negative and $|A\beta| \ll 1$? What is the relationship between A and A_f for $A\beta = -0.9$? What happens to A_f when $A\beta = -1.0$? Operation is unstable under the latter condition and the behavior of the circuit depends on amplifier saturation and the existence of energy-storage elements (L, C) as will be seen in Chapter 14.

*12.47 Show that the circuit in Fig. P12.47 incorporates positive feedback and find the loop gain. Assume an ideal amplifier with a gain of $+10$. Use the feedback method to find the closed-loop gain V_o/V_s and the input resistance R'_{if}.

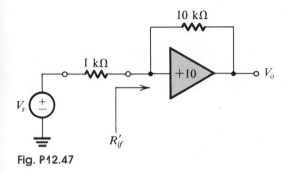

Fig. P12.47

12.48 For the circuit in Fig. P12.48 assume the op amp to be ideal except for having a finite open-loop gain μ. Use the feedback method to find an expression for the closed-loop gain A_f. For $\mu = 1$, what is A_f when (a) $R_1 = R_2 = R$, (b) $R_2 = 0$, (c) $R_1 = \infty$?

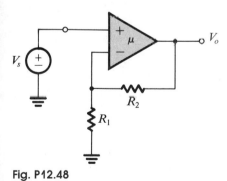

Fig. P12.48

12.49 Repeat Problem 12.48 for the case in which the input terminals of the op amp are interchanged. Reconsider for $\mu = 0.99$.

**12.50 The op amp in the circuit of Fig. P12.50 has an open-loop gain of 10^5 and a single-pole rolloff with $f_{3dB} = 10$ rad/s.

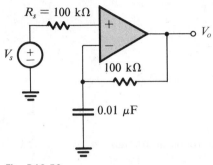

Fig. P12.50

(a) Sketch a Bode plot for the loop gain.
(b) Find the frequency at which $|A\beta| = 1$ and the corresponding phase margin.
(c) Find the closed-loop transfer function, including its zero and poles. Sketch the magnitude of the transfer function versus frequency and label the important parameters on your sketch.

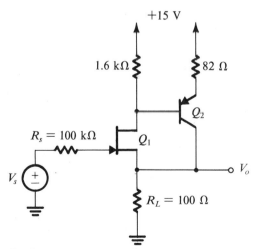

Fig. P12.51

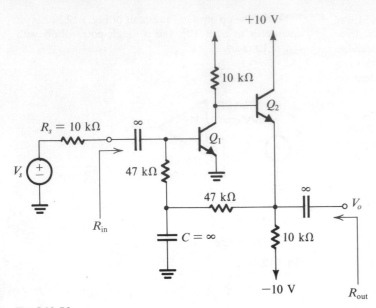

Fig. P12.52

***12.51** For the amplifier in Fig. P12.51 the FET has $V_P = -2$ V and $I_{DSS} = 4$ mA and the BJT has $|V_{BE}| = 0.7$ V and $\beta = 100$. Find the dc operating point of each device assuming that the dc component of V_s is zero. Find A, β, and A_f. Also, calculate the output resistance ex-

cluding R_L. Verify all results using direct analysis (that is, without making use of feedback methods).

***12.52** For the circuit in Fig. P12.52 let the BJTs have $V_{BE} = 0.7$ V and $\beta = 100$. Find V_o/V_s, R_{in} and R_{out} for **(a)** C in the circuit and **(b)** C removed.

Chapter 13

Analog Integrated Circuits

INTRODUCTION

Analog ICs include operational amplifiers, analog multipliers, analog-to-digital (A/D) and digital-to-analog (D/A) converters, phase-locked loops, and a variety of other, more specialized functional blocks. All of these analog subsystems are internally constructed using the basic building blocks we have studied in previous chapters, including differential pairs, current mirrors, MOS switches, and others.

In this chapter we shall study the internal circuitry of the most important analog IC—namely, the operational amplifier and data converters. The terminal characteristics and circuit applications of op amps have already been covered in Chapter 3.

Here, our objective is to expose the reader to some of the ingenious, techniques that have been evolved over the years for combining elementary analog circuit building blocks so as to realize a complete op amp. Specifically, we shall study in some detail the circuit of the most popular analog IC in production today, the 741 internally compensated op amp. This op amp was introduced in 1966 and is currently produced by almost every manufacturer of analog semiconductors. We shall also study a popular circuit of a CMOS op amp. Although the performance of CMOS op amps does not match that available in bipolar units, it is more than adequate for their application in VLSI systems.

In addition to introducing the reader to some of the ideas that make analog IC design an exciting topic, this chapter should serve to tie together many of the concepts and methods studied in the previous chapters.   □

13.1 THE 741 OP-AMP CIRCUIT

We shall begin with a qualitative study of the 741 op-amp circuit, which is shown in Fig. 13.1. Note that in keeping with the IC design philosophy the circuit uses a large number of transistors but relatively few resistors and only one capacitor. This philosophy is dictated by the economics (silicon area, ease of fabrication, quality of realizable components) of the fabrication of active and passive components in IC form (see Appendix A).

As is the case with most modern IC op amps, the 741 requires two power supplies, $+V_{CC}$ and $-V_{EE}$. Normally, $V_{CC} = V_{EE} = 15$ V, but the circuit operates satisfactorily with the power supplies reduced to much lower values (such as ± 5 V). It is important to observe that no circuit node is connected to ground, the common terminal of the two supplies.

With a relatively large circuit such as that shown in Fig. 13.1, the first step in the analysis is the identification of its recognizable parts and their functions. This can be done as follows.

Bias Circuit

The reference bias current of the 741 circuit, I_{REF}, is generated in the branch at the extreme left, consisting of the two diode-connected transistors Q_{11} and Q_{12} and the resistance R_5. Using a Widlar current source formed by Q_{11}, Q_{10}, and R_4, bias current for the first stage is generated in the collector of Q_{10}. Another current mirror formed by Q_8 and Q_9 takes part in biasing the first stage.

The reference bias current I_{REF} is used to provide two proportional currents in the collectors of Q_{13}. This double-collector *lateral*[1] *pnp* transistor can be thought of as two transistors whose base-emitter junctions are connected in parallel. Thus Q_{12} and Q_{13} form a two-output current mirror: one output, the collector of Q_{13B}, provides bias current for Q_{17}, and the other output, the collector of Q_{13A}, provides bias current for the output stage of the op amp.

Two more transistors, Q_{18} and Q_{19}, take part in the dc bias process. The purpose of Q_{18} and Q_{19} is to establish two V_{BE} drops between the bases of the output transistors Q_{14} and Q_{20}.

---

[1] See Appendix A for a description of lateral *pnp* transistors.

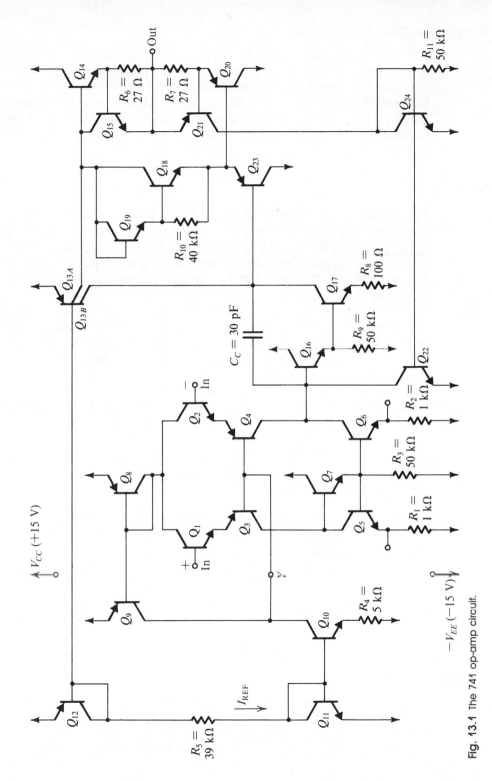

Fig. 13.1 The 741 op-amp circuit.

Short-Circuit Protection Circuitry

The 741 circuit includes a number of transistors that are normally off and that conduct only in the event that one attempts to draw a large current from the op-amp output terminal. This would happen, for example, if the output terminal is short-circuited to one of the two supplies. The short-circuit protection network consists of R_6, R_7, Q_{15}, Q_{21}, Q_{24}, and Q_{22}. In the following we shall assume that these transistors are off. Operation of the short-circuit protection network will be explained in Section 13.5.

The Input Stage

The 741 circuit consists of three stages: an input differential stage, an intermediate single-ended high-gain stage, and an output-buffering stage. The input stage consists of transistors Q_1 through Q_7, with biasing performed by Q_8, Q_9, and Q_{10}. Transistors Q_1 and Q_2 act as emitter followers, causing the input resistance to be high and delivering the differential input signal to the differential common-base amplifier formed by Q_3 and Q_4.

Transistors Q_5, Q_6, and Q_7 and resistors R_1, R_2, and R_3 form the load circuit of the input stage. This is an elaborate current mirror load circuit, which we will analyze in detail in Section 13.3. It will be shown that this load circuit not only provides a high-resistance load but also converts the signal from differential to single-ended with no loss in gain or common-mode rejection. The output of the input stage is taken single-endedly at the collector of Q_6.

As mentioned in Section 9.8, every op-amp circuit includes a *level shifter* whose function is to shift the dc level of the signal so that the signal at the op-amp output can swing positive and negative. In the 741, level shifting is done in the first stage using the lateral *pnp* transistors Q_3 and Q_4. Although lateral *pnp* transistors have poor high-frequency performance, their use in the common-base configuration (which is known to have good high-frequency response) does not seriously impair the op-amp frequency response.

The use of the lateral *pnp* transistors Q_3 and Q_4 in the first stage results in an added advantage: protection of the input-stage transistors Q_1 and Q_2 against emitter–base junction breakdown. Since the emitter–base junction of an *npn* transistor breaks down at about 7 V of reverse bias (see Section 8.12), regular *npn* differential stages would suffer such a breakdown if, say, the supply voltage is accidently connected between the input terminals. Lateral *pnp* transistors, however, have high emitter–base breakdown voltages (about 50 V) and because they are connected in series with Q_1 and Q_2, they provide protection of the 741 input transistors, Q_1 and Q_2,

The Second Stage

The second or intermediate stage is composed of Q_{16}, Q_{17}, Q_{13B}, and the two resistors R_8 and R_9. Transistor Q_{16} acts as an emitter follower, thus giving the second stage a high input resistance. This minimizes the loading on the input stage and avoids loss of gain. Transistor Q_{17} acts as a common-emitter amplifier with a 100-Ω resistor in the emitter. Its load is composed of the high output resistance of the *pnp* current source Q_{13B} in parallel with the input resistance of the output stage (seen looking into the base of Q_{23}). Using a transistor current source as a load resistance is a technique called *active load* (Section 9.5). It enables one to obtain high gain without resorting to the use of high load resistances, which would occupy a large chip area.

The output of the second stage is taken at the collector of Q_{17}. Capacitor C_C is connected in the feedback path of the second stage to provide frequency compensation using the Miller compensation technique studied in Section 12.11. It will be shown in Section 13.6 that the relatively small capacitor C_C gives the 741 a dominant pole at about 4 Hz. Furthermore, pole splitting causes other poles to be shifted to much higher frequencies, giving the op amp a uniform -20 dB/decade gain rolloff with a unity-gain bandwidth of about 1 MHz. It should be pointed out that although C_C is small in value, the chip area that it occupies is about 13 times that of a standard *npn* transistor!

The Output Stage

The purpose of the output stage (Chapter 10) is to provide the amplifier with a low output resistance. In addition, the output stage should be able to supply relatively large load currents without dissipating an unduly large amount of power in the IC. The 741 uses an efficient class AB output stage, which we shall study in detail in Section 13.5.

The output stage consists of the complementary pair Q_{14} and Q_{20}, where Q_{20} is a *substrate pnp* (see Appendix A). Transistors Q_{18} and Q_{19} are fed by current source Q_{13A} and bias the output transistors Q_{14} and Q_{20}. Transistor Q_{23} (which is another substrate *pnp*) acts as an emitter follower, thus minimizing the effect of loading of the output stage on the second stage.

Device Parameters

In the following sections we shall carry out a detailed analysis of the 741 circuit. For the standard *npn* and *pnp* transistors, the following parameters will be used:

npn: $I_S = 10^{-14}$ A, $\beta = 200$, $V_A = 125$ V
pnp: $I_S = 10^{-14}$ A, $\beta = 50$, $\ \ V_A = 50$ V

In the 741 circuit the nonstandard devices are Q_{13}, Q_{14}, and Q_{20}. Transistor Q_{13} will be assumed to be equivalent to two transistors, Q_{13A} and Q_{13B}, with parallel base-emitter junctions and with the following saturation currents:

$$I_{SA} = 0.25 \times 10^{-14} \text{ A} \qquad I_{SB} = 0.75 \times 10^{-14} \text{ A}$$

Transistors Q_{14} and Q_{20} will be assumed to each have an area three times that of a standard device. Output transistors usually have relatively large areas in order to be able to supply large load currents and dissipate relatively large amounts of power with only a moderate increase in the device temperature.

13.1 For the standard *npn* transistor whose parameters are given above, find approximate values for the following parameters if $I_C = 1$ mA: V_{BE}, g_m, r_e, r_π, r_o, r_μ. (*Note:* Assume $r_\mu = 10\beta r_o$.)

Ans. 633 mV; 40 mA/V; 25 Ω; 5 kΩ; 125 kΩ; 250 MΩ

13.2 For the circuit in Fig. E13.2, neglect base currents and use the exponential i_C-v_{BE} relationship to show that

$$I_3 = I_1 \sqrt{\frac{I_{S3}I_{S4}}{I_{S1}I_{S2}}}$$

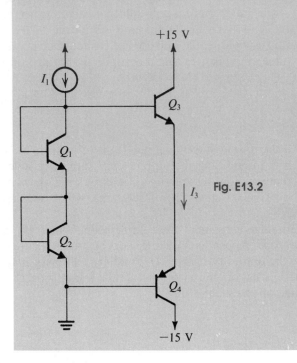

Fig. E13.2

13.2 DC ANALYSIS OF THE 741

In this section we shall carry out a dc analysis of the 741 circuit to determine the bias point of each device. For the dc analysis of an op-amp circuit the input terminals are grounded. Theoretically speaking, this should result in zero dc voltage at the output. However, because the op amp has very large gain, any slight approximation in the analysis will show that the output voltage is far from being zero and is close to either $+V_{CC}$ or $-V_{EE}$. In actual practice an op amp left open-loop will have an output voltage saturated close to one of the two supplies. To overcome this problem in the dc analysis, it will be assumed that the op amp is connected in a negative-feedback loop that stabilizes the output dc voltage to zero volts.

Reference Bias Current

The reference bias current I_{REF} is generated in the branch composed of the two diode-connected transistors Q_{11} and Q_{12} and resistor R_5. With reference to Fig. 13.1, we can write

$$I_{REF} = \frac{V_{CC} - V_{EB12} - V_{BE11} - (-V_{EE})}{R_5}$$

For $V_{CC} = V_{EE} = 15$ V and $V_{BE11} = V_{EB12} \simeq 0.7$ V we have $I_{REF} = 0.73$ mA.

Input Stage Bias

Transistor Q_{11} is biased by I_{REF}, and the voltage developed across it is used to bias Q_{10}, which has a series emitter resistance R_4. This part of the circuit is redrawn in Fig. 13.2 and can be recognized as the Widlar current source studied in Section 9.4. From the circuit we have

$$V_{BE11} - V_{BE10} = I_{C10}R_4$$

Thus

$$V_T \ln \frac{I_{REF}}{I_{C10}} = I_{C10}R_4 \tag{13.1}$$

where it has been assumed that $I_{S10} = I_{S11}$. Substituting the known values for I_{REF} and R_4, this equation can be solved by trial and error to determine I_{C10}. For our case the result is $I_{C10} = 19 \ \mu A$.

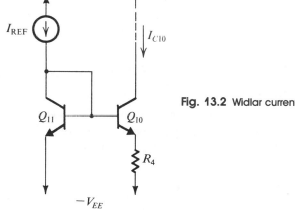

Fig. 13.2 Widlar current source.

Exercise

13.3 Design the Widlar current source of Fig. 13.2 to generate a current $I_{C10} = 10 \ \mu A$ given that $I_{REF} = 1$ mA. If at a collector current of 1 mA $V_{BE} = 0.7$ V, find V_{BE11} and V_{BE10}.

Ans. $R_4 = 11.5$ kΩ; $V_{BE11} = 0.7$ V; $V_{BE10} = 0.585$ V

Having determined I_{C10}, we proceed to determine the dc current in each of the input-stage transistors. Part of the input stage is redrawn in Fig. 13.3. From symmetry we see that

$$I_{C1} = I_{C2}$$

Denote this current by I. We see that if the *npn* β is high, then

$$I_{E3} = I_{E4} \simeq I$$

and the base currents of Q_3 ad Q_4 are equal, with a value of $I/(\beta_P + 1) \cong I/\beta_P$, where β_P denotes β of the *pnp* devices.

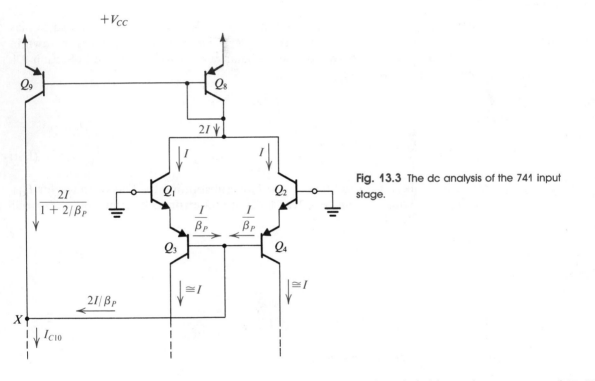

Fig. 13.3 The dc analysis of the 741 input stage.

The current mirror formed by Q_8 and Q_9 is fed by an input current of $2I$. Using the result in Eq. (9.65), we can express the output current of the mirror as

$$I_{C9} = \frac{2I}{1 + 2/\beta_P}$$

We can now write a node equation for node X in Fig. 13.3 and thus determine the value of I. If $\beta_P \gg 1$, then this node equation gives

$$2I \simeq I_{C10}$$

For the 741, $I_{C10} = 19 \ \mu\text{A}$; thus $I \simeq 9.5 \ \mu\text{A}$. We have thus determined that

$$I_{C1} = I_{C2} \simeq I_{C3} = I_{C4} = 9.5 \ \mu\text{A}$$

At this point we should note that transistors Q_1 through Q_4, Q_8, and Q_9 form a *negative-feedback loop*, which works to stabilize the value of I at approximately $I_{C10}/2$. To appreciate this fact, assume that for some reason the current I in Q_1 and Q_2 increases. This will cause the current pulled from Q_8 to increase, and the output current of the Q_8-Q_9 mirror will correspondingly increase. However, since I_{C10} remains constant, node X forces the combined base currents of Q_3 and Q_4 to decrease. This in turn will cause the emitter currents of Q_3 and Q_4, and hence the collector currents of Q_1 and Q_2, to decrease. This is opposite in direction to the change originally assumed. Hence the feedback is negative, and it stabilizes the value of I.

Figure 13.4 shows the remainder of the 741 input stage. If we neglect the base current of Q_{16}, then

$$I_{C6} \simeq I$$

Similarly, neglecting the base current of Q_7 we obtain

$$I_{C5} \simeq I$$

The bias current of Q_7 can be determined from

$$I_{C7} \simeq I_{E7} = \frac{2I}{\beta_N} + \frac{V_{BE6} + IR_2}{R_3} \tag{13.2}$$

where β_N denotes β of the *npn* transistors. To determine V_{BE6} we use the transistor exponential relationship and write

$$V_{BE6} = V_T \ln \frac{I}{I_S}$$

Substituting $I_S = 10^{-14}$ A and $I = 9.5$ μA results in $V_{BE6} = 517$ mV. Then substituting in Eq. (13.2) yields $I_{C7} = 10.5$ μA. Note that the base current of Q_7 is indeed negligible compared to the value of I, as has been assumed.

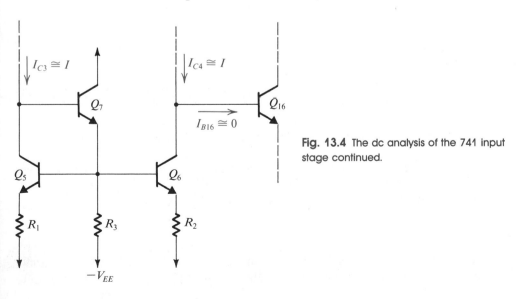

Fig. 13.4 The dc analysis of the 741 input stage continued.

Input Bias and Offset Currents

The *input bias current* of an op amp is defined (Chapters 3 and 9)

$$I_B = \frac{I_{B1} + I_{B2}}{2}$$

For the 741 we obtain

$$I_B = \frac{I}{\beta_N}$$

Using $\beta_N = 200$ yields $I_B = 47.5$ nA. Note that this value is reasonably small and is typical of general-purpose op amps that use BJTs in the input stage. Much lower input bias currents (in the picoamp range) can be obtained using an FET input stage. Also, there exist techniques for reducing the input bias current of bipolar-input op amps.

Because of possible mismatches in the β values of Q_1 and Q_2, the input base currents will not be equal. Given the value of the β mismatch, one can use Eq. (9.62) to calculate the *input offset current*, defined as

$$I_{\text{off}} = |I_{B1} - I_{B2}|$$

Input Offset Voltage

From Chapter 9 we know that the input offset voltage is determined primarily by mismatches between the two sides of the input stage. In the 741 op amp the input offset voltage is due to mismatches between Q_1 and Q_2, between Q_3 and Q_4, between Q_5 and Q_6, and between R_1 and R_2. Evaluation of the components of V_{off} corresponding to the various mismatches follows the method outlined in Section 9.3. Basically, we find the current that results at the output of the first stage due to the particular mismatch being considered. Then we find the differential input voltage that must be applied to reduce the output current to zero.

Input Common-Mode Range

The *input common-mode range* is the range of input common-mode voltages over which the input stage remains in the linear active mode. Refer to Fig. 13.1. We see that in the 741 circuit the input common-mode range is determined at the upper end by saturation of Q_1 and Q_2 and at the lower end by saturation of Q_3 and Q_4.

Exercise

13.4 Neglect the voltage drops across R_1 and R_2 and assume that $V_{CC} = V_{EE} = 15$ V. Show that the input common-mode range of the 741 is approximately -12.6 to $+14.4$ V. (Assume that $V_{BE} \simeq 0.6$ V.)

Second-Stage Bias

If we neglect the base current of Q_{23} then we see from Fig. 13.1 that the collector current of Q_{17} is approximately equal to the current supplied by current source Q_{13B}.

Because Q_{13B} has a scale current 0.75 times that of Q_{12}, its collector current will be

$$I_{C13B} \simeq 0.75 I_{REF}$$

where we have assumed that $\beta_P \gg 1$. Thus $I_{C13B} = 550\ \mu A$ and $I_{C17} \simeq 550\ \mu A$. At this current level the base–emitter voltage of Q_{17} is

$$V_{BE17} = V_T \ln \frac{I_{C17}}{I_S} = 618\ \text{mV}$$

The collector current of Q_{16} can be determined from

$$I_{C16} \simeq I_{E16} = I_{B17} + \frac{I_{E17}R_8 + V_{BE17}}{R_9}$$

This calculation yields $I_{C16} = 16.2\ \mu A$. Note that the base current of Q_{16} will indeed be negligible compared to the input-stage bias I, as we have previously assumed.

Output-Stage Bias

Figure 13.5 shows the output stage of the 741 with the short-circuit protection circuitry omitted. Current source Q_{13A} delivers a current of $0.25I_{REF}$ (because I_S of Q_{13A} is 0.25 times the I_S of Q_{12}) to the network composed of Q_{18}, Q_{19}, and R_{10}. If

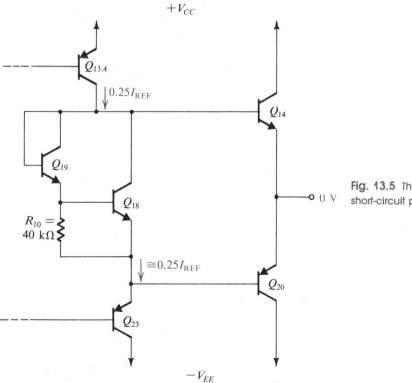

Fig. 13.5 The 741 output stage without the short-circuit protection devices.

we neglect the base currents of Q_{14} and Q_{20}, then the emitter current of Q_{23} will also be equal to $0.25I_{REF}$. Thus

$$I_{C23} \simeq I_{E23} \simeq 0.25I_{REF} = 180 \ \mu A$$

Thus we see that the base current of Q_{23} is only $180/50 = 3.6 \ \mu A$, which is negligible compared to I_{C17}, as we have previously assumed.

If we assume that V_{BE18} is approximately 0.6 V, we can determine the current in R_{10} as 15 μA. The emitter current of Q_{18} is therefore

$$I_{E18} = 180 - 15 = 165 \ \mu A$$

Also,

$$I_{C18} \simeq I_{E18} = 165 \ \mu A$$

At this value of current we find that $V_{BE18} = 588$ mV, which is quite close to the value assumed. The base current of Q_{18} is $165/200 = 0.8 \ \mu A$, which can be added to the current in R_{10} to determine the Q_{19} current as

$$I_{C19} \simeq I_{E19} = 15.8 \ \mu A$$

The voltage drop across the base–emitter junction of Q_{19} can now be determined as

$$V_{BE19} = V_T \ln \frac{I_{C19}}{I_S} = 530 \ \text{mV}$$

As mentioned in Section 13.1, the purpose of the Q_{18}-Q_{19} network is to establish two V_{BE} drops between the bases of the output transistors Q_{14} and Q_{20}. This voltage drop, V_{BB}, can be now calculated as

$$V_{BB} = V_{BE18} + V_{BE19} = 588 + 530 = 1.118 \ \text{V}$$

Since V_{BB} appears across the series combination of the base–emitter junctions of Q_{14} and Q_{20}, we can write

$$V_{BB} = V_T \ln \frac{I_{C14}}{I_{S14}} + V_T \ln \frac{I_{C20}}{I_{S20}}$$

Using the calculated value of V_{BB} and substituting $I_{S14} = I_{S20} = 3 \times 10^{-14}$ A, we determine the collector currents as

$$I_{C14} = I_{C20} = 154 \ \mu A$$

Table 13.1 DC COLLECTOR CURRENTS OF THE 741 CIRCUIT (μA)

Q_1	9.5	Q_8	19	Q_{13B}	550	Q_{19}	15.8
Q_2	9.5	Q_9	19	Q_{14}	154	Q_{20}	154
Q_3	9.5	Q_{10}	19	Q_{15}	0	Q_{21}	0
Q_4	9.5	Q_{11}	730	Q_{16}	16.2	Q_{22}	0
Q_5	9.5	Q_{12}	730	Q_{17}	550	Q_{23}	180
Q_6	9.5	Q_{13A}	180	Q_{18}	165	Q_{24}	0
Q_7	10.5						

Summary

For future reference, Table 13.1 provides a listing of the values of the collector bias currents of the 741 transistors.

Exercise

13.5 If in the circuit of Fig. 13.5 the Q_{18}-Q_{19} network is replaced by two diode-connected transistors, find the current in Q_{14} and Q_{20}. (*Hint:* Use the result of Exercise 13.2.)

Ans.   540 μA

13.3 SMALL-SIGNAL ANALYSIS OF THE 741 INPUT STAGE

Figure 13.6 shows part of the 741 input stage for the purpose of performing small-signal analysis. Note that since the collectors of Q_1 and Q_2 are connected to a constant dc voltage, they are shown grounded. Also, the constant-current biasing of the bases of Q_3 and Q_4 is equivalent to having the common base terminal open-circuited.

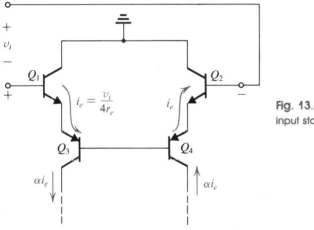

Fig. 13.6 Small-signal analysis of the 741 input stage.

The differential signal v_i applied between the input terminals effectively appears across four equal emitter resistances connected in series—those of Q_1, Q_2, Q_3, and Q_4. As a result, emitter signal currents flow as indicated in Fig. 13.6 with

$$i_e = \frac{v_i}{4r_e} \tag{13.3}$$

where r_e denotes the emitter resistance of each of Q_1 through Q_4. Thus

$$r_e = \frac{V_T}{I} = \frac{25\,\text{mV}}{9.5\,\mu\text{A}} = 2.63\,\text{k}\Omega$$

Thus the four transistors Q_1 through Q_4 supply the load circuit with a pair of complementary current signals αi_e, as indicated in Fig. 13.6.

The input differential resistance of the op amp can be obtained from Fig. 13.6 as

$$R_{id} = 4r_\pi = 4(\beta_N + 1)r_e \tag{13.4}$$

For $\beta_N = 200$ we obtain $R_{id} = 2.1$ MΩ.

Proceeding with the input-stage analysis, we show in Fig. 13.7 the load circuit fed with the complementary pair of current signals found above. Neglecting the signal current in the base of Q_7, we see that the collector signal current of Q_5 is approximately equal to the input current αi_e. Now, since Q_5 and Q_6 are identical and their bases are tied together, and since equal resistances are connected in their emitters, it follows that their collector signal currents must be equal. Thus the signal current in the collector of Q_6 is forced to be equal to αi_e. In other words, the load circuit functions as a *current mirror*.

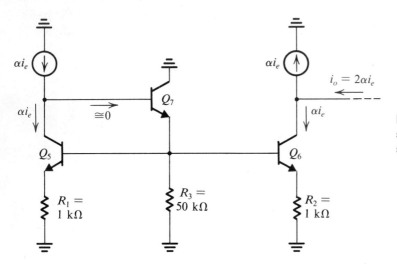

Fig. 13.7 The load circuit of the input stage fed by the two complementary current signals generated by Q_1 through Q_4 in Fig. 13.6.

Now consider the output node of the input stage. The output current i_o is given by

$$i_o = 2\alpha i_e \tag{13.5}$$

The factor of two in this equation indicates that conversion from differential to single-ended is performed without losing half the signal. The trick, of course, is the use of the current mirror to invert one of the current signals and then add the result to the other current signal (see Section 9.5).

Equations (13.3) and (13.5) can be combined to obtain the transconductance of the input stage, G_{m1}:

$$G_{m1} \equiv \frac{i_o}{v_i} = \frac{\alpha}{2r_e} \tag{13.6}$$

Substituting $r_e = 2.63$ kΩ and $\alpha \simeq 1$ yields $G_{m1} = 1/5.26$ mA/V.

Exercise

To complete our modeling of the 741 input stage we must find its output resistance R_{o1}. This is the resistance seen "looking back" at the collector terminal of Q_6 in Fig. 13.7. Thus R_{o1} is the parallel equivalent of the output resistance of the current source supplying the signal current αi_e and the output resistance of Q_6. The first component is the resistance looking into the collector of Q_4 in Fig. 13.6. Finding this resistance is considerably simplified if we assume that the common bases of Q_3 and Q_4 are at a *virtual ground*. This of course happens only when the input signal v_i is applied in a complementary fashion. Nevertheless, this assumption does not result in a large error.

Assuming that the base of Q_4 is at virtual ground, the resistance we are after is R_{o4}, indicated in Fig. 13.8a. This is the output resistance of a common-base transistor that has a resistance (r_e of Q_2) in its emitter. To find R_{o4} we may use the expression developed in Chapter 9 [Eq. (9.78)]:

$$R_o = r_o[1 + g_m(R_E \| r_\pi)] \tag{13.7}$$

Substituting $R_E = r_e \equiv 2.63 \text{ k}\Omega$ and $r_o = V_A/I$, where $V_A = 50$ V and $I = 9.5 \ \mu\text{A}$, (thus $r_o = 5.26 \text{ M}\Omega$) results in $R_{o4} = 10.5 \text{ M}\Omega$.

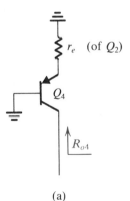

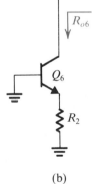

Fig. 13.8 Simplified circuit for finding the two components of the output resistance R_{o1} of the first stage.

(a)   (b)

The second component of the output resistance is that seen looking into the collector of Q_6 in Fig. 13.7. Although the base of Q_6 is not at signal ground, we shall assume that the signal voltage at the base is sufficiently small to make this approximation valid. The circuit then takes the form in Fig. 13.8b, and R_{o6} can be determined using Eq. (13.7) with $R_E = R_2$. Thus $R_{o6} \simeq 18.2 \text{ M}\Omega$.

Finally, we combine R_{o4} and R_{o6} in parallel to obtain the output resistance of the input stage, R_{o1}, as $R_{o1} = 6.7 \text{ M}\Omega$.

Figure 13.9 shows the equivalent circuit that we have derived for the input stage. This is a simplified version of the y-parameter model of a two-port network with y_{12} assumed negligible (see Appendix B).

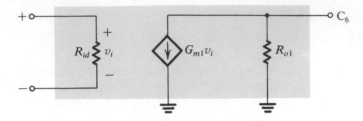

Fig. 13.9 Small-signal equivalent circuit for the input stage of the 741 op amp.

EXAMPLE 13.1

We wish to find the input offset voltage resulting from a 2% mismatch between the resistances R_1 and R_2 in Fig. 13.1.

Solution

Consider first the situation when both input terminals are grounded and assume that $R_1 = R$ and $R_2 = R + \Delta R$, where $\Delta R/R = 0.02$. From Fig. 13.10 we see that while Q_5 still conducts a current equal to I, the current in Q_6 will be smaller by ΔI. The value of ΔI can be found from

$$V_{BE5} + IR = V_{BE6} + (I - \Delta I)(R + \Delta R)$$

Thus

$$V_{BE5} - V_{BE6} = I\,\Delta R - \Delta I(R + \Delta R) \tag{13.8}$$

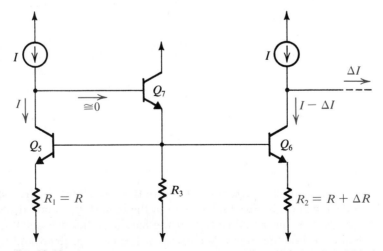

Fig. 13.10 Input stage with both inputs grounded and a mismatch ΔR between R_1 and R_2.

The quantity on the left-hand side is in effect the change in V_{BE} due to a change in I_E of ΔI. We may therefore write

$$V_{BE5} - V_{BE6} \simeq \Delta I r_e \tag{13.9}$$

Equations (13.8) and (13.9) can be combined to obtain

$$\frac{\Delta I}{I} = \frac{\Delta R}{R + \Delta R + r_e} \tag{13.10}$$

Substituting $R = 1\ \text{k}\Omega$ and $r_e = 2.63\ \text{k}\Omega$ shows that a 2% mismatch between R_1 and R_2 gives rise to an output current $\Delta I = 5.5 \times 10^{-3} I$. To reduce this output current to zero we have to apply an input voltage V_{off} given by

$$V_{\text{off}} = \frac{\Delta I}{G_{m1}} = \frac{5.5 \times 10^{-3} I}{G_{m1}} \tag{13.11}$$

Substituting $I = 9.5\ \mu\text{A}$ and $G_{m1} = 1/5.26\ \text{mA/V}$ results in the offset voltage $V_{\text{off}} \simeq 0.3\ \text{mV}$.

It should be pointed out that the offset voltage calculated is only one component of the input offset voltage of the 741. Other components arise because of mismatches in transistor characteristics. The 741 offset voltage is specified to be typically 2 mV.

Exercises

The purpose of the following series of exercises is to determine the finite common-mode gain that results from a mismatch in the load circuit of the input stage of the 741 op amp. Figure E13.7 shows the input stage with an input common-mode signal v_{icm} applied and with a mismatch ΔR between the two resistances R_1 and R_2. Note that to simplify matters we have opened the common-mode feedback loop and included a resistance R_o, which is the resistance seen looking to the left of node Y in the circuit of Fig. 13.1. Thus R_o is the parallel equivalent of R_{o9} (the output resistance of Q_9) and R_{o10} (the output resistance of Q_{10}).

13.7 Show that the current i (Fig. E13.7) is given by

$$i = \frac{v_{\text{icm}}}{r_{e1} + r_{e3} + [2R_o/(\beta_P + 1)]}$$

13.8 Show that

$$i_o = i \frac{\Delta R}{R + r_{e5} + \Delta R}$$

13.9 Using the results of Exercises 13.7 and 13.8, and assuming that $\Delta R \ll (R + r_e)$ and $R_o/(\beta_P + 1) \gg (r_{e1} + r_{e3})$, show that the common-mode transconductance G_{mcm} is given approximately by

$$G_{\text{mcm}} \equiv \frac{i_o}{v_{\text{icm}}} \simeq \frac{\beta_P}{2R_o} \frac{\Delta R}{R + r_{e5}}$$

13.10 Refer to Fig. 13.1 and assume that the bases of Q_9 and Q_{10} are at approximately constant voltages (signal ground). Find R_{o9}, R_{o10} and hence R_o. Use $V_A = 125\ \text{V}$ for npn and $50\ \text{V}$ for pnp transistors and neglect r_μ.

Ans. $R_{o9} = 2.63\ \text{M}\Omega$; $R_{o10} = 31.1\ \text{M}\Omega$; $R_o = 2.43\ \text{M}\Omega$

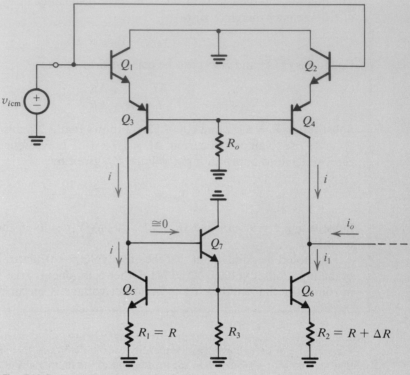

Fig. E13.7

13.11 For $\beta_P = 50$ and $\Delta R/R = 0.02$ evaluate G_{mcm} obtained in Exercise 13.9.

Ans. $0.057 \ \mu\text{A/V}$

13.12 Use the value of G_{mcm} obtained in Exercise 13.11 and the value of G_{m1} obtained from Eq. (13.6) to find the CMRR that is the ratio of G_{m1} to G_{mcm}, expressed in decibels.

Ans. 70.5 dB

13.13 Noting that with the common-mode negative-feedback loop in place the common-mode gain will decrease by the amount of feedback, and noting that the loop gain is approximately equal to β_P (see Problem 13.9), find the CMRR with the feedback loop in place.

Ans. 104.6 dB

13.4 SMALL-SIGNAL ANALYSIS OF THE 741 SECOND STAGE

Figure 13.11 shows the 741 second stage prepared for small-signal analysis. In this section we shall analyze the second stage to determine the values of the parameters of the equivalent circuit shown in Fig. 13.12. Again, this is a smplified y-parameter equivalent circuit with y_{12} neglected.

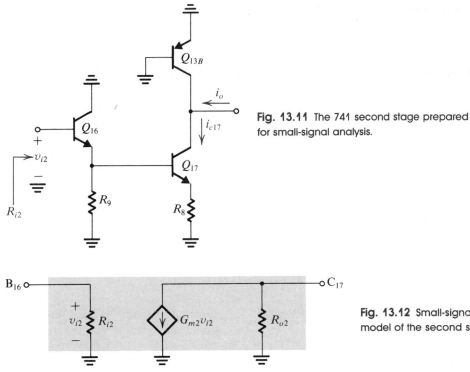

Fig. 13.11 The 741 second stage prepared for small-signal analysis.

Fig. 13.12 Small-signal equivalent circuit model of the second stage.

Input Resistance

The input resistance R_{i2} can be found by inspection to be

$$R_{i2} = (\beta_{16} + 1)[r_{e16} + R_9 \| (\beta_{17} + 1)(r_{e17} + R_8)] \tag{13.12}$$

Substituting the appropriate parameter values yields $R_{i2} \simeq 4\ \text{M}\Omega$.

Transconductance

From the equivalent circuit of Fig. 13.12 we see that the transconductance G_{m2} is the ratio of the *short-circuit output current* to the input voltage. Short-circuiting the output terminal of the second stage (Fig. 13.11) to ground makes the signal current through the output resistance of Q_{13B} zero, and the output short-circuit current becomes equal to the collector signal current of Q_{17} (i_{c17}). This latter current can be easily related to v_{i2} as follows:

$$i_{c17} = \frac{\alpha v_{b17}}{r_{e17} + R_8} \tag{13.13}$$

$$v_{b17} = v_{i2} \frac{(R_9 \| R_{i17})}{(R_9 \| R_{i17}) + r_{e16}} \tag{13.14}$$

$$R_{i17} = (\beta_{17} + 1)(r_{e17} + R_8) \tag{13.15}$$

These equations can be combined to obtain

$$G_{m2} \equiv \frac{i_{c17}}{v_{i2}} \tag{13.16}$$

which, for the 741 parameter values, is found to be $G_{m2} = 6.5$ mA/V.

Output Resistance

To determine the output resistance R_{o2} of the second stage in Fig. 13.11, we ground the input terminal and find the resistance looking back into the output terminal. It follows that R_{o2} is given by

$$R_{o2} = (R_{o13B} \| R_{o17}) \tag{13.17}$$

where R_{o13B} is the resistance looking into the collector of Q_{13B} while its base and emitter are connected to ground. It can be easily shown that

$$R_{o13B} = r_{o13B} \tag{13.18}$$

For the 741 component values we obtain $R_{o13B} = 90.9$ kΩ.

The second component in Eq. (13.17), R_{o17}, is the resistance seen looking into the collector of Q_{17}, as indicated in Fig. 13.13. Since the resistance between the base of Q_{17} and ground is relatively small, one can considerably simplify matters by assuming that the base is grounded. Doing this, we can use Eq. (13.7) to determine R_{o17}. For our case the result is $R_{o17} \cong 787$ kΩ. Combining R_{o13B} and R_{o17} in parallel yields $R_{o2} = 81$ kΩ.

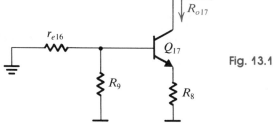

Fig. 13.13 Definition of R_{o17}.

Thévenin Equivalent Circuit

The second-stage equivalent circuit can be converted to the Thévenin form, as shown in Fig. 13.14. Note that the stage open-circuit voltage gain is $-G_{m2}R_{o2}$.

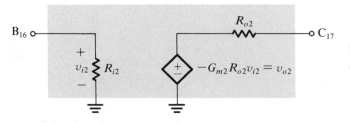

Fig. 13.14 Thévenin form of the small-signal model of the second stage.

Exercises

13.5 ANALYSIS OF THE 741 OUTPUT STAGE

The 741 output stage is shown in Fig. 13.15 without the short-circuit protection circuit. The stage is shown driven by the second-stage transistor Q_{17} and loaded with a 2-kΩ resistance. The circuit is of the AB class (Chapter 10), with the network composed of Q_{18}, Q_{19}, and R_{10} providing the bias of the output transistors Q_{14} and Q_{20}. The use of this network rather than two diode-connected transistors in series enables biasing the output transistors at a low current (0.15 mA) in spite of the fact that the output devices are about three times as large as the standard devices. This is obtained by arranging that the current in Q_{19} is very small and thus its V_{BE} is also small. We analyzed the dc bias in Section 13.2.

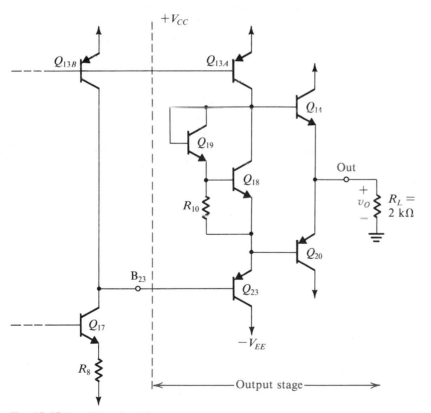

Fig. 13.15 The 741 output stage.

Another feature of the 741 output stage worth noting is that the stage is driven by an emitter follower Q_{23}. As will be shown, this emitter follower provides added buffering, which makes the op-amp gain almost independent of the parameters of the output transistors.

Output Voltage Limits

The maximum positive output voltage is limited by the saturation of current-source transistor Q_{13A}. Thus

$$v_{O\text{max}} = V_{CC} - V_{CE\text{sat}} - V_{BE14} \tag{13.19}$$

which is about 1 V below V_{CC}. The minimum output voltage (that is, maximum negative amplitude) is limited by the saturation of Q_{17}. Neglecting the voltage drop across R_8 we obtain

$$v_{O\text{min}} = -V_{CC} + V_{CE\text{sat}} + V_{EB23} + V_{EB20} \tag{13.20}$$

which is about 1.5 V above $-V_{CC}$.

Small-Signal Model

We shall now carry out a small-signal analysis of the output stage for the purpose of determining the values of the parameters of the equivalent circuit model shown in Fig. 13.16. The model is shown fed by v_{o2}, which is the open-circuit output voltage of the second stage. From Fig. 13.14, v_{o2} is given by

$$v_{o2} = -G_{m2}R_{o2}v_{i2} \tag{13.21}$$

and G_{m2} and R_{o2} were previously determined as $G_{m2} = 6.5$ mA/V and $R_{o2} = 81$ kΩ. Resistance R_{i3} is the input resistance of the output stage determined with the amplifier loaded with R_L. Although the effect of loading an amplifier stage on its input resistance is negligible in the input and second stages, this is not the case in general in an output stage. Defining R_{i3} in this manner enables correct evaluation of the voltage gain of the second stage, A_2, as

$$A_2 \equiv \frac{v_{i3}}{v_{i2}} = -G_{m2}R_{o2}\frac{R_{i3}}{R_{i3} + R_{o2}} \tag{13.22}$$

To determine R_{i3} assume that one of the two output transistors—say, Q_{20}—is conducting a current of, say, 5 mA. It follows that the input resistance looking into

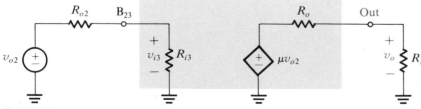

Fig. 13.16 Model for the 741 output stage.

the base of Q_{20} is approximately $\beta_{20}R_L$. Assuming $\beta_{20} = 50$, for $R_L = 2\ \text{k}\Omega$ the input resistance of Q_{20} is 100 kΩ. This resistance appears in parallel with the series combination of the output resistance of Q_{13A} ($r_{o13A} \simeq 2.5\ \text{M}\Omega$) and the resistance of the Q_{18}-Q_{19} network. This series combination will be much greater than 100 kΩ, and we may assume that the total resistance in the emitter of Q_{23} is approximately 100 kΩ. Thus the input resistance R_{i3} is given by

$$R_{i3} \simeq \beta_{23} \times 100\ \text{k}\Omega$$

which for $\beta_{23} = 50$ is $R_{i3} \simeq 5\ \text{M}\Omega$. Since $R_{o2} = 81\ \text{k}\Omega$, we see that $R_{i3} \gg R_{o2}$, and the value of R_{i3} will have little effect on the performance of the op amp. We can use the value obtained for R_{i3} to determine the gain of the second stage in Eq. (13.22) as $A_2 = -518$. The value of A_2 will be needed in Section 13.6 in connection with frequency-response analysis.

Continuing with the determination of the equivalent circuit-model-parameters, we note from Fig. 13.16 that μ is the *open-circuit voltage gain* of the output stage,

$$\mu = \left.\frac{v_o}{v_{o2}}\right|_{R_L = \infty} \tag{13.23}$$

With $R_L = \infty$ the gain of the emitter-follower output transistor (Q_{14} or Q_{20}) will be nearly unity. Also, with $R_L = \infty$ the resistance in the emitter of Q_{23} will be very large. This means that the gain of Q_{23} will be nearly unity and the input resistance of Q_{23} will be very large. We thus conclude that $\mu \simeq 1$.

Next we shall find the value of the output resistance of the op amp, R_o. For this purpose refer to the circuit shown in Fig. 13.17. In accordance with the definition of R_o, the input source feeding the output stage is grounded but its resistance (which

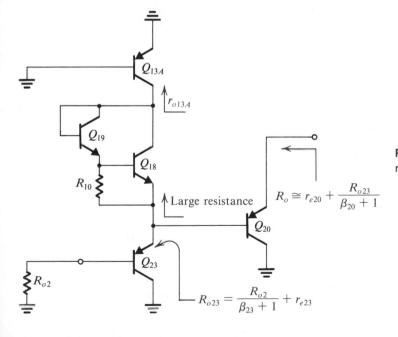

Fig. 13.17 Circuit for finding the output resistance R_o.

is the output resistance of the second stage, R_{o2}) is included. We have assumed that the output voltage v_O is negative, and thus Q_{20} is conducting most of the current; transistor Q_{14} has therefore been eliminated. The exact value of the output resistance will of course depend on which transistor (Q_{14} or Q_{20}) is conducting and on the value of load current. Nevertheless, we wish to find an estimate of R_o.

As indicated in Fig. 13.17, the resistance seen looking into the emitter of Q_{23} is

$$R_{o23} = \frac{R_{o2}}{\beta_{23} + 1} + r_{e23} \tag{13.24}$$

Substituting $R_{o2} = 81$ kΩ, $\beta_{23} = 50$, and $r_{e23} = 25/0.18 = 139$ Ω yields $R_{o23} = 1.73$ kΩ. This resistance appears in parallel with the series combination of r_{o13A} and the resistance of the Q_{18}-Q_{19} network. Since r_{o13A} alone (2.5 MΩ) is much larger than R_{o23}, the effective resistance between the base of Q_{20} and ground is approximately equal to R_{o23}. Now we can find the output resistance R_o as

$$R_o = \frac{R_{o23}}{\beta_{20} + 1} + r_{e20} \tag{13.25}$$

For $\beta_{20} = 50$ the first component of R_o is 34 Ω. The second component depends critically on the value of output current. For an output current of 5 mA, r_{e20} is 5 Ω and R_o is 39 Ω. To this value we must add the resistance R_7 (27 Ω) (see Fig. 13.1), which is included for short-circuit protection. The output resistance of the 741 is specified to be typically 75 Ω.

Exercises

13.18 Using a simple (r_π, g_m) model for each of the two transistors Q_{18} and Q_{19} in Fig. E13.18, find the small-signal resistance between A and A'. (*Note:* From Table 13.1, $I_{C18} = 165$ μA and $I_{C19} \simeq 16$ μA.

Ans.   163 Ω

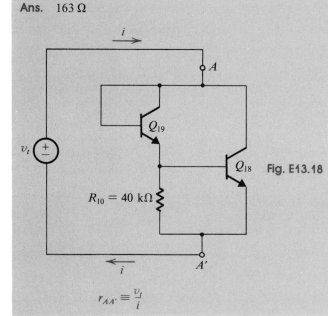

Fig. E13.18

13.19 Figure E13.19 shows the circuit for determining the op-amp output resistance when v_O is positive and Q_{14} is conducting most of the current. Using the resistance of the Q_{18}-Q_{19} network calculated in Exercise 13.18 and neglecting the output resistance of Q_{13A}, find R_o.

Ans.   14.4 Ω

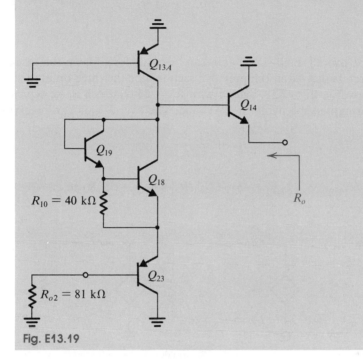

Fig. E13.19

Output Short-Circuit Protection

If the op-amp output terminal is short-circuited to one of the power supplies, one of the two output transistors could conduct a large amount of current. Such a large current can result in sufficient heating to cause burnout of the IC. To guard against this possibility, the 741 op amp is equipped with a special circuit for short-circuit protection. The function of this circuit is to limit the current in the output transistors in the event of a short circuit.

Refer to Fig. 13.1. Resistance R_6 together with transistor Q_{15} limits the current that would flow out of Q_{14} in the event of a short circuit. Specifically, if the current in the emitter of Q_{14} exceeds about 20 mA, the voltage drop across R_6 exceeds 540 mV, which turns Q_{15} on. As Q_{15} turns on, its collector robs some of the current supplied by Q_{13A}, thus reducing the base current of Q_{14}. This mechanism thus limits the maximum current that the op amp can source (that is, supply from the output terminal in the outward direction) to about 20 mA.

Limiting of the maximum current that the op amp can sink, and hence the current through Q_{20}, is done by a mechanism similar to the one discussed above. The relevant circuit is composed of R_7, Q_{21}, Q_{24}, and Q_{22}.

13.6 GAIN AND FREQUENCY RESPONSE OF THE 741

In this section we shall evaluate the overall small-signal voltage gain of the 741 op amp. We shall then consider the op amp's frequency response and its slew-rate limitation.

Small-Signal Gain

The overall small-signal gain can be easily found from the cascade of the equivalent circuits derived in the previous sections for the three op-amp stages. This cascade is shown in Fig. 13.18, loaded with $R_L = 2 \text{ k}\Omega$, which is the typical value used in measuring and specifying the 741 data. The overall gain can be expressed as

$$\frac{v_o}{v_i} = \frac{v_{i2}}{v_i} \frac{v_{o2}}{v_{i2}} \frac{v_o}{v_{o2}} \qquad (13.26)$$

$$= -G_{m1}(R_{o1} \parallel R_{i2})(-G_{m2}R_{o2})\mu \frac{R_L}{R_L + R_o} \qquad (13.27)$$

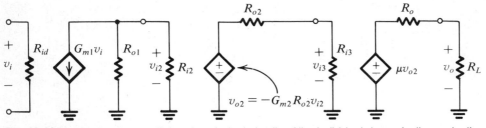

Fig. 13.18 Cascading the small-signal equivalent circuits of the individual stages for the evaluation of the overall voltage gain.

Using the values found in previous sections yields

$$\frac{v_o}{v_i} = -476.1 \times (-526.5) \times 0.97 = 243,147 \text{ V/V}$$

$$= 107.7 \text{ dB} \qquad (13.28)$$

Frequency Response

The 741 is an internally compensated op amp. It employs the Miller compensation technique, studied in Section 12.11, to introduce a dominant low-frequency pole. Specifically, a 30-pF capacitor (C_C) is connected in the negative-feedback path of the second stage. An approximate estimate of the frequency of the dominant pole can be obtained as follows:

Using Miller's theorem (Section 2.5) the effective capacitance due to C_C between the base of Q_{16} and ground is (see Fig. 13.1)

$$C_i = C_C(1 + |A_2|) \qquad (12.29)$$

where A_2 is the second-stage gain. Use of the value calculated for A_2 in Section 13.6, $A_2 = -518$, results in $C_i = 15,570$ pF. Since this capacitance is quite large, we shall neglect all other capacitances between the base of Q_{16} and signal ground. The total resistance between this node and ground is

$$R_t = (R_{o1} \parallel R_{i2})$$
$$= (6.7 \text{ M}\Omega \parallel 4 \text{ M}\Omega) = 2.5 \text{ M}\Omega \qquad (13.30)$$

Thus the dominant pole has a frequency f_P given by

$$f_P = \frac{1}{2\pi C_i R_t} = 4.1 \text{ Hz} \qquad (13.31)$$

It should be noted that this approach is equivalent to using the approximate formula in Eq. (12.64).

As discussed in Section 12.11, Miller compensation produces an additional advantageous effect, namely pole splitting. As a result, the other poles of the circuit are moved to very high frequencies. This has been confirmed by computer-aided analysis [see Gray and Meyer (1984)].

Assuming that all nondominant poles are at high frequencies, the calculated values give rise to the Bode plot shown in Fig. 13.19. The unity-gain bandwidth f_t can be calculated from

$$f_t = A_0 f_{3dB} \qquad (13.32)$$

Thus

$$f_t = 243,147 \times 4.1 \cong 1 \text{ MHz} \qquad (13.33)$$

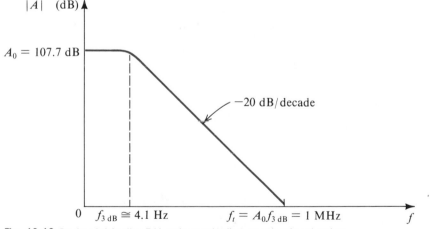

Fig. 13.19 Bode plot for the 741 gain, neglecting nondominant poles.

Although this Bode plot implies that the phase shift at f_t is $-90°$ and thus that the phase margin is $90°$, in practice a phase margin of about $80°$ is obtained. The excess phase shift (about $10°$) is due to the nondominant poles. This phase margin is sufficient to provide stable operation of closed-loop amplifiers with any value of feedback factor β. This convenience of use of the internally compensated 741 is achieved at the expense of a great reduction in open-loop gain and hence in the amount of negative feedback. In other words, if one requires a closed-loop amplifier with a gain of 1,000, then the 741 is overcompensated for such an application and one would be much better off designing one's own compensation (assuming, of course, the availability of an op amp that is not internally compensated).

A Simplified Model

Figure 13.20 shows a simplified model of the 741 op amp in which the high-gain second stage, with its feedback capacitance C_C, is modeled by an ideal integrator. In this model the gain of the second stage is assumed sufficiently large that a virtual ground appears at its input. For this reason the output resistance of the input stage and the input resistance of the second stage have been omitted. Furthermore, the output stage is assumed to be an ideal unity-gain follower. The reader will recall that this model was used in our study of the op-amp terminal characteristics in Chapter 3.

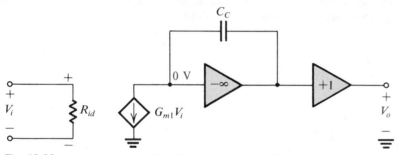

Fig. 13.20 A simple model for the 741 based on modeling the second stage as an integrator.

Analysis of the model in Fig. 13.20 gives

$$A(s) \equiv \frac{V_o(s)}{V_i(s)} = \frac{G_{m1}}{sC_C} \tag{13.34}$$

Thus

$$A(j\omega) = \frac{G_{m1}}{j\omega C_C} \tag{13.35}$$

and the magnitude of gain becomes unity at $\omega = \omega_t$, where

$$\omega_t = \frac{G_{m1}}{C_C} \tag{13.36}$$

Substituting $G_{m1} = 1/5.26$ mA/V and $C_C = 30$ pF yields

$$f_t = \frac{\omega_t}{2\pi} \simeq 1 \text{ MHz} \tag{13.37}$$

which is equal to the value calculated before. It should be pointed out, however, that this model is valid only at frequencies $f \gg f_{3dB}$. At such frequencies the gain falls off with a slope of -20 dB/decade, just like an integrator.

Slew Rate

The slew-rate limitation of op amps is discussed in Chapter 3. Here we shall illustrate the origin of the slewing phenomenon in the context of the 741 circuit.

Consider the unity-gain follower of Fig. 13.21 with a step of, say, 10 V applied at the input. Because of amplifier dynamics, its output will not change in zero time. Thus immediately after the input is applied, almost the entire value of the step will appear as a differential signal between the two input terminals. This large input voltage causes the input stage to be *overdriven*, and its small-signal model no longer applies. Rather, half the stage cuts off and the other half conducts all the current. Specifically, reference to Fig. 13.1 shows that a large differential input voltage causes Q_1 and Q_3 to conduct all the available bias current ($2I$) while Q_2 and Q_4 will be cut off. The current mirror Q_5, Q_6, and Q_7 will still function, and Q_6 will produce a collector current of $2I$.

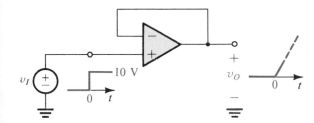

Fig. 13.21 A unity-gain follower with a large step input. Since the output voltage cannot change immediately, a large differential voltage appears between the op-amp input terminals.

Using the above observations, and modeling the second stage as an ideal integrator, results in the model of Fig. 13.22. From this circuit we see that the output voltage will be a ramp with a slope of $2I/C_C$:

$$v_O(t) = \frac{2I}{C_C} t \tag{13.38}$$

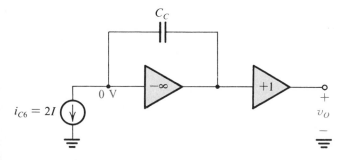

Fig. 13.22 Model for the 741 op-amp when a large differential signal is applied.

Thus the slew rate SR is given by

$$SR = \frac{2I}{C_C} \tag{13.39}$$

For the 741, $I = 9.5\ \mu A$ and $C_C = 30$ pF, resulting in SR $= 0.63$ V/μs.

It should be pointed out that this is a rather simplified model of the slewing process. More detail can be found in Gray and Meyer (1984).

Exercise

13.20 Use the value of the slew rate calculated above to find the full-power bandwidth f_M of the 741 op amp. Assume that the maximum output is ± 10 V.

Ans.   10 kHz

Relationship Between f_t and SR

A simple relationship exists between the unity-gain bandwidth f_t and the slew rate SR. This relationship is obtained from Eqs. (13.36) and (13.39) together with

$$G_{m1} = 2\frac{1}{4r_e}$$

where r_e is the emitter resistance of each of Q_1 through Q_4. Thus

$$r_e = \frac{V_T}{I}$$

and

$$G_{m1} = \frac{I}{2V_T} \tag{13.40}$$

Substituting in Eq. (13.36) results in

$$\omega_t = \frac{I}{2C_c V_T} \tag{13.41}$$

Substituting for I/C_C from Eq. (13.39) gives

$$\omega_t = \frac{SR}{4V_T} \tag{13.42}$$

which can be expressed in the alternative form

$$SR = 4\omega_t V_T \tag{13.43}$$

As a check, for the 741 we have

$$SR = 4 \times 2\pi \times 10^6 \times 25 \times 10^{-3} = 0.63\,V/\mu s$$

which is the result obtained previously.

Exercises

13.21 Consider the integrator model of the op amp in Fig. 13.20. Find the value of the resistor that, when connected across C_C, provides the correct value of the dc gain?

Ans. 1279 MΩ

13.22 If a resistance R_E is included in each of the emitter leads of Q_3 and Q_4 show that SR $= 4\omega_t(V_T + IR_E/2)$. Hence find the value of R_E that would double the 741 slew rate while keeping ω_t and I unchanged. What are the new values of C_C, the dc gain, and the 3-dB frequency?

Ans. 5.26 kΩ; 15 PF; 101.7 dB (a 6-dB decrease); 8.2 Hz

(*Note:* This is a viable technique in general for increasing slew rate. It is referred to as the G_m-reduction method.)

13.7 CMOS OP AMPS

Unlike the 741 op amp, which is a general-purpose operational amplifier intended for a variety of applications, most CMOS op amps are designed to be used as part of a VLSI circuit. This constrained application environment implies that the amplifier specifications can be relaxed in return for a simpler circuit that occupies a relatively small silicon area. The most significant specification relaxation is in the load-driving capability of the op amp. Most CMOS op amps are required to drive-on-chip capacitive loads of few picofarads. It follows that such an op amp does not need a sophisticated output stage. In fact, most CMOS op amps do not have a low-impedance output stage. On a VLSI chip, however, some of the amplifiers are required to drive off-chip loads, and these few amplifiers are usually equipped with an output stage of a classical type.

Two-Stage Topology

Figure 13.23 shows a popular two-stage CMOS op-amp configuration. The circuit utilizes two power supplies, which are usually $+5$ V but can be as low as ± 2.5 V for advanced, reduced-feature-size technologies. A reference bias current I_{REF} is generated either externally or using on-chip circuits [see Gray and Meyer (1984)]. The current mirror formed by Q_8 and Q_5 supplies the differential pair Q_1-Q_2 with bias current. The W/L ratio of Q_5 is selected to yield the desired input-stage bias. The input differential pair is actively loaded with the current mirror formed by Q_3 and Q_4. Thus the input stage is identical to that studied in Section 9.7.

The second stage consists of Q_6, which is actively loaded with the current-source transistor Q_7. As in the 741, frequency compensation is implemented using a Miller feedback capacitor C_C. Here, however, there is an additional resistor included in series with C_C. The function of this resistor, which is usually implemented using one or two MOS transistors, will be explained shortly.

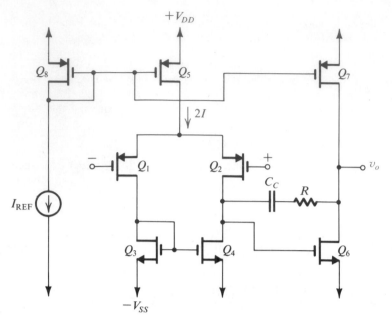

Fig. 13.23 Two-stage CMOS op-amp configuration.

Voltage Gain

The voltage gain of the first stage was found in Section 9.7 to be given by

$$A_1 = -g_{m1}(r_{o2} \parallel r_{o4}) \tag{13.44}$$

The second stage is an actively loaded common-source amplifier whose voltage gain is given by

$$A_2 = -g_{m6}(r_{o6} \parallel r_{o7}) \tag{13.45}$$

The dc open-loop gain of the op amp is the product of A_1 and A_2.

Exercise

13.23 Consider the circuit in Fig. 13.23 with the following device geometries.

Transistor	Q_1	Q_2	Q_3	Q_4	Q_5	Q_6	Q_7	Q_8
W/L	120/8	120/8	50/10	50/10	150/10	100/10	150/10	150/10

Let $I_{REF} = 25\ \mu\text{A}$, $|V_t|$ (for all devices) $= 1$ V, $\mu_n C_{OX} = 20\ \mu\text{A/V}^2$, $\mu_p C_{OX} = 10\ \mu\text{A/V}^2$, $|V_A|$ (for all devices) $= 25$ V, $V_{DD} = V_{SS} = 5$ V. For all devices evaluate I_D, $|V_{GS}|$, g_m, and r_o. Also find A_1, A_2, the dc open-loop voltage gain, the input common-mode range, and the output voltage range. Neglect the effect of V_A on bias current.

Ans.

	Q_1	Q_2	Q_3	Q_4	Q_5	Q_6	Q_7	Q_8
I_D (μA)	12.5	12.5	12.5	12.5	25	25	25	25
$|V_{GS}|$ (V)	1.4	1.4	1.5	1.5	1.6	1.5	1.6	1.6
g_m (μA/V)	62.5	62.5	50	50	83.3	100	83.3	83.3
r_o (MΩ)	2	2	2	2	1	1	1	1

$A_1 = -62.5$ V/V; $A_2 = -50$ V/V; $A = 3125$ V/V; -4.5 V to 3 V; -4.5 V to 4.4 V

Input Offset Voltage

The inevitable device mismatches in the input stage give rise to an input offset voltage. The components of this input offset voltage can be calculated using the methods developed in Chapter 9 and applied in previous sections for the 741 op amp. Because device mismatches are random in nature, the resulting offset voltage is referred to as *random offset*. This is to distinguish it from another type of input offset voltage that can be found in CMOS op amps even if all appropriate devices are perfectly matched. This predictable or *systematic offset* can be minimized by careful design. It does not occur in BJT op amps because of the large gain per stage.

To see how systematic offset can occur, consider the circuit of Fig. 13.23 with the two input terminals grounded. If the input stage is perfectly balanced, then the voltage appearing at the drain of Q_4 will be equal to that at the drain of Q_3, which is $(-V_{SS} + V_{GS4})$. Now this is also the voltage that is fed to the gate of Q_6. In other words, a voltage equal to V_{GS4} appears between gate and source of Q_6. If this voltage is different from the value of V_{GS6} that will make $I_6 = I_7$, an output current, and hence an output offset voltage, results. It can be shown that this output offset can be eliminated by selecting device geometries that satisfy the constraint

$$\frac{K_4}{K_6} = \frac{1}{2}\frac{K_5}{K_7}$$

(13.46)

where K is the conductivity parameter.

Exercise

13.24 Derive Eq. (13.46).

Frequency Response

To appreciate the need for the resistor R placed in series with the Miller compensation capacitor C_C in the circuit of Fig. 13.23, consider first the situation without R. Figure 13.24a shows the small-signal equivalent circuit of the op amp with only C_C included. Note that G_{m1} is the transconductance of the input stage ($G_{m1} = g_{m1} = g_{m2}$), R_1 is the output resistance of the first stage [$R_1 = r_{o2} \| r_{o4}$], C_1 is the total capacitance at the interface between the first and second stages, G_{m2} is the transconductance of the second stage ($G_{m2} = g_{m6}$), R_2 is the output resistance of the second

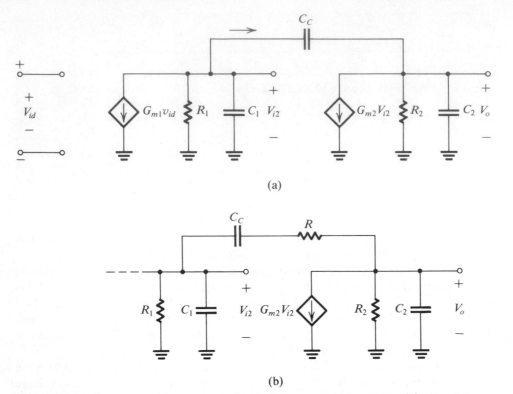

Fig. 13.24 Small-signal equivalent circuit of the CMOS op amp in Fig. 13.23: (a) without the resistance R; (b) with R included.

stage ($R_2 = r_{o6} \parallel r_{o7}$), and C_2 is the total capacitance at the output node of the op amp. Since C_2 includes the load capacitance, it is usually much larger than C_1.

A circuit similar to that in Fig. 13.24a was analyzed in Section 12.11 and the two poles were found to be as follows:

$$\omega_{P1} \simeq \frac{1}{G_{m2}R_2C_CR_1} \tag{13.47}$$

$$\omega_{P2} \simeq \frac{G_{m2}C_C}{C_1C_2 + C_C(C_1 + C_2)} \tag{13.48}$$

We observe that the first pole is due to the Miller capacitance $(1 + G_{m2}R_2)C_C \simeq G_{m2}R_2C_C$ (which is much larger than C_1) interacting with R_1. To make ω_{P1} the dominant pole, we select a value for C_C that will result in a value for ω_{P1} that, when multiplied by the dc gain A_0, gives the desired unity-gain frequency ω_t. The value of ω_t is usually selected to be lower than the frequencies of non-dominant poles and zeros. Thus, for our case,

$$A_0\omega_{P1} = \omega_t \tag{13.49}$$

$$(G_{m1}R_1G_{m2}R_2)\left(\frac{1}{G_{m2}R_2C_CR_1}\right) = \omega_t$$

which yields

$$\omega_t = \frac{G_{m1}}{C_C} \qquad (13.50)$$

The Miller capacitance C_C also introduces a right-half-plane zero in the amplifier transfer function. We paid no attention to this zero in the case of the 741 because it was at a very high frequency. Unfortunately, in the CMOS amplifier this is not the case. The zero location can be most easily determined directly from the circuit in Fig. 13.24a. We wish to find the value of s at which $V_o = 0$. Setting $V_o = 0$, the current in C_C becomes sC_CV_{i2} in the direction indicated. Now because $V_o = 0$, there will be no current in R_2 and C_2. Thus a node equation at the output provides

$$sC_CV_{i2} = G_{m2}V_{i2}$$

In other words the zero is at

$$s = \frac{G_{m2}}{C_C} \qquad (13.51)$$

Since G_{m2} for CMOS amplifiers is of the same order of magnitude as G_{m1}, the zero frequency will be close to ω_t given by Eq. (13.50). Since the zero is in the right half-plane, the phase shift it introduces will decrease the phase margin and thus impair the amplifier stability. Once again we note that this problem is not encountered in BJT op amps because G_{m2} is usually much greater than G_{m1} and thus the zero is at a much higher frequency than ω_t.

The above problem can be solved by including the resistance R in series with the feedback capacitor C_C, as shown in Fig. 13.24b. To find the new location of the transfer function zero, set $V_o = 0$. Then the current through C_C will be $V_{i2}/(R + 1/sC_C)$, and a node equation at the output yields

$$\frac{V_{i2}}{R + 1/sC_C} = G_{m2}V_{i2}$$

Thus the zero is at

$$s = \frac{1}{C_C(1/G_{m2} - R)} \qquad (13.52)$$

We observe that by selecting $R = 1/G_{m2}$, the zero can be placed at infinite frequency. An even better choice would be to select R greater than $1/G_{m2}$, thus placing the zero at a negative real-axis location where the phase it introduces *adds* to the phase margin.

Even with the inclusion of the resistor R, another problem still remains: The frequency of the second pole [Eq. (13.48)] is not very far from ω_t. Thus the second pole introduces appreciable phase shift at ω_t, which reduces the phase margin. To see this more clearly, consider the case in which C_2 and C_C are much greater than C_1. Equation (13.48) can be approximated by

$$\omega_{P2} \simeq \frac{G_{m2}}{C_2} \qquad (13.53)$$

Now, comparing (13.50) with (13.53), we see that for C_2 of the order of C_C (which can happen in the case of a relatively large load capacitance), ω_{P2} will be close to ω_t. This difficulty can be alleviated by increasing C_C and thus decreasing ω_t (see Problem 13.32).

Exercises

The following exercises refer to the amplifier analyzed in Exercise 13.23.

13.25 Find the value of C_C that will result in $f_t = 1$ MHz.

Ans. 10 pF

13.26 Find the value of R that will place the transfer-function zero at infinity.

Ans. 10 kΩ

13.27 Find the frequency of the second pole under the condition that the total capacitance at the output is 10 pF. Hence find the excess phase introduced by the second pole at $\omega = \omega_t$ and the resulting phase margin, assuming that the zero is at $\omega = \infty$.

Ans. 1.59 MHz; 32.2°; 57.8°

Slew Rate

The slew rate of the CMOS op amp of Fig. 13.23 is given by

$$SR = \frac{2I}{C_C} \tag{13.54}$$

Using $\omega_t = g_{m1}/C_C = 2I/[(|V_{GS}| - V_t)C_C]$, where $|V_{GS}|$ is the magnitude of the gate-to-source voltage of Q_1 and Q_2, we relate SR and ω_t by

$$SR = (V_{GS} - V_t)\omega_t \tag{13.55}$$

Comparing this relationship with the corresponding one for BJT amplifiers [Eq. (13.43)], we see that since $(V_{GS} - V_t)$ is usually much greater than $(4V_T)$, CMOS op amps exhibit higher slew rates than BJT units (for the same ω_t). We also observe that higher SR can be obtained by operating the input-stage devices at a higher V_{GS}. This, however, increases the input offset voltage (see Section 9.7).

Exercise

13.28 Calculate the slew rate of the op amp in Fig. 13.23, whose parameters are specified in Exercise 13.23, for the compensation designed in Exercise 13.25 ($C_C = 10$ pF, $f_t = 1$ MHz).

Ans. 2.5 V/μs

13.8 SUMMARY

● The internal circuit of the 741 op amp embodies many of the design techniques employed in bipolar analog integrated circuits.

● The 741 circuit consists of an input differential stage, a high-gain single-ended second stage, and a class AB output stage. This structure is typical of modern op amps and is known as the two-stage design (not counting the output stage). The same structure is used in CMOS op amps.

● To obtain low input offset voltage and current and high CMRR, the 741 input stage is designed to be perfectly balanced. The CMRR is increased by common-mode feedback, which also stabilizes the dc operating point.

● To obtain high input resistance and low input bias current the input stage of the 741 is operated at a very low current level.

● In the 741, output short-circuit protection is accomplished by turning on a transistor that takes away most of the base current drive of the output transistor.

● The use of Miller frequency compensation in the 741 circuit enables locating the dominant pole at a very low frequency, while utilizing a relatively small compensating capacitance.

● Two-stage op amps can be modeled as a transconductance amplifier feeding an ideal integrator with C_C as the integrating capacitor.

● The slew rate of a two-stage op amp is determined by the first-stage bias current and the frequency-compensation capacitor.

● Most CMOS op amps are designed to operate as part of a VLSI circuit and thus are required to drive only small capacitive loads. Therefore, most do not have a low output-resistance stage.

● In CMOS op amps, approximately equal gains are realized in the two stages.

● The threshold mismatch ΔV_t and the low transconductance of the input stage result in a larger input offset voltage for CMOS op amps as compared to bipolar units.

● Miller compensation is employed in CMOS op amps also, but a series resistor is required in order to place the transmission zero at either $s = \infty$ or on the negative real axis.

● CMOS op amps have higher slew rates than their bipolar counterparts with comparable f_t values.

BIBLIOGRAPHY

J. A. Connely (Ed.), *Analog Integrated Circuits*, New York: Wiley-Interscience, 1975.

P. R. Gray, D. A. Hodges, and R. W. Brodersen, *Analog MOS Integrated Circuits*, New York: IEEE Press, 1980.

P. R. Gray and R. G. Meyer, *Analysis and Design of Analog Integrated Circuits*, 2nd ed. New York: Wiley, 1984.

A. B. Grebene (Ed.), *Analog Integrated Circuits*, New York: IEEE Press, 1978.

A. B. Grebene, *Bipolar and MOS Analog Integrated Circuit Design*, New York: Wiley, 1984.

IEEE Journal of Solid-State Circuits. The December issue of each year has been devoted to analog ICs.

R. G. Meyer (Ed.), *Integrated-Circuit Operational Amplifiers*, New York: IEEE Press, 1978.

J. E. Solomon, "The monolithic op amp: A tutorial study," *IEEE Journal of Solid-State Circuits*, vol. SC-9, no. 6, pp. 314–332, Dec. 1974.

R. J. Widlar, "Some circuit design techniques for linear integrated circuits," *IEEE Transactions on Circuit Theory*, vol. CT-12, pp. 586–590, Dec. 1965.

R. J. Widlar, "Design techniques for monolithic operational amplifiers," *IEEE Journal of Solid-State Circuits*, vol. SC-9, no. 6, pp. 314-322, Dec. 1974.

PROBLEMS

13.1 In the circuit of Fig. 13.1, what current flows through R_5 for supply voltages of **(a)** ± 15 V and **(b)** ± 5 V? Ignore base currents.

13.2 For the circuit of Fig. 13.1 find the total current drawn from each power supply and hence estimate the quiescent power dissipation in the circuit. Use the data in Table 13.1.

13.3 Transistor Q_{13} in Fig. 13.1 consists, in effect, of two transistors whose emitter–base junctions are connected in parallel and for which $I_{SA} = 0.25 \times 10^{-14}$ A, $I_{SB} = 0.75 \times 10^{-4}$ A, $\beta = 50$ and $V_A = 50$ V. For operation at a total emitter current of 1.0 mA, find values for the parameters V_{EB}, g_m, r_e, r_π, r_o, and r_μ. Assume that $r_\mu \simeq 10\,\beta r_o$.

13.4 If in the circuit of Fig. 13.1, Q_1, Q_2, Q_3, and Q_4 all exhibit emitter–base breakdown at 7 V, what differential input voltage would result in this type of breakdown?

13.5 Repeat Exercise 13.1 for the standard *pnp* transistor in the 741 op amp circuit.

13.6 Repeat Exercise 13.1 for the standard *npn* transistor operating at 0.1 mA.

13.7 Figure P13.7 shows the CMOS version of the circuit in Fig. E13.2. Find the relationship between I_1 and I_3 in terms of K_1, K_2, K_3, and K_4 of the four devices, assuming the threshold voltages of all devices to be equal. In the event that $K_1 = K_2$ and $K_3 = K_4$, find I_3 for $I_1 = 100\ \mu$A and $K_3 = 16K_1$.

13.8 In the reference-bias part of the circuit of Fig. 13.1, consider the replacement of the resistor R_5 by an *n*-channel JFET suitably connected.
(a) If the JFET has an infinite output resistance in pinch-off, find its required I_{DSS} so that $I_{REF} = 0.73$ mA.
(b) If the JFET has an Early voltage $V_A = 50$ V, $I_{DSS} = 0.5$ mA, and $V_P = -2$ V, find I_{REF} for supply voltages of ± 5 V and ± 15 V.
(c) What is the minimum supply voltage below which I_{REF} begins to deviate rapidly from the design value?

*13.9 Consider the common-mode feedback loop in the 741 circuit comprised of transistors Q_1, Q_2, Q_3, Q_4, Q_8, Q_9, and Q_{10}. We wish to find the loop gain. This can be conveniently done by breaking the loop between the collectors of Q_1 and Q_2, and the diode-connected transistor Q_8. Apply a test current signal I_t to Q_8 and find the returned current signal I_r in the combined collector connection of Q_1 and Q_2. Thus determine the loop gain. Assume that Q_9 and Q_{10} act as ideal current sources. If Q_3 and Q_4 have $\beta = 50$, find the amount of common-mode feedback in dB.

13.10 Consider the input circuit of the 741 op amp of Fig. 13.1 under the conditions that the emitter current of Q_8 is about 19 μA. If β of Q_1 is 175 and that of Q_2 is 225, find the input bias current I_B and the input offset current I_{off} of the op amp.

13.11 The solution to Exercise 13.4 was based on the assumption that a BJT ceases linear operation as soon as its collector–base junction becomes forward-biased. If linear operation continues for as much as 0.2 V of collector–base forward bias, what does the input common-mode range become?

13.12 In Fig. 13.1 what is the voltage at the base of Q_{16}? Use the data in Table 13.1.

13.13 In the circuit of Fig. 13.5, in which the voltage between the bases of Q_{14} and Q_{20} is maintained at 1.118 V, what is the effect on their current if **(a)** the scale current of each is doubled, **(b)** the scale current of one is doubled, **(c)** the scale current of one is halved while that of the other is doubled?

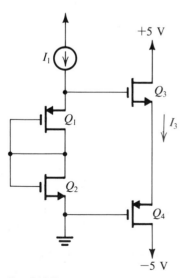

Fig. P13.7

*13.14 In Fig. 13.5 for the conditions calculated (and given in Table 13.1), what is the incremental resistance of the two-terminal network formed by Q_{18}, Q_{19}, and R_{10}? (Use the small-signal equivalent circuit for each transistor.)

**13.15 An alternative approach to providing the diode function of the Q_{18}-Q_{19}-R_{10} network of Fig. 13.5 is the circuit shown in Fig. P13.15. Design two versions of the circuit, for $\beta = 200$ and $I_S = 10^{-14}$ A: (a) one in which half the current flows in R_1, and   (b) one in which one-

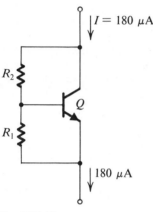

Fig. P13.15

tenth the current flows in R_1. In each case the overall voltage drop should be 1.118 V. In each case find the incremental resistance of the two-terminal circuit (using the BJT small-signal model). Now if β reduces to 100 in each case, what do the terminal voltage and incremental resistance become?

*13.16 Show that the maximum voltage developed across the Q_{18}-Q_{19} network in Fig. 13.5 is obtained when the 180-μA current splits about equally between Q_{18} and R_{10}. Find the value of R_{10} required and the resulting terminal voltage. Assume $\beta = \infty$.

*13.17 Through a processing imperfection, the β of Q_4 in Fig. 13.1 is reduced to 25, while the β of Q_3 remains at its regular value of 50. Find the input offset voltage that this mismatch introduces. (*Hint:* Follow the general procedure outlined in Example 13.1.)

*13.18 Referring to Problem 13.17, what mismatch between R_1 and R_2 would be necessary to compensate for the β difference indicated, thus reducing the offset voltage to zero? [*Hint:* Use Eq. (13.10).]

*13.19 What is the maximum input offset voltage that can be compensated by completely short-circuiting either R_1 or R_2 in Fig. 13.1? [*Hint:* Use Eq. (13.10).]

13.20 In Exercises 13.9 and 13.12, what is the effect on CMRR of using external components to reduce $\Delta R/R$ by a factor of 10?

*13.21 Use the circuit of Fig. E13.7 to find the CMRR obtained with $R_1 = R_2$ (that is, $\Delta R = 0$), but under the condition that β of Q_3 is 55 whereas that of Q_4 is 50. Note that the signal currents in the collectors of Q_3 and Q_4 will no longer be equal. Recall that the input-stage transistors are biased at 9.5 μA and that for differential input signals, the input-stage transconductance is 0.19 mA/V. Also, $R_o = 2.43$ MΩ.

**13.22 What is the effect on the differential gain of the 741 op amp of short-circuiting one, or the other, or both, of R_1 and R_2 in Fig. 13.1? (Refer to Fig. 13.7.) For simplicity, assume $\beta = \infty$.

*13.23 Figure P13.23 shows the equivalent common-mode half-circuit of the input stage of the 741. Here R_o is the resistance seen looking to the left of node Y in Fig. 13.1; its value is approximately 2.4 MΩ. Transistors Q_1 and Q_3 operate at a bias current of 9.5 μA. Find the input resistance of the common-mode half circuit using $\beta_N = 200$, $\beta_P = 50$, $r_\mu = 10\beta r_o$, $V_A = 125$ V for *npn* and 50 V for *pnp* transistors. To find the common-mode input

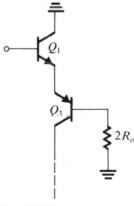

Fig. P13.23

resistance of the 741 note that it has common-mode feedback that increases the input common-mode resistance. The loop gain is approximately equal to β_P. Find the value of R_{icm}.

13.24 In the analysis of the 741 second stage, note that R_{o2} is affected most strongly by the low value of R_{o13B}. Consider the effect of placing appropriate resistors in the emitters of Q_{12}, Q_{13A} and Q_{13B} on this value. What resistor in the emitter of Q_{13B} would be required to make R_{o13B} equal to R_{o17} and thus R_{o2} half as great? What resistors in each of the other emitters would be required?

13.25 An internally compensated op amp having an f_t of 5 MHz and dc gain of 10^6 utilizes Miller compensation around an inverting amplifier stage with a gain of 1,000. If space exists for at most a 50-pF capacitor, what resistance level must be reached at the input of the Miller amplifier for compensation to be possible?

13.26 Consider the integrator op-amp model shown in Fig. 13.20. For $G_{m1} = 10$ mA/V, $C_C = 50$ pF, and a resistor of $10^8\ \Omega$ shunting C_C, sketch and label a Bode plot for the magnitude of the open-loop gain. If G_{m1} is related to the first-stage bias current via Eq. (13.40), find the slew rate of this op amp.

13.27 For an amplifier with a slew rate of 10 V/μs, what is the full-power bandwidth for outputs of ± 10 V? What unity-gain bandwidth, ω_t, would you expect if the topology was similar to that of the 741?

**13.28 Figure P13.28 shows a circuit suitable for op amp applications. For all transistors, $\beta = 100$, $V_{BE} = 0.7$ V, and $r_o = \infty$.
(a) For inputs grounded and output held at 0 V (by negative feedback) find the emitter currents of all transistors.
(b) Calculate the gain of the amplifier with a load of 10 kΩ.
(c) With load as in (b) calculate the capacitor C required for a 3-dB frequency of 1 kHz.

13.29 Consider the amplifier of Fig. 13.23, whose parameters are specified in Exercise 13.23. If a manufacturing error results in the W/L ratio of Q_6 being 120/10, find the current that Q_6 will now conduct. Thus find the systematic offset voltage that will appear at the output. (Use the results of Exercise 13.23.) Assuming that the open-loop gain will remain approximately unchanged from the value found in Exercise 13.23, find the corresponding value of input offset voltage.

*13.30 Consider the input stage of the CMOS op amp in Fig. 13.23 with both inputs grounded. Assume that the two sides of the input stage are perfectly matched except that the threshold voltages of Q_3 and Q_4 show a mismatch ΔV_t. Show that a current $g_{n3}\ \Delta V_t$ appears at the output of the first stage. What is the corresponding input

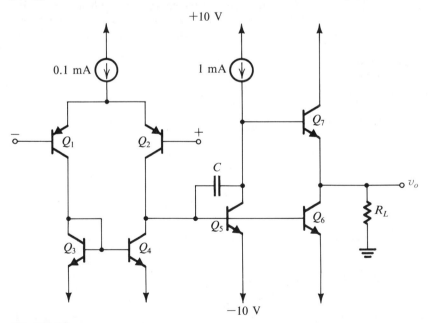

Fig. P13.28

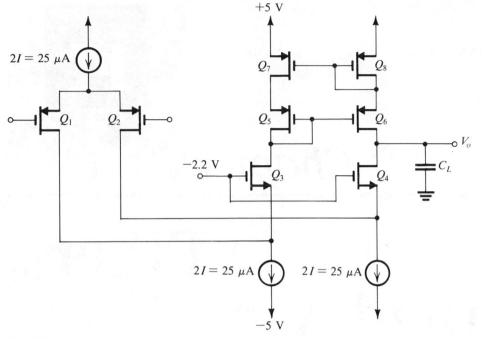

Fig. P13.34

offset voltage? Evaluate this offset voltage for the circuit specified in Exercise 13.23 for $\Delta V_t = 2$ mV. (Use the results of Exercise 13.23.)

13.31 What is the open-loop output resistance of the amplifier of Fig. 13.23, whose parameters are specified in Exercise 13.23? (Use the results of Exercise 13.23.)

***13.32** It is required to design the frequency-compensation network for the amplifier in Fig. 13.23, whose parameters are specified in Exercise 13.23. The transmission zero is to be placed at infinite frequency and the amplifier is to have 80° of phase margin when the total capacitance at the output is 10 pF. What are the required values of C_C and R? What are the resulting values of f_t and SR?

***13.33** Redesign the compensation network of Problem 13.32, this time placing the transmission zero on the negative real axis. In this way it introduces a positive phase that increases the phase margin. Design the network so that $f_t = 1$ MHz and the phase margin is 80° with a total capacitance at the output of 10 pF. Find C_C and R and the resulting slew rate.

*****13.34** Figure P13.34 shows an alternative CMOS op-amp configuration. Here Q_1 and Q_2 form the input differential stage, Q_3 and Q_4 form a common-gate stage that is loaded by the modified Wilson mirror formed by Q_5, Q_6, Q_7, and Q_8. Assuming that all devices have W/L of 120/8, $|V_t| = 1$ V, and $|V_A| = 25$ V, and that $\mu_n C_{\text{OX}} = 2\mu_p C_{\text{OX}} = 20$ μA/V^2, find the dc bias current and $|V_{GS}|$ for each transistor. Also find g_m and r_o for each device. Ignore the effect of V_A on bias calculations.

Observe that Q_1, Q_2, Q_3, and Q_4 in effect form a cascode stage. Since this cascode stage uses complementary devices, it is known as a "folded cascode." Using the results from Chapter 9 on the output resistance of a cascode circuit [see Eq. (9.133)], which also apply for the Wilson current mirror, find the dc open-loop gain of this amplifier.

This amplifier uses the load capacitance C_L for frequency compensation. Assuming that C_L denotes the total capacitance at the output, find its value so that $f_t = 1$ MHz. Also, find the slew rate. Finally, what is the range of allowable output-voltage swing?

Chapter

Filters, Tuned Amplifiers, and Oscillators

INTRODUCTION

Except for the frequency compensation of feedback amplifiers, any shaping of the frequency response of the amplifiers studied in previous chapters was, in general, not deliberate; it was caused by the frequency limitations of the transistors and by the coupling and bypass capacitors. In contrast, the circuits studied in this chapter have deliberate *frequency-selective* characteristics.

The first functional block to be considered, the filter, has as its main objective the selection of signals based on their frequency spectra. Filters are used extensively in communications systems, and there is hardly an electronic instrument that does not contain a filter of some kind.

The second topic studied is the tuned amplifier, which is a bandpass amplifier whose gain peaks at

a certain frequency and falls off outside a band of frequencies (usually a narrow band) centered around the peak. Such amplifiers are employed, for example, in radio and TV receivers where their tuned nature enables one to select a particular radio or TV station. Although tuned amplifiers are in effect bandpass filters, they are studied separately because their design is based on somewhat different techniques.

In addition to filters and tuned amplifiers, this chapter is also concerned with the design of *linear sinusoidal oscillators*. The word "linear" is used to distinguish this type of sinusoidal generator from the ones discussed in Chapter 5, which create sine waves by nonlinear shaping of triangular waveforms. The sinusoidal oscillators we shall study employ an amplifier (an op amp or a transistor) in a feedback loop with a frequency selective network (an RC network, an RLC network, or a piezoelectric crystal) that determines the frequency of oscilla-tion. Sinusoidal oscillators are employed in a wide variety of communications and instrumentation systems. □

14.1 FILTER TECHNOLOGIES

A filter is a two-port network that passes signals with frequencies within a specified band (the filter *passband*) and attenuates signals whose frequencies lie outside this band (in the filter *stopband*). Figure 14.1 shows the transmission (in decibels) of a third-order low-pass filter with a 3-dB frequency at $\omega = 1$ rad/s. (This response is for a type of filter known as a Butterworth filter.) Also shown for comparison is the transmission of a first-order RC low-pass network of the type studied in Chapter 2. Observe that the third-order filter has a flatter passband and much greater attenuation outside the passband. Also, its transmission falls off at a much greater rate (compared to that of the first-order filter), and thus the third-order filter is said to have greater *selectivity*.

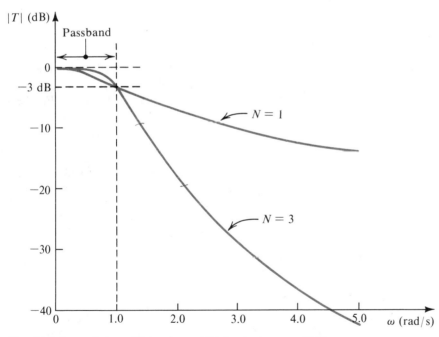

Fig. 14.1 Transmission of first-order and third-order low-pass filters.

The selectivity of RC networks is limited because their poles are restricted to lie on the negative real axis in the s plane. As an example, Fig. 14.2 shows the first-order RC low-pass filter together with its pole location. Much greater selectivity can be obtained by utilizing the resonance properties of LC circuits. As an example, we show in Fig. 14.3 an RLC circuit having the third-order low-pass response of Fig.

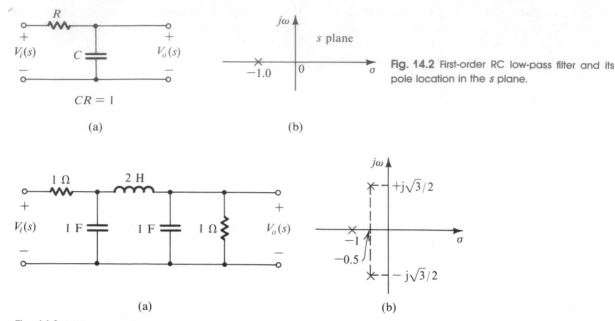

Fig. 14.2 First-order RC low-pass filter and its pole location in the s plane.

Fig. 14.3 A third-order LC low-pass filter together with its pole locations in the s plane.

14.1, together with the locations of its poles. Observe that this third-order filter has a pair of complex conjugate poles. In general, selective filter responses are obtained when the poles are complex (except that odd-order filters have one real pole). Furthermore, the closer the poles are to the $j\omega$ axis, the more selective the filter response becomes.

Exercises

14.1 Find the transfer function $T(s)$ of the third-order low-pass filter having the poles shown in Fig. 14.3b. Assume that all three zeros are at $s = \infty$, and that the transmission is unity at dc.

Ans. $T(s) = 1/(s + 1)(s^2 + s + 1)$

14.2 For the filter of Exercise 14.1 show that $|T(j\omega)| = 1/\sqrt{1 + \omega^6}$. Hence find ω_{3dB} and the attenuation at $\omega = 3$ rad/s.

Ans.   1 rad/s; 28.6 dB

14.3 Analyze the RLC network of Fig. 14.3a to determine its voltage transfer function $T(s) = V_o(s)/V_i(s)$. Show that the transfer function is identical to that in Exercise 14.1 except for a multiplicative constant of 0.5. (*Hint:* Begin the analysis at the output and work your way back to the input.)

The oldest technology for realizing filters makes use of inductors and capacitors, and the resulting filters are called *passive LC filters*. An example of a passive LC filter is the network shown in Fig. 14.3a.

The problem with LC filters is that in low-frequency applications (dc to 100 kHz) the required inductors are large and physically bulky, and their characteristics are quite nonideal. Furthermore, such inductors are impossible to manufacture in monolithic form and are incompatible with any of the modern techniques for assembling electronic systems. Therefore, there has been considerable interest in finding filter realizations that do not require inductors. Of the various possible types of *inductorless filters*, we shall study *active-RC filters* and *switched-capacitor filters*.

In its simplest form, the active-RC filter utilizes an op amp together with an RC network in a feedback loop that is arranged so as to shift the poles from the negative real axis (where the poles of the RC network lie) to complex conjugate locations. In this way, highly selective filter responses can be obtained. Thus the active-RC filter effectively utilizes the pole-shifting property of feedback (see Section 12.9).

In the following sections we shall study op-amp–RC circuits that realize second-order filter functions. Filters of higher order can be realized by cascading second-order sections. Of course, if the filter is of odd order, a first-order section will be required but this can be realized using a simple RC circuit.

Active-RC filters are fabricated using discrete, hybrid thin-film, or hybrid thick-film technology. However, for large-volume production, such technologies do not yield the economies achieved by monolithic IC fabrication. At the present time it appears that the most viable approach for realizing fully integrated monolithic filters is the switched-capacitor technique discussed in Section 14.6.

The active-RC and switched-capacitor filters studied in this chapter are readily applicable in the frequency range up to about 100 kHz. This limit is due to the finite bandwidth of op amps. Utilization of high-frequency op amps and special circuit techniques allows the extension of the useful frequency range of application to the low-MHz band.

14.2 SECOND-ORDER FILTER FUNCTIONS

The general second-order, biquadratic, or simply *biquad*, transfer function can be written in the form

$$T(s) = \frac{n_2 s^2 + n_1 s + n_0}{s^2 + s(\omega_0/Q) + \omega_0^2} \tag{14.1}$$

Here ω_0 and Q are parameters that determine the location of the poles. Figure 14.4 shows a pair of complex conjugate poles and illustrates the definition of ω_0 and Q. As indicated, the *pole frequency* ω_0 is the radial distance of the poles from the origin, and the pole Q factor, or simply *pole-Q*, determines how far the poles are from the $j\omega$ axis. The pole locations are given by

$$P_1, P_2 = -\frac{\omega_0}{2Q} \pm j\omega_0 \sqrt{1 - 1/4Q^2}$$

It can be easily shown that Q less than 0.5 means that the poles are on the negative real axis, $Q = 0.5$ means that the poles are coincident and real, $Q > 0.5$ means that the poles are complex conjugate, and $Q = \infty$ means the poles are on the $j\omega$ axis.

The numerator coefficients, n_0, n_1, and n_2, determine the transmission zeros and thus the shape of the magnitude response and the type of filter (low pass, high pass,

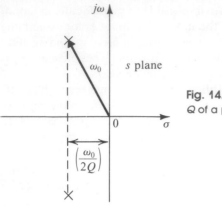

Fig. 14.4 Definition of the parameters ω_0 and Q of a pair of complex conjugate poles.

etc.). Special cases of interest are as follows:

1. *Low-Pass (LP) Filter.* In this case $n_1 = n_2 = 0$ and the transfer function becomes

$$T(s) = \frac{n_0}{s^2 + s(\omega_0/Q) + \omega_0^2} \qquad (14.2)$$

Thus the two transmission zeros are at $s = \infty$. Figure 14.5a shows the magnitude response, $20 \log|T(j\omega)|$, for a second-order low-pass filter with a unity dc gain ($n_0 = \omega_0^2$). Note that the response exhibits a peak. It can be shown that this peak occurs if Q is greater than 0.707. The case $Q = 0.707$ results in the *maximally flat* magnitude response. Also note that at high frequencies the gain falls off with an asymptotic slope of -40 dB/decade.

2. *High-Pass (HP) Filter.* In this case $n_0 = n_1 = 0$ and the transfer function becomes

$$T(s) = \frac{n_2 s^2}{s^2 + s(\omega_0/Q) + \omega_0^2} \qquad (14.3)$$

Thus the two transmission zeros are at $s = 0$ (dc). Figure 14.5b shows the magnitude response for a second-order high-pass filter with unity high-frequency gain ($n_2 = 1$). The effect of the value of Q on the response shape is similar to that in the low-pass case. Note that at low frequencies the transmission falls off with an asymptotic slope of 40 dB/decade.

3. *Bandpass (BP) Filter.* In this case $n_0 = n_2 = 0$ and the transfer function becomes

$$T(s) = \frac{n_1 s}{s^2 + s(\omega_0/Q) + \omega_0^2} \qquad (14.4)$$

Thus there is one transmission zero at $s = 0$ (dc) and one at $s = \infty$. Figure 14.5c shows the magnitude response of a second-order bandpass filter with unity center-frequency gain ($n_1 = \omega_0/Q$). Note that the response peaks at $\omega = \omega_0$; thus ω_0 is also called the *center frequency*. The pole-Q determines how narrow (selective) the bandpass filter is. Specifically, the two frequencies ω_1 and ω_2 at which the transmission drops by 3 dB below the value at ω_0

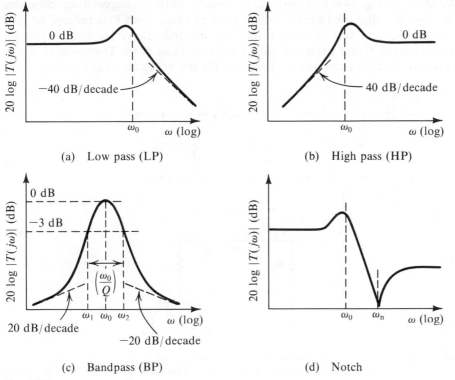

(a)   Low pass (LP)

(b)   High pass (HP)

(c)   Bandpass (BP)

(d)   Notch

Fig. 14.5 Second-order filter responses.

are separated by ω_0/Q. This distance is called the 3-dB bandwidth. As Q is increased, the bandwidth decreases and the filter becomes more selective. Finally, note that the gain falls off at very low and high frequencies with a slope of 20 dB/decade.

4. *Notch Filter.*   In this case $n_1 = 0$ and the transfer function can be written in the form

$$T(s) = n_2 \frac{s^2 + \omega_n^2}{s^2 + s(\omega_0/Q) + \omega_0^2} \tag{14.5}$$

Thus the zeros are at $s = \pm j\omega_n$, and the transmission becomes zero at $\omega = \omega_n$. Figure 14.5d shows the response of a second-order notch filter for which $\omega_n > \omega_0$. This type of response is referred to as a *low-pass notch* (LPN).

5. *All-Pass Filter.*   In this case the transfer function is given by

$$T(s) = n_2 \frac{s^2 - s(\omega_0/Q) + \omega_0^2}{s^2 + s(\omega_0/Q) + \omega_0^2} \tag{14.6}$$

Thus the zeros are in the right half-plane at locations which are mirror images of the locations of the poles. The magnitude response is constant independent of frequency; hence the name all-pass. Such networks are used as phase shifters or to modify the phase response of a system in a desired manner.

We show in Fig. 14.6 RLC circuit realizations of the five special second-order transfer functions discussed above. Note that in all cases when V_i is reduced to zero (that is, when the input voltage source is short-circuited) the circuit reduces to a parallel LCR tuned circuit. Hence all circuits shown have the same poles. The reader is urged, as an exercise, to derive the transfer functions for the networks of Fig. 14.6.

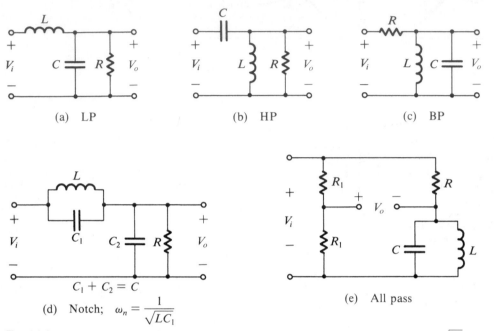

(a) LP   (b) HP   (c) BP

(d) Notch; $\omega_n = \dfrac{1}{\sqrt{LC_1}}$

(e) All pass

Fig. 14.6 Passive realizations of special second-order filter functions. For all circuits $\omega_0 = 1/\sqrt{LC}$ and $Q = \omega_0 CR$.

In the following sections we shall study circuits that use op amps and RC networks to realize second-order transfer functions. Such circuits are usually referred to as biquadratic circuits or simply biquads.

Exercises

14.4 The third-order low-pass filter whose poles are shown in Fig. 14.3b (see also Exercise 14.1) can be realized as the cascade of a second-order section and a first-order (RC) section. Find the type, ω_0, and Q of the second-order section.

Ans.   LP; 1 rad/s; 1

14.5 Give the transfer function of a second-order bandpass filter having a center frequency of 10^4 rad/s, a 3-dB bandwidth of 10^3 rad/s, and a center-frequency gain of 10.

Ans. $T(s) = \dfrac{10^4 s}{s^2 + 10^3 s + 10^8}$

14.6 What is the center-frequency gain of the circuit in Fig. 14.6c? Use this circuit to realize the bandpass function of Exercise 14.5 (except for a different gain). For $R = 10\ \text{k}\Omega$ find L and C.

Ans.   1 (0 dB); $L = 0.1$ H; $C = 0.1\ \mu\text{F}$

14.3 SINGLE-AMPLIFIER BIQUADRATIC FILTERS

The simplest realization of second-order filter functions utilizes a single op amp and is known as a *single-amplifier biquad* (SAB). In this section we shall study the synthesis of SABs following a two-step approach:

1. Synthesis of a feedback loop that realizes a pair of complex conjugate poles characterized by a frequency ω_0 and a Q factor Q.

2. Injecting the input signal in a way that realizes the desired transmission zeros.

Synthesis of the Feedback Loop

Consider the circuit shown in Fig. 14.7a, which consists of a two-port RC network n placed in the negative-feedback path of an op amp. We shall assume that, except for having a finite gain A, the op amp is ideal. We shall denote by $t(s)$ the open-circuit voltage transfer function of the RC network n, where the definition of $t(s)$ is illustrated in Fig. 14.7b. The transfer function $t(s)$ can in general be written as the ratio of two polynomials $N(s)$ and $D(s)$:

$$t(s) = \frac{N(s)}{D(s)}$$

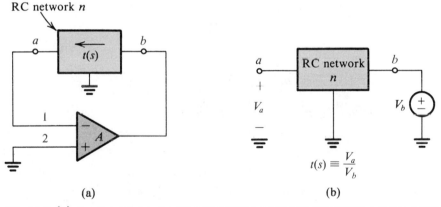

Fig. 14.7 (a) Feedback loop obtained by placing a two-port RC network n in the feedback path of an op amp. **(b)** Definition of the open-circuit transfer function $t(s)$ of the RC network.

The roots of $N(s)$ are the transmission zeros of the RC network, and the roots of $D(s)$ are its poles. Study of network theory shows that while the poles of an RC network are restricted to lie on the negative real axis, the zeros can in general lie anywhere in the s plane.

The loop gain $L(s)$ of the feedback circuit in Fig. 14.7a can be determined using the method of Section 12.7. It is simply the product of the op amp gain A and the transfer function $t(s)$,

$$L(s) - At(s) = \frac{AN(s)}{D(s)} \tag{14.7}$$

Substituting for $L(s)$ into the characteristic equation

$$1 + L(s) = 0 \tag{14.8}$$

results in the poles s_P of the closed-loop circuit obtained as solutions to the equation

$$t(s_P) = -\frac{1}{A} \tag{14.9}$$

In the ideal case, $A = \infty$ and the poles are obtained from

$$N(s_P) = 0 \tag{14.10}$$

That is, the poles are identical to the zeros of the RC network.

Since our objective is to realize a pair of complex conjugate poles, we should select an RC network that has complex conjugate transmission zeros. The simplest such networks are the bridged-T networks shown in Fig. 14.8 together with their transfer function $t(s)$ from b to a, with a open-circuited. As an example, consider the circuit generated by placing the bridged-T network of Fig. 14.8a in the negative-feedback path of an op amp, as shown in Fig. 14.9. The pole polynomial of the active-filter circuit will be equal to the numerator polynomial of the bridged-T network; thus

$$s^2 + s\frac{\omega_0}{Q} + \omega_0^2 = s^2 + s\left(\frac{1}{C_1} + \frac{1}{C_2}\right)\frac{1}{R_3} + \frac{1}{C_1 C_2 R_3 R_4}$$

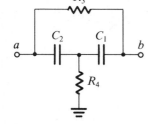

$$t(s) = \frac{s^2 + s\left(\dfrac{1}{C_1} + \dfrac{1}{C_2}\right)\dfrac{1}{R_3} + \dfrac{1}{C_1 C_2 R_3 R_4}}{s^2 + s\left(\dfrac{1}{C_1 R_3} + \dfrac{1}{C_2 R_3} + \dfrac{1}{C_1 R_4}\right) + \dfrac{1}{C_1 C_2 R_3 R_4}}$$

(a)

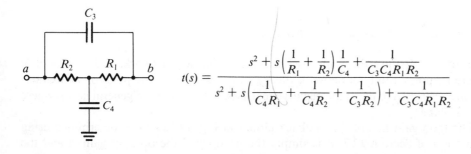

$$t(s) = \frac{s^2 + s\left(\dfrac{1}{R_1} + \dfrac{1}{R_2}\right)\dfrac{1}{C_4} + \dfrac{1}{C_3 C_4 R_1 R_2}}{s^2 + s\left(\dfrac{1}{C_4 R_1} + \dfrac{1}{C_4 R_2} + \dfrac{1}{C_3 R_2}\right) + \dfrac{1}{C_3 C_4 R_1 R_2}}$$

(b)

Fig. 14.8 Two RC networks (called bridged-T networks) that have complex transmission zeros. The transfer functions given are from b to a with a open-circuited.

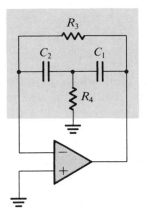

Fig. 14.9 An active filter feedback loop generated using the bridged-T network of Fig. 14.8a.

which enables us to obtain ω_0 and Q as

$$\omega_0 = \frac{1}{\sqrt{C_1 C_2 R_3 R_4}} \tag{14.11}$$

$$Q = \left[\frac{\sqrt{C_1 C_2 R_3 R_4}}{R_3} \left(\frac{1}{C_1} + \frac{1}{C_2} \right) \right]^{-1} \tag{14.12}$$

If we are designing this circuit, ω_0 and Q are given and Eqs. (14.11) and (14.12) can be used to determine C_1, C_2, R_3, and R_4. It follows that there are two degrees of freedom. Let us exhaust one of these by selecting $C_1 = C_2 = C$. Let us also denote $R_3 = R$ and $R_4 = R/m$. By substituting in Eqs. (14.11) and (14.12) and with some manipulation, we obtain

$$m = 4Q^2 \tag{14.13}$$

$$CR = \frac{2Q}{\omega_0} \tag{14.14}$$

Thus if we are given the value of Q, Eq. (14.13) can be used to determine the ratio of the two resistances R_3 and R_4. Then the given values of ω_0 and Q can be substituted in Eq. (14.14) to determine the time constant CR. There remains one degree of freedom—the value of C or R can be arbitrarily chosen. In an actual design this value, which sets the *impedance level* of the circuit, should be chosen so that the resulting component values are practical.

Exercises

14.7 Design the circuit of Fig. 14.9 to realize a pair of poles with $\omega_0 = 10^4$ rad/s and $Q = 1$. Select $C_1 = C_2 = 1$ nF.

Ans. $R_3 = 200$ kΩ; $R_4 = 50$ kΩ

14.8 For the circuit designed in Exercise 14.7, find the location of the poles of the RC network in the feedback loop.

Ans. -0.382×10^4 and -2.618×10^4 rad/s

Injecting the Input Signal

Having synthesized a feedback loop that realizes a given pair of poles, we now consider connecting the input signal source to the circuit. We wish to do this, of course, without altering the poles.

Since, for the purpose of finding the poles of a circuit, an ideal voltage source is equivalent to a short-circuit, it follows that any circuit node that is connected to ground can instead be connected to the input voltage source without causing the poles to change. Thus the method of injecting the input signal into the feedback loop is simply to disconnect a component (or several components) which is (are) connected to ground and connect it (them) to the input source. Depending on the component(s) through which the input signal is injected, different transmission zeros are obtained.

As an example, consider the feedback loop of Fig. 14.9. Here we have two grounded nodes (one terminal of R_4 and the positive input terminal of the op amp) that can serve for injecting the input signal. Figure 14.10 shows the circuit with the input signal injected through part of the resistance R_4. Note that the two resistances R_4/α and $R_4/(1-\alpha)$ have a parallel equivalent of R_4. It can be shown that this circuit realizes the second-order bandpass function and that the value of α $(0 < \alpha \leq 1)$ can be used to obtain the desired center-frequency gain.

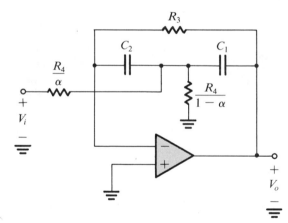

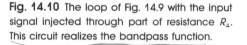

Fig. 14.10 The loop of Fig. 14.9 with the input signal injected through part of resistance R_4. This circuit realizes the bandpass function.

Exercise

14.9 Assume the op amp to be ideal and analyze the circuit in Fig. 14.10 to obtain its transfer function. Thus show that the circuit realizes the bandpass function and that its poles are identical to the zeros of $t(s)$ in Fig. 14.8a. Using the component values of Exercise 14.7 find the values of R_4/α and $R_4/(1-\alpha)$ to obtain unity center-frequency gain.

Ans. 100 kΩ; 100 kΩ

Generation of Equivalent Feedback Loops

The *complementary transformation* of feedback loops is based on the property of linear networks illustrated in Fig. 14.11 for the case of the two-port (three-terminal) network

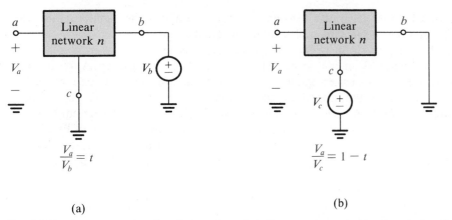

(a)   (b)

Fig. 14.11 Interchanging input and ground results in the complement of the transfer function.

n. In part (a) of the figure, terminal *c* is grounded and a signal V_b is applied to terminal *b*. The transfer function from *b* to *a* with *c* grounded is denoted *t*. Then, in part (b) of the figure, terminal *b* is grounded and the input signal is applied to terminal *c*. The transfer function from *c* to *a* with *b* grounded can be shown to be the complement of *t*—that is, $1 - t$.

Application of the complementary transformation to a feedback loop to generate an equivalent feedback loop is a two-step process:

1. Nodes of the feedback network and any of the op-amp inputs that are connected to ground should be disconnected from ground and connected to the op-amp output. Conversely, those nodes that were connected to the op-amp output should be now connected to ground. That is, we simply interchange the op-amp output terminal with ground.
2. The two input terminals of the op amp should be interchanged.

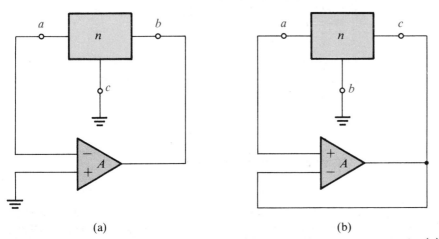

(a)   (b)

Fig. 14.12 Application of the complementary transformation to the feedback loop in **(a)** results in the equivalent (same poles) loop in **(b)**.

The feedback loop generated by this transformation has the same characteristic equation, and hence the same poles, as the original loop.

To illustrate, we show in Fig. 14.12a the feedback loop formed by connecting a two-port RC network in the negative-feedback path of an op amp. Application of the complementary transformation to this loop results in the feedback loop of Fig. 14.12b. Note that in the latter loop the op amp is used in the unity-gain follower configuration. We shall now show that the two loops of Fig. 14.12 are equivalent.

If the op amp has an open-loop gain A, the follower in the circuit of Fig. 14.12b will have a gain of $A/(A + 1)$. This together with the fact that the transfer function of network n from c to a is $1 - t$ (see Fig. 14.11) enables us to write for the circuit in Fig. 14.12b the characteristic equation

$$1 - \frac{A}{A + 1}(1 - t) = 0$$

This equation can be manipulated to the form

$$1 + At = 0$$

which is the characteristic equation of the loop in Fig. 14.12a. As an example, consider the application of the complementary transformation to the feedback loop of Fig. 14.9: the feedback loop of Fig. 14.13a results. Injecting the input signal through C_1 results in the circuit in Fig. 14.13b, which can be shown to realize a second-order high-pass function.

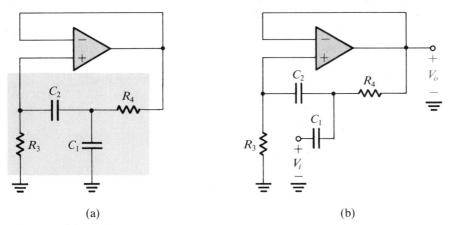

(a)   (b)

Fig. 14.13 (a) Feedback loop obtained by applying the complementary transformation to the loop in Fig. 14.9. **(b)** Injecting the input signal through C_1 realizes the high-pass function.

Exercise

14.10 Use the bridged-T network of Fig. 14.8b in the negative-feedback path of an op amp to realize a pair of complex poles. Then apply the complementary transformation to this circuit and show that the resulting circuit can be used to realize a second-order low-pass function.

14.4 SENSITIVITY

Because of the tolerances in component values and because of the finite op-amp gain, the response of the actual assembled filter will deviate from the ideal response. As a means for predicting such deviations, the filter designer employs the concept of *sensitivity*. Specifically, for second-order filters one is normally interested in finding how *sensitive* their poles are relative to variations (both initial tolerances and future changes) in RC component values and amplifier gain. These sensitivities can be quantified using the *classical sensitivity function* S_x^y, defined

$$S_x^y \equiv \lim_{\Delta x \to 0} \frac{\Delta y/y}{\Delta x/x} \qquad (14.15)$$

Thus

$$S_x^y = \frac{\partial y}{\partial x} \frac{x}{y} \qquad (14.16)$$

Here x denotes the value of a component (a resistor, a capacitor, or an amplifier gain) and y denotes an output parameter of interest (say, ω_0 or Q). For small changes

$$S_x^y \cong \frac{\Delta y/y}{\Delta x/x} \qquad (14.17)$$

Thus we can use the value of S_x^y to determine the per-unit change in y due to a given per-unit change in x. For instance, if the sensitivity of Q relative to a particular resistance R_1 is 5, then a 1% increase in R_1 results in a 5% increase in the value of Q.

EXAMPLE 14.1

For the feedback loop of Fig. 14.9 find the sensitivities of ω_0 and Q relative to all the passive components and the op-amp gain. Evaluate these sensitivities for the design considered in the previous section for which $C_1 = C_2$.

Solution
To find the sensitivities with respect to the passive components, called *passive sensitivities*, we assume that the op-amp gain is infinite. In this case ω_0 and Q are given by Eqs. (14.11) and (14.12). Thus for ω_0 we have

$$\omega_0 = \frac{1}{\sqrt{C_1 C_2 R_3 R_4}}$$

which can be used together with the sensitivity definition of Eq. (14.16) to obtain

$$S_{C_1}^{\omega_0} = S_{C_2}^{\omega_0} = S_{R_3}^{\omega_0} = S_{R_4}^{\omega_0} = -\tfrac{1}{2}$$

For Q we have

$$Q = \left[\sqrt{C_1 C_2 R_3 R_4} \left(\frac{1}{C_1} + \frac{1}{C_2} \right) \frac{1}{R_3} \right]^{-1}$$

on which we apply the sensitivity definition to obtain

$$S_{C_1}^Q = \frac{1}{2}\left(\sqrt{\frac{C_2}{C_1}} - \sqrt{\frac{C_1}{C_2}}\right)\left(\sqrt{\frac{C_2}{C_1}} + \sqrt{\frac{C_1}{C_2}}\right)^{-1}$$

For the design with $C_1 = C_2$ we see that $S_{C_1}^Q = 0$. Similarly, we can show that

$$S_{C_2}^Q = 0, \qquad S_{R_3}^Q = \tfrac{1}{2}, \qquad S_{R_4}^Q = -\tfrac{1}{2}$$

It is important to remember that the sensitivity expression should be derived *before* substituting values corresponding to a particular design.

Next we consider the sensitivities relative to the amplifier gain. Assuming the op amp to have a finite gain A, the characteristic equation for the loop becomes

$$1 + At(s) = 0 \tag{14.18}$$

where $t(s)$ is given in Fig. 14.8a. To simplify matters we can substitute for the passive components by their design values. This causes no errors in evaluating sensitivities, since we are now finding the sensitivity with respect to the amplifier gain. Using the design values previously obtained—namely, $C_1 = C_2 = C$, $R_3 = R_4 = R/4Q^2$, and $CR = 2Q/\omega_0$—gives

$$t(s) = \frac{s^2 + s(\omega_0/Q) + \omega_0^2}{s^2 + s(\omega_0/Q)(2Q^2 + 1) + \omega_0^2} \tag{14.19}$$

where ω_0 and Q denote the nominal or design values of the pole frequency and Q factor. The actual values are obtained by substituting for $t(s)$ in Eq. (14.18):

$$s^2 + s\frac{\omega_0}{Q}(2Q^2 + 1) + \omega_0^2 + A\left(s^2 + s\frac{\omega_0}{Q} + \omega_0^2\right) = 0$$

Assuming the gain A to be real and dividing both sides by $A + 1$ gives

$$s^2 + s\frac{\omega_0}{Q}\left(1 + \frac{2Q^2}{A + 1}\right) + \omega_0^2 = 0 \tag{14.20}$$

From this equation we see that the actual pole frequency, ω_{0a}, and pole-Q, Q_a, are

$$\omega_{0a} = \omega_0 \tag{14.21}$$

$$Q_a = \frac{Q}{1 + 2Q^2/(A + 1)} \tag{14.22}$$

Thus,

$$S_A^{\omega_{0a}} = 0$$

$$S_A^{Q_a} = \frac{A}{A + 1}\frac{2Q^2/(A + 1)}{1 + 2Q^2/(A + 1)}$$

For $A \gg 2Q^2$ and $A \gg 1$ we obtain

$$S_A^{Q_a} \cong \frac{2Q^2}{A}$$

It is usual to drop the subscript a in this expression and write

$$S_A^Q \cong \frac{2Q^2}{A} \tag{14.23}$$

Note that if Q is high ($Q \geq 5$) its sensitivity relative to the amplifier gain can be quite high.[1]

Exercise

14.11 In a particular filter utilizing the feedback loop of Fig. 14.9, with $C_1 = C_2$, find the expected percentage change in ω_0 and Q under the conditions that:
(a) R_3 is 2% high.
(b) R_4 is 2% high.
(c) Both R_3 and R_4 are 2% high.
(d) Both capacitors are 2% high and both resistors are 2% low.

Ans. (a) -1%, $+1\%$; (b) -1%, -1%; (c) -2%, 0%; (d) 0%, 0%

14.5 MULTIPLE-AMPLIFIER BIQUADRATIC FILTERS

The results of Example 14.1 indicate a serious disadvantage of single-amplifier biquads—the sensitivity of Q relative to the amplifier gain is quite high. Although there exists a technique for reducing S_A^Q in SABs, this is done at the expense of increased passive sensitivities. Nevertheless, the resulting SABs are used extensively in many applications. In the following, however, we shall present two circuits that utilize two and three op amps, respectively, to realize second-order filter functions with reduced sensitivities and increased versatility.

GIC-Based Biquads

In Example 3.6 we introduced the generalized impedance converter (GIC) and considered its use in realizing an inductance. (The reader is urged to review Example 3.6 at this point.) Figure 14.14 shows the GIC circuit with the components selected according to the first case of Example 3.6. Assume that $A_1 = A_2 = \infty$. It can be shown that the input impedance Z_{in} between node 1 and ground is given by

$$Z_{in}(s) = sC_2 \frac{R_1 R_3 R_5}{R_4} \tag{14.24}$$

Thus between node 1 and ground we have an inductance of value L,

$$L = \frac{C_2 R_1 R_3 R_5}{R_4} \tag{14.25}$$

---

[1] Because the open-loop gain A of op amps usually has wide tolerance, it is important to keep $S_A^{\omega_o}$ and S_A^Q very small.

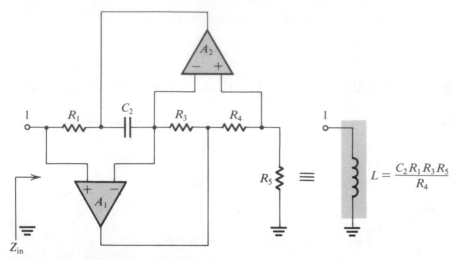

Fig. 14.14 The GIC used to realize an inductance.

Resonance can be obtained by connecting a capacitor C_6 between node 1 and ground. The input signal can then be fed to this *resonator* through a resistor R_7. Figure 14.15a shows the resulting circuit, and Fig. 14.15b shows the equivalent circuit. This resonator has a pole frequency ω_0 given by

$$\omega_0 = \frac{1}{\sqrt{LC_6}}$$

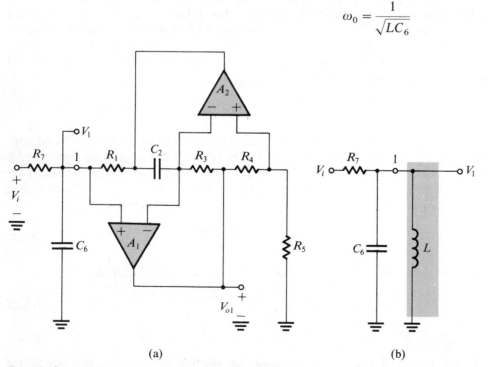

Fig. 14.15 Using the inductance of Fig. 14.14 to obtain resonance.

Thus

$$\omega_0 = \frac{1}{\sqrt{C_2 C_6 R_1 R_3 R_5 / R_4}} \tag{14.26}$$

The Q factor is determined by the resistance R_7 according to

$$Q = \omega_0 C_6 R_7 = R_7 \sqrt{\frac{C_6 R_4}{C_2 R_1 R_3 R_5}} \tag{14.27}$$

If the output is taken across the tuned circuit—that is, between node 1 and ground—the function realized is a bandpass one. In fact, this circuit is equivalent to the passive realization shown in Fig. 14.6c. Node 1, however, is a high-impedance node and is not suitable as an output node. In other words, connecting a load impedance between node 1 and ground changes the function realized. Fortunately, however, a low-impedance node exists where the voltage is directly proportional to that at node 1. This is the output terminal of op amp A_1, where, as the reader can easily verify, the voltage is

$$V_{o1} = V_1 \frac{R_4 + R_5}{R_5}$$

Thus the filter output is usually taken as V_{o1}.

The analysis of this circuit to determine its sensitivity relative to the gains of the op amps is straightforward but quite tedious (see Sedra & Brackett, 1978). The results indicate that S_{A1}^Q and S_{A2}^Q are proportional to the value of Q rather than to Q^2, as in the case of SABs.

Finally, it should be mentioned that a complete set of circuits based on the GIC has been developed for the realization of the various second-order filter functions (see Sedra & Brackett, 1978).

Exercises

14.12 Design the biquad circuit of Fig. 14.15a to realize a bandpass function with center frequency $f_0 = 10$ kHz and $Q = 20$. Select $C_2 = C_6 = C$ and $R_1 = R_3 = R_4 = R_5 = 10$ kΩ. Find the values of C and R_7. Also find the value of the center-frequency gain.

Ans.   1.59 nF; 200 kΩ; 2

14.13 What filter function is realized by interchanging R_7 and C_6 in the circuit of Fig. 14.15?

Ans.   High-pass

Two-Integrator-Loop Biquads

The circuits discussed next utilize two integrators in a feedback loop. To develop this two-integrator-loop realization consider the high-pass transfer function

$$\frac{V_{hp}}{V_i} = \frac{n_2 s^2}{s^2 + s(\omega_0/Q) + \omega_0^2} \tag{14.28}$$

Cross multiplying and dividing both sides by s^2 gives

$$V_{hp} = -\frac{1}{Q} \frac{\omega_0}{s} V_{hp} - \frac{\omega_0^2}{s^2} V_{hp} + n_2 V_i \tag{14.29}$$

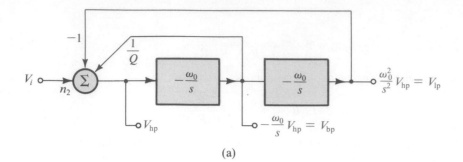

(a)

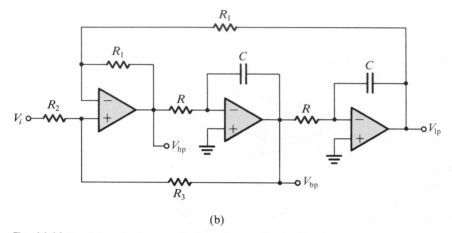

(b)

Fig. 14.16 Two-integrator-loop realization of second-order filter functions.

Utilizing two inverting integrators and a summer, Fig. 14.16a shows a block-diagram realization of this equation. An actual circuit is shown in Fig. 14.16b. The reader is urged to verify the correspondence between the circuit and the block diagram and to use this correspondence to show that

$$ CR = \frac{1}{\omega_0}, \qquad \frac{R_3}{R_2} = 2Q - 1, \qquad \text{and} \qquad n_2 = 2 - \frac{1}{Q}. $$

It should be noted that the function available at the output of the summer is high pass, that at the output of the first integrator is bandpass, and that at the output of the second integrator is low-pass. The simultaneous availability of three different filtering functions makes this circuit suitable for application as a universal filter. Other filtering functions can be obtained by employing an additional summer that forms a weighted sum of the outputs of the three op amps.

An alternative two-integrator-loop realization in which all three op amps are used in a single-ended mode can be developed as follows: Rather than using the input summer to add signals with positive and negative coefficients, we can introduce an additional inverter, as shown in Fig. 14.17a. Now the summer has all positive coefficients, and we may dispense with the summer amplifier altogether and perform

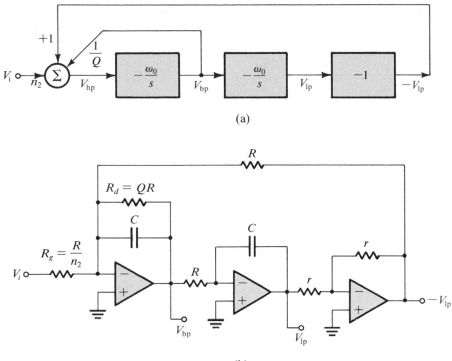

Fig. 14.17 An alternative two-integrator-loop filter circuit.

the summation at the virtual-ground input of the first integrator. The resulting circuit is shown in Fig. 14.17b, from which we observe that the high-pass function is no longer available. This is the price paid for obtaining a circuit that utilizes the op amps in a single-ended mode.

Two integrator-loop circuits display low sensitivities and considerable flexibility and versatility. Although the circuits shown here suffer drastically from the effects of finite op-amp bandwidth, simple modifications desensitize the circuits to these effects. Such a topic, however, is beyond the scope of this book.

Exercise

14.6 Use the circuit of Fig. 14.17b to realize a bandpass filter with $f_0 = 10$ kHz, $Q = 20$, and unity center-frequency gain. If $R = 10$ kΩ, give the values of C, R_d, and R_g.

Ans. 1.59 nF; 200 kΩ; 200 kΩ

14.6 SWITCHED-CAPACITOR FILTERS

The active-RC filter circuits presented above have two properties that make their production in monolithic IC form difficult, if not practically impossible; these are the need for large-valued capacitors and the requirement of accurate RC time constants. The search therefore continued for a method of filter design that would lend itself more naturally to IC implementation. In this section we shall introduce one such method. At the time of this writing it appears to be the best contender for the task.

The Basic Principle

The switched-capacitor filter technique is based on the realization that a capacitor switched between two circuit nodes at a sufficiently high rate is equivalent to a resistor connecting these two nodes. To be specific, consider the active-RC integrator of Fig. 14.18a. This is the familiar Miller integrator, which we used in the two-integrator-loop biquad in Section 14.5. In Fig. 14.18b we have replaced the input resistor R_1 by a grounded capacitor C_1 together with two MOS transistors acting as switches. In some circuits more elaborate switch configurations are used, but such details are beyond our present need.

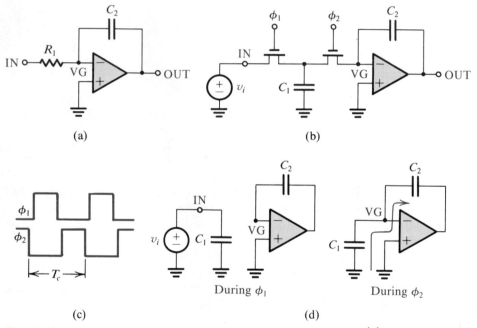

Fig. 14.18 Basic principles of the switched-capacitor filter technique. **(a)** Active-RC integrator. **(b)** Switched-capacitor integrator. **(c)** Two-phase clock (nonoverlapping). **(d)** During ϕ_1, C_1 charges up to the current value of v_i and then, during ϕ_2, discharges into C_2.

The two MOS switches in Fig. 14.18b are driven by a *nonoverlapping* two-phase clock. Figure 14.18c shows the clock waveforms. We shall assume in this introductory exposition that the clock frequency f_c ($f_c = 1/T_c$) is much higher than the frequency of the signal being filtered. Thus during clock phase ϕ_1, when C_1 is connected across the input signal source v_i, the variations in the input signal are negligibly small. It follows that during ϕ_1 capacitor C_1 charges up to the voltage v_i,

$$q_{C1} = C_1 v_i$$

Then, during clock phase ϕ_2, capacitor C_1 is connected to the virtual-ground input of the op amp, as indicated in Fig. 14.18d. Capacitor C_1 is thus forced to discharge, and its previous charge q_{C1} is transferred to C_2, in the direction indicated in Fig. 14.18d.

From the above description we see that during each clock period T_c an amount of charge $q_{C1} = C_1 v_i$ is extracted from the input source and supplied to the integrator capacitor C_2. Thus the average current flowing between the input node (IN) and the virtual ground node (VG) is

$$i_{av} = \frac{C_1 v_i}{T_c}$$

If T_c is sufficiently short, one can think of this process as almost continuous and define an equivalent resistance R_{eq} that is in effect present between nodes IN and VG:

$$R_{eq} \equiv \frac{v_i}{i_{av}}$$

Thus

$$R_{eq} = \frac{T_c}{C_1} \tag{14.30}$$

Using R_{eq} we obtain an equivalent time constant for the integrator:

$$\text{Time constant} = C_2 R_{eq} = T_c \frac{C_2}{C_1} \tag{14.31}$$

Thus the time constant that determines the frequency response of the filter is determined by the clock period T_c and the capacitor ratio C_2/C_1. Both of these parameters can be well controlled in an IC process. Specifically, note the dependence on capacitor ratios rather than on absolute values of capacitors. The accuracy of capacitor ratios in MOS technology can be controlled to within 0.1%.

Another point worth observing is that with a reasonable clocking frequency (such as 100 kHz) and not-too-large capacitor ratios (say, 10) one can obtain reasonably large time constants (such as 10^{-4} s) suitable for audio applications. Since capacitors typically occupy relatively large areas on the IC chip, it is important to mention that the ratio accuracies quoted above are obtainable with the smaller capacitor value as low as 0.2 pF.

Practical Circuits

The switched-capacitor (SC) circuit in Fig. 14.18b realizes a negative integrator. As we saw in Section 14.5, a two-integrator-loop active filter is composed of one inverting and one noninverting integrator.[2] To realize a switched-capacitor biquad filter we

---

[2] In the two-integrator loop of Fig. 14.17 the noninverting integrator is realized by the cascade of a Miller integrator and an inverter.

therefore need a pair of complementary switched-capacitor integrators. Figure 14.19a shows a noninverting, or positive, integrator circuit. The reader is urged to follow the operation of this circuit during the two clock phases and thus show that it operates much the same way as the basic circuit of Fig. 14.18b, except for a sign reversal.

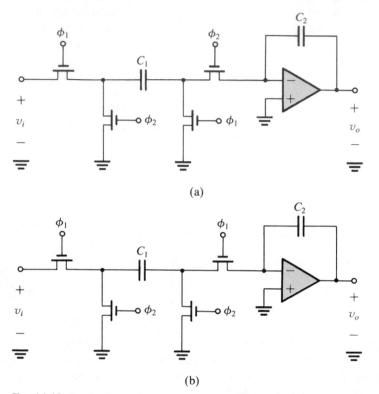

(a)

(b)

Fig. 14.19 A pair of complementary stray-insensitive switched-capacitor integrators. **(a)** Noninverting switched-capacitor integrator. **(b)** Inverting switched-capacitor integrator.

In addition to realizing a noninverting integrator function, the circuit in Fig. 14.19a is insensitive to stray capacitances; however, we shall not explore this point any further. The interested reader is referred to Ghausi and Laker (1981). By reversal of the clock phases on two of the switches, the circuit in Fig. 14.19b is obtained. This latter circuit realizes the inverting integrator function, like the circuit of Fig. 14.18b, but is insensitive to stray capacitances (which the original circuit of Fig. 14.18b is not). At the time of this writing, the pair of complementary integrators of Fig. 14.19 has become the standard building block in the design of switched-capacitor filters.

Let us now consider the realization of a complete biquad circuit. Figure 14.20a shows the active-RC two-integrator-loop circuit previously studied. By considering the cascade of integrator 2 and the inverter as a positive integrator, and then simply replacing each resistor by its switched-capacitor equivalent, we obtain the circuit in Fig. 14.20b. Ignore the damping around the first integrator (that is, the switched capacitor C_5) for the time being and note that the feedback loop indeed consists of

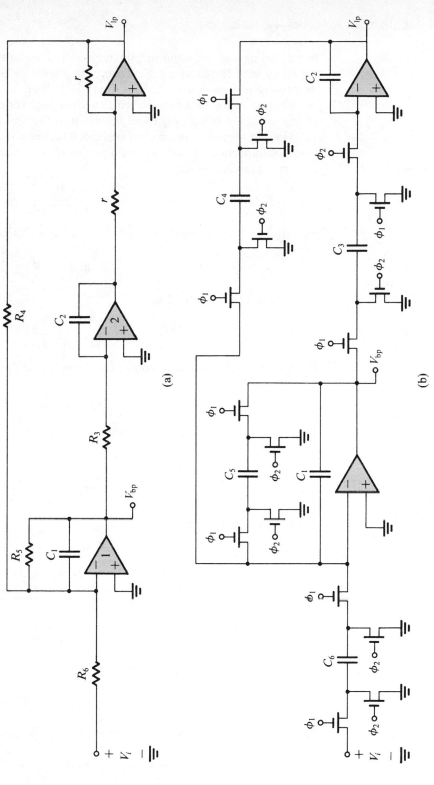

Fig. 14.20 A two-integrator-loop active-RC biquad and its switched-capacitor counterpart.

one inverting and one noninverting integrator. Then note the phasing of the switched capacitor used for damping. Reversing the phases here would convert the feedback to positive and move the poles to the right half of the s plane. On the other hand, the phasing of the feed-in switched capacitor (C_6) is not that important; a reversal of phases would result only in an inversion in the sign of the function realized.

Having identified the correspondences between the active-RC biquad and the switched-capacitor biquad, we can now derive design equations. Analysis of the circuit in Fig. 14.20a yields

$$\omega_0 = \frac{1}{\sqrt{C_1 C_2 R_3 R_4}} \tag{14.32}$$

Substituting for R_3 and R_4 by their SC equivalent values, that is,

$$R_3 = \frac{T_c}{C_3} \quad \text{and} \quad R_4 = \frac{T_c}{C_4}$$

gives ω_0 of the SC biquad as

$$\omega_0 = \frac{1}{T_c} \sqrt{\frac{C_3}{C_2} \frac{C_4}{C_1}} \tag{14.33}$$

It is usual to select the time constants of the two integrators to be equal; that is,

$$\frac{T_c}{C_3} C_2 = \frac{T_c}{C_4} C_1 \tag{14.34}$$

If we further select the two integrating capacitors C_1 and C_2 to be equal,

$$C_1 = C_2 = C \tag{14.35}$$

then

$$C_3 = C_4 = KC, \tag{14.36}$$

where from Eq. (14.33)

$$K = \omega_0 T_c \tag{14.37}$$

For the case of equal time-constants the Q factor of the circuit in Fig. 14.20a is given by R_5/R_4. Thus the Q factor of the corresponding SC circuit in Fig. 14.20b is given by

$$Q = \frac{T_c/C_5}{T_c/C_4} \tag{14.38}$$

Thus C_5 should be selected from

$$C_5 = \frac{C_4}{Q} = \frac{KC}{Q} = \omega_0 T_c \frac{C}{Q} \tag{14.39}$$

Finally, the center-frequency gain of the bandpass function is given by

$$\text{Center-frequency gain} = \frac{C_6}{C_5} = Q \frac{C_6}{\omega_0 T_c C} \tag{14.40}$$

Exercise

14.15 Use $C_1 = C_2 = 20$ pF and design the circuit in Fig. 14.20b to realize a bandpass function with $f_0 = 10$ kHz, $Q = 20$, and unity center-frequency gain. Use a clock frequency $f_c = 200$ kHz. Find the values of C_3, C_4, C_5, and C_6.

Ans. 6.283 pF; 6.283 pF; 0.314 pF; 0.314 pF

A Final Remark

We have attempted only to provide an introduction to switched-capacitor filters. We have made many simplifying assumptions, the most important being the switched-capacitor–resistor equivalence [Eq. (14.30)]. This equivalence is correct only at $f_c = \infty$ and is approximately correct for $f_c \gg f$. Switched-capacitor filters are, in fact, sampled-data networks whose analysis and design can be carried out exactly using z-transform techniques. The interested reader is referred to the bibliography at the end of this chapter.

14.7 TUNED AMPLIFIERS

In this section we study a special kind of frequency-selective network, the LC-tuned amplifier. Figure 14.21 shows the general frequency-response shape of a tuned amplifier. The techniques to be discussed apply to amplifiers with center frequencies in the range of few hundred kHz to few hundred MHz. Tuned amplifiers find application in the radio-frequency (RF) and intermediate-frequency (IF) sections of communications receivers and in a variety of other systems. It should be noted that the tuned-amplifier response of Fig. 14.21 is similar to that of the bandpass filter discussed in Section 14.2.

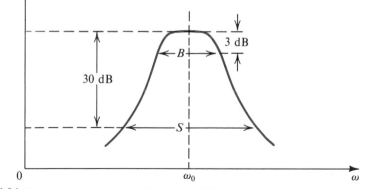

Fig. 14.21 Frequency response of a tuned amplifier.

As indicated in Fig. 14.21, the response is characterized by the center frequency ω_0, the 3-dB bandwidth B, and the *skirt selectivity*, which is usually measured as the ratio of the 30-dB bandwidth to the 3-dB bandwidth. In many applications, the 3-dB bandwidth is less than 5% of ω_0. This *narrow-band* property makes possible certain approximations that can simplify the design process, as will be explained later.

The tuned amplifiers studied in this section are small-signal voltage amplifiers in which the transistors operate in the class A mode. Tuned power amplifiers based on class C and other switching modes of operation are not studied in this book.

The Basic Principle

The basic principle underlying the design of tuned amplifiers is the use of a parallel RLC circuit as the load, or at the input, of a BJT or a FET amplifier. This is illustrated in Fig. 14.22 with a MOSFET amplifier having a tuned-circuit load. For simplicity, the bias details are not included. Since this circuit uses a single tuned circuit, it is known as a *single-tuned* amplifier. The amplifier equivalent circuit is shown in Fig. 14.22b. Here R denotes the parallel equivalent of R_L and the output resistance r_o of the FET, and C is the parallel equivalent of C_L and the FET output capacitance (usually very small). From the equivalent circuit we can write

$$V_o = \frac{-g_m V_i}{Y_L}$$

$$= \frac{-g_m V_i}{sC + 1/R + 1/sL}$$

Thus the voltage gain can be expressed as

$$\frac{V_o}{V_i} = -\frac{g_m}{C} \frac{s}{s^2 + s(1/CR) + 1/LC} \tag{14.41}$$

which is a second-order bandpass function of the form given in Eq. (14.4). Thus the tuned amplifier has a center frequency of

$$\omega_0 = \frac{1}{\sqrt{LC}} \tag{14.42}$$

a 3-dB bandwidth of

$$B = \frac{1}{CR} \tag{14.43}$$

a Q factor of

$$Q \equiv \frac{\omega_0}{B} = \omega_0 CR \tag{14.44}$$

and a center-frequency gain of

$$\frac{V_o(j\omega_0)}{V_i(j\omega_0)} = -g_m R \tag{14.45}$$

Note that the expression for the center-frequency gain could have been written by inspection; at resonance the reactances of L and C cancel out and the impedance of the parallel LCR circuit reduces to R.

EXAMPLE 14.2

It is required to design a tuned amplifier of the type shown in Fig. 14.22, having $f_0 = 1$ MHz, 3-dB bandwidth $= 10$ kHz, and center-frequency gain $= -10$ V/V. The FET available has at the bias point $g_m = 5$ mA/V and $r_o = 10$ kΩ. The output capacitance is negligibly small. Determine the values of R_L, C_L, and L.

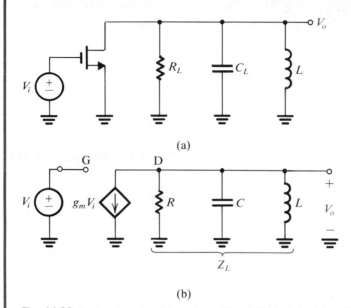

(a)

(b)

Fig. 14.22 Basic principle of tuned amplifiers is illustrated using a MOSFET with a tuned-circuit load. Bias details are not shown.

Solution

Center-frequency gain $= -10 = -5$ R. Thus $R = 2$ kΩ. Since $R = R_L \| r_o$, then $R_L = 2.5$ kΩ.

$$B = 2\pi \times 10^4 = \frac{1}{CR}$$

Thus

$$C = \frac{1}{2\pi \times 10^4 \times 2 \times 10^3} = 7{,}958 \text{ pF}$$

Since $\omega_0 = 2\pi \times 10^6 = 1/\sqrt{LC}$, we obtain

$$L = \frac{1}{4\pi^2 \times 10^{12} \times 7{,}958 \times 10^{-12}} = 3.2 \ \mu\text{H}.$$

Inductor Losses

The power loss in the inductor is usually represented by a series resistance r_s as shown in Fig. 14.23a. However, rather than specifying the value of r_s, the usual

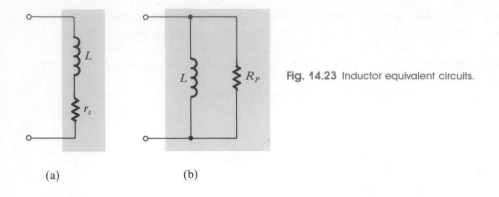

Fig. 14.23 Inductor equivalent circuits.

(a)   (b)

practice is to specify the inductor Q factor at the frequency of interest,

$$Q_0 \equiv \frac{\omega_0 L}{r_s} \tag{14.46}$$

Typically, Q_0 is in the range of 50 to 200.

The analysis of a tuned amplifier is greatly simplified by representing the inductor loss by a parallel resistance R_p as shown in Fig. 14.23b. The relationship between R_p and Q_0 can be found by writing, for the admittance of the circuit in Fig. 14.23a,

$$Y(j\omega_0) = \frac{1}{r_s + j\omega_0 L}$$

$$= \frac{1}{j\omega_0 L} \frac{1}{1 - j(1/Q_0)} = \frac{1}{j\omega_0 L} \frac{1 + j(1/Q_0)}{1 + (1/Q_0^2)}$$

For $Q_0 \gg 1$,

$$Y(j\omega_0) \simeq \frac{1}{j\omega_0 L}\left(1 + j\frac{1}{Q_0}\right) \tag{14.47}$$

Equating this to the admittance of the circuit in Fig. 14.23b gives

$$Q_0 = \frac{R_p}{\omega_0 L} \tag{14.48}$$

or, equivalently,

$$R_p = \omega_0 L Q_0 \tag{14.49}$$

Finally, it should be noted that the coil Q factor poses an upper limit on the value of Q achieved by the tuned circuit.

Exercise

14.16 If the inductor in Example 14.2 has $Q_0 = 150$, find R_p and hence find the value to which R_L should be changed so as to keep the overall Q, and hence the bandwidth, unchanged.

Ans. 3 kΩ; 15 kΩ

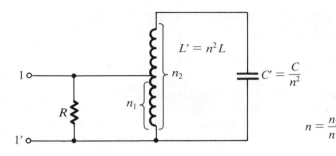

Fig. 14.24 A tapped inductor is used as an impedance transformer to allow using a higher inductance, L', and a smaller capacitance, C'.

$$L' = n^2 L$$

$$C' = \frac{C}{n^2}$$

$$n = \frac{n_2}{n_1}$$

Use of Transformers

In many cases it is found that the required value of inductance is not practical, in the sense that coils with the required inductance might not be available with the required high values of Q_0. A simple solution is to use a transformer to effect an impedance change. Alternatively, a tapped coil, known as an *autotransformer*, can be used, as shown in Fig. 14.24. Provided the two parts of the inductor are tightly coupled, which can be achieved by winding it on a ferrite core, the transformation relationships shown hold. The result is that the tuned circuit seen between terminals 1 and 1′ is equivalent to that in Fig. 14.22b. For example, if a turns ratio $n = 3$ is used in the amplifier of Example 14.2, then a coil with inductance $L' = 9 \times 3.2 = 28.8\ \mu\text{H}$ and a capacitance $C' = 7{,}958/9 = 884$ pF will be required. Both of these values are more practical than the original ones.

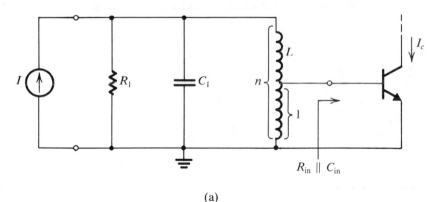

(a)

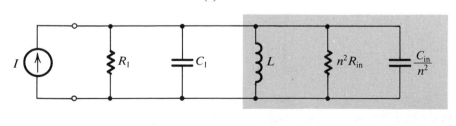

(b)

Fig. 14.25 (a) The output of a tuned amplifier is coupled to the input of another amplifier via a tapped coil and **(b)** equivalent circuit.

In applications that involve coupling the output of a tuned amplifier to the input of another amplifier, the tapped coil can be used to raise the effective input resistance of the latter amplifier stage. In this way, one can avoid reduction of the overall Q. This point is illustrated in Fig. 14.25 and in the following exercises.

Exercises

14.17 Consider the circuit in Fig. 14.25a, first without tapping the coil. Let $L = 5\ \mu H$ and assume that R_1 is fixed at 1 kΩ. We wish to design a tuned amplifier with $f_0 = 455$ kHz and a 3-dB bandwidth of 10 kHz (this is the IF amplifier of an AM radio). If the BJT has $R_{in} = 1$ kΩ and $C_{in} = 200$ pF, find the actual bandwidth obtained, and the required value of C_1.

Ans. 13 kHz; 24.27 nF

14.18 Since the bandwidth realized in Exercise 14.17 is greater than desired, find an alternative design utilizing a tapped coil as in Fig. 14.25a. Find the value of n that allows the specifications to be just met. Also find the new required value of C_1 and the current gain I_c/I at resonance. Assume that at the bias point the BJT has $g_m = 40$ mA/V.

Ans. 1.36; 24.36 nF; 19.1 A/A

Amplifiers with Multiple Tuned Circuits

The selectivity achieved with the single-tuned circuit of Fig. 14.22 is not sufficient in many applications—for instance, in the IF amplifier of a radio or a TV receiver. Greater selectivity is obtained by using additional tuned stages. Figure 14.26 shows a BJT with tuned circuits at both the input and the output.[3] In this circuit the bias

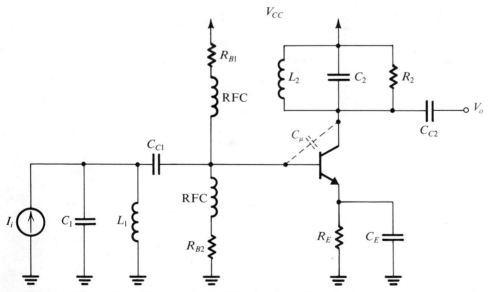

Fig. 14.26 A BJT amplifier with tuned circuits at the input and the output.

---

[3] Note that because the input circuit is a parallel resonant circuit, an input current source (rather than voltage source) signal is utilized.

details are shown, from which we note that biasing is quite similar to the classical arrangement employed in discrete-circuit design. However, in order to avoid the loading effect of the bias resistors R_{B1} and R_{B2} on the input tuned circuit, a *radio-frequency choke* (RFC) is inserted in series with each resistor. These chokes have high impedances at the frequencies of interest. The use of RFCs in biasing tuned RF amplifiers is common practice.

The analysis and design of the double-tuned amplifier of Fig. 14.26 is complicated by the Miller effect[4] due to capacitance C_μ. Since the load is not simply resistive, as was the case in the amplifiers studied in Chapter 11, the Miller impedance at the input will be complex. This reflected impedance will cause detuning of the input circuit as well as "skewing" of the response of the input circuit. Needless to say, the coupling introduced by C_μ makes tuning or aligning the amplifier quite difficult. Worse still, the capacitor C_μ can cause oscillations to occur (see Gray and Searle, 1969 and Problem 14.36).

Methods exist for *neutralizing* the effect of C_μ, using additional circuits so arranged as to feed back a current equal and opposite to that through C_μ. An alternative, and preferred, approach is to use circuit configurations that do not suffer from the Miller effect. These will be discussed below. Before leaving this section, however, we wish to point out that circuits of the type shown in Fig. 14.26 are usually designed utilizing the y-parameter model of the BJT (see Appendix B). This is done because here, in view of the fact that C_μ plays a significant role, the y-parameter model makes the analysis simpler (as compared to that using the hybrid-π model). Also, the y parameters can easily be measured at the particular frequency of interest, ω_0. For narrow-band amplifiers, the assumption is usually made that the y parameters remain approximately constant over the passband.

The Cascode and the CC-CB Cascade

From our study of amplifier frequency response in Chapter 11 we know that two amplifier configurations do not suffer from the Miller effect. These are the cascode configuration and the common-collector common-base cascade. Figure 14.27 shows tuned amplifiers based on these two configurations. The CC-CB cascade is usually preferred in IC implementations because its differential structure makes it suitable for IC biasing techniques. (Note that the biasing details of the cascode circuit are not shown in Fig. 14.27. Biasing can be done using arrangements similar to those discussed in earlier chapters).

Synchronous Tuning

In the design of a tuned amplifier with multiple tuned circuits the question arises as to the frequency to which each circuit should be tuned. The object, of course, is for the overall response to exhibit high passband flatness and skirt selectivity. To investigate this question we shall assume that the overall response is the product of the

---

[4] Here we use "Miller effect" to refer to the effect of the feedback capacitance C_μ in reflecting an input impedance that is a function of the amplifier load impedance.

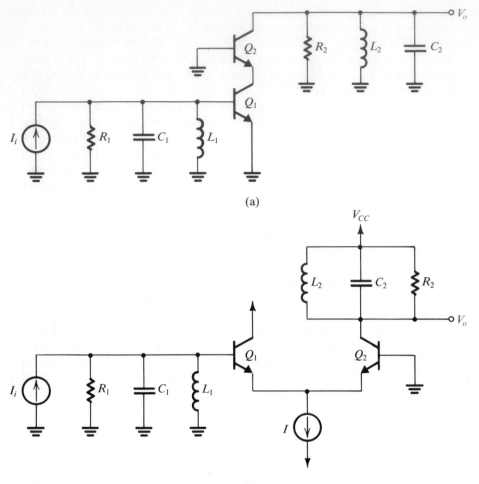

Fig. 14.27 Two tuned-amplifier configurations that do not suffer from the Miller effect: **(a)** cascode; **(b)** common-collector common-base cascade. (Note that bias details of the cascode circuit are not shown.)

individual responses; in other words, the stages do not interact. This can be easily achieved using circuits such as those in Fig. 14.27.

Consider first the case of N identical resonant circuits, known as the *synchronously tuned* case. Figure 14.28 shows the response of an individual stage and that of the cascade. Observe the bandwidth "shrinkage" of the overall response. The 3-dB bandwidth B of the overall amplifier is related to that of the individual tuned circuits, ω_0/Q, by (see Problem 14.38)

$$B = \frac{\omega_0}{Q} \sqrt{2^{1/N} - 1} \tag{14.50}$$

The factor $\sqrt{2^{1/N} - 1}$ is known as the bandwidth-shrinkage factor. Given B and N, Eq. (14.50) can be used to determine the bandwidth required of the individual stages.

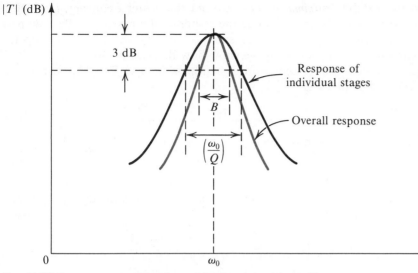

Fig. 14.28 Frequency response of a synchronously tuned amplifier.

Exercise

14.19 Consider the design of an IF amplifier for an FM radio receiver. Using two synchronously tuned stages with $f_0 = 10.7$ MHz, find the 3-dB bandwidth of each stage so that the overall bandwidth is 200 kHz. Using 3-μH inductors find C and R for each stage.

Ans.   310.8 kHz; 73.7 pF; 6.95 kΩ

Stagger Tuning

A much better overall response is obtained by stagger-tuning the individual stages, as illustrated in Fig. 14.29 Stagger-tuned amplifiers are usually designed so that the

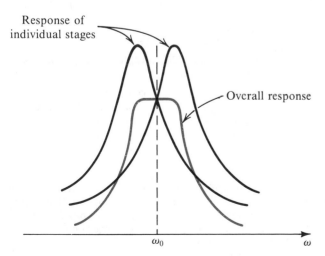

Fig. 14.29 Stagger-tuning the individual resonant circuits can result in an overall response with a passband flatter than that obtained with synchronous tuning (Fig. 14.28).

overall response exhibits *maximal flatness* around the center frequency f_0. Such a response can be obtained by transforming the response of a maximally flat low-pass filter (known as a Butterworth response), such as that shown in Fig. 14.1, up the frequency axis to ω_0. In the following we show how this can be done.

The transfer function of a second-order bandpass filter in Eq. (14.4) can be expressed in terms of its poles as

$$T(s) = \frac{n_1 s}{\left(s + \dfrac{\omega_0}{2Q} - j\omega_0 \sqrt{1 - \dfrac{1}{4Q^2}}\right)\left(s + \dfrac{\omega_0}{2Q} + j\omega_0 \sqrt{1 - \dfrac{1}{4Q^2}}\right)} \tag{14.51}$$

For a narrow-band filter $Q \gg 1$, and for values of s in the neighborhood of $+ j\omega_0$ (see Fig. 14.30b) the second factor in the denominator is approximately $(2j\omega_0)$. Hence Eq. (14.51) can be approximated by

$$T(s) \simeq \frac{n_1/2}{s + \omega_0/2Q - j\omega_0}$$

$$= \frac{n_1/2}{(s - j\omega_0) + \omega_0/2Q} \tag{14.52}$$

This is known as the *narrow-band approximation.*[5] Note that the magnitude response, for $s = j\omega$, has a peak value of $n_1 Q/\omega_0$ at $\omega = \omega_0$, as expected.

Now consider a first-order low-pass network with a single pole at $p = -\omega_0/2Q$ (we use p to denote the complex frequency variable for the low-pass filter). Its transfer function is

$$T(p) = \frac{K}{p + \omega_0/2Q} \tag{14.53}$$

where K is a constant. Comparing Eqs. (14.52) and (14.53) we note that they are identical for $p = s - j\omega_0$ or, equivalently,

$$s = p + j\omega_0 \tag{14.54}$$

This result implies that the response of the second-order bandpass filter *in the neighborhood of its center frequency* $s = j\omega_0$ is identical to the response of a first-order low-pass filter with a pole at $(-\omega_0/2Q)$ *in the neighborhood of* $p = 0$. Thus the bandpass response can be obtained by shifting the pole of the low-pass prototype and adding the complex conjugate pole, as illustrated in Fig. 14.30. This is called a *low-pass to bandpass transformation* for *narrow-band* filters.

---

[5] The bandpass response is *geometrically symmetric* around the center frequency ω_0. That is, each two frequencies ω_1 and ω_2 at which the magnitude response is equal are related by $\omega_1\omega_2 = \omega_0^2$. For high Q, the symmetry becomes almost *arithmetic* for frequencies close to ω_0. That is, two frequencies with the same magnitude response are almost equally spaced from ω_0. The same is true for higher-order bandpass filters designed using the transformation presented in this section.

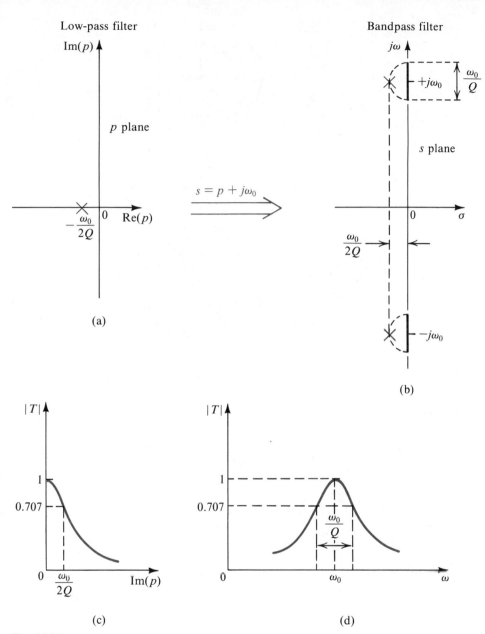

Fig. 14.30 Obtaining a second-order narrow-band bandpass filter by transforming a first-order low-pass filter.

The transformation $p = s - j\omega_0$ can be applied to low-pass filters of order greater than one. For instance, we can transform a maximally flat second-order low-pass filter ($Q = 1/\sqrt{2}$) to obtain a maximally flat bandpass filter. If the 3-dB bandwidth of the bandpass filter is to be B rad/s, then the low-pass filter should have a pole frequency of $(B/2)$ rad/s, as illustrated in Fig. 14.31. The resulting fourth-order bandpass

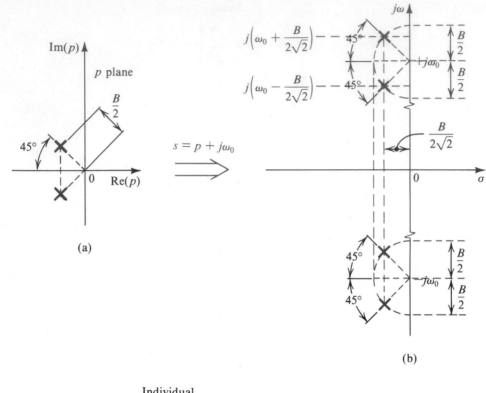

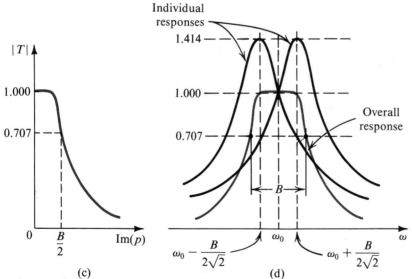

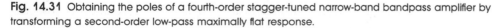

Fig. 14.31 Obtaining the poles of a fourth-order stagger-tuned narrow-band bandpass amplifier by transforming a second-order low-pass maximally flat response.

filter will be a stagger-tuned one, with its two tuned circuits (refer to Fig.14.31) having

$$\omega_{01} = \omega_0 + \frac{B}{2\sqrt{2}} \qquad B_1 = \frac{B}{\sqrt{2}} \qquad Q_1 \simeq \frac{\sqrt{2}\omega_0}{B} \tag{14.55}$$

$$\omega_{02} = \omega_0 - \frac{B}{2\sqrt{2}} \qquad B_2 = \frac{B}{\sqrt{2}} \qquad Q_2 \simeq \frac{\sqrt{2}\omega_0}{B} \tag{14.56}$$

Note that in order for the overall response to have a normalized center-frequency gain of unity, the individual responses are shown in Fig. 14.31d to have equal center-frequency gains of $\sqrt{2}$. In practice, however, the individual responses need not have equal center-frequency gains, as illustrated in the following exercises.

Exercises

14.20 A stagger-tuned design for the IF amplifier specified in Exercise 14.19 is required. Find f_{01}, B_1, f_{02} and B_2. Also give the value of C and R for each of the two stages. (Recall that 3-μH inductors are to be used.)

Ans.   10.77 MHz; 141.4 kHz; 10.63 MHz; 141.4 kHz; 72.8 pF; 15.5 kΩ; 74.7 pF; 15.1 kΩ

14.21 Using the fact that the voltage gain at resonance is proportional to the value of R, find the ratio of the gain at 10.7 MHz of the stagger-tuned amplifier designed in Exercise 14.20 and the synchronously tuned amplifier designed in Exercise 14.19. (*Hint:* For the stagger-tuned amplifier note that the gain at ω_0 is equal to the product of the gains of the individual stages at their 3-dB frequencies.)

Ans.   2.42

14.8 BASIC PRINCIPLES OF SINUSOIDAL OSCILLATORS

The generation of sine waves is an important task that electronics engineers are often required to undertake. Basically there are two approaches to the design of sine-wave generators. The first is to design a *nonlinear oscillator*, which generates square and triangular waveforms, and apply the triangular wave to a *sine-wave shaper*, which usually consists of diodes and resistors. Nonlinear oscillators, or *function generators* as they are usually called, are studied in Chapter 5.

The second approach, which is the subject of the remainder of this chapter, employs a *positive-feedback loop* that contains a *frequency-selective network*. The loop is designed to have a gain of unity at a single frequency determined by the frequency-selective network. In this type of oscillator, called a *linear oscillator*, sine waves are generated essentially by a resonance phenomenon.

In spite of the name linear oscillator, some form of nonlinearity has to be employed to provide control of the amplitude of the output sine wave. In fact, all oscillators are essentially nonlinear circuits. This complicates the task of analysis and design of oscillators; no longer is one able to apply transform methods (*s* plane) directly. Nevertheless, techniques have been developed by which the design of sinusoidal oscillators can be performed in two steps. The first step is a linear one, and frequency-domain methods of feedback circuit analysis can be readily employed. Subsequently, a nonlinear mechanism for amplitude control can be provided.

The Oscillator Feedback Loop

The basic structure of a sinusoidal oscillator consists of an amplifier and a frequency-selective network connected in a positive-feedback loop, such as that shown in block-diagram form in Fig. 14.32. Although in an actual oscillator circuit no input signal will be present, we include an input signal here to help explain the principle of operation. It is important to note that unlike the negative-feedback loop of Fig. 12.1, here the feedback signal x_f is summed with a *positive* sign. Thus the gain with feedback is given by

$$A_f(s) = \frac{A(s)}{1 - A(s)\beta(s)} \tag{14.57}$$

where we note the negative sign in the denominator.

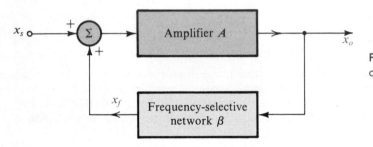

Fig. 14.32 The basic structure of a sinusoidal oscillator.

According to the definition of loop gain in Chapter 12, the loop gain of the circuit in Fig. 14.32 is $-A(s)\beta(s)$. However, for our purposes here it is more convenient to drop the minus sign and define the loop gain $L(s)$ as

$$L(s) = A(s)\beta(s) \tag{14.58}$$

The characteristic equation thus becomes

$$1 - L(s) = 0 \tag{14.59}$$

Note that this new definition of loop gain corresponds directly to the actual gain seen around the feedback loop of Fig. 14.32.

The Oscillation Criterion

If at a specific frequency f_0 the loop gain $A\beta$ is equal to unity, it follows from Eq. (14.57) that A_f will be infinite. That is, at this frequency the circuit will have a finite output for zero input signal. Such a circuit is by definition an oscillator. Thus the condition for the feedback loop of Fig. 14.32 to provide sinusoidal oscillations of frequency ω_0 is that

$$L(j\omega_0) \equiv A(j\omega_0)\beta(j\omega_0) = 1 \tag{14.60}$$

That is, at ω_0 the phase of the loop gain should be zero and the magnitude of the loop gain should be unity. This is known as the Barkhausen criterion. Note that for the circuit to oscillate at one frequency the oscillation criterion should be satisfied at one frequency only (that is, ω_0); otherwise the resulting waveform will not be a simple sinusoid.

An intuitive feeling for the Barkhausen criterion can be gained by considering the feedback loop of Fig. 14.32. For this loop to produce and sustain an output x_o with no input applied ($x_s = 0$), the feedback signal x_f,

$$x_f = \beta x_o$$

should be sufficiently large so that when multiplied by A it produces x_o,

$$A x_f = x_o$$

that is,

$$A \beta x_o = x_o$$

which results in

$$A\beta = 1$$

It should be noted that the frequency of oscillation ω_0 is determined solely by the phase characteristics of the feedback loop; the loop oscillates at the frequency for which the phase is zero. It follows that the stability of the frequency of oscillation will be determined by the manner in which the phase $\phi(\omega)$ of the feedback loop varies with frequency. A "steep" function $\phi(\omega)$ will result in a more stable frequency. This can be seen if one imagines a change in phase $\Delta\phi$ due to a change in one of the circuit components. If $d\phi/d\omega$ is large, the resulting change in ω_0 will be small, as illustrated in Fig. 14.33.

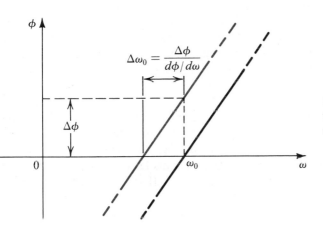

Fig. 14.33 Dependence of the oscillator frequency stability on the slope of the phase response.

An alternative approach to the study of oscillator circuits consists of examining the circuit poles, which are the zeros of the characteristic equation [Eq. (14.59)]. For the circuit to produce sustained oscillations at a frequency ω_0 the characteristic equation has to have zeros at $s = \pm j\omega_0$. Thus $1 - A(s)\beta(s)$ should have a factor of the form $s^2 + \omega_0^2$.

Nonlinear Amplitude Control

The oscillation condition discussed above guarantees sustained oscillations in a mathematical sense. It is well known, however, that the parameters of any physical system cannot be maintained constant for any length of time. In other words, suppose we work hard to make $A\beta = 1$ at $\omega = \omega_0$; then the temperature changes and $A\beta$ becomes slightly less than unity. Obviously, oscillations will cease in this case. Conversely, if $A\beta$ exceeds unity, oscillations will grow in amplitude. We therefore need a mechanism for forcing $A\beta$ to remain equal to unity at the desired value of output amplitude. This task is accomplished by providing a nonlinear circuit for gain control.

Basically, the function of the gain-control mechanism is as follows: First, to ensure that oscillations will start, one designs the circuit such that $A\beta$ is slightly greater than unity. This corresponds to designing the circuit so that the poles are in the right half of the s plane. Thus as the power supply is turned on, oscillations will grow in amplitude. When the amplitude reaches the desired level, the nonlinear network comes into action and causes the loop gain to be reduced to exactly unity. In other words, the poles will be "pulled back" to the $j\omega$ axis. This action will cause the circuit to sustain oscillations at this desired amplitude. If, for some reason, the loop gain is reduced below unity, the amplitude of the sine wave will diminish. This will be detected by the nonlinear network, which will cause the loop gain to increase to exactly unity.

As will be seen, there are two basic approaches to the implementation of the nonlinear amplitude-stabilization mechanism. The first approach makes use of a limiter circuit (see Chapter 5). Oscillations are allowed to grow until the amplitude reaches the level to which the limiter is set. Once the limiter comes into operation, the amplitude remains constant. Obviously, the limiter should be "soft" in order to minimize nonlinear distortion. Such distortion, however, is reduced by the filtering action of the frequency-selective network in the feedback loop. In fact, in one of the oscillator circuits studied in Section 14.9, the sine waves are hard-limited, and the resulting square waves are applied to a bandpass filter present in the feedback loop. The "purity" of the output sine waves will be a function of the selectivity of this filter. That is, the higher the Q of the filter, the less the harmonic content of the sine-wave output.

The other mechanism for amplitude stabilization is more elaborate. The amplitude of the sine wave output is detected and converted to a dc level, which is compared with a preset value. The output of the comparator is then used to adjust the value of a resistance that determines the loop gain. This latter resistance, a voltage-controlled one, may be implemented for example by a JFET operated in the triode region (see Chapter 6).

14.9 OP-AMP–RC OSCILLATOR CIRCUITS

In this section we shall study some practical oscillator circuits utilizing op amps and RC networks.

The Wien-Bridge Oscillator

One of the simplest oscillator circuits is based on the Wien bridge. Figure 14.34 shows a Wien-bridge oscillator without the nonlinear gain-control network. The circuit consists of an op amp connected in the noninverting configuration, with a closed-loop gain of $1 + R_2/R_1$. In the feedback path of this positive-gain amplifier an RC network

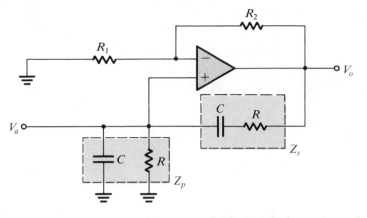

Fig. 14.34 Wien-bridge oscillator without amplitude stabilization.

is connected. The loop gain can be easily obtained by multiplying the transfer function $V_a(s)/V_o(s)$ of the feedback network by the amplifier gain,

$$L(s) = \left[1 + \frac{R_2}{R_1}\right]\frac{Z_p}{Z_p + Z_s}$$

Thus

$$L(s) = \frac{1 + R_2/R_1}{3 + sCR + 1/sCR} \tag{14.61}$$

Substituting $s = j\omega$ results in

$$L(j\omega) = \frac{1 + R_2/R_1}{3 + j(\omega CR - 1/\omega CR)} \tag{14.62}$$

The loop gain will be a real number (that is, the phase will be zero) at one frequency given by

$$\omega_0 CR = \frac{1}{\omega_0 CR}$$

That is,

$$\omega_0 = 1/CR \tag{14.63}$$

To obtain sustained oscillations at this frequency, one should set the magnitude of the loop gain to unity. This can be achieved by selecting

$$\frac{R_2}{R_1} = 2 \tag{14.64}$$

To ensure that oscillations will start, one chooses R_2/R_1 slightly greater than 2. The reader can easily verify that if $R_2/R_1 = 2 + \delta$, where δ is a small number, the roots of the characteristic equation $1 - L(s) = 0$ will be in the right half of the s plane.

The amplitude of oscillation can be determined and stabilized by using a nonlinear control network. Two different implementations of the amplitude control are shown in Figs. 14.35 and 14.36. The circuit in Fig. 14.35 employs a symmetrical feedback

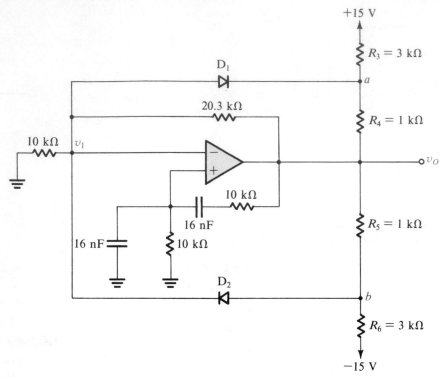

Fig. 14.35 A Wien-bridge oscillator with a limiter used for amplitude control.

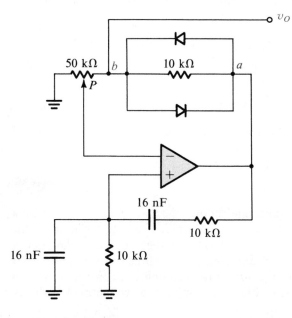

Fig. 14.36 A Wien-bridge oscillator with an alternative method for amplitude stabilization.

limiter (see Section 5.9) formed by diodes D_1 and D_2 together with resistors R_3, R_4, R_5, and R_6. The limiter operates in the following manner: At the positive peak of the output voltage v_O, the voltage at node b will exceed the voltage v_1 (which is about $\frac{1}{3}v_O$), and diode D_2 conducts. This will clamp the positive peak to a value determined by R_5, R_6, and the negative power supply. The value of the positive output peak can be calculated by writing a node equation at node b and neglecting the current through D_2. Similarly, the negative peak of the output sine wave will be clamped to the value that causes diode D_1 to conduct. The value of the negative peak can be determined from the node equation for node a while neglecting the current through D_1. Finally, note that in order to obtain a symmetrical output waveform, R_3 is chosen equal to R_6 and R_4 equal to R_5.

Exercise

14.22 For the circuit in Fig. 14.35,
(a) Disregard the limiter circuit and find the location of the closed-loop poles.
(b) Find the frequency of oscillation.
(c) Find the amplitude of the output sine wave (assume that the diode drop is 0.7 V).

Ans. $(10^5/16)(0.015 \pm j)$; 1 kHz; 21.36 V (peak-to-peak)

The circuit of Fig. 14.36 employs an inexpensive implementation of the parameter variation mechanism of amplitude control. Potentiometer P is adjusted until oscillations just start to grow. As the oscillations grow, the diodes start to conduct, causing the effective resistance between a and b to decrease. Equilibrium will be reached at the output amplitude that causes the loop gain to be exactly unity. The output amplitude can be varied by adjusting potentiometer P.

As indicated in Fig. 14.36, the output is taken at point b rather than at the op-amp output terminal because the signal at b has lower distortion than that at a (why?). Node b, however, is a high-impedance node, and a buffer will be needed if a load is to be connected.

Exercise

14.23 For the circuit in Fig. 14.36 find the following.
(a) The setting of potentiometer P at which oscillations start.
(b) The frequency of oscillation.

Ans.   **(a)** 20 kΩ to ground; **(b)** 1 kHz

The Phase-Shift Oscillator

The basic structure of the phase-shift oscillator is shown in Fig. 14.37. It consists of a negative-gain amplifier $(-K)$ with a three-section (third-order) RC ladder network in the feedback. The circuit will oscillate at the frequency for which the phase shift of the RC network is 180°. Only at this frequency will the total phase shift around the loop be 0 or 360°. Here we should note that the reason for using a three-section RC network is that three is the minimum number of sections (that is, lowest order) that is capable of producing a 180° phase shift at a finite frequency.

For oscillations to be sustained, the value of K should be equal to the inverse of the magnitude of the RC network transfer function at the frequency of oscillation.

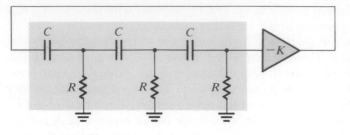

Fig. 14.37 Phase-shift oscillator.

However, in order to ensure that oscillations start, the value of K has to be chosen slightly higher than the value that satisfies the unity-loop-gain condition. Oscillations will then grow in magnitude until limited by some nonlinear control mechanism.

Figure 14.38 shows a practical phase-shift oscillator with a feedback limiter, consisting of diodes D_1 and D_2 and resistors R_1, R_2, R_3, and R_4 for amplitude stabilization. To start oscillations, R_f has to be made slightly greater than the minimum required value. Although the circuit stabilizes more rapidly, and provides sine waves with more stable amplitude, if R_f is made much larger than this minimum, the price paid is an increased output distortion.

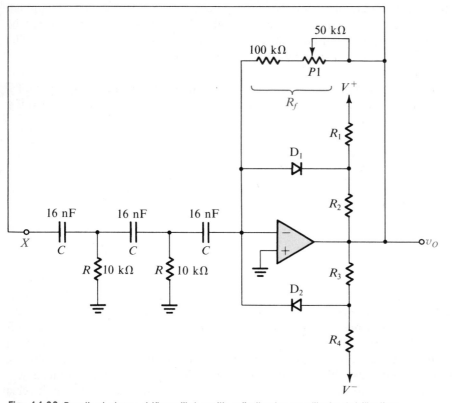

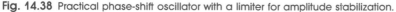

Fig. 14.38 Practical phase-shift oscillator with a limiter for amplitude stabilization.

Exercises

14.24 Consider the circuit of Fig. 14.38 *without* the limiter. Break the feedback loop at X and find the loop gain $A\beta \equiv V_o(j\omega)/V_x(j\omega)$. To do this it is easier to start at the output and work backward, finding the various currents and voltages, and eventually V_x in terms of V_o.

Ans. $\dfrac{\omega^2 C^2 R R_f}{4 + j(3\omega CR - 1/\omega CR)}$

14.25 Use the expression derived in Exercise 14.24 to find the frequency of oscillation f_0 and the minimum required value of R_f for oscillations to start in the circuit of Fig. 14.38.

Ans. 574.3 Hz; 120 kΩ

The Quadrature Oscillator

The *quadrature oscillator* is based on the two-integrator loop studied in Section 14.6. As an active filter the loop is damped so as to locate the poles in the left half of the s plane. Here no such damping will be used, since we wish to locate the poles on the $j\omega$ axis in order to provide sustained oscillations. In fact, to ensure that oscillations start, the poles are initially located in the right half-plane and then "pulled back" by the nonlinear gain control.

Figure 14.39 shows a practical quadrature oscillator. Amplifier 1 is connected as an inverting Miller integrator with a limiter in the feedback for amplitude control. Amplifier 2 is connected as a noninverting integrator (thus replacing the cascade

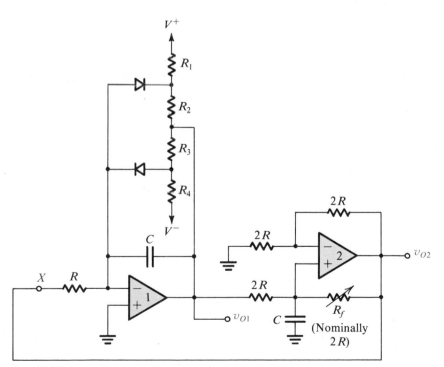

Fig. 14.39 A quadrature oscillator circuit.

connection of the Miller integrator and the inverter in the two-integrator loop of Fig. 14.17b). This noninverting integrator circuit is studied in Example 3.5.

The resistance R_f in the positive-feedback path of op amp 2 is made variable, with a nominal value of $2R$. Decreasing the value of R_f moves the poles to the right half-plane (Problem 14.46) and ensures that the oscillations start. Too much positive feedback, although it results in better amplitude stability, also results in higher output distortion (because the limiter has to operate "harder"). In this regard, note that the output v_{O2} will be "purer" than v_{O1} because of the filtering action provided by the second integrator on the peak-limited output of the first integrator.

If we disregard the limiter and break the loop at X, the loop gain can be obtained as

$$L(s) \equiv \frac{V_{o2}}{V_x} = -\frac{1}{s^2 C^2 R^2} \tag{14.65}$$

Thus the loop will oscillate at frequency ω_0, given by

$$\omega_0 = \frac{1}{CR} \tag{14.66}$$

Finally, it should be pointed out that the name *quadrature oscillator* is used because the circuit provides two sinusoids with 90° phase difference. This should be obvious, since v_{O2} is the integral of v_{O1}. There are many applications for which quadrature sinusoids are required.

The Active-Filter Tuned Oscillator

The last oscillator circuit that we shall discuss is quite simple both in principle and in design. Nevertheless, the approach is general and versatile and can result in high-quality (that is, low-distortion) output sine waves. The basic principle is illustrated in Fig. 14.40. The circuit consists of a high-Q bandpass filter connected in a positive-feedback loop with a hard limiter. To understand how this circuit works, assume

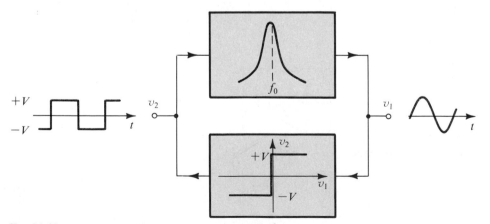

Fig. 14.40 Block diagram of the active-filter tuned oscillator.

that oscillations have already started. The output of the bandpass filter will be a sine wave whose frequency is equal to the center frequency of the filter, f_0. The sine-wave signal v_1 is fed to the limiter, which produces at its output a square wave whose levels are determined by the limiting levels, and whose frequency is f_0. The square wave in turn is fed to the bandpass filter, which filters out the harmonics and provides a sinusoidal output v_1 at the fundamental frequency f_0. Obviously, the purity of the output sine wave will be a direct function of the selectivity (or Q factor) of the bandpass filter.

The simplicity of this approach to oscillator design should be apparent. We have independent control of frequency and amplitude as well as of distortion of the output sinusoid. Any filter circuit with positive gain can be used to implement the bandpass filter. The frequency stability of the oscillator will be directly determined by the frequency stability of the bandpass-filter circuit. Also, a variety of limiter circuits (see Chapter 5) with different degrees of sophistication can be used to implement the limiter block.

Figure 14.41 shows one possible implementation of the active-filter tuned oscillator. This circuit uses the GIC-based bandpass filter studied in Section 14.6. The limiter used is a very simple one consisting of a resistance R_1 and two diodes.

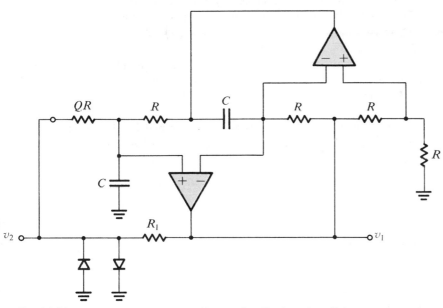

Fig. 14.41 Practical implementation of the active-filter tuned oscillator.

Exercise

14.26 Use $C = 16$ nF and find the value of R such that the circuit of Fig. 14.41 produces 1-kHz sine waves. If the diode drop is 0.7 V, find the peak-to-peak amplitude of the output sine wave. (*Hint:* A square wave with peak-to-peak amplitude of V volts has a fundamental component with $4V/\pi$ volts peak-to-peak amplitude.)

Ans. 10 kΩ; 3.6 V

A Final Remark

The op-amp–RC oscillator circuits studied are useful for operation in the range 10 Hz to 100 kHz (or perhaps 1 MHz at most). Whereas the lower frequency limit is dictated by the size of passive components required, the upper limit is governed by the frequency-response and slew-rate limitations of op amps. For higher frequencies, circuits that employ transistors together with LC tuned circuits or crystals are frequently used.[6] These are discussed in the next section.

14.10 LC AND CRYSTAL OSCILLATORS

Oscillators utilizing transistors (FETs or BJTs) and LC tuned circuits or crystals as feedback elements, are used in the frequency range of 100 kHz to hundreds of MHz. They exhibit higher Q than the RC types. However, LC oscillators are difficult to tune over wide ranges, and crystal oscillators operate at a single frequency.

LC Tuned Oscillators

Figure 14.42 shows two commonly used configurations of LC tuned oscillators. They are known as (a) the Colpitts oscillator and (b) the Hartley oscillator. Both utilize a parallel LC circuit connected between collector and base (or between drain and gate if an FET is used) with a fraction of the tuned-circuit voltage fed to the emitter (the source in a FET). This feedback is achieved via a capacitive divider in the Colpitts oscillator and an inductive divider in the Hartley circuit. Note that the bias details are not shown in order to focus attention on the oscillator structure.

If the frequency of operation is sufficiently low that we can neglect the transistor capacitances, the frequency of oscillation will be determined by the resonance frequency of the parallel-tuned circuit (also known as a *tank circuit* because it behaves

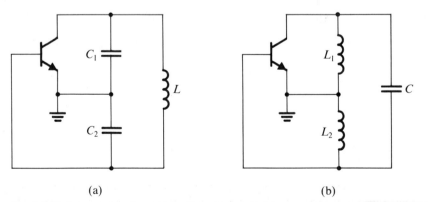

(a)   (b)

Fig. 14.42 Two commonly used configurations of LC-tuned oscillators: **(a)** Colpitts; **(b)** Hartley.

[6] Of course, transistors can be used in place of the op amps in the circuits just studied. At higher frequencies, however, better results are obtained with LC tuned circuits and crystals.

as a reservoir for energy storage). Thus, for the Colpitts oscillator we have

$$\omega_0 = 1 \Big/ \sqrt{L\left(\frac{C_1 C_2}{C_1 + C_2}\right)} \tag{14.67}$$

and for the Hartley oscillator

$$\omega_0 = 1/\sqrt{(L_1 + L_2)C} \tag{14.68}$$

The divider ratio determines the feedback factor and thus must be adjusted in conjunction with the transistor gain to ensure that oscillations will start. To determine the oscillation condition of the Colpitts oscillator we replace the transistor with its low-frequency equivalent circuit, as shown in Fig. 14.43. To find the loop gain we break the loop at the transistor base, apply an input voltage V_π and find the returned voltage that appears across the input terminals of the transistor. We then equate the loop gain to unity. An alternative approach is to analyze the circuit and eliminate all current and voltage variables and thus obtain one equation that governs circuit operation. Oscillations will start if this equation is satisfied. In other words, the resulting equation will give us the conditions for oscillation.

Fig. 14.43 Equivalent circuit of the Colpitts oscillator of Fig. 14.42a at low frequencies.

Analysis of the circuit provides

$$\left(\frac{1}{r_\pi} + sC_2\right)V_\pi + g_m V_\pi + sC_1 V_\pi\left[1 + sL\left(sC_2 + \frac{1}{r_\pi}\right)\right] = 0$$

which can be arranged in the form

$$s^3 L C_1 C_2 + s^2\left(\frac{LC_1}{r_\pi}\right) + s(C_1 + C_2) + \left(g_m + \frac{1}{r_\pi}\right) = 0 \tag{14.69}$$

Substituting $s = j\omega$ gives

$$\left(g_m + \frac{1}{r_\pi} - \frac{\omega^2 L C_1}{r_\pi}\right) + j[\omega(C_1 + C_2) - \omega^3 L C_1 C_2] = 0 \tag{14.70}$$

For oscillations to start, both the real and imaginary parts must be zero. Equating the imaginary part to zero gives the frequency of oscillation as

$$\omega_0 = 1 \Big/ \sqrt{L\left(\frac{C_1 C_2}{C_1 + C_2}\right)} \tag{14.71}$$

which is the resonance frequency of the tank circuit, as anticipated. Equating the real part to zero together with using Eq. (14.71) and substituting $g_m r_\pi = \beta_0$ gives

$$\frac{C_1}{C_2} = \beta_0 \qquad (14.72)$$

This is a limiting condition and is usually stated as

$$\frac{C_1}{C_2} \leq \beta_0$$

to ensure that the loop gain at ω_0 is at least unity. As oscillations grow in amplitude, the transistor nonlinear characteristics reduce the loop gain to unity, thus sustaining the oscillations.

Analysis similar to that above can be carried out for the Hartley circuit. At high frequencies, the transistor internal capacitances can no longer be ignored and the analysis must use the complete hybrid-π model. Alternatively, the y parameters of the BJT can be measured at the intended frequency ω_0, and the analysis carried out using the y-parameter model (see Appendix B). This is usually simpler and more accurate, especially at frequencies above about 30% of the transistor f_T.

As an example of a practical LC oscillator we show in Fig. 14.44 the circuit of a Colpitts oscillator, complete with bias details. Here the radio-frequency choke (RFC) provides a high reactance at ω_0 but a low dc resistance.

Finally, a few words are in order on the mechanism that determines the amplitude of oscillations in the LC-tuned oscillators discussed above. Unlike the op-amp oscillators which incorporate special amplitude-control circuitry, LC-tuned oscillators utilize the nonlinear $i_C - v_{BE}$ characteristics of the BJT for amplitude control,

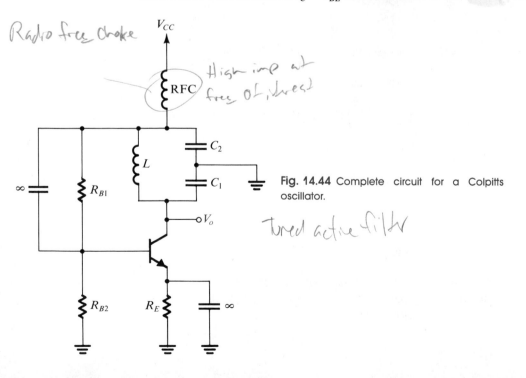

Fig. 14.44 Complete circuit for a Colpitts oscillator.

and are thus known as *self-limiting* oscillators. Specifically, as the oscillations grow in amplitude, the effective gain of the transistor is reduced below its small-signal value. Eventually, an amplitude is reached at which the effective gain is reduced to the point that the Barkhausen criterion is satisfied exactly. The amplitude then remains constant at this value.

Reliance on the nonlinear characteristics of the BJT implies that the collector current waveform will be nonlinearly distorted. Nevertheless, the output voltage signal will still be a sinusoid of high purity because of the filtering action of the LC-tuned circuit. Detailed analysis of amplitude control makes use of nonlinear circuit techniques and is beyond the scope of this book. The interested reader is referred to Clarke and Hess, 1971.

Exercises

14.27 Show that for the Hartley oscillator of Fig. 14.42b the frequency of oscillation is given by Eq. (14.68) and that for oscillations to start $L_2/L_1 = \beta_0$.

14.28 Using a BJT specified to have a minimum β of 50 and an inductor of 100 μH, design a Colpitts oscillator to operate at 1 Mrad/s. Find C_1 and C_2 so that oscillations are just guaranteed to start.

Ans. 0.5 μF; 0.01 μF

Crystal Oscillators

A piezoelectric crystal, such as quartz, exhibits electromechanical-resonance characteristics that are very stable (with time and temperature) and highly selective (having very-high Q factors). The circuit symbol of a crystal is shown in Fig. 14.45a, and the equivalent circuit model is given in Fig. 14.45b. The resonance properties are characterized by a large inductance L (as high as hundreds of henrys), a very small series

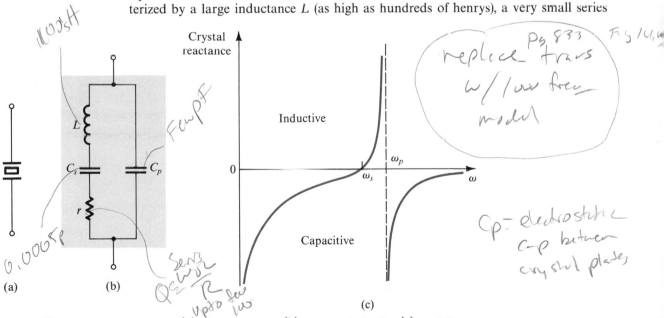

Fig. 14.45 A piezoelectric crystal: **(a)** circuit symbol; **(b)** equivalent circuit; **(c)** crystal reactance versus frequencies [note that, neglecting the small resistance r, $Z_{\text{crystal}} = jX(\omega)$].

capacitance C_s (as small as 0.0005 pF), a series resistance r representing a Q factor $\omega_0 L/r$ that can be as high as a few hundred thousand, and a parallel capacitance C_p (a few picofarads). Capacitor C_p represents the electrostatic capacitance between the two parallel plates of the crystal. Note that $C_p \gg C_s$.

Since the Q factor is very high, we may neglect the resistance r and express the crystal impedance as

$$Z(s) = 1 \left/ \left[sC_p + \frac{1}{sL + 1/sC_s} \right] \right.$$

which can be manipulated to the form

$$Z(s) = \frac{1}{sC_p} \frac{s^2 + (1/LC_s)}{s^2 + [(C_p + C_s)/LC_sC_p]} \tag{14.73}$$

From Eq. (14.73) and from Fig. 14.45b we see that the crystal has two resonance frequencies: a series resonance at ω_s,

$$\omega_s = \frac{1}{\sqrt{LC_s}} \tag{14.74}$$

and a parallel resonance at ω_p,

$$\omega_p = 1 \left/ \sqrt{L \left(\frac{C_sC_p}{C_s + C_p} \right)} \right. \tag{14.75}$$

Thus for $s = j\omega$ we can write

$$Z(j\omega) = -j \frac{1}{\omega C_p} \left(\frac{\omega^2 - \omega_s^2}{\omega^2 - \omega_p^2} \right) \tag{14.76}$$

From Eqs. (14.74) and (14.75) we note that $\omega_p > \omega_s$. However, since $C_p \gg C_s$, the two resonance frequencies are very close. Expressing $Z(j\omega) = jX(\omega)$, the crystal reactance $X(\omega)$ will have the shape shown in Fig. 14.45c. We observe that the crystal reactance is inductive over the very narrow frequency band between ω_s and ω_p. For a given crystal, this frequency band is well defined. Thus, we may use the crystal to replace the inductor of the Colpitts oscillator (Fig. 14.42). The resulting circuit will oscillate at the resonance frequency of the crystal inductance L with the series equivalent of C_s and

$$\left(C_p + \frac{C_1C_2}{C_1 + C_2} \right)$$

Since C_s is much smaller than the three other capacitances, it will be dominant and

$$\omega_0 \simeq \frac{1}{\sqrt{LC_s}} = \omega_s.$$

In addition to the basic Colpitts oscillator, a variety of configurations exist for crystal oscillators. Figure 14.46 shows a popular configuration (called the Pierce oscillator) utilizing a CMOS inverter (see Chapter 15) as amplifier. Resistor R_f determines a dc operating point in the high-gain region of the CMOS inverter. Note that this circuit also is based on the Colpitts configuration.

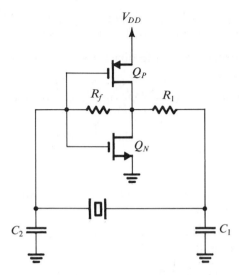

Fig. 14.46 A Colpitts (or Pierce) crystal oscillator utilizing a CMOS inverter as an amplifier.

The extremely stable resonance characteristics and the very high Q factors of quartz crystals result in oscillators with very accurate and stable frequencies. Crystals are available with resonance frequencies in the range of few kHz to hundreds of MHz. Temperature coefficients of ω_0 of 1 or 2 parts per million (ppm)/°C are achievable. Unfortunately, however, crystal oscillators are fixed-frequency circuits.

Exercise

14.29 A 2-MHz quartz crystal is specified to have $L = 0.52$ H, $C_s = 0.012$ pF, $C_p = 4$ pF, and $r = 120\ \Omega$. Find f_s, f_p, and Q.

Ans. 2.015 MHz; 2.018 MHz; 55,000

14.11 SUMMARY

● Active-RC filters utilize feedback to shift the poles of the RC network from their natural locations on the negative real axis (of the s plane) to complex conjugate locations, thereby realizing selective responses.

● The poles of a second-order filter are characterized by the pole frequency (ω_0) and the pole-Q (Q).

● The numerator coefficients of the second-order transfer function determine the filter type (that is, low pass, bandpass, etc.).

● The 3-dB bandwidth of a second-order bandpass filter is ω_0/Q.

● The parallel RLC resonator has $\omega_0 = 1/\sqrt{LC}$ and $Q = \omega_0 CR$.

● Placing a bridged-T RC network in the negative-feedback path of an infinite-gain op amp results in an active filter with poles identical to the transmission zeros of the bridged-T network.

- In a linear circuit, if the connection to the input voltage source is interchanged with ground, the complement of the transfer function (that is, $1 - t$) is realized.

- Application of the complementary transformation to a feedback loop yields another loop with the same transfer-function poles.

- The sensitivity function is utilized to evaluate the effects of changes in component values on filter response.

- The single-amplifier filters studied have low sensitivities relative to the passive components (that is, the resistors and the capacitors). However, the pole Q can be quite sensitive to the amplifier gain.

- An excellent second-order filter circuit is based on using the generalized impedance converter (GIC) to simulate an inductance.

- A popular and versatile second-order filter realization employs two integrators, one inverting and one noninverting, in a feedback loop.

- Switched-capacitor filters can be fabricated on a monolithic integrated circuit using either NMOS or CMOS technology.

- Switched-capacitor filters are based on the fact that a capacitor, C, switched between two circuit nodes at a high rate, f_c, is equivalent to a resistance $R = 1/Cf_c$, connecting the two circuit nodes.

- Tuned amplifiers utilize LC tuned circuits as loads, or at the input, of transistor amplifiers.

- The inductor power loss can be modeled by either a series resistance r_s or a parallel resistance R_p. The Q factor is given by $Q_0 = \omega_0 L/r_s = R_p/\omega_0 L$.

- The cascode and the CC-CB cascade configurations are frequently used in the design of tuned amplifiers.

- Stagger tuning results in a tuned amplifier with a flatter passband response (as compared to that obtained with synchronous tuning).

- An oscillator can be realized by placing a frequency-selective network in the feedback path of an amplifier (an op amp or a transistor). The circuit will oscillate at the frequency at which the total phase shift around the loop is zero, provided that at this frequency the magnitude of loop gain is equal to, or greater than, unity.

- If in an oscillator the magnitude of loop gain is greater than unity, the amplitude will increase until a nonlinear amplitude-control mechanism is activated.

- The Wien-bridge oscillator, the phase-shift oscillator, the quadrature oscillator, and the active-filter-tuned oscillator, are popular configurations for frequencies up to about 1 MHz. These circuits employ RC networks together with op amps or transistors. For higher frequencies, LC-tuned or crystal-tuned oscillators are utilized. A popular configuration is the Colpitts circuit.

- Crystal oscillators provide the highest possible frequency accuracy and stability.

BIBLIOGRAPHY

P. E. Allen and E. Sanchez-Sinencio, *Switched-Capacitor Circuits*, New York: Van Nostrand Reinhold, 1984

R. W. Brodersen, P. R. Gray, and D. A. Hodges, "MOS switched-capacitor filters," *Proceedings of the IEEE*, vol. 67, pp. 61–74, Jan. 1979.

K. K. Clarke and D. T. Hess, *Communication Circuits: Analysis and Design*, Chap. 6, Reading, Mass.: Addison Wesley, 1971.

M. S. Ghausi, *Electronic Devices and Circuits: Discrete and Integrated*, New York: Holt, Rinehart and Winston, 1985.

M. S. Ghausi and K. Laker, *Modern Filter Design*, Chap. 6, Englewood Cliffs, N.J.: Prentice-Hall, 1981.

J. G. Graeme, G. E. Tobey, and L. P. Huelsman, *Operational Amplifiers: Design and Applications*, New York: McGraw-Hill, 1971.

P. E. Gray and C. L. Searle, *Electronic Principles*, Chap. 17, New York: Wiley, 1969.

A. B. Grebene, *Bipolar and MOS Analog Integrated Circuit Design*, New York: Wiley, 1984.

W. Jung, *IC Op Amp Cookbook*, Indianapolis, Ind.: Howard Sams, 1974.

K. Martin, "Improved circuits for the realization of switched-capacitor filters," *IEEE Transactions on Circuits and Systems*, vol. CAS-27, no. 4, pp. 237–244, April 1980.

J. K. Roberge, *Operational Amplifiers: Theory and Practice*, New York: Wiley, 1975.

R. Schaumann, M. Soderstrand, and K. Laker (Eds.), *Modern Active Filter Design*, New York: IEEE Press, 1981.

A. S. Sedra, "Switched-Capacitor Filter Synthesis," in *MOS VLSI Circuits for Telecommunications*, Y. Tsividis and P. Antognetti, (Eds.), Englewood Cliffs, N.J.,: Prentice-Hall, 1985.

A. S. Sedra and P. O. Brackett, *Filter Theory and Design: Active and Passive*, Portland, Ore.: Matrix, 1978.

A. S. Sedra, M. Ghorab, and K. Martin, "Optimum configurations for single-amplifier biquadratic filters", *IEEE Transactions on Circuits and Systems*, vol. CAS-27, no. 12, pp. 1155–1163, Dec. 1980.

J. I. Smith, *Modern Operational Circuit Design*, New York: Wiley-Interscience, 1971.

L. Strauss, *Wave Generation and Shaping*, 2nd ed., New York: McGraw-Hill, 1970.

M. E. Van Valkenburg, *Analog Filter Design*, New York: Holt, Rinehart and Winston, 1981.

J. V. Wait, L. P. Huelsman, and G. A. Korn, *Introduction to Operational Amplifier Theory and Applications*, New York: McGraw-Hill, 1975.

PROBLEMS

14.1 The third-order filter of Fig. 14.3 has $\omega_{3\text{dB}} = 1$ rad/s. The 3-dB frequency can be shifted to $\omega_{3\text{dB}} = \omega_H$, by dividing all inductances and capacitances by ω_H, a process known as *frequency denormalization*. Also, if the network is to be terminated in source and load resistances of R ohms (instead of 1 Ω), all inductances are multiplied by R and all capacitances are divided by R, a process known as *impedance denormalization*. Use the circuit of Fig. 14.3 to design a third-order low-pass filter having the same response shape as in Fig. 14.1 but with $\omega_{3\text{dB}} = 10^4$ rad/s and terminated in source and load resistances of 10 kΩ.

14.2 A Butterworth low-pass filter of order N has the magnitude response $|T(j\omega)| = 1/\sqrt{1 + \omega^{2N}}$. Plot on Fig. 14.1 the transmission of a fifth-order filter. Note that its passband is flatter than that of the third-order filter. Also note the increased selectivity obtained with a fifth-order filter.

14.3 Consider the biquad transfer function represented by $T(s)$ in Eq. (14.1). Indicate by a simple sketch, the location of the poles corresponding to each of the following. Also, state the pole values.
(a) $\omega_0 = 100$ rad/s, $Q = \infty$

(b) $\omega_0 = 50$ rad/s, $Q = 1$
(c) $\omega_0 = 10$ rad/s, $Q = 0.5$
(d) $\omega_0 = 100$ rad/s, $Q = 0.1$

14.4 Write the transfer function of a second-order low-pass filter having unity dc gain, a pole frequency of 10 rad/s, and a maximally flat magnitude response.

14.5 Repeat Problem 14.4 for a high-pass filter with unity high-frequency gain.

*__14.6__ Substituting $s = j\omega$ in Eq. (14.4), show that the two frequencies ω_1, ω_2 at which the magnitude of the transfer function drops by 3 dB below the value at ω_0 are given by

$$\omega_{1,2} = \sqrt{\omega_0^2 + \left(\frac{\omega_0}{2Q}\right)^2} \pm \frac{\omega_0}{2Q}$$

Hence show that

$$\omega_1 - \omega_2 = \frac{\omega_0}{Q} \qquad \text{and} \qquad \omega_1\omega_2 = \omega_0^2$$

14.7 Write the transfer function of a second-order notch filter for which the dc gain is unity, the pole frequency is 10 rad/s, the pole Q is 0.5, and the transmission is zero at 100 rad/s.

*__14.8__ For the second-order all-pass filter described by Eq. (14.6) show that the phase shift ϕ is given by

$$\phi = -2 \tan^{-1}\left[\frac{\omega\omega_0}{Q(\omega_0^2 - \omega^2)}\right]$$

For $\omega_0 = 1$ rad/s and $Q = 1$, sketch ϕ versus ω. Find the frequencies at which the phase shift is $-90°$, $-180°$, and $-270°$.

14.9 Consider a parallel RLC circuit fed with a constant-current source I_s. Find the voltage $V(s)$ across the circuit and express $V(s)/I(s)$ in the form of Eq. (14.4) Hence show that $\omega_0 = 1/\sqrt{LC}$ and $Q = \omega_0 CR = R/\sqrt{L/C}$.

14.10 Consider the notch circuit of Fig. 14.6d. For what ratio of C_1 and C_2 does the notch occur at $1.1\omega_0$? For this case, what is the magnitude of the transmission at frequencies $\ll \omega_0$? at frequencies $\gg \omega_0$?

14.11 Consider the connection of the bridged-T network of Fig. 14.8b in the negative-feedback path of an infinite-gain op amp. Let $R_1 = R_2 = R$, $C_4 = C$, and $C_3 = C/m$. Find m and CR so that the closed loop has a pair of complex conjugate poles characterized by ω_0 and Q.

14.12 Design the circuit of Fig. 14.9 to realize a pair of poles with $\omega_0 = 10^4$ rad/s and $Q = 1/\sqrt{2}$. Use $C_1 = C_2 = 1$ nF.

14.13 Consider the bridged-T network of Fig. 14.8a with $R_1 = R_2 = R$ and $C_1 = C_2 = C$, and denote $CR = \tau$. Find the zeros and poles of the bridged-T network. If the network is placed in the negative-feedback path of an infinite-gain op amp, as in Fig. 14.9, find the poles of the closed-loop amplifier.

**__14.14__ Consider the bridged-T network of Fig. 14.8b with $R_1 = R_2 = R$, $C_4 = C$, and $C_3 = C/16$. Let the network be placed in the negative-feedback path of an infinite-gain op amp and let C_4 be disconnected from ground and connected to the input signal source V_i. Analyze the resulting circuit to determine its transfer function $V_o(s)/V_i(s)$, where $V_o(s)$ is the voltage at the op-amp output. Show that the filter realized is a bandpass and find its ω_0, Q, and center-frequency gain.

__14.15__ Consider the bandpass circuit shown in Fig. 14.10. Let $C_1 = C_2 = C$, $R_3 = R$, $R_4 = R/4Q^2$, $CR = 2Q/\omega_0$, and $\alpha = 1$. Disconnect the positive input terminal of the op amp from ground and apply V_i through a voltage divider R_1, R_2 to the positive input terminal. Analyze the circuit to find its transfer function V_o/V_i. Find the voltage-divider ratio $R_2/(R_1 + R_2)$ so that the circuit realizes **(a) an all-pass function and **(b)** a notch function. Assume the op amp to be ideal.

*__14.16__ Derive the transfer function of the circuit in Fig. 14.13b assuming the op amp to be ideal. Thus show that the circuit realizes a high-pass function. What is the high-frequency gain of the circuit? Design the circuit for a maximally flat response with a 3-dB frequency of 10^3 rad/s. Use $C_1 = C_2 = 10$ nF. (*Hint:* For a maximally flat response, $Q = 1/\sqrt{2}$ and $\omega_{3dB} = \omega_0$.)

*__14.17__ This is a continuation of Problem 14.11 and Exercise 14.10. Design the resulting low-pass circuit to obtain a dc gain of unity and maximally flat response with a 3-dB frequency of 10 kHz. Use $R_1 = R_2 = 10$ kΩ. (*Hint:* For a maximally flat response, $Q = 1/\sqrt{2}$ and $\omega_{3dB} = \omega_0$.)

14.18 The process of obtaining the complement of a transfer function by interchanging input and ground, as illustrated in Fig. 14.11, applies to any general network (not just RC networks as shown). Show that if the network n is a bandpass with center-frequency gain of unity then the complement obtained is a notch. Verify this by using the RLC circuits of Figs. 14.6c and d (with C_2 in the latter set to zero).

14.19 Evaluate the sensitivities of ω_0 and Q relative to R, L, and C of the bandpass circuit in Fig. 14.6c.

*14.20 Verify the following sensitivity identities:
(a) If $y = uv$, then $S_x^y = S_x^u + S_x^v$.
(b) If $y = u/v$, then $S_x^y = S_x^u - S_x^v$.
(c) If $y = ku$, where k is a constant, then $S_x^y = S_x^u$.
(d) If $y = u^n$, where n is a constant, then $S_x^y = nS_x^u$.
(e) If $y = f_1(u)$ and $u = f_2(x)$, then $S_x^y = S_u^y \cdot S_x^u$.

14.21 For the high-pass filter of Fig. 14.13b, what are the sensitivities of ω_0 and Q to amplifier gain A?

*14.22 For the feedback loop generated by placing the bridged-T of Fig. 14.8b in the negative-feedback path of an op amp, find the sensitivities of ω_0 and Q relative to each passive component, for the design in which $R_1 = R_2$.

14.23 For the GIC-based bandpass filter of Fig. 14.15, find the sensitivities of ω_0 and Q relative to all passive components. [*Hint:* Utilize Eqs. (14.26) and (14.27).]

14.24 Use the fact illustrated in Fig. 14.11 to show that if $t(s)$ is a bandpass function with a center-frequency gain of 2, then an all-pass function can be obtained by interchanging the input and ground terminals. Apply this process to the two-amplifier bandpass filter of Fig. 14.15a (with $R_1 = R_3 = R_4 = R_5 = R$, $C_2 = C_6 = C$, and $R_7 = QR$). Sketch the circuit of the resulting all-pass filter.

14.25 Use the circuit of Fig. 14.16b to realize a bandpass filter with $f_0 = 10$ kHz and $Q = 50$. Utilize $R = 10$ kΩ and find suitable values for C, R_1, R_2, and R_3. What is the value of the center-frequency gain?

14.26 Use the circuit of Fig. 14.17b to realize a low-pass filter with $f_0 = 10$ kHz, $Q = 1/\sqrt{2}$, and unity low-frequency gain. If $R = 10$ kΩ, give the values of C, R_d, and R_g.

14.27 Consider the circuit of Fig. 14.16b augmented as follows: The outputs V_{1p}, V_{bp}, and V_{hp} are fed to a summing op amp through resistors R_4, R_5, and R_6, respectively. The feedback resistance of the summing op amp is R_7. Derive an expression for the transfer function from the circuit input to the output of the summing op amp. Show that to obtain a realization for the notch function of Eq. (14.5) the following resistance values may be used: R_6 = arbitrary, $R_5 = \infty$, $R_4 = R_6(\omega_0/\omega_n)^2$, and $R_7 = R_6$ $[n_2/(2 - 1/Q)]$. Note that the high-pass function realized in the circuit of Fig. 14.16b has a high-frequency gain of $(2 - 1/Q)$.

14.28 For the switched-capacitor input circuit of Fig. 14.18b, in which a clock frequency of 100 kHz is used, what input resistances correspond to C_1 capacitance values of 1 pF and 10 pF?

14.29 For a dc voltage of 1 V applied to the input of the circuit of Fig. 14.18b, in which the input capacitance is 1 pF, what charge is transferred for each cycle of the two-phase clock? For a 100-kHz clock, what is the average current drawn from the input source? For a feedback capacitance of 10 pF, what change would you expect in the output for each cycle of the clock? For an amplifier that saturates at ± 10 V and the feedback capacitor initially discharged, how many clock cycles would it take to saturate the amplifier? What is the average slope of the staircase output voltage produced?

14.30 Repeat Exercise 14.15 for a clock frequency of 400 kHz.

14.31 Repeat Exercise 14.15 for $Q = 40$.

14.32 Design the circuit of Fig. 14.20b to realize, at the output of the second (noninverting) integrator, a maximally flat low-pass function with $\omega_{3dB} = 10^4$ rad/s and unity dc gain. Use a clock frequency $f_c = 100$ kHz and select $C_1 = C_2 = 10$ pF. Give the values of C_3, C_4, C_5 and C_6. (*Hint:* For a maximally flat response, $Q = 1/\sqrt{2}$ and $\omega_{3dB} = \omega_0$.)

*14.33 A voltage signal source with a resistance $R_s = 10$ kΩ is connected to the input of a common-emitter BJT amplifier. Between base and emitter is connected a tuned circuit with $L = 1$ μH and $C = 200$ pF. The transistor is biased at 1 mA and has $\beta = 200$, $C_\pi = 10$ pF, and $C_\mu = 1$ pF. The transistor load is a resistance of 5 kΩ. Find ω_0, Q, the 3-dB bandwidth, and the center-frequency gain of this single-tuned amplifier.

14.34 A coil having an inductance of 10 μH is intended for applications around 1-MHz frequency. Its Q is specified to be 200. Find the equivalent parallel resistance R_p. What is the value of the capacitor required to produce resonance at 1 MHz? What additional parallel resistance is required to produce a 3-dB bandwidth of 10 kHz?

14.35 An inductance of 36 μH is resonated with a 1,000-pF capacitor. If the inductor is tapped at one-third of its turns and a 1-kΩ resistor is connected across the third of the coil turns, find f_0 and Q of the resonator.

*14.36 Consider a common-emitter transistor amplifier loaded with an inductance L. Ignoring r_o and r_x, show that for $\omega C_\mu \ll 1/\omega L$, the input admittance is given by

$$Y_{in} \simeq \left(\frac{1}{r_\pi} - \omega^2 C_\mu L g_m \right) + j\omega(C_\pi + C_\mu)$$

Note: The real part of the input admittance can be negative. This can lead to oscillations.

**14.37 (a) Substituting $s = j\omega$ in Eq. (14.4), find $|T(j\omega)|$. For ω in the vicinity of ω_0 [that is, $\omega = \omega_0 + \delta\omega = \omega_0(1 + \delta\omega/\omega_0)$, where $\delta\omega/\omega_0 \ll 1$ so that $\omega^2 \simeq \omega_0^2 (1 + 2\delta\omega/\omega_0)$] show that, for $Q \gg 1$.

$$|T(j\omega)| \simeq \frac{|T(j\omega_0)|}{\sqrt{1 + 4Q^2 \left(\dfrac{\delta\omega}{\omega_0}\right)^2}}$$

(b) Use the result obtained in (a) to show that the 3-dB bandwidth B, of N synchronously tuned sections connected in cascade, is

$$B = \frac{\omega_0}{Q} \sqrt{2^{1/N} - 1}$$

***14.38 (a) Using the fact that the second-order bandpass response in the neighborhood of ω_0 is the same as the response of a first-order low-pass with 3-dB frequency of $(\omega_0/2Q)$, show that the bandpass response at $\omega = \omega_0 + \delta\omega$, $\delta\omega \ll \omega_0$, is given by

$$|T(j\omega)| \simeq \frac{|T(j\omega_0)|}{\sqrt{1 + 4Q^2(\delta\omega/\omega_0)^2}}$$

(b) Use the relationship derived in (a) together with Eq. (14.50) to show that a bandpass amplifier with a 3-dB bandwidth B, designed using N synchronously tuned stages, has an overall transfer function given by

$$|T(j\omega)|_{\text{overall}} = \frac{|T(j\omega_0)|_{\text{overall}}}{\left[1 + 4(2^{1/N} - 1)\left(\dfrac{\delta\omega}{B}\right)^2\right]^{N/2}}$$

(c) Use the relationship derived in (b) to find the attenuation (in decibels) obtained at a bandwidth $2B$ for $N = 1$ to 5. Also find the ratio of the 30-dB bandwidth to the 3-dB bandwidth for $N = 1$ to 5.

**14.39 This problem investigates the selectivity of maximally flat stagger-tuned amplifiers derived in the manner illustrated in Fig. 14.31.

(a) The low-pass maximally flat filter of bandwidth $(B/2)$ and order N has the magnitude response

$$|T| = 1 \bigg/ \sqrt{1 + \left(\frac{\Omega}{B/2}\right)^{2N}}$$

where $\Omega = \text{Im}(p)$, is the frequency in the low-pass domain. Use this expression to obtain for the corresponding bandpass filter at $\omega = \omega_0 + \delta\omega$, $\delta\omega \ll \omega_0$,

$$|T| = 1 \bigg/ \sqrt{1 + \left(\frac{\delta\omega}{B/2}\right)^{2N}}$$

(b) Use the transfer function of (a) to find the attenuation (in decibels) obtained at a bandwidth of $2B$ for $N = 1$ to 5. Also find the ratio of the 30-dB bandwidth to the 3-dB bandwidth for $N = 1$ to 5.

**14.40 Consider a sixth-order stagger-tuned bandpass amplifier with center-frequency ω_0 and 3-dB bandwidth B. The poles are to be obtained by shifting those of the third-order maximally flat low-pass filter, given in Fig. 14.3. (Note that you must scale the poles so that the 3-dB frequency is $B/2$, not unity as shown). For the three resonant circuits find ω_0, the 3-dB bandwidth, and Q.

14.41 In a particular oscillator characterized by the structure of Fig. 14.32, the frequency-selective network exhibits a minimum loss of 20 dB and a phase shift of $180°$ at frequency ω_0. What is the minimum gain and the phase shift that the amplifier must have for oscillations to begin?

14.42 For the Wien-bridge oscillator of Fig. 14.34, let the components of Z_s and Z_p be denoted R_s, C_s and R_p, C_p, respectively. Find expressions for ω_0 and R_2/R_1 for oscillation to occur. Apply these expressions in the following special cases.
(a) $C_s = C$, $R_s = R$, $C_p = C/10$, $R_p = 10 R$
(b) $C_s = C/2$, $R_s = R$, $C_p = 2C$, $R_p = R$
(c) $C_s = C$, $R_s = 2R$, $C_p = C$, $R_p = R/2$
(d) $C_s = C$, $R_s = 10R$, $C_p = C$, $R_p = R/10$

*14.43 Reconsider Exercise 14.22 with R_3 and R_6 increased to reduce the output voltage. What values are required for an output of 10 V peak-to-peak? What results if R_3 and R_6 are open-circuited?

**14.44 Redesign the circuit of Fig. 14.36 for operation at 10 kHz using the same values of resistance. If at 10 kHz the op amp provides an excess phase shift of $5.7°$, what will the frequency of oscillation be? To restore operation to 10 kHz what change must be made in the shunt resistor in the Wien bridge?

*14.45 Repeat Exercise 14.24 with a resistor $R = 10$ kΩ connected in series with the rightmost capacitor C. What is the frequency of oscillation and what minimum R_f is required?

14.46 Consider the quadrature-oscillator circuit of Fig. 14.39 without the limiter. Let the resistance R_f be equal to $2R/(1 + \Delta)$ where $\Delta \ll 1$. Show that the poles of the characteristic equation are in the right-half s-plane and given by $s \simeq (1/CR)[(\Delta/4) \pm j]$.

14.47 Assuming that the diode-clipped waveform in Exercise 14.26 is nearly an ideal square wave and that

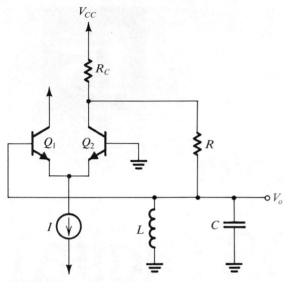

Fig. P14.48

the resonator Q is 20, provide an estimate of the distortion in the output sine wave by calculating the magnitude (relative to the fundamental) of (a) the second harmonic, (b) the third harmonic, (c) the fifth harmonic, (d) the root mean square of harmonics to the tenth. Note that a square wave of amplitude V and frequency ω is represented by the series

$$\frac{4V}{\pi}\left(\cos \omega t - \frac{1}{3}\cos 3\omega t + \frac{1}{5}\cos 5\omega t - \frac{1}{7}\cos 7\omega t + \cdots\right)$$

*14.48 Consider the oscillator circuit in Fig. P14.48 and assume for simplicity that transistor β is very high $(\beta = \infty)$.

(a) Find the frequency of oscillation and the minimum value of R_C (in terms of the transistor r_e) for oscillations to start.

(b) If $R_C = 1/I$ ohms and oscillations grow to the point that V_o is sufficiently large to turn the BJTs on and off, find the peak-to-peak amplitude of V_o.

MOS Digital Circuits

INTRODUCTION

This chapter and the next are devoted to the study of digital circuits. In a digital circuit, signals take on a limited number of values. The most common digital systems employ two values and are said to be *binary* systems. In such a system, voltage signals are either "high" or "low" and the symbols 1 and 0 are used to denote the two possible levels.

Digital circuits operate on binary-valued input signals and produce binary-valued output signals. The operation of digital circuits can be described by a special kind of algebra called *Boolean algebra*. It is not our intention here to study Boolean algebra, a topic normally covered in digital systems texts. We shall in fact assume that the reader is familiar with the basic concepts of digital systems, and concentrate here on digital electronics.

Digital circuits play a very important role in today's electronic systems. They are employed in almost every facet of electronics, including communications, control, instrumentation, and, of course,

computing. This widespread usage is due mainly to the availability of inexpensive integrated-circuit packages that contain powerful digital circuitry. The circuit complexity of a digital IC chip ranges from a small number of logic gates to a complete computer (a *microprocessor*) or a million bits of memory.

The conventional approach to designing digital systems consists of assembling the system using standard IC packages of various levels of complexity (and hence integration). As an alternative to using "off-the-shelf" components, the designer might opt for implementing part or all of the system using one or more customized very-large-scale-integrated (VLSI) circuits. However, custom IC design is usually justified only when the production volume is large (greater than about 100,000 parts). An intermediate approach, known as *semicustom design*, utilizes *gate array* chips. These are integrated circuits containing up to 10,000 unconnected logic gates. Their interconnection can be achieved by a final metallization step according to a pattern specified by the user so as to implement the user's particular functions.

Whatever approach is taken in digital design, some familiarity with the various digital circuit technologies and design techniques is essential. The material to follow is intended to provide the reader with a basic understanding of digital electronics. It should also serve as an introduction to VLSI circuit design. □

15.1 LOGIC CIRCUITS—SOME BASIC CONCEPTS

In this section we consider some basic concepts that underlie the design and specification of logic gate circuits. The material presented applies to both MOS and bipolar digital circuits.

Digital Signals

In binary digital circuits, two distinct voltage levels can represent the two values of binary variables. However, in order to allow for the inevitable component tolerances and miscellaneous other effects that change the signal voltage levels, two distinct voltage ranges are usually defined. As shown in Fig. 15.1, if the signal voltage lies in the range V_{L1} to V_{L2}, the signal is interpreted (by the digital circuit) as a logic 0. If the signal amplitude falls in the V_{H1} to V_{H2} range, it is interpreted as a logic 1. The two voltage bands are separated by a range in which the signal amplitude is not supposed to lie. This forbidden band represents an undefined or excluded region.

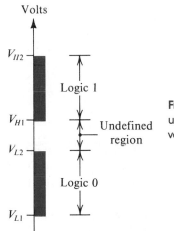

Fig. 15.1 Two distinct voltage ranges are used to represent the two values of binary variables.

Since the logic-1 voltages are higher than those that represent logic 0, the system illustrated in Fig. 15.1 is said to use *positive logic*. We can of course reverse our voltage band assignment, thus obtaining a *negative logic* system. Throughout this book we shall assume the positive-logic convention. Furthermore, on many occasions we shall use the words "high" and "low" interchangeably with 1 and 0, respectively.

Logic Circuit Families

Integrated logic circuits are classified into a number of different families. Members of each family are made with the same technology, have similar circuit structure, and exhibit the same basic features. In this chapter we shall study two MOS families: NMOS and CMOS. The first utilizes *n*-channel MOSFETs exclusively; the latter uses both *n*- and *p*-channel transistors in complementary symmetric circuit configurations. In the next chapter we shall study two BJT logic families: transistor-transistor logic (TTL) and emitter-coupled logic (ECL). CMOS, TTL, and ECL are available as off-the-shelf components for conventional logic design, as well as being utilized in the design of special VLSI circuits. NMOS, on the other hand, is used only in the design of VLSI circuits such as microprocessors and memory chips.

It will be shown that each of the four logic circuit families offers a unique set of advantages and disadvantages. In designing a digital system one selects an appropriate logic family and attempts to implement as much of the system as possible using packages that belong to this family. In this way interconnection of the various packages is relatively straightforward. If, on the other hand, packages from more than one family have to be used, one has to design suitable *interface circuits*. The selection of a logic family is based on such considerations as logic flexibility, speed of operation, availability of complex functions, noise immunity, operating-temperature range, power dissipation, and cost. We will discuss some of these considerations in this chapter and the next.

Scale of Integration

Many types of logic functions are usually available within each logic family. Depending on the complexity of the circuit on the IC chip, the package can be classified as one of four types:

1. small-scale integrated (SSI) circuit

2. medium-scale integrated (MSI) circuit

3. large-scale integrated (LSI) circuit

4. very-large-scale integrated (VLSI) circuit

Although the boundaries between the different levels of integration are not very sharp, a rough guide, based on the number of "equivalent logic gates" on the chip, is as follows: SSI, 1 to 10 gates; MSI, 10 to 100 gates; LSI, 100 to 1,000 gates; and VLSI, >1,000 gates.

The basic circuit properties of a logic family can be established by studying the basic inverter circuit of that family. In the following we shall consider the general characteristics and structure of the logic inverter circuit.

The Basic Inverter

The logic inverter is basically a voltage-controlled switch such as that represented schematically in Fig. 15.2. As shown, the switch connected between terminals 2 and

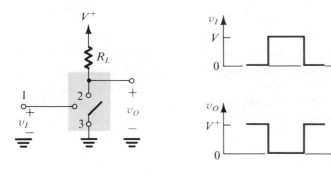

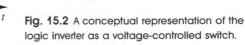

Fig. 15.2 A conceptual representation of the logic inverter as a voltage-controlled switch.

3 is controlled by the input signal v_I applied between terminals 1 and 3 (terminal 3 is connected to the reference or ground point). The inverter output voltage v_O is taken across the switch—that is, between 2 and ground. When v_I is low (around 0 V) the switch is open and the output voltage v_O is high (equal to the supply voltage V^+). When v_I is high (above a specified threshold voltage, as explained below) the switch is closed and the output voltage is low (0 V). It should be obvious that this circuit realizes the logic inversion operation.

Practical logic inverter circuits differ from the conceptual circuit in Fig. 15.2 in a number of ways. First, the input terminal of the inverter usually draws some current from the driving source. Second, the switch is not ideal; specifically, when the switch is closed it does not behave as a short circuit but rather has a finite closure resistance (called *on resistance*) and sometimes an additional voltage drop (called *offset voltage*). Figure 15.3 shows an equivalent circuit of the switch in the closed position. As a result of these imperfections, the voltage v_O will not be zero in the on state. Third, the inverter switch may not switch instantaneously; rather, there may be a delay time between the application of the input change and the appearance of the output change. Even if the inverter switches instantaneously, the capacitance inevitably present between the output terminals will cause the output waveform to have finite rise and fall times. This point will be discussed in some detail shortly. Fourth, actual inverters may not exhibit a well-defined switching threshold as implied by the discussion of the conceptual circuit of Fig. 15.2. Transfer characteristics of logic inverters are discussed next.

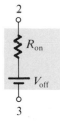

Fig. 15.3 Equivalent circuit for a typical switch in the on position.

Practical implementations of the logic inverter utilize a transistor (a MOSFET or a BJT) as the switching element and a resistor or another transistor for the load resistor R_L. Since MOS transistors require much smaller chip areas than medium- and high-valued resistors, MOS logic circuits almost always utilize MOSFETs as load elements.

The Inverter Transfer Characteristic

Figure 15.4 shows the ideal transfer characteristic for a logic inverter operated from a power supply V^+. As indicated, the inverter exhibits a threshold voltage $V_{th} = \frac{1}{2}V^+$. Input signals below this threshold voltage are interpreted as being low, and the inverter output is equal to the supply voltage V^+. Input signals above the threshold are interpreted as high, and the inverter output is equal to 0 V. Note that this inverter is quite tolerant of errors in the value of the input signal. Also, with the threshold voltage at half the power supply, this tolerance is equally distributed between the two domains of input signal (low and high). Finally, the inverter characteristics are ideal in the sense that, independent of the exact value of input signal, the output is either V^+ or 0 V.

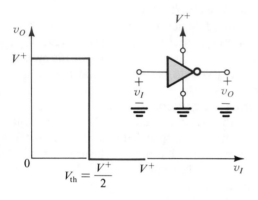
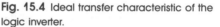

Fig. 15.4 Ideal transfer characteristic of the logic inverter.

Actual logic inverters have transfer characteristics that only approximate the ideal one of Fig. 15.4. Figure 15.5 shows a typical inverter transfer characteristic. Note that the threshold voltage is no longer well defined and that there exists a *transition region* between the high and low states. Also, the high output (V_{OH}) and low output (V_{OL}) are no longer equal to V^+ and 0 V, respectively.

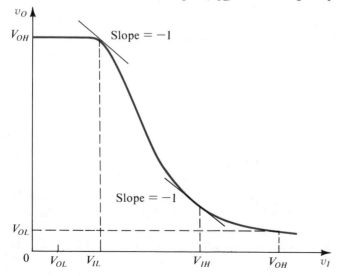

Fig. 15.5 Typical transfer characteristic of a logic inverter, illustrating the definition of the critical points.

Examining the inverter transfer characteristic in Fig. 15.5 in more detail, we note that there are three distinct regions:

1. The low-input region: $v_I < V_{IL}$

2. The transition (or uncertain) region: $V_{IL} \leq v_I \leq V_{IH}$

3. The high-input region: $v_I > V_{IH}$

Since the transition between one region and the next might not be sharp, it is customary to define V_{IL} and V_{IH} as the points at which the slope of the voltage transfer curve is -1, as indicated in Fig. 15.5. Of course, the slope in the transition region is much greater than unity. Input voltages less than V_{IL} are acknowledged by the gate as representing logic 0. Thus, V_{IL} *is the maximum allowable logic-0 value*. Similarly, input voltages greater than V_{IH} are acknowledged by the gate as representing logic 1. Thus, V_{IH} *is the minimum allowable logic-1 value*.

We should note in passing that the inverter transfer characteristic indicates that it is a grossly nonlinear device. In fact, digital circuits are extreme cases of the nonlinear circuits discussed in Chapter 5. If the inverter whose transfer characteristic is depicted in Fig. 15.5 were to be used as a linear amplifier, it would be biased at a point somewhere in the transition region and the input signal swing would be kept small to restrict operation to a short, almost-linear segment in the transition region of the characteristic.

Noise Margins

A great advantage of binary digital circuits is their tolerance to variations in value of input signals. As long as the input signal is correctly interpreted as low or high, the accuracy of operation is not affected. This tolerance to variation in signal level can be considered as immunity to noise superimposed on the input signal. Here *noise* refers to extraneous signals that can be capacitively or inductively coupled into the digital circuit from other parts of the system or from outside the system. We shall now quantify the ability of a logic gate to reject noise.[1]

Consider once more the inverter transfer characteristic of Fig. 15.5, where the four parameters V_{IL}, V_{IH}, V_{OL}, and V_{OH} are indicated. On the horizontal axis we have also marked the points corresponding to the output levels V_{OL} and V_{OH}. In a logic system, one gate usually drives another. Thus a gate whose output is high, at V_{OH}, drives an identical gate whose specified minimum input logic-1 level is V_{IH}. Furthermore, we see that the difference $V_{OH} - V_{IH}$ represents a margin of safety, for if noise were superimposed on the output signal of the driving gate (V_{OH}), the driven gate would not be bothered as long as the amplitude of the noise voltage was lower than ($V_{OH} - V_{IH}$). This difference is therefore called the *logic-1 or "high" noise margin*

---

[1] Unlike analog circuits where injected noise propagates throughout the system, in digital circuits once a gate rejects noise at the input, the gate output will be correct and we need no longer concern ourselves with the input noise.

and is denoted NM_H; that is

$$NM_H \equiv V_{OH} - V_{IH} \tag{15.1}$$

The *logic-0 or "low" noise margin*, NM_L, is similarly defined, as

$$NM_L \equiv V_{IL} - V_{OL} \tag{15.2}$$

Due to the unavoidable variabilities in the values of circuit components and power-supply voltage, the manufacturer usually specifies *worst-case* values for the four parameters V_{OH}, V_{IH}, V_{OL}, and V_{IL}. These values are defined as follows:

V_{OH}	the minimum voltage that will be available at a gate output when the output is supposed to be logic 1 (high)
V_{IH}	the minimum gate input voltage that will be unambiguously recognized by the gate as corresponding to logic 1
V_{OL}	the maximum voltage that will be available at a gate output when the output is supposed to be logic 0 (low)
V_{IL}	the maximum gate input voltage which will be unambiguously recognized by the gate as corresponding to logic 0

Instead of drawing the complete transfer characteristic, one usually is satisfied with the logic band diagram of Fig. 15.6. From this diagram it can be inferred that to maximize and equalize NM_L and NM_H, one ideally desires that $V_{IL} = V_{IH} =$ a value midway in the *logic swing* V_{OL} to V_{OH}. This implies that the transfer characteristic should switch abruptly; that is, it should exhibit high gain in the transition region. It also implies that switching should occur in the middle of the logic swing. To maximize this swing, V_{OL} should be as low as possible (ideally 0 V) and V_{OH} should be as high as possible (ideally equal to the power supply voltage V^+).

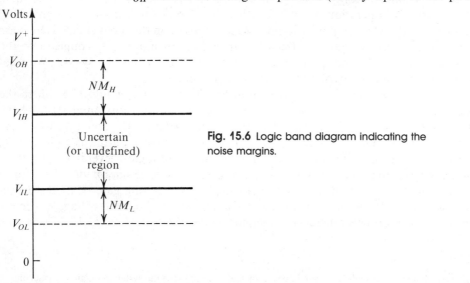

Fig. 15.6 Logic band diagram indicating the noise margins.

Such an idealized transfer characteristic is shown in Fig. 15.4. As will be seen in Section 15.5, the CMOS logic family provides an excellent approximation to the ideal performance.

Power Dissipation

Of interest to a logic circuit designer is the amount of power consumed by the circuit. Knowledge of power dissipation enables the designer to determine the current that the power supply for a digital system must provide. The power dissipated in a logic circuit is composed of two components: static and dynamic. The *static power* is the power dissipated while the circuit is not changing states. Thus in the idealized inverter of Fig. 15.2 we see that when the output is high, the static power is zero; however, when the output is low, the static power is $(V^+/R_L)V^+$. If one assumes that on average a gate spends half the time in either state, then the average static power dissipation will be $(V^+)^2/2R_L$.

To visualize *dynamic-power dissipation*, consider the idealized inverter of Fig. 15.2 when driving a load capacitance C_L. This could be the input capacitance of another logic gate or the capacitance of the wire interconnection between the inverter and another part of the system. Let us assume that initially, at $t = 0-$, the input was high and the switch was closed. Thus the capacitor was initially discharged. Now let v_I change to low at $t = 0$. It follows that the switch will open, but since the capacitor voltage cannot change instantaneously, v_O will rise exponentially toward V^+. The charging current will be supplied through R_L, and thus power will be dissipated in R_L. If the instantaneous supply current is denoted by i, then the energy drawn from the supply will be $\int V^+ i \, dt = V^+ \int i \, dt = V^+ Q$, where Q is the charge supplied to the capacitor; that is, $Q = C_L V^+$. Thus the energy drawn from the supply is $C_L(V^+)^2$. Now, since the capacitor initially had zero energy and will finally have a stored energy of $\frac{1}{2}C_L(V^+)^2$, it follows that the energy dissipated in R_L is $\frac{1}{2}C_L(V^+)^2$. If we next let v_I go high, the switch will close and C_L will discharge through the switch resistance R_{on}, thus dissipating power through the switch. If we assume that V_{off} is zero, then the energy dissipated in R_{on} will be equal to the energy stored on the capacitor, $\frac{1}{2}C_L(V^+)^2$. It follows that in one cycle of operation the gate dissipates $C_L(V^+)^2$ watts. Thus if the inverter is switched on and off f times per second, the dynamic power dissipation will be

$$\text{Dynamic power dissipation} = f C_L(V^+)^2 \qquad (15.3)$$

which is the sum of the power dissipation in R_L and R_{on}.

Fan-in and Fan-out

Fan-in of a gate is the number of its inputs. Thus a four-input NOR gate has a fan-in of 4. Fan-out is the maximum number of similar gates that a gate can drive while remaining within the guaranteed specifications. More will be said about fan-in and fan-out in the context of studying the various logic families.

Propagation Delay

Because of the dynamics involved in the action of a switching device such as a bipolar transistor, the switch in the inverter of Fig. 15.2 may not respond instantaneously to the control signal v_I. In addition, the inevitable load capacitance at the inverter output causes the waveform of v_O to depart from an ideal pulse. Figure 15.7 illustrates the typical response of an inverter to an input pulse with finite rise and fall times. The output pulse, of course, exhibits finite rise and fall times. In addition, there is a delay time between the input and output pulses. There are various ways for expressing this *propagation delay*. The usual way is to specify the times between the 50% points of the input and output waveforms at both the leading and trailing edges. These times are referred to as t_{PHL} (where HL indicates the high-to-low transition of the output) and t_{PLH} (where LH indicates the low-to-high transition of the output) in Fig. 15.7. The propagation time delay t_P can then be defined as the average of these two times,

$$t_P = \tfrac{1}{2}(t_{PHL} + t_{PLH}) \tag{15.4}$$

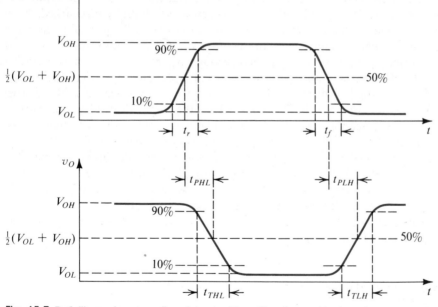

Fig. 15.7 Definitions of propagation delay and transition times of logic gates.

Delay-Power Product

One is usually interested in high-speed performance (low t_P) combined with low power dissipation. Unfortunately these two requirements are in conflict since, generally when designing a gate, if one attempts to reduce power dissipation by decreasing supply current, the gate delay will increase. It follows that a figure of merit for comparing logic families is the delay-power product DP, defined as

$$DP \equiv t_P P_D \tag{15.5}$$

where P_D is the power dissipation of the gate. Note that DP has the unit of joules. The lower the DP figure for a family, the more effective this logic family is.

Physical Packaging of Logic Circuits

Figure 15.8a shows a common physical package used to house IC logic circuits. This package is made either of plastic or ceramic and is called a *dual-in-line* (DIP) package. It has 14 leads, 7 brought out to each side. Other packages with 16, 24, and 40 pins exist.

A functional diagram of a quad two-input NAND package is shown in Fig. 15.8b.

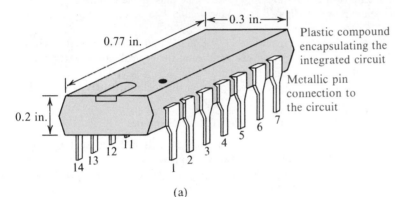

(a)

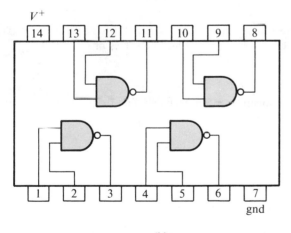

(b)

Fig. 15.8 A 14-pin integrated circuit. **(a)** Physical appearance. **(b)** Schematic of an integrated circuit providing 4 two-input NAND gates.

Exercises

15.1 The data sheet of the SN7400 quad 2-input NAND gate of the TTL family provides the following.

> Logic-1 input voltage required at both input terminals to ensure a logic-0 level at the output: MIN (minimum) 2 V.
> Logic-0 input voltage required at either input terminals to ensure a logic-1 level at the output: MAX (maximum) 0.8 V.
> Logic-1 output voltage: MIN 2.4 V, TYP (typical) 3.3 V.
> Logic-0 output voltage: TYP 0.22 V, MAX 0.4 V.
> Logic-0 level supply current: TYP 12 mA, MAX 22 mA (for the entire package).
> Logic-1 level supply current: TYP 4 mA, MAX 8 mA (for the entire package).
> Propagation delay time to logic-0 level: TYP 7 ns, MAX 15 ns.
> Propagation delay time to logic-1 level: TYP 11 ns, MAX 22 ns.

(a) Find the noise margin in both the 0 and 1 states.
(b) Assuming that the gate is in the 1 state 50% of the time and in the 0 state 50% of the time, find the average static power dissipated in a typical gate. The power supply voltage is +5 V.
(c) Assuming that the gate drives a capacitance $C_L = 45$ pF and is switched at 1 MHz frequency, find the dynamic power dissipation per gate using the typical values of the logic 1 and 0 levels at the output.
(d) Find the typical value of the gate delay-power product (neglecting the dynamic power dissipation).

Ans. **(a)** 0.4 V, 0.4 V; **(b)** 10 mW; **(c)** 0.7 mW; **(d)** 90 picojoules

15.2 A logic inverter having a negligible static power dissipation is switched at the rate of 1 MHz. If the inverter is operated from a 10-V power supply, and drives a 50-pF load capacitance find the dynamic power dissipation and the average current drawn from the power supply

Ans. 5 mW; 0.5 mA

15.3 Consider the inverter of Fig. 15.2 under the conditions that $V^+ = 5.5$ V, $R_L = 10$ kΩ and the switch on-resistance, $R_{on} = 1$ kΩ. Let $V_{off} = 0$. Find the values of V_{OH} and V_{OL}.

Ans. 5.5 V; 0.5 V

15.4 Let the inverter specified in Exercise 15.3 be fed with an ideal pulse having zero rise and fall times. Assuming that the switch operates instantaneously, find the propagation delay t_{PLH}, t_{PHL}, and t_P that result from a 50-pF load capacitance.

Ans. 347 ns; 32 ns; 189 ns

15.2 NMOS INVERTER WITH ENHANCEMENT LOAD

As mentioned above, the basic building block of a digital circuit family is the logic inverter. For NMOS, there are two basic inverter circuits: the enhancement-load inverter and the depletion-load inverter. In this section we study the first.

Static Characteristics

Figure 15.9a shows the enhancement-load inverter. The application of this circuit as an amplifier was studied in Section 7.8, where we derived the voltage transfer characteristic shown in Fig. 15.9b. As a logic inverter, the circuit operates as follows: Logic-0 input signals, represented by voltages lower than the threshold voltage of Q_1, V_{t1}, cause Q_1 to be off and thus the output voltage will be high at

$$V_{OH} = V_{DD} - V_{t2} \qquad (15.6)$$

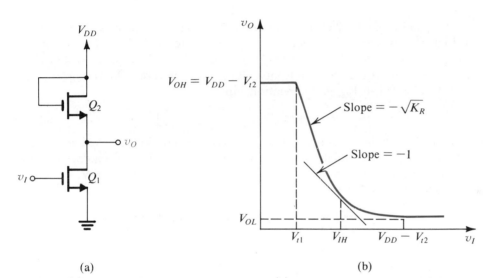

(a)   (b)

Fig. 15.9 (a) The enhancement-load inverter and **(b)** its voltage transfer characteristic.

We note that the high output is lower than V_{DD} by the threshold voltage V_{t2}. This is a serious drawback of the enhancement-load inverter, resulting in reduced signal swing and noise margin.

With v_I high at the logic-1 level of $(V_{DD} - V_{t2})$, Q_1 will be in the triode region while Q_2 remains in the pinch-off region. The output will be V_{OL}.

In the transition region the voltage transfer characteristic is linear with a slope of $-\sqrt{K_1/K_2}$ [see Eqs. (7.60)–(7.62)]. The ratio of the conductance parameters K_1 and K_2 is denoted K_R, and is as follows:

$$K_R \equiv \frac{K_1}{K_2} = \frac{(W/L)_1}{(W/L)_2} \qquad (15.7)$$

where we have substituted $K_1 = \frac{1}{2}\mu_n C_{ox}(W/L)_1$ and $K_2 = \frac{1}{2}\mu_n C_{ox}(W/L)_2$. The constant K_R is known as the geometry ratio or the aspect ratio of the inverter. In order to obtain a reasonably sharp voltage transfer characteristic, and hence acceptable noise margins, K_R is usually greater than 8. Note, however, that the larger the value of K_R the greater is the silicon area occupied by the inverter, since there is a lower limit to the usable dimensions of a device in a given technology.

The Body Effect

The derivation in Chapter 7 of the inverter transfer characteristic neglected the body effect. Figure 15.10 shows the enhancement-load inverter with the substrate connections explicitly indicated. Note that while the source-to-body voltage (V_{SB}) of Q_1 is zero, that of Q_2 is equal to the output voltage v_O. The threshold voltage V_t of a MOSFET is related to V_{SB} via the relationship

$$V_t = V_{t0} + \gamma[\sqrt{V_{SB} + 2\phi_F} - \sqrt{2\phi_F}] \tag{15.8}$$

where V_{t0} is the threshold voltage at $V_{SB} = 0$, γ is a constant for the given fabrication process, and $2\phi_F$ is the equilibrium electrostatic potential of the p material forming the body. Typically, $V_{t0} = 1$ to 1.5 V, $\gamma = 0.3$ to 1 V$^{1/2}$, and $2\phi_F \simeq 0.6$ V. Using Eq. (15.8) we find that $V_{t1} = V_{t0}$ and

$$V_{t2} = V_{t0} + \gamma(\sqrt{v_O + 2\phi_F} - \sqrt{2\phi_F}) \tag{15.9}$$

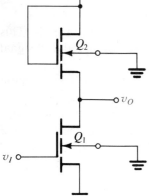

Fig. 15.10 The enhancement-load inverter with the body connections explicitly shown.

This relationship can be used together with the i-v relationship of the load transistor Q_2,

$$i_{D2} = K_2(V_{DD} - v_O - V_{t2})^2 \tag{15.10}$$

and the i_D-v_{DS} characteristics of the inverting transistor Q_1, to obtain a more accurate transfer characteristic for the enhancement-load inverter. This process, however, is quite tedious. Fortunately, such a detailed analysis is rarely necessary. It is sufficient to note that the body effect is significant only when v_O is high and results simply in a reduction in the value of V_{OH}.

EXAMPLE 15.1

Consider an enhancement-load inverter having $V_{t0} = 1$ V, $(W/L)_1 = 3$, $(W/L)_2 = \frac{1}{3}$, $\mu_n C_{OX} = 20$ $\mu A/V^2$, $2\phi_F = 0.6$ V, $\gamma = 0.5$ $V^{1/2}$, and $V_{DD} = 5$ V.

(a) Neglecting the body effect, find the critical points of the voltage transfer characteristic, and hence find the noise margins.

(b) Taking the body effect into account, find the modified values of V_{OH} and NM_H.

(c) Find the inverter current in both states, and hence find the average static power dissipation.

Solution

(a) Neglecting the body effect, $V_{t1} = V_{t2} = 1$ V. By reference to Fig. 15.9 we find that

$$V_{OH} = V_{DD} - V_t = 5 - 1 = 4 \text{ V}$$

The transfer characteristic in the transition region is a straight line with a slope of $-\sqrt{K_R} = -\sqrt{9} = -3$ V/V. Thus the slope changes abruptly from 0 to -3 at $v_I = V_t$. It follows that

$$V_{IL} = V_t = 1 \text{ V}$$

To find V_{OL} we assume that Q_1 is operating in the triode region and that Q_2 is operating in the pinch-off region, and we equate their drain currents to obtain

$$K_1[2(v_I - V_{t1})v_O - v_O^2] = K_2(V_{DD} - v_O - V_{t2})^2 \tag{15.11}$$

where we have neglected the finite MOSFET output resistance in pinch-off. Substituting $v_I = V_{OH} = V_{DD} - V_t$, $v_O = V_{OL}$, and $K_1 - 9K_2$ yields a quadratic equation in V_{OL} whose solution is

$$V_{OL} \simeq 0.3 \text{ V}$$

At this value of v_O, Q_1 will indeed be in the triode region, as assumed.

The value of V_{IH} is determined by assuming that Q_1 is in the triode region. Thus Eq. (15.11) applies. Differentiating both sides of Eq. (15.11) relative to v_I gives

$$K_1\left[2(v_I - V_{t1})\frac{dv_O}{dv_I} + 2v_O - 2v_O\frac{dv_O}{dv_I}\right] = -2K_2(V_{DD} - v_O - V_{t2})\frac{dv_O}{dv_I}$$

Substituting $v_I = V_{IH}$, $dv_O/dv_I = -1$ and $K_1 = K_R K_2$ results in

$$K_R[-(V_{IH} - V_{t1}) + 2v_O] = V_{DD} - v_O - V_{t2} \tag{15.12}$$

Substituting $v_I = V_{IH}$ in Eq. (15.11) and solving the resulting equation simultaneously with (15.12) results in

$$V_{IH} \simeq 2.2 \text{ V} \quad \text{and} \quad v_O \simeq 0.8 \text{ V}$$

From these values we can verify that Q_1 is indeed operating in the triode region.

The noise margins can now be found as

$$NM_H = V_{OH} - V_{IH} \qquad\qquad NM_L = V_{IL} - V_{OL}$$
$$= 4 - 2.2 = 1.8 \text{ V} \qquad\qquad = 1 - 0.3 = 0.7 \text{ V}$$

(b) Taking the body effect into account, for V_{OH} we write

$$V_{OH} = V_{DD} - V_{t2}|_{V_{SB} = V_{OH}}$$
$$= V_{DD} - V_{t0} - \gamma(\sqrt{V_{OH} + 2\phi_F} - \sqrt{2\phi_F})$$

which for the given numerical values becomes

$$V_{OH} = 3.4 \text{ V}$$

Thus the body effect increases the threshold voltage of Q_2 to 1.6 V, and thus decreases V_{OH} to 3.4 V and NM_H to 1.2 V.

(c) With v_O high, the current in the inverter is negligibly small, and hence the static power dissipation is negligible. For $v_O = V_{OL} = 0.3$ V, the inverter current can be determined from

$$I_{D2} = K_2(V_{DD} - V_{OL} - V_t)^2$$
$$= \tfrac{1}{2} \times 20 \times \tfrac{1}{3} \times (5 - 0.3 - 1)^2 = 46 \text{ } \mu\text{A}$$

Thus in the low state, the static power dissipation is

$$P_{DL} = 46 \times 5 = 230 \text{ } \mu\text{W}$$

The average power dissipation of the inverter is

$$P_D = \tfrac{1}{2}(230 + 0) = 115 \text{ } \mu\text{W}$$

Dynamic Operation

We next consider the dynamic operation of the enhancement-load NMOS inverter. Figure 15.11a shows the inverter with a capacitive load C. It is assumed that C includes all the relevant MOSFET capacitances, the total input capacitance of all gates that are driven by the inverter under study, and the associated wiring (routing) capacitance. Lumping the various capacitive effects into a single load capacitance is, of course, an approximation, the purpose of which is to make the problem tractable for pencil-and-paper analysis. More accurate results can be obtained using more elaborate models together with a circuit analysis computer program such as SPICE (see Appendix C). Nevertheless, for gaining insight into circuit operation, there is no substitute for manual analysis.

Further simplification of the analysis is achieved by assuming that the input pulse is ideal, having zero rise and fall times, as shown in Fig. 15.11b. This figure defines the propagation delay times that we wish to calculate. Figure 15.11c shows a sketch of i_{D2} versus v_O, which is the load curve, and a sketch of i_{D1} versus v_O for $v_I = V_{OH}$.

Before the application of the input pulse, $v_I = V_{OL}$ and $v_O = V_{OH}$. The inverter is operating at point B in Fig. 15.11c and the load capacitor C is charged to V_{OH}.

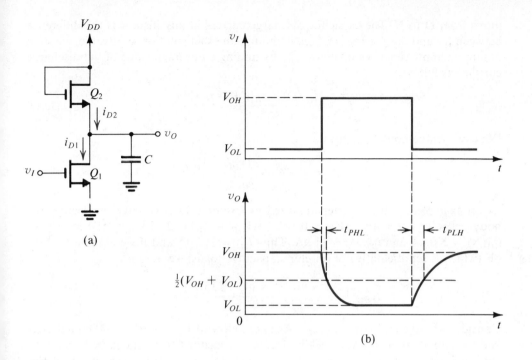

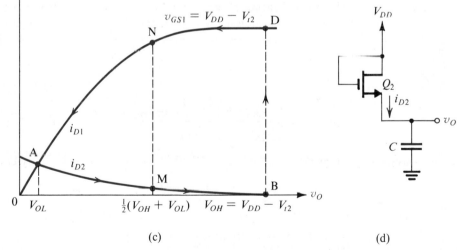

Fig. 15.11 Dynamic operation of the enhancement-load NMOS inverter.

When v_I goes high (to V_{OH}), transistor Q_1 turns on. However, since capacitor C cannot discharge instantaneously, the operating point jumps from B to D in Fig. 15.11c. Transistor Q_1 will sink a relatively large current, thus discharging C. The operating point moves along the i_{D1} curve until it finally reaches point A, where $v_O = V_{OL}$. The propagation delay time t_{PHL} is the time for the operating point to move from D to N. The capacitor discharge current at any instant is the difference between i_{D1} and i_{D2} (see Fig. 15.11c). Although an exact solution is possible, we shall obtain an approximate estimate of t_{PHL} by finding an average value of the discharge current, as follows:

$$I_{HL} = \frac{i_{D1}(D) + i_{D1}(N) - i_{D2}(M)}{2} \tag{15.13}$$

We can then compute t_{PHL} from

$$t_{PHL} = \frac{C[V_{OH} - \frac{1}{2}(V_{OH} + V_{OL})]}{I_{HL}} \tag{15.14}$$

As an example, for the inverter analyzed in Example 15.1 we have (neglecting the body effect) $V_{OH} = 4$ V, $V_{OL} = 0.3$ V, $\frac{1}{2}(V_{OH} + V_{OL}) = 2.15$ V, $i_{D1}(D) = 270$ μA, $i_{D1}(N) = 250$ μA and $i_{D2}(M) = 11$ μA. Thus $I_{HL} = 255$ μA and if we assume $C = 0.1$ pF (which is representative of on-chip capacitive loads) we obtain

$$t_{PHL} = \frac{0.1 \times 10^{-12} \times (4 - 2.15)}{255 \times 10^{-6}} = 0.7 \text{ ns}$$

Consider next the evaluation of t_{PLH}. As v_I goes low to V_{OL}, Q_1 turns off immediately and the circuit reduces to that in Fig. 15.11d. Capacitor C is then charged up by the current supplied by Q_2, and the operating point moves along the load curve of Fig. 15.11c from A to, eventually, B. Because the available current is low, we expect t_{PLH} to be much longer than t_{PHL}. Although t_{PLH} can be found by solving the differential equation that describes the operation of the circuit in Fig. 15.11d (see Problem 15.15) we shall obtain an approximate value using an average charging current, computed as

$$I_{LH} = \frac{1}{2}[i_{D2}(A) + i_{D2}(M)] \tag{15.15}$$

Thus

$$t_{PLH} = \frac{C[\frac{1}{2}(V_{OH} + V_{OL}) - V_{OL}]}{I_{LH}} \tag{15.16}$$

For example, for the inverter of Example 15.1 we have (neglecting the body effect)

$$i_{D2}(A) = 46 \text{ } \mu\text{A} \quad \text{and} \quad i_{D2}(M) = 11 \text{ } \mu\text{A}$$

Thus, $I_{LH} \simeq 29$ μA and

$$t_{PLH} = \frac{0.1 \times 10^{-12} \times [\frac{1}{2}(4 + 0.3) - 0.3]}{29 \times 10^{-6}} = 6.4 \text{ ns}$$

The propagation delay of the inverter can be now computed as the average of t_{PHL} and t_{PLH}. For the inverter of Example 15.1 we obtain

$$t_P = \frac{1}{2}(0.7 + 6.4) \simeq 3.6 \text{ ns}$$

Here it should be noted that if the inverter is used to drive off-chip circuitry, then the load capacitance C will be two orders of magnitude larger than the value used in the above example. The propagation delay will increase proportionately and become hundreds of nanoseconds. This explains why enhancement-load NMOS is not used in SSI and MSI logic design.

Delay-Power Product

The delay-power product (DP) for enhancement-load NMOS can be found by multiplying the propagation delay by the static power dissipation. For the inverter in Example 15.1,

$$DP = 3.6 \times 10^{-9} \times 115 \times 10^{-6} = 0.4 \text{ pJ}$$

If we use capacitance values representative of those encountered in conventional discrete component design we obtain DP of the order of 10 to 100 pJ.

An approximate but useful expression for DP can be derived as follows: Assuming that $V_{OL} \simeq 0$ we can show that I_{LH} of Eq. (15.15) is approximately

$$I_{LH} \simeq \tfrac{5}{8} K_2 (V_{DD} - V_t)^2 \tag{15.17}$$

Using this expression in Eq. (15.16) gives

$$t_{PLH} \simeq \frac{0.8 \; C}{K_2 (V_{DD} - V_t)} \tag{15.18}$$

Now, since $t_{PHL} \ll t_{PLH}$, we may evaluate t_P as

$$t_P \simeq \tfrac{1}{2} t_{PLH} = \frac{0.4 C}{K_2 (V_{DD} - V_t)} \tag{15.19}$$

For $V_{OL} \simeq 0$, the average static power dissipation is approximately

$$P_D \simeq \tfrac{1}{2} K_2 (V_{DD} - V_t)^2 V_{DD} \tag{15.20}$$

Equations (15.19) and (15.20) can be combined to obtain

$$DP \simeq 0.2 C V_{DD} (V_{DD} - V_t) \tag{15.21}$$

This expression shows that DP can be reduced by either reducing C or V_{DD}. Reducing V_{DD}, however, reduces the signal swing and the noise margins.

Exercises

15.6 For the enhancement-load inverter of Fig. 15.10, find K_R that results in $V_{OL} = 0.1$ V. Let $V_{t1} = V_{t2} = 1$ V and $V_{DD} = 5$ V, and neglect the body effect.

Ans. 26

15.7 Use the approximate expressions in Eqs. (15.19) and (15.21) to determine t_P and DP for the inverter of Example 15.1. Let $C = 0.1$ pF.

Ans. 3 ns; 0.4 pJ

15.8 Use the expression in Eq. (15.21) to determine DP for an enhancement-load NMOS inverter operated from $V_{DD} = 5$ V and having $V_t = 1$ V. Let $C = 10$ pF.

Ans. 40 pJ

15.3 NMOS INVERTER WITH DEPLETION LOAD

The use of a depletion MOSFET as the load element results in an inverter with higher gain having a sharper voltage transfer characteristic, and hence increased noise margins. Furthermore, the improved noise margins can be obtained while using a smaller geometry ratio, K_R, and hence smaller silicon area than required for the enhancement-load inverter. The price paid for the performance improvement is an extra fabrication step required to implant the channel of the depletion device.

Figure 15.12a shows the depletion-load inverter. We have studied the amplifier application of this circuit in Section 7.9. However, in deriving the voltage transfer

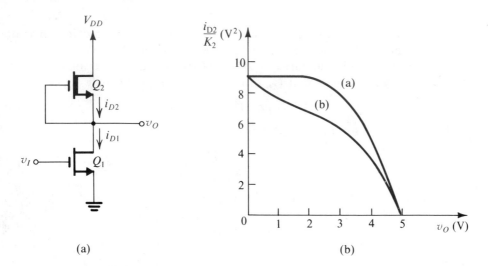

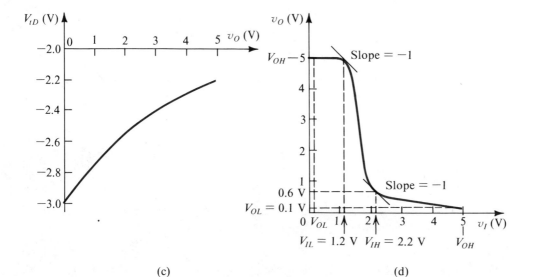

Fig. 15.12 The depletion-load inverter and its characteristics.

characteristic in Section 7.9, we did not emphasize the body effect, which plays a significant role in the operation of the circuit as a logic inverter.[2] To be specific, we show in Fig. 15.12b plots of i_{D2} (normalized relative to K_2) versus v_O without [curve (a)] and with [curve (b)] the body effect taken into account. These are the curves representing the load of the inverting transistor Q_1. We note that without the body effect the depletion device behaves as a constant-current source over a wide range of v_O. This implies that the inverter transfer characteristic will be very sharp in the transition region. It also implies that a relatively large current is available to charge the load capacitance, resulting in a short t_{PLH}. Unfortunately, however, the body effect causes the depletion load to depart from constant-current operation. Thus, the inverter characteristics, though superior to those obtained with enhancement load, are not as good as was thought to be possible in the early stages of the development of this technology.

Analytically, the current in the depletion load is given by

$$i_{D2} = K_2|V_{tD}|^2 \qquad \text{for } v_O \leq V_{DD} - |V_{tD}| \tag{15.22}$$

and

$$i_{D2} = K_2[2|V_{tD}|(V_{DD} - v_O) - (V_{DD} - v_O)^2] \qquad \text{for } v_O \geq V_{DD} - |V_{tD}| \tag{15.23}$$

where V_{tD} is the threshold voltage of the depletion device. For $V_{SB}|_{Q_2} = 0$, $V_{tD} = V_{tD0}$, which is typically -3 V. However, $V_{SB}|_{Q_2} = v_O$ and V_{tD} is given by

$$V_{tD} = V_{tD0} + \gamma[\sqrt{v_O + 2\phi_F} - \sqrt{2\phi_F}] \tag{15.24}$$

Figure 15.12c shows a plot of V_{tD} versus v_O for $V_{tD0} = -3$ V, $\gamma = 0.5$ V$^{1/2}$, and $2\phi_F = 0.6$ V. The plots in Fig. 15.12b are obtained using Eqs. (15.22)–(15.24) with the power supply $V_{DD} = 5$ V.

Static Characteristics

The characteristic curve of the depletion load [curve (b) in Fig. 15.12b] can be used in conjunction with the i_D-v_{DS} characteristics of the enhancement inverting transistor Q_1 to obtain the inverter transfer characteristic shown in Fig. 15.12d. Observe that $V_{OH} - V_{DD}$, which is 1 to 1.5 V higher than the logic-1 level of the enhancement-load inverter. The following example illustrates the calculation of the critical points of the voltage transfer characteristic.

EXAMPLE 15.2

Consider a depletion-load inverter with $V_{tE} = 1$ V, $K_R = 9$, $V_{tD0} = -3$ V, $\mu_n C_{OX} = 20$ μA/V^2, $2\phi_F = 0.6$ V, $\gamma = 0.5$ V$^{1/2}$, and $V_{DD} = 5$ V. (Note that since this inverter has the same geometry ratio as the enhancement-load inverter of Example 15.1, the performance results can be easily compared.)

---

[2] In Section 7.9 we did take the body effect into account, however, in determining the small-signal voltage gain of the depletion-load amplifier.

(a) Find the critical points of the voltage transfer charactertistic, taking into account, where necessary, the body effect. Hence calculate the noise margins.

(b) Find $(W/L)_2$ so that the inverter current in the low-output state equals the corresponding value in the enhancement-load inverter of Example 15.1. Also find $(W/L)_1$.

(c) Find the average static power dissipation.

Solution

(a) $V_{OH} = V_{DD} = 5$ V. From Fig. 15.12c we find that at $v_O = 5$ V, $V_{tD} = -2.2$ V. Thus Q_2 with $v_{GS} = 0$ can conduct a load current, if required. Of course, because Q_1 is off, the current in the inverter is zero.

To find V_{OL}, we let $v_I = V_{OH} = 5$ V and assume Q_1 to be in the triode region and Q_2 to be in the pinch-off region. Since we expect V_{OL} to be small, we may neglect the body effect and assume that $V_{tD} \simeq -3$ V. Thus,

$$i_{D1} = K_1[2(5-1)V_{OL} - V_{OL}^2]$$
$$i_{D2} = K_2|V_{tD}|^2 = 9K_2$$

Equating i_{D1} and i_{D2}, substituting $K_1/K_2 = 9$, and solving the resulting quadratic equation, we get

$$V_{OL} \simeq 0.1 \text{ V}$$

To determine V_{IH} we assume that Q_1 is in the triode region while Q_2 is in pinch-off and equate their drain currents to obtain

$$K_1[2(v_I - 1)v_O - v_O^2] = 9K_2$$

Substituting $K_2 = K_1/9$ gives

$$2(v_I - 1)v_O - v_O^2 = 1 \tag{15.25}$$

Differentiating with respect to v_I yields

$$2(v_I - 1)\frac{dv_O}{dv_I} + 2v_O - 2v_O\frac{dv_O}{dv_I} = 0$$

Substituting $v_I = V_{IH}$ and $dv_O/dv_I = -1$ results in

$$v_O = \tfrac{1}{2}(V_{IH} - 1) \tag{15.26}$$

Substituting $v_I = V_{IH}$ in Eq. (15.25) and replacing v_O with the value given in Eq. (15.26) results in a quadratic equation in V_{IH} whose solution is $V_{IH} = 2.2$ V and from Eq. (15.26) the corresponding value of $v_O = 0.6$ V. Note that Q_1 is indeed in the triode region and Q_2 is in the pinch-off region, as assumed. Also note that the values obtained are approximate since we have assumed the body effect to be negligible.

To determine V_{IL} we assume that Q_1 is in pinch-off and Q_2 in the triode region and equate their currents. Since we expect v_O to be close to 5 V, we use $V_{tD} \simeq -2.2$ V (see Fig. 15.12c). The resulting equation is

$$(v_I - 1)^2 = (0.49)(5 - v_O) - \tfrac{1}{9}(5 - v_O)^2 \tag{15.27}$$

Differentiating relative to v_I, we obtain

$$2(v_I - 1) = -0.49 \frac{dv_O}{dv_I} + \frac{2}{9}(5 - v_O)\frac{dv_O}{dv_I}$$

Substituting $v_I = V_{IL}$ and $dv_O/dv_I = -1$ gives

$$v_O = 9(V_{IL} - 0.69) \tag{15.28}$$

Substituting $v_I = V_{IL}$ in Eq. (15.27) and for v_O from Eq. (15.28) results in a quadratic equation whose solution yields V_{IL} and the corresponding v_O as

$$V_{IL} = 1.2 \text{ V} \qquad v_O = 4.8 \text{ V}$$

The noise margins can now be calculated as follows:

$$NM_L = V_{IL} - V_{OL} = 1.2 - 0.1 = 1.1 \text{ V}$$
$$NM_H = V_{OH} - V_{IH} = 5 - 2.2 = 2.8 \text{ V}$$

Both of these values are larger than the corresponding values obtained for the enhancement-load inverter in Example 15.1 (0.7 V and 1.8 V, respectively).

(b) The enhancement-load inverter of Example 15.1 has at $v_O = V_{OL}$ a current of 46 μA. To establish an equal current in the depletion-load inverter we need to choose K_2 so that

$$K_2 |V_{tD}|^2 = 46 \ \mu\text{A}$$

Thus,

$$K_2 = \tfrac{46}{9} = 5.1 \ \mu\text{A/V}^2$$

$$\tfrac{1}{2}(\mu_n C_{OX})\left(\frac{W}{L}\right)_2 = 5.1$$

which leads to

$$\left(\frac{W}{L}\right)_2 \simeq 0.5$$

With $K_R = 9$,

$$\left(\frac{W}{L}\right)_1 = 4.5$$

(c) Since the inverter is designed to have a current equal to that of the enhancement-load inverter of Example 15.1, the average static power dissipation will be the same; that is,

$$P_D = 115 \ \mu\text{W}$$

The results of the above example indicate that a depletion-load inverter has markedly increased noise margins as compared to an enhancement-load inverter with the same geometry ratio, K_R. This implies that K_R of the depletion-load inverter can be reduced, thus reducing device sizes and silicon area while still maintaining good noise margins.

Exercise

15.9 If K_R of the inverter in Example 15.2 is reduced to 4 by reducing $(W/L)_1$ from 4.5 to 2, calculate the new values of V_{OH}, V_{OL}, V_{IH}, V_{IL}, NM_H, and NM_L.

Ans. 5 V; 0.3 V; 2.7 V; 1.5 V; 2.3 V; 1.2 V

Dynamic Operation

Figure 15.13 illustrates the dynamic operation of the depletion-load inverter in the presence of a load capacitance C. The propagation delay times t_{PLH} and t_{PHL} can be calculated by finding the average currents available to charge and discharge C, in the same manner used in the enhancement-load case. We shall not show the details of this analysis (see Exercise 5.10) but point out that the depletion load provides a relatively higher current over a wider range of v_O than does the enhancement load. This results in a somewhat faster charging of the load capacitance, and correspondingly a slightly[3] shorter t_{PLH}.

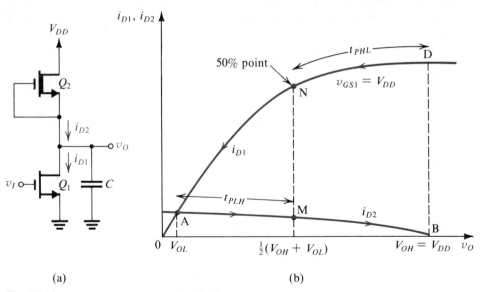

(a)   (b)

Fig. 15.13 Dynamic operation of the depletion-load inverter.

---

[3] The reduction in t_{PLH} is not as great because the voltage swing is higher in the depletion-load inverter.

Exercise	**15.10** Consider the depletion-load inverter of Exercise 15.9 with $(W/L)_1 = 2$, $(W/L)_2 = 0.5$, and all the other parameters as specified in Example 15.2. Find the average values of the currents I_{HL} and I_{LH}. Also, for $C = 0.1$ pF calculate t_{PHL}, t_{PLH}, and t_P. **Ans.** 300 μA; 38 μA; 0.8 ns; 6.2 ns; 3.5 ns

Delay-Power Product

An approximate expression can be derived for the delay-power product DP in a manner similar to that used for the enhancement-load inverter. The result (see Problem 15.21) is as follows:

$$DP \simeq \frac{1}{8\alpha} CV_{DD}^2 \qquad (15.29)$$

where α is a fraction smaller than, but close to, 1, accounting for the variation of V_{tD} with v_O. This expression yields values somewhat smaller than those for enhancement-load circuits.

Exercise	**15.11** For the depletion-load inverter of Exercises 15.9 and 15.10 let $\alpha \simeq 0.9$. Find the delay-power product first by multiplying P_D by t_P and then using the approximate expression in Eq. (15.29). **Ans.** 0.4 pJ; 0.35 pJ

A Final Remark

Although it occupies a smaller chip area, the depletion-load NMOS inverter exhibits higher noise margins and a somewhat higher speed of operation than the enhancement-load inverter. For this reason, virtually all modern NMOS logic and memory circuits utilize the depletion-load technology.

15.4 NMOS LOGIC CIRCUITS

Figure 15.14a shows a two-input NOR gate in depletion-load NMOS technology. The logic gate operates as follows: If the voltage at any of the input terminals is high (at V_{DD}), then the corresponding transistor will be on and the output voltage will be low (at V_{OL}). The output voltage will be high only if the two inputs are simultaneously low. In this case both input transistors will be off and $v_Y = V_{DD}$. Thus the operation can be described by the Boolean expression

$$Y = \overline{A}\overline{B}$$

or, equivalently,

$$Y = \overline{A + B}$$

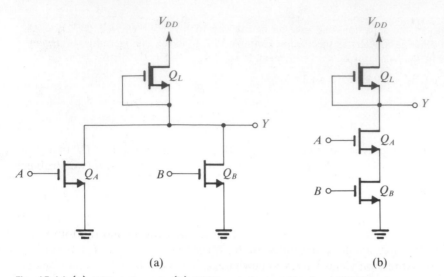

Fig. 15.14 (a) NOR gate and **(b)** NAND gate in depletion-load NMOS logic.

The input transistors are identical and have the same dimensions as the inverter transistor. Thus if either input is high, the output will be at V_{OL}. If both inputs are simultaneously high, then the output voltage will be lower than V_{OL}. The gate fan-in can be increased by adding additional input transistors.

Exercise

15.12 For the NOR gate in Fig. 15.14a let Q_A and Q_B be identical with $W = 12\ \mu$m and $L = 6\ \mu$m. The load transistor Q_L has $W = 6\ \mu$m and $L = 12\ \mu$m. Also let $V_{tE} = 1$ V, $V_{tD0} = -3$ V and $V_{DD} = 5$ V. Find the output voltage when **(a)** one input is high and **(b)** both inputs are high.

Ans.   0.29 V; 0.14 V

Whereas the NOR gate is formed by connecting input transistors in parallel, the NAND function is obtained by placing the input transistors in series. A two-input NAND gate is shown in Fig. 15.14b. Here the output will be low only when both Q_A and Q_B are on, a situation obtained when both inputs are simultaneously high. Thus

$$\bar{Y} = AB$$

or

$$Y = \overline{AB} = \bar{A} + \bar{B}$$

When both Q_A and Q_B are conducting, the effective channel length between the output node and ground is double that of the inverter transistor. It follows that to keep the output voltage at the value of V_{OL} obtained in the inverter, each of the input transistors in the NAND gate should have double the width of the inverter transistor. In this way, the two series conducting transistors will have the same effective W/L ratio as the single inverter transistor. Of course, if we have N inputs, the width of each of the input transistors should be N times that of the inverter transistor. As a result, the silicon area required by a NAND gate is greater than that required by a NOR gate having the same number of inputs. This limits the application of the NAND implementation.

Exercise

15.13 If the optimum dimensions of a depletion-load NMOS inverter are $W/L = 12\ \mu m/6\ \mu m$ for the inverting transistor and $6\ \mu m/12\ \mu m$ for the load transistor, find the dimensions of the input transistors in a 3-input NAND gate.

Ans. $36\ \mu m/6\ \mu m$

As a final remark we note that NMOS logic circuits are simple to fabricate and require small chip area, and thus can be very densely packed on an IC chip. This permits very high levels of integration. Indeed, the most important application of NMOS is in the design of VLSI circuits such as microprocessors and random-access memory. In such applications, the low load-driving capability of NMOS is not a serious drawback. It is this low load-driving capability, however, that makes NMOS impractical for conventional digital system design. Thus NMOS logic is not available as SSI or MSI "off-the-shelf" components as is the case, for example, with CMOS or TTL.

15.5 THE CMOS INVERTER

Complementary MOS or CMOS is currently the most popular digital circuit technology. CMOS logic circuits are available as standard SSI and MSI packages for use in conventional digital system design. CMOS is also used in the design of general-purpose VLSI circuits such as memory and microprocessors. For custom and semi-custom VLSI, CMOS is the technology of choice. Furthermore, CMOS is currently popular in analog-circuit applications as well (see Chapters 7, 9, and 13).

In this section we study the basic CMOS inverter, shown in Fig. 15.15a. The inverter utilizes two matched enhancement-type MOSFETs: one, Q_N, with an n channel

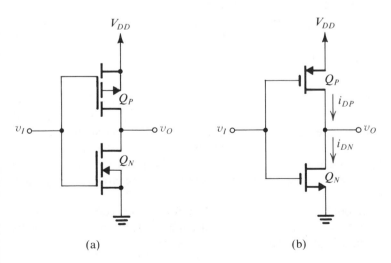

Fig. 15.15 (a) CMOS inverter. **(b)** Simplified circuit schematic for the inverter.

(a)   (b)

and the other, Q_P, with a p channel. As indicated, the body of each device is connected to its source and thus no body effect arises. We shall therefore use the simplified circuit schematic diagram shown in Fig. 15.15b.

Circuit Operation

We first consider the two extreme cases: when v_I is at logic-0 level, which is approximately 0V, and when v_I is at logic-1 level, which is approximately V_{DD} volts. In both cases, we shall consider the n-channel device Q_N to be the driving transistor and the p-channel device Q_P to be the load. However, since the circuit is completely symmetric, this assumption is obviously arbitrary, and the reverse would lead to identical results.

Figure 15.16 illustrates the case when $v_I = V_{DD}$, showing the i_D-v_{DS} characteristic curve for Q_N with $v_{GSN} = V_{DD}$. (Note that $i_D = i$ and $v_{DSN} = v_O$.) Superimposed on the Q_N characteristic curve is the load curve, which is the i_D-v_{SD} curve of Q_P for the case $v_{SGP} = 0V$. Since $v_{SGP} < |V_t|$, the load curve will be a horizontal straight line at almost zero current level. The operating point will be at the intersection of the two curves, where we note that the output voltage is nearly zero (typically less than 10 mV) and the current through the two devices is also nearly zero. This means that the power dissipation in the circuit is very small (typically a fraction of a microwatt).

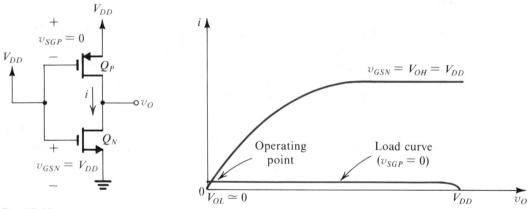

Fig. 15.16 Operation of the CMOS inverter when v_I is high.

The other extreme case, when $v_I = 0V$, is illustrated in Fig. 15.17. In this case Q_N is operating at $v_{GSN} = 0$; hence its i_D-v_{DS} characteristic is almost a horizontal straight line at zero current level. The load curve is the i_D-v_{SD} characteristic of the p-channel device with $v_{SGP} = V_{DD}$. As shown, at the operating point the output voltage is almost equal to V_{DD} (typically less than 10 mV below V_{DD}) and the current in the two devices is still nearly zero. Thus the power dissipation in the circuit is very small in both extreme states.

From the above we conclude that the basic CMOS logic inverter behaves as an "ideal" logic element: the output voltage is almost equal to zero volts or V_{DD} volts and the power dissipation is almost zero.

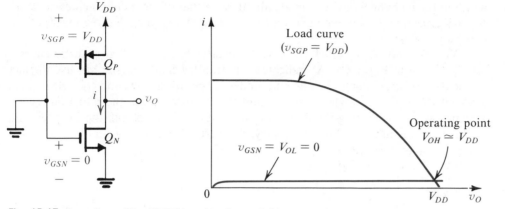

Fig. 15.17 Operation of the CMOS inverter when v_I is low.

It should be noted, however, that in spite of the fact that the quiescent current is zero, the load-driving capability of the CMOS inverter is high. For instance, with the input high, as in the circuit of Fig. 15.16, transistor Q_N can sink a relatively large load current. This current can quickly discharge the load capacitance, as will be seen shortly. Because of its action in sinking load current and thus pulling the output voltage down toward ground, transistor Q_N is known as the "pull-down" device. Similarly, with the input low, as in the circuit of Fig. 15.17, transistor Q_P can source a relatively large load current. This current can quickly charge up a load capacitance, thus pulling the output voltage up toward V_{DD}. Hence, Q_P is known as the pull-up device.

The Voltage Transfer Characteristic

The complete voltage transfer characteristic of the CMOS inverter can be obtained by repeating the graphical procedure, used above in the two extreme cases, for all intermediate value of v_I. In the following, we shall calculate the critical points of the resulting voltage transfer curve. For this we need the i-v relationships of Q_N and Q_P. For Q_N,

$$i_{DN} = K_n[2(v_I - V_{tn})v_O - v_O^2] \quad \text{for } v_O \leq v_I - V_{tn} \tag{15.30}$$

and

$$i_{DN} = K_n(v_I - V_{tn})^2 \quad \text{for } v_O \geq v_I - V_{tn} \tag{15.31}$$

For Q_P,

$$i_{DP} = K_p[2(V_{DD} - v_I - |V_{tp}|)(V_{DD} - v_O) - (V_{DD} - v_O)^2] \quad \text{for } v_O \geq v_I + |V_{tp}| \tag{15.32}$$

and

$$i_{DP} = K_p(V_{DD} - v_I - |V_{tp}|)^2 \quad \text{for } v_O \leq v_I + |V_{tp}| \tag{15.33}$$

The CMOS inverter is usually designed to have $V_{tn} = |V_{tp}| = V_t$ and $K_n = K_p = K$. It should be noted that since μ_p is about half the value of μ_n, to make $K_p = K_n$ the width of the *p*-channel device is made about twice that of the *n*-channel device. With $K_n = K_p$, the inverter has equal current-driving capability in both directions (pull-up and pull-down).

With Q_N and Q_P matched, the CMOS inverter has the voltage transfer characteristic shown in Fig. 15.18. As indicated, the transfer characteristic has five distinct segments corresponding to the different modes of operation of Q_N and Q_P. The vertical segment BC is obtained when both Q_N and Q_P are operating in the pinch-off region. Because we are neglecting the finite output resistance in pinch-off, the inverter gain in this region is infinite. From symmetry, this vertical segment occurs at $v_I = V_{DD}/2$ and is bounded by $v_O(\text{B}) = V_{DD}/2 + V_t$ and $v_O(\text{C}) = V_{DD}/2 - V_t$.

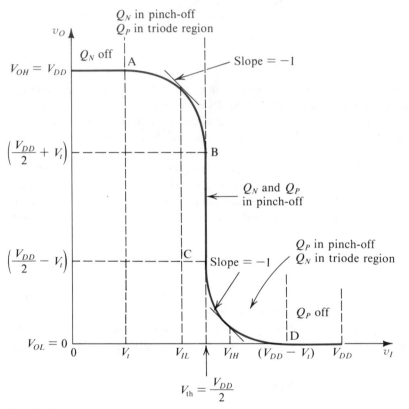

Fig. 15.18 The voltage transfer characteristic of the CMOS inverter.

To determine V_{IH} we note that Q_N is in the triode region, and thus its current is given by Eq. (15.30), while Q_P is in pinch-off and its current is given by Eq. (15.33). Equating i_{DN} and i_{DP}, and assuming matched devices, gives

$$2(v_I - V_t)v_O - v_O^2 = (V_{DD} - v_I - V_t)^2 \tag{15.34}$$

Differentiating both sides relative to v_I results in

$$2(v_I - V_t)\frac{dv_O}{dv_I} + 2v_O - 2v_O\frac{dv_O}{dv_I} = -2(V_{DD} - v_I - V_t)$$

in which we substitute $v_I = V_{IH}$ and $dv_O/dv_I = -1$ to obtain

$$v_O = V_{IH} - \frac{V_{DD}}{2} \tag{15.35}$$

Substituting in Eq. (15.34) $v_I = V_{IH}$ and for v_O from Eq. (15.35) gives

$$V_{IH} = \tfrac{1}{8}(5V_{DD} - 2V_t) \tag{15.36}$$

V_{IL} can be determined in a manner similar to that used to find V_{IH}. Alternatively, we can use the symmetry relationship

$$V_{IH} - \frac{V_{DD}}{2} = \frac{V_{DD}}{2} - V_{IL}$$

together with V_{IH} from Eq. (15.36) to obtain

$$V_{IL} = \tfrac{1}{8}(3V_{DD} + 2V_t) \tag{15.37}$$

The noise margins can now be determined as follows:

$$\begin{aligned} NM_H &= V_{OH} - V_{IH} \\ &= V_{DD} - \tfrac{1}{8}(5V_{DD} - 2V_t) \\ &= \tfrac{1}{8}(3V_{DD} + 2V_t) \end{aligned} \tag{15.38}$$

$$\begin{aligned} NM_L &= V_{IL} - V_{OL} \\ &= \tfrac{1}{8}(3V_{DD} + 2V_t) - 0 \\ &= \tfrac{1}{8}(3V_{DD} + 2V_t) \end{aligned} \tag{15.39}$$

As expected, the symmetry of the voltage transfer characteristic results in equal noise margins. Of course, if Q_N and Q_P are not matched, the voltage transfer characteristic will no longer be symmetric and the noise margins will not be equal (see Problems 15.30 and 15.31).

Exercises

15.14 For a CMOS inverter with matched MOSFETs having $V_t = 1$ V find V_{IL}, V_{IH}, and the noise margins if $V_{DD} = 5$ V.

Ans. 2.1 V; 2.9 V; 2.1 V

15.15 Consider a CMOS inverter with $V_{tn} = |V_{tp}| = 2$ V, $(W/L)_N = 20$, $(W/L)_P = 40$, $\mu_n C_{OX} = 2\mu_p C_{OX} = 20\mu A/V^2$, and $V_{DD} = 10$ V. For $v_I = V_{DD}$ find the maximum current that the inverter can sink while v_O remains ≤ 0.5 V.

Ans. 1.55 mA

Current Flow and Power Dissipation

As the CMOS inverter is switched, current flows through the series connection of Q_N and Q_P. Figure 15.19 shows the inverter current as a function of v_I. We note that the current peaks at the switching point, $v_I = V_{DD}/2$. This current gives rise to dynamic power dissipation in the CMOS inverter. However, a more significant component of dynamic power dissipation results from the current that flows in Q_N and Q_P when

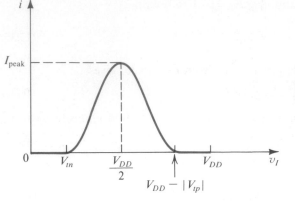

Fig. 15.19 The current in the CMOS inverter versus the input voltage.

the inverter is loaded by a capacitor C. This latter component of power dissipation, as given by Eq. (15.3), is

$$P_D = fCV_{DD}^2$$

where f is the frequency at which the inverter is switched.

Exercises

15.16 For the inverter specified in Exercise 15.15, find the peak current drawn from V_{DD} during switching.

Ans. 1.8 mA

15.17 Let the inverter specified in Exercise 15.15 be loaded by a 15-pF capacitance. Find the dynamic power dissipation that results when the inverter is switched at a frequency of 2 MHz. What is the average current drawn from the power supply?

Ans. 3 mW; 0.3 mA

Dynamic Operation

Unlike the NMOS inverters, the CMOS inverter features equally fast turn-on and turn-off times in the presence of capacitive loads. To illustrate, consider the capacitively loaded CMOS inverter of Fig. 15.20a driven by the ideal pulse (zero rise and fall times) shown in Fig. 15.20b. Since the circuit is symmetric (assuming matched MOSFETs), the rise and fall times of the output waveform should be equal. It is sufficient, therefore, to consider either the turn-on or the turn-off process. In the following we consider the first.

Figure 15.20c shows the trajectory of the operating point obtained when the input pulse goes from $V_{OL} = 0$ to $V_{OH} = V_{DD}$ at time $t = 0$. Just prior to the leading edge of the input pulse (that is, at $t = 0-$) the output voltage equals V_{DD} and capacitor C is charged to this voltage. At $t = 0$, v_I rises to V_{DD}, causing Q_P to turn off immediately. From there on, the circuit is equivalent to that shown in Fig. 15.20d with the initial value of $v_O = V_{DD}$. Thus the operating point at $t = 0+$ is point E, at which it is seen that Q_N will be in the pinch-off region and conducting a large current. As C discharges, the current of Q_N remains constant until $v_O = V_{DD} - V_t$ (point F). Denoting this

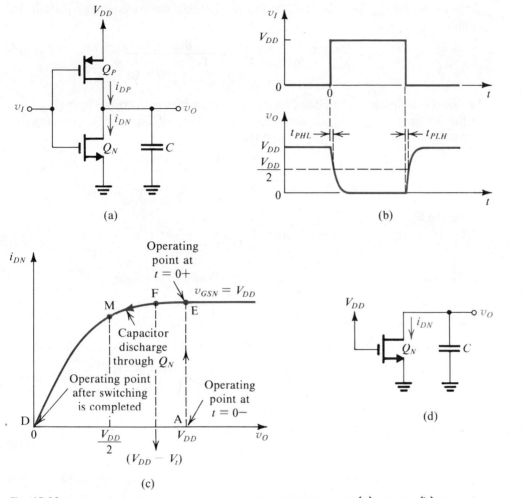

Fig. 15.20 Dynamic operation of a capacitively loaded CMOS inverter: **(a)** circuit; **(b)** input and output waveforms; **(c)** trajectory of operating point as the input goes high and C discharges through Q_N; **(d)** equivalent circuit during the capacitor discharge.

portion of the discharge interval t_{PHL1} we can write

$$t_{PHL1} = \frac{C[V_{DD} - (V_{DD} - V_t)]}{K_n(V_{DD} - V_t)^2}$$

$$= \frac{CV_t}{K_n(V_{DD} - V_t)^2} \tag{15.40}$$

Beyond point F, transistor Q_N operates in the triode region and thus its current is given by Eq. (15.30). This portion of the discharge interval can be described by

$$i_{DN}\, dt = -C\, dv_O$$

Substituting for i_{DN} from Eq. (15.30) and rearranging the differential equation, we obtain

$$-\frac{K_n}{C}\,dt = \frac{1}{2(V_{DD} - V_t)}\frac{dv_O}{\dfrac{1}{2(V_{DD} - V_t)}v_O^2 - v_O} \tag{15.41}$$

To find the component of the delay time t_{PHL} during which v_O decreases from $(V_{DD} - V_t)$ to the 50% point, $v_O = V_{DD}/2$, we integrate both sides of Eq. (15.41). Denoting this component of delay time t_{PHL2} we find that

$$-\frac{K_n}{C}\,t_{PHL2} = \frac{1}{2(V_{DD} - V_t)}\int_{v_O = V_{DD} - V_t}^{v_O = V_{DD}/2}\frac{dv_O}{\dfrac{1}{2(V_{DD} - V_t)}v_O^2 - v_O} \tag{15.42}$$

Using the fact that

$$\int\frac{dx}{ax^2 - x} = \ln\left(1 - \frac{1}{ax}\right)$$

enables us to evaluate the integral in Eq. (15.42) and thus obtain

$$t_{PHL2} = \frac{C}{2K_n(V_{DD} - V_t)}\ln\left(\frac{3V_{DD} - 4V_t}{V_{DD}}\right) \tag{15.43}$$

The two components of t_{PHL} in Eqs. (15.40) and (15.43) can be added to obtain

$$t_{PHL} = \frac{C}{K_n(V_{DD} - V_t)}\left[\frac{V_t}{V_{DD} - V_t} + \frac{1}{2}\ln\left(\frac{3V_{DD} - 4V_t}{V_{DD}}\right)\right] \tag{15.44}$$

For the usual case of $V_t \simeq 0.2V_{DD}$ this equation reduces to

$$t_{PHL} = \frac{0.8C}{K_nV_{DD}} \tag{15.45}$$

Similar analysis of the turn-off process yields an expression for t_{PLH} identical to that in Eq. (15.44) except for K_p replacing K_n.

Exercises

15.18 A CMOS inverter in a VLSI circuit operating from a 5-V supply has $(W/L)_N = 10\ \mu\text{m}/5\ \mu\text{m}$, $(W/L)_P = 20\ \mu\text{m}/5\ \mu\text{m}$, $V_{tn} = |V_{tp}| = 1$ V, $\mu_nC_{OX} = 2\mu_pC_{OX} = 20\ \mu\text{A/V}^2$. If the total effective load capacitance is 0.1 pF, find t_{PHL}, t_{PLH}, and t_P.

Ans. 0.8 ns; 0.8 ns; 0.8 ns

15.19 For the CMOS inverter of Exercise 15.15, which is intended for SSI and MSI circuit applications, find t_P if the load capacitance is 15 pF.

Ans. 6 ns

Delay-Power Product

The delay-power product of CMOS is obtained by multiplying the dynamic power dissipation by the propagation delay time. The resulting values range from much less than 1 pJ for VLSI circuits to about 10 pJ for commercially available SSI. The value of *DP* is directly proportional to the rate of switching and thus can be considerably reduced by operating at low rates. Also, like NMOS, *DP* for CMOS can be reduced by reducing the load capacitance and/or the power-supply voltage.

Exercise

15.20 Calculate the delay-power product of the CMOS inverter of Exercise 15.18 when it is operating at a switching rate of 50 MHz.

Ans. 0.1 pJ

15.6 CMOS GATE CIRCUITS

CMOS gate circuits are formed by extending the basic inverter circuit. To illustrate, we show a two-input NOR gate in Fig. 15.21a. At the heart of the gate is the inverter Q_1, Q_2. The NOR gate circuit is obtained by adding a parallel *n*-channel device Q_3 and a series *p*-channel device Q_4. This process is repeated for each additional input. Since each additional input requires an additional pair of complementary MOSFETs, CMOS consumes more silicon area than NMOS, where, as we have seen in Section 15.4, each additional gate input requires the addition of only one transistor. The NAND gate of Fig. 15.21b is formed by a *dual* process: for each additional input we add a series *n*-channel device and a parallel *p*-channel device. It is interesting to note that if in one gate (say the NOR gate) we change each *n*-channel device to a *p*-channel device and vice versa, and exchange the connections to V_{DD} and ground we obtain the other gate.

Operation of the gate circuits in Fig. 15.21 is straightforward. The output of the circuit in Fig. 15.21a will be high (at V_{DD}) if both Q_2 and Q_4 are simultaneously on.

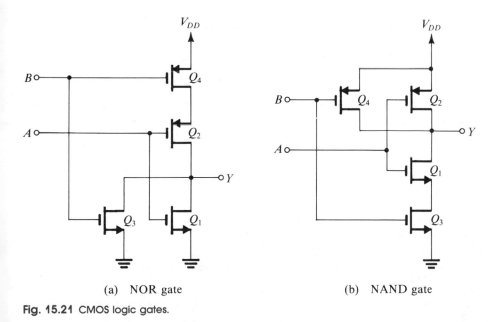

(a)   NOR gate   (b)   NAND gate

Fig. 15.21 CMOS logic gates.

This will be achieved only when A and B are simultaneously low. Thus

$$Y = \bar{A}\bar{B}$$

or, equivalently,

$$Y = \overline{A + B}$$

which is the NOR logic function. On the other hand, the output of the circuit in Fig. 15.21b will be low (at 0 V) only when Q_1 and Q_3 are simultaneously on. This will occur only when A and B are simultaneously high. Thus,

$$\bar{Y} = AB$$

or, equivalently,

$$Y = \overline{AB}$$

which is the NAND logic function.

Device Dimensions

It is desirable to design the CMOS gate circuits so as to provide equal output current-driving capability in both directions (pull-down and pull-up). For the two-input NOR gate of Fig. 15.21a we see that if the two inputs are tied together, then the pull-down current is the sum of the currents provided by Q_1 and Q_3. Since Q_1 and Q_3 are identical, the pull-down current is double that of the n-channel device. On the other hand, the pull-up current is provided by Q_2 and Q_4 in series and is thus equal to that of a single p-channel device. It follows that the output currents will be equal if each p-channel device is designed to have double the value of K of each of the n-channel devices; that is,

$$K_2 = K_4 = 2K_1 = 2K_3 \qquad (15.46)$$

Now since $\mu_p \simeq \frac{1}{2}\mu_n$, we see that to satisfy the above condition the W/L ratio of each of the p-channel devices must be about four times the W/L ratio of each of the n-channel devices. In general, for an N-input NOR gate,

$$(W/L)_p \simeq 2N(W/L)_n \qquad (15.47)$$

Applying the same reasoning to the NAND gate, we find that for the two-input NAND of Fig. 15.21b we should have $(W/L)_p \simeq (W/L)_n$. It follows that a two-input NAND requires smaller silicon area than a two-input NOR. In general, for an N-input NAND, one should use

$$\left(\frac{W}{L}\right)_n = \frac{N}{2}\left(\frac{W}{L}\right)_p \qquad (15.48)$$

Exercise

15.21 Consider a CMOS technology in which the channel length of all devices is 5 μm and in which the minimum desired output current capability of the basic inverter is achieved with a $(W/L)_n = 2$. For a two-input NOR gate find the width of each of the n-channel and p-channel devices and the total area occupied by all devices. Repeat for a two-input NAND gate.

Ans. **(a)** NOR: $W_n = 10$ μm, $W_p = 40$ μm, area $= 500$ μm^2; **(b)** NAND: $W_n = 20$ μm, $W_p = 20$ μm, area $= 400$ μm^2

Gate Threshold

Even if the devices in a CMOS gate are designed according to the above considerations, the observed gate threshold will differ from the ideal value of $V_{DD}/2$ for a number of reasons. First, there are unavoidable mismatches. Second, the series devices (except for one) suffer from the body effect. Third, the gate threshold will depend on the signal values at the other inputs to the gate. The latter effect is illustrated in the following example.

EXAMPLE 15.3

Consider the two-input NOR gate of Fig. 15.21a. Assume that $K_2 = K_4 = 2K_1 = 2K_3$ and that all devices have a V_t of 2 V. Calculate the value of the switching threshold of the gate V_{th} if (a) input terminal B is connected to ground and (b) input terminals A and B are joined together. Let the power supply voltage $V_{DD} = 10$ V and neglect the body effect in Q_2.

Solution

(a) Figure 15.22a shows a simplified circuit diagram of the NOR circuit with terminal B connected to ground. Since Q_3 will be off, we have eliminated it from the circuit altogether. The switching threshold V_{th} is the value of input voltage v_I at which both Q_1 and Q_2 are in the pinch-off region. (Figure 15.18 shows the threshold voltage obtained in the ideal situation.) For the no-load condition, the currents in Q_1 and Q_2 will be equal and given by

$$I = K_1(V_{th} - V_t)^2 = K_2(V_1 - V_{th} - V_t)^2 \qquad (15.49)$$

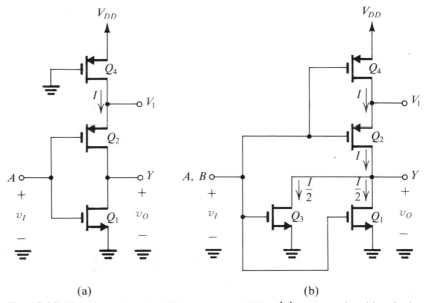

(a)   (b)

Fig. 15.22 Circuits for Example 15.3: a two-input NOR **(a)** with one input terminal grounded and **(b)** with the input terminals tied together.

where V_1 denotes the voltage at the source of Q_2. Since the value of V_{th} will be close to $V_{DD}/2$, V_1 will be higher than this value by at least V_t. Hence it is reasonable to expect that Q_4 will be operating in the triode region. We shall make this assumption and later check its validity. Thus we can write for Q_4

$$I = K_4[2(V_{DD} - V_t)(V_{DD} - V_1) - (V_{DD} - V_1)^2] \qquad (15.50)$$

Eliminating I from Eqs. (15.49) and (15.50) and substituting $K_2 = K_4 = 2K_1$ gives two equations whose simultaneous solution yields $V_{th} = 5.31$ V and $V_1 = 9.65$ V. Thus Q_4 is indeed in the triode region, as assumed. Note that the value of V_{th} is slightly greater than the ideal value of $V_{DD}/2$, a result of the additional series device Q_4. At the switching threshold voltage V_{th} of 5.31 V the output voltage abruptly changes from $(5.31 + V_t)$ to $(5.31 - V_t)$—that is, from 7.31 V to 3.31 V.

(b) Next we consider the case where the two inputs are tied together as indicated in the circuit of Fig. 15.22b. Again, we shall assume that with $v_I = V_{th}$ the value of V_1 will be such that Q_4 will be operating in the triode region. The other three transistors will be assumed to operate in pinch-off. Straightforward analysis gives $V_1 \simeq 9$ V and $V_{th} \simeq 4.5$ V. Thus we see that the gate threshold is 0.5 V below the ideal value of $V_{DD}/2 = 5$ V.

Transmission Gates

In addition to the basic CMOS gate circuits discussed above, the CMOS transmission gate, shown in Fig. 15.23, is an important building block in both digital and analog systems. We have studied the transmission gate in Section 7.13. An application of CMOS transmission gates will be presented in the next section.

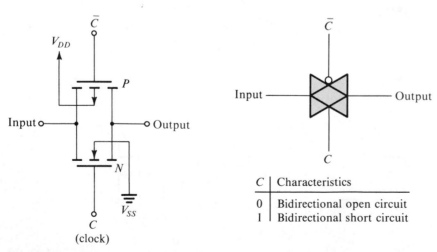

Fig. 15.23 Basic CMOS transmission gate.

C	Characteristics
0	Bidirectional open circuit
1	Bidirectional short circuit

Commercially Available CMOS Logic

As well as being used in VLSI circuit design, CMOS is available commercially in SSI, MSI and LSI logic circuit packages for conventional digital system design. The oldest family of CMOS logic circuits, the 4000 series, utilizes metal gate technology (see Appendix A). These standard CMOS parts can operate over a supply voltage range of 3 to 18 V. (This should be contrasted to the typical allowed range for TTL of 4.75 to 5.25 V; see Chapter 16.)

Manufacturers of CMOS circuits usually specify minimum and maximum transfer characteristics in the manner of Fig. 15.24. All supplied CMOS gates are guaranteed to have transfer characteristics in the shaded area. The figure illustrates the definition of worst-case values for V_{IL} and V_{IH}—namely, as the input values at which v_O deviates from the corresponding ideal value (V_{DD} and 0) by a specified maximum, ΔV. (Note that these definitions differ from those we have been using and which are currently the standard.) Normally the deviation in output voltage ΔV is taken as $0.1V_{DD}$. As an example, Motorola specifies for standard CMOS logic:

$$V_{DD} = 5 \text{ V}: \quad V_{IL} = 1 \text{ V}, V_{IH} = 4 \text{ V}$$

$$V_{DD} = 10 \text{ V}: \quad V_{IL} = 2 \text{ V}, V_{IH} = 8 \text{ V}$$

$$V_{DD} = 15 \text{ V}: \quad V_{IL} = 2.5 \text{ V}, V_{IH} = 12.5 \text{ V}$$

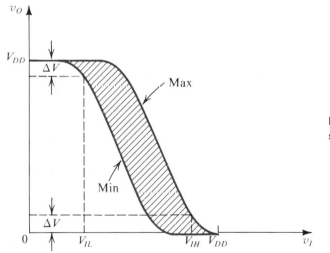

Fig. 15.24 The two limit transfer characteristics specified for a CMOS gate.

These values can be used together with $\Delta V = 0.1V_{DD}$ to find the worst-case noise margins NM_H and NM_L:

$$V_{DD} = 5 \text{ V}: \quad NM_L = NM_H = 0.5 \text{ V}$$

$$V_{DD} = 10 \text{ V}: \quad NM_L = NM_H = 1 \text{ V}$$

$$V_{DD} = 15 \text{ V}: \quad NM_L = NM_H = 1 \text{ V}$$

The dynamic response of CMOS gates is specified in terms of the rise time, fall time, and propagation delay of the output pulse (see Fig. 15.7). As an example, for standard CMOS logic operated at $V_{DD} = 10$ V and driven with a pulse having 20 ns rise and fall times, Motorola specifies

$$t_{TLH} = (1.5 \text{ ns/pF})C_L + 15 \text{ ns}$$

$$t_{THL} = (0.75 \text{ ns/pF})C_L + 12.5 \text{ ns}$$

$$t_{PLH}, t_{PHL} = (0.66 \text{ ns/pF})C_L + 22 \text{ ns}$$

Note that the rise time is longer than the fall time because in these circuits the p device has a lower K than that of the n device. These formulas can be used to compute the switching times for a given load capacitance C_L. A standard SSI CMOS gate has an input capacitance of about 5 pF. Maximum fan-out of a CMOS gate is limited by the deterioration in its dynamic response due to the increased load capacitance. This should be contrasted to the case of TTL (Chapter 16), where maximum fan-out is limited by the degradation in noise margins.

An example of specifying dynamic power dissipation in CMOS logic is as follows: Motorola states that a standard CMOS gate operating with $V_{DD} = 10$ V and loaded with $C_L = 50$ pF draws an average dc supply current of 0.6 μA/kHz. Thus at a frequency of 1 MHz such a gate dissipates 6 mW. Finally, we note that although higher operating speeds are possible at higher supply voltages, these are obtained at the expense of increased dynamic power dissipation.

Improved-Performance CMOS

A variety of CMOS subfamilies exist with varying degrees of improved performance, including increased noise margins, higher output-driving capability, and higher speed of operation. Of special note is the buffered or B-series CMOS which includes two cascaded CMOS inverters placed between the output node of the logic gate circuit and the external output terminal. The MOSFETs in the buffer inverters have W/L ratios that are much greater than those of the gate circuit proper. Thus the buffer provides an increased output drive capability. Also, the buffer provides additional voltage gain in the transition region. Since this results in a much sharper voltage transfer characteristic, the noise margins are increased.

Exercise

15.22 The buffered CMOS family is specified to have the following worst-case values.
(a) At $V_{DD} = 5$ V, $V_{IL} = 1.5$ V and $V_{IH} = 3.5$ V.
(b) At $V_{DD} = 10$ V, $V_{IL} = 3$ V and $V_{IH} = 7$ V.
(c) At $V_{DD} = 15$ V, $V_{IL} = 4$ V and $V_{IH} = 11$ V.
By reference to Fig. 15.24 and using $\Delta V = 0.1\ V_{DD}$, calculate the corresponding worst-case noise margins.

Ans. **(a)** 1 V; **(b)** 2 V; **(c)** 2.5 V

(*Note:* Compare these values with the corresponding ones given in the text above for standard CMOS.)

Further improved versions of CMOS utilize devices with channel lengths as short as 1.25 μm. The use of *small-feature-size* devices results in increased speed of operation.

Some Practical Considerations

As mentioned in Chapter 7, because of the very high input resistance of MOS devices, static electricity can cause a charge to accumulate on the input capacitances during handling. Such charges can cause large voltages to appear at the MOSFET gate, which in turn can cause breakdown of the thin gate oxide. CMOS circuits include input diode networks, together with appropriate series resistors, for the purpose of limiting the voltages that may appear at the MOS device inputs (see Hodges and Jackson, 1983).

Since the input resistances of a CMOS gate are very high, a gate input left unconnected will float at an unspecified voltage. Usually, however, leakage currents of the input protection diodes are such that the input devices enter the active mode, allowing large currents to flow that cause overheating. Accordingly it is important that spare gate inputs be connected to an appropriate local power-supply pin or paralleled with another input (keeping in mind the effect of this on the gate switching threshold).

A Final Remark

CMOS logic circuits are highly versatile, finding application in diverse areas ranging from micropower circuits for watches and calculators to circuits that require high noise immunity, such as those used in automobiles and home appliances.

15.7 LATCHES AND FLIP-FLOPS

The logic circuits considered thus far are called *combinational* (or *combinatorial*). Their output depends only on the current value of the input. Thus these circuits do *not* have memory.

Memory is a very important part of digital systems. Its availability in digital computers allows for storing programs and data. Furthermore, it is important for temporary storage of the output produced by a combinational circuit for use at a later time in the operation of a digital system.

Logic circuits that incorporate memory are called *sequential circuits;* that is, their output depends not only on the present value of the input but also on the input's previous values. Such circuits require a timing generator (a *clock*) for their operation.[4]

In this section we shall study the basic memory element, the latch. We shall also examine two applications of latches, that of an NMOS *set/reset* (SR) flip/flop and a CMOS D (or data) flip-flop.

---

[4] Some combinational logic circuits also require a clock. Such circuits are called *dynamic logic* (see Hodges and Jackson, 1983). All the logic circuits we have studied fall into the static-logic category.

The Latch

The basic memory element, the latch, is shown in Fig. 15.25a. It consists of two cross-coupled logic inverters, G_1 and G_2. The inverters form a positive-feedback loop. To investigate the operation of the latch we break the feedback loop at the input of one of the inverters, say G_1, and apply an input signal, v_W in Fig. 15.25b. Assuming that the input impedance of G_1 is large, breaking the feedback loop will not change the loop voltage transfer characteristic, which can be determined from the circuit of Fig. 15.25b by plotting v_Z versus v_W. This is the voltage transfer characteristic of two cascaded inverters and thus takes the shape shown in Fig. 15.25c. Observe that the transfer characteristic consists of three segments, with the middle segment corresponding to the transition region of the inverters.

Also shown in Fig. 15.25c is a straight line with unity slope. This straight line represents the relationship $v_W = v_Z$ which is realized by reconnecting Z to W to close

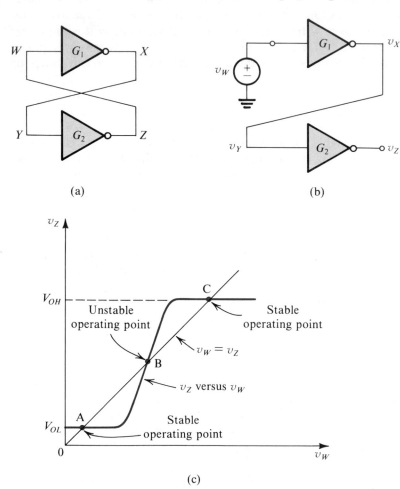

(a) (b)

(c)

Fig. 15.25 (a) Basic latch. **(b)** The latch with the feedback loop opened. **(c)** Determining the operating point of the latch.

the feedback loop. As indicated, the straight line intersects the loop transfer curve at three points, A, B, and C. Thus any of these three points can serve as the operating point for the latch. We shall now show that while points A and C are stable operating points in the sense that the circuit can remain at either indefinitely, point C is an unstable operating point; the latch cannot operate at B for any period of time.

The reason point B is unstable can be seen by considering the latch circuit in Fig. 15.25a to be operating at point B, and taking account of the interference or noise that is inevitably present in any circuit. Let the voltage v_W increase by a small increment v_w. The voltage at X will increase (in magnitude) by a larger increment, equal to the product of v_w and the incremental gain of G_1 at point B. The resulting signal v_x is applied to G_2 and gives rise to an even larger signal at node Z. The voltage v_z is related to the original increment v_w by the loop gain at point B which is the slope of the v_Z versus v_W curve at point B. This gain is usually much greater than unity. Since v_z is coupled to the input of G_1, it will be further amplified by the loop gain. This regenerative process continues, shifting the operating point from B upwards to point C. Since at C the loop gain is zero (or almost zero) no regeneration can take place.

In the description above, we assumed an initial positive voltage increment at W. Had we instead assumed a negative voltage increment we would have seen that the operating point moves downward from B to A. Again, since at point A the slope of the transfer curve is zero (or almost zero), no regeneration can take place. In fact for regeneration to occur the loop gain must be greater than unity, which is the case at point B.

The above discussion leads us to conclude that the latch has two stable operating points, A and C. At point C, v_W is high, v_X is low, v_Y is low, and v_Z is high. The reverse is true at point A. If we consider X and Z as the latch outputs, we see that in one of the stable states (say that corresponding to operating point A) v_X is high (at V_{OH}) and v_Z is low (at V_{OL}). In the other state (corresponding to operating point C) v_X is low (at V_{OL}) and v_Z is high (at V_{OH}). Thus the latch is a *bistable* circuit having two complementary outputs. In which of its two stable states the latch operates depends on the external excitation that forces it to the particular state. The latch then *memorizes* this external action by staying indefinitely in the acquired state. As a memory element the latch is capable of storing one bit of information. For instance, we can arbitrarily designate the state in which v_X is high and v_Z is low as corresponding to a stored logic 1. The other complementary state then designates a stored logic 0.

It now remains to devise a mechanism by which the latch can be *triggered* to change state. The latch together with the triggering circuitry forms a *flip-flop*. This will be discussed next. At this point, however, we wish to remind the reader that bistable circuits utilizing op amps were presented in Chapter 5.

The SR Flip-Flop

The simplest type of flip-flop is the set-reset (SR) flip-flop shown in Fig. 15.26a. It is formed by cross-coupling two NOR gates, and thus it incorporates a latch. The second input of each NOR gate together serve as the trigger inputs of the flip-flop.

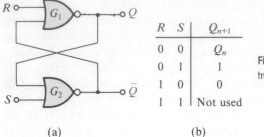

R	S	Q_{n+1}
0	0	Q_n
0	1	1
1	0	0
1	1	Not used

Fig. 15.26 The set-reset (SR) flip-flop and its truth table.

(a)   (b)

These two inputs are labeled S (for set) and R (for reset). The outputs are labeled Q and $\bar{Q}$, emphasizing the fact that these are complementary. The flip-flop is considered set (that is, storing a logic 1) when Q is high and $\bar{Q}$ is low. When the flip-flop is in the other state (Q low, $\bar{Q}$ high), it is considered reset (storing a logic 0). In the rest or memory state (that is, when we do not wish to change the state of the flip-flop) both the S and R inputs should be low.

Consider the case when the flip-flop is storing a logic zero. Q will be low and thus both inputs to the NOR gate G_2 will be low. Its output will therefore be high. This high is applied to the input of G_1, causing its output Q to be low. To set the flip-flop we raise S to the logic-1 level while leaving R at 0. The 1 at the S terminal will force the output of G_2, $\bar{Q}$, to 0. Thus the two inputs to G_1 will be 0 and its output Q will go to 1. Now even if S returns to 0, the flip-flop remains in the newly acquired set state. Obviously, if we raise S to 1 again (with R remaining at 0) no change will occur. To reset the flip-flop we need to raise R to 1 while leaving $S = 0$. We can readily show that this forces the flip-flop into the reset state and that the flip-flop remains in this state even after R returns to 0. It should be observed that the trigger signal merely starts the regenerative action of the positive-feedback loop of the latch.

Finally, we inquire into what happens if both S and R are simultaneously raised to 1. The two NOR gates will cause both Q and $\bar{Q}$ to become 0 (note that in this case the complementary labeling of these two variables is incorrect). However, if R and S return to the rest state ($R = S = 0$) simultaneously, the state of the flip-flop will be undefined. In other words it will be impossible to predict the final state of the flip-flop. For this reason, this input combination is usually disallowed (that is, not used). Note, however, that this situation arises only in the idealized case, when both R and S return to 0 precisely simultaneously. In actual practice one of the two will return to 0 first, and the final state will be determined by the input that remains high longest.

The operation of the flip-flop is summarized by the truth table in Fig. 15.26b, where Q_n denotes the value of Q at time t_n just before the application of the R and S signals, and Q_{n+1} denotes the value of Q at time t_{n+1} after the application of the input signals.

Rather than using two NOR gates, one can also implement an SR flip-flop by cross-coupling two NAND gates.

An SR flip-flop constructed using two two-input depletion-load NMOS NOR gates is shown in Fig. 15.27. This circuit is a direct implementation of the flip-flop shown in logic form in Fig. 15.26, and thus its operation should be self-evident.

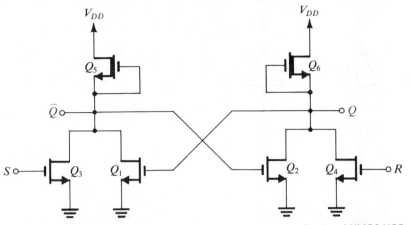

Fig. 15.27 An SR flip-flop formed by cross-coupling two depletion-load NMOS NOR gates.

A CMOS D Flip-Flop

A variety of flip-flop types exist. Most can be synthesized in terms of logic gates. The gates can then be replaced with their circuit implementations using the desired technology (NMOS, CMOS, TTL, etc.). For this reason we shall not present these flip-flop circuits here. An exception is a CMOS data or D flip-flop that utilizes transmission gates together with NOR gates in an interesting fashion.

The D flip-flop is shown in block diagram form in Fig. 15.28. It has two inputs, the data input D and the clock input C. The complementary outputs are labeled Q and $\bar{Q}$. When the clock is low, the flip-flop is in the memory or rest state; signal changes on the D input line have no effect on the state of the flip-flop. As the clock goes high, the flip-flop acquires the logic level that existed on the D line just before the rising edge of the clock. Such a flip-flop is said to be *edge-triggered*. Some implementations of the D flip-flop include also direct set and reset inputs that override the clocked operation just described.

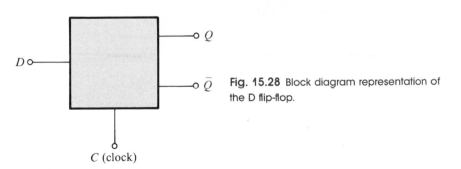

Fig. 15.28 Block diagram representation of the D flip-flop.

Figure 15.29 shows a CMOS implementation of the D flip-flop including direct set and reset inputs. The flip-flop is composed of two SR flip-flops called *master* and *slave*. Each consists of two NOR gates with a transmission gate inserted in the feedback loop. In addition, a transmission gate TG1 connects the D input to the master

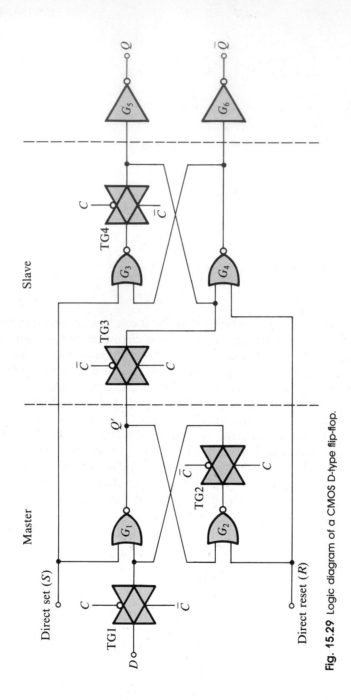

Fig. 15.29 Logic diagram of a CMOS D-type flip-flop.

flip-flop, and another, TG3, connects the output of the master flip-flop to the input of the slave flip-flop. The outputs of the D flip-flop are buffered versions of the outputs of the slave section.

Triggering the flip-flop from the set and reset inputs is straightforward and needs no further discussion. Operation as an edge-triggered D flip-flop is best understood by considering the transmission gates. When the clock is low, the transmission gates are in the following states.

$$TG1 \rightarrow ON, TG2 \rightarrow OFF, TG3 \rightarrow OFF, TG4 \rightarrow ON$$

Thus the slave is isolated from the master, and the feedback loop of the slave flip-flop is closed, making it retain its previous state. Meanwhile the feedback loop of the master flip-flop is open and the output of the master, Q', simply follows the complement of the input D.

When the clock input goes high, the transmission gates change to the following states

$$TG1 \rightarrow OFF, TG2 \rightarrow ON, TG3 \rightarrow ON, TG4 \rightarrow OFF$$

The result is that the master is disconnected from input D and its feedback loop is closed. Thus its output Q' adopts the complement of the signal that existed on input line D just prior to the positive transition of the clock. Meanwhile the feedback loop of the slave is opened and its output Q adopts the complement of Q', which is the value of input D. The overall effect is that on the positive transition of the clock the output Q adopts the value of input D that existed just prior to the transition.

Exercise

15.23 For the SR flip-flop of Fig. 15.27 let $V_{DD} = 5$ V, $(W/L)_{1,2,3,4} = 2$, $(W/L)_{5,6} = 0.5$, $V_{tE} = 1$ V, $\mu_n C_{OX} = 20$ μA/V^2, and $V_{tD} = -3$ V, and ignore the body effect. Assume that the total capacitance between each of the two output nodes and ground is 0.1 pF. Estimate the minimum required width of pulse at S or R to set or reset the flip-flop. (*Hint*: Use the results of Exercise 15.10.)

Ans. 7 ns

15.8 MULTIVIBRATOR CIRCUITS

As mentioned before, the flip-flop has two stable states and is called a bistable multivibrator. There are two other types of multivibrators: monostable and astable. The monostable multivibrator has one stable state in which it can remain indefinitely. It has another *quasi-stable* state to which it can be triggered. The monostable can remain in the quasi-stable state for a predetermined interval T, after which it automatically reverts to the stable state. In this way the monostable generates an output pulse of duration T. This pulse duration is in no way related to the details of the triggering pulse, as is indicated schematically in Fig. 15.30. The monostable can therefore be used as a *pulse stretcher* or, more appropriately, a *pulse standardizer*. A monostable is also referred to as a *one-shot*.

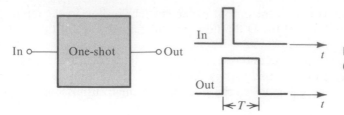

Fig. 15.30 The monostable multivibrator (one-shot) as a functional block.

The astable multivibrator has no stable states. Rather, it has two quasi-stable states, and it remains in each for predetermined intervals T_1 and T_2. Thus after T_1 seconds in one of the quasi-stable states the astable switches to the other quasi-stable state and remains there for T_2 seconds, after which it reverts back to the original state, and so on. The astable thus oscillates with a period $T = T_1 + T_2$ or a frequency $f = 1/T$, and it can be used to generate periodic pulses such as those required for clocking.

In Chapter 5 we studied astable and monostable multivibrator circuits that use op amps. In the following we shall discuss monostable and astable circuits using logic gates.

A CMOS Monostable Circuit

Figure 15.31 shows a simple and popular circuit for a monostable multivibrator. It is composed of two two-input CMOS NOR gates, G_1 and G_2, a capacitor of capacitance C, and a resistor of resistance R. The input source v_I supplies the triggering pulses for the monostable.

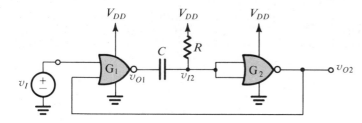

Fig. 15.31 Monostable circuit using CMOS NOR gates.

CMOS gates have a special arrangement of diodes connected at their input terminals, as indicated in Fig. 15.32a. The purpose of these diodes is to prevent the input voltage signal from rising above the supply voltage V_{DD} (by more than one diode drop) and from falling below ground voltage (by more than one diode drop). These clamping diodes have an important effect on the operation of the monostable circuit. Specifically, we shall be interested in the effect of these diodes on the operation of the inverter connected gate G_2. In this case each pair of corresponding diodes appears in parallel, giving rise to the equivalent circuit in Fig. 15.32b. While the diodes provide a low-resistance path to the power supply for voltages exceeding the power supply limits, the input current for intermediate voltages is essentially zero.

To simplify matters we shall use the approximate equivalent output circuit of the gate, illustrated in Fig. 15.33. Figure 15.33a indicates that when the gate output is

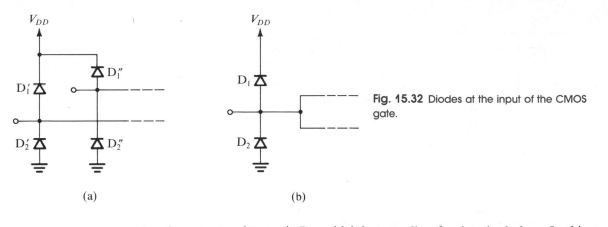

Fig. 15.32 Diodes at the input of the CMOS gate.

low, its output resistance is R_{on}, which is normally a few hundred ohms. In this state, current can flow from the external circuit into the output terminal of the gate; the gate is said to be *sinking* current. Similarly, the equivalent output circuit in Fig. 15.33b applies when the gate output is high. In this state, current can flow from V_{DD} through the output terminal of the gate into the external circuit; the gate is said to be *sourcing* current.

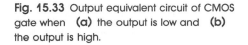

Fig. 15.33 Output equivalent circuit of CMOS gate when **(a)** the output is low and **(b)** the output is high.

To see how the monostable circuit of Fig. 15.31 operates, consider the timing diagram given in Fig. 15.34. Here a short triggering pulse of duration τ is shown in Fig. 15.34a. In the following we shall neglect the propagation delays through G_1 and G_2. These delays, however, set a lower limit on the pulse width τ, $\tau > (t_{P1} + t_{P2})$.

Consider first the stable state of the monostable—that is, the state of the circuit before the trigger pulse is applied. The output of G_1 is high at V_{DD}, the capacitor is discharged, and the input voltage to G_2 is high at V_{DD}. Thus the output of G_2 is low, at ground voltage. This low voltage is fed back to G_1; since v_I is low, the output of G_1 is high, as initially assumed.

Next consider what happens as the trigger pulse is applied. The output voltage of G_1 will go low. However, because G_1 will be sinking some current and because of its finite output resistance R_{on}, its output will not go all the way to 0 V. Rather, the output of G_1 drops by a value ΔV_1, which we shall shortly evaluate.

The drop ΔV_1 is coupled through C (which acts as a short circuit during the transient) to the input of G_2. Thus the input voltage of G_2 drops by an identical amount ΔV_1. Here we note that during the transient there will be an instantaneous current that flows from V_{DD} through R and C and into the output terminal of G_1 to ground. We thus have a voltage divider formed by R and R_{on} (note that the

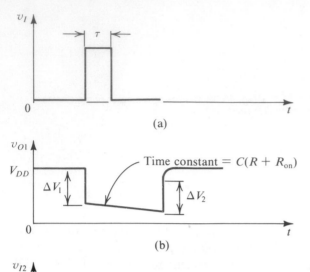

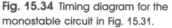

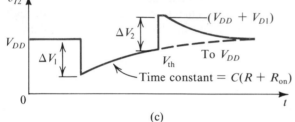

Fig. 15.34 Timing diagram for the monostable circuit in Fig. 15.31.

instantaneous voltage across C is zero) from which we can determine ΔV_1 as

$$\Delta V_1 = V_{DD} \frac{R}{R + R_{on}} \tag{15.51}$$

Returning to G_2, we see that the drop of voltage at its input causes its output to go high (to V_{DD}). This signal keeps the output of G_1 low even after the triggering pulse has disappeared. The circuit is now in the quasi-stable state.

We next consider operation in the quasi-stable state. The current through R, C, and R_{on} causes C to charge, and the voltage v_{I2} rises exponentially toward V_{DD} with a time constant $C(R + R_{on})$, as indicated in Fig. 15.34c. The voltage v_{I2} will continue to rise until it reaches the value of the threshold voltage V_{th} of inverter G_2. At this time G_2 will switch and its output v_{O2} will go to 0 V, which will in turn cause G_1 to switch. The output of G_1 will attempt to rise to V_{DD} but, as will become obvious shortly, its instantaneous rise will be limited to an amount ΔV_2. This rise in v_{O1} is

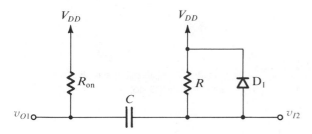

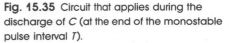

Fig. 15.35 Circuit that applies during the discharge of C (at the end of the monostable pulse interval T).

coupled faithfully through C to the input of G_2. Thus the input of G_2 will rise by an equal amount ΔV_2. Note here that because of diode D_1, between the input of G_1 and V_{DD}, the voltage v_{I2} can rise only to $V_{DD} + V_{D1}$, where V_{D1} (approximately 0.7 V) is the drop across D_1. Thus from Fig. 15.46c we see that

$$\Delta V_2 = V_{DD} + V_{D1} - V_{th} \tag{15.52}$$

Thus it is diode D_1 that limits the size of the increment ΔV_2.

Because now v_{I2} is higher that V_{DD} (by V_{D1}) current will flow from the output of G_1 through C and then through the parallel combination of R and D_1. This current discharges C until v_{I2} drops to V_{DD} and v_{O1} rises to V_{DD}. The charging circuit is depicted in Fig. 15.35, from which we note that the existence of the diode causes the discharging to be a nonlinear process. Although the details of the transient at the end of the pulse are not of immense interest, it is important to note that the monostable should not be retriggered until the capacitor has been discharged, since otherwise the output obtained will not be the standard pulse which the one-shot is intended to provide. The capacitor discharge interval is known as the *recovery time*.

Exercise

15.24 Derive an expression for the pulse interval T of the monostable circuit in Fig. 15.43. (*Hint:* Use the information given in the timing diagram of Fig. 15.34c and the value of ΔV_1 given in Eq. (15.51).)

Ans.

$$T = C(R + R_{on}) \ln\left(\frac{R}{R + R_{on}} \frac{V_{DD}}{V_{DD} - V_{th}}\right)$$

An Astable Circuit

Figure 15.36 shows a popular astable circuit composed of two inverter-connected NOR gates, a resistor, and a capacitor. We shall consider its operation, assuming that the NOR gates are of the CMOS family. However, to simplify matters we shall make some further approximations: The finite output resistance of the CMOS gate will be neglected. Also, the clamping diodes will be assumed ideal (thus have zero voltage drop when conducting).

With these simplifying assumptions, the waveforms of Fig. 15.37 are obtained. The reader is urged to consider the operation of this circuit in a step-by-step manner and verify that the waveforms shown indeed apply.[5]

---

[5] Practical circuits often use a large resistance in series with the input to G_1. This limits the effect of diode conduction and allows v_{I1} to rise to a voltage greater than V_{DD} and, as well, fall below zero.

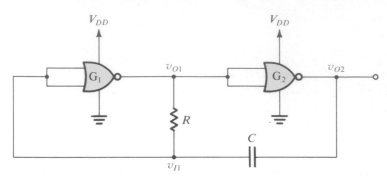

Fig. 15.36 A simple astable multivibrator circuit using CMOS gates.

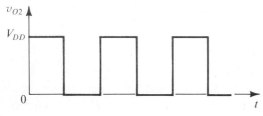

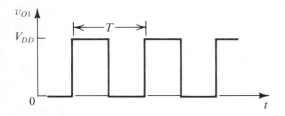

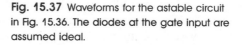

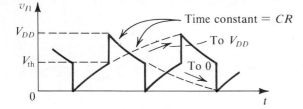

Fig. 15.37 Waveforms for the astable circuit in Fig. 15.36. The diodes at the gate input are assumed ideal.

Exercise

15.25 Using the waveforms in Fig. 15.37, derive an expression for the period T of the astable multivibrator of Fig. 15.36.

Ans.

$$T = CR \ln \left(\frac{V_{DD}}{V_{DD} - V_{th}} + \frac{V_{DD}}{V_{th}} \right)$$

15.9 SUMMARY

● The characteristics of a logic circuit family are determined by analyzing the circuit of its basic inverter.

● The static performance of a logic inverter is characterized by its voltage transfer characteristic, which in turn is characterized by the four parameters V_{OL}, V_{OH}, V_{IL}, and V_{IH} (see Fig. 15.5 for definitions). The noise margins are $NM_H = V_{OH} - V_{IH}$ and $NM_L = V_{IL} - V_{OL}$. An ideal inverter has $V_{OH} = V_{DD}$, $V_{OL} = 0$, $V_{IL} = V_{IH} = V_{DD}/2$, and thus $NM_H = NM_L = V_{DD}/2$.

● The dynamic performance of a logic inverter is characterized by the propagation delay t_P and the rise and fall times of the output pulse waveform (see Fig. 15.7 for definitions).

● The two performance measures of a logic inverter—the average propagation delay t_P and the average power dissipation P_D—are combined to obtain the delay-power product $DP = t_P P_D$ which serves as a figure of merit in comparing logic circuit families. Modern logic circuits have DP in the range 1 to 100 pJ.

● There are two MOS logic circuit families: NMOS and CMOS. NMOS has low output-driving capability and thus its application is limited to VLSI circuit design. CMOS, on the other hand, is used in VLSI as well as in conventional design. At the present time CMOS is perhaps the most popular logic circuit technology. NMOS, however, occupies the least silicon area and thus allows higher levels of integration than CMOS.

● Of the two NMOS types, depletion-load circuits offer higher performance and smaller silicon area than enhancement-load circuits.

● Performance of depletion-load NMOS circuits is seriously degraded by the body effect on the load device.

● Although the static power dissipation of CMOS is negligibly small, dynamic power dissipation can be significant at high operating speeds. It is given by $f C_L V_{DD}^2$ where f is the switching frequency and C_L is the load capacitance.

● Flip-flops employ one or more latches. The basic latch is a bistable circuit (bistable multivibrator) implemented using two inverters connected in a positive-feedback loop. The latch can remain in either stable state indefinitely.

● A monostable multivibrator has one stable state, in which it can remain indefinitely, and one quasi-stable state, which it enters upon triggering and in which it remains for a predetermined interval T. Monostable circuits can be used to generate pulse signals of predetermined height and width.

● An astable multivibrator has no stable states. It has two quasi-stable states, between which it oscillates. The astable circuit is, in effect, a square-wave generator.

BIBLIOGRAPHY

M. I. Elmasry (Ed), *Digital MOS Integrated Circuits*, New York: IEEE Press, 1981.

D. A. Hodges and H. G. Jackson, *Analysis and Design of Digital Integrated Circuits*, New York: McGraw-Hill, 1983.

IEEE Journal of Solid-State Circuits. The October issue of each year has been devoted to digital circuits.

C. Mead and L. Conway, *Introduction to VLSI Systems*, Reading, Mass.: Addison-Wesley, 1980.

Motorola, *McMOS Handbook*, Phoenix, Ariz.: Motorola Inc., 1974.

RCA, *COS/MOS Digital Integrated Circuits*, Publication No. SSD-203B, Somerville, N. J.: RCA Solid-State Division, 1974.

N. Weste and K. Eshraghian, *Principles of CMOS VLSI Design*, Reading, Mass.: Addison-Wesley, 1985.

PROBLEMS

15.1 Consider the conceptual inverter of Fig. 15.2 for which $V^+ = 5$ V, $R_L = 1$ kΩ, $R_{on} = 50$ Ω, and $V_{off} = 0.2$ V, connected to a load circuit which can be represented by a 2-kΩ resistor connected to $+2$ V. Find the output voltage levels of the inverter.

***15.2** An inverter that can be characterized by Fig. 15.2 has $V^+ = 5$ V, $R_L = 1$ kΩ, and $R_{on} = 100$ Ω. It is loaded by a similar inverter whose switching threshold is at 2 V, by means of a connection whose capacitance to ground is 50 pF. If both switches exhibit a pure delay of 10 ns from the moment their input signal threshold is crossed, how long does it take for an input step to open the switch of the second inverter? To close it?

15.3 Consider a logic inverter having a voltage transfer characteristic consisting of three straight-line segments: two horizontal ones at $v_o = V_{OH} = 4$ V and at $v_o = V_{OL} = 0.5$ V, and the third joining the points $v_1 = V_{IL} = 1.5$ V and $v_I = V_{IH} = 2.5$ V. Find **(a)** the noise margins, **(b)** the value at which v_I and v_o are equal, and **(c)** the voltage gain in the transition region.

15.4 For a particular logic family for which the supply voltage is V^+, $V_{OL} = 0.1V^+$, $V_{OH} = 0.8V^+$, $V_{IL} = 0.3V^+$, $V_{IH} = 0.6V^+$, what are the noise margins? What is the width of the transition region? For a minimum noise margin of 1 V what value of V^+ is required?

****15.5** An inverter modeled by Fig. 15.2 for which $V^+ = 10$ V, $R_L = 1$ kΩ, and $R_{on} = 100$ Ω is operated at 1 MHz by an input that is high 75% of the time and with a load of 100 pF. Find the average power drawn from the power supply, the average supply current, and the power dissipated in the switch.

15.6 A logic inverter for which $t_{PHL} = 20$ ns, $t_{PLH} = 30$ ns, $t_{THL} = 30$ ns, $t_{TLH} = 40$ ns is driven by a signal for which the rise and fall times (10 to 90%) are 20 ns. Assuming linear (ramp) transitions, find the time intervals from the 10% level of the input to the 90% level of the output, for both transitions. What is the propagation delay through four of these gates connected in cascade?

***15.7** Consider the situation in which three identical inverting gates are connected in a closed ring. Using the ideas expressed in Fig. 15.7 and assuming that a rising edge occurs at the input of one of the inverters, sketch the signals that result at this and all other inputs. For simplicity, assume the rise and fall times to be zero, but that the propagation times are not zero. For a uniform propagation delay of 50 ns, what is the frequency of the resulting oscillator? What is it if $t_{PLH} = 60$ ns and $t_{PHL} = 40$ ns?

15.8 For an NMOS inverter with enhancement load for which $K_R = 9$, $V_t = 1$ V, $V_{DD} = 5$ V, and the body effect is ignored, find the input voltage for which **(a)** $v_O = V_{DD} - 2V_t$, **(b)** $v_O = v_I$, **(c)** $v_O = V_t$, and **(d)** $dv_O/dv_I = -1$.

15.9 Repeat Problem 15.8 for the situation in which $K_R = 36$.

15.10 Two NMOS enhancement-load inverter designs are being considered, both of which use the minimum-size load structure available for which $K_2 = 1$ μA/V². The supply voltage $V_{DD} = 5$ V and the threshold voltage $V_t = 1$ V. For one inverter, $K_R = 9$, and for the other, $K_R = 36$. Neglecting the body effect, find the current available for driving a capacitor load downward when $v_I = V_{OH}$ and $v_O = 4$ V, 3 V, and 1 V.

15.11 An n-channel enhancement MOS device, used in an inverter load operating from a 5-V supply, has V_{t0} ranging from 1 to 1.5 V and γ ranging from 0.3 to 1 V$^{1/2}$ and $2\phi_F = 0.6$ V. What is the lowest value of V_{OH} that will be found using transistors of this kind?

***15.12** Reconsider Example 15.1. Note that the calculations that result in $V_{IH} = 2.2$ V at $v_O = 0.8$ V and $V_{OL} = 0.3$ V are performed with the body effect neglected. Noting that with the body effect $V_{OH} = 3.4$ V, what does V_{OL} become? Now using the stated result for V_{IH}, iterate once by finding the corresponding V_{t2} and thus V_{IH} and the corresponding v_O. Calculate the noise margins.

15.13 An enhancement-load inverter operated from a 5-V supply is found to have $V_{OH} = 3$ V. If $V_{t0} = 1$ V and $2\phi_F = 0.6$ V, what must γ be?

15.14 A particular NMOS logic inverter having a 5-V supply utilizes devices for which $V_{t0} = 0$ V but whose substrate is connected to a -5-V supply (generated locally within the IC). $2\phi_F = 0.6$ V and $\gamma = 0.5$ $V^{1/2}$. For this arrangement find V_{OH} and V_{OL} for $K_R = 9$.

*****15.15** Reconsider the analysis of t_{PLH} for the capacitively loaded enhancement-load inverter approximated by Eqs. (15.15) and (15.16) by writing and solving the differential equation that describes the operation of the circuit in Fig. 15.11d. Ignore the body effect in your analysis. Contrast your result for the inverter of Example 15.1 and a 0.1-pF load with the value of 6.4 ns obtained by the approximate method.

15.16 For the circuit of Example 15.1, for which t_{PHL} is approximately 0.7 ns and t_{PLH} is 6.4 ns with a 0.1 pF load, it is required to reduce t_{PLH} as much as possible by increasing the width of the load device. This, however, has the undesirable effect of raising V_{OL}. If the width is increased so that $V_{OL} = 1$ V, what does t_{PLH} become? Ignore the body effect.

15.17 Derive the result given in Eq. (15.17).

15.18 Use Eqs. (15.19), (15.20), and (15.21) to calculate the average propagation delay, the average power dissipation and the delay-power product of an inverter for the following conditions.

K_2 ($\mu A/V^2$)	V_t (V)	V_{DD} (V)	C (pF)
5	1	5	0.1
1	1	5	0.1
5	0.5	5	0.1
5	1	10	0.1
5	1	5	1.0
1	0.5	5	0.1

*****15.19** For three depletion load transistors subject to body effect, consider operation at $v_O = 0$, $V_{DD}/2$, and $0.8V_{DD}$ for $V_{DD} = 5$ V, $2\phi_F = 0.6$ V, and $\gamma = 0.5$ $V^{1/2}$. Determine the values of i_D that result for
(a) $K = 10/4 = 2.50$ $\mu A/V^2$ and $V_{tD0} = -2$ V.
(b) $K = 10/9 = 1.11$ $\mu A/V^2$ and $V_{tD0} = -3$ V.
(c) $K = 10/16 = 0.625$ $\mu A/V^2$ and $V_{tD0} = -4$ V.

15.20 For the inverter in Example 15.2, find the resistance of the depletion load at $v_O = V_{OH}$. Note that $K_2 = 5.1$ $\mu A/V^2$.

*****15.21** For the depletion-load inverter, repeat an analysis similar to the one leading to Eqs. (15.19), (15.20), and (15.21) for an enhancement-load inverter. In particular, verify Eq. (15.29) and indicate the origin of the constant α.

15.22 For a technology in which the optimum dimensions of a depletion-load NMOS inverter are $W/L = 12$ $\mu m/6$ μm for the inverting transistor and 6 $\mu m/12$ μm for the load, estimate the areas of a 3-input NOR and a 3-input NAND.

15.23 Find the logic function implemented by the circuit shown in Fig. P15.23. Give device sizes if the technology is that indicated in Problem 15.22.

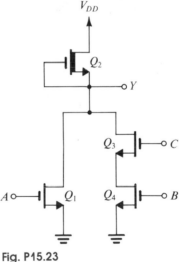

Fig. P15.23

15.24 For the circuit in Fig. P15.24, find F as a Boolean function of A, B, C, D, and E. If X is joined to Y, find the modified expression for F.

*****15.25** The circuit shown in Fig. P15.25 can be used in applications such as output buffers and clock drivers where capacitive loads require high current-drive capability. Let $V_{tE} = 1$ V, $V_{tD0} = -3$ V, $V_{DD} = 5$ V, $\mu_n C_{ox} = 20$ $\mu A/V^2$, $W_1 = W_2 = 12$ μm, $L_1 = L_2 = L_3 = L_4 = 6$ μm, $W_3 = 120$ μm, and $W_4 = 240$ μm. To simplify matters we shall assume the body effect to be negligible.
(a) Find V_{AH}, V_{OH}, V_{AL}, and V_{OL} for $V_{IH} = 5$ V. (Note that V_{AH} denotes the high level at node A, etc.)

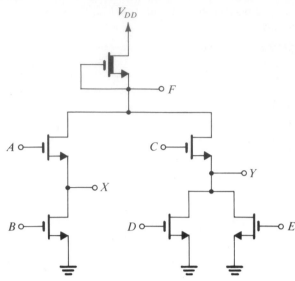

Fig. P15.24

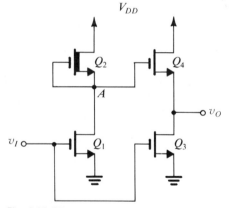

Fig. P15.25

(b) Find the total static current with output high and with output low.

(c) As v_O goes low, find the peak current available to discharge a load capacitance $C = 10$ pF. Also find the discharge current for v_O at the 50% output swing point and thus find the average discharge current and t_{PHL}. Assume the capacitance at node A to be zero.

(d) Repeat (c) for v_O going high, thus finding t_{PLH}.

***15.26** The circuit shown in Fig. P15.26, called a bootstrap driver, is intended to provide a large output voltage swing and high output drive current with a reasonable standing current. Let $V_{DD} = 5$ V, $V_t = 1$ V, $K_1 =$

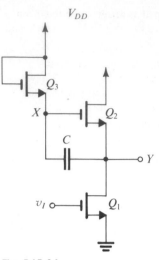

Fig. P15.26

$40 \ \mu\text{A/V}^2$, $K_2 = 10 \ \mu\text{A/V}^2$, and $K_3 = 1 \ \mu\text{A/V}^2$, and assume that C is large compared to any capacitances at nodes X and Y. Also, ignore the body effect.

(a) Find the voltages at nodes X and Y when v_I has been high (at 4 V) for some time.

(b) When v_I goes low to 0 V and Q_1 turns off, to what voltages do nodes X and Y rise?

(c) Find the current available to charge a load capacitance at Y as soon as v_I goes low and Q_1 turns off.

(d) As v_I goes high again, what current is immediately available to discharge the load capacitance at Y?

***15.27** Consider the circuit shown in Fig. P15.27 for which $V_{DD} = 5$ V, $V_{tE} = 1$ V, $V_{tD} = -3$ V, $K_1 = K_3 = K_5 = 9 \ \mu\text{A/V}^2$, and $K_2 = K_4 = 1 \ \mu\text{A/V}^2$. Express D and E as logic (Boolean) functions of A, B, and C. What are

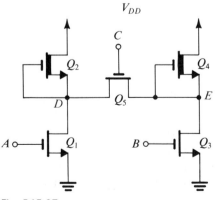

Fig. P15.27

the voltage levels to be found at nodes D and E? Ignore the body effect.

***15.28** For the circuit in Fig. P15.28, express Y as a logic (Boolean) function of A and B. For a technology in which the basic inverter has $K_R = 4$ and the depletion load devices have $W = 6$ μm and $L = 12$ μm, suggest W and L values for Q_1, Q_3, Q_5, and Q_6 (assumed to be identical) so that the value of V_{OL} is not higher than that obtained in the basic inverter.

*15.29** Consider the CMOS transfer characteristic of Fig. 15.18, for which $\mu_n C_{OX} = 2\,\mu_p C_{OX} = 20\,\mu A/V^2$, $W_p = 2W_n = 24\,\mu m$, $L_p = L_n = 6\,\mu m$, $V_{DD} = 5$ V, $V_t = 1$ V and the Early voltage (see Chapter 7) $|V_A| = 100$ V.

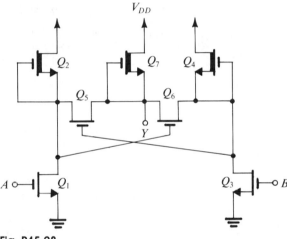

Fig. P15.28

What is the slope of the segment BC? (Use small-signal analysis.) The inverter is biased by connecting a resistor $R_G = 10$ MΩ between input and output. What is the dc voltage at input and output? What is the small-signal voltage gain of the resulting amplifier? Find the input resistance.

*15.30** A particular CMOS inverter uses n- and p-channel devices of identical size for which $K_n = 2K_p$, $|V_t| = 1$ V, and $V_{DD} = 5$ V. Sketch its voltage transfer characteristic (in the manner of Fig. 15.18) and find the coordinates of points A, B, C, and D. Determine V_{IL} and V_{IH} and thus the noise margins.

15.31 Show that for a CMOS inverter in which the two MOSFETs are characterized by K_n, V_{tn}, K_p, and V_{tp}, the threshold voltage V_{th} (that is, the voltage that defines the segment BC of the voltage transfer characteristic in Fig.

15.18) is given by

$$V_{th} = \frac{V_{DD} + \sqrt{K_n/K_p}\,V_{tn} - |V_{tp}|}{1 + \sqrt{K_n/K_p}}$$

If $K_n = 2K_p$, $|V_{tp}| = 1$ V and $V_{DD} = 5$ V, find V_{tn} that will result in $V_{th} = V_{DD}/2$.

15.32 Repeat Exercise 15.14 for $V_{DD} = 10$ V and 15 V.

15.33 Repeat Exercise 15.14 for $V_t = 0.5$ V, 1.5 V, and 2.0 V.

15.34 Reconsider the gate in Exercise 15.15 for operation with a 5-V supply. What are the maximum source and sink currents for which V_{OH} and V_{OL} deviate from the ideal values of V_{DD} and zero, respectively, by at most $0.1V_{DD}$? What must be done to the width of each device to raise this current to 1.55 mA? If the enlarged inverter is used at 10 V, what current results? What is the current with a supply voltage of 15 V?

15.35 For the inverter specified in Exercise 15.15, find the peak current drawn from the supply, during switching, for $V_{DD} = 5$, 10, and 15 V.

15.36 Repeat Exercise 15.17 for $V_{DD} = 5$ and 15 V

*15.37** The inverter specified in Exercise 15.15 is driven by a ramp input that rises from 0 to 10 V in 1 ms. Sketch the waveforms of the inverter current and the output voltage. Calculate the total charge that flows in the switch. If the signal is part of a 1-kHz sawtooth waveform, what is the average inverter current? What is the average power dissipation in the inverter? What is the peak power dissipation in either one of the devices?

15.38 Using Eq. (15.44), explore the effect of variation of V_t on t_{PHL}. Derive expressions for $V_t = 0.1V_{DD}$, $0.2V_{DD}$ and $0.3V_{DD}$.

15.39 Extend the analysis of the dynamic operation of the CMOS inverter to derive an expression for the fall time, t_{THL}, which is the time for the output waveform to decrease from $0.9V_{DD}$ to $0.1V_{DD}$. Assume that the input is an ideal pulse. Evaluate t_{THL} for the inverter whose parameters are given in Exercise 15.18.

15.40 Repeat Exercise 15.21 for three-input NOR and NAND gates.

15.41 For standard CMOS gates whose propagation delay times are specified in the text, find the time of propagation through four levels of inverting (NAND) logic. Assume that each stage has a total fanout of 4 (each with input capacitance of 5 pF) and that the wiring capacitance is 10 pF.

15.42 As specified in the text, a standard CMOS gate operating with $V_{DD} = 10$ V and loaded with $C_L = 50$ pF draws an average dc supply current of 0.6 μA/kHz. What is the component of this current that arises because of the repeated charge and discharge of C_L? Estimate the supply current for an unloaded gate ($C_L = 0$) in μA/kHz. Find the total gate dissipation for $V_{DD} = 10$ V, $f = 10$ MHz, and $C_L = 100$ pF.

15.43 A three-stage buffered CMOS inverter uses p-channel devices that are twice as wide as their n-channel counterparts. The output stage is the same size as the unbuffered gate, and each of the preceding stages has a W/L ratio that is one-tenth that of the stage that it drives. The input stage uses the smallest available n-channel device of unit area. What is the relative size of the buffered and unbuffered inverters? Note that all the channel lengths are the same.

*15.44 For the situation described in Problem 15.43, input capacitance is dominated by the gate-to-drain capacitance and Miller multiplication. If the gain per stage is about 50 and the input capacitance of the unbuffered gate is 5 pF, what is the input capacitance of the buffered inverter likely to be? If, instead, the unbuffered input capacitance has a 1-pF part due to wiring and connection, what would you expect the input capacitance of the buffered gate to be?

**15.45 Consider the latch of Fig. 15.25 as implemented in CMOS technology. Let $\mu_n C_{OX} = 2\,\mu_p C_{OX} = 20\ \mu$A/V^2, $W_p = 2W_n = 24\ \mu$m, $L_p = L_n = 6\ \mu$m, $|V_t| = 1$ V, and $V_{DD} = 5$ V.
(a) Plot the transfer characteristic of each inverter—that is, v_X versus v_W and v_Z versus v_Y. From the curve, determine the output of each inverter at input voltages of 1, 1.5, 2, 2.25, 2.5, 2.75, 3, 3.5, 4, and 5 volts.
(b) Use the characteristics in (a) to determine the loop voltage transfer curve of the latch—that is, v_Z versus v_W. Find the coordinates of points A, B, and C as defined in Fig. 15.25c.
(c) If the finite pinch-off output resistance of the MOSFET is taken into account, with $|V_A| = 100$ V, find the slope of the loop transfer characteristic at point B. What is the approximate width of the transition region?

15.46 Sketch the logic gate symbolic representation of an SR flip-flop using NAND gates. Give the truth table that describes its operation.

15.47 Sketch the circuit of the NAND SR flip-flop in

Problem 15.46 using depletion-load NMOS. What is the rest state of the S and R inputs?

15.48 Sketch the circuit of the NOR SR flip-flop of Fig. 15.26 using CMOS.

*15.49 Consider the SR flip-flop circuit shown in Fig. P15.49. Let $V_{DD} = 5$ V, $|V_t| = 1$ V, and $K_1 = K_2 = K_3 = $

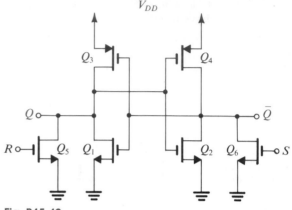

Fig. P15.49

$K_4 = K$. Find the value of $K_5 = K_6$ so that the flip-flop switches state when a set or reset signal of $V_{DD}/2$ volts is applied.

15.50 For the monostable circuit of Fig. 15.31, find an approximation to the result of Exercise 15.24 under the conditions that $R_{on} \ll R$ and $V_{th} \simeq V_{DD}/2$.

*15.51 For the monostable circuit of Fig. 15.31 whose waveforms are given in Fig. 15.34 and an expression for its period T is given in Exercise 15.24, let $V_{DD} = 10$ V, $V_{th} = V_{DD}/2$, $R = 10$ kΩ, $C = 0.001\ \mu$F, and $R_{on} = 200\ \Omega$. Find the values of T, ΔV_1, and ΔV_2. By how much does v_{O1} change during the quasi-stable state? What is the peak current that G_1 is required to sink?

15.52 Using the circuit of Fig. 15.31 (for which an expression for T is given in Exercise 15.24) design a monostable circuit with CMOS logic for which $R_{on} = 100\ \Omega$, $V_{DD} = 5$ V, and $V_{th} = 0.4V_{DD}$. Use $C = 1\ \mu$F to generate an output pulse of duration $T = 1$ s. What value of R should be used?

15.53 Consider the astable circuit of Fig. 15.36 under the conditions that $V_{DD} = 10$ V, $V_{th} = V_{DD}/2$, $C = 0.001\ \mu$F and $R = 10$ kΩ. Find the frequency of oscillation. (You may use the result of Exercise 15.25.)

Chapter

Bipolar Digital Circuits

INTRODUCTION

This is the second chapter of the two-chapter sequence devoted to digital circuits: In Chapter 15 we studied MOS digital circuits; here we shall study circuits implemented with bipolar junction transistors. A prerequisite for this material is a thorough familiarity with the BJT (Chapter 8). Also, it will be assumed that the reader is familiar with the general digital circuit concepts introduced in Section 15.1.

Our study of BJT digital circuits will begin with the Ebers-Moll model. This is a large-signal model whose application yields considerable insight into the operation of the BJT in the saturation region. We shall then discuss the dynamic operation of the BJT, relating its response times to the charge stored in its base.

Following a brief overview of early bipolar logic-circuit families, we study in detail two contemporary families: transistor-transistor logic

(TTL) and emitter-coupled logic (ECL). For many years TTL has been the most popular circuit technology for implementing digital systems using SSI, MSI, and LSI packages. At the present time, TTL continues to be popular and is rivaled only by CMOS (Chapter 15). An important factor that contributed to the longevity of TTL is the continual performance improvement that this circuit technology has undergone over the years. Modern forms of TTL feature gate delays as low as 1.5 ns. As will be seen, in these improved circuits, the BJTs are prevented from saturation. This is done to avoid the time delay required to bring a transistor out of saturation. The other popular bipolar family, ECL, also avoids transistor saturation.

Emitter-coupled logic is the fastest digital circuit technology available, featuring SSI and MSI gate delays of less than 1 ns, and even shorter delays in VLSI implementations. ECL finds application in digital communications circuits as well as in the high-speed circuits utilized in supercomputers.

A bipolar digital circuit technology that was popular a few years ago in VLSI applications is integrated injection logic (I^2L). It has, however, lost application ground to CMOS and will not be studied here.

Although the emphasis in this chapter is on logic circuits, other digital system building blocks such as flip-flops and multivibrators can be implemented in TTL and ECL following conventional approaches. Very-high-density memory chips, however, remain the exclusive domain of MOS technology. □

16.1 THE BJT AS A DIGITAL CIRCUIT ELEMENT

We shall begin our study of BJT logic circuits with a summary of pertinent BJT characteristics. In addition, a popular large-signal model for the BJT will be introduced. Before proceeding with this material the reader is advised to review Chapter 8.

Saturating and Nonsaturating Logic

The most common usage of the BJT in digital circuits is to employ its two extreme modes of operation: cutoff and saturation. The resulting logic circuits are called *saturated* (or *saturating*) logic. The advantages of this mode of application include relatively large, well-defined logic swings and reasonably low power dissipation. The main disadvantage is the relatively slow response due to the long turnoff times of saturated transistors.

To obtain faster logic, one has to arrange the design such that the BJT does not saturate. We shall study two forms of nonsaturating BJT logic—namely, emitter-coupled logic (ECL),which is based on the differential pair studied in Section 9.1, and Schottky TTL, which is based on the use of special low-voltage-drop diodes called Schottky diodes.

The Ebers-Moll (EM) Model

Although the simple large-signal transistor model developed in Chapter 8 is usually quite adequate for the approximate analysis of BJT digital circuits, more insight can be obtained from a more formal approach using a popular large-signal model of the BJT known as the *Ebers-Moll* (EM) model.

The EM model is a low-frequency (static) model based on the fact that the BJT is composed of two *pn* junctions, the emitter-base junction and the collector-base junction. One can therefore express the terminal currents of the BJT as the super-position of the currents due to the two *pn* junctions, as shown in the following.

Figure 16.1 shows an *npn* transistor together with its EM model. The model consists of two diodes and two controlled sources. The diodes are D_E, the emitter-base junction diode, and D_C, the collector-base junction diode. The diode currents i_{DE} and i_{DC} are given by the diode equation:

$$i_{DE} = I_{SE}(e^{v_{BE}/V_T} - 1) \tag{16.1}$$

$$i_{DC} = I_{SC}(e^{v_{BC}/V_T} - 1) \tag{16.2}$$

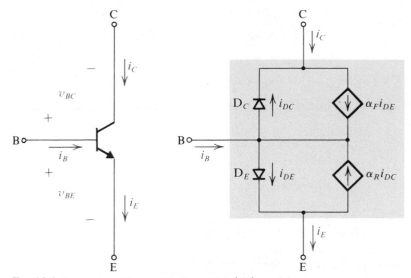

Fig. 16.1 An *npn* transistor and its Ebers-Moll (EM) model.

where I_{SE} and I_{SC} are the saturation or scale currents of the two diodes. Since the collector-base junction is usually of larger area than the emitter-base junction, I_{SC} is usually larger than I_{SE} (by a factor of 2 to 50).

As explained in Chapter 8, part of the emitter-base junction current i_{DE} reaches the collector and registers as collector current. It is this component that gives rise to the current source $\alpha_F i_{DE}$ in the model of Fig. 16.1. Here α_F denotes the *forward* α of the transistor (which is the parameter we simply called α previously). The value of α_F is usually very close to unity. Similarly, part of the collector-base junction current i_{DC} is transported across the base region and reaches the emitter. This component is represented in the EM model by the current source $\alpha_R i_{DC}$, where α_R denotes the *reverse* α of the transistor. Since the transistor structure is not physically symmetric but rather is optimized to have a large forward α, α_R is usually small (0.02 to 0.5).

A relationship exists (see Harris, Gray, & Searle, 1966) between the four parameters of the EM model and the transistor current scale I_S (see Chapter 8):

$$\alpha_F I_{SE} = \alpha_R I_{SC} = I_S \tag{16.3}$$

Since $\alpha_F \cong 1$, we see that

$$I_{SE} \cong I_S \tag{16.4}$$

Recall that for low-power (small-signal) transistors I_S is of the order of 10^{-14} to 10^{-15} A and is proportional to the area of the emitter-base junction.

The Transistor Terminal Currents

Having provided a qualitative physical justification for the EM model, we shall now use it to express the BJT terminal currents in terms of the junction voltages. From

Fig. 16.1 we can write

$$i_E = i_{DE} - \alpha_R i_{DC} \tag{16.5}$$

$$i_C = -i_{DC} + \alpha_F i_{DE} \tag{16.6}$$

$$i_B = (1 - \alpha_F)i_{DE} + (1 - \alpha_R)i_{DC} \tag{16.7}$$

Substituting for i_{DE} and i_{DC} from Eqs. (16.1) and (16.2) and using the relationship in Eq. (16.3) gives

$$i_E = \frac{I_S}{\alpha_F}(e^{v_{BE}/V_T} - 1) - I_S(e^{v_{BC}/V_T} - 1) \tag{16.8}$$

$$i_C = I_S(e^{v_{BE}/V_T} - 1) - \frac{I_S}{\alpha_R}(e^{v_{BC}/V_T} - 1) \tag{16.9}$$

$$i_B = \frac{I_S}{\beta_F}(e^{v_{BE}/V_T} - 1) + \frac{I_S}{\beta_R}(e^{v_{BC}/V_T} - 1) \tag{16.10}$$

where β_F is the forward β and β_R is the reverse β,

$$\beta_F = \frac{\alpha_F}{1 - \alpha_F} \tag{16.11}$$

$$\beta_R = \frac{\alpha_R}{1 - \alpha_R} \tag{16.12}$$

While β_F is usually large, β_R is very small.

Exercise

16.1 A particular transistor is said to have $\alpha_F \simeq 1$ and $\alpha_R = 0.02$. Its emitter scale current is about 10^{-14} A. What is its collector scale current? What is the size of the collector junction relative to the emitter junction? What is the value of β_R?

Ans. 50×10^{-14} A; 50 times as large; 0.02

Application of the EM Model

We shall now consider the application of the EM model to characterize transistor operation in various modes.

The Normal Active Mode.   Here the emitter–base junction is forward-biased and the collector–base junction is reverse-biased. The word *normal* is used to distinguish this mode from that in which the roles of the two junctions are interchanged (the *reverse* or *inverse* active mode). Since v_{BC} is negative and its magnitude is usually much greater than V_T, Eqs. (16.8) through (16.10) can be approximated as

$$i_E \simeq \frac{I_S}{\alpha_F} e^{v_{BE}/V_T} + I_S\left(1 - \frac{1}{\alpha_F}\right) \tag{16.13}$$

$$i_C \simeq I_S e^{v_{BE}/V_T} + I_S\left(\frac{1}{\alpha_R} - 1\right) \tag{16.14}$$

$$i_B \simeq \frac{I_S}{\beta_F} e^{v_{BE}/V_T} - I_S\left(\frac{1}{\beta_F} + \frac{1}{\beta_R}\right) \tag{16.15}$$

In each of these three equations one can normally neglect the second term on the right-hand side. This results in the familiar current–voltage relationships that characterize the active mode of operation.

Exercise

16.2 Use Eq. (16.8) to show that the *i-v* characteristic of the diode-connected transistor of Fig. E16.2 is given by

$$i = \frac{I_S}{\alpha_F}(e^{v/V_T} - 1) \cong I_S e^{v/V_T}$$

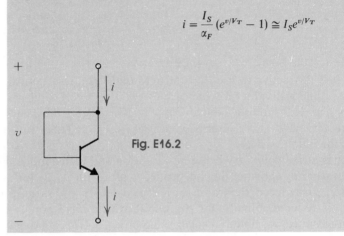

Fig. E16.2

The Saturation Mode. Consider first the normal (as opposed to reverse) saturation mode, as can be obtained in the circuit of Fig. 16.2. Assume that a current I_B is pushed into the base and that its value is sufficient to drive the transistor into saturation. Thus the collector current will be $\beta_{\text{forced}}I_B$, where $\beta_{\text{forced}} < \beta_F$. We wish to use the EM equations to derive an expression for $V_{CE\text{sat}}$.

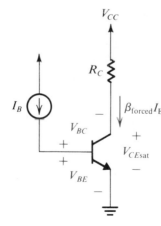

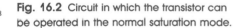

Fig. 16.2 Circuit in which the transistor can be operated in the normal saturation mode.

In saturation both junctions are forward-biased. Thus V_{BE} and V_{BC} are both positive, and their values are much greater than V_T. Thus in Eqs. (16.9) and (16.10) we can assume that $e^{V_{BE}/V_T} \gg 1$ and $e^{V_{BC}/V_T} \gg 1$. Making these approximations and substituting $i_B = I_B$ and $i_C = \beta_{\text{forced}}I_B$ results in two equations that can be solved to obtain V_{BE} and V_{BC}. The saturation voltage $V_{CE\text{sat}}$ can be then obtained as the difference

Table 16.1 NUMERICAL VALUES

β_{forced}	50	48	45	40	30	20	10	1	0
V_{CEsat} (mV)	∞	235	211	191	166	147	123	76	60

between these two voltage drops:

$$V_{CEsat} = V_T \ln \frac{1 + (\beta_{forced} + 1)/\beta_R}{1 - \beta_{forced}/\beta_F} \qquad (16.16)$$

It is instructive to use Eq. (16.16) to find V_{CEsat} in a typical case. Table 16.1 provides numerical values for the case $\beta_F = 50$, $\beta_R = 0.1$, and various values of β_{forced}. Also, Fig. 16.3 shows a sketch of V_{CEsat} versus β_{forced}. This curve is simply the v_{CE}-i_C characteristic for a constant base current I_B. From Table 16.1 and Fig. 16.3 we note that the infinite value of V_{CEsat} obtained at $\beta_{forced} = \beta_F$ is an indication that the transistor is at the boundary between saturation and active mode. Figure 16.3 illustrates this further by showing the independence of v_{CE} on β_{forced} (or i_C) in the active mode. As β_{forced} is reduced, the transistor is driven deeper into saturation, V_{BC} increases, and V_{CEsat} is reduced. Finally, for $\beta_{forced} = 0$, which corresponds to the collector being open-circuited, we obtain a small value of V_{CEsat}. This small value is almost equal to the *offset voltage* of the BJT switch, as defined in Fig. 8.57.

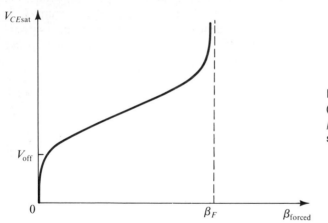

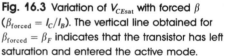

Fig. 16.3 Variation of V_{CEsat} with forced β ($\beta_{forced} = I_C/I_B$). The vertical line obtained for $\beta_{forced} = \beta_F$ indicates that the transistor has left saturation and entered the active mode.

The numerical values of Table 16.1 suggest that for a saturated transistor $V_{CEsat} \cong$ 0.1–0.3 V. For approximate calculations we shall henceforth assume that for a transistor on the verge of saturation $V_{CEsat} = 0.3$ V; for a transistor "comfortably" saturated $V_{CEsat} = 0.2$ V; and for a transistor deep into saturation $V_{CEsat} = 0.1$ V.

Exercise

16.3 Use the entries for $\beta_{forced} = 1$ and 10 in Table 16.1 to calculate an approximate value for the collector–emitter saturation resistance of a transistor having $I_B = 1$ mA.

Ans. 5.2 Ω

The Inverse Mode. We shall next consider the operation of the BJT in the inverse or reverse mode. Figure 16.4 shows a simple circuit in which the transistor is used with its collector and emitter interchanged. Note that the currents indicated—namely, I_B, I_1, and I_2—have positive values. Thus since $i_C = -I_2$ and $i_E = -I_1$, both i_C and i_E will be negative.

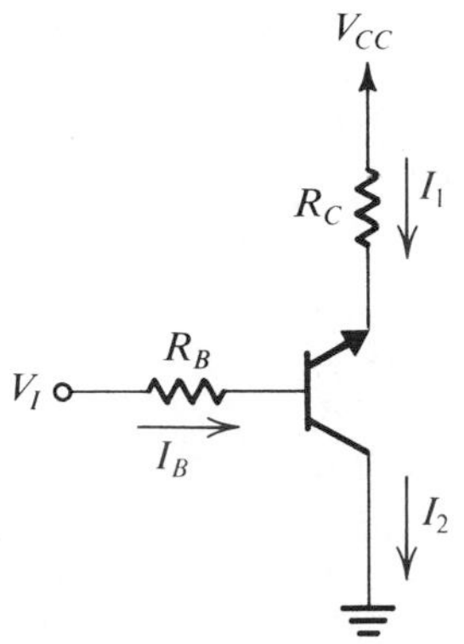

Fig. 16.4 Circuit in which the transistor is used in the reverse (or inverse) mode.

Since the roles of the emitter and collector are interchanged, the transistor in the circuit of Fig. 16.4 will operate in the active mode (called the *reverse active mode* in this case) when the emitter-base junction is reverse-biased. In this case

$$I_1 = \beta_R I_B$$

Since β_R is usually very low, it makes little sense to operate the BJT in the reverse active mode.

The transistor in the circuit of Fig. 16.4 will saturate (that is, operate in the reverse saturation mode) when the emitter-base junction becomes forward-biased. In this case

$$\frac{I_1}{I_B} < \beta_R$$

We can use the EM equations to find an expression for $V_{EC\text{sat}}$ in this case. Such an expression can be directly obtained from Eq. (16.16) as follows: Replace β_{forced} by $-I_2/I_B$ and then replace I_2 by $I_1 + I_B$. The result is

$$V_{EC\text{sat}} = V_T \ln \frac{1 + \dfrac{1}{\beta_F} + \left(\dfrac{I_1}{I_B}\right)\left(\dfrac{1}{\beta_F}\right)}{1 - \left(\dfrac{I_1}{I_B}\right)\left(\dfrac{1}{\beta_R}\right)} \qquad (16.17)$$

From this equation it can be seen that the minimum $V_{EC\text{sat}}$ is obtained when $I_1 = 0$. This minimum is very close to zero. Furthermore, we observe that the condition $I_1/I_B < \beta_R$ has to be satisfied in order that the denominator remain positive. This, of course, is the condition for the transistor to operate in the reverse saturation mode. Finally, note that since β_R is usually very low, I_1 has to be much smaller than I_B,

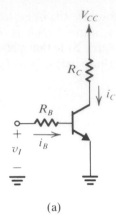

(a)

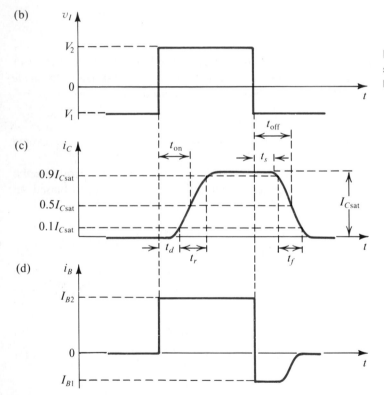

Fig. 16.5 Switching times of the BJT in the simple inverter circuit of (a) when the input v_I has the pulse waveform in (b).

with the result that $V_{EC\text{sat}}$ will be very small. This indeed is the reason for operating the BJT in the reverse saturation mode. Saturation voltages as low as a fraction of a millivolt have been reported. The disadvantage of the reverse saturation mode of operation is a relatively long turnoff time.

16.4 For the circuit in Fig. 16.4, let $R_B = 1$ kΩ and $V_{CC} = V_I = +5$ V. Assume that $V_{BC} = 0.6$ V, $\beta_R = 0.1$, and $\beta_F = 50$. Calculate approximate values for the emitter voltage in the following cases: $R_C = 1$ kΩ; $R_C = 10$ kΩ; and $R_C = 100$ kΩ.

Ans. $+4.56$ V; $+0.6$ V; $+3.5$ mV

Transistor Switching Times

Because of their internal capacitive effects, transistors do not switch in zero time. Figure 16.5 illustrates this point by displaying the waveform of the collector current i_C of a simple transistor inverter together with the waveforms for the input voltage v_I and the base current i_B. As indicated, when the input voltage v_I rises from the negative (or zero) level V_1 to the positive level V_2, the collector current does not respond immediately. Rather, a delay time t_d elapses before any appreciable collector current begins to flow. This delay time is required mainly for the EBJ depletion capacitance[1] to charge up to the forward-bias voltage V_{BE} (approximately 0.7 V). After this charging process is completed the collector current begins an exponential rise toward a final value of βI_{B2}, where I_{B2} is the current pushed into the base,[2] and is given as follows:

$$I_{B2} = \frac{V_2 - V_{BE}}{R_B} \qquad (16.18)$$

The time constant of the exponential rise is determined by the junction capacitances. In fact, it is during the interval of the rising edge of i_C that the excess minority carrier charge is being stored in the base region (see Chapter 8).

Although the exponential rise of i_C is heading toward βI_{B2}, this value will never be reached, since the transistor will saturate and the collector current will be limited to $I_{C\text{sat}}$. A measure of the BJT switching speed is the *rise time* t_r indicated in Fig. 16.5c. Another, measure is the *turn-on time*, also indicated in Fig. 16.5c.

Figure 16.6a shows the profile of excess minority carrier charge stored in the base of a saturated transistor. Unlike the active mode case, the excess minority concentration is not zero at the edge of the CBJ. This is because the CBJ is now forward-biased. Of special interest here is the extra charge stored in the base and represented by the colored area in Fig. 16.6a. Because this charge does not contribute to the slope of the concentration profile, it does not result in a corresponding collector current component. Rather, this extra stored charge arises from the pushing of more current into the base than is required to saturate the transistor. The higher the overdrive factor used,

----

[1] Since during t_d the current is zero, the diffusion capacitance will be zero.
[2] We use β and β_F interchangeably.

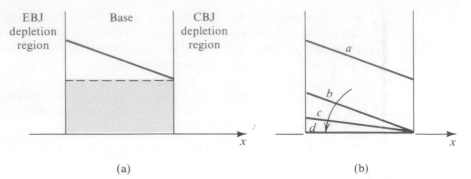

Fig. 16.6 (a) Profile of excess minority carriers in the base of a saturated transistor. The colored area represents the extra (or saturating) charge. **(b)** As the transistor is turned off, the extra stored charge has to be removed first. During this interval the profile changes from line *a* to line *b*. Then the profile decreases toward zero (line *d*), and the collector current falls exponentially to zero.

the greater the amount of extra charge stored in the base. In fact the extra charge Q_x, called *saturating charge* or *excess charge*, is proportional to the excess base drive $I_{B2} - I_{Csat}/\beta$; that is,

$$Q_x = \tau_s(I_{B2} - I_{Csat}/\beta) \tag{16.19}$$

where τ_s is a transistor parameter known as the *storage time constant*.

Let us now consider the turn-off process. When the input voltage v_I returns to the low level V_1, the collector current does not respond but remains almost constant for a time t_s (Fig. 16.5c). This is the time required to remove the saturating charge from the base. During the time t_s, called the *storage time*, the profile of stored minority carriers will change from that of line *a* to that of line *b* in Fig. 16.6b. As indicated in Fig. 16.5d, the base current reverses direction because v_{BE} remains approximately 0.7 V while v_I is at the negative (or zero) level V_1. The reverse current I_{B1} helps to "discharge the base" and remove the extra stored charge; in the absence of the reverse base current I_{B1}, the saturating charge has to be removed entirely by recombination. It can be shown (see Millman & Taub, 1965) that the storage time t_s is given by

$$t_s = \tau_s \frac{I_{B2} - I_{Csat}/\beta}{I_{B1} + I_{Csat}/\beta} \tag{16.20}$$

Once the extra stored charge has been removed, the collector current begins to fall exponentially with a time constant determined by the junction capacitances. During the fall time, the slope of excess charge profile decreases toward zero, as indicated in Fig. 16.6b. Finally, note that the reversed base current eventually decreases to zero as the EBJ capacitance charges up to the reverse-bias voltage V_1.

Typically t_d, t_r, and t_f are of the order of a few nanoseconds to a few tens of nanoseconds. The storage time t_s, however, is larger and usually constitutes the limiting factor on the switching speed of the transistor. As mentioned before, t_s increases with the overdrive factor (that is, with how deep the transistor is driven into saturation). It follows that if one desires high-speed digital circuits, then operation in the saturation region should be avoided. This is the idea behind the two nonsaturating forms of logic circuits we shall study—namely, Schottky TTL and ECL.

Exercise

16.5 We wish to use the transistor equivalent circuit of Fig. 8.32 to find an expression for the delay time t_d of a simple transistor inverter fed by a step voltage source with a resistance R_B. Since during the delay time the transistor is not conducting, the resistance r_π is infinite and C_π will consist entirely of the depletion capacitance C_{je}. As an approximation, C_{je} can be a assumed to remain constant during t_d. Also, since the collector voltage does not change during t_d, the collector can be considered grounded. Assume that the two levels of v_I are V_1 and V_2 and that the end of t_d can be taken as the time at which $v_\pi = 0.7$ V.

Ans. $t_d = R_B(C_{je} + C_\mu) \ln[(V_2 - V_1)/(V_2 - 0.7)]$

16.2 EARLY FORMS OF BJT DIGITAL CIRCUITS

In order to place the material of this chapter in proper perspective we shall examine briefly two early forms of bipolar logic circuit families.

The Basic BJT Inverter

Figure 16.7 shows the basic BJT logic inverter together with its voltage transfer characteristic. We have studied this circuit in detail in Chapter 8. Furthermore, in the previous section we studied BJT models that help in the analysis of the static and dynamic operation of the BJT inverter. Note that for a logic-0 input, $v_I \leq V_{IL}$, the transistor will be cut off and the output voltage will be equal to V_{CC}; that is, $V_{OH} = V_{CC}$. For a logic-1 input, $v_I \geq V_{IH}$, the BJT will be saturated and the output voltage will be equal to V_{CEsat}; that is, $V_{OL} = V_{CEsat} = 0.1$ to 0.2 V.

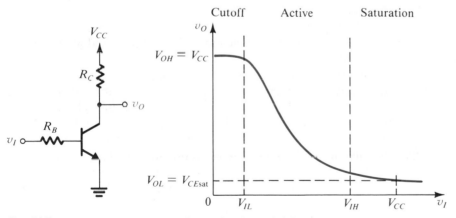

Fig. 16.7 The basic BJT inverter and its transfer characteristic.

Resistor-Transistor Logic (RTL)

By paralleling the outputs of two or more basic inverters we obtain the basic gate circuit of an early logic circuit family known as resistor-transistor logic (RTL). Figure 16.8 shows such a two-input NOR gate. The circuit works as follows: If one of the inputs—say, A—is high (logic 1), then the corresponding transistor (Q_A) will be on

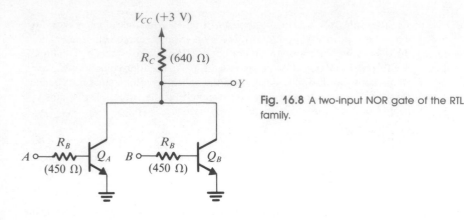

Fig. 16.8 A two-input NOR gate of the RTL family.

and saturated. This will result in $v_Y = V_{CEsat}$, which is low (logic 0). If the other input (B) is also high (logic 1), then the corresponding transistor (Q_B) will be on and saturated, thus keeping the output low. It can be seen that for the output to be high ($v_Y = V_{CC}$), both Q_A and Q_B have to be off simultaneously. Clearly, this is obtained if both A and B are simultaneously low. That is, a logic 1 will appear at the output only in one case: A and B are low. We may therefore write the Boolean expression

$$Y = \bar{A}\bar{B}$$

which can be also written

$$Y = \overline{A + B}$$

which is a NOR function.

The fan-in of the RTL NOR gate can be increased by adding more input transistors. The resistor values and supply voltage indicated are those used in integrated-circuit RTL (circa the 1960s).

Although the high output level (V_{OH}) of a single gate is V_{CC}, this is not the case when the RTL gate is driving another similar gate. Since the input transistor of the driven gate will be turned on, its base current will be supplied through resistor R_C of the driving gate. Thus the value of V_{OH} will be considerably lowered, to a value closer to 1 V. Furthermore, this value will be reduced as the gate fan-out is increased. As a result the noise margins of the RTL gate are rather narrow. This, together with the fact that RTL gates dissipate a rather large amount of power (the delay-power product is about 140 pJ) has resulted in the demise of RTL.

An RTL SR Flip-Flop

Before leaving RTL we wish to show the SR flip-flop circuit shown in Fig. 16.9. This circuit can be formed by cross-coupling two two-input NOR gates. However, the circuit was in fact quite popular in discrete-circuit design before the advent of the integrated circuit in the early 1960s. Operation of the circuit is straightforward and follows closely the logic description of the SR flip-flop given in Section 15.7.

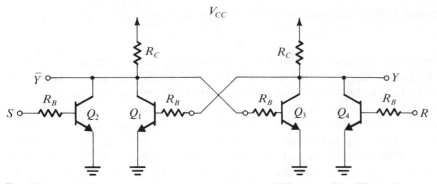

Fig. 16.9 An SR flip-flop formed by cross-coupling two NOR gates of the RTL family.

Diode-Transistor Logic (DTL)

Another early BJT logic circuit family is diode-transistor logic or DTL, exemplified by the two-input NAND gate shown in Fig. 16.10. DTL is of particular interest to us because, as we shall see in the next section, it is the circuit from which TTL has evolved.

The DTL circuit operates as follows: Let input B be left open. If a logic-0 signal ($\simeq 0$ V) is applied to A, diode D_1 will conduct and the voltage at node X will be one diode drop (0.7 V) above the logic-0 value. The two diodes D_3 and D_4 will be conducting, thus causing the base of transistor Q to be two diode drops below the voltage at node X. Thus the base will be at a small negative voltage, and hence Q will be off and $v_Y = V_{CC}$ (logic 1).

Consider now increasing the voltage v_A. It can be seen that diode D_1 will remain conducting and node X will keep rising in potential. Diodes D_3 and D_4 will remain conducting, and hence the base will also rise in potential. This situation will continue until the voltage at the base reaches about 0.5 V, at which point the transistor will

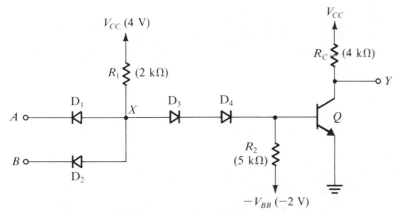

Fig. 16.10 A two-input NAND gate of the DTL family.

start to conduct. This will occur when the voltage at A is

$$v_A \simeq 0.5 + V_{D4} + V_{D3} - V_{D1} \simeq 1.2 \text{ V}$$

Small increases in v_A above this threshold value will appear as increases in v_{BE} and hence in i_C. In this range the transistor will be in the active region. Eventually the voltage at the base will reach 0.7 V and the transistor will be fully conducting. At this point the voltage at X will be clamped to two diode drops above V_{BE}, and any further increases in v_A will reverse-bias D_1. It can be seen that the current in D_1 will begin to decrease when v_A reaches about 1.4 V. As D_1 stops conducting, all the current through R_1 will be diverted through D_3 and D_4 into the base of the transistor. The circuit is normally designed such that the current into the base will be sufficient to drive the transistor into saturation. Thus when A is at logic 1 level the transistor will be saturated and v_Y will be equal to V_{CEsat} ($\simeq 0.2$ V), which is a logic 0 output.

Extrapolating from the above discussion, it can be seen that if either or both of the inputs is low the corresponding diode (D_1, D_2, or both) will be conducting, the transistor will be off, and the output Y will be high. The output will be low if the transistor is on, which will happen for only one particular input combination, when all the inputs are simultaneously high. We may therefore write the Boolean expression

$$\bar{Y} = AB$$

which can be rewritten

$$Y = \overline{AB}$$

which is a NAND function. This should come as no surprise, since the DTL circuit consists of a diode AND gate formed by D_1, D_2, and R_1 (see Section 5.12), followed by a transistor inverter. Finally, we note that because of their function in steering current into either R_2 or the transistor base, diodes D_3 and D_4 are known as "steering diodes".

DTL was popular in the 1960s and was implemented first using discrete components and subsequently in IC form. However, it was eventually replaced with transistor-transistor logic (TTL).

Exercises

16.6 Consider the RTL gate of Fig. 16.8 when driving N identical gates. Let both inputs to the driving gate be low. Convince yourself that the output voltage V_{OH} can be determined

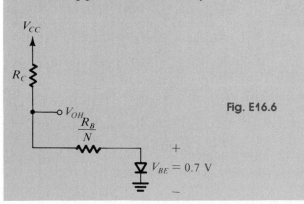

Fig. E16.6

using the equivalent circuit shown in Fig. E16.6. Hence show that

$$V_{OH} = V_{CC} - R_C \frac{V_{CC} - V_{BE}}{R_C + R_B/N}$$

For $N = 5$, use the values given in Fig. 16.8, together with $V_{BE} = 0.7$ V to obtain the value of V_{OH}.

Ans. 1 V

16.7 For the DTL gate of Fig. 16.10 assume all conducting junctions to have a voltage drop of 0.7 V:
(a) Find the current through D_1 when $v_A = 0.2$ V and $v_B = +4$ V. Also find the voltage at the base.
(b) With $v_A = v_B = +4$ V find the transistor base current. If $V_{CEsat} = 0.2$ V find the value of β_{forced}.

Ans. (a) 1.25 mA, -0.5 V; (b) 0.41 mA, 2.3

16.3 TRANSISTOR-TRANSISTOR LOGIC (TTL OR T²L)

For almost two decades TTL has enjoyed immense popularity. Indeed, for the bulk of digital systems applications employing SSI and MSI packages, TTL is rivaled only by CMOS (Chapter 15).

We shall begin this section with a study of the evolution of TTL from DTL. In this way we shall explain the function of each of the stages of the complete TTL gate circuit. Characteristics of standard TTL gates will be studied in Section 16.4. Standard TTL, however, has now been virtually replaced with more advanced forms of TTL that feature improved performance. These will be discussed in Section 16.5.

Evolution of TTL From DTL

The basic DTL gate circuit in discrete form was discussed in the previous section (see Fig. 16.10). The integrated-circuit form of the DTL gate is shown in Fig. 16.11 with only one input indicated. As a prelude to introducing TTL, we have drawn the input diode as a diode-connected transistor (Q_1), which corresponds to how diodes are made in IC form.

This circuit differs from the discrete DTL circuit of Fig. 16.10 in two important aspects. First, one of the steering diodes is replaced by the base-emitter junction of a transistor (Q_2) that is either cut off (when the input is low) or in the active mode (when the input is high). This is done to increase the fan-out capability of the gate. A detailed explanation of this point, however, is not relevant to our study of TTL. Second the resistance R_B is returned to ground rather than to a negative supply, as was done in the earlier discrete circuit. An obvious advantage of this is the elimination of the additional power supply. The disadvantage, however, is that the reverse base current available to remove the excess charge stored in the base of Q_3 is rather small. We shall elaborate on this point below.

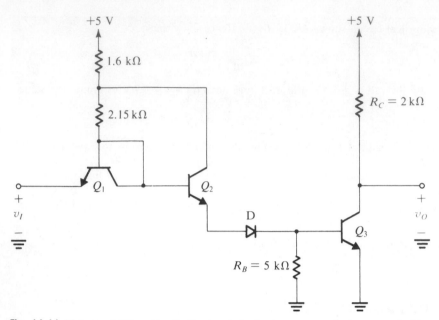

Fig. 16.11 IC form of DTL gate with the input diode shown as a diode-connected transistor (Q_1). Only one input terminal is shown.

Exercise

16.8 Consider the DTL gate circuit shown in Fig. 16.11 and assume that $\beta(Q_2) = \beta(Q_3) = 50$.
(a) When $v_I = 0.2$ V, find the input current.
(b) When $v_I = +5$ V, find the base current of Q_3.

Ans. **(a)** 1.1 mA; **(b)** 1.6 mA

Reasons for the Slow Response of DTL

The DTL gate has relatively good noise margins and reasonably good fan-out capability. Its response, however, is rather slow. This due to two reasons. First, when the input goes low and Q_2 and D turn off, the charge stored in the base of Q_3 has to leak through R_B to ground. The initial value of the reverse base current that accomplishes this "base discharging" process is approximately 0.7 V/R_B, which is about 0.14 mA. Because this current is quite small in comparison to the forward base current, the time required for the removal of base charge is rather long, which contributes to lengthening the gate delay.

The second reason for the relatively slow response of DTL derives from the nature of the output circuit of the gate, which is simply a common-emitter transistor. Figure 16.12 shows the output transistor of a DTL gate driving a capacitive load C_L. The capacitance C_L represents the input capacitance of another gate and/or the wiring and parasitic capacitances that are inevitably present in any circuit. When Q_3 is turned on, its collector voltage cannot instantaneously fall because of the existence of C_L. Thus Q_3 will not immediately saturate but rather will operate in the active region. The collector of Q_3 will therefore act as a constant-current source and will sink a rel-

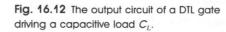

Fig. 16.12 The output circuit of a DTL gate driving a capacitive load C_L.

atively large current (βI_B). This large current will rapidly discharge C_L. We thus see that the common-emitter output stage features a short turn-on time. However, turnoff is another matter.

Consider next the operation of the common-emitter output stage when Q_3 is turned off. The output voltage will not rise immediately to the high level (V_{CC}). Rather, C_L will charge up to V_{CC} through R_C. This is a rather slow process, and it results in lengthening the DTL gate delay (and similarly the RTL gate delay).

Having identified the two reasons for the slow response of DTL, we shall see in the following how these problems are remedied in TTL.

Input Circuit of the TTL Gate

Figure 16.13 shows a conceptual TTL gate with only one input terminal indicated. The most important feature to note is that the input diode has been replaced by a transistor. One can think of this simply as if the short circuit between base and collector of Q_1 in Fig. 16.12 has been removed.

To see how the conceptual TTL circuit of Fig. 16.13 works, let the input v_I be high (say, $v_I = V_{CC}$). In this case current will flow from V_{CC} through R, thus forward-biasing the base–collector junction of Q_1. Meanwhile, the base–emitter junction of Q_1 will be reverse-biased. Therefore Q_1 will be operating in the *inverse active mode*–that is, in the active mode but with the roles of emitter and collector interchanged. The voltages

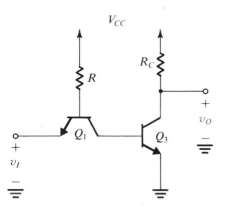

Fig. 16.13 Conceptual form of TTL gate. Only one input terminal is shown.

and currents will be as indicated in Fig. 16.14, where the current I can be calculated from

$$I = \frac{V_{CC} - 1.4}{R}$$

In actual TTL circuits Q_1 is designed to have a very low reverse β ($\beta_R \cong 0.02$). Thus the gate input current will be very small, and the base current of Q_3 will be approximately equal to I. This current will be sufficient to drive Q_3 into saturation, and the output voltage will be low (0.1 to 0.2 V).

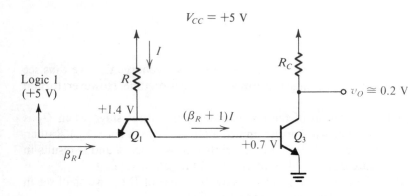

Fig. 16.14 Analysis of the conceptual TTL gate when the input is high.

Next let the gate input voltage be brought down to the logic-0 level (say, $v_I \cong$ 0.2 V). The current I will then be diverted to the emitter of Q_1. The base–emitter junction of Q_1 will become forward-biased, and the base voltage of Q_1 will therefore drop to 0.9 V. Since Q_3 *was* in saturation, its base voltage will remain at $+0.7$ V pending the removal of the excess charge stored in the base region. Figure 16.15 indicates the various voltage and current values immediately after the input is lowered. We see that Q_1 will be operating in the normal active mode[3] and its collector will carry a large current ($\beta_F I$). This large current rapidly discharges the base of Q_3 and drives it into cutoff. We thus see the action of Q_1 in speeding up the turnoff process.

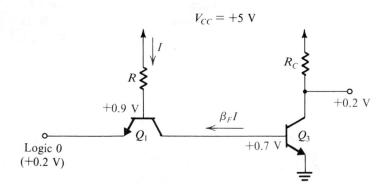

Fig. 16.15 Voltage and current values in the conceptual TTL circuit immediately after the input is lowered.

---

[3] Although the collector voltage of Q_1 is lower than its base voltage by 0.2 V, the collector-base junction will in effect be cutoff and Q_1 will be operating in the active mode.

As Q_3 turns off, the voltage at its base is reduced, and Q_1 enters the saturation mode. Eventually the collector current of Q_1 will become negligibly small, which implies that its V_{CEsat} will be approximately 0.1 V and the base of Q_3 will be at about 0.3 V, which keeps Q_3 in cutoff.

Output Circuit of the TTL Gate

The above discussion illustrates how one of the two problems that slow down the operation of DTL is solved in TTL. The second problem, the long rise time of the output waveform, is solved by modifying the output stage, as we shall now explain.

First, recall that the common-emitter output stage provides fast discharging of load capacitance but rather slow charging. The opposite is obtained in the emitter-follower output stage shown in Fig. 16.16. Here, as v_I goes high, the transistor turns on and provides a low output resistance (characteristic of emitter followers), which results in fast charging of C_L. On the other hand, when v_I goes low, the transistor turns off and C_L is then left to discharge slowly through R_E.

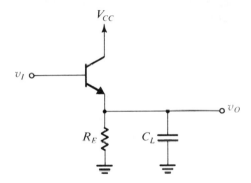

Fig. 16.16 An emitter-follower output stage with capacitive load.

It follows that an optimum output stage would be a combination of the common-emitter and the emitter-follower configurations. Such an output stage, shown in Fig. 16.17, has to be driven by two *complementary* signals v_{I1} and v_{I2}. When v_{I1} is high v_{I2} will be low, and in this case Q_3 will be on and saturated, and Q_4 will be off. The

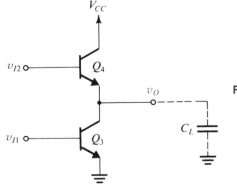

Fig. 16.17 The totem-pole output stage.

common-emitter transistor Q_3 will then provide the fast discharging of load capacitance and in steady state provide a low resistance (R_{CEsat}) to ground. Thus when the output is low, the gate can *sink* substantial amounts of current through the saturated transistor Q_3.

When v_{I1} is low and v_{I2} is high, Q_3 will be off and Q_4 will be conducting. The emitter follower Q_4 will then provide fast charging of load capacitance. It also provides the gate with a low output resistance in the high state and hence with the ability to *source* a substantial amount of load current.

Because of the appearance of the circuit in Fig. 16.17, with Q_4 stacked on top of Q_3, the circuit has been given the name *totem-pole output stage*. Also, because of the action of Q_4 in *pulling up* the output voltage to the high level, Q_4 is referred to as the *pull-up transistor*. Since the pulling up is achieved here by an active element (Q_4), the circuit is said to have an *active pull-up*. This is in contrast to the *passive pull-up* of RTL and DTL gates. Finally, note that a special *driver circuit* is needed to generate the two complementary signals v_{I1} and v_{I2}.

EXAMPLE 16.1

We wish to analyze the circuit shown together with its driving waveforms in Fig. 16.18 to determine the waveform of the output signal v_O. Assume that Q_3 and Q_4 have $\beta = 50$.

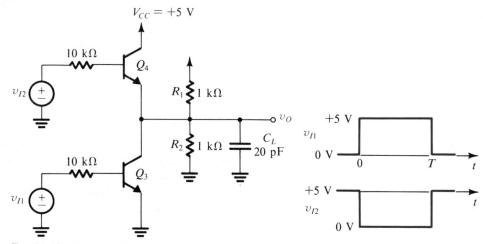

Fig. 16.18 Circuit and input waveforms for Example 16.1.

Solution

Consider first the situation before v_{I1} goes high—that is, at time $t < 0$. In this case Q_3 is off and Q_4 is on, and the circuit can be simplified to that shown in Fig. 16.19. In this simplified circuit we have replaced the voltage divider (R_1, R_2) by its Thévenin equivalent. In the steady state, C_L will be charged to the output voltage v_O, whose value can be obtained as follows:

$$5 = 10 \times I_B + V_{BE} + I_E \times 0.5 + 2.5$$

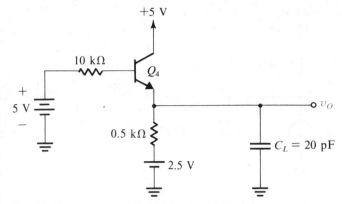

Fig. 16.19 The circuit of Fig. 16.18 when Q_3 is off.

Substituting $V_{BE} \cong 0.7$ V and $I_B = I_E/(\beta + 1) = I_E/51$ gives $I_E = 2.59$ mA. Thus the output voltage v_O is given by

$$v_O = 2.5 + I_E \times 0.5 = 3.79 \text{ V}$$

We next consider the circuit as v_{I1} goes high and v_{I2} goes low. Transistor Q_3 turns on and transistor Q_4 turns off, and the circuit simplifies to that shown in Fig. 16.20. Again we have used the Thévenin equivalent of the divider (R_1, R_2). We shall also assume that the switching times of the transistors are negligibly small. Thus at $t = 0+$ the base current of Q_3 becomes

$$I_B = \frac{5 - 0.7}{10} = 0.43 \text{ mA}$$

Since at $t = 0$ the collector voltage of Q_3 is 3.79 V, and since this value cannot change instantaneously because of C_L, we see that at $t = 0+$ transistor Q_3 will be in the active mode. The collector current of Q_3 will be βI_B, which is 21.5 mA, and the circuit will have the equivalent shown in Fig. 16.21a. A simpler version of this equivalent circuit, obtained using Thévenin's theorem, is shown in Fig. 16.21b.

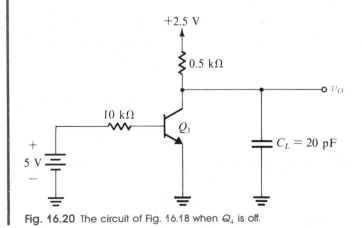

Fig. 16.20 The circuit of Fig. 16.18 when Q_4 is off.

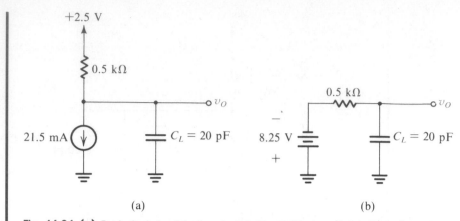

(a)   (b)

Fig. 16.21 (a) Equivalent circuit for the circuit in Fig. 16.20 when Q_3 is in the active mode. **(b)** Simpler version of the circuit in (a) obtained using Thévenin's theorem.

The equivalent circuit of Fig. 16.21 applies as long as Q_3 remains in the active mode. This condition persists while C_L is being discharged and until v_O reaches about $+0.3$ V, at which time Q_3 enters saturation. This is illustrated by the waveform in Fig. 16.22. The time for the output voltage to fall from $+3.79$ V to $+0.3$ V, which can be considered the *fall time* t_f, can be obtained from

$$-8.25 - (-8.25 - 3.79)e^{-t_f/\tau} = 0.3$$

which results in

$$t_f \cong 0.34\tau$$

where

$$\tau = C_L \times 0.5 \text{ k}\Omega = 10 \text{ ns}$$

Thus $t_f = 3.4$ ns.

After Q_3 enters saturation, the capacitor discharges further to the final steady-state value of $V_{CE\text{sat}}$ ($\cong 0.2$ V). The transistor model that applies during this interval

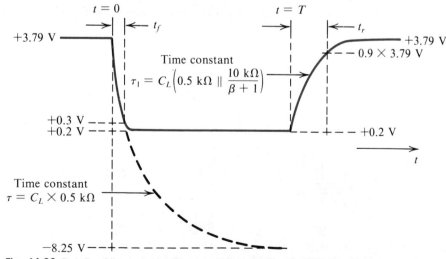

Fig. 16.22 Details of the output voltage waveform for the circuit in Fig. 16.18.

is more complex; since the interval in question is quite short, we shall not pursue the matter further.

Consider next the situation as v_{I1} goes low and v_{I2} goes high at $t = T$. Transistor Q_3 turns off as Q_4 turns on. We shall assume that this occurs immediately and thus at $t = T+$ the circuit simplifies to that in Fig. 16.19. We have already analyzed this circuit in the steady state and thus know that eventually v_O will reach $+3.79$ V. Thus v_O rises exponentially from $+0.2$ V toward $+3.79$ V with a time constant of $C_L\{0.5 \text{ k}\Omega \,\|\, [10 \text{ k}\Omega/(\beta + 1)]\}$, where we have neglected the emitter resistance r_e. Denoting this time constant τ_1 we obtain $\tau_1 = 2.8$ ns. Defining the rise time t_r as the time for v_O to reach 90% of the final value, we obtain $3.79 - (3.79 - 0.2)e^{-t_r/\tau_1} = 0.9 \times 3.79$, which results in $t_r = 6.4$ ns. Figure 16.22 illustrates the details of the output voltage waveform.

The Complete Circuit of the TTL Gate

Figure 16.23 shows the complete TTL gate circuit. It consists of three stages: the input transistor Q_1, whose operation has already been explained, the driver stage Q_2, whose function is to generate the two complementary voltage signals required to drive the totem-pole circuit, which is the third (output) stage of the gate. The totem-pole circuit in the TTL gate has two additional components: the 130-Ω resistance in the collector circuit of Q_4 and the diode D in the emitter circuit of Q_4. The function of these two additional components will be explained shortly. Notice that the TTL gate is shown with only one input terminal indicated. Inclusion of additional input terminals will be considered in Section 16.5.

Because the driver stage Q_2 provides two complementary (that is, out-of-phase) signals, it is known as a *phase splitter*.

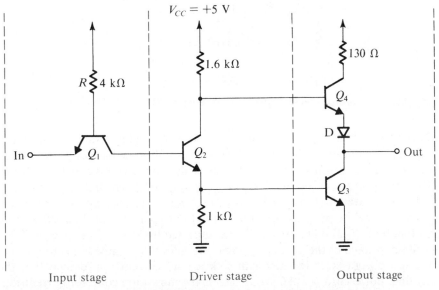

Fig. 16.23 The complete TTL gate circuit with only one input terminal indicated.

We shall now provide a detailed analysis of the TTL gate circuit in its two extreme states: one with the input high and one with the input low.

Analysis when the Input Is High

When the input is high (say, $+5$ V), the various voltages and currents of the TTL circuit will have the values indicated in Fig. 16.24. The analysis illustrated in Fig. 16.24 is quite straightforward. As expected, the input transistor is operating in the inverse active mode, and the input current, called the *input high current* I_{IH}, is small; that is,

$$I_{IH} = \beta_R I \cong 15 \ \mu A$$

where we assume that $\beta_R \cong 0.02$.

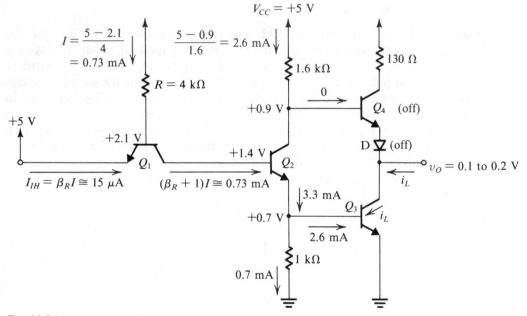

Fig. 16.24 Analysis of the TTL gate with the input high.

The collector current of Q_1 flows into the base of Q_2, and its value is sufficient to saturate the phase-splitter transistor Q_2. The latter supplies the base of Q_3 with sufficient current to drive it into saturation and lower its output voltage to V_{CEsat} (0.1 to 0.2 V). The voltage at the collector of Q_2 is $V_{BE3} + V_{CEsat} (Q_2)$, which is approximately $+0.9$ V. If diode D were not included, this voltage would be sufficient to turn Q_4 on, which is contrary to the proper operation of the totem-pole circuit. Including diode D ensures that both Q_4 and D remain off. The saturated transistor Q_3 then establishes the low output voltage of the gate (V_{CEsat}) and provides a low impedance to ground.

In the low-output state the gate can sink a load current i_L provided that the value of i_L does not exceed $\beta \times 2.6$ mA, which is the maximum collector current that

Q_3 can sustain while remaining in saturation. Obviously the greater the value of i_L, the greater the output voltage will be. To maintain the logic-0 level below a certain specified limit, a corresponding limit has to be placed on the load current i_L. As will be seen shortly, it is this limit that determines the maximum fan-out of the TTL gate.

Figure 16.25 shows a sketch of the output voltage v_O versus the load current i_L of the TTL gate when the output is low. This is simply the i_C-v_{CE} characteristic curve of Q_3 measured with a base current of 2.6 mA. Note that at $i_L = 0$, v_O is the offset voltage, which is about 100 mV.

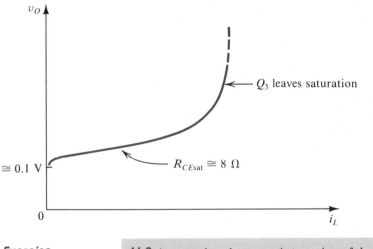

Fig. 16.25 The v_O-i_L characteristic of the TTL gate when the output is low.

Exercise

16.9 Assume that the saturation portion of the v_O-i_L characteristic shown in Fig. 16.25 can be approximated by a straight line (of slope = 8 Ω) that intersects the v_O axis at 0.1 V. Find the maximum load current that the gate is allowed to sink if the logic 0 level is specified to be ≤ 0.3 V.

Ans. 25 mA

Analysis when the Input Is Low

Consider next the operation of the TTL gate when the input is at the logic 0 level ($\cong 0.2$ V). The analysis is illustrated in Fig. 16.26, from which we see that the base–emitter junction of Q_1 will be forward-biased and the base voltage will be approximately $+0.9$ V. Thus the current I can be found to be approximately 1 mA. Since 0.9 V is insufficient to forward-bias the series combination of the collector–base junction of Q_1 and the base–emitter junction of Q_2 (at least 1.2 V would be required), the latter will be off. Therefore the collector current of Q_1 will be almost zero and Q_1 will be saturated, with $V_{CEsat} \cong 0.1$ V. Thus the base of Q_2 will be at approximately $+0.3$ V, which is indeed insufficient to turn Q_2 on.

The gate input current in the low state, called *input-low current* I_{IL}, is approximately equal to the current I ($\cong 1$ mA) and flows out of the emitter of Q_1. If the TTL gate is driven by another TTL gate, the output transistor Q_3 of the driving gate should sink this current I_{IL}. Since the output current that a TTL gate can sink is limited to

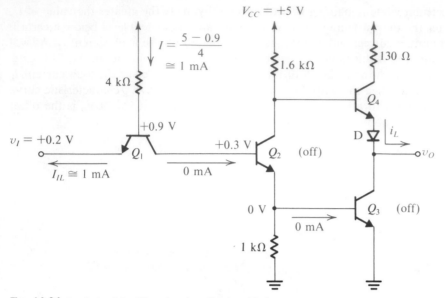

Fig. 16.26 Analysis of the TTL gate when the input is low.

a certain maximum value, the maximum fan-out of the gate is directly determined by the value of I_{IL}.

Exercises

16.10 Consider the TTL gate analyzed in Exercise 16.9. Find its maximum allowable fan-out using the value of I_{IL} calculated above.

Ans.   25

16.11 Use Eq. (16.16) to find V_{CEsat} of transistor Q_1 when the input of the gate is low (0.2 V). Assume that $\beta_F = 50$ and $\beta_R = 0.02$.

Ans.   98 mV

Let us continue with our analysis of the TTL gate. When the input is low, we see that both Q_2 and Q_3 will be off. Transistor Q_4 will be on and will supply (source) the load current i_L. Depending on the value of i_L, Q_4 will be either in the active mode or in the saturation mode.

With the gate output terminal open, the current i_L will be very small (mostly leakage) and the two junctions (base-emitter junction of Q_4 and diode D) will be barely conducting. Assuming that each junction has a 0.65-V drop and neglecting the voltage drop across the 1.6-kΩ resistance, we find that the output voltage will be

$$v_O \cong 5 - 0.65 - 0.65 = 3.7 \text{ V}$$

As i_L is increased, Q_4 and D conduct more heavily, but for a range of i_L, Q_4 remains in the active mode, and v_O is given by

$$v_O = V_{CC} - \frac{i_L}{\beta + 1} \times 1.6 \text{ k}\Omega - V_{BE4} - V_D \tag{16.21}$$

If we keep increasing i_L, a value will be reached at which Q_4 saturates. Then the output voltage becomes determined by the 130-Ω resistance according to the approximate relationship

$$v_O \cong V_{CC} - i_L \times 130 - V_{CE_{\text{sat}}}(Q_4) - V_D \tag{16.22}$$

Function of the 130-Ω Resistance

At this point the reason for including the 130-Ω resistance should be evident: It is simply to limit the current that flows through Q_4, especially in the event that the output terminal is accidentally short-circuited to ground. This resistance also limits the supply current in another circumstance, namely, when Q_4 turns on while Q_3 is still in saturation. To see how this occurs, consider the case where the gate input was high and then is suddenly brought down to the low level. Transistor Q_2 will turn off relatively fast because of the availability of a large reverse current supplied to its base terminal by the collector of Q_1. On the other hand, the base of Q_3 will have to discharge through the 1-kΩ resistance, and thus Q_3 will take some time to turn off. Meanwhile Q_4 will turn on, and a large current pulse will flow through the series combination of Q_4 and Q_3. Part of this current will serve the useful purpose of charging up any load capacitance to the logic-1 level. The magnitude of the current pulse will be limited by the 130-Ω resistance to about 30 mA.

The occurrence of these current pulses of short duration (called current spikes) raises another important issue. The current spikes have to be supplied by the V_{CC} source and, because of its finite source resistance, will result in voltage spikes (or "glitches") superimposed on V_{CC}. These voltage spikes could be coupled to other gates and flip-flops in the digital system and thus might produce false switching in other parts of the system. This effect, which might loosely be called *crosstalk*, is a problem in TTL systems. To reduce the size of the voltage spikes, capacitors (called bypass capacitors) should be connected to ground at frequent locations on the supply rail. These capacitors lower the impedance of the supply voltage source and hence reduce the magnitude of the voltage spikes. Alternatively, one can think of the bypass capacitors as supplying the impulsive current spikes.

Exercises

16.12 Assuming that Q_4 has $\beta = 50$ and that at the verge of saturation $V_{CE_{\text{sat}}} - 0.3$ V, find the value of i_L at which Q_4 saturates.

Ans.   4.16 mA

16.13 Assuming that at a current of 1 mA the voltage drops across the emitter-base junction of Q_4 and the diode D are each 0.7 V, find v_O when $i_L = 1$ mA and 10 mA. (Note the result of the previous exercise.)

Ans.   3.6 V; 2.74 V

16.14 Find the maximum current that can be sourced by a TTL gate while the output high level (V_{OH}) remains greater than the minimum guaranteed value of 2.4 V.

Ans.   12.3 mA

16.4 CHARACTERISTICS OF STANDARD TTL

Because of its popularity and importance, TTL will be studied further in this and the next sections. In this section we shall consider some of the important characteristics of standard TTL gates. Special improved forms of TTL will be dealt with in Section 16.5.

Transfer Characteristic

Figure 16.27 shows the TTL gate together with a sketch of its voltage transfer characteristic drawn in a piecewise-linear fashion. The actual characteristic is, of course, a smooth curve. We shall now explain the transfer characteristic and calculate the various break-points and slopes. It will be assumed that the output terminal of the gate is open.

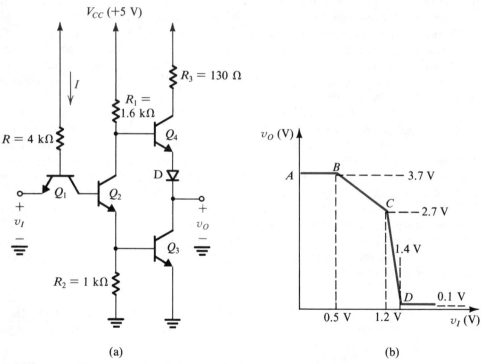

(a)   (b)

Fig. 16.27 The TTL gate and its voltage transfer characteristic.

Segment AB is obtained when transistor Q_1 is saturated, Q_2 and Q_3 are off, and Q_4 and D are on. The output voltage is approximately two diode drops below V_{CC}. At point B the phase splitter (Q_2) begins to turn on because the voltage at its base reaches 0.6 V (0.5 V + V_{CEsat} of Q_1).

Over segment BC transistor Q_1 remains saturated, but more and more of its base current I gets diverted to its base-collector junction and into the base of Q_2, which operates as a linear amplifier. Transistor Q_4 and diode D remain on, with Q_4 acting as an emitter follower. Meanwhile the voltage at the base of Q_3, although increasing, remains insufficient to turn Q_3 on (less than 0.6 V).

Let us now find the slope of segment BC of the transfer characteristic. Let the input v_I increase by an increment Δv_I. This increment appears at the collector of Q_1, since the saturated Q_1 behaves (approximately) as a three-terminal short circuit as far as signals are concerned. Thus at the base of Q_2 we have a signal Δv_I. Neglecting the loading of emitter follower Q_4 on the collector of Q_2, we can find the gain of the phase splitter from

$$\frac{v_{c2}}{v_{b2}} = \frac{-\alpha_2 R_1}{r_{e2} + R_2} \tag{16.23}$$

The value of r_{e2} will obviously depend on the current in Q_2. This current will range from zero (as Q_2 begins to turn on) to the value that results in a voltage of about 0.6 V at the emitter of Q_2 (the base of Q_3). This latter value is about 0.6 mA and corresponds to point C on the transfer characteristic. Assuming an average current in Q_2 of 0.3 mA, we obtain $r_{e2} \cong 83 \ \Omega$. For $\alpha = 0.98$, Eq. (16.23) results in a gain value of 1.45. Since the gain of the output follower Q_4 is close to unity, the overall gain of the gate, which is the slope of the BC segment, is about -1.45.

As already implied, breakpoint C is determined by Q_3 starting to conduct. The corresponding input voltage can be found from

$$v_I(C) = V_{BE3} + V_{BE2} - V_{CEsat}(Q_1)$$
$$= 0.6 + 0.7 - 0.1 = 1.2 \text{ V}$$

At this point the emitter current of Q_2 is approximately 0.6 mA. The collector current of Q_2 is also approximately 0.6 mA; neglecting the base current of Q_4, the voltage at the collector of Q_2 is

$$v_{C2}(C) = 5 - 0.6 \times 1.6 \cong 4 \text{ V}$$

Thus Q_2 is still in the active mode. The corresponding output voltage is

$$v_O(C) = 4 - 0.65 - 0.65 = 2.7 \text{ V}$$

As v_I is increased past the value of $v_I(C) = 1.2$ V, Q_3 begins to conduct and operates in the active mode. Meanwhile, Q_1 remains saturated, and Q_2 and Q_4 remain in the active mode. The circuit behaves as an amplifier until Q_2 and Q_3 saturate and Q_4 cuts off. This occurs at point D on the transfer characteristic, which corresponds to an input voltage $v_I(D)$ obtained from

$$v_I(D) = V_{BE3} + V_{BE2} + V_{BC1} - V_{BE1}$$
$$= 0.7 + 0.7 + 0.7 - 0.7 = 1.4 \text{ V}$$

Note that we have in effect assumed that at point D transistor Q_1 is still saturated, but with $V_{CEsat} \cong 0$. To see how this comes about, note that from point B on, more and more of the base current of Q_1 is diverted to its base–collector junction. Thus while the drop across the base–collector junction increases, that across the base–emitter junction decreases. At point D these drops become almost equal. For

$v_I > v_I(D)$ the base–emitter junction of Q_1 cuts off; thus Q_1 leaves saturation and enters the inverse active mode.

Calculation of gain over the segment CD is a relatively complicated task. This is due to the fact that there are two paths from input to output: one through Q_3 and one through Q_4. A simple but gross approximation for the gain of this segment can be obtained from the coordinates of points C and D in Fig. 16.27b, as follows:

$$\text{Gain} = -\frac{v_O(C) - v_O(D)}{v_I(D) - v_I(C)}$$

$$= -\frac{2.7 - 0.1}{1.4 - 1.2} = -13 \text{ V/V}$$

From the transfer curve of Fig. 16.27b we can determine the critical points and the noise margins as follows: $V_{OH} = 3.7$ V; V_{IL} is somewhere in the range 0.5 V to 1.2 V, and thus a conservative estimate would be 0.5 V; $V_{OL} = 0.1$ V; $V_{IH} = 1.4$ V; $NM_H = V_{OH} - V_{IH} = 2.3$ V; and $NM_L = V_{IL} - V_{OL} = 0.4$ V. It should be noted that these values are computed assuming that the gate is not loaded and without taking into account power-supply or temperature variations.

Exercise

16.15 Taking into account the fact that the voltage across a forward-biased pn junction changes by about -2 mV/°C, find the coordinates of points A, B, C, and D of the gate transfer characteristic at -55°C and at $+125$°C. Assume that the characteristic in Fig. 16.27b applies at 25°C, and neglect the small temperature coefficient of $V_{CE\text{sat}}$.

Ans. At -55°C: (0, 3.38), (0.66, 3.38), (1.52, 2.16), (1.72, 0.1); at $+125$°C: (0, 4.1), (0.3, 4.1), (0.8, 3.46), (1.0, 0.1)

Manufacturers' Specifications

Manufacturers of TTL usually provide curves for the gate transfer characteristics, the input i-v characteristics, and the output i-v characteristics, measured at the limits of the specified operating temperature range. In addition, guaranteed values are usually given for the parameters V_{OL}, V_{OH}, V_{IL}, and V_{IH}. For standard TTL (known as the 74 series) these values are $V_{OL} = 0.4$ V, $V_{OH} = 2.4$ V, $V_{IL} = 0.8$ V, and $V_{IH} = 2$ V. These limit values are guaranteed for a specified tolerance in power-supply voltage and for a maximum fan-out of 10. From our discussion in Section 16.3 we know that the maximum fan-out is determined by the maximum current that Q_3 can sink while remaining in saturation and while maintaining a saturation voltage lower than a guaranteed maximum ($V_{OL} = 0.4$ V). Calculations performed in Section 16.3 indicate the possibility of a maximum fan-out of 20 to 30. Thus the figure specified by the manufacturer is appropriately conservative.

The parameters V_{OL}, V_{OH}, V_{IL}, and V_{IH} can be used to compute the noise margins as follows:

$$NM_H = V_{OH} - V_{IH} = 0.4 \text{ V}$$
$$NM_L = V_{IL} - V_{OL} = 0.4 \text{ V}$$

Exercises

16.16 In Section 16.3 we found that when the gate input is high, the base current of Q_3 is approximately 2.6 mA. Assume that this value applies at 25°C and that at this temperature $V_{BE} \cong 0.7$ V. Taking into account the -2-mV/°C temperature coefficient of V_{BE} and neglecting all other changes, find the base current of Q_3 at -55°C and at $+125$°C.

Ans.   2.2 mA; 3 mA

16.17 Figure E16.17 shows sketches of the i_L-v_O characteristics of a TTL gate when the output is low. Use these characteristics together with the results of Exercise 16.16 to calculate the value of β of transistor Q_3 at -55°C, $+25$°C, and $+125$°C.

Ans.   16; 25; 28

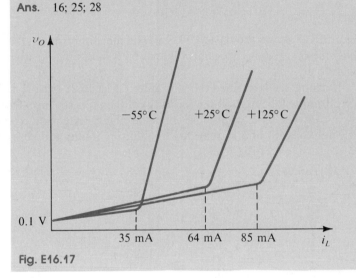

Fig. E16.17

Propagation Delay

The propagation delay of TTL gates is defined conventionally as the time between the 1.5-V points of corresponding edges of the input and output waveforms. For standard TTL (also known as *medium-speed* TTL) t_P is typically about 10 ns.

As far as power dissipation is concerned it can be shown (see Exercise 16.18 below) that when the gate output is high the gate dissipates 5 mW, and when the output is low the dissipation is 16.7 mW. Thus the average dissipation is 11 mW, resulting in a delay-power product of about 100 pJ.

Exercise

16.18 Calculate the value of the supply current (I_{CC}), and hence the power dissipated in the TTL gate, when the output terminal is open and the input is **(a)** low at 0.2 V (see Fig. 16.26) and **(b)** high at $+5$ V (see Fig. 16.24).

Ans.   **(a)** 1 mA, 5 mW;   **(b)** 3.33 mA, 16.7 mW

Dynamic Power Dissipation

In Section 16.3 the occurrence of supply current spikes was explained. These spikes give rise to additional power drain from the V_{CC} supply. This *dynamic power* is also dissipated in the gate circuit. It can be evaluated by multiplying the average current due to the spikes by V_{CC}, as illustrated by the solution of Exercise 16.19.

Exercise

16.19 Consider a TTL gate that is switched on and off at the rate of 1 MHz. Assume that each time the gate is turned off (that is, the output goes high) a supply-current pulse of 30-mA amplitude and 2-ns width occurs. Also assume that no current spike occurs when the gate is turned on. Calculate the dynamic power dissipation.

Ans.   0.3 mW

The TTL NAND Gate

Figure 16.28 shows the basic TTL gate. Its most important feature is the *multiemitter transistor* Q_1 used at the input. Figure 16.29 shows the structure of the multiemitter transistor.

It can be easily verified that the gate of Fig. 16.28 performs the NAND function. The output will be high if one (or both) of the inputs is (are) low. The output will be low in only one case: when both inputs are high. Extension to more than two inputs is straightforward and is achieved by diffusing additional emitter regions.

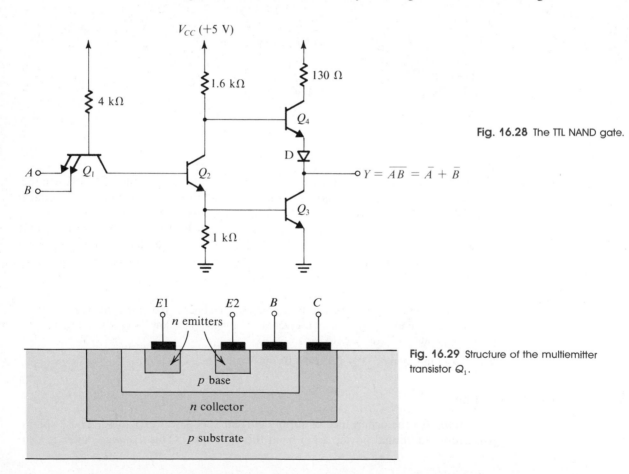

Fig. 16.28 The TTL NAND gate.

Fig. 16.29 Structure of the multiemitter transistor Q_1.

Although theoretically an unused input terminal may be left open-circuited, this is generally not a good practice. An open-circuit input terminal could act as an "antenna" that "picks up" interfering signals and thus could cause erroneous gate switching. An unused input terminal should therefore be connected to the positive power supply *through a resistance* (of, say, 1 kΩ). In this way the corresponding base–emitter junction of Q_1 will be reverse-biased and thus will have no effect on the operation of the gate. The series resistance is included in order to limit the current in case of breakdown of the base-emitter junction due to transients on the power supply.

Other TTL Logic Circuits

On a TTL MSI chip there are many cases in which logic functions are implemented using "stripped-down" versions of the basic TTL gate. As an example we show in Fig. 16.30 the TTL implementation of the AND-OR-INVERT function. As shown, the phase-splitter transistors of two gates are connected in parallel and a single output stage is used. The reader is urged to verify that the logic function realized is as indicated.

$$\overline{AB + CD} = \overline{AB} \cdot \overline{CD}$$

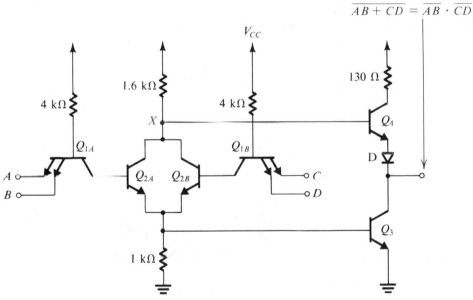

Fig. 16.30 A TTL AND-OR-INVERT gate.

At this point it should be noted that the totem-pole output stage of TTL does *not* allow connecting the output terminals of two gates to realize the AND function of their outputs (known as the wired-AND connection). To see the reason for this, consider two gates whose outputs are connected together and let one gate have a high output and the other have a low output. Current will flow from Q_4 of the first gate through Q_3 of the second gate. The current value will fortunately be limited by the 130-Ω resistance. Obviously, however, no useful logic function is realized by this connection.

The lack of wired-AND capability is a drawback of TTL. Nevertheless, the problem is solved in a number of ways, including doing the paralleling at the phase-splitter stage, as illustrated in Fig. 16.30. Another solution consists of deleting the emitter-follower transistor altogether. The result is an output stage consisting solely of the common-emitter transistor Q_3 without even a collector resistance. Obviously, one can connect the outputs of such gates together to a common collector resistance and achieve a wired-AND capability. TTL gates of this type are known as *open-collector TTL*. The obvious disadvantage is the slow rise time of the output waveform.

Another useful variant of TTL is the *tristate* output arrangement explored in Exercise 16.20.

Exercise

16.20 The circuit shown in Fig. E16.20 is called tristate TTL. Verify that when the terminal labeled Third state is high, the gate functions normally and that when this terminal is low, transistor Q_4 is cut off and the output of the gate is an open circuit. This latter state is the third state, or the high-output-impedance state.

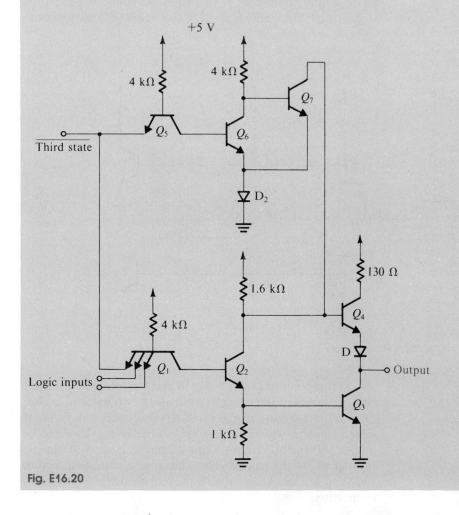

Fig. E16.20

Tristate TTL enables the connection of a number of TTL gates to a common output line (or *bus*). At any particular time the signal on the bus will be determined by the one TTL gate that is *enabled* (by raising its third-state input terminal). All other gates will be in the third state and thus will have no control of the bus.

16.5 TTL FAMILIES WITH IMPROVED PERFORMANCE

The standard TTL circuits studied in the two previous sections were introduced in the mid-1960s. Since then several improved versions have been developed. In this section we shall discuss some of these improved TTL subfamilies. As will be seen the improvements are in two directions: increasing speed and reducing power dissipation.

The speed of the standard TTL gate of Fig. 16.28 is limited by two mechanisms. First, transistors Q_1, Q_2, and Q_3 saturate, and hence we have to contend with their finite storage time. Although Q_2 is discharged reasonably quickly because of the active mode of operation of Q_1, as already explained, this is not true for Q_3, whose base charge has to leak out through the 1-kΩ resistance in its base circuit. Second, the resistances in the circuit, together with the various transistor and wiring capacitances, form relatively long time constants, which contribute to the gate delay.

It follows that there are two approaches to speeding up the operation of TTL. The first is to prevent transistor saturation and the second is to reduce the values of all resistances. Both approaches are utilized in the Schottky TTL circuit family.

Schottky TTL

In Schottky TTL, transistors are prevented from saturation by connecting a low-voltage-drop diode between base and collector, as shown in Fig. 16.31. These diodes, formed as a metal to semiconductor junction, are called Schottky diodes and have a forward voltage drop of about 0.5 V. Schottky diodes are easily fabricated and do not increase chip area. In fact, the Schottky TTL fabrication process has been designed to yield transistors with smaller areas than those produced by the standard TTL process. Thus Schottky transistors have higher β and f_T. Figure 16.31 also shows the symbol used to represent a Schottky transistor.

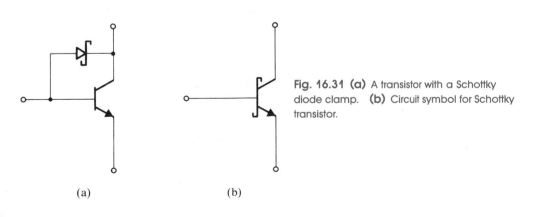

Fig. 16.31 (a) A transistor with a Schottky diode clamp. **(b)** Circuit symbol for Schottky transistor.

(a)   (b)

The Schottky transistor does not saturate since some of its base current drive is shunted away from the base by the Schottky diode. The latter then conducts and clamps the base–collector junction voltage to about 0.5 V. This voltage is smaller than the value required to forward-bias the collector–base junction of these small-sized transistors. In fact, a Schottky transistor begins to conduct when its v_{BE} is about 0.7 V, and the transistor is fully conducting when v_{BE} is about 0.8 V. With the Schottky diode clamping v_{BC} to about 0.5 V, the collector-to-emitter voltage is about 0.3 V and the transistors are still operating in the active mode. By avoiding saturation, the Schottky-clamped transistor exhibits a very short turnoff time.

Figure 16.32 shows a Schottky-clamped, or simply Schottky, TTL NAND gate. Comparing this circuit to that of standard TTL shown in Fig. 16.28 reveals a number of variations. First and foremost, Schottky clamps have been added to all transistors except Q_4. As will be seen shortly, transistor Q_4 can never saturate and thus does not need a Schottky clamp. Second, all resistances have been reduced to almost half the values in the standard circuit. These two changes result in a much shorter gate delay. The reduction in resistance values, however, increases the gate power dissipation (by a factor of about 2).

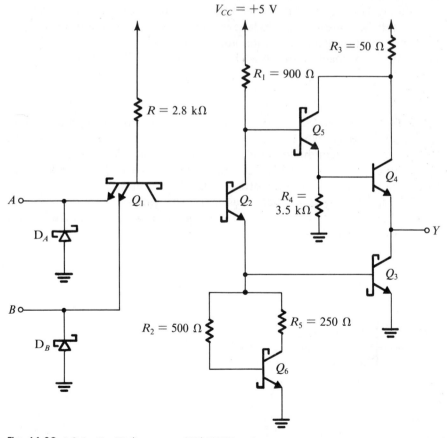

Fig. 16.32 A Schottky TTL (known as STTL) NAND gate.

The Schottky TTL gate features three other circuit techniques that further improve performance. These are as follows:

1. The diode D needed to prevent Q_4 from conducting when the gate output is low is replaced by transistor Q_5, which together with Q_4 forms a Darlington pair. This Darlington stage provides increased current gain and hence increased current-sourcing capability. This together with the lower output resistance of the gate (in the output-high state) yields a reduction in the time required to charge the load capacitance to the high level. Note that transistor Q_4 never saturates because

$$V_{CE4} = V_{CE5} + V_{BE4}$$
$$\simeq 0.3 + 0.8 = 1.1 \text{ V}$$

2. Input clamping diodes, D_A and D_B, are included.[4] These diodes conduct only when the input voltages go below ground level. This could happen due to "ringing" on the wires connecting the input of the gate to the output of another gate. Ringing occurs because such connecting wires behave as *transmission lines* that are not properly terminated. Without the clamping diodes, ringing can cause the input voltage to transiently go sufficiently negative so as to cause the substrate-to-collector junction of Q_1 (see Fig. 16.29) to become forward-biased. This in turn would result in improper gate operation. Also, ringing can cause the input voltage to go sufficiently positive to result in false gate switching. The input diodes clamp the negative excursions of the input ringing signal (to -0.5 V). Their conduction also provides a power loss in the transmission line, which results in *damping* of the ringing waveform and thus a reduction in its positive-going part.

3. The resistance between the base of Q_3 and ground has been replaced by a nonlinear resistance realized by transistor Q_6 and two resistors, R_2 and R_5. This nonlinear resistance is known as an *active pull-down*, in analogy to the active pull-up provided by the emitter-follower part of the totem pole output stage. This feature is an ingenious one, and we shall discuss it in more detail.

Active Pull-Down.   Figure 16.33 shows a sketch of the *i-v* characteristic of the nonlinear network composed of Q_6 and its two associated resistors, R_2 and R_5. Also shown for comparison is the linear *i-v* characteristic of a resistor that would be connected between the base of Q_3 and ground if the active pull-down were not used.

The first characteristic to note of the active pull-down is that it conducts negligible current and thus behaves as a high resistance until the voltage across it reaches a V_{BE} drop. Thus the gain of the phase splitter (as a linear amplifier) will remain negligibly small until a V_{BE} drop develops between its emitter and ground, which is the onset of conduction of Q_3. In other words, Q_2, Q_3, and Q_6 will turn on almost simultaneously. Thus segment BC of the transfer characteristic (see Fig. 16.27b) will be absent and the gate transfer characteristic will become much sharper, as shown in Fig. 16.34. Since the active pull-down circuit causes the transfer characteristic to

---

[4] Some standard TTL circuits also include input clamping diodes. Note, however, that in Schottky TTL, the input clamping diodes are of the Schottky type.

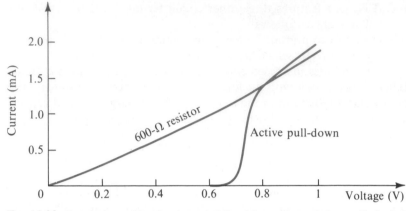

Fig. 16.33 Comparison of the *i-v* characteristic of the active pull-down with that of a 600-Ω resistor.

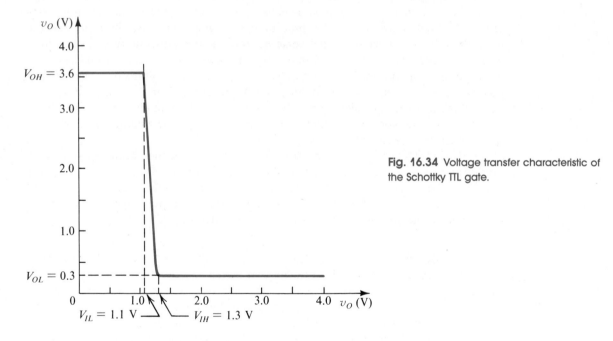

Fig. 16.34 Voltage transfer characteristic of the Schottky TTL gate.

become "squarer," it is also known as a "squaring circuit." The result is an increase in noise margins.

The active pull-down circuit also speeds up the turn-on and turnoff of Q_3. To see how this comes about observe that the active pull-down draws a negligible current over a good part of its characteristic. Thus the current supplied by the phase splitter will initially be diverted into the base of Q_3, thereby speeding up the turn-on of Q_3. On the other hand, during turnoff the active pull-down transistor Q_6 will be operating in the active mode. Its large collector current will flow through the base of Q_3 in the reverse direction, thus quickly discharging the base–emitter capacitance of Q_3 and turning it off fast.

Finally, the function of the 250-Ω collector resistance of Q_6 should be noted; without it the base–emitter voltage of Q_3 would be clamped to V_{CE6}, which is about 0.3 V.

16.21 Show that the values of V_{OH}, V_{IL}, V_{IH} and V_{OL} of the Schottky TTL gate are as given in Fig. 16.34. Assume that the gate output current is small and that a Schottky transistor conducts at $v_{BE} = 0.7$ V and is fully conducting at $v_{BE} = 0.8$ V.

16.22 Calculate the input current of a Schottky TTL gate with the input voltage low (at 0.3 V).

Ans. 1.4 mA

16.23 For the Schottky TTL gate calculate the current drawn from the power supply with the input low at 0.3 V (remember to include the current in the 3.5-kΩ resistor) and with the input high at 3.6 V. Hence calculate the gate power dissipation in both states and the average power dissipation.

Ans. 2.6 mA; 5.3 mA; 13 mW; 26.5 mW; 20 mW

Performance Characteristics. Schottky TTL (known as Series 74S) is specified to have the following worst-case parameters.

$$V_{OH} = 2.7 \text{ V} \qquad V_{OL} = 0.5 \text{ V}$$
$$V_{IH} = 2.0 \text{ V} \qquad V_{IL} = 0.8 \text{ V}$$
$$t_P = 3 \text{ ns} \qquad P_D = 20 \text{ mW}$$

Note that the delay-power product is 60 pJ, as compared with about 100 pJ for standard TTL.

16.24 Use the values given above to calculate the noise margins NM_H and NM_L for a Schottky TTL gate.

Ans. 0.7 V; 0.3 V

Low-Power Schottky TTL

Although Schottky TTL achieves a very low propagation delay, the gate power dissipation is rather high. This limits the number of gates that can be included per package. The need therefore arose for a modified version that achieves a lower gate dissipation, possibly at the expense of an increase in gate delay. This is obtained in the low-power Schottky (LS) TTL subfamily, represented by the two-input NAND gate shown in Fig. 16.35.

The low-power Schottky circuit of Fig. 16.35 differs from that of the regular Schottky TTL of Fig. 16.32 in several ways. Most important, note that the resistances used are about ten times larger, with the result that the power dissipation is only about a tenth of that of the Schottky circuit (2 mW versus 20 mW). However, as expected, the use of larger resistances is accompanied by a reduction in the speed of operation. To help compensate for this speed reduction some circuit design innovations are employed: First, the input multiemitter transistor has been eliminated

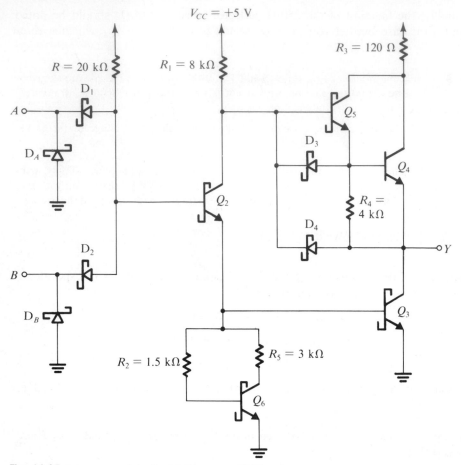

Fig. 16.35 A low-power Schottky TTL (known as LSTTL) gate.

in favor of Schottky diodes, which occupy a smaller silicon area and hence have smaller parasitic capacitances. In this regard it should be recalled that the main advantage of the input multiemitter transistor is that it rapidly removes the charge stored in the base of Q_2. In Schottky-clamped circuits, however, transistor Q_2 does not saturate, eliminating this aspect of the need for Q_1.

Second, two Schottky diodes, D_3 and D_4, have been added to the output stage to help speed up the turnoff of Q_4 and the turn-on of Q_3 and hence the transition of the output from high to low. Specifically, as the gate input is raised and Q_2 begins to turn on, some of its collector current will flow through diode D_3. This current constitutes a reverse base current for Q_4 and thus aids in the rapid turnoff of Q_4. Simultaneously, the emitter current of Q_2 will be supplied to the base of Q_3, thus causing it to turn on faster. Some of the collector current of Q_2 will also flow through D_4. This current will help discharge the gate load capacitance and thus shorten the transition time from high to low. Both D_3 and D_4 will be cut off under static conditions.

Finally, observe that the other terminal of the emitter resistor of Q_5 is now connected to the gate output. It follows that when the gate output is high, the output current will be first supplied by Q_5 through R_4. Transistor Q_4 will turn on only when a 0.7-V drop develops across R_4, and thereafter it sources additional load current. However, when the gate is supplying a small load current, Q_4 will be off and the output voltage will be approximately

$$V_{OH} = V_{CC} - V_{BE5}$$

which is higher than the value obtained in the regular Schottky TTL gate.

Exercises

16.25 The voltage transfer characteristic of the low-power Schottky TTL gate of Fig. 16.35 has the same shape as that of the regular Schottky TTL gate, shown in Fig. 16.34. Calculate the values of V_{OH}, V_{IL}, V_{IH}, and V_{OL}. Assume that a Schottky transistor conducts at $v_{BE} = 0.7$ V and is fully conducting at $v_{BE} = 0.8$ V, and that a Schottky diode has a 0.5-V drop. Also assume that the gate output current is very small.

Ans. 4.3 V; 0.9 V; 1.1 V; 0.3 V

16.26 For the low-power Schottky TTL gate, using the specifications given in Exercise 16.25, calculate the supply current in both states. Hence calculate the average power dissipation.

Ans. 0.2 mA; 0.66 mA; 2 mW

Although the guaranteed voltage levels and noise margins of low-power Schottky TTL (known as Series 74LS) are similar to those of the regular Schottky TTL, the gate delay and power dissipation are

$$t_P = 10 \text{ ns} \qquad P_D = 2 \text{ mW}$$

Thus although the power dissipation has been reduced by a factor of 10, the delay has been increased by only a factor of 3. The result is a delay-power product of only 20 pJ.

Further Improved TTL Families

There are other TTL families with further improved characteristics. Of particular interest is the advanced Schottky (Series 74AS) and the advanced low-power Schottky (Series 74ALS). We shall not discuss the circuit details of these families here. Table 16.2 provides a comparison of the TTL subfamilies based on gate delay and power dissipation.

Table 16.2 PERFORMANCE COMPARISON OF TTL FAMILIES

	Standard TTL (Series 74)	Schottky TTL (Series 74S)	Low-Power Schottky TTL (Series 74LS)	Advanced Schottky TTL (Series 74AS)	Advanced Low-Power Schottky TTL (Series 74ALS)
t_P, ns	10	3	10	1.5	4
P_D, mW	10	20	2	20	1
DP, pJ	100	60	20	30	4

Observe that advanced low-power Schottky offers a very small delay-power product. In conclusion, we note that at the present time TTL is still a very popular logic circuit family. Although standard TTL is no longer used in new designs, the advanced circuit types are frequently employed. The speed of the advanced Schottky family is rivaled only by that achieved in emitter-coupled logic (ECL).

16.6 EMITTER-COUPLED LOGIC (ECL)

Emitter-coupled logic (ECL) is the fastest logic circuit family. High speed is achieved by operating all transistors out of saturation, thus avoiding storage time delays, and by keeping the logic signal swings relatively small (about 0.8 V), thus reducing the time required to charge and discharge the various load and parasitic capacitances.

Unlike Schottky TTL, where saturation is prevented by diverting the excess base current drive into the Schottky clamp, saturation in ECL is avoided by using the BJT differential pair as a current switch. An introduction to the BJT differential pair was given in Sections 9.1 and 9.2, which we urge the reader to review before proceeding with the study of ECL.

ECL Families

Currently there are two popular forms of ECL—namely, ECL 10K and ECL 100K. The ECL 100K Series features gate delays of the order of 0.75 ns and dissipates about 40 mW/gate, for a delay-power product of 30 pJ. Although its power dissipation is relatively high, the 100K Series provides the shortest available gate delay.

The ECL 10K Series is slightly slower; it features a gate propagation delay of 2 ns and a power dissipation of 25 mW for a delay-power product of 50 pJ. Although the value of DP is higher than that obtained in the 100K Series, the 10K Series is easier to use. This is due to the fact that the rise and fall times of the pulse signals are deliberately made longer, thus reducing signal coupling, or crosstalk, between adjacent signal lines. ECL 10K has an "edge speed" of about 3.5 ns, as compared with the approximately 1 ns of ECL 100K. In the following we shall study the popular ECL 10K in some detail.

The Basic Gate Circuit

The basic gate circuit of the ECL 10K family is shown in Fig. 16.36. The circuit consists of three parts. The network composed of Q_1, D_1, D_2, R_1, R_2, and R_3 generates a reference voltage V_R whose value at room temperature is -1.32 V. As will be shown below, the value of this reference voltage is made to change with temperature in a predetermined manner so as to keep the noise margins almost constant. Also, the reference voltage V_R is made relatively insensitive to variations in the power supply voltage V_{EE}.

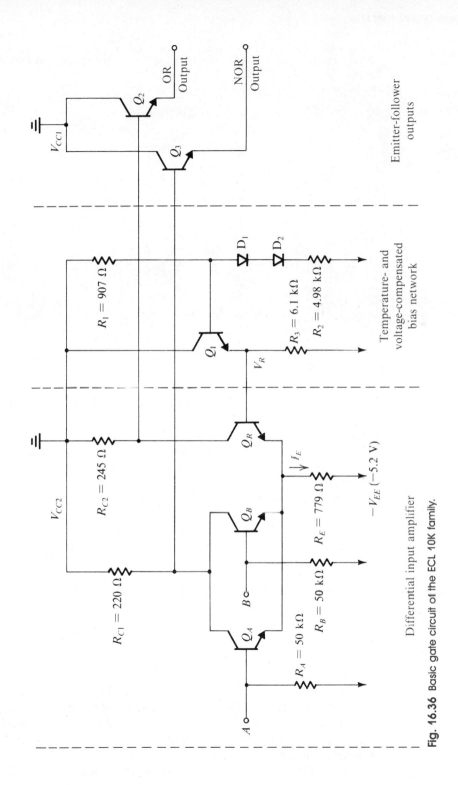

Fig. 16.36 Basic gate circuit of the ECL 10K family.

16.27 Figure E16.27 shows the circuit that generates the reference voltage V_R. Assuming that the voltage drop across each of D_1, D_2, and the base-emitter junction of Q_1 is 0.75 V, calculate the value of V_R.

Ans. -1.32 V

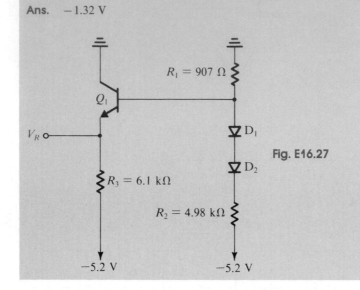

Fig. E16.27

The second part, and the heart of the gate, is the differential amplifier formed by Q_R and either Q_A or Q_B. This differential amplifier is biased not by a constant-current source, as was done in the circuits of Chapter 9, but with a resistance R_E connected to the negative supply $-V_{EE}$. One side of the differential amplifier consists of the reference transistor Q_R, whose base is connected to the reference voltage V_R. The other side consists of a number of transistors (two in the case shown) connected in parallel, with separated bases connected to the gate inputs. If the voltages applied to A and B are at the logic-0 level, which, as we will soon find out, is about 0.4 V below V_R, both Q_A and Q_B will be off and the current I_E in R_E will flow through the reference transistor Q_R. The resulting voltage drop across R_{C2} will cause the collector voltage of Q_R to be low.

On the other hand, when the voltage applied to A or B is at the logic-1 level, which, as we will show shortly, is about 0.4 V above V_R, transistor Q_A or Q_B, or both, will be on and Q_R will be off. Thus the current I_E will flow through Q_A or Q_B, or both, and an almost equal current flows through R_{C1}. The resulting voltage drop across R_{C1} will cause the collector voltage to drop. Meanwhile, since Q_R is off, its collector voltage rises. We thus see that the voltage at the collector of Q_R will be high if A or B, or both, is high, and thus at the collector of Q_R the OR logic function, $A + B$, is realized. On the other hand, the common collector of Q_A and Q_B will be high only when A and B are simultaneously low. Thus, at the common collector of Q_A and Q_B the logic function $\overline{AB} = \overline{A + B}$ is realized. We therefore conclude that the two-input gate of Fig. 16.36 realizes the OR function and its complement, the NOR function. The availability of complementary outputs is an important advantage of ECL; it simplifies logic design and avoids the use of additional inverters with associated time delay.

It should be noted that the resistance connecting each of the gate input terminals to the negative supply enables the user to leave an unused input terminal open. An open input terminal will then be *pulled down* to the negative supply voltage, and its associated transistor will be off.

Exercise

16.28 With input terminals A and B left open, find the current I_E through R_E. Also find the voltages at the collector of Q_R and at the common collector of the input transistors Q_A and Q_B. Use $V_R = -1.32$ V, V_{BE} of $Q_R \simeq 0.75$ V, and assume that β of Q_R is very high.

Ans. 4 mA; -1 V; 0 V

The third part of the ECL gate circuit is composed of the two emitter followers, Q_2 and Q_3. The emitter followers do not have on-chip loads, since in most applications of high-speed logic circuits the gate output drives a transmission line terminated at the other end, as indicated in Fig. 16.37.

The emitter followers have two purposes. First, they shift the level of the output signals by one V_{BE} drop. Thus, using the results of Exercise 16.28 above, we see that the output levels become approximately -1.75 V and -0.75 V. These shifted levels are centered approximately around the reference voltage ($V_R = -1.32$ V), which means that one gate can drive another. This compatibility of logic levels at input and output is always an essential requirement in the design of gate circuits.

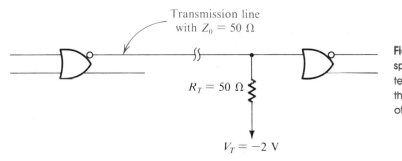

Transmission line with $Z_0 = 50\ \Omega$

$R_T = 50\ \Omega$

$V_T = -2$ V

Fig. 16.37 Proper way for connecting high-speed logic gates such as ECL. Properly terminating the transmission line connecting the two gates eliminates "ringing" that would otherwise corrupt the logic signals.

The second function of the output emitter followers is that they provide the gate with low output resistances and with the large output currents required for charging load capacitances. Since these large transient currents can cause spikes on the power-supply line, the collectors of the emitter followers are connected to a power-supply terminal V_{CC1} separate from that of the differential amplifier and the reference-voltage circuit, V_{CC2}. Here we note that the supply current of the differential amplifier and the reference circuit remains almost constant. The use of separate power-supply terminals prevents the coupling of power-supply spikes from the output circuit to the gate circuit, and thus lessens the likelihood of false gate switching. Both V_{CC1} and V_{CC2} are of course connected to the same system ground external to the chip.

Voltage Transfer Characteristics

Having provided a qualitative description of the operation of the ECL gate we shall now derive its voltage transfer characteristics. This will be done under the conditions that the outputs are terminated in the manner indicated in Fig. 16.37. Assuming that the B input is low and thus Q_B is off, the circuit simplifies to that shown in Fig. 16.38. We wish to analyze this circuit to determine v_{OR} versus v_I and v_{NOR} versus v_I (where $v_I \equiv v_A$).

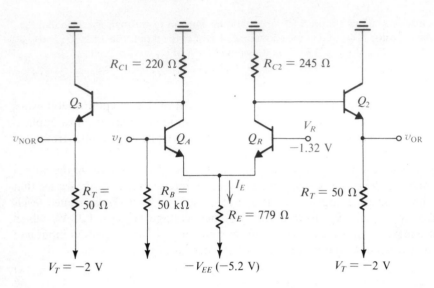

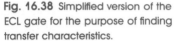

Fig. 16.38 Simplified version of the ECL gate for the purpose of finding transfer characteristics.

In the analysis to follow we shall make use of the exponential i_C-v_{BE} characteristic of the BJT. Since the BJTs used in ECL circuits have small areas (in order to have small capacitances and hence high f_T), their scale currents I_S are small. We will therefore assume that at an emitter current of 1 mA an ECL transistor has a V_{BE} drop of 0.75 V.

The OR Transfer Curve

Figure 16.39 shows a sketch of the OR transfer characteristic, v_{OR} versus v_I, with the parameters V_{OL}, V_{OH}, V_{IL}, and V_{IH} indicated. However, in order to simplify the calculation of V_{IL} and V_{IH} we shall use an alternative to the unity-gain definition. Specifically, we shall assume that at point x, transistor Q_A is conducting 1% of I_E while Q_R is conducting 99% of I_E. The reverse will be assumed for point y. Thus at point x we have

$$\frac{I_E|_{Q_R}}{I_E|_{Q_A}} = 99$$

Using the exponential i_E-v_{BE} relationship we obtain

$$V_{BE}|_{Q_R} - V_{BE}|_{Q_A} = V_T \ln 99 = 115 \text{ mV}$$

which gives

$$V_{IL} = -1.32 - 0.115 = -1.435 \text{ V}$$

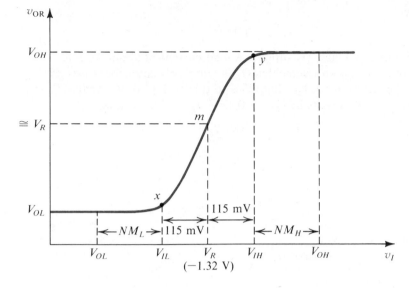

Fig. 16.39 The OR transfer characteristic, v_{OR} versus v_I, for the circuit in Fig. 16.38.

Assuming Q_A and Q_R to be matched, we can write

$$V_{IH} - V_R = V_R - V_{IL}$$

which can be used to find V_{IH} as

$$V_{IH} = -1.205 \text{ V}$$

To obtain V_{OL} we note that Q_A is off and Q_R carries the entire current I_E, given by

$$I_E = \frac{V_R - V_{BE}|_{Q_R} + V_{EE}}{R_E}$$

$$= \frac{-1.32 - 0.75 + 5.2}{0.779}$$

$$\simeq 4 \text{ mA}$$

(If we wish, we can iterate to determine a better estimate of $V_{BE}|_{Q_n}$ and hence of I_E.) Assuming that Q_R has a high β so that its $\alpha \simeq 1$, then its collector current will be approximately 4 mA. If we neglect the base current of Q_2, we obtain for the collector voltage of Q_R

$$V_C|_{Q_R} \simeq -4 \times 0.245 = -0.98 \text{ V}$$

Thus a first approximation for the value of the output voltage V_{OL} is

$$V_{OL} = V_C|_{Q_R} - V_{BE}|_{Q_2}$$
$$\simeq -0.98 - 0.75 = -1.73 \text{ V}$$

We can use this value to find the emitter current of Q_2 and then iterate to determine a better estimate of its base–emitter voltage. The result is $V_{BE2} \simeq 0.79$ V and, correspondingly,

$$V_{OL} \simeq -1.77 \text{ V}$$

At this value of output voltage, Q_2 supplies a load current of about 4.6 mA.

To find the value of V_{OH} we assume that Q_R is completely cutoff (because $v_I > V_{IH}$). Thus the circuit for determining V_{OH} simplifies to that in Fig. 16.40. Analysis of this circuit assuming $\beta_2 = 100$ results in $V_{BE2} \simeq 0.83$ V, $I_{E2} = 22.4$ mA, and

$$V_{OH} \simeq -0.88 \text{ V}$$

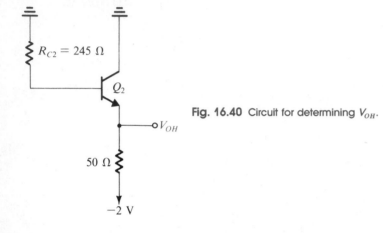

Fig. 16.40 Circuit for determining V_{OH}.

Noise Margins

The results of Exercise 16.29 indicate that the bias current I_E remains approximately constant. Also, the output voltage corresponding to $v_I = V_R$ is approximately equal to V_R. Notice further that this is also approximately the midpoint of the logic swing; specifically,

$$\frac{V_{OL} + V_{OH}}{2} = -1.325 \simeq V_R$$

Thus the output logic levels are centered around the midpoint of the input transition band. This is an ideal situation from the point of view of noise margins, and it is one of the reasons for selecting the rather arbitrary-looking numbers $V_R = -1.32$ V and $V_{EE} = 5.2$ V.

The noise margins can now be evaluated as follows:

$$NM_H = V_{OH} - V_{IH} \qquad\qquad NM_L = V_{IL} - V_{OL}$$
$$= -0.88 - (-1.205) = 0.325 \text{ V} \qquad = -1.435 - (-1.77) = 0.335 \text{ V}$$

Note that these values are approximately equal.

The NOR Transfer Curve

The NOR transfer characteristic, which is v_{NOR} versus v_I for the circuit in Fig. 16.38, is sketched in Fig. 16.41. The values of V_{IL} and V_{IH} are identical to those found above for the OR characteristic. To emphasize this we have labeled the threshold points x and y, the same letters used in Fig. 16.39.

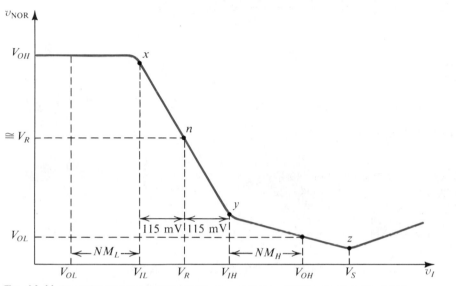

Fig. 16.41 The NOR transfer characteristic, v_{NOR} versus v_I, for the circuit in Fig. 16.38.

For $v_I < V_{IL}$, Q_A is off and the output voltage v_{NOR} can be found by analyzing the circuit composed of R_{C1}, Q_3, and its 50-Ω termination. Except that R_{C1} is slightly smaller than R_{C2}, this circuit is identical to that in Fig. 16.40. Thus the output voltage will be only slightly greater than the value V_{OH} found previously. In the sketch of Fig. 16.41 we have assumed that the output voltage is approximately equal to V_{OH}.

For $v_I > V_{IH}$, Q_A is on and is conducting the entire bias current. The circuit then simplifies to that in Fig. 16.42. This circuit can be easily analyzed to obtain v_{NOR} versus v_I for the range $v_I \geqq V_{IH}$. A number of observations are in order. First, note that $v_I = V_{IH}$ results in an output voltage slightly higher than V_{OL}. This is because R_{C1} is smaller than R_{C2}. In fact, R_{C1} is chosen lower in value than R_{C2} so that with v_I equal to the normal logic 1 value (that is, V_{OH}, which is approximately -0.88 V) the output will be equal to the V_{OL} value previously found for the OR output.

Second, note that as v_I exceeds V_{IH}, transistor Q_A operates in the active mode and the circuit of Fig. 16.42 can be analyzed to find the gain of this amplifier, which is the slope of the segment yz of the transfer characteristic. At point z transistor Q_A saturates. Further increments in v_I (beyond the point $v_I = V_S$) cause the collector voltage and hence v_{NOR} to increase. The slope of the segment of the transfer characteristic beyond point z, however, is not unity but is about 0.5 because as Q_A is driven deeper into saturation a portion of the increment in v_I appears as an increment in the base–collector forward-bias voltage. The reader is urged to solve Exercise 16.30, which is concerned with the details of the NOR transfer characteristic.

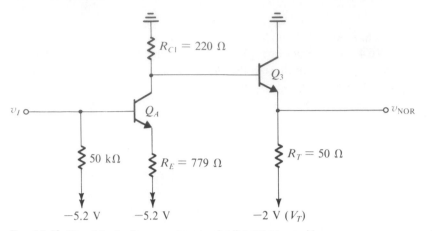

Fig. 16.42 Circuit for finding v_{NOR} versus v_I for the range $v_I > V_{IH}$.

Exercise

16.30 Consider the circuit in Fig. 16.42.
(a) For $v_I = V_{IH} = -1.205$ V, find v_{NOR}.
(b) For $v_I = V_{OH} = -0.88$ V, find v_{NOR}.
(c) Find the slope of the transfer characteristic at the point $v_I = V_{OH} = -0.88$ V.
(d) Find the value of v_I at which Q_A saturates (that is, V_S). Assume that $V_{CEsat} \simeq 0.3$ V and $\beta = 100$.

Ans.   **(a)** -1.71 V;   **(b)** -1.79 V;   **(c)** -0.24 V/V;   **(d)** -0.58 V

Manufacturer's Specifications

ECL manufacturers supply gate transfer characteristics of the form shown in Figs. 16.39 and 16.41. The manufacturer usually provides such curves measured at a number of temperatures. In addition, at each relevant temperature, worst-case values for the parameters V_{IL}, V_{IH}, V_{OL}, and V_{OH} are given. These worst-case values are specified with the inevitable component tolerances taken into account. As an example, Motorola specifies that for MECL 10,000 at 25°C the following worst-case values apply:[5]

$$V_{ILmax} = -1.475 \text{ V} \qquad V_{IHmin} = -1.105 \text{ V}$$
$$V_{OLmax} = -1.630 \text{ V} \qquad V_{OHmin} = -0.980 \text{ V}$$

These values can be used to determine worst-case noise margins

$$NM_L = 0.155 \text{ V} \qquad NM_H = 0.125 \text{ V}$$

which are about half the *typical* values previously calculated.

For additional information on MECL specifications the interested reader is referred to the Motorola (1972, 1978) publications listed in the bibliography at the end of this chapter.

---

[5] MECL is the trade name used by Motorola for its ECL.

Fan-Out

When the input signal to an ECL gate is low, the input current is equal to the current that flows in the 50-kΩ pull-down resistor. Thus

$$I_{IL} = \frac{-1.77 + 5.2}{50} \cong 69 \ \mu A$$

When the input is high, the input current is greater because of the base current of the input transistor. Thus, assuming a transistor β of 100, we obtain

$$I_{IH} = \frac{-0.88 + 5.2}{50} + \frac{4}{101} \cong 126 \ \mu A$$

Both of these current values are quite small, which, coupled with the fact that the output resistance of the ECL gate is very small, ensures that little degradation of logic signal levels results from the input currents of fan-out gates. It follows that the fan-out of ECL gates is not limited by logic-level considerations but rather by the degradation of the circuit speed (rise and fall times). This latter effect is due to the capacitance that each fan-out gate presents to the driving gate (approximately 3 pF). Thus while the *dc fan-out* can be as high as 90 and thus does not represent a design problem, the *ac fan-out* is limited by considerations of circuit speed to 10 or so.

Speed

The speed of operation of a logic family is measured by the delay of its basic gate and by the rise and fall times of the output waveforms. Typical values of these parameters for ECL have already been given. Here we should note that because the output circuit is an emitter follower the rise time of the output signal is shorter than its fall time, since on the rising edge of the output pulse the emitter follower functions and provides the output current required to charge up the load and parasitic capacitances. On the other hand, as the signal at the base of the emitter follower falls, the emitter follower cuts off and the load capacitance discharges through the combination of load and pull-down resistances. This point was explained in detail in Section 16.3.

Signal Transmission

In order to take full advantage of the very high speed of operation possible with ECL, special attention should be paid to the method of interconnecting the various logic gates in a system. To appreciate this point we shall briefly discuss the problem of signal transmission.

ECL deals with signals whose rise times may be 1 ns or even less, the time it takes for light to travel only 30 cm or so. For such signals a wire and its environment becomes a relatively complex circuit element along which signals propagate with finite speed (perhaps half the speed of light—that is, 30 cm/ns). Unless special care is taken, energy that reaches the end of such a wire is not absorbed but rather returns as a *reflection* to the transmitting end, where (without special care) it may be re-reflected. The result of this process of reflection is what can be observed as *ringing*, a damped oscillatory excursion of the signal about its final value.

Unfortunately ECL is particularly sensitive to ringing because the signal levels are so small. Thus it is important that transmission of signals be well controlled and surplus energy absorbed to prevent reflections. The accepted technique is to limit the nature of connecting wires in some way. One way is to insist that they be very "short", where short is taken with respect to the signal rise time. The reason for this is that if the wire connection is so short that reflections return while the input is still rising, the result becomes only a somewhat slowed and "bumpy" rising edge.

If, however, the reflection returns *after* the rising edge, it produces not simply a modification of the initiating edge but an *independent second event*. This is clearly bad! The restriction is thus made that the time taken for a signal to go from one end of line and back should be less than the rise time of the driving signal by some factor—say, 5. Thus for a signal with a 1-ns rise time and for propagation at the speed of light (30 cm/ns), a double path of only 0.2 ns equivalent length, or 6 cm, would be allowed, representing in the limit a wire only 3 cm from end to end.

Such is the restriction on ECL 100K. However, ECL 10K has intentionally slower rise time of about 3.5 ns. Using the same rules, wires can accordingly be as long as about 10 cm for ECL 10K.

If greater lengths are needed, then transmission lines must be used. These are simply wires in a controlled environment in which the distance to a ground reference plane or second wire is highly controlled. Thus they might simply be twisted pairs of wires, one of which is grounded, or parallel ribbon wires every second of which is grounded, or so-called microstrip lines on a printed-circuit (PC) board. The last are simply copper strips of controlled geometry on one side of a printed-circuit board, the other side of which consists of a grounded plane.

Such transmission lines have a *characteristic impedance* R_0 which ranges from a few tens of ohms to 1 kΩ or so. Signals propagate on such lines somewhat slower than the speed of light, perhaps half as fast. When a transmission line is terminated at its receiving end in a resistance equal to its characteristic impedance R_0, all the energy sent on the line is absorbed at the receiving end, and no reflections occur. Thus signal integrity is maintained. Such transmission lines are said to be *properly terminated*. A properly terminated line appears at its sending end as a resistor of value R_0. The followers of ECL 10K with their open emitters and low output resistances (specified to be 7 Ω maximum) are ideally suited for driving transmission lines. ECL is also good as a line receiver. The simple gate with its high (50 kΩ) pull-down input resistor represents a very high resistance to the line. Thus a few such gates can be connected to a terminated line with little difficulty.

Much more on the subject of logic-signal transmission in ECL can be found in Taub and Schilling (1977) and Motorola (1972).

Power Dissipation

Because of the differential-amplifier nature of ECL, the gate current remains approximately constant and is simply steered from one side of the gate to the other depending on the input logic signals. Thus, unlike TTL, the supply current and hence the gate power dissipation of unterminated ECL remain relatively constant independent of the logic state of the gate. It follows that no voltage spikes are introduced on the supply line. Such spikes are a dangerous source of noise in a digital system, as explained in connection with TTL. It follows that in ECL the need for supply-line bypassing is not as great as in TTL. This is another advantage of ECL.

At this point we should mention that although an ECL gate would operate with $V_{EE} = 0$ and $V_{CC} = +5.2$ V, the selection of $V_{EE} = -5.2$ V and $V_{CC} = 0$ V is recommended because all signal levels are referenced to V_{CC}, and ground is certainly an excellent reference.

Exercise

16.31 For the ECL gate in Fig. 16.36 calculate an approximate value for the power dissipated in the circuit under the condition that all inputs are low and that the emitters of the output followers are left open. Assume that the reference circuit supplies four identical gates, and hence only a quarter of the power dissipated in the reference circuit should be attributed to a gate.

Ans. 22.3 mW

Thermal Effects

In our analysis of the ECL gate of Fig. 16.36 it was found that at room temperature the reference voltage V_R is -1.32 V. We have also shown that the midpoint of the output logic swing is approximately equal to this voltage, which is an ideal situation in that it results in equal high and low noise margins. In Example 16.2 below we shall derive expressions for the temperature coefficients of the reference voltage and of the output low and high voltages. In this way, it will be shown that the midpoint of the output logic swing varies with temperature at the same rate as the reference voltage. As a result, although the magnitudes of the 1 and 0 noise margins change with temperature, their values remain equal. This is an added advantage of ECL and is a demonstration of the degree of design optimization of this gate circuit.

EXAMPLE 16.2

We wish to determine the temperature coefficient of the reference voltage V_R and of the midpoint between V_{OL} and V_{OH}.

Solution

To determine the temperature coefficient of V_R, consider the circuit in Fig. E16.27 and assume that the temperature changes by $+1°C$. Denoting the temperature coefficient of the diode and transistor voltage drops by δ, where $\delta \cong -2$ mV/°C, we obtain the equivalent circuit shown in Fig. 16.43. In this latter circuit the changes in device voltage drops are considered as signals, and hence the power supply is shown as a signal ground.

In the circuit of Fig. 16.43 we have two signal generators, and we wish to analyze the circuit to determine ΔV_R, the change in V_R. We shall do so using the principle of superposition. Consider first the branch R_1, D_1, D_2, 2δ, and R_2 and neglect the signal base current of Q_1. The voltage signal at the base of Q_1 can be easily obtained from

$$v_{b1} = \frac{2\delta \times R_1}{R_1 + r_{d1} + r_{d2} \mid R_2}$$

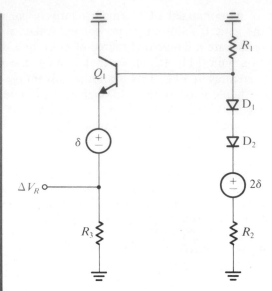

Fig. 16.43 Equivalent circuit for determining the temperature coefficient of the reference voltage V_R.

where r_{d1} and r_{d2} denote the incremental resistances of diodes D_1 and D_2, respectively. The dc bias current through D_1 and D_2 is approximately 0.64 mA, and thus $r_{d1} = r_{d2} = 39.5\ \Omega$. Hence $v_{b1} \cong 0.3\delta$. Since the gain of the emitter-follower Q_1 is approximately unity, it follows that the component of ΔV_R due to the generator 2δ is approximately equal to v_{b1}, that is, $\Delta V_{R1} = 0.3\delta$.

Consider next the component of ΔV_R due to the generator δ. Reflection of the total resistance of the base circuit, $[R_1 \parallel (r_{d1} + r_{d2} + R_2)]$, into the emittter circuit by dividing it by $\beta + 1$ ($\beta \cong 100$) results in the following component of ΔV_R.

$$\Delta V_{R2} = -\frac{\delta \times R_3}{[R_B/(\beta + 1)] + r_{e1} + R_3}$$

where R_B denotes the total resistance in the base circuit and r_{e1} denotes the emitter resistance of Q_1 ($\cong 40\ \Omega$). This calculation yields $\Delta V_{R2} \cong -\delta$. Adding this value to that due to the generator 2δ gives $\Delta V_R \cong -0.7\delta$. Thus for $\delta = -2$ mV/°C the temperature coefficient of V_R is $+1.4$ mV/°C.

We next consider determination of the temperature coefficient of V_{OL}. The circuit for accomplishing this analysis is shown in Fig. 16.44. Here we have three generators whose contributions can be considered separately and the resulting components of ΔV_{OL} summed. The result is

$$\Delta V_{OL} \cong \Delta V_R \frac{-R_{C2}}{r_{eR} + R_E} \frac{R_T}{R_T + r_{e2}}$$

$$- \delta \frac{-R_{C2}}{r_{eR} + R_E} \frac{R_T}{R_T + r_{e2}}$$

$$- \delta \frac{R_T}{R_T + r_{e2} + R_{C2}/(\beta + 1)}$$

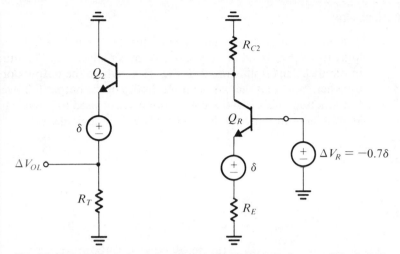

Fig. 16.44 Equivalent circuit for determining the temperature coefficient of V_{OL}.

Substituting the values given and those obtained throughout the analysis of this section, we find $\Delta V_{OL} \cong -0.43\delta$.

The circuit for determining the temperature coefficient of V_{OH} is shown in Fig. 16.45, from which we obtain

$$\Delta V_{OH} = -\delta \frac{R_T}{R_T + r_{e2} + R_{C2}/(\beta + 1)} \cong -0.93\delta$$

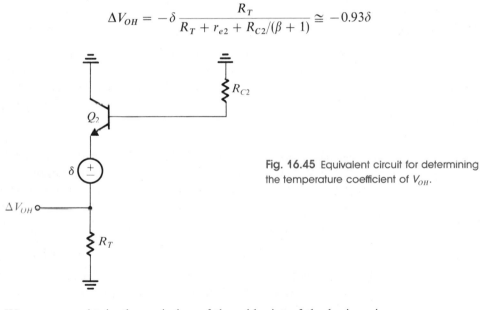

Fig. 16.45 Equivalent circuit for determining the temperature coefficient of V_{OH}.

We now can obtain the variation of the midpoint of the logic swing as

$$\frac{\Delta V_{OL} + \Delta V_{OH}}{2} = -0.68\delta$$

which is approximately equal to that of the reference voltage $V_R(-0.7\delta)$.

The Wired-OR Capability

The emitter-follower output stage of the ECL family allows an additional level of logic to be performed at very low cost by simply wiring the outputs of several gates in parallel. This is illustrated in Fig. 16.46 where the outputs of two gates are wired together. Note that the base-emitter diodes of the output followers provide a positive OR function: this *wired-OR* connection may be used to provide gates with high fan-in as well as to increase the flexibility of ECL in logic design.

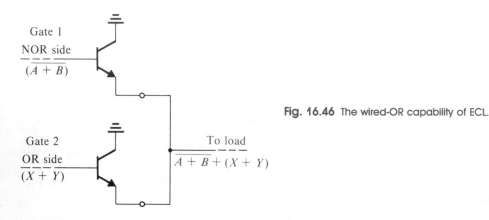

Gate 1
NOR side
$\overline{(A + B)}$

Gate 2
OR side
$(X + Y)$

To load
$\overline{A + B} + (X + Y)$

Fig. 16.46 The wired-OR capability of ECL.

A Final Remark

ECL is an important logic family that has been successfully applied in the design of high-speed digital communications systems as well as computer systems. It is also currently applied in LSI and VLSI circuit design.

16.7 SUMMARY

● The Ebers-Moll model is based on the fact that the BJT is composed of two *pn* junctions: the emitter–base junction, having a scale current I_{SE}, and the collector–base junction, having a scale current I_{SC}. The current I_{SC} is usually 2 to 50 times greater than I_{SE}.

● Although the forward α, α_F, is close to unity, the reverse α, α_R, is 0.02 to 0.5.

● For small collector currents, a saturated transistor has $V_{CE_{sat}} \simeq 0.1$ V.

● Saturation voltages as low as a fraction of a millivolt can be obtained when the transistor is operated in the reverse saturation mode.

● Before a saturated transistor begins to turn off, the extra charge stored in its base must be removed. The storage time can be shortened by avoiding deep saturation and by arranging for a reverse base current to flow. For high-speed operation, saturation should be avoided altogether.

● Transistor-transistor logic (TTL) evolved from diode-transistor logic (DTL).

● The TTL gate consists of three sections: the input stage, which implements the AND function and utilizes either a miltiemitter transistor (in standard TTL) or Schottky diodes (in modern forms of TTL); the phase splitter, which generates a pair of complementary signals to drive the output stage; and the output stage, which utilizes the totem-pole configuration and provides logic inversion. The basic gate implements the NAND function.

● In standard TTL, when the gate input goes low, the input multiemitter transistor operates in the active mode and supplies a large collector current to rapidly discharge the base of the phase splitter.

● The totem-pole output stage consists of a common-emitter transistor, which can sink large load currents and thus rapidly discharge the load capacitance, and an emitter follower, which can source large load currents and thus rapidly charge the load capacitance.

● To increase the speed of TTL, transistors are prevented from saturation. This is achieved by connecting a Schottky diode between the base and collector. Schottky diodes are formed as metal-to-semiconductor junctions and exhibit low forward voltage drops. The Schottky diode shunts some of the base current drive of the BJT and thus keeps it out of saturation.

● The state of the art in TTL circuits is represented by Advanced Schottky TTL, having $t_P = 1.5$ ns, $P_D = 20$ mW, and $DP = 30$ pJ; and Advanced Low-Power Schottky TTL, having $t_P = 4$ ns, $P_D = 1$ mW, and $DP = 4$ pJ.

● Emitter-coupled logic (ECL) is the fastest logic circuit family. It achieves its high speed of operation by avoiding transistor saturation and by utilizing small logic-signal swings.

● In ECL the input signals are used to steer a constant bias current between a reference transistor and the input transistors. The basic gate configuration is that of a differential amplifier.

● There are two popular ECL types: ECL 10K, having $t_P = 2$ ns, $P_D = 25$ mW and $DP = 50$ pJ; and ECL 100K, having $t_P = 0.75$ ns, $P_D = 40$ mW, and $DP = 30$ pJ. ECL 10K is easier to use because the rise and fall times of its signals are deliberately made long (about 3.5 ns).

● Because of the very high operating speeds of ECL, care should be taken in connecting the output of one gate to the input of another. Transmission line techniques are usually employed.

● The design of the ECL gate is optimized so that the noise margins are equal and remain equal as temperature changes.

● The ECL gate provides two complementary outputs, realizing the OR and NOR functions.

● The outputs of ECL gates can be wired together to realize the OR function of the individual output variables.

BIBLIOGRAPHY

Fairchild, *TTL Data Book*, Mountain View, Calif.: Fairchild Camera and Instrument Corp., Dec. 1978.

L. S. Garrett, "Integrated-circuit digital logic families," a three-part article published in *IEEE Spectrum*, Oct., Nov., and Dec. 1970.

I. Getreu, *Modeling the Bipolar Transistor*, Beaverton, Ore.: Tektronix Inc., 1976.

J. N. Harris, P. E. Gray, and C. L. Searle, *Digital Transistor Circuits*, Vol. 6 of the SEEC Series, New York: Wiley, 1966.

D. A. Hodges and H. G. Jackson, *Analysis and Design of Digital Integrated Circuits*, New York: McGraw-Hill, 1983.

IEEE Journal of Solid-State Circuits. The October issue of each year has been devoted to digital circuits.

J. Millman and H. Taub, *Pulse, Digital, and Switching Waveforms*, chap. 20, New York: McGraw-Hill, 1965.

Motorola, *MECL System Design Handbook*, Phoenix, Ariz.: Motorola Semiconductor Products Inc., 1972.

Motorola, *MECL High-Speed Integrated Circuits*, Phoenix, Ariz.: Motorola Semiconductor Products Inc., 1978.

L. Strauss, *Wave Generation and Shaping*, 2nd ed., New York: McGraw-Hill, 1970.

H. Taub and D. Schilling, *Digital Integrated Electronics*, New York: McGraw-Hill, 1977.

Texas Instruments Staff, *Designing with TTL Integrated Circuits*, New York: McGraw-Hill, 1971.

PROBLEMS

16.1 Repeat Exercise 16.1 for a transistor for which $\alpha_R = 0.5$.

16.2 A transistor characterized by the Ebers-Moll model shown in Fig. 16.1 is operated with both emitter and collector grounded and a base current of 1 mA. If the collector junction is 9 times larger than the emitter junc-

16.3 An *npn* BJT for which $\alpha_F = 0.99$ and $I_S = 10^{-15}$ A, operates in the normal active mode with a base current of 10 μA. Neglecting the voltage-independent terms in Eq. (16.15), calculate the corresponding base–emitter voltage. If α_F is reduced to 0.98 by some means, what does v_{BE} become?

16.4 Repeat Problem 16.3 for the same device operating in a circuit for which the emitter current is constant at 1 mA. What is v_{BE} for $\alpha_F = 0.99$? For $\alpha_F = 0.98$?

16.5 Find an expression for the *i-v* characteristic of a transistor connected as a diode by joining its base to its emitter.

16.6 A transistor operating in the diode-connected configuration of Fig. E16.2 at a current I is found to have a voltage drop of 0.7 V. When the configuration is changed by connecting the emitter to the base (the case considered in Problem 16.5), the voltage drop becomes 0.6 V. Find the relative sizes of the two junctions.

16.7 Use Eq. (16.16) to contrast the operation of three types of transistors in a circuit for which the base current is large and the collector current is small, such that $\beta_{\text{forced}} \simeq 0$, if

Q_1 has $\beta_F = 100$, $\beta_R = 1$ (i.e., a normal transistor).
Q_2 has $\beta_F = \beta_R = 20$ (i.e., a symmetric transistor).
Q_3 has $\beta_F = 1$, $\beta_R = 100$ (i.e., it could be Q_1 in the inverted mode).

Find $V_{CE\text{sat}}$ in the three cases.

16.8 A BJT for which I_B is 1 mA has $V_{CE\text{sat}} = 100$ mV at $I_C = 2$ mA, and $V_{CE\text{sat}} = 150$ mV at $I_C = 20$ mA. What is its saturation resistance? What are β_F and β_R? What is its offset voltage?

16.9 For a grounded-emitter transistor for which $\beta_F = 100$ and $\beta_R = 1$ in a circuit in which $I_B = 1$ mA and $\beta_{\text{forced}} = 10$, label all currents in the branches of the EM model shown in Fig. 16.1b.

16.10 Repeat Problem 16.9 for the situation in which the collector lead is cut while the base connection remains. Estimate the expected change in base–emitter voltage which results.

***16.11** Measurements taken on the circuit of Fig. P16.11 show $V_E = -700$ mV and $V_C = -650$ mV. Estimate a value for β_R. (Assume β_F to be large.)

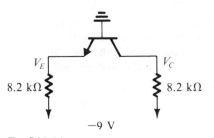

8.2 kΩ   8.2 kΩ

V_E V_C

−9 V

Fig. P16.11

***16.12** A BJT with fixed base current has $V_{CE\text{sat}}$ of -1.0 mV with the emitter open and $V_{CE\text{sat}}$ of 60 mV with the collector open. Estimate values of β_R and β_F for this transistor.

16.13 A transistor whose β is 100 and whose storage time constant is 20 ns is operated with a collector current of 10 mA and a forced base current of 1 mA. What is the excess saturating base charge under these conditions? What does it become if the base drive is reduced to 0.11 mA? In both cases find the storage time, assuming that the turnoff base current is 0.1 mA?

16.14 When well saturated at a low forced β, a particular class of BJT is known to have a storage delay time of 10 ns when its forward base current (I_{B2} in Fig. 16.5) is exchanged for a reverse current (I_{B1} in Fig. 16.5) of equal magnitude. It is to be used in a circuit in which the forward base current is 1 mA while the reverse current is limited to 0.25 mA. What is the storage delay time to be expected?

16.15 The transistor described in the previous problem, for which V_{BE} is 0.7 V, is used to switch a small collector load current. Its base is controlled via a 1-kΩ base resistor from a source which falls from $+1$ to $+0.2$ V. Estimate the storage delay time to be observed.

16.16 Someone unaware of the subtleties of BJT storage delay proposes to use the switch described in Problem 16.16 under control of a signal that he believes to be more than adequate. This signal falls from $+10$ to 0 V. Estimate the storage delay he will experience.

16.17 If the BJT switch described in Problems 16.14 and 16.15 is required to switch a collector current of 20 mA, estimate the resulting storage delay. $\beta_F = 100$.

***16.18** In this problem we wish to find the critical points of the voltage transfer characteristic, and hence the noise margins, of the RTL NOR gate of Fig. 16.8. Consider the situation where $v_B = 0$V, and thus Q_B is cut off. The voltage transfer characteristic is v_Y versus v_A. For this purpose note that:
(a) V_{OH} was calculated in Exercise 16.6, under the conditions of a fan-out $N = 5$, and was found to be 1 V.
(b) V_{IL} is the value of v_A at which Q_A begins to conduct. Let it be approximately 0.6 V.
(c) V_{IH} is the value of v_A at which Q_A saturates with $V_{CE\text{sat}} = 0.2$ V. Let Q_A have $\beta_F = 50$, $\beta_R = 0.1$, and $V_{BE} = 0.7$ V and use Eq. (16.16) to calculate β_{forced} and hence I_B and V_{IH}.
(d) $V_{OL} = V_{CE\text{sat}} = 0.2$ V.
(e) Sketch the voltage transfer characteristic.
(f) Calculate NM_H and NM_L.

16.19 Consider the RTL gate shown in Fig. 16.8.
(a) Find the current drawn from the dc supply when $v_Y = V_{OL} = 0.2$ V. Hence find the power dissipation in the gate in this state (neglect the power dissipated due to the base drive of the BJTs.)
(b) With the transistor cut off and the gate driving other gates so that $v_Y = V_{OH} = 1$ V, find the current drawn from the dc supply and hence the gate power dissipation in this state.
(c) Use the results of (a) and (b) above to compute the average power dissipation of the RTL gate.
(d) If the propagation delay of the RTL gate is 10 ns, find its delay-power product.

16.20 For the DTL gate of Fig. 16.10 calculate the total supply current and hence the gate power dissipation for the two cases: v_Y high and v_Y low. Hence find the average power dissipation in the DTL gate.

16.21 Consider the circuit in Fig. 16.9. If $V_{CC} = 5$ V, $R_C = R_B = 1$ kΩ, $V_{CE\text{sat}} = 0.2$ V, and $V_{BE} = 0.7$ V, what are the voltage levels at Y and $\bar{Y}$ when the set and reset inputs are inactive.

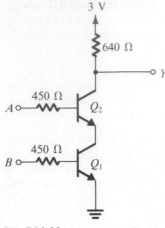

Fig. P16.22

16.22 What is the logic function implemented by the circuit shown in Fig. P16.22?

*16.23 For the DTL gate of Fig. 16.11 let $\beta = 100$ and $V_{BE} = 0.7$ V. Calculate the base current supplied to Q_3 as v_I goes high. As v_I goes low, what is the value of reverse current that flows through the base of Q_3 to remove the saturating charge? If $\tau_s = 10$ ns, use Eq. (16.20) to compute the storage delay time t_s.

16.24 For the DTL gate in Fig. 16.11, what is the total supply current with input high? With input low? Assume $\beta = 100$ and $V_{CEsat} = 0.2$ V.

*16.25 For the circuit in Fig. 16.11, what is the minimum β of Q_2 and Q_3 (assumed matched) that ensures that Q_3 saturates with a base overdrive factor of 4? Assume $V_{BE} = V_D = 0.7$ V and $V_{CEsat} = 0.2$ V. (Recall that the base overdrive factor is the ratio of the base current to the minimum base current required to saturate the transistor.)

*16.26 Consider the output circuit shown in Fig. 16.12 for which $R_C = 2$ kΩ, $C_L = 10$ pF, $V_{CC} = 5$ V, $I_{B3} = 1$ mA, and $\beta_3 = 50$. What times are required for the output to move high from 0.2 V and low from +5 V to the gate threshold level of 1.4 V?

16.27 Consider the circuit of Fig. 16.13 with $V_{CC} = 3$ V, $R = 3$ kΩ, and $R_C = 1$ kΩ as the input rises slowly from 0 V. If V_{CEsat} of Q_1 is 0.1 V and if Q_3 turns on when its V_{BE} reaches 0.6 V, at what value of input voltage does Q_3 begin to conduct? This is a good estimate of V_{IL}.

*16.28 For the circuit of Fig. 16.16 the input voltage rises linearly from 0 to +5 V in 50 ns. Let $V_{CC} = 5$ V, $R_E = 1$ kΩ, $C_L = 100$ pF and assume that the BJT con-

ducts when its v_{BE} reaches 0.7 V, and that thereafter v_{BE} remains constant at this value. What is the delay before the output voltage begins to rise? What is the maximum output voltage? What is the peak emitter current?

*16.29 A variant of the T²L gate shown in Fig. 16.23 is being considered in which all resistances are tripled. For input high, estimate all node voltages and branch currents with $\beta_F = 30$, $\beta_R = 0.01$, $V_{BE} = 0.7$ V, and a load to +5 V of 1 kΩ.

*16.30 Repeat the analysis of the circuit suggested in Problem 16.29 with input low (at +0.2 V) and a resistor of 1 kΩ connected from the output to ground.

*16.31 Two TTL gates of the type described in Problem 16.29, one with input high and one with input low, have their outputs accidentally joined. What output voltage result? What current flows in the short circuit?

*16.32 A transistor for which $\beta_F = 50$ and $\beta_R = 5$ is used for Q_3 in Fig. 16.24. For a base current of 2.5 mA, what is V_{CEsat} for $i_L = 0$, 1, 10, and 100 mA? Estimate R_{CEsat} at 0.5, 5, and 50 mA.

16.33 Consider the output circuit of the gate in Fig. 16.26. What is the output voltage when a current of 1 mA is extracted? What is the (small-signal) output resistance? Use $\beta = 50$.

16.34 Consider the output circuit of the gate in Fig. 16.26. For $\beta = \infty$ and $V_{CEsat} = 0.2$ V, at what output current does Q_4 saturate? For $\beta = 20$, at what current does saturation occur?

16.35 If the output of the circuit in Fig. 16.26 is short-circuited to ground, what current flows? Assume high β. What is the minimum value of β for which your analysis holds?

16.36 Using the data provided in the answers to Exercise 16.15, find the noise margins of the T²L gate at $-55°C$ and $+125°C$.

16.37 For a particular TTL implementation, the current from the 5-V supply is 5 mA with inputs high and 3 mA with inputs low. The propagation delay for output rising is 18 ns, whereas for output falling it is 12 ns. Estimate the delay-power product.

*16.38 For an 8-input TTL NAND gate resembling that in Fig. 16.28, β_R of Q_1 is as large as 0.04. For all inputs high, what is the additional current supplied to Q_2 as a result of Q_1 operating in the inverted mode?

16.39 Consider a variant of the circuit in Fig. 16.30 which utilizes Q_{1A}, Q_{1B}, Q_{1C}, each with a single input A,

B, *C*, in conjunction with Q_{2A}, Q_{2B}, Q_{2C}. What is the logic function realized?

*16.40 Consider the tristate input of the tristate gate shown in Fig. E16.20. What is the input voltage just re-

quired to turn off Q_7 and thus release the 1.6-kΩ resistor to drive Q_4 or Q_3 via Q_2?

*16.41 The BJT in the circuit shown in Fig. P16.41 begins to conduct at $v_{BE} = 0.7$ V and is fully conducting at $v_{BE} = 0.8$ V. The same applies to D_7. The Schottky diodes have a voltage drop of 0.5 V.
(a) What is the logic function performed?
(b) Find V_{OL} and V_{OH}.
(c) Find V_{IL} and V_{IH}.
(d) Find the noise margins.
(e) Find the largest and the smallest currents drawn from the power supply.

*16.42 For the ECL circuit in Fig. P16.42, the transistors exhibit V_{BE} of 0.75 V at an emitter current I and have very high β.
(a) Find V_{OH} and V_{OL}.
(b) For the input at B sufficiently negative for Q_B to be cut off, what voltage at A causes a current of $I/2$ to flow in Q_R?
(c) Repeat (b) for a current in Q_R of 0.99I.
(d) Repeat (c) for a current in Q_R of 0.01I.
(e) Use the results of (c) and (d) to specify V_{IL} and V_{IH}.
(f) Find NM_H and NM_L.
(g) Find the value of IR that makes the noise margins equal to the width of the transition region, $V_{IH} - V_{IL}$.
(h) Using the IR value obtained in (g) give numerical values for V_{OH}, V_{OL}, V_{IH}, V_{IL}, and V_R for this ECL gate.

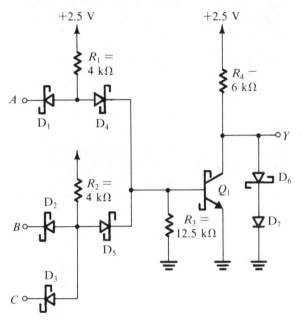

Fig. P16.41

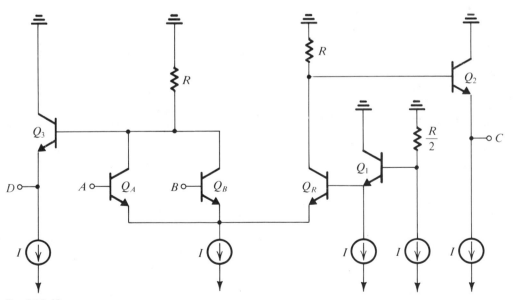

Fig. P16.42

*16.43 Three logic inverters are connected in a ring Specifications for this family of gates indicates a typical propagation delay of 3 ns for high-to-low output transitions and 7 ns for low-to-high transitions. Assume that for some reason the input to one of the gates undergoes a low-to-high transition. By sketching the waveforms at the outputs of the three gates and keeping track of their relative positions, show that the circuit functions as an oscillator. What is the frequency of oscillation of this ring oscillator?

*16.44 Following the idea of a ring oscillator introduced in Problem 16.45, consider an implementation using a ring of five ECL 100K inverters. Assume that the inverters have linearly rising and falling edges (and thus the waveforms are trapezoidal in shape). Let the 0 to 100% rise and fall times be equal to 1 ns. Also, let the propagation delay (in both directions) be equal to 1 ns. Provide a labeled sketch of the five output signals, taking care that relevant phase information is provided. What is the frequency of oscillation?

*16.45 Using the logic and circuit flexibility of ECL indicated by Figs. 16.36 and 16.46, sketch an ECL circuit that realizes the Exclusive OR function, $Y = \bar{A}B + A\bar{B}$. (*Hint:* $\bar{A}B + A\bar{B} = \overline{A + \bar{B} + \bar{A} + B}$.)

*16.46 For the circuit in Fig. 16.38, whose transfer characteristic is shown in Fig. 16.39, calculate the incremental voltage gain from input to the OR output at points x, m, and y of the transfer characteristic. Assume $\beta = 100$. Use the results of Exercise 16.29.

16.47 For the circuit in Fig. 16.38, whose transfer characteristic is shown in Fig. 16.39, find V_{IL} and V_{IH} if x and y are defined as the points at which
(a) 90% of the current I_E is switched.
(b) 99.9% of the current I_E is switched.

16.48 For the symmetrically loaded circuit of Fig. 16.38 and for typical output signal levels ($V_{OH} = -0.88$ V and $V_{OL} = -1.77$ V), calculate the power lost in both load resistors R_T and both output followers. What then is the total power dissipation of a single ECL gate including its symmetrical output terminations?

16.49 Considering the circuit of Fig. 16.40, what is the value of β of Q_2, for which the high noise margin (NM_H) is reduced by 50%?

*16.50 Consider an ECL gate whose output is terminated in a 50-Ω resistance and connected to a load capacitance C. As the input of the gate rises, the output emitter follower cuts off and the load capacitance C discharges through the 50-Ω load (until the emitter follower conducts again). Find the value of C that will result in a discharge time of 1 ns. Assume that the two output levels are -0.88 V and -1.77 V.

16.51 For signals whose rise and fall times are 3.5 ns, what length of unterminated gate-to-gate wire interconnect can be used if a ratio of rise time to return time of 5 to 1 is required? Assume the environment of the wire to be such that the signal propagates at two-thirds the speed of light (which is 30 cm/ns).

*16.52 For the circuit in Fig. P16.52 express E as a logic function of A, B, C, and D. For all inputs low at 0 V what is the voltage at E? If A and C are raised to $+5$ V, what is the voltage at E? Assume $|V_{BE}| = 0.7$ V and $\beta = 50$.

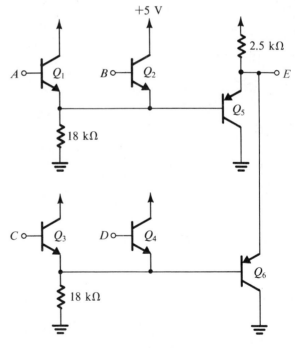

Fig. P16.52

Integrated-Circuit Technology

INTRODUCTION

The purpose of this appendix is to familiarize the reader with integrated-circuit terminology. An explanation is given of standard integrated-circuit processes. The characteristics of the devices available in integrated-circuit form are also presented. This is done to aid the reader in understanding those aspects of integrated-circuit design that are distinct from discrete-circuit design. In order to take proper advantage of the economics of integrated circuits, designers have had to overcome some serious device limitations (such as poor resistor tolerances) while exploiting device advantages (such as good resistor matching). An understanding of device characteristics is therefore essential in designing good integrated circuits, and also helps when applying commercial integrated circuits to system design.

This appendix will consider only silicon tech-nology. Although germanium and gallium arsenide are also used to make semiconducting devices, silicon is still the most popular material and will remain so for some time. The physical properties of silicon make it suitable for fabricating active devices with good electrical characteristics. In addition, silicon can easily be oxidized to form an excellent insulating layer (glass). This insulator is used to make capacitor structures and allows the construction of field-controlled devices. It also serves as a good mask against foreign impurities which could diffuse into the high-purity silicon material. This masking property allows the formation of integrated circuits; active and passive circuit elements can be built together on the same piece of material (substrate), at the same time, and inter-connected to form a complete circuit function. □

INTEGRATED-CIRCUIT PROCESSES

The basic processes involved in the fabrication of integrated circuits will be described in the following sections.

Wafer Preparation

The starting material for modern integrated circuits is very-high-purity silicon. The material is grown as a single crystal. It takes the shape of a solid cylinder 10 to 12.5 cm in diameter and 1 m in length, which is steel gray in color. This crystal is then sawed (like a loaf of bread) to produce wafers 10 to 12.5 cm in diameter and 200 μm thick (a micrometer or micron is a millionth of a meter). The surface of the wafer is then polished to a mirror finish.

The basic electrical and mechanical properties of the wafer depend on the direction in which the crystal was grown (crystal orientation) and the number and type of impurities present. Both variables are strictly controlled during crystal growth. Impurities can be added on purpose to the pure silicon in a process known as *doping*. Doping allows controlled alteration of the electrical properties of the silicon, in particular the resistivity. It is also possible to control the type of carrier used to produce electrical conduction, being either holes (in *p*-type silicon) or electrons (in *n*-type silicon). If a large number of impurity atoms is added, then the silicon is said to be heavily doped. When designating relative doping concentrations on diagrams of devices, it is common to use + and − symbols. Thus a heavily doped (low-resistivity) *n*-type silicon wafer would be referred to as n^+ material. This ability to control the doping of silicon permits the formation of diodes, transistors, and resistors in integrated circuits.

Oxidation

Oxidation refers to the chemical process of silicon reacting with oxygen to form silicon dioxide. To speed up the reaction, it is necessary to heat up the wafers to the 1000 to 1200°C range. The heating is performed in special high-temperature furnaces. To avoid the introduction of even small quantities of contaminants (which could significantly alter the electrical properties of the silicon), it is necessary to maintain an ultraclean environment for the processing. This is true for all processing steps involved in the fabrication of an integrated circuit. Specially filtered air is circulated in the processing area, and all personnel must wear special lint-free clothing.

The oxygen used in the reaction can be introduced either as a high-purity gas (referred to as a "dry oxide") or as water vapor (forming a "wet oxide"). In general, a wet oxide has a faster growth rate, but a dry oxide has better electrical characteristics. The oxide layer grown has excellent electrical insulation properties. It has a dielectric constant of about 3.5, and it can be used to form excellent capacitors. It also serves as a good mask against many impurities. It can therefore be used to protect the silicon surface from contaminants. It can also be used as a masking layer, allowing the introduction of dopants into the silicon only in regions that are not covered with oxide. This masking property is what permits the convenient fabrication of integrated circuits.

The silicon dioxide layer is a thin, transparent film and the silicon surface is highly reflective. If white light is incident on the oxidized wafer, constructive and destructive interference effects occur in the oxide, causing certain colors to be absorbed strongly. The wavelengths absorbed depend on the thickness of the oxide layer. This absorption produces different colors in the different regions of a processed wafer. The colors can be quite vivid, ranging from blue to greens and reds, and are immediately obvious when a finished chip is viewed under the microscope. It should be remembered though that the color is due to an optical effect. The oxide layer is transparent and the silicon underneath it is a steel-gray color.

Diffusion

Diffusion is the process by which atoms move through the crystal lattice. In fabrication, it relates to the introduction of impurity atoms (dopants) into silicon to change its doping. The rate at which dopants diffuse in silicon is a strong function of temperature. This allows us to introduce the impurities at a high temperature (1000 to 1200°C) to obtain the desired doping. The slice is then cooled to room temperature and the impurities are essentially "frozen" in position. The diffusion process is performed in furnances similar to those used for oxidation. The depth to which the impurities diffuse depends both on the temperature and time allowed.

The two most common impurities used as dopants are boron and phosphorus. Boron is a *p*-type dopant and phosphorus is an *n*-type dopant. Both dopants are effectively masked by thin silicon dioxide layers. By diffusing boron into an *n*-type substrate, a *pn* junction is formed (diode). A subsequent phosphorus diffusion will produce an *npn* structure (transistor). In addition, if the doping concentration is heavy, the diffused layer can be used as a conductor.

Ion Implantation

In addition to diffusion, impurities can be introduced into silicon by the use of an ion implanter. An ion implanter produces ions of the desired impurity, accelerates them by an electric field, and allows them to strike the silicon surface. The ions become embedded in the silicon. The depth of penetration is related to the energy of the ion beam, which can be controlled by the accelerating-field voltage. The quantity of ions implanted can be controlled by varying the beam current (flow of ions). Since both voltage and current can be accurately measured and controlled, ion implantation results in much more accurate and reproducible impurity profiles than can be obtained by diffusion. In addition, ion implantation can be performed at room temperature. Ion implantation normally is used when accurate control of the dopant is essential for device operation.

Chemical Vapor Deposition

Chemical-vapor deposition (CVD) is a process by which gases or vapors are chemically reacted, leading to the formation of a solid on a substrate. In our case, CVD can be used to deposit silicon dioxide on a silicon substrate. For instance, if silane gas and oxygen are mixed above a silicon substrate, silicon dioxide deposits as a solid on the silicon. The oxide layer formed is not as good as a thermally grown oxide, but it is good enough to act as an electrical insulator. The advantage of a CVD layer is that the oxide deposits at a faster rate and a lower temperature (below 500°C).

If silane gas alone is used, then a silicon layer deposits on the wafer. If the reaction temperature is high enough (above 1000°C), then the layer is deposited as a crystalline layer (assuming the substrate is crystalline silicon). This is because the atoms have enough energy to align themselves in the proper crystal directions. Such a layer is said to be an epitaxial layer, and the deposition process is referred to as *epitaxy* instead of CVD. At lower temperatures, or if the substrate is not single-crystal silicon, the atoms are not all aligned along the same crystal direction. Such a layer is called *polycrystalline silicon*, since it consists of many small crystals of silicon aligned in various directions. Such layers are normally doped very heavily to form a high conductivity region that can be used for interconnecting devices.

Metallization

The purpose of metallization is to interconnect the various components of the integrated circuit (transistors, resistors, etc.) to form the desired circuit. Metallization involves the deposition of a metal (aluminum) over the entire surface of the silicon. The required interconnection pattern is then selectively etched. The aluminum is deposited by heating it in vacuum until it vaporizes. The vapors then contact the silicon surface and condense to form a solid aluminum layer.

Packaging

A finished silicon wafer may contain from 100 to 1,000 finished circuits or chips. Each chip contains between 10 and 10^6 transistors and is rectangular in shape, typically between 1 and 10 mm on each edge. The circuits are first tested electrically (while still in wafer form) using an automatic probing station. Bad circuits are marked for later identification. The circuits are then separated from each other (by dicing) and the good circuits (dies) are mounted in packages (headers). Fine gold wires are then used to interconnect the pins of the package to the metallization pattern on the die. Finally, the package is sealed under vacuum or in an inert atmosphere.

Photolithography

The surface geometry of the various integrated-circuit components is defined photographically: The silicon surface is coated with a photosensitive layer and then exposed to light through a master pattern on a photographic plate. The layer is then developed to reproduce the pattern on the wafer. Very fine surface geometries can be reproduced accurately by this technique. The resulting layer is not attacked by the chemical etchants used for silicon dioxide or aluminum and so forms an effective mask. This allows "windows" to be etched in the oxide layer in preparation for subsequent diffusion processes. This process is used to define transistor regions and isolate one transistor from another.

INTEGRATED-CIRCUIT STRUCTURES

The structure of the basic integrated-circuit components available to the circuit designer will be presented in this section. Only a standard bipolar junction-isolated process will be considered initially. The processing steps involved are outlined in Fig. A.1 along with the corresponding transistor cross section. The process requires six masking steps, four diffusions, and one epitaxial layer growth. The possible components produced include *npn* and *pnp* bipolar transistors, *p*-channel junction field-effect transistors, resistors, and capacitors. Each of the components will be described in the following sections.

Resistors

Resistors can be made from the collector layer (*n*-type), base region (*p*-type), emitter region (n^+-type), or the "pinched-base" region (base region between the n^+ diffused region and the *n*-type epitaxial region). The value of resistance obtained is related to the surface geometry and the region resistivity (doping). Figure A.2 gives the surface and cross-sectional views of the possible resistor structures. Table A.1 gives typical characteristics of the resulting resistors. Note that it is difficult to obtain high resistor values. Also note that while the tolerance of the resistor value is very poor (20% to 50%), matching of two similar resistor values is quite good (5%). Thus circuit designers should design circuits which exploit resistor matching, and should avoid all circuits which require a specific resistor value. Also note that diffused resistors have a significant temperature coefficient.

Capacitors

Two types of capacitor structure are available in bipolar integrated circuits. Figure A.3 shows cross sections of those structures. The junction capacitor makes use of one of the junctions available by diffusion. Either the collector–base or emitter–base junction can be used, but the low breakdown voltage of the emitter–base junction (6 to 7 V) makes it less popular. The capacitance is determined by geometry and doping levels, but normally is restricted to values less than 20 pF. The tolerance is 20%. Junction capacitors must always be reverse-biased and have a large voltage dependence.

A capacitor can also be formed using the oxide as a dielectric. The aluminum metallization and the emitter region form a parallel-plate structure. This structure makes an excellent capacitor with very low voltage and temperature coefficients, low leakage, and high breakdown voltage. Although the tolerance is 10%, capacitors on the same chip can be matched to better than 1%. Again the values are restricted to a few ten's of picofarads.

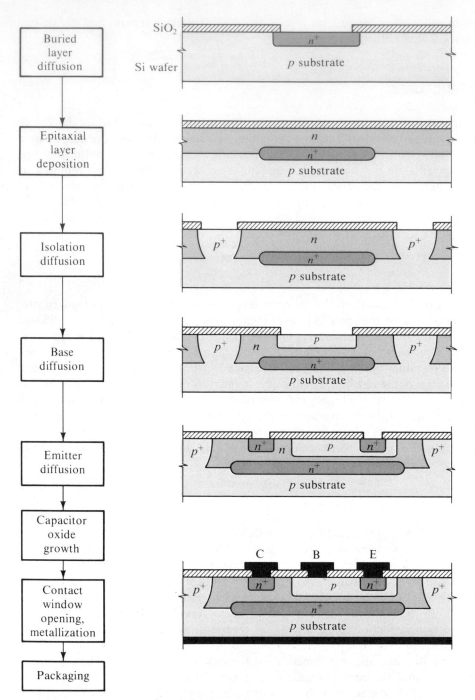

Fig. A.1 Standard bipolar integrated-circuit process.

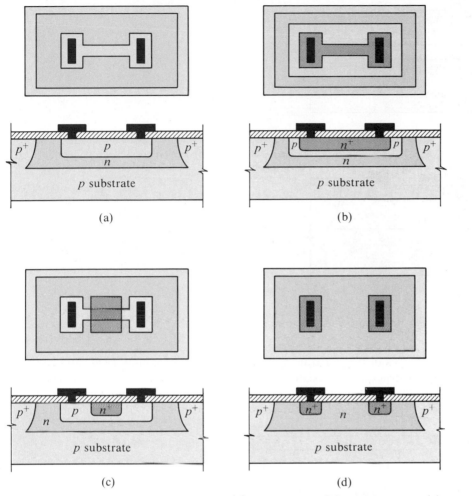

Fig. A.2 Integrated-circuit resistor structures. **(a)** Base resistor. **(b)** Emitter resistor. **(c)** Pinched base resistor. **(d)** Collector resistor.

Table A.1 DIFFUSED RESISTOR CHARACTERISTICS

	Resistor Type			
	Base	Emitter	Pinched Base	Collector
Range (ohms)	50–50 k	5–100	10 k–500 k	1 k–10 k
Tolerance	20%	20%	50%	50%
Matching	5%	5%	5%	5%
Temperature coefficient	0.1%/°C	0.2%/°C	0.5%/°C	0.8%/°C
Breakdown voltage (V)	40	6	6	70

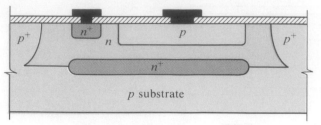

(a)

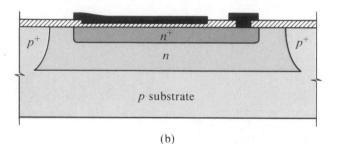

(b)

Fig. A.3 Integrated-circuit capacitor structures. **(a)** Collector-base junction capacitor. **(b)** Oxide capacitor.

Bipolar Junction Transistors

Three basic types of bipolar transistors are available: *npn, lateral pnp,* and *substrate pnp.* The *npn* structure is repeated in cross-section form in Fig. A.4. It has a beta typically of 100 to 500, a collector breakdown voltage of 40 V, and a cutoff frequency of 500 MHz. The normal operating-current range is from a few microamperes to

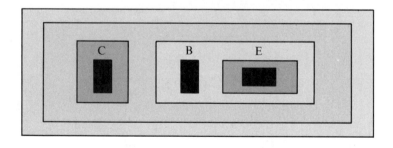

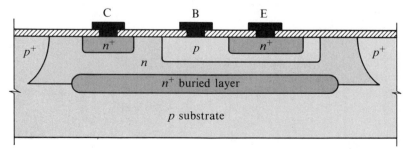

Fig. A.4 An *npn* bipolar transistor.

a few tens of milliamperes. Higher operating currents are obtained by increasing the surface area of the device.

The lateral *pnp* is shown in Fig. A.5. The base region in this case is the lateral area between the two *p* diffusions, hence the name (the other transistor structures are vertical devices). This is the most commonly used *pnp* structure. It suffers from very low beta, being typically about 20 at low collector currents (a few microamperes) and diminishing rapidly as current levels are increased. At milliampere current levels, its gain may be only slightly greater than unity. It also has a low cutoff frequency, around 1 MHz. Its breakdown voltage is over 40 V. Despite its poor characteristics, the lateral *pnp* is the best *pnp* transistor available in integrated circuits.

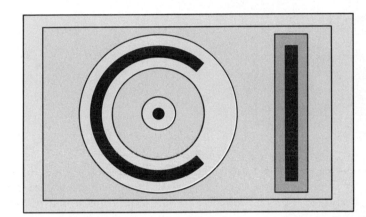

Fig. A.5 Lateral *pnp* bipolar transistor.

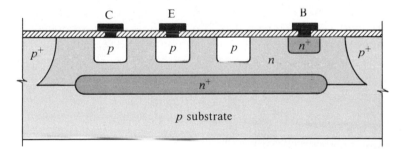

The substrate *pnp* structure is shown in Fig. A.6. In this case, the collector is electrically connected to the substrate, which is normally a signal ground. This limits application of the device to common-collector configurations. It does, however, have a slightly better current gain than the lateral *pnp* and exhibits somewhat better performance at higher currents. The characteristics are only slightly improved, so integrated-circuit designers avoid using *pnp* transistors, if at all possible. If needed, they are best used in unity-gain configurations to avoid serious degradation of bandwidth due to their low cutoff frequency.

For digital integrated circuits, the storage time of the *npn* transistor is most important if the device is operated in the saturation region. One technique for reducing the storage time is to allow gold to diffuse into the silicon. Gold acts as an impurity

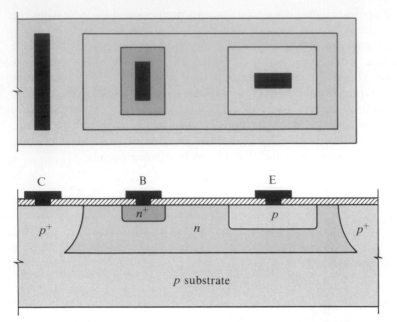

Fig. A.6. Substrate *pnp* bipolar transistor.

in the silicon and serves to reduce the carrier lifetime, hence the storage time. It also reduces the current gain and breakdown voltage, but these reductions are not important in digital circuits. It is also not possible to make *pnp* transistor structures with gains greater than unity due to the low lifetime, but again the *pnp* is not needed in digital circuits. An alternative to gold doping is to avoid saturation region operation by Schottky clamping the collector–base junction. This will not effect any transistor parameters, and yet the storage time can be eliminated. Fortunately, a suitable Schottky diode can be formed by using aluminum as the metal and the collector *n* region as the semiconductor. This leads to the very simple structure shown in Fig. A.7.

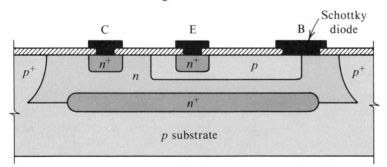

Fig. A.7. Schottky clamped *npn* bipolar transistor.

Junction Field-Effect Transistors

A *p*-channel JFET can be formed using the bipolar process. The corresponding cross section is shown in Fig. A.8. This structure suffers from a low gate-drain breakdown

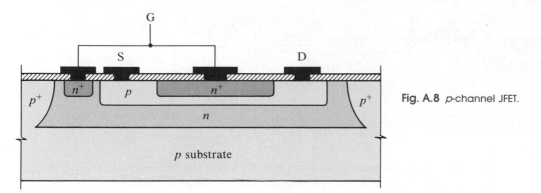

Fig. A.8 *p*-channel JFET.

voltage of 6 V, and poor matching of pinch-off voltages. The use of ion implantation as an extra processing step, however, permits much improvement in pinch-off matching and control. The structure can be used to functionally replace *pnp* bipolar devices in circuits. It is capable of providing increased gain and cutoff frequency, with values similar to those of the *npn* device. Thus the *p*-channel JFET is very useful in analog integrated-circuit design, and is to be preferred over the use of lateral *pnp* devices.

MOS Technology

We conclude this appendix with a brief discussion of MOS IC technology. Though originally employed for the fabrication of digital circuits, MOS technology is currently used also for analog circuits, and for VLSI chips containing both digital and analog circuits. Initially, MOS IC technology was based on the PMOS transistor, whose cross section is shown in Fig. A.9. The simplicity of the structure should be noted. Indeed, the PMOS process is simple and economical. The performance of the resulting circuits, however, falls considerably short of that currently available from modern NMOS and CMOS processes. The PMOS process, exemplified by the transistor of Fig. A.9, utilizes aluminum gates. Most modern NMOS and CMOS processes employ polycrystalline silicon (poly) to form the gate electrodes.

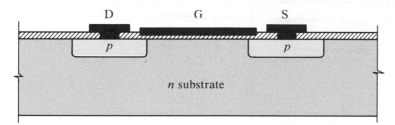

Fig. A.9 PMOS transistor (aluminum gate).

Figure A.10 shows a simplified cross section of an NMOS transistor fabricated using the silicon-gate process. Here also the structure is simple, with the result that a large number of devices can be fabricated on the same chip. Indeed, NMOS technology provides the highest possible integration density.

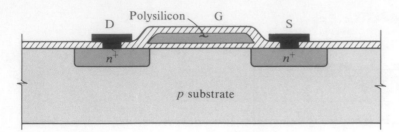

Fig. A.10 NMOS transistor (silicon gate).

The NMOS transistor shown in Fig. A.10 is of the enhancement type. Depletion devices can be formed by using ion implantation to implant a channel at the substrate surface. Modern NMOS technology employs depletion transistors as load devices.

Compared to NMOS, CMOS technology requires additional processing steps but allows much greater flexibility in circuit design. Also, the resulting circuits, both analog and digital, exhibit improved performance over those fabricated in NMOS technology.

Although CMOS technology can utilize either the metal gate or the silicon gate technique, most modern CMOS processes are of the silicon gate variety. Figure A.11 shows a somewhat simplified cross section of silicon gate CMOS transistors. As indicated, the p-channel transistor is formed in an n well. For this reason, the process illustrated is known as an n-well process. (p-well processes are possible and are equally popular.) The gate electrodes are formed of high-conductivity n^+ polycrystalline silicon that is deposited using a CVD process. The poly is then covered with a thick oxide layer that is deposited also using CVD. For the n-well CMOS structure shown, the p substrate is tied to ground while the n well is tied to V_{DD}.

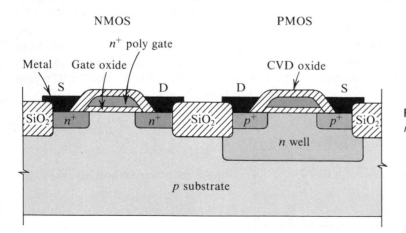

Fig. A.11 CMOS transistors in silicon-gate, n-well technology.

The CMOS process illustrated in Fig. A.11 utilizes a single layer of polysilicon and is thus known as a single-poly process. More advanced processes employ two layers of polysilicon. These double-poly processes provide increased flexibility. For instance, high-quality capacitors for analog circuit applications can be implemented using the two poly layers as electrodes and the silicon dioxide as the dielectric. Also, because of their high conductivity, the polysilicon layers can be used for forming interconnection patterns, thus enabling greater levels of integration.

BIBLIOGRAPHY

A. B. Grebene, *Bipolar and MOS Analog Integrated Circuit Design*, New York: Wiley, 1984.

D. J. Hamilton and W. G. Howard, *Basic Integrated Circuit Engineering*, New York: McGraw-Hill, 1975.

Appendix

Two-Port Network Parameters

INTRODUCTION

At various points throughout the text, we made use of some of the different possible ways to character- ize linear two-port networks. A summary of this topic is presented in this appendix. □

CHARACTERIZATION OF LINEAR TWO-PORT NETWORKS

A two-port network (Fig. B.1) has four port variables: V_1, I_1, V_2, and I_2. If the two-port network is linear, we can use two of the variables as excitation and the other two as response variables. For instance, the network can be excited by a voltage V_1 at port 1 and a voltage V_2 at port 2, and the two currents, I_1 and I_2, measured to represent the network response. In this case V_1 and V_2 are independent variables and I_1 and I_2 are dependent variables, and the network operation can be described by the two equations

$$I_1 = y_{11}V_1 + y_{12}V_2 \tag{B.1}$$

$$I_2 = y_{21}V_1 + y_{22}V_2 \tag{B.2}$$

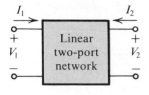

Fig. B.1 The reference directions of the four port variables in a linear two-port network.

Here the four parameters y_{11}, y_{12}, y_{21}, and y_{22} are admittances, and their values completely characterize the linear two-port network.

Depending on which two of the four port variables are used to represent the network excitation, a different set of equations (and a correspondingly different set of parameters) is obtained for characterizing the network. In the following we shall present the four parameter sets commonly used in electronics.

y Parameters

The short-circuit admittance or y-parameter characterization is based on exciting the network by V_1 and V_2, as shown in Fig. B.2a. The describing equations are Eqs. (B.1) and (B.2) above. The four admittance parameters can be defined according to their roles in Eqs. (B.1) and (B.2).

Specifically, from Eq. (B.1) we see that y_{11} is defined as follows:

$$y_{11} = \frac{I_1}{V_1}\bigg|_{V_2 = 0} \tag{B.3}$$

Thus y_{11} is the input admittance at port 1 with port 2 short-circuited. This definition is illustrated in Fig. B.2b, which also provides a conceptual method for measuring the input short-circuit admittance y_{11}.

The definition of y_{12} can be obtained from Eq. (B.1) as

$$y_{12} = \frac{I_1}{V_2}\bigg|_{V_1 = 0} \tag{B.4}$$

Thus y_{12} represents transmission from port 2 to port 1. Since in amplifiers, port 1 represents the input port and port 2 the output port, y_{12} represents internal *feedback* in the network. Figure B.2c illustrates the definition and the method for measuring y_{12}.

The definition of y_{21} can be obtained from Eq. (B.2) as

$$y_{21} = \frac{I_2}{V_1}\bigg|_{V_2 = 0} \tag{B.5}$$

Thus y_{21} represents transmission from port 1 to port 2. If port 1 is the input port and port 2 the output port of an amplifier, then y_{21} provides a measure of the forward gain or transmission. Figure B.2d illustrates the definition and the method for measuring y_{21}.

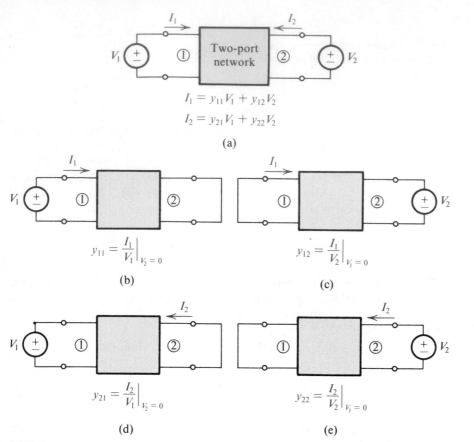

$$I_1 = y_{11}V_1 + y_{12}V_2$$
$$I_2 = y_{21}V_1 + y_{22}V_2$$

(a)

$$y_{11} = \left.\frac{I_1}{V_1}\right|_{V_2=0}$$

(b)

$$y_{12} = \left.\frac{I_1}{V_2}\right|_{V_1=0}$$

(c)

$$y_{21} = \left.\frac{I_2}{V_1}\right|_{V_2=0}$$

(d)

$$y_{22} = \left.\frac{I_2}{V_2}\right|_{V_1=0}$$

(e)

Fig. B.2 Definition and conceptual measurement circuits for y parameters.

The parameter y_{22} can be defined, based on Eq. (B.2), as

$$y_{22} = \left.\frac{I_2}{V_2}\right|_{V_1=0} \tag{B.6}$$

Thus y_{22} is the admittance looking into port 2 while port 1 is short-circuited. For amplifiers y_{22} is the output short-circuit admittance. Figure B.2e illustrates the definition and the method for measuring y_{22}.

z Parameters

The open-circuit impedance (or z-parameter) characterization of two-port networks is based on exciting the network by I_1 and I_2, as shown in Fig. B.3a. The describing

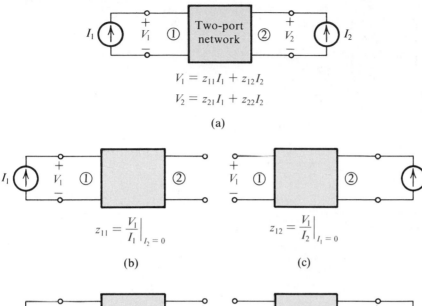

$$V_1 = z_{11}I_1 + z_{12}I_2$$
$$V_2 = z_{21}I_1 + z_{22}I_2$$

(a)

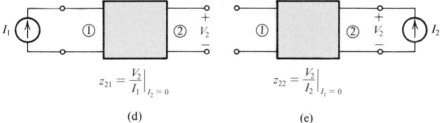

(b)   (c)

(d)   (e)

Fig. B.3 Definition and conceptual measurement circuits for z parameters.

equations are

$$V_1 = z_{11}I_1 + z_{12}I_2 \qquad \text{(B.7)}$$
$$V_2 = z_{21}I_1 + z_{22}I_2 \qquad \text{(B.8)}$$

Owing to the duality between the z- and y-parameter characterizations we shall not give a detailed discussion of z parameters. The definition and method of measuring each of the four z parameters is given in Fig. B.3.

h Parameters

The hybrid (or h-parameter) characterization of two-port networks is based on exciting the network by I_1 and V_2, as shown in Fig. B.4a (note the reason behind the name *hybrid*). The describing equations are

$$V_1 = h_{11}I_1 + h_{12}V_2 \qquad \text{(B.9)}$$
$$I_2 = h_{21}I_1 + h_{22}V_2 \qquad \text{(B.10)}$$

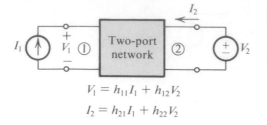

$$V_1 = h_{11}I_1 + h_{12}V_2$$
$$I_2 = h_{21}I_1 + h_{22}V_2$$

(a)

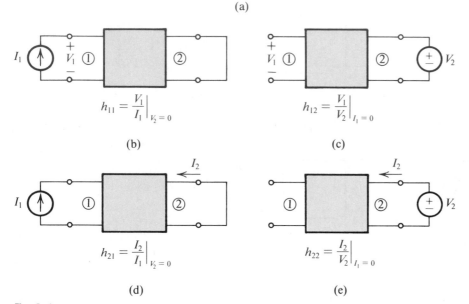

$$h_{11} = \left.\frac{V_1}{I_1}\right|_{V_2=0}$$

(b)

$$h_{12} = \left.\frac{V_1}{V_2}\right|_{I_1=0}$$

(c)

$$h_{21} = \left.\frac{I_2}{I_1}\right|_{V_2=0}$$

(d)

$$h_{22} = \left.\frac{I_2}{V_2}\right|_{I_1=0}$$

(e)

Fig. B.4 Definition and conceptual measurement circuits for h parameters.

from which the definition of the four h parameters can be obtained as

$$h_{11} = \left.\frac{V_1}{I_1}\right|_{V_2=0} \qquad h_{21} = \left.\frac{I_2}{I_1}\right|_{V_2=0}$$

$$h_{12} = \left.\frac{V_1}{V_2}\right|_{I_1=0} \qquad h_{22} = \left.\frac{I_2}{V_2}\right|_{I_1=0}$$

Thus h_{11} is the input impedance at port 1 with port 2 short-circuited. The parameter h_{12} represents the reverse or feedback voltage ratio of the network, measured with the input port open-circuited. The forward-transmission parameter h_{21} represents the current gain of the network with the output port short-circuited; for this reason h_{21} is called the *short-circuit current gain*. Finally, h_{22} is the output admittance with the input port open-circuited.

The definitions and conceptual measuring setups of the h parameters are given in Fig. B.4.

g Parameters

The inverse-hybrid (or g-parameter) characterization of two-port networks is based on excitation of the network by V_1 and I_2, as shown in Fig. B.5a. The describing equations are

$$I_1 = g_{11}V_1 + g_{12}I_2 \tag{B.11}$$

$$V_2 = g_{21}V_1 + g_{22}I_2 \tag{B.12}$$

The definitions and conceptual measuring setups are given in Fig. B.5.

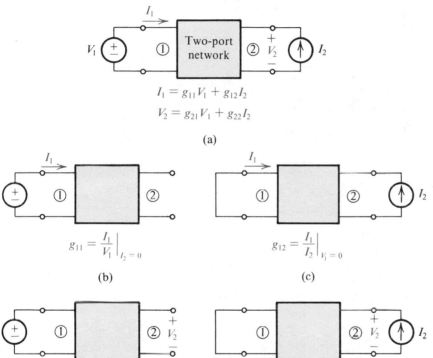

$$I_1 = g_{11}V_1 + g_{12}I_2$$
$$V_2 = g_{21}V_1 + g_{22}I_2$$

(a)

$$g_{11} = \left.\frac{I_1}{V_1}\right|_{I_2 = 0}$$

(b)

$$g_{12} = \left.\frac{I_1}{I_2}\right|_{V_1 = 0}$$

(c)

$$g_{21} = \left.\frac{V_2}{V_1}\right|_{I_2 = 0}$$

(d)

$$g_{22} = \left.\frac{V_2}{I_2}\right|_{V_1 = 0}$$

(e)

Fig. B.5 Definition and conceptual measurement circuits for *g* parameters.

Equivalent Circuit Representation

A two-port network can be represented by an equivalent circuit based on the set of parameters used for its characterization. Figure B.6 shows four possible equivalent circuits corresponding to the four parameter types discussed above. Each of these equivalent circuits is a direct pictorial representation of the corresponding two equations describing the network in terms of the particular parameter set.

Finally, it should be mentioned that other parameter sets exist for characterizing two-port networks, but these are not discussed or used in this book.

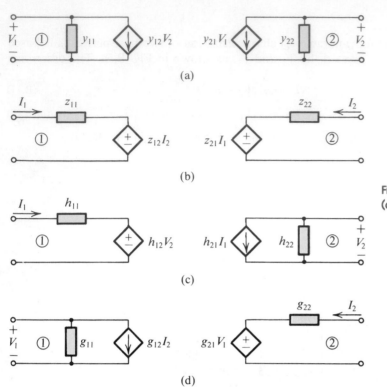

(a)

(b)

Fig. B.6 Equivalent circuits for **(a)** y, **(b)** z, **(c)** h, and **(d)** g parameters.

(c)

(d)

Exercise

B.1 Figure EB.1 shows the small-signal equivalent circuit model of a transistor. Calculate the values of the h parameters.

Ans. $h_{11} \simeq 2.6 \text{ k}\Omega$; $h_{12} \simeq 2.5 \times 10^{-4}$; $h_{21} \simeq 100$; $h_{22} \simeq 2 \times 10^{-5} \text{ Ʊ}$

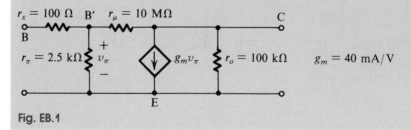

Fig. EB.1

PROBLEMS

B.1 (a) An amplifier characterized by the h-parameter equivalent circuit of Fig. B.6c is fed with a source having a voltage V_s and a resistance R_s, and is loaded in a resistance R_L. Show that its voltage gain is given by

$$\frac{V_2}{V_s} = \frac{-h_{21}}{(h_{11} + R_s)(h_{22} + 1/R_L) - h_{12}h_{21}}$$

(b) Use the expression derived in (a) to find the voltage gain of the transistor in Exercise B.1 for $R_s = 1 \text{ k}\Omega$ and $R_L = 10 \text{ k}\Omega$.

B.2 The terminal properties of a two-port network are measured with the following results: With the output short-circuited and an input current of 0.01 mA, the out-

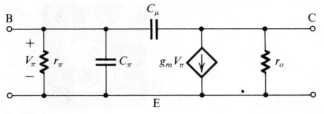

Fig. PB.3

put current is 1.0 mA and the input voltage is 26 mV. With the input open-circuited and a voltage of 10 V applied to the output, the current in the output is 0.2 mA and the voltage measured at the input is 2.5 mV. Find values for the h parameters of this network.

B.3 Figure PB.3 shows the high-frequency equivalent circuit of a BJT. (For simplicity, r_x has been omitted.) Find the y parameters.

Appendix

Computer Aids for Electronic Circuit Design

INTRODUCTION

Earlier in the book we have indicated situations where computer aids may be appropriate in the process of circuit design. In this appendix we wish to expand upon these comments, describe more generally the role which we see for such aids and, finally, by means of an example, show the benefits which can be derived from their use. □

THE CURRENT SITUATION

At the present time, computer aids for those working with electronic circuits, whether linear or digital, are becoming increasingly popular. Although design and analysis aids were once available only on relatively large machines on a time-sharing basis, the present trend is toward the provision of such tools at lower and lower cost on more accessible machines. Many such tools are already available (often in somewhat limited form) for use with the more expensive personal computers. Development of computer aids is continuing at a rapid pace.

Computer aids for the process of development of electronic products take many forms: some relatively simple and low-cost, but useful, such as printed-circuit-board (PCB) layout programs, and some quite sophisticated, such as parameterized circuit simulators, a class of programs which have not yet reached their potential even on large machines, and certainly not in a low-cost form. Since the number and variety of such aids is quite large, particularly if design is contemplated to the level of custom VLSI, our approach here will be to focus on one of the most important classes of programs—circuit simulation.

The simulator we have chosen for illustrative purposes is SPICE 2, an evolved (intermediate) circuit-simulator program developed at the Electronics Research Laboratory of the University of California, Berkeley, as part of a long-term research effort in electronic circuit design. While there are other more sophisticated, more complete, and more parameterized proprietary products available (some of which have origins in earlier SPICE developments), none are as accessible as SPICE. As a result of the generous distribution policies of the Berkeley Group, SPICE 2 is installed in a large number of academic and medium-to-large-scale industrial organizations throughout the world. Thus circuit simulation by SPICE 2 at modest (but not negligible) cost is a reality for everyone interested in electronics.

WHEN TO USE SPICE

Before we proceed with the details of SPICE data specification and application, it is important to review its place among the tools and techniques related to the study of electronics.

In spite of the fact that SPICE and similar tools are popularly associated with design, as implied by the popular acronym CAD (Computer-Aided Design), this connection is really somewhat indirect. At present SPICE and other such tools are simply circuit simulators, and thus are actually tools for performing analysis. While we have already seen that analysis is an essential part of the design process, it is certainly not all there is, nor even the most important part. But most critically, as this text has demonstrated again and again, the process of analysis is not unique: there are many choices, each representing a particular balance in a cost-benefit trade-off.

Thus we have seen that at the very early stages of a design a very simple, perhaps approximate, analysis is appropriate. The challenge then is to rapidly discover whether a conjectured solution to a circuit's problem is likely to work at all. Thus, very simple and rapid analysis is the key to "good design," particularly at the stage of selection of circuit topology. For without some degree of this ability the designer is not likely to find the circuit whose intrinsic strengths match the specifications of the problem to be solved.

On the other hand, at later stages of the design process, a more thorough analysis is justified. Often it is the only means by which one of several competing designs may be selected; frequently it is the final step before a commitment to production is made. Thorough analysis is particularly important in the process of design for VLSI, since the potential loss of time and money resulting from a design fault can be very great indeed.

So a circuit simulator such as SPICE is best used very near the end of the design process, both as an independent verification of operation and as a means by which relatively subtle parameter optimization can be performed. In general it is *not* a technique for selection of a basic circuit topology.

Certainly at present there is no design aid capable of the conceptualization, winnowing, and selection of topologies which characterize the techniques used by the best circuit designers. Nor in a general sense is there likely to be. What *is* becoming available, however, are parameterized design aids in which, for a given topology, ranges of component values are tested and even optimized. In this regard, in order to avoid the combinatorial explosion (and corresponding cost of computer time) implied by simultaneous variation of many parameters, random sampling techniques are employed to "cover" the corresponding multidimensional parameter space. (In view of the implied wager placed on the adequacy of coverage, these techniques are referred to as Monte Carlo!) While certainly of great importance in the past for special purposes, and likely to be more generally applied in the future, such techniques are beyond our present need and will not be discussed further here.

SOME SPECIFICS ON SPICE

"SPICE is a general-purpose circuit simulation program for nonlinear dc, nonlinear transient, and linear ac analysis. Circuits may contain resistors, capacitors, inductors, mutual inductors, independent voltage and current sources, four types of dependent sources, transmission lines, and the four most common semiconductor devices: diodes, BJTs, JFETs, and MOSFETs." (See Vladimirescu, Newton, and Pederson, 1980.)

SPICE provides a hierarchy of built-in models for semiconductor devices. If model-parameter data is available, the more sophisticated models can be invoked; otherwise a simpler model is used by default. Thus for BJTs, although the integral-charge model of Gummel and Poon (which we have not studied in this book) is potentially available, the Ebers-Moll model is used in default mode. Correspondingly there is a hierarchy of three MOS models which can include second-order effects such as channel-length modulation (short-channel effects) and subthreshold conduction.

TYPES OF ANALYSIS

DC Analysis

The dc-analysis part of SPICE determines the dc operating point of the circuit with inductors shorted and capacitors open. It is performed automatically prior to transient analysis and small-signal analysis to determine the operating point of each device, and thus to establish appropriate device-model parameters.

The dc-analysis part of SPICE can also be used to generate dc transfer curves as an independent source is stepped over a range, and as corresponding dc output variables are calculated. Also available are dc small-signal sensitivities of specified output variables with respect to circuit parameters.

AC Small-Signal Analysis

The ac small-signal part of SPICE computes ac output variables over a specifiable range of frequency on the basis of linearized small-signal models established from dc bias-point data. Typical transfer functions (gain, transimpedance, etc.) are available. Techniques are also available for evaluating noise and distortion characteristics.

Transient Analysis

The transient analysis part of SPICE computes transient output variables as a function of time over a specifiable interval. Such an analysis is useful for determining, for example, the step and pulse responses of logic-gate circuits.

Convergence

It is interesting and important to note that SPICE uses an iterative process to obtain both dc and transient solutions. Iteration is terminated when branch currents and node voltages converge to within a tolerance of 0.1% or 1 pA or 1 μV, whichever is larger. Although SPICE algorithms are found to be quite reliable, convergence is not guaranteed. In the event of failure to converge, the program terminates with no guarantee of relevance of data provided. This may be the case, for example, when dc analysis of regenerative circuits, such as flip-flops and Schmitt triggers, is attempted.

Input Format

The input format for SPICE is of the (relatively) free-format type in which labeled entries are separated by blanks, a comma, an equal sign, or a parenthesis. Extra spaces are ignored. Number fields may be either integer or floating point, using either decimal or scientific notation or mnemonic scale factors (for example, G for 10^9, MEG for 10^6, M for 10^{-3}, U for 10^{-6}, N for 10^{-9}, and so on). Letters following a number are ignored if not scale factors, as are other letters after scale factors. Thus 10, 10V, 10 VOLTS, and 10 HZ all represent the same values and M, MA, MSEC, and MMHOS represent the same scale factor.

CIRCUIT DESCRIPTION

The circuit to be analyzed is described to SPICE by a sequence of lines entered at a computer terminal or by element cards in a noninteractive system. There are element lines which define the topology and element values, and control lines which set model parameters and measurement nodes. The first line must be a title and the last .END. In between, the order is arbitrary, except that continued lines must follow immediately.

Each element in the circuit is specified by an element line containing the element name, connected nodes, and electrical parameter value(s). The first letter of an element name denotes element type (for example, R for resistor) and the name can be from one to eight characters. Nodes are specified by nonnegative integers, but need not be numbered sequentially. The datum node (ground) must be numbered 0. Every node must have two connections, except for MOSFET substrate nodes (and unterminated transmission lines).

Comments can be interspersed by prefacing them with an asterisk(*). Comments are *very important* for later understanding (both by the originating designer and other users) and should be used liberally.

Basic (but incomplete) specification formats are given in the table below:

Component	Name			Nodes and value
Resistor	Rxxxxxxx	N+	N−	VALUE
Capacitor	Cxxxxxxx	N+	N−	VALUE
Inductor	Lxxxxxxx	N+	N−	VALUE
VCCS	Gxxxxxxx	N+	N−	NC + NC − VALUE
VCVS	Exxxxxxx	N+	N−	NC + NC − VALUE
CCCS	Fxxxxxxx	N+	N−	VNAM VALUE
CCVS	Hxxxxxxx	N+	N−	VNAM VALUE
Voltage Source	Vxxxxxxx	N+	N−	QUAL
Current Source	Ixxxxxxxx	N+	N−	QUAL

where

1. The component name begins with a particular letter as indicated, and is from one to eight alphanumeric characters long.

2. N+ and N− indicate the nodes connected, the first being positive (if that matters).

3. VALUE is in units of ohms, farads, henries, mhos, V/V, A/A, ohms, respectively, for the first seven components above.

4. NC+ and NC− are nodes across which the controlling voltage appears.

5. VNAM is the voltage source through which the controlling current flows.

6. QUAL is a set of qualifiers of the source, whether DC or transient (including pulse, sinusoid, exponential or piecewise-linear) with amplitudes and other qualifiers, or AC with magnitude and phase.

7. Note that a voltage source of zero volts is used in SPICE as a means to indicate the location of a current measurement.

While the previous elements are simple enough to allow for relatively rudimentary descriptions, often on a single line, that is not the case for semiconductor devices for which named separate definitions using .MODEL are required. Subsequent semiconductor device specifications refer to the named model definition (for example,

MName). Thus for the four semiconductor devices, we have

Device	Name			Nodes and models
Diode	Dxxxxxxx	N+	N−	MNAME AREA
BJT	Qxxxxxxx	NC	NB	NE NS MNAME AREA
JFET	Jxxxxxxx	ND	NG	NS MNAME AREA
MOSFET	Mxxxxxxx	ND	NG	NS NB MNAME L W

where

1. MNAME refers to a user- or system-defined device model.

2. AREA is an (optional) area-scaling factor.

3. NC, NB, NE, NS are nodes to which the collector, base, emitter, and (optionally) substrate are connected.

4. ND, NG, NS are nodes to which the drain, gate, and source are connected.

5. ND, NG, NS, NB are nodes to which the drain, gate, source, and substrate are connected.

6. L, W are the channel length and width in meters.

A description of model creation and the MODEL construct is beyond our present need and will not be included here. *Note again* that the preceding description of specification-line formats, while essentially correct, is not complete and is intended only to provide an introductory appreciation of the example which follows.

EXAMPLE C.1

The requirement is to provide a SPICE analysis of the two-stage CMOS op amp presented originally in Fig. 13.23, for two sets of device models: a simple one and a considerably more completely specified one characteristic of the 5-micron process at Bell Northern Research (BNR). The simple model was presented originally in Exercise 13.23 where (in SPICE notation) VTO corresponds to V_t, KP to (μC_{OX}) and LAMBDA to $1/|V_A|$ with values of 1 V, 20 μA/V^2, 0.04 V^{-1}, and -1 V, 10 μA/V^2 for n- and p-channel devices, respectively. For the BNR model, the devices are similar, though not identical. In particular $|V_A|$ is higher as we shall see when comparing gain and output-resistance results.

Solution

Figure C.1 repeats Fig. 13.23 but with nodes labeled for SPICE input purposes. It also includes the W/L specification for each of the devices employed.

Figure C.2 is an exact version of the input presented via an interactive terminal to a version of SPICE, namely SPICE 2G.5 (10 Aug 81), installed in a DEC VAX 11/780 at the University of Toronto.

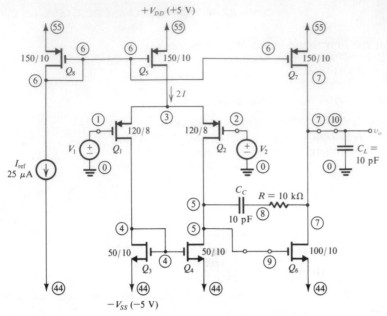

Fig. C.1 Two-stage CMOS op amp for Example C.1.

```
p-channel input cmos opamp    * opamp.spi *
.options gmin=1e-8 nomod
.width out=80
* transistor models
.model  mn  nmos (level=2  vto=1 kp=20u lambda=0.04)
* ciruit description
.model  mp  pmos (level=2  vto=-1 kp=10u lambda=0.04)
m1 4 1 3 55 mp l=8u w=120u
m2 5 2 3 55 mp l=8u w=120u
m3 4 4 44 44 mn l=10u w=50u
m4 5 4 44 44 mn l=10u w=50u
m5 3 6 55 55 mp l=10u w=150u
m6 7 9 44 44 mn l=10u w=100u
m7 7 6 55 55 mp l=10u w=150u
m8 6 6 55 55 mp l=10u w=150u
* frequency compensation
cc 5 8 10pF
rc 8 7 10k
*
* load
cl 10 0 10pf
*
*
vdd 55 0 dc 5
vss 44 0 dc -5
v2 2 0   dc 0 ac
v1 1 0 dc 0
vtmp 9 5 dc 0
ibias 6 44 dc 25u
*
* ammeters
vampo 7 10 dc 0
*
* initialize node voltages
.nodeset v(3)=1.2 v(4)=-3.2 v(5)=-3.2 v(6)=3.5 v(7)=0 v(8)=0
*
* small signal gain
.tf v(7) v2
*
.end
```

Fig. C.2 SPICE input for the circuit of Fig. C.1 utilizing the simple MOS device model.

The following list is a commentary on the input on nearly a line-by-line basis. Note that input is in lowercase letters; it will be reproduced in uppercase format in the output.

1. Line 1 is the title which will appear directly on the output.

2. Line 2 is an options-control list where gmin sets the value of minimum conductance to 1×10^{-8} mhos and nomod suppresses the output of model parameters.

3. Line 3 is a (printer) width control which includes 80-column output.

4. Line 4 is a comment.

5. Line 5 is a model specification for mn, an NMOS device using the MOS2 model with $V_t = 1$ V, $\mu C_{OX} = 20\ \mu A/V^2$, and $V_A = 1/\lambda = 25$ V.

6. Line 6 is a premature comment with a typo.

7. Line 7 is a model specification for mp, a PMOS device using the MOS2 model with $V_t = 1$ V, $\mu C_{OX} = 10\ \mu A/V^2$, and $V_A = 1/\lambda = 25$ V.

8. Lines 8 through 15 provide a device specification and wiring list for eight MOS devices where in terms of line 8, m1 is Q_1, connected with drain at node 4, gate at node 1, source at node 3, and substrate at node 55, for which model mp applies, with length of 8 microns and width of 120 microns.

9. Line 16 is a comment.

10. Line 17 describes a capacitor cc connected from node 5 to node 8 of value 10 pF. Note the typo F, (only the p is used).

11. Line 18 describes a resistor rc connected from node 8 to node 7 of value 10 kΩ.

12. Line 19 is a (blank) comment.

13. Line 20 is a comment.

14. Line 21 describes a capacitor cl connected from node 10 to node 0 (ground) of value 10 pF.

15. Lines 22 and 23 are (blank) comments.

16. Line 24 describes an independent voltage source vdd connected between node 55 (and thus defining it) and node 0 (ground) of type dc and value +5 V.

17. Line 25 establishes −5V on node 44.

18. Line 26 describes an independent voltage source v2 connected from node 2 [the (+)input] to node 0 (ground) with a dc component of 0 and an ac component of unit magnitude and 0 phase (by default).

19. Line 27 describes an independent voltage source v1 connected from node 1 [the (−)input] to node 0 (ground) of type dc and value 0.

20. Line 28 describes an independent voltage source vtmp connected from node 9 to node 5 of type dc and value 0 (to be used for current measurement).

21. Line 29 describes an independent current source ibias connected from node 6 to node 44 of type dc and value 25 μA.

22. Lines 29 and 30 are comments.

23. Line 31 describes an independent voltage source vampo connected from node 7 to node 10 of type dc and value 0 (to be used for current measurement).

24. Lines 32 and 33 are comments.

25. Line 34 is an initializing control to establish voltages at nodes 3, 4, 5, 6, 7, 8 prior to iteration.

26. Lines 35 and 36 are comments.

27. Line 37 is an output-control which requests a small-signal analysis with output at node 7 and input at node 2.

28. Line 38 is a comment.

29. Line 39 is an end-control signifying the end of the process in line 1.

Figure C.3 provides the output results. Note that the input data are first reproduced complete with typos, then a small-signal bias solution is presented. This includes dc node voltages and supply currents with total power dissipation. Then operating-point information for each of the transistors is presented. Note that in the latter part of the table all capacitances except C_{gs} are zero, a consequence of the simple model specified. Finally note that the gain and input and output resistances are provided at the end.

For the complex MOS model, Fig. C.4 shows the input model data (which replaces lines 5 and 7 of Fig. C.2). Figure C.5 provides the output corresponding to Fig. C.3 for the simple model. Note here the relative completeness of the operating-point information, including more nonzero capacitances. Finally note that the computer CPU time taken has increased to 4.32 units.

Finally, in the sense of a summary of all we now know about this op amp, Table C.1 presents a tabulation of results. While a lot can be said about these results, only a few points will be made here. Possibly the most reassuring is that the results are quite similar. Probably the most interesting is that the hand-calculated results lie somewhere between the extremes of SPICE using the simple model and the more complex one. That the results are strongly dependent on the detail of the diverse models is noteworthy as well.

Now, having analyzed the op amp by several means, we are in a particularly good position to find anomalous behavior, and possibly to correct for it in a redesign. We see, for example, that the slew rate is somewhat less than desired and asymmetric as well. Motivated by this result we reconsider the slew rate computation and realize that SR+ is essentially $I_{D7}/C_C + C_L$ where C_C and C_L are about the same size, and in particular that the result is a slew rate of about

$$\frac{I}{C} = \frac{30 \times 10^{-6}}{(10 + 10) \times 10^{-12}}$$

or 1.5 V/μs, about as the SPICE results show.

```
1*******07/04/86 ********  SPICE 2G.5 (10AUG81)  *******00:06:51*****

0P-CHANNEL INPUT CMOS OPAMP   * OPAMP.SPI *

0****     INPUT LISTING                 TEMPERATURE =   27.000 DEG C

0**********************************************************************


    .OPTIONS GMIN=1E-8 NOMOD
    .WIDTH OUT=80
    * TRANSISTOR MODELS
    .MODEL   MN NMOS (LEVEL=2  VTO=1 KP=20U LAMBDA=0.04)
    * CIRUIT DESCRIPTION
    .MODEL   MP PMOS (LEVEL=2  VTO=-1 KP=10U LAMBDA=0.04)
    M1 4 1 3 55 MP L=8U W=120U
    M2 5 2 3 55 MP L=8U W=120U
    M3 4 4 44 44 MN L=10U W=50U
    M4 5 4 44 44 MN L=10U W=50U
    M5 3 6 55 55 MP L=10U W=150U
    M6 7 9 44 44 MN L=10U W=100U
    M7 7 6 55 55 MP L=10U W=150U
    M8 6 6 55 55 MP L=10U W=150U
    * FREQUENCY COMPENSATION
    CC 5 8 10PF
    RC 8 7 10K
    *
    * LOAD
    CL 10 0 10PF
    *
    *
    VDD 55 0 DC 5
    VSS 44 0 DC -5
    V2 2 0  DC 0 AC
    V1 1 0 DC 0
    VTMP 9 5 DC 0
    IBIAS 6 44 DC 25U
    *
    * AMMETERS
    VAMPO 7 10 DC 0
    *
    * INITIALIZE NODE VOLTAGES
    .NODESET V(3)=1.2 V(4)=-3.2 V(5)=-3.2 V(6)=3.5 V(7)=0 V(8)=0
    *
    * SMALL SIGNAL GAIN
    .TF V(7) V2
    *
    .END
1*******07/04/86 ********  SPICE 2G.5 (10AUG81)  *******00:06:51*****

0P-CHANNEL INPUT CMOS OPAMP   * OPAMP.SPI *

0****     SMALL SIGNAL BIAS SOLUTION     TEMPERATURE =   27.000 DEG C

0**********************************************************************

   NODE   VOLTAGE    NODE   VOLTAGE    NODE   VOLTAGE    NODE   VOLTAGE


  ( 1)     .0000   ( 2)     .0000   ( 3)    1.3841   ( 4)   -3.4904

  ( 5)   -3.4904   ( 6)    3.4411   ( 7)   -1.0221   ( 8)   -1.0221

  ( 9)   -3.4904   ( 10)  -1.0221   ( 44)  -5.0000   ( 55)   5.0000
```

Fig. C.3 SPICE output obtained in response to the input of Fig. C.2.

```
VOLTAGE SOURCE CURRENTS

NAME       CURRENT

VDD        -8.359e-05

VSS         8.359e-05

V2          0.000e+00

V1          0.000e+00

VTMP        0.000e+00

VAMPO       0.000e+00

   TOTAL POWER DISSIPATION   6.25e-04  WATTS
1*******07/04/86 ******** SPICE 2G.5 (10AUG81)  ********00:06:51*****

0P-CHANNEL INPUT CMOS OPAMP   * OPAMP.SPI *

0****     OPERATING POINT INFORMATION     TEMPERATURE =   27.000 DEG C

0************************************************************************

0
0**** MOSFETS

0            M1        M2        M3        M4        M5        M6        M7
0MODEL       MP        MP        MN        MN        MP        MN        MP
 ID       -1.37e-05 -1.37e-05  1.38e-05  1.38e-05 -2.74e-05  3.09e-05 -3.09e-05
 VGS        -1.384    -1.384     1.510     1.510    -1.559     1.510    -1.559
 VDS        -4.875    -4.875     1.510     1.510    -3.616     3.978    -6.022
 VBS         3.616     3.616      .000      .000     3.616      .000      .000
 VTH        -1.000    -1.000     1.000     1.000    -1.000     1.000    -1.000
 VDSAT       -.384     -.384      .510      .510     -.559      .510     -.559
 GM        7.16e-05  7.16e-05  5.42e-05  5.42e-05  9.80e-05  1.21e-04  1.10e-04
 GDS       6.83e-07  6.83e-07  5.88e-07  5.88e-07  1.28e-06  1.47e-06  1.63e-06
 GMB       0.00e+00  0.00e+00  0.00e+00  0.00e+00  0.00e+00  0.00e+00  0.00e+00
 CBD       0.00e+00  0.00e+00  0.00e+00  0.00e+00  0.00e+00  0.00e+00  0.00e+00
 CBS       0.00e+00  0.00e+00  0.00e+00  0.00e+00  0.00e+00  0.00e+00  0.00e+00
 CGSOVL    0.00e+00  0.00e+00  0.00e+00  0.00e+00  0.00e+00  0.00e+00  0.00e+00
 CGDOVL    0.00e+00  0.00e+00  0.00e+00  0.00e+00  0.00e+00  0.00e+00  0.00e+00
 CGBOVL    0.00e+00  0.00e+00  0.00e+00  0.00e+00  0.00e+00  0.00e+00  0.00e+00
 CGS       2.21e-13  2.21e-13  1.15e-13  1.15e-13  3.45e-13  2.30e-13  3.45e-13
 CGD       0.00e+00  0.00e+00  0.00e+00  0.00e+00  0.00e+00  0.00e+00  0.00e+00
 CGB       0.00e+00  0.00e+00  0.00e+00  0.00e+00  0.00e+00  0.00e+00  0.00e+00

0            M8
0MODEL       MP
 ID       -2.50e-05
 VGS        -1.559
 VDS        -1.559
 VBS         .000
 VTH        -1.000
 VDSAT       -.559
 GM        8.94e-05
 GDS       1.07e-06
 GMB       0.00e+00
 CBD       0.00e+00
 CBS       0.00e+00
 CGSOVL    0.00e+00
 CGDOVL    0.00e+00
 CGBOVL    0.00e+00
 CGS       3.45e-13
 CGD       0.00e+00
 CGB       0.00e+00

0****     SMALL-SIGNAL CHARACTERISTICS

0    V(7)/V2                          =  2.145e+03
0    INPUT RESISTANCE AT V2           =  1.000e+20
0    OUTPUT RESISTANCE AT V(7)        =  3.210e+05
0
     JOB CONCLUDED
0    TOTAL JOB TIME            2.57
```

Fig. C.3 (Continued)

```
* transistor models
*
.model   mn nmos (level=2  vto=1  nsub=1e16  tox=8.5e-8  uo=750
+                  cgso=4e-10  cgdo=4e-10  cgbo=2e-10
+                  ucrit=5e4  uexp=.14  utra=0  vmax=5e4  rsh=15
+                  cj=4e-4  mj=2  pb=.7  cjsw=8e-10  mjsw=2
+                  js=1e-6  xj=1u  ld=.7u)
.model   mp pmos (level=2  vto=-1  nsub=2e15  tox=8.5e-8  uo=250
+                  cgso=4e-10  cgdo=4e-10  cgbo=2e-10
+                  ucrit=1e4  uexp=.03  utra=0  vmax=3e4  rsh=75
+                  cj=1.8e-4  mj=2  pb=.7  cjsw=6e-10  mjsw=2
+                  js=1e-6  xj=.9u  ld=.6u)
*
```

Fig. C.4 Input model data (to replace lines 5 and 7 of Fig. C.2) for the more elaborate MOS model.

```
 * SMALL SIGNAL GAIN
 .TF V(7) V2
 *
 .END
1*******07/04/86 ********  SPICE 2G.5 (10AUG81)  ********00:06:04*****

0P-CHANNEL INPUT CMOS OPAMP   * OPAMP.SPI *

0****     SMALL SIGNAL BIAS SOLUTION        TEMPERATURE =   27.000 DEG C

0**********************************************************************


   NODE   VOLTAGE    NODE   VOLTAGE    NODE   VOLTAGE    NODE   VOLTAGE


  (  1)    .0000   (  2)    .0000   (  3)   1.8834   (  4)  -3.5384

  (  5)  -3.5384   (  6)   3.4244   (  7)   -.9032   (  8)   -.9032

  (  9)  -3.5384   ( 10)   -.9032   ( 44)  -5.0000   ( 55)   5.0000



     VOLTAGE SOURCE CURRENTS

     NAME       CURRENT


     VDD      -8.173e-05

     VSS       8.173e-05

     V2        0.000e+00

     V1        0.000e+00

     VIMP      1.059e-22

     VAMPO     0.000e+00


     TOTAL POWER DISSIPATION   6.07e-04  WATTS
1*******07/04/86 ********  SPICE 2G.5 (10AUG81)  ********00:06:04*****

0P-CHANNEL INPUT CMOS OPAMP   * OPAMP.SPI *

0****     OPERATING POINT INFORMATION     TEMPERATURE =   27.000 DEG C

0**********************************************************************
```

Fig. C.5 SPICE output for the circuit in Example C.1 utilizing the more elaborate MOS device model specified in Fig. C.4.

0	M1	M2	M3	M4	M5	M6	M7
0MODEL	MP	MP	MN	MN	MP	MN	MP
ID	-1.35e-05	-1.35e-05	1.35e-05	1.35e-05	-2.69e-05	2.96e-05	-2.96e-05
VGS	-1.883	-1.883	1.462	1.462	-1.576	1.462	-1.576
VDS	-5.422	-5.422	1.462	1.462	-3.117	4.097	-5.903
VBS	3.117	3.117	.000	.000	.000	.000	.000
VTH	-1.515	-1.515	.952	.952	-.959	.941	-.952
VDSAT	-.318	-.318	.284	.284	-.456	.292	-.462
GM	7.18e-05	7.18e-05	5.37e-05	5.37e-05	8.64e-05	1.15e-04	9.43e-05
GDS	7.87e-07	7.87e-07	6.08e-07	6.08e-07	1.09e-06	7.76e-07	8.92e-07
GMB	8.55e-06	8.55e-06	3.90e-05	3.90e-05	2.63e-05	8.25e-06	2.83e-05
CBD	0.00e+00	0.00e+00	0.00e+00	0.00e+00	0.00e+00	0.00e+00	0.00e+00
CBS	0.00e+00	0.00e+00	0.00e+00	0.00e+00	0.00e+00	0.00e+00	0.00e+00
CGSOVL	4.80e-14	4.80e-14	2.00e-14	2.00e-14	6.00e-14	4.00e-14	6.00e-14
CGDOVL	4.80e-14	4.80e-14	2.00e-14	2.00e-14	6.00e-14	4.00e-14	6.00e-14
CGBOVL	1.36e-15	1.36e-15	1.72e-15	1.72e-15	1.76e-15	1.72e-15	1.76e-15
CGS	2.21e-13	2.21e-13	1.16e-13	1.16e-13	3.58e-13	2.33e-13	3.58e-13
CGD	0.00e+00	0.00e+00	0.00e+00	0.00e+00	0.00e+00	0.00e+00	0.00e+00
CGB	0.00e+00	0.00e+00	0.00e+00	0.00e+00	0.00e+00	0.00e+00	0.00e+00

0	M8
0MODEL	MP
ID	-2.50e-05
VGS	-1.576
VDS	-1.576
VBS	.000
VTH	-.964
VDSAT	-.451
GM	8.08e-05
GDS	1.42e-06
GMB	2.48e-05
CBD	0.00e+00
CBS	0.00e+00
CGSOVL	6.00e-14
CGDOVL	6.00e-14
CGBOVL	1.76e-15
CGS	3.58e-13
CGD	0.00e+00
CGB	0.00e+00

0****   SMALL-SIGNAL CHARACTERISTICS

0	V(7)/V2	=	3.435e+03
0	INPUT RESISTANCE AT V2	=	1.000e+20
0	OUTPUT RESISTANCE AT V(7)	=	5.960e+05

0

    JOB CONCLUDED
0   TOTAL JOB TIME   4.32

Fig. C.5 (Continued)

Table C.1 COMPARISON OF THE RESULTS OF THE ANALYSIS OF THE CIRCUIT OF FIG. C.1 BY HAND AND USING SPICE WITH TWO DIFFERENT MODELS

	Units	Hand Calculation	SPICE 2	
			Simple	Complex
First stage gain	V/V	−62.5	−55.1	−50.4
First stage output resistance	kΩ	—	779	714
Second stage gain	V/V	−50.0	−38.9	−68.2
Second stage output resistance	kΩ	—	321	596
Open-loop gain	V/V	3125	2145	3435
Output resistance	kΩ	—	321	596
Input resistance	Ω	—	1×10^{20}	1×10^{20}
Unity-gain frequency	MHz	1.0	0.972	0.955
Phase margin	degrees	57.8	68.3	63.4
Positive slew rate	V/μs	—	1.37	1.25
Negative slew rate	V/μs	−2.5	−3.07	−2.90

To improve this, our only option is to raise I_{D7}, by perhaps a factor of 2, expecting that SR$-$, controlled primarily by I_{D1}, will remain the same. We note, however regrettably, that the gain of stage 2 will reduce somewhat. Our logical next step is to evaluate this by hand computation and then select a modified design which we would check by SPICE analysis.

CONCLUDING REMARKS

SPICE is an extremely valuable circuit-simulation program. But it does not provide an alternative to hand analysis; it is usually used at a later stage in the design process to help in design optimization.

We hope that the reader has access to SPICE. If so, considerable benefit can be gained by using it to analyze some of the circuits described in the examples, exercises, and problems in this text.

BIBLIOGRAPHY

A. Vladimirescu, A. R. Newton, and D. O. Pederson, "SPICE Version 2G.1 User's Guide," Berkeley: University of California, Department of Electrical Engineering and Computer Science, 1980.

Appendix

Answers to Selected Problems

Chapter 1
1.2 81%   **1.4** 2%   **1.6** (a) 9 V; (b) 3.86 V; 42.9%
1.7 (a) 40 mV; (b) 0.4 V; (c) 2.5%; (d) 1.25%

Chapter 2
2.2 (a) 31.8%; (b) 25%; (c) 50%   **2.4** $i(t) = 10 \sin(2\pi \times 10^3 t)$
2.5 100 V/V (40 dB); 100 A/A (40 dB); 10^4 W/W (40 dB)   **2.7** 11 mW; 69.4%
2.9 0.691 V; $0.04v_d$; 25 Ω; 2.5 Ω
2.10 2,000 A/A; 200 V/V; 4×10^5 W/W; 10 kΩ; 100 Ω; 220 V/V

2.13 22 V/V (27 dB); 1.936×10^8 W/W (83 dB)   **2.15** $1\left/\left(g_m + \dfrac{1}{R_i}\right)\right.$

2.17 400 V/V (52 dB); 100 A/A (40 dB); 4×10^4 W/W (46 dB)   **2.18** -33.3 V/V
2.22 666.7 V/V (56.5 dB); 3.33×10^3 A/A (70.5 dB); 2.24×10^6 W/W (63.5 dB)

2.27 (a) $1/(1 + sCR)$; (b) $sCR/(1 + sCR)$; (c) $\left(sCR + \dfrac{1}{k}\right)\left/\left[sCR + \left(1 + \dfrac{1}{k}\right)\right]\right.$

2.29 $0.64\omega_0$; $0.51\omega_0$; $0.43\omega_0$ **2.32** $0.707 \sin\left(10^3 t - \dfrac{\pi}{4}\right)$

2.33 $V_{oc} = V_t$; $I_{sc} = V_t/Z_t$; $Z_t = V_{oc}/I_{sc}$ **2.35** 1 V; 900 Ω; 0.526 V   **2.36** 0.588 V; 470 Ω

2.38 20 pF; 1 μF; 11 pF; 0; 1 pF   **2.39** $v_o = g_m R(v_1 - v_2)$; 0 V; 10 V
2.40 2.5 V; 9 V; 0.56 mA; 0.3 V   **2.43** $R_1 = 9$ MΩ, $C_1 = 3.33$ pF; 10 M$\Omega \parallel 3$ pF
2.44 10^5 rad/s   **2.45** 20 Hz and 5 kHz; 0.17 dB; 20 Hz and 5 kHz
2.46 $-50/(1 + 0.555 \times 10^{-6}s)$ **2.48** HP; 10 rad/s   **2.49** 11.1 MΩ
2.51 0.69τ; 2.3τ; 4.6τ; 6.9τ **2.52** 3.5 ns   **2.54** 4.67 V   **2.57** 20 μF, 10 rad/s
2.59 15.76 kΩ

Chapter 3

3.1 (a) -10 V/V, 1 kΩ; (b) -1 V/V, 1 kΩ; (c) -0.1 V/V, 10 kΩ; (d) 0 V/V, 1 kΩ;
(e) -10 V/V, 100 Ω; (f) $-\infty$ V/V, 0 Ω
3.3 $R_1 = 100$ kΩ, $R_2 = 10$ MΩ; $R_{in} = 100$ kΩ

3.6 $\dfrac{R_2}{R_1} \dfrac{sC_1R_1 + 1}{sC_2R_2 + 1}$; $C_1R_1 = C_2R_2$; 1 MΩ; 10 MΩ, 30 pF, 3 pF

3.8 $R_1 = 10$ kΩ, $R_2 = 100$ kΩ, $C_2 = 0.01$ μF; 0 dB; $-10(1 - e^{-10^3 t})$
3.10 100 pulses; -1 V   **3.13** $sCR/(1 + sCR_1)$; 20 dB; 40 dB   **3.14** 13
3.16 56 V peak-to-peak; 20 V rms   **3.18** $v_O = v_1 + v_2$
3.21 $-x/(1 - x)$; $1/(1 - x)$; $-2(2 - x)/(1 - x)$ **3.22** $i_O = v/R$; $i_O = -v/R$
3.24 0.999; 0.05%   **3.25** $Z_{in} = -1/\omega^2 C^2 R$ (a frequency-dependent negative resistance)
3.28 1,000; 1 kHz; 1 MHz   **3.29** 2 MHz; 1 MHz; 0.643 MHz   **3.31** 99; -0.98
3.32 10 kHz; 58.5 kHz   **3.34** 10 kHz; 35 μs
3.35 (b) $\beta = 0.5$, $v_o/v_i = 2$; (c) $\beta = \frac{1}{5}$, $v_o/v_i = 5$; (d) $\beta = \frac{1}{13}$, $v_o/v_i = 13$
3.37 200,000; 8 Hz; 1.6 MHz   **3.39** 2 V/μs; 24.5 kHz   **3.41** 6 V peak   **3.43** 40 V
3.48 (a) 0.9999 Ω; (b) 99 Ω; (c) 834 Ω **3.50** 0.159 H; 1.26 kHz
3.52 ±0.25 V; ∓0.5 V; of opposite polarities
3.54 (a) 0.1 V; (b) 0.2 V; (c) 9.9 kΩ, 10 mV; (d) 0.111 V   **3.57** 1 mV; equal at 0.1 V/s; 60 s
3.58 $R_1 = 100$ kΩ, $R_2 = 3.9$ MΩ, $C_2 = 0.1$ μF; 10 ms; $\simeq 0.7$ s
3.61 $+1$ V; when S is opened, v_B ramps up to $+13$ V with a slope of 1 V/s; 10 s
3.62 0 V; -10 mV; 1 Ω
3.64 $R_1 = 10$ kΩ, $R_2 = 1$ MΩ, $C_1 = 0.159$ μF, $C_2 = 15.9$ pF; $f_t = 10$ MHz

Chapter 4

4.2 0 V, 1 mA; 0 V, 0 mA; 10 V, 0 mA; 0 V, 0 mA; 5 V, 0 mA; 0 V, 0.5 mA; 0 V, 0 mA; 0 V, 1 mA
4.3 15 V, 0.75 mA; 10 V, 0 mA; 2 V, 0 mA; -2 V, 1.2 mA; 0 V, 0.45 mA; 0 V, 0.125 mA
4.4 14.14 V; 8.3 ms; 8.3 ms; 4.5 V; 45 mA   **4.7** $\frac{1}{2}$ cycle; 120 mA; 60 mA   **4.9** 6.7 kΩ
4.11 1 V peak   **4.14** $\Delta v = -17.32$ mV   **4.15** 697.4 mV; 684.4 mV; -1.3 mV/°C
4.18 740 mV; 540 mV   **4.19** 2.67 kΩ; 4.94 mA; 33.6 mW   **4.21** $R = 7.45$ kΩ
4.24 19.6 ppm/°C   **4.27** 0.186 mA, 616 mW
4.29 (a) 0.514 mA, 989 mV; (b) 1.5 mA; (c) 0.1 mA; (d) 14 V
4.31 573 mV, 6.6 Ω; 609.7 mV; 699.6 mV
4.34 Two parallel branches, each consisting of three series diodes; decreases by 15 mV
(approximately), 15.8 mV using exponential model.

Chapter 5

5.4 For $-1 \le v_I \le +1$, $v_O = 0$; for $v_I > 1$, $v_O = \frac{2}{3}(v_I - 1)$; for $v_I < -1$, $v_O = \frac{2}{3}(v_I + 1)$
5.11 $+2$ V; -2 V   **5.15** $0.636V_p$; $1.32V_p$ **5.16** 11.1 V; 30 V; 0 V; 18.6 V
5.18 4.5 mA; sine wave of $5\sqrt{2}$ V peak; 7.14 V   **5.19** $v_O(t_1) = 9.2$ V; 9.23 V
5.21 0.0995 V; 0.198 V; 0.952 V; 100; 230; 461
5.23 0.01 V; 0.02 V, 0.10 V; 50; 100; impossible to reach 2 V

5.25 833 μF; 3.2%; 3.13 A; 6.26 A   **5.27** $\left(2V_p - \dfrac{V_p}{fCR}\right)$; V_p/fCR

5.29 70 MΩ; $R_3 = 10$ MΩ, $R_4 = 1$ kΩ, $R_5 = 6$ kΩ **5.30** 833 μF; 8.17 V

5.34 $v_I = 0.51, 0.70, 1.7, 10.8, 0, -0.51, -0.70, -1.7, -10.8$ V; soft limiter; $K = 0.98$; $L_+ = 0.8$ V; $L_- = -0.8$ V

5.35 2 V; $+12$ V or -12 V; square wave of frequency f and ± 12 V levels, lagging the input by an angle of $65.4°$; 0.1 V

5.41 10 MΩ   **5.42** $R_2 = 200$ kΩ; $V_R = V/21$ **5.43** $R_2 = 173.3$ kΩ; $R_3 = 200$ kΩ

5.47 ± 0.7 V; ± 7.7 V; As R_1 is reduced, the hysteresis width is reduced, reaching zero at $R_1 = 10$ kΩ. For R_1 less than 10 kΩ, the circuit does not exhibit bistability.

5.48 ± 0.93 V   **5.49** $R_1 = 10$ kΩ, $R_2 = 90$ kΩ, $R_3 = 10$ kΩ, $C_1 = 0.249$ μF; 18.2%

5.52 $V = 1.1$ V, $R = 400$ Ω   **5.53** ± 2.5 V

Chapter 6

6.1 n **6.4** $-V_P/2I_{DSS}$; $-V_P/I_{DSS}$; $-5V_P/I_{DSS}$ **6.6** 17 V   **6.8** 400 kΩ

6.10 0.225 V; -0.293 V   **6.12** $\frac{20}{3}$ mA; $\frac{4}{3}$ V   **6.14** 1 mA; 2 V   **6.15** 0.515 V; 0.05 V

6.16 1.22 mA; -1.22 V; -3.9 V

6.18 (a) $I_1 = 2.2$ mA; (b) $I_2 = 4.4$ mA; (c) $I_3 = 4.4$ mA; (d) $V_4 = 0.658$ V; (e) $V_5 = 5$ V; $I_5 = 2.1$ mA; (f) $V_6 = 5$ V

6.19 (a) $V_1 = 6$ V; (b) $V_2 = 1$ V, $V_3 = 5$ V; (c) $V_4 \simeq 0.87$ V, $V_5 \simeq 1.24$ V; (d) $V_6 = 1$ V, $V_7 = 5$ V; (e) $V_8 = 0$ V, $V_9 = 6$ V; (f) $V_{10} = 0.586$ V, $V_{11} = 6$ V; (g) $V_{12} = 0.585$ V, $V_{13} = 6$ V; (h) $V_{14} = 9.12$ V; (i) $V_{15} = 0.882$ V

6.22 $V_X = +0.05$ V, $V_Y = -0.05$ V; -50 mV; -7.7 V; $+5$ V, -5 V   **6.23** 4 mA

6.26 (a) $R_D = 5$ kΩ, $R_S = 1$ kΩ; (b) $R_D = 22$ kΩ, $R_S = 6$ kΩ

6.27 (a) 1.62 mA; (b) 0.45 mA   **6.30** g_m and voltage gain are doubled in value.

6.32 $R_1 = 10$ MΩ, $R_2 = 5$ MΩ   **6.33** 1 kΩ   **6.34** 12.5%; 40 mV   **6.36** $R_S = 18/g_{m1}$

6.37 22.1×10^3 (43.4 dB); 25×10^3 (44 dB)   **6.38** 353 Ω; 353 Ω   **6.40** 10 mV; 1 V

6.46 $0.095 \sin \omega t$; 7.7 V   **6.47** 100 μF   **6.48** -2.3 V

6.49 For $v_I = 0, \pm 8$ V, ± 9 V, $v_O = 0, \pm 8$ V, ± 8.41 V; 31.25 Ω; -62.5 mV

6.53 $V_P = -2$ V, $I_{DSS} = 2$ mA, $R_S = 22$ kΩ; 306 Ω

6.54 (a) $1 \Big/ \left(500 + \dfrac{1}{10^{-5}s} \right)$; (b) $1 \Big/ \left(1{,}500 + \dfrac{1}{10^{-5}s} \right)$; (c) $1 \Big/ \left(1{,}500 + \dfrac{1}{10^{-5}s} \right)$;

(d) $1 \Big/ \left[1{,}500 + \left(10 \text{ k}\Omega \Big\| \dfrac{1}{10^{-5}s} \right) \right]$ **6.56** 1 V; 1.697 V; 2.210 V; 2.598 V

6.57 0 V; -5 V; -10 V/V

Chapter 7

7.1 (a) 2 V, 8 mA; (b) 6 V, 72 mA   **7.2** 6 mA; 8 mA; 10 mA/V   **7.3** 1.66 V

7.6 10 kΩ; 2.5 kΩ   **7.7** 1 to 4 V   **7.9** 20 mA; 150 Ω; 75 Ω

7.11 1 V; 0.3 mA; 4 V; 0.365 mA; 2.7 V   **7.12** 134 μA; 169 μA; 26% versus 100%

7.15 12 pF   **7.16** (a) 0.1 mA/V; (b) 0.5 mA/V   **7.18** -8.33 V/V; 1.07 MΩ

7.19 2.24 V; 258 μA/V; 25.8 μA/V; 500 kΩ   **7.22** 0 V; 1 kΩ   **7.23** 0 V, 6 kΩ

7.25 $+1$ V, -0.5 V, $+1$ V, $+2.5$ V, $+2.5$ V; -1 V/V; -10 V/V; 0.6 MΩ

7.26 (a) 0 mA, 10 V; (b) 0.5 mA; 7.5 V; (c) 1.46 mA, 2.7 V; (d) 1.03 mA, 4.86 V; (e) 0.125 mA, 5.5 V; (f) 0.5 mA, 5 V; (g) 0.5 mA, 4 V; (h) 0.125 mA, 4 V

7.28 (a) 1.056 V, 1.12 kΩ; (b) 0.111 V, 0.111 kΩ; (c) 0.011 V, 0.011 kΩ

7.30 $V_O = 2.5, 5$ and 7.5 V; $I = 0.1125, 0.800, 2.11$ mA   **7.31** -10 V/V

7.33 (a) 8 V, 4 V, 20 μA; (b) 8.59 V, 3.41 V, 10 μA   **7.35** 0.0625

7.37 (a) 2.5 V, -1 V/V; (b) 2.24 V, -3.16 V/V; (c) 2.08 V, -10 V/V

7.39 (a) 6.67 kΩ; (b) 1 MΩ, 0.25 MΩ; (c) 10.1 kΩ, 5 kΩ, 38.5 kΩ   **7.42** 100 kΩ

7.44 10 μm; 50 μm   **7.46** 0.3 V; 0.212 V; 0 V

7.47 225 μA, 225 μA; $-3{,}000$ V/V; -115.4 V/V, 85.9 kΩ; -9.13 V/V; ± 2 V

7.49 2.5 kΩ; 3.33 kΩ; 333 kΩ   **7.51** 47.8 kHz

Chapter 8

8.1 0.691 V   **8.2** 144.6 mA   **8.4** 667   **8.5** 5.05 mA; 100; 0.990   **8.6** 10.24 μA
8.9 -0.558 V; 0.5 mA   **8.11** 0.8; 4
8.12 (a) $V_E = -0.7$ V, $I_E = 0.93$ mA, $I_B = 9.2$ μA, $V_C = 0.792$ V; (b) $V_E = +0.7$ V,
$I_E = 1.86$ mA, $I_C = 1.842$ mA, $V_C = -5.792$ V
8.14 0.3 V; 0.015 mA; 0.8 mA; 0.785 mA; 52.3, 0.98   **8.15** $+0.4$ V; 0 V   **8.17** 50 kΩ; 50 V
8.18 (a) active, 1.3 V, 0.99 mA; (b) saturated, 1.7 V; (c) active, -4.3 V, 0.492 mA;
(d) active, -20.7 V, 1.86 mA
8.21 $V_E = 1.802$ V; $V_B = 2.502$ V; $I_E = 0.060$ mA; $I_C = 0.059$ mA; $V_C = 6.06$ V
8.24 0.09; 5 mV; 0.5 mV   **8.25** 40 mA/V, 2.5 kΩ **8.28** 400 V/V; 200 V/V
8.30 185.9 V/V; 169 V/V   **8.33** 160 V/V; 0.976 V/V
8.34 $R_1 - 10$ kΩ, $R_2 = 5$ kΩ, $R_E = 430$ Ω, $R_C = 430$ Ω; 9.3 mA, 7.05 V; 5.87 mA, 10.18 V
8.37 5 kΩ; 2.46 kΩ; 24.6 kΩ; 49.2 kΩ; 1.34 mA; 1.26 mA; 1.78 kΩ; -125 V/V; -18.9 V/V
8.38 1.18 mA; 15.52 kΩ; -4.6 V/V; 0.247 V; 1.136 V   **8.40** 2 V; 80 mV; 2.08 V; 10.62 V
8.41 0.5 mA; 50 Ω; 69.8 V/V; the transistor leaving the active region   **8.43** -37 V/V
8.44 (a) 7.18 mA; (b) 51.1 kΩ, 0.997 V/V, 3.47 Ω; (c) 33.7 kΩ, 0.994 V/V; (d) 0.767 V/V
8.46 109 Ω; 1.000, 0.902; 2.5 V; 2.5 V
8.48 $\beta = \infty$: 9.3 V; 9.3 mA; 0.997; 100 kΩ.
   $\beta = 100$: 4.67 V; 4.67 mA; 0.995; 50.4 kΩ
8.49 $\beta = \infty$: 5.97 V; 5.97 mA; 0.9956; 98 kΩ.
   $\beta = 100$: 3 V; 3 mA; 0.9918; 50 kΩ
8.51 1.13 V; -9.5 V/V   **8.52** (a) 1.3 V, 2; (b) 6.9 V, 4.1, 3.1, 2.9 (for Q_4), 1.3 (for Q_3)
8.57 90; 100   **8.58** 6.8 V

Chapter 9

9.1 4.51 V   **9.2** 4.02 V; 5 V   **9.4** 54.9 mV   **9.6** $v_d/4V_T$; $\frac{1}{2}$; $\frac{1}{4}$; 10 mW   **9.8** 100 V/V
9.10 1 mA; 100   **9.12** (a) 100 V/V, 0, ∞; (b) 50 V/V, 0.0025 V/V, 20,000   **9.14** 1.01
9.16 149 MΩ; 201 kΩ **9.17** 1.09×10^{-3} V/V
9.19 $I/3$; $2I/3$; $IR_C/3$; small-signal analysis: $v_d = 0.75V_T$; large-signal analysis: $v_d = 0.693V_T$;
1 μA; 0.33 μA
9.22 0.99%   **9.24** 25 Ω **9.26** 18; 198   **9.27** $R = 9.3$ kΩ; 0.093 mA
9.33 $R - 10$ kΩ; $R_E = 48$ kΩ; $R_{out} = 24.2$ MΩ; 0.384 μA   **9.34** 2.8 MΩ
9.36 $5/(1 + 6/\beta)$
9.38 (a) 2 mA, -0.7 V, 5 V, 0.7 V, -0.7 V, -5.7 V; (b) 0.2 mA, -0.7 V, $+5$ V, $+0.7$ V,
-0.7 V, -5.7 V
9.39 $i = v/R$; $v_2 \le v$; $v_2 \le 10v$ **9.43** 1 V   **9.44** 14.14 V/V; 10^{-3} V/V; 14,000; 14.1 mV
9.46 150.5 kΩ **9.47** (a) ∞; (b) 2.6 MΩ **9.48** (a) 16 μA; (b) 80 μA/V; (c) 0.28 V;
(d) 3 MΩ; (e) 4 mV; (f) 240 V/V
9.53 58.4 kΩ **9.54** 3.84 V/V

Chapter 10

10.1 -9.3 V   **10.2** 2.86 kΩ; 2.47 kΩ; 20 mA; 20.8 mA   **10.4** $V_{CC}I$ (in all cases)
10.5 10%; 13.3%; 16%; 20%
10.7 For 5-V output: 1.25 W; 1.59 W; 0.17 W; 78.6%. For 1-V output: 0.05 W; 0.318 W;
0.13 W; 15.7%
10.8 For 5-V output: 2.5 W; 2.5 W; 100%. For 1-V output: 0.1 W; 0.5 W; 0.2 W; 20%
10.10 ± 5 V; ± 6.67 V; 721.3 μF   **10.11** (a) 1.4 V; (b) 1.285 V; (c) 1.266 V
10.13 1 mA; a negative step of 1.058 V; 4 V, 6 V
10.14 0.5 mA; -10 V to $+2.55$ V; 199; 1.96 mA
10.17 78.3; 0.99   **10.18** 50; 4.48 V; 4.4 V   **10.19** (a) 1.2 V; (b) 1.212 V; (c) 1.31 V
10.20 140°C; 570 mV   **10.23** $P_{Dmax} = 0.8$ W; $P_{D0} = 1.2$ W   **10.25** 84°C; $\frac{2}{3}$ W/°C; 3 cm

10.27 $T_J = 152.5°C$; 18.6 W

10.28 $v_I = 0$: $i_I = 0$; $v_I = +10$ V: 20 μA for $R_L = \infty$ and 22.5 μA for $R_L = 100$ Ω; $v_I = -10$ V: same but with directions reversed

10.30 $i_C = [\beta_1 + \beta_2(\beta_1 + 1)]i_B$; $i_E = (\beta_1 + 1)(\beta_2 + 1)i_B$;

$$v_{BE} = V_T \ln\left[\frac{\beta_1\beta_2}{(\beta_1 + 1)(\beta_2 + 1)^2} \frac{i_E}{I_{S1}I_{S2}}\right]; \ 1.264 \text{ V}; \ 0.98 \ \mu\text{A}; \ 9.999 \text{ mA}$$

10.33 $R_1 = 58$ kΩ; $R_2 = 3.64$ kΩ; 2.1 nA; 129°C (approx.)

10.34 18 μA; 18 μA; 0.36 mA; 0.36 mA; 10.4 V   **10.36** 14.37 kΩ; 13.07 Ω

10.37 25 kΩ; 20 kΩ **10.39** 250 cm^2/V·s   **10.41** 463 Ω; 92.6 kΩ; 14 kΩ; 28 kΩ; 14 kΩ

Chapter 11

11.1 (a) $\dfrac{s + 1/R_2C}{s + 1/(R_1 \parallel R_2)C}$; (b) $\dfrac{s + 1/(L/R_1)}{s + 1/[L/(R_1 + R_2)]}$

11.3 20 dB, 84°; 37 dB, 45°; 40 dB, 0°; 37 dB, $-45°$; 20 dB, $-84°$; 0 dB, $-90°$

11.5 Peak gain of 80 dB at 3×10^7 rad/s; $\phi = 0°$ **11.7** 0.0175

11.8 (a) $n = 1{:}1$; $n = 2{:}0.642$; $n = 3{:}0.509$; $n = 4{:}0.434$; (b) $n = 1{:}1$; $n = 2{:}0.414$; $n = 3{:}0.268$; $n = 4{:}0.199$

11.9 $T(s) = \dfrac{-2(1 + s)(1 + s/10)}{(1 + s/5)(1 + s/100)}$; -2 V/V; 99.1 rad/s   **11.11** 111 rad/s

11.12 $\dfrac{-100(1 + s/10^6)}{(1 + s/10^5)(1 + s/10^7)}$; 10^5 rad/s; 1.01×10^5 rad/s; 1.01×10^5 rad/s; $\dfrac{-100}{1 + s/10^7}$; 10^7 rad/s.

11.13

	(i)	(ii)	(iii)
(b)	-261, -38.2	-137, -73	-1192, -8.4
(c)	285	192	1196
(d)	261	137	1192
(e)	264	155	1192
(f)	300	210	1200

11.14 838 pF; 1,000 pF; 83.6 Hz   **11.16** 83.4 nF

11.18 -18.2 V/V; 38.9 kHz; 318 MHz; 239.3 MHz (212.2 MHz using the approximate formula)

11.20 21.2 kHz; 22.5 MHz   **11.22** 400 mA/V; 2.5 Ω; 300 Ω; 120 Ω; 30 Ω

11.24 (a) Approximately: 5 kΩ and 200; more accurately: 5.26 kΩ and 212.6; (b) 5210 Ω, 24.4 Ω, 40.8 mA/V, 1.025 mA; (c) 0.2×10^{-3}, low frequency to avoid effect of C_μ shunting r_μ; (d) 26 MΩ, 6.12 kHz; (e) 15.9 μA/V, 129.5 kΩ.

11.26 3.64 Grad/s; 24.2 Mrad/s   **11.28** 0.32 pF   **11.30** $100\omega_\beta$

11.34 $r_e/[1 + s(C_\pi + C_\mu)r_e]$; 404 MHz

11.35 $\dfrac{g_m - sC_\mu}{\dfrac{1}{r_e} + \dfrac{1}{r_\pi} + s(2C_\pi + C_\mu)}$; 213.6 MHz; 318 MHz   **11.37** 107 MHz; 127 MHz

11.39 (a) 1.06 GHz, 31.8 MHz, 31.7 MHz, and $f_H = 22.4$ MHz; (b) 1.06 GHz, 31.8 MHz, 11.27 MHz, and $f_H = 10.62$ MHz

11.40 (a) 0.7 V, 0.7 V, 5 V; (b) -52 V/V; (c) 243 Hz   **11.44** -48 V/V; 648 kHz

11.49 (a) -66.7 V/V, 117 kHz; (b) -66.7 V/V, 3.8 MHz; (c) 50 V/V, 4.6 MHz; (d) -194 V/V, 1.6 MHz; (e) -66.7 V/V, 3.8 MHz; (f) 50 V/V, 4.6 MHz

Chapter 12

12.1 (a) 9,900; (b) 99,900   **12.2** 1; 0.909 V/V; 20.8 dB; 0.909 V, 0.091 V; 0.909%

12.5 10; (a) approximately 990, exactly 989; (b) approximately 950, exactly 909.1

12.7 10 Hz   **12.8** 90.9 V/V; $12.42\omega_H$; $88.6\omega_H$ **12.11** ± 7 mV; 0.990 V/V

12.13 10.9 kΩ; 10 MΩ; 159 pF   **12.15** 1 kΩ; 1 Ω; 1.59 mH

12.17 (a) 85.7 V/V, 0.1 V/V, 8.96 V/V, 191 kΩ, 19.1 Ω; (b) 62.5 V/V, 1 V/V, 0.984 V/V, 1.84 MΩ, 2.91 Ω; pole at 896 rad/s

12.18 1.01 MΩ **12.20** 40 A/V; 50 V/A; 20 mA/V; -12 V/V; 18.5 MΩ **12.22** 10.1 Ω

12.24 -7.9 V/V; 433 Ω; 1 kΩ **12.25** -5.58 V/V; 143.8 Ω; 5.67 kΩ

12.27 2.74 μF; 19.6 Hz   **12.29** -192 V/V; 29.1 Ω; 29.5 Ω **12.33** 10^4 rad/s; 0.002; 500

12.35 8×10^{-3} **12.37** 7.86 kHz; 51.9°   **12.39** 10 Hz

12.40 23,256 V/V; 6,542 V/V   **12.44** 10 Hz; 0.0159 μF

12.46 Positive feedback; negative feedback; positive feedback; $|A_f| > |A|$; $A_f = 10A$; $A_f = \infty$

12.47 $A\beta = -10/11$; 100 V/V; -1.11 kΩ

12.48 $A_f = \mu \Big/ \left(1 + \mu \dfrac{R_1}{R_1 + R_2}\right)$; (a) 0.667; (b) 0.5; (c) 0.5

Chapter 13

13.2 1.7 mA; 51 mW

13.3 634 mV; 10 mA/V and 30 mA/V; 100 Ω and 33.3 Ω for a total of 25 Ω; 1275 Ω (total); 200 kΩ and 68 kΩ; 100 MΩ and 34 MΩ

13.4 15.4 V   **13.7** $\dfrac{I_3}{I_1} = \left(\dfrac{\dfrac{1}{\sqrt{K_1}} + \dfrac{1}{\sqrt{K_2}}}{\dfrac{1}{\sqrt{K_3}} + \dfrac{1}{\sqrt{K_4}}}\right)^2$; 1.6 mA   **13.9** Loop gain $= \beta_P = 50$; 34 dB

13.10 48.2 nA; 12.1 nA   **13.12** -13.8 V   **13.14** 165 Ω **13.16** 6.38 kΩ; 1.146 V

13.18 7%   **13.19** 10.8 mV   **13.21** 46.3 dB   **13.24** 318 Ω, 953 Ω; 239 Ω

13.25 637 kΩ **13.27** 159 kHz; 15.9 MHz   **13.29** 30 μA; -2.5 V; 0.8 mV

13.31 500 kΩ **13.33** 10 pF; 16.7 kΩ; 2.5 V/μs

Chapter 14

14.1 10 kΩ, 0.01 μF; 2 H; 0.01 μF; 10 kΩ **14.4** $100/(s^2 + 14.14s + 100)$

14.7 $\dfrac{(s/2)^2 + 100}{s^2 + 20s + 100}$ **14.10** $C_1 = 4.76 \, C_2$; 1; 0.826

14.12 $R_3 = 141.4$ kΩ; $R_4 = 70.7$ kΩ **14.14** $\dfrac{-s/RC_3}{s^2 + \dfrac{2}{RC}s + \dfrac{1}{R^2CC_3}}$, $\omega_0 = 4/RC$; $Q = 2$; 8

14.16 $s^2 \Big/ \left[s^2 + s\dfrac{2}{C_1 R_3} + \dfrac{1}{C_1 C_2 R_3 R_4}\right]$; 1, $R_3 = 141.4$ kΩ; $R_4 = 70.7$ kΩ

14.17 $C_3 = 1{,}125$ pF; $C_4 = 2{,}251$ pF   **14.19** $S_R^{\omega_0} = 0$; $S_L^{\omega_0} = S_C^{\omega_0} = -\frac{1}{2}$, $S_R^Q - 1$; $S_L^Q = -\frac{1}{2}$; $S_C^Q = \frac{1}{2}$

14.23 $S_{C_2}^{\omega_0} = S_{C_6}^{\omega_0} = S_{R_1}^{\omega_0} = S_{R_3}^{\omega_0} = S_{R_5}^{\omega_0} = -\frac{1}{2}$; $S_{C_2}^Q = S_{R_1}^Q = S_{R_3}^Q = S_{R_5}^Q = -\frac{1}{2}$; $S_{C_6}^Q = S_{R_4}^Q = \frac{1}{2}$; $S_{R_7}^Q = +1$

14.25 $C = 1591$ pF, $R_1 = 10$ kΩ, $R_2 = 10$ kΩ, $R_3 = 490$ kΩ, $R = 10$ kΩ; gain $= -99$ V/V

14.28 10 MΩ, 1 MΩ **14.29** 1 pC; 0.1 μA; 0.1 V; 100; 10^6 V/s

14.32 $C_3 = C_4 = C_6 = 1$ pF, $C_5 = 1.414$ pF   **14.33** 49.4 Mrad/s; 67.7; 729.4 krad/s; -66.6 V/V   **14.35** 838.8 kHz; 47.4

14.40 $\omega_{01} = \omega_0$, $B_1 = B$, $Q_1 = \omega_0/B$; $\omega_{02} = \omega_0 + \sqrt{3}B/4$, $B_2 = B/2$, $Q_2 \simeq 2\omega_0/B$; $\omega_{03} = \omega_0 - \sqrt{3}B/4$, $B_3 = B/2$, $Q_3 \simeq 2\omega_0/B$

14.43 6.6 kΩ; 2.1 V peak-to-peak   **14.45** 406.1 Hz; 290 kΩ

14.47 (a) 0; (b) 6.25×10^{-3}; (c) 2.08×10^{-3}; (d) 6.7×10^{-3}

14.48 (a) $1/\sqrt{LC}$, $R_C = 2r_e = 4V_T/I$; (b) 1.27 V

Chapter 15

15.1 4 V; 0.465 V   **15.2** 24.9 ns; 40.8 ns   **15.4** $NM_L = NM_H = 0.2V^+$; $0.3V^+$; 5 V

15.5 77.3 mW; 7.73 mA; 10.8 mW   **15.7** 3.33 MHz; frequency remains unchanged.

15.8 (a) 1.33 V; (b) 1.75 V; (c) 2 V; (d) 2.2 V

15.10 $v_O = 4$ V: 81 μA, 324 μA; $v_O = 3$ V: 80 μA, 323 μA; $v_O = 1$ V: 36 μA, 171 μA

15.12 $V_{OL} = 0.334$ V; $V_{IH} = 2.09$ V; $NM_L = 0.666$ V, $NM_H = 1.31$ V

15.14 $V_{OH} = 3.85$ V; $V_{OL} \simeq 0.3$ V   **15.15** 8.1 ns   **15.16** 1.6 ns

15.19 (a) 10 μA, 5.63 μA, 4.05 μA; (b) 10 μA, 6.94 μA, 4.02 μA; (c) 10 μA, 7.03 μA, 3.51 μA

15.22 288 μm^2; 720 μm^2

15.24 Separate: $\bar{F} = AB + CD + CE$; joined: $\bar{F} = (A + C)(B + D + E)$

15.25 (a) $+5$ V, 4 V, 1.35 V, 0.03 V; (b) 0, 229 μA; (c) 3.2 mA, 2.4 mA, 2.8 mA, 7.14 ns;
(d) 6.4 mA, 1.6 mA, 4 mA, 5 ns

15.27 $\bar{E} = B + AC$; $\bar{D} = A + BC$; 0.13 V, 0.26 V, 0.4 V and 5 V

15.29 -133.3 V/V; 2.5 V; -133.3 V/V; 74.5 kΩ **15.31** 1.44 V

15.34 0.55 mA, increase width by a factor of 2.82; 4.37 mA; 7.17 mA

15.35 50 μA, 1.8 mA; 6.05 mA   **15.37** 0.36 μC; 0.36 mA; 3.6 mW; 12.6 mW

15.40 NOR: $W_n = 10$ μm, $W_p = 60$ μm, area $= 1,050$ μm^2;
NAND: $W_p = 20$ μm, $W_n = 30$ μm, area $= 750$ μm^2

15.42 0.5 μA; 0.1 μA/kHz; 110 mW   **15.43** Unbuffered: 300 units; buffered: 333 units

15.51 6.9 μs; 9.8 V; 5.7 V; -0.1 V; 1 mA   **15.53** 72.1 kHz

Chapter 16

16.2 1.125 mA; 0.125 mA; 0.125 mA; 0.125 mA   **16.3** 690.5 mV; 672.9 mV

16.5 $i = \dfrac{I_S}{\alpha_R}(e^{v/V_T} - 1)$ **16.7** 17.3 mV; 1.22 mV; 0.25 mV

16.8 2.78 Ω; 268.5, 0.056; 94.4 mV   **16.10** V_{BE} drops by 44.8 mV   **16.11** 0.078

16.13 18 pC; 0.2 pC; 180 ns; 2 ns   **16.15** 6 ns   **16.17** 1.43 ns

16.18 (c) $V_{IH} = 0.746$ V; (f) $NM_H = 0.254$ V, $NM_L = 0.4$ V

16.20 Average power $= 7.5$ mW   **16.22** NAND   **16.23** 1.65 mA; 0.14 mA; 98.5 ns

16.26 5.75 ns, 0.73 ns   **16.28** 7 ns; 4.3 V; 14.3 mA

16.32 4.56, 6.4, 19.4, 95.7 mV; 1.84 Ω, 1.44 Ω, 0.85 Ω **16.34** 3.85 mA, 10.5 mA

16.35 31.5 mA; 14   **16.38** 0.232 mA   **16.39** NOR

16.41 (a) $Y = \overline{A + BC}$; (b) 0.3 V, 1.3 V; (c) 0.7 V, 0.8 V; (d) $NM_H = 0.5$ V, $NM_L = 0.4$ V;
(e) 1.092 mA, 0.967 mA

16.42 (a) -0.75 V, $-0.75 - IR$; (b) $-0.75 - IR/2$; (c) $-0.65 - IR/2$; (d) $-0.635 - IR/2$;
(e) $V_{IL} = -0.865 - IR/2$, $V_{IH} = -0.635 - IR/2$; (f) $NM_H = NM_L = IR/2 - 0.115$;
(g) $IR = 0.69$ V; (h) -0.75 V, -1.44 V, -0.98 V, -1.21 V, -1.095 V

16.44 100 MHz   **16.47** (a) -1.375, -1.265 V; (b) -1.493, -1.147 V   **16.49** 20.5

16.50 12.6 pF   **16.52** $E = (A + B)(C + D)$; 0.984 V; $+5$ V

Appendix B

B.1 -245.7 V/V   **B.2** $h_{11} = 2,600$ Ω; $h_{21} = 100$; $h_{12} = 2.5 \times 10^{-4}$, $h_{22} = 2 \times 10^{-5}$ $\mho$

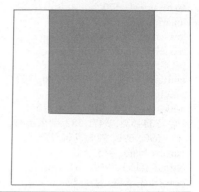

Index

The boldface page numbers signify either the presence of a definition or of an important concept.